Modeling and Optimization in Science and Technologies

Volume 3

For further volumes:
http://www.springer.com/series/10577

About this Series

The book series *Modeling and Optimization in Science and Technologies (MOST)* publishes basic principles as well as novel theories and methods in the fast-evolving field of modeling and optimization. Topics of interest include, but are not limited to: methods for analysis, design and control of complex systems, networks and machines; methods for analysis, visualization and management of large data sets; use of supercomputers for modeling complex systems; digital signal processing; molecular modeling; and tools and software solutions for different scientific and technological purposes. Special emphasis is given to publications discussing novel theories and practical solutions that, by overcoming the limitations of traditional methods, may successfully address modern scientific challenges, thus promoting scientific and technological progress. The series publishes monographs, contributed volumes and conference proceedings, as well as advanced textbooks. The main targets of the series are graduate students, researchers and professionals working at the forefront of their fields.

Constandinos X. Mavromoustakis · Evangelos Pallis
George Mastorakis

Resource Management in Mobile Computing Environments

Springer

Constandinos X. Mavromoustakis
Department of Computer Science
University of Nicosia
Nicosia
Cyprus

George Mastorakis
Department of Commerce and Marketing
Technological Educational Institute of Crete
Crete
Greece

Evangelos Pallis
Department of Informatics Engineering
Technological Educational Institute of Crete
Heraklion
Greece

ISSN 2196-7326 ISSN 2196-7334 (electronic)
ISBN 978-3-319-06703-2 ISBN 978-3-319-06704-9 (eBook)
DOI 10.1007/978-3-319-06704-9
Springer Cham Heidelberg New York Dordrecht London

Library of Congress Control Number: 2014938112

Printed on acid-free paper

Springer is part of Springer Science+Business Media (www.springer.com)

Dedication

To my wife Afrodyte, and our daughters Thelma, Aristi and Despina, for their continuous love, patience and support (Constandinos X. Mavromoustakis)

To my wife Athina, for her constant support and love (George Mastorakis)

To my son Meletios and my wife Vasiliki, for their support and understanding (Evangelos Pallis)

Preface

Resource manipulation in dynamically changing environments has emerged as a critical part of the information technology infrastructure. The need for efficient resource manipulation stems from the increasing growth of wireless infrastructures and mobile networks. The need for a reliable management of resources in mobile computing environments, facilitating ubiquitous availability and efficient access to large quantities of distributed resources, has become apparent, as the number of people that communicate and collaborate computationally over the Internet, via different accessing systems and mobile devices, has increased. Mobile computing paradigm is set to drive technology over the next decade and integrate resources availability through the 3As (Anywhere, Anything, Anytime). Notwithstanding, there are a lot of challenges to meet, in order to have mobile computing paradigm applicable in all aspects and in an efficiently utilized manner. In this context, this book aims at presenting state-of-the-art research and future trends on resource management in mobile and heterogeneous networking systems and applications. It combines mobile communications and resource management field, in a common research ground to present various research concepts that contribute to enable highly efficient management of networking resources. The major subjects of the book cover resource management methodologies, modeling, analysis and efficient resource management of mobile computing environments, newly introduced technologies, facing the scarceness of resources and model formulation for resources management in wireless networking systems.

This book is an excellent source of new scheme developed in the field for students, researchers and practitioners. It may be used in undergraduate and graduate courses on design and development of mobile computing systems and their applications. Researchers will find the book useful as it provides a state-of-the-art guide to resource management in mobile computing environments and technology and offers an unfolding perspective on current and future trends in these environments. Practitioners will find the book useful as a means of updating their knowledge on particular topics such as mobile and Ad-hoc wireless networks, peer-to-peer systems for mobile computing, novel resource management in cognitive radio networks and power management in mobile computing systems. Mainly the book captures the current state-of-the-art in resource management in wireless environments and is a solid source of comprehensive reference material on the related research field.

The book contains 26 refereed chapters from prominent researchers working in this area around the world. It is organized along five themes (sections) in the field of resource management issues for different mobile computing environments.

Section I—Mobile and Ad Hoc Wireless Networks: Chapter 1 by Petre, Chilipirea and Dobre elaborates on delay tolerant networks for disaster scenarios. Chapter 2 by Carmona-Murillo *et al.* provides an analytical study on Quality of Service (QoS) issues in next generation mobile networks and chapter 3 by Vale Pinheiro and Boavida discusses performance evaluation issues of network mobility paradigms. Chapter 4 by Palma and Curado proposes scalable routing mechanisms for mobile ad-hoc networks and chapter 5 by Stratakis *et al.* introduces novel methods for measurements on modern wireless communication technologies and estimation of human exposure. Chapter 6 by Pirmoradian, Adigun and Politis discusses issues related with radio spectrum and cognitive mobile computing and chapter 7 by Basholli and Lagkas presents resource request mapping techniques for OFDMA networks.

Section II—Peer-to-Peer systems for Mobile Computing: Chapter 8 by Mirtchev *et al.* provides traffic management issues and chapter 9 by Bleda *et al.* introduces ambient assisted living tools for a sustainable aging society. Chapter 10 by Andreou *et al.* focuses on adaptive heuristic-based peer-to-peer network connectivity issues and chapter 11 by Sousa *et al.* studies evaluation issues of ubiquity in mobile computing systems. Chapter 12 by Markakis *et al.* reviews peer-to-peer constellations in interactive broadcasting networks enhanced with advanced management mechanisms.

Section III—Resource management in mobile cognitive radio networks: Chapter 13 by Karyotis, Anifantis and Papavassiliou introduces novel cross-layer based resource management frameworks for mobile cognitive radio networks and chapter 14 by Mastorakis *et al.* elaborates on energy-efficient routing protocols for cognitive radio networks. Chapter 15 by Kormentzas studies challenges and solutions on spectrum trading in cognitive radio networks and chapter 16 by Mastorakis *et al.* focuses on energy and resource consumption evaluation issues of mobile cognitive radio devices. Chapter 17 by Mastorakis *et al.* proposes a novel network architecture for efficient resource management, utilizing content-aware multipath routing.

Section IV—Resource and Power management in Mobile Computing Systems: Chapter 18 by Kouskouli and Koutsakis presents novel approaches on scheduling for efficient telemedicine traffic transmission over next generation cellular networks and chapter 19 by Mesodiakaki, Adelantado, Alonso and Verikoukis presents an energy-efficient contention-aware algorithm for channel selection in cognitive radio networks. Chapter 20 by Saleem *et al.* elaborates on resource management issues in mobile sink based wireless sensor networks through cloud computing and chapter 21 by Narayanan presents research approaches on mobile video streaming resource management. Chapter 22 by Mavromoustakis *et al.* elaborates on energy harvesting issues in wireless networks devices and chapter 23 by Zotos *et al.* introduces a unified platform for machine-to-machine virtual object interoperability.

Section V—Performance Evaluation of Mobile Computing Systems: Chapter 24 by Correa *et al.* reviews computational resource management issues for video coding in mobile environments and chapter 25 by Khan and Martini studies resource allocation and scheduling issues for video transmission over mobile computing systems. Finally, chapter 26 by Wang and Haas provides a performance analysis of epidemic routing for delay-tolerant networks by proposing a mathematical framework to model such systems.

Although the covered topics may not be an exhaustive representation of all the resource management issues in mobile computing environments, they do represent a rich and useful sample of the related research approaches and contents.

This book has been made possible by the great efforts and contributions of many people. First of all, we would like to thank all the contributors for putting together excellent chapters that are very comprehensive and informative. Second, we would like to thank all the reviewers for their valuable suggestions and comments, which have greatly enhanced the quality of this book. Third, we would like to thank the staff members from Springer, for putting this book together and assisting us throughout the process. Finally, we would like to dedicate this book to our families.

Contents

Part IV: Resource and Power Management in Mobile Computing Systems

Part V: Performance Evaluation of Mobile Computing Systems

Part I

Mobile and Ad Hoc Wireless Networks

Delay Tolerant Networks for Disaster Scenarios

Andreea-Cristina Petre, Cristian Chilipirea, and Ciprian Dobre*

University Politehnica of Bucharest, Computer Science Department
Splaiul Independentei 313, Bucharest 060042, Romania
{andreea.petre,cristian.chilipirea}@cti.pub.ro,
ciprian.dobre@cs.pub.ro

Abstract. Disaster and emergency management refers to a range of activities designed to maintain control over crisis situations, providing the rescue and assistance equipment with a framework for helping victims and reducing its impact. The range of activities include prevention, advance warning, early detection, analysis of the problem and assessment of scope, notification of the public and appropriate authorities, mobilization of a response, containment of damage, and relief and medical care for those affected. One of the challenges in emergency scenarios is the fact that communications can be interrupted, cutting the information flow. This lack of communication infrastructure makes an appropriate response to the disaster more challenging, and leads to reduced quality of services experienced by vulnerable civilians. As an example, emergency scenarios with big agglomerations of people or traffic jams following accidents demand a unified communication infrastructure to optimize the response and decision making. This can be overcome using self-configured wireless networks, because they do not require any pre-existing infrastructure to be established, and are easy to deploy and fast to operate. The continuous use of modern smartphones facilitates the accessibility to wireless technologies. However, when incorporating mobile smartphones into disaster assisting networks, the biggest challenge is that such wireless networks need to be specifically designed and used for supporting victims, people and assistance equipment in crisis scenarios. Because of this, in future mobile networks designed for disaster management, there is a need for new architectures and protocols, capable to adapt existing and available wireless technologies for smart data capturing and decision making. This chapter analyses specific challenges and requirements related to supporting communication in such challenged situations. We present an extensive analysis of networking solutions designed to support situations such the ones described.

Keywords: Disaster Scenario, Networking, Delay Tolerant Networks, Mobile Devices, Communication and mobility support.

1 Introduction

In the past few years our dependency on technologies, such as the personal computer or the Internet, has grown to a point where we cannot sustain our normal day to day

* Corresponding author.

C.X. Mavromoustakis et al., *Resource Management in Mobile Computing Environments*, Modeling and Optimization in Science and Technologies 3,
DOI: 10.1007/978-3-319-06704-9_1, © Springer International Publishing Switzerland 2014

activities if we were to be denied access to such advancements. More and more businesses migrate part of their infrastructure to the Internet [1, 2], either to construct stronger business-to-client relations, or to simplify the process of acquiring new clients by having a more efficient infrastructure and communication process inside the company. Many e-commerce web sites, like Amazon, have managed to dominate the market by having their service available 24/7, with easy access to anyone interested in acquiring their products. The same can be said of many other e-commerce platforms that have managed to beat the competition, platforms which range from fast-food home delivery to electronics.

Social Networks, such as Facebook [3, 4], on the other hand, have become ubiquitous in the lives of many people. The number of users and time spent on such social networks increased with the emergence of new similar services, such as Google+ [5] and others dedicated to a certain activity (such as finding a job or sharing experience in programming). Furthermore, technologies such as Twitter [6], a micro-blogging platform, are used every day by millions of users, and have recently stood at the center of major political events, such as the ones in Libya [7].

The Internet is not only a tool represented by the various websites that are part of the web, but it is also an infrastructure that supports other services such as communication, like in the case of Skype [8], a Voice-over-IP service that only requires an Internet connection and is otherwise completely free that offers voice, text and video communication as well as file transfer.

Emerging technologies such as Internet of Things [9] or Smart Cities [10, 11, 12] will further increase our (already high) need for continuous Internet access. These emerging trends will connect more and more devices to the Internet, and will require high-availability connections to function correctly. An air conditioner will probably need to access a weather service to predetermine its usage and increase or lower the temperature of a room autonomously, before the user arrives home. Such technologies will probably be able to function without a permanent connection to the Internet, but their functionality and efficiency would dramatically be reduced if the connection is not available for longer periods of time.

This dependency continues with the centralization of user data, especially in the case of government centralized data. Considering the example of a hospital [13] that uses a government issued system to track patient confidential information; if an unconscious patient arrives at the hospital in a critical condition and the hospital does not have access to his data because of an Internet outage, mistakes can be made that can jeopardize his well-being. Such mistakes can be caused by the lack of patient history or the list of medication the patient is using.

To top all of these methods that the Internet is used in a day to day basis, cloud computing [14] (or, in a larger sense, utility computing) makes individual and businesses heavily dependent on good Internet connections. Without an Internet connection available, companies will not have access to data that would otherwise be always available on site. Most technologies, already available, or emerging ones, all increase our need for a stable Internet connection. Because of this, it is important to develop new technologies that guarantee Internet access for most people even in extreme scenarios. Furthermore we need to create technologies that give assurances that no one individual or organization can restrict access to a group of people.

The problem of managing access to communication resources becomes even more critical in case of disasters. The Great East Japan Earthquake in 2011 showed that disaster management services supported by communication between rescuers and victims are mandatory to be fast put in place to support efficient evacuation guidance procedures. Such services could use opportunistic communication. In recent years, as a practical use of Delay Tolerant Network and Mobile Opportunistic Network, researches on disaster evacuation guidance effective against situations of large-scale urban disasters have been undertaken. In such disaster evacuation guidance procedures based on opportunistic communication, evacuees collect location information of impassable and congested roads due to disaster in their smartphones by themselves, and share it with each other opportunistically by short-range wireless communication between nearby smartphones in order to not only navigate evacuating crowds to refuges, but also rapidly aggregate the disaster information.

In this paper we study existing and emerging technologies that provide an Internet connection or at least a communication method even in extreme circumstances such as natural disasters. In the following section we present why there is a need for such a technology and then we present the available networking technologies that try to solve this problem.

The rest of this chapter is structured as follows. We first present an analysis of extreme networking scenarios. Section 3 presents the set of requirements for networks in the case of disaster scenarios, and in Section 4 we present an analysis of current networking alternatives for disaster scenarios. This is followed in Section 5 by ad-hoc networking alternatives, and experimental results. Section 6 concludes and presents future work.

2 Extreme Networking Scenarios

High Price/Low Availability – There are a number of regions in the world, such as sub-Saharan Africa [15], where the disparity between the price of an Internet connection and the income of most citizens is so high that most people cannot afford to own an Internet connection. This is not a problem that is confined to 3rd world countries like the ones in Africa, but are also common in rural area of more advanced countries.

This problem is introduced by the fact that the cheapest way to spread the Internet network is currently given by wired technology. Wired technology has severe disadvantages: the infrastructure needs to be deployed, usually requiring digging holes or raising posts; because of technologies such as fiber optic, it is usually enough to get one cable to a densely populated area and offer high speed bandwidth to a large number of users, this translates in lower prices only for densely populated areas; there are range-related limits for both fiber optic and cooper based networking, there is a need for repeaters which can dramatically raise price; usually deployment is not enough, there is need for administration and redundancy.

For sparse populated areas, or populated areas that are extremely far apart, like the case of cities separated by a desert, or villages built on top of mountains, the costs of constructing a wired networking infrastructure can be so high that it is not even worth considering. Furthermore because only a small number of people live in each of this villages, the price of constructing the initial infrastructure is split only by a small number, thus keeping price per person extremely high. This is opposite to a dense populated city, with huge buildings shared by several families, where the price per person to build the infrastructure can be extremely low.

The further we move from the large cities and the metropolises that many of us have grown accustomed to, the more we can feel the lack of stable Internet connections, or even their complete availability.

Because of this any networking technology that wants to provide Internet infrastructure to such areas should not have its price dependent on the number of users in an area that need to adopt it.

Unstable Electricity – The same areas we presented earlier, in which the people have low income and the costs of infrastructure are extremely high, suffer from problems with stable, always available electricity. This areas usually have no or a small number of redundancies to their main line that provides electricity. If this main line fails for whatever reason it needs to be repaired before the electric service can again be available. The problem is further complicated by the fact that cables and infrastructure in these areas is not changed often enough and they are more likely to fail because of this.

To top all of this, repairs to any infrastructure, be it Internet, electrical or of any other type can take huge periods of time as access to such areas tends to be limited, or difficult.

To build a network that can provide high stability and functionality even in such environments it needs to require very little administration or change of infrastructure equipment and it needs to have a way to work even if there is no electricity through the use of batteries or alternative energy sources, like the use of solar panels.

Natural Disasters – these represent problems that not only affect the regions presented earlier, but also affect densely populated regions. There are a number of natural disasters that can affect infrastructure, starting from earthquakes, volcanoes and all the way to tsunamis.

The problem with these is even greater if we consider the chaos they induce and the higher need for communication to save lives in such scenarios. People want to contact their loved ones and make sure they are safe and can overload even the remaining part of any infrastructure. Search and rescue teams need to communicate between themselves, to the central command and control and to the individuals they intend to rescue.

Search and rescue teams can also have huge benefits from any data that they have about a location. This data can be of multiple types, from recent maps to pictures from the area they are investigating to sensor data such as gas leakage detectors. There is an extremely high amount of data that can help search and rescue personnel to asses risk and to decide what the best course of action is.

Because of the potentially high number of people affected, the necessity of information, the potential to use data and communication to save life, and the reduction in stress in individuals given by the opportunity to communicate with close ones, we believe that this scenario is one of the most important one to take into account when considering developing technologies for extreme networking.

Building such technologies requires either a way to assure that the infrastructure will resist after the natural disaster event, which is really difficult, or the infrastructure can be replaced in a simple and fast manner. There is also the case where no infrastructure is required, this is the case of ad-hoc networks.

Oppression – after the incident in Libya, also known as the "Twitter Revolution" governments that oppress their citizens have tried to prevent the start of a revolution by limiting or stopping the access to Internet to its citizens. This was seen in January 2011 when Egyptian government [16] has shutdown Internet access in an attempt to stop civil unrest against President Mubarak.

There are other cases where the Internet is not stopped entirely but censored. This is the case of the "Great firewall of China" [17] which limit the access of Chinese citizens to the information on the Internet, both for in country websites and other countries.

Attempts of censorship have even crossed country borders in cases such as the one in which Pakistani government has tried to block access to YouTube [18], and in doing so has blocked access to the site for the entire world for several hours by adding a BGP route that sent all request to a server that dropped them. A mistake caused the route to spread in the entire Internet infrastructure with a claim that the shortest path to YouTube servers is through the Pakistani server.

Governments are not the only ones that try to censor its citizens, but various criminal organizations can be behind such activities. An attempt was recorded in March, 2013 when a group of Egyptian military has stopped several individuals in their attempt to cut an underwater cable [19]. The event would have caused the loss of Internet in 50% of the Egyptian infrastructure. This shows that Internet availability can be threatened by a targeted attack to the existing infrastructure. There are countries which only have a small number of connections with other countries and if this connections are disabled it can take large amounts of time for them to be rebuilt, thus cutting entire countries Internet access.

There are other forms of oppression present, like the modification of DNS entries by governments that have control of these servers. This types of oppression are common even in advanced and democratic countries such as the U.S.A..

Building the technology or the infrastructure in such a way that no one can block the access of a different individual to the Internet is extremely difficult. It requires a large number of redundancies and putting the control of the infrastructure in the hands of the citizens. This is usually not possible. However in the case of ad-hoc networks, this can be achieved on a smaller scale. Communication between devices can still be guaranteed, even if the infrastructure that provides access to the Internet is stopped.

To make sure the Internet is available, and accessible, at high quality rates, we need to consider technologies that take into account all the scenarios presented above.

We need consider that probably no technology can bring 100% guarantees; even wireless connections can be jammed for instance, and all devices are currently dependent on electricity. To top all of this, we always need to consider the cost. A high cost directly translates in low adoption.

One would ask if the cost is really such a big problem, and argue that a person that cannot afford an Internet connection can probably also not afford a computer with which to utilize such a connection. This is not necessarily true, the cost of computing hardware has dropped dramatically in the past few years. Projects like the Raspberry Pi [20], have brought the size and the price of a computer down significantly. A unit is now not bigger than a credit card and it costs only 30$. This is an extremely small one time investment compared to the cost of an Internet connection that requires monthly recurring payments.

In the following Sections we will further present a discussion of the particular case of communication support for natural disasters. Such solutions could easily be adapted further to other cases (such as the ones presented here).

3 Requirements for Networks in the Case of Disaster Scenarios

In case of a disaster, a network needs to be able to operate effective even when parts or its entire infrastructure is no longer available. This requirement is the single most important one; the network could not be very fast, but it is this requirement which must hold in order to truly benefit the users. By users, we distinguish several categories. Also, we distinguish several periods after the disaster in which the network is used (which yields specific requirements on how the network needs to behave). The normal user wants to communicate with his close friends or family. This kind of usage is not as important as the communication between search and rescue teams. As such we can identify a few periods after a disaster in which we need to consider network needs separately.

The Search and Rescue Phase – This is the main phase after any disaster. Search and rescue teams are deployed, and the wreckage is searched to find and identify survivors. This is a critical period. The communication between search and rescue teams is essential, as well as communication with a center responsible with the command and control of coordination between rescue teams. If we consider a network that is shared between search and rescue teams as well as individuals trying to communicate between themselves, there is a need for special protocols that would give priority to the traffic created by the search and rescue teams. We also need to consider the data that needs to pass over the network. For example, text takes a smaller amount of bandwidth. Building a network that can transfer small amounts of data is simpler and cheaper, but the majority of today's communication protocols insert an additional header which might prove to be inadequate for tomorrow's small data chunks needing to be transfer (and this is also true for machine-to-machine control communication, in a more generic sense). However, using voice and even video can help the search and rescue dramatically. But such data pose different requirements on the communication protocols. So, could both types of data co-exist efficiently over the communication channels in place for disaster management? Do we need separate solutions to route control / text messages (a control plane) completely separated from a multimedia plane? This is a question many researchers are trying to answer today.

Data availability is also important. Sending the teams in dangerous environments can be extremely risky, and everything needs to be considered when making such a decision. For instance: video feeds from individuals that require extraction, but are stuck, can help identify the priority targets; having sensors can help determine anything from stability of a building to gas or fire levels to other potential risks or hazards. There are two things to be considered here: first is how to acquire the data, usually networks of sensors need to be spread before the incident; second is how to sustain the flow and filtration of all available data. Filtering the data is important as multiple sensors can indicate the same thing and will only overload the network without providing any extra information.

In this phase any infrastructure is most likely to be down. There are a number of reasons for this to happen. Wired connections can be broken. Repeaters or any devices that help extend the network can be down because of lack of electricity or even destroyed by electrical spikes. We can also consider a lack of available personnel to manage the infrastructure or to do any repairs.

It is also important to consider the destruction of Internet Service Providers server centers. These are probably the most expensive and hardest parts to replace. This process might also take a high amount of time.

The Reconstruction Phase – after the search and rescue phase, there is another period in which infrastructure is not yet available. The reconstruction of critical infrastructure is under way, but it can take a large period of time until everything becomes fully operational. The search and rescue teams have mostly finished their tasks and communication between them is not as big a priority as it was.

In recent years, as a practical use of Delay Tolerant Network and Mobile Opportunistic Network, disaster evacuation guidance effective against situations of large-scale urban disasters have been studied. In disaster situations, this navigation becomes disaster evacuation guidance navigating crowds of evacuees to the nearest refuges. Also, disaster information sharing is also available by opportunistic communication. For example, authors of [52, 53] have previously proposed the use of opnets for autonomously navigating crowds of evacuees to refuges, but also rapidly aggregating real-time disaster information. They propose the opportunistic collaboration between individual mobile nodes, using wireless communication, to assist with the reconstruction phase through guidance: evacuees naturally collect location information of impassable and congested roads in their smartphones by themselves and share it with nearby smartphones by opportunistic communication. Using the collected information, the guidance proposes an effective shortest-path based evacuation route to the nearest refuge avoiding known impassable and congested roads beforehand. By simulating a simple mathematical model of urban disaster scenarios, authors have actually shown numerically that the guidance reduces the average and maximum evacuation times even when the effects of congestion by evacuees is applied.

However, in the reconstruction phase there are several issues remaining to be solved as well. Everyone needs to communicate and reach as many other individuals as they can, make assurances to families and friends that everything is alright and start rebuilding what they lost. In this phase, hospitals and other centers can be overcrowded (network in such locations can be overused).

These are several places where, with a higher priority, the infrastructure needs to be restored fast. As infrastructure is restored, offering higher bandwidth, most people will try to use it to share whatever information they can, be it text, video or voice.

Finally, after this phase is completed, the infrastructure should be completely restored and the entire area struck by the disaster should go back to its standard functionality.

It is however important to minimize the time an area stays in reconstruction phase. The faster a network can be rebuilt the more content will the individuals using it be.

In summary, there should be a way to classify traffic and to offer QoS assurances to a number of individuals, like the search and rescue teams; the network should work with 0 or very little infrastructure and be easily extendable; reconstructing the original network should be as simple and as time and cost effective as possible.

To add to the above mentioned features we should also add that security can prove to be a huge concern. Individuals can try and disable the remaining network. Individuals can use information they gather on the network to do more harm. A network should be able to withstand any cyber-attack that can be used against it. This means the use of a number of security protocols and security considerations. The ability to remotely remove malicious individuals from the network could prove a huge benefit.

It is important to add that some individuals can create a high load on the network without even realizing they are doing a great decrement to everyone else. Take for instance a network with very low band-width and an individual that streams video that he considers is relevant for a search and rescue. This individual needs to be stopped without permanently blocking his access to the network.

4 Current Networking Alternatives for Disaster Scenarios

Wired networking solutions are the most popular to support disaster management operations at the moment, because of the high bandwidth they provide. Also, the security is higher, because the medium provides low direct access (it is more difficult to connect to the network simply because the access requires a physical connection). Furthermore, wired networking gives certain assurances to the user. Once connected, it is difficult for instance to lose connection, like the case with a Wi-Fi network in which movement inside a house can affect the performance of the connection and there can even be a complete loss of signal.

The price of a wired network is a complicated topic however. Deploying a temporary small area (inside a house or a room) network is extremely cheap because the devices and technology used have their price lowered by the popularity of the medium. However, deploying such a network on a large scale requires taking certain assurances that the cables used to form the connections are always secure and safe, regardless of changes in the location in which they have been deployed. Cables need to be put up on posts or underground to assure no one has direct access to them. Some need even more extensive protection against rodents or other such animals that can damage the cable.

We have to keep in mind that most wireless connection requires a wired backbone. If such a backbone is not available the price for a large scale wireless network rises because the technology to support large distance or to position directed antennas is nor as widely used. Wireless networks also have signals that overlap each other, for instance Wi-Fi 802.11b has only 4 non-overlapping frequencies available. This means that an entire Wi-Fi 802.11b network can only support 4 distinct connections before suffering a loss of performance. This means that building a large scale Wi-Fi Network requires either a strong wired backbone or special technology that can sustain high bandwidth connection between the center devices.

If bandwidth is not a strict requirement the deployment of a wireless only network is possible, this is proved by projects such as Loon [36]. Project Loon plans to put a fleet of balloons in the sky with wireless equipment and solar panels to provide power. Each balloon has a theoretical area limit of 40 km's. Deploying a large number of these devices can provide a cheap networking alternative even in places with sparse population. However these devices do not offer high bandwidth or low latency and require specialized equipment for every connecting node on the ground.

Further on, we will not discuss wired networks as they do not provide a cost effective way of deployment in the scenarios we are interested in. Furthermore, deployment can take a lot of time. We do mention these networks as they stand as backbone for most of the wireless alternatives we present.

The Public Safety/Security (PSS) sector currently relies on Professional Mobile Radio (PMR)/TETRA networks due to its benefits on high security and resilience. As a result, PSS organizations lack advanced communication functionality because TETRA is currently lagging behind the commercially available networks (3G, LTE, WIMAX etc.) in terms of data rate, speed, and coverage (see Table 1). According to Gartner, the trend towards a more data centric set of operational services in PSS services (e.g. image/video communication) will follow that of the wider commercial market, where data traffic has increased significantly. However, the current TETRA network is unsuitable to host most data centric applications and services [21] and would need to adopt wireless and mobile broadband networks.

Moreover, with more industries becoming reliant on mobile communications (as part of their operations and/or service offerings), there will be an increased demand for more complex mobile device functionality and application (e.g. GPS and other remote sensing applications). Apart from increased pressure on the network providers (as mentioned above), this will also put more pressure on the processing/energy resource of the mobile device which has a memory capacity and processing power limitation. This increases the need to intelligently manage mobile device resources and ensure resources are available when needed.

There is a need for a future communication network to cope with the projected industry growth of mobile and fixed connected devices and satisfy the needs and requirements across all industry sectors (including PSS). This need is in line with the

Digital Agenda Europe (DAE) [22], which aims to help Europe's citizens and businesses to get the most out of digital technologies. It particularly aligns with pillars II (Interoperability & Standards), IV (Fast and ultra-fast Internet access), and VII (ICT-enabled benefits for EU society); address the specific actions around telecoms and the Internet [23]. Hence, this future network will need to be:

- High-speed and secure, with high quality-of-service (QoS) and quality-of-experience (QoE) for users.
- Resilient, robust, flexible across spectrum and always available.
- Energy-efficient and capable of managing mobile device and network resource intelligently.

Public Safety and Security (PSS) organizations, such as law enforcement, ambulance services, fire services, and other civil emergency/disaster management services, are tasked with providing public safety and security service. Due to the nature of the services, mobile communication is a main requirement and Professional Mobile Radio (PMR) communication systems are extensively relied on to conduct critical operations.

Table 1. Characteristics of the various Communication Standards

Specification	Frequency	Spectrum	Coverage (Urban - Rural)	Data rate	Latency
TETRA	400, 800 MHz	5 - 20 MHz	5 - 15 Km	13 Kbps	250 ms
TEDS	400, 800 MHz	20 MHz	2 - 7 Km	20 - 80 Kbps	200 ms
GSM	900 MHz	35 MHz	40 - 100 Km	11.4 - 22.8 Kbps	800- 3000 ms
WI-FI	2.4, 5 GHz	20 - 40 MHz	5-20m (indoor) to 2 Km (Outdoor)	50 Mbps - 1 Gbps	5 - 20 ms
3G/UMTS	900, 1800, 2100 MHz	30 - 50 MHz	1.5 – 4 Km	64 - 384 Kbps	170 ms
HSDPA	1800, 2100 MHz	10 - 15 MHz	500m - 1.5 Km	1.8 - 7.2 Mbps	60 ms
WIMAX	700 MHz, (2.4, 3.5, 5.8) GHz	200 MHz	400m - 1Km	4 - 20 Mbps	30 ms
LTE	700, 1800, 2100 MHz	18 - 60 MHz	300 - 800 m	10 - 100 Mbps	10 ms

TETRA is the widely accepted communication network choice in Europe. PSS organizations cannot afford the risk of failures in their 'mission critical' communications (voice, data or video), hence they require: resilient and highly available infrastructure; reliable and secure communication; point-to-multipoint communication; large geographical coverage; and (recently) interoperability between different PSS organizations locally and across borders. This can only be ensured by a robust, secure and resilient mobile and fixed communication network infrastructure. Also, for these organizations to be adequately prepared to tackle any future event like those previously witnessed (September 11th World Trade Centre attack, the Atocha

(Madrid) bombing, the London underground attack, the major earthquake in Van, Turkey), they need to be properly equipped.

There are needs for new advanced applications envisioned in the next generation PSS communication such as:

- Remote sensor networks (personnel monitoring, forest fire tracking or water/flood level monitoring).
- Multi-functional mobile terminals (real-time video, biometric data, ID verification, image transfer).
- Remote database access, device control, etc.

However, with the growing demand for resilient, reliable and secure high-speed data communication, the current capacity for PSS communication network (i.e. TETRA) will be exceeded. It is considered likely that an upgrade or replacement in current network will be required across Europe in the next 5-10 years. TETRA is constantly being evolved by ETSI (European Telecommunications Standards Institute) and new features are being introduced to fulfill the growing PSS requirements. Like GSM moving to GPRS, EDGE, 3G/UMTS, and now LTE, TETRA attempted to evolve to satisfy increasing user demand for new data services by upgrading the original TETRA standard (TETRA 1, which had less emphasis on data) to TETRA 2 (TETRA Enhanced Date Service - TEDS). However, demand growth for frequencies and more data intensive applications is likely to go beyond TEDS's capacity and will not satisfy future needs for the essential services highlighted.

After TETRA 2, an attempt has been made to continue working on TETRA 3 (also called TETRA Broadband, DAWS and MESA), but the development of this standard has officially been cancelled [24]. This is because PSS network manufacturers and operators cannot afford the several billion euro research and development budgets to develop next generation PMR mobile radio network in parallel to or even ahead of commercial networks. Even if they could, it may take as long as 10 years to plan and deploy such network [25]. Hence, there is the need to extend the capabilities of PSS communication by interfacing with commercial communication networks.

A new ETSI working group has started working on the standardization of mission-critical broadband communications, whereby LTE has been opted for as the basis for this standard. This is because of the increasingly rapid progress in the capability of communication technologies deployed in the commercial network, particularly with regard to over-the-air data rates and the spectrum efficiency that can be achieved. This has seen the rise of an increasing gulf between the capabilities of commercial networks and dedicated PSS networks. However, replacing TETRA with LTE is not a short term option, for many reasons. First of all, LTE does not yet support the voice communication features of TETRA and secondly, there is a huge TETRA infrastructure that cannot be replaced rapidly. This infrastructure needs to be used effectively.

There has been strong research effort in the last decade on the development and integration of new wireless access technologies for mobile Internet access. This is considered by many experts to be the de-facto direction for future researches to support disaster management, as mobile phones and tabletPCs become more commonly encountered everywhere, and their communication capabilities could be used to sustain mobile wireless networks in support for rescue procedures. But, still, such wireless network will need to work with wired backbones, so researchers investigate various means to create next-generation heterogeneous networks. Among the main research concepts for taking advantage of the availability of various heterogeneous networking technologies in place, Always Best Connected (ABC) and Quality of Experience (QoE) Bandwidth Aggregation concepts have been at the center of attention.

- *Always Best Connected* implies that end-users expect to be able to connect anytime, anywhere – also when on the move – by their terminal of choice. End-users also expect to be able to specify in each situation whether "best" is defined by price or capability. However, the current state-of-the-art solutions, such as IETF Mobile IPv6 (MIP) or the emerging Host Identity Protocol (HIP), mainly focus on mobility management, instead of considering additional user-related issues, such as user preferences, associated cost, access-network operator reputation, and trust and mainly application-related issues like (Quality of Service) QoS and failure recovery in conjunction with mobility.
- *Quality of Experience (QoE)* reflects the collective effect of service performances that determines the degree of satisfaction of the end-user, e.g. what the user really perceives in terms of usability, accessibility, retain-ability and integrity of the service. Seamless communications is mostly based on technical Network QoS parameters so far, but a true end-user view of QoS is needed to link between QoS and QoE. While existing 3GPP or IETF specifications describe procedures for QoS negotiation, signalling and resource reservation for multimedia applications (such as audio/video communication and multimedia messaging, support for more advanced services, involving interactive applications with diverse and interdependent media components) is not specifically addressed. Additionally, although the QoS parameters required by multimedia applications are well known, there is no standard QoS specification enabling to deploy the underlying mechanisms in accordance with the application QoS needs.

One of the early attempts to provide all-IP architecture and integrate different access technologies for public safety communications was by the project MESA (Mobility for Emergency and Safety Applications), an international partnership project by ETSI and TIA dating back to 2000 [26]. A.K. Salkintzis proposed a solution for integrating WLAN and TETRA networks that fits to the all-IP architecture of MESA and allows TETRA terminals to interface the TETRA infrastructure over a broadband WLAN radio access network instead of the conventional narrowband TETRA radio network, while remaining fully interoperable with conventional TETRA terminals and services. Chiti et al. [27] proposed a wireless network that aims to interconnect several heterogeneous systems and provide multimedia access to groups of people for disaster management. The authors address the issues of heterogeneous network interconnection, full and fault tolerant coverage of the disaster area, localization to enable an efficient coordination of the rescue operations, and security. The focus of this work is on the use of WiMAX-based wireless network as a backbone to provide reliable and secure multimedia communications to operators during the disaster management. Also, Durantini et al. [28] present a solution for interoperability and integration among Professional Mobile Radio systems (TETRA and Simulcast), public systems (GSM/GPRS/UMTS), and broadband wireless technologies, such as WiMAX, with the aim of enabling distributed service provisioning while guaranteeing always best connection to bandwidth demanding applications provided by an IP-based core network. Furthermore, the authors address the issue of optimizing the quality of service management in a multi-network environment, and propose a QoS mapping between WiMAX QoS classes and TETRA service typologies.

Alcatel-Lucent recently demonstrated at TETRA World Congress a very interesting use case for TETRA communications over an LTE network. A Standardized Digital Professional Mobile Radio systems user with a TETRA client application running on an LTE rugged terminal, can communicate with other TETRA users over an LTE network, while using all of the TETRA services. This option opens the way to TETRA/LTE hybrid solutions combining the best features of the two technologies to provide broadband overlay services to existing TETRA networks. There is a multitude of other similar work focusing on the integration of various network technologies in and out of the scope of public safety communications. However, solutions available to date are fragmented and each considers only a subset of the ideal QoE-aware and autonomous connectivity solution that can also simultaneously exploit all available network interfaces. During large scale emergencies and disasters, it is crucial to aggregate the scarce communication resources of multiple technologies and be able to use simultaneously, since the left-over capacity of a single technology may suffer due to infrastructural damages.

An alternative consists in the incorporation of Multipath TCP into the wireless communication world. The transmission control protocol (TCP), which serves as the data transport basis of many telecommunication services of today, was designed to work on single links and does not cope well with the simultaneous use of multiple links at the same time. A survey of TCP performance in heterogeneous networks [29] shows the existing solutions to date and their problems. Magalhaes et al. present a solution for channel aggregation at the transport layer, called R-MTP (Reliable Multiplexing Transport Protocol), which multiplexes data from a single application data stream across multiple network interfaces [30]. The recent EU-funded 'Trilogy' project introduced the Multipath TCP (MPTCP) solution, towards enabling the simultaneous use of several paths by a modification of TCP that presents a normal TCP interface to applications, while in fact spreading data across several subflows [31]. An IETF working group has been formed to develop the MPTCP protocol, which is an on-going effort. However, through extensive evaluation studies over MPTCP, some authors [32] report that heterogeneous network environment (Ethernet, Wi-Fi and 3G) has a great impact on MPTCP throughput and reveals the need of an intelligent algorithm for interface selection in MPTCP.

In terms of security, the Terrestrial Trunked Radio (TETRA) supports two types of security: air-interface security and end-to-end security. Air-interface security [33] protects user's identity, signaling, voice and data between mobile station (MS) and base station (BS). It specifies air-interface encryption, (mutual) authentication, key management (OTAR: over-the-air-rekeying) and enable/disable functionality. End-to-end security [34] encrypts the voice from MS to MS. Current candidates as encryption algorithms are IDEA (owned by MediaCrypt AG) and AES as the encryption schemes.

One of the main challenge for multi-technology communication is the compatibility problem between the security mechanisms (encryption, authentication, integrity and key management) supported by these technologies. Wireless LAN supports various security mechanisms, uses of which are mostly optional. MAC address filtering and hidden service set identifier (SSID) are the simplest techniques. Today very few access points use Wired Equivalent Privacy (WEP) because many cracking tools are publicly available on Internet. Wi-Fi Protected Access (WPA and WPA2 based on 802.11i) are introduced to overcome this problem but weak passwords are still a problem. 802.1x defines the encapsulation of the Extensible Authentication Protocol (EAP), and enables authentication through third party authentication servers such as Radius and Diameter. End-to-end security can be provided by use of Internet Protocol Security (IPSEC), Transport Layer Security (TLS), Secure Sockets Layer (SSL), Secure Shell (SSH), pretty Good Privacy (PGP), etc. Security of GSM and 3G suffers from similar compatibility problems with TETRA. GSM security defines Subscriber Identity Module (SIM), the MS, and the GSM network. SIM hosts subscribe authentication key (K), Personal Identification Number (PIN), key generation algorithm (A8) and authentication algorithm (A3). MS contains the encryption algorithm (A5) for air interface. Encryption is only provided for the air-interface. 3G security builds upon the security of GSM. It addresses the weaknesses in 2G systems with integrity and enhanced authentication as well as with enhanced encryption using longer keys and stronger algorithms.

Another challenge in multi-technology communication is that most of the security mechanisms are optional, and they are maintained based on the policies of different administrative domains. An end-to-end connection between two MS may go through an unsecure public network which may permit in variety of attacks including denial-of-service and man-in-the-middle. Cost of mitigating these attacks on MS side may be higher than the benefit of the connection in terms of Quality of Service (QoS) and Quality of Experience (QoE) metrics. Therefore, QoS and QoE mechanisms must involve related metrics to provide predictable security service levels to the end users [35].

5 Ad-Hoc Networking Alternatives

Worldwide communication networks are growing exponentially, with close to 80 per cent of the world's population now enjoying access to a mobile phone [37]. And with more mobile devices than people in 97 countries around the world, the mobile communication industry in particular is constantly evolving and growing at a rapid pace. This is due to the impact made by exciting new devices including iPhone, Android, Windows phone and tablets [38]. Besides the volume of devices in the market, the data consumption is also surging, with a recent Ericsson Mobility Report [39] estimating that *mobile data traffic doubled in just one year between 2011 and 2012*. It also estimated that *by 2018, demand for mobile data will increase by a factor* *of 12.*While this exceptional pace of growth is exciting, it also presents a whole new set of challenges as network service providers need to invest in improving network speed, flexibility, availability, utilization and efficiency.

From mobile phones, tablets, laptops and devices such as smart appliances/meters, it is predicted that *more than 50 billion devices will be connected to the web by 2020* [40] as the 'Internet of Things' concept develops further. It is estimated that there are currently approximately 14 billion devices connected to the Internet. Hence, the forecast is predicting this number to almost quadruple by 2020 with *social network, commerce, transport, public safety/security, entertainment and utilities industries being dependent on wireless and mobile broadband services and having increased data usage* [41].

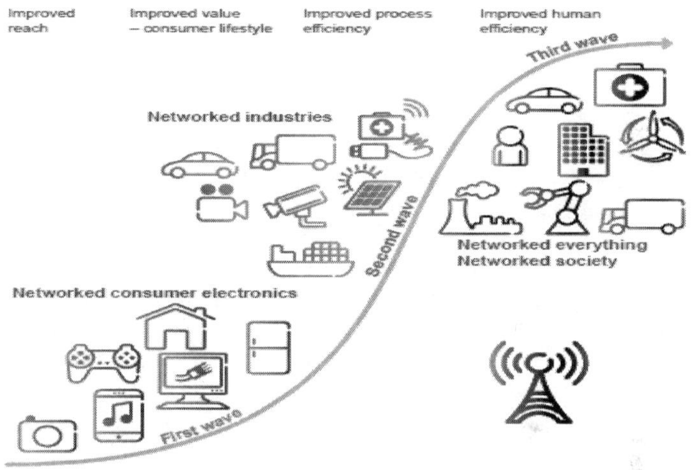

Fig. 1. The Development Waves of Connected Devices

Mobile Ad Hoc Networks are multi hop networks where nodes can be stationary or mobile and they are formed on a dynamic basis. They allow people to perform tasks efficiently by offering unprecedented levels of access to information. In mobile ad-hoc networks, topology is highly dynamic and random and in addition, the distribution of nodes and their capability of self-organizing play an important role. Their main characteristics can be summarized as follows:

- Topology is highly dynamic and frequent changes in the topology may be hard to predict.
- Based on wireless links, this may have a lower capacity than their wired counterparts.
- Physical security is limited due to the wireless transmission.
- Affected by higher loss rates and can present higher delays/jitter than fixed networks due to wireless links.

- Nodes rely on batteries or other exhaustible means for their energy. As a result, energy savings are an important system design criterion.
- Furthermore, nodes have to be power-aware: the set of functions offered by a node depends on its available power (CPU, memory, etc.).

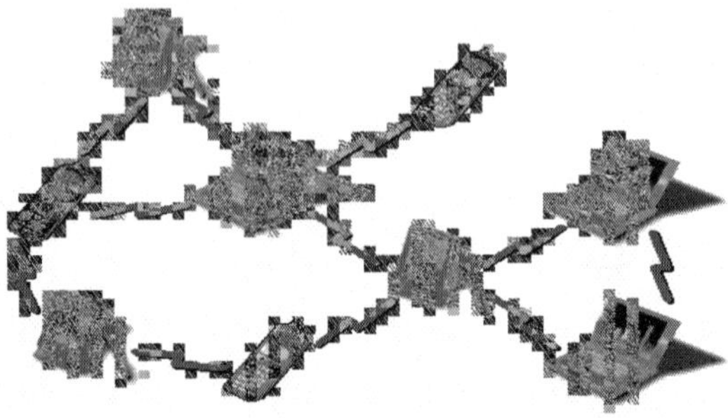

Fig. 2. Mobile Ad-Hoc Network

A well-designed architecture for mobile ad-hoc networks involves all networking layers, ranging from the physical to the application layer. Power management is of paramount importance and general strategies for saving power need to be addressed, as well as adaptation to the specifics of nodes of general channel and source coding methods of radio resource management and multiple accesses. In mobile ad-hoc networks, with the unique characteristic of being totally independent from any authority and infrastructure, there is a great potential for a user-centric network (users in control). Two or more users can become a mobile ad-hoc network simply by being close enough to meet the radio constraints, without any external intervention.

Fig. 3. Infrastructure Based Wireless Network

Delay Tolerant Networks also known as DTNs are an important branch of Ad-Hoc Networks. These networks work on the assumption that nodes in the network are sparse and very mobile. Here we use nodes based on a probability that the other node will be able to transmit the package, without any assurances. As such a store-and-forward policy is needed.

A node stores messages that do not have itself as a destination and collects messages for the destinations it has a high probability to encounter in the near future. When a destination or a node with a bigger probability of success is reached the packages are forwarded to this other node.

Fig. 4. Delay Tolerant Network

It is important to not see DTN networks as a standalone network. At any point in the process a node can choose to forward the packets to an existing backbone. This is usually done on a node by node basis, as the cost of forwarding to a backbone can vary greatly. Because of this property a DTN network can even be used to join two or more distinct pieces of Network that have lost connection between them because of a natural disaster or a similar incident.

A DTN has also an extremely important property of being very low cost. Being based on devices that are already ubiquitous in most areas and having the possibility to be deployed using inexpensive computers like the Raspberry Pi the network can function without any other infrastructure. There are papers like the case of TRIAGE [49] that try to offer a delay tolerant solution specifically for disaster scenarios. Furthermore DTN networks can be used to deliver important sensor data in case of a disaster scenario [51] even if the infrastructure normally used to gather this data is down.

There are already a large number of proposals for algorithms that work in this manner and studying the complex requirements of each of them is beyond the scope of this paper. **Epidemic** [42] tries to get the biggest number of packets to their destination, to achieve this it sends each packet to all the nodes that it encounters if the node has not yet received that packet. This assures that Epidemic has the highest cost and delivery rate from all the DTN algorithms. **Wait** [43] does the complete opposite of Epidemic, it sends packets only to the destination. As such the cost of delivering all packets is equal to the delivery ratio and they are both low. Both Epidemic and Wait are algorithms that are not meant to be used in real life scenarios: Epidemic has a very high cost and assumes that devices have unlimited memory resources while Wait has a delivery ratio that is unacceptably low in most scenarios.

Multiple-Copy-multiple-hoP [44] is a version of Epidemic in which the packets are sent only for a number of hops and only a limited number of copies are created every time. This algorithm can provide a well enough comparison for any others as its performance is acceptable for real-life scenarios.

dLife [45] assumes a pattern in the way nodes encounter each other. Using this assumption daily statistics are calculated and nodes receive a social weight. **Rank** [44] makes the assumption that some nodes are more popular than others and packets are forwarded to these nodes in hope that it has a higher chance of meeting the destination. In our experience Rank has extremely good results in both delivery ratio and cost.

Label [46] assumes that nodes are grouped in communities and to forward a packet it is first important to have it reach the community of the destination. Label does require a way to determine the community, before the packets are forwarded, this can be done in a static manner or calculated while the network is being used. **BUBBLE Rap** [44] makes a merger between rank and label and tries to forward packets so that they both reach the community they are intended for and the most popular nodes in each community. There are 2 versions the initial one, A, and the modified version, B, that removes a packet after it was forwarded.

PRoPHET [47] uses contact history to compute the probability of encountering another node. This metric is called delivery predictability and one such value should exist for any 2 node combination in the network, although not all values need to be stored. This value is then used to send a packet on an increasing probability path to the destination.

PROPICMAN and **CiPRO** [48] use context information to determine the path, any data is useful like information of where an individual carrying a node lives or where he works. This information is not always available and people might be reluctant to share it.

To evaluate the feasibility of such opportunistic routing algorithms, for disaster management, we developed the CCPAC simulator (http://ccpac.hpc.pub.ro/). Furthermore, we used the simulator in a series of realistic experiments, where we compared all previously mentioned algorithms.

The experimental scenario used for this case was a Random waypoint simulation, in which nodes where moving randomly. We have conducted similar tests on real life traces and the results are very similar, with only small variations in algorithms performance.

These figures should offer a comparison view of the presented algorithms, they all have their individual stats and performance issues and choosing one of them should take into account the scenario in which they should be used.

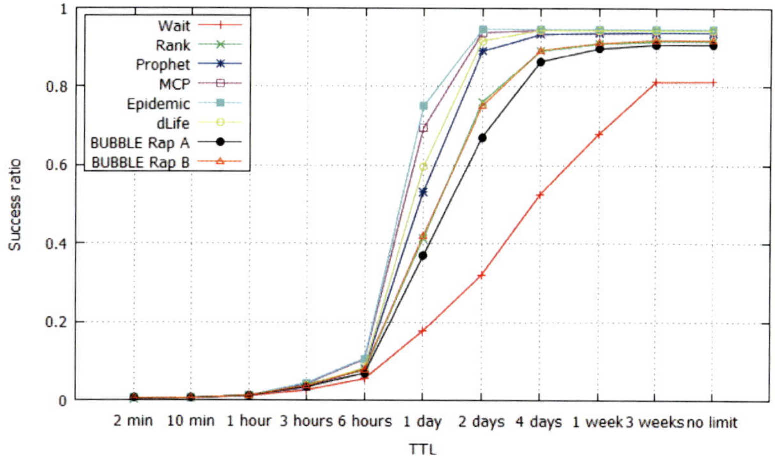

Fig. 5. Random Waypoint Delivery Ratio

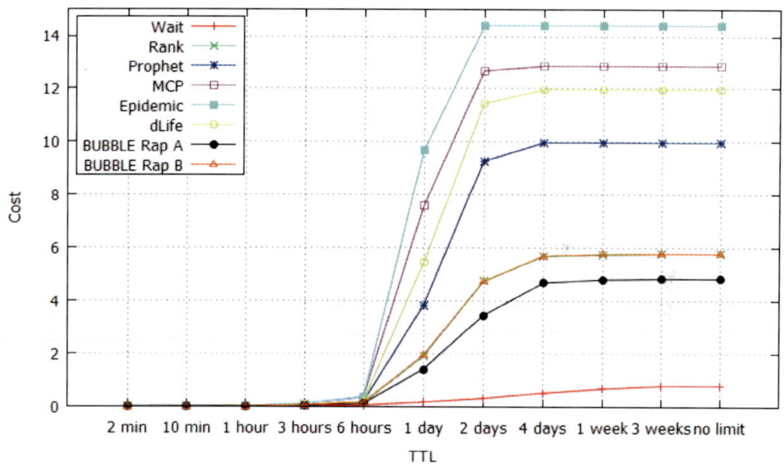

Fig. 6. Random Waypoint Cost

6 Conclusion

In this paper we have presented a number of scenarios in which building and sustaining networking infrastructure can be a complex matter. We have presented the existing types of networking solutions available at the moment and we have discussed their advantages/disadvantages in these scenarios.

We strongly believe that the best way to insure functionality in extreme scenarios and especially after natural disasters technologies that assure operability between different networking systems [51] are needed. This is probably the best way to assure simple, fast and efficient deployment of an infrastructure.

DTNs have proven to be an emerging trend that is stimulated by the mass usage of mobile devices such as smart phones. These networks are extremely resilient because they do not require any infrastructure. Furthermore one can use a DTN network to insure connection between different networks that, during a disaster, or from some other cause have lost connectivity.

One can envision many scenarios in which these networks and interoperability between them can lead to a safer world.

References

1. Chen, L.-D., et al.: Small business Internet commerce: a case study. Information Resources Management Journal (IRMJ) 3, 17–41 (2003)
2. Chen, K., Tarn, J.M., Han, B.T.: Internet dependency: its impact on online behavioral patterns in e-commerce. Human Systems Management 23(1), 49–58 (2004)
3. Böhmer, M., et al.: Falling asleep with angry birds, facebook and kindle: a large scale study on mobile application usage. In: Proceedings of the 13th International Conference on Human Computer Interaction with Mobile Devices and Services, ACM (2011)
4. Ulusu, Y.: Determinant factors of time spent on Facebook: brand community engagement and usage types. Journal of Yasar University 18(5), 2949–2957 (2010)
5. Curran, K., Morrison, S., Cauley, S.M.: Google+ v Facebook: The Comparison. TELKOMNIKA (Telecommunication, Computing, Electronics and Control) 10(2), 379–388 (2012)
6. Java, A., et al.: Why we twitter: understanding microblogging usage and communities. In: Proceedings of the 9th WebKDD and 1st SNA-KDD 2007 Workshop on Web Mining and Social Network Analysis. ACM (2007)
7. Duncombe, C.: The twitter revolution? Social media, representation and crisis in Iran and Libya. In: Australian Political Science Association Conference (APSA 2011), School of Politics & International Relations (2011)
8. Guha, S., Daswani, N.: An experimental study of the skype peer-to-peer voip system. Cornell University (2005)
9. Atzori, L., Iera, A., Morabito, G.: The Internet of things: A survey. Computer Networks 54(15), 2787–2805 (2010)
10. Shapiro, J.M.: Smart cities: quality of life, productivity, and the growth effects of human capital. The Review of Economics and Statistics 88(2), 324–335 (2006)
11. Caragliu, A., Del Bo, C., Nijkamp, P.: Smart cities in Europe. Vrije Universiteit, Faculty of Economics and Business Administration (2009)
12. Mahizhnan, A.: Smart cities: the Singapore case. Cities 16(1), 13–18 (1999)
13. Fuller, S.S.: Internet connectivity for hospitals and hospital libraries: strategies. Bulletin of the Medical Library Association 83(1), 32 (1995)
14. Armbrust, M., et al.: A view of cloud computing. Communications of the ACM 53(4), 50–58 (2010)
15. Oyelaran-Oyeyinka, B., Lal, K.: Internet diffusion in sub-Saharan Africa: A cross-country analysis. Telecommunications Policy 29(7), 507–527 (2005)
16. Woodcock, B.: Packet Clearing House. Overview of the Egyptian Internet Shutdown (2011)
17. Clayton, R.C., Murdoch, S.J., Watson, R.N.M.: Ignoring the great firewall of china. In: Danezis, G., Golle, P. (eds.) PET 2006. LNCS, vol. 4258, pp. 20–35. Springer, Heidelberg (2006)

18. Bailey, M., Labovitz, C.: Censorship and Co-option of the Internet Infrastructure, Ann Arbor, vol. 1001, p. 48104 (2011)
19. (June 21, 2013) http://english.ahram.org.eg/NewsContent/1/64/67894/Egypt/Politics-/Egypts-military-stops-attempt-to-cut-Internet-cabl.aspx (last accessed June 20, 2013)
20. Upton, E., Halfacree, G.: Meet the Raspberry Pi. Wiley (2012)
21. Feature: Airwave 'Could Be Replaced', Police Oracle (February 19, 2013), http://www.policeoracle.com/news/Police+IT+and+Technology/2013/Feb/19/Feature-Airwave-Could-Be-Replaced_61319.html (last accessed June 20, 2013)
22. Digital Agenda Europe, http://ec.europa.eu/digital-agenda/digital-agenda-europe (last accessed June 20, 2013)
23. Telecoms Internet, Europe, http://ec.europa.eu/digital-agenda/telecoms-Internet-0 (last accessed June 20, 2013)
24. TETRA standard and applications, http://www.tetra-applications.com/item.html&objID=15774 (last accessed June 20, 2013)
25. The evolution of TETRA, white paper, http://www.p3-group.com/downloads/4/1/7/5/P3_-_Evolution_of_TETRA_-_White_Paper_-_v1.0.pdf (last accessed June 20, 2013)
26. The MESA project, http://www.projectmesa.org (last accessed June 20, 2013)
27. Chiti, F., Fantacci, R., Maccari, L., Marabissi, D., Tarchi, D.: A broadband wireless communications system for emergency management. Wireless Communications 15(3), 8–14 (2008)
28. Durantini, A., Petracca, M., Vatalaro, F., Civardi, A., Ananasso, F.: Integration of Broadband Wireless Technologies and PMR Systems for Professional Communications. In: Fourth International Conference on Networking and Services, ICNS 2008, March 16-21, pp. 84–89 (2008)
29. Barakat, C., Altman, E., Dabbous, W., "On, T.C.P.: performance in a heterogeneous network: a survey. IEEE Communications Magazine 38(1), 40–46 (2000)
30. Magalhaes, L., Kravets, R.: Transport level mechanisms for bandwidth aggregation on mobile hosts. In: Ninth International Conference on Network Protocols 2001, November 11-14, pp. 165–171 (2001)
31. Barré, S., Paasch, C., Bonaventure, O.: MultiPath TCP: From Theory to Practice. In: Domingo-Pascual, J., Manzoni, P., Palazzo, S., Pont, A., Scoglio, C. (eds.) NETWORKING 2011, Part I. LNCS, vol. 6640, pp. 444–457. Springer, Heidelberg (2011)
32. Nguyen, S.C., Nguyen, T.M.T.: Evaluation of multipath TCP load sharing with coupled congestion control option in heterogeneous networks. In: Global Information Infrastructure Symposium (GIIS 2011), August 4-6, pp. 1–5 (2011)
33. Terrestrial Trunked Radio (TETRA), Voice plus Data (V+D), Part 7: Security, ETSI EN 300 392-7 V3.2.1 (2010-06)
34. Terrestrial Trunked Radio (TETRA), Direct Mode Operation (DMO), Part 6: Security, ETSI EN 300 396-6 V1.4.1 (2010-07)
35. Spyropoulou, E., Agar, C., Levin, T., Irvine, C.: IPsec Modulation for Quality of Security Service. In: Proceedings of the International Systems Security Engineering Conference (2002)
36. The Loon project (June 24, 2013), http://www.google.com/loon (last accessed June 20, 2013)
37. International Telecommunication Union (October 2010)

38. The new mobile world order, http://www.cisco.com/web/about/ac79/docs/sp/New-Mobile-World-Order.pdf (last accessed June 20, 2013)
39. Erricson, Mobility report, http://www.ericsson.com/res/docs/2012/ericsson-mobility-report-november-2012.pdf (last accessed June 20, 2013)
40. 50 billion devices online by 2020, The Telegraph (January 31, 2012), http://www.telegraph.co.uk/technology/Internet/9051590/50-billion-devices-online-by-2020.html (last accessed June 20, 2013)
41. Erricson, http://www.ericsson.com/res/docs/whitepapers/wp-50-billions.pdf (last accessed June 20, 2013)
42. Vahdat, A., Becker, D.: Epidemic routing for partially connected ad hoc networks. Technical Report CS-200006, Duke University (2000)
43. Spyropoulos, T., Psounis, K., Raghavendra, C.S.: Single-copy routing in intermittently connected mobile networks. In: Proc. of First Annual IEEE Communications Society Conference on Sensor and Ad Hoc Communications and Networks (SECON 2004), pp. 235–244 (October 2004)
44. Hui, P., Crowcroft, J.: BUBBLE Rap: Social-Based Forwarding in Delay-Tolerant Networks. IEEE Transactions on Mobile Computing 10(11), 1576–1589 (2011)
45. Moreira, W., Mendes, P., Sargento, S.: Opportunistic routing based on daily routines. In: Proc. of IEEE Int. Symp. World of Wireless, Mobile and Multimedia Networks (WoWMoM 2012), pp. 1–6 (June 2012)
46. Hui, P., Crowcroft, J.: How Small Labels Create Big Improvements. In: Proc. of the 5th IEEE International Conference on Pervasive Computing and Communications Workshops (PERCOMW 2007), Washington, DC, USA, pp. 65–70 (2007)
47. Lindgren, A., Doria, A., Schelen, O.: Probabilistic Routing in Intermittently Connected Networks. SIGMOBILE Mob. Comput. Commun. Rev. 7(3), 19–20 (2003)
48. Anh Nguyen, H., Giordano, S., Puiatti, A.: Probabilistic Routing Protocol for Intermittently Connected Mobile Ad hoc Network (PROPICMAN). In: Proc. of IEEE Int. Symp. on World of Wireless, Mobile and Multimedia Networks (WoWMoM 2007), pp. 1–6 (June 2007)
49. Luqman, F.: TRIAGE: Applying context to improve timely delivery of critical data in mobile ad hoc networks for disaster response. In: Pervasive Computing and Communications 2011 IEEE International Conference on Workshops (PERCOM Workshops), March 21-25, pp. 407–408 (2011)
50. Yu, F.R., Jie, Z., Tang, H., Chan, H.C.B., Leung, V.C.M.: Enhancing interoperability in heterogeneous mobile wireless networks for disaster response. IEEE Transactions on Wireless Communications 8(5), 2424–2433 (2009)
51. Cayirci, E., Coplu, T.: SENDROM: sensor networks for disaster relief operations management. Wireless Networks 13(3), 409–423 (2007)
52. Fujihara, A., Miwa, H.: Real-time Disaster Evacuation Guidance using Opportunistic Communications. In: The 2012 IEEE/IPSJ International Symposium on Applications and the Internet (SAINT 2012), pp. 326–331 (2012)
53. Fujihara, A., Miwa, H.: Effect of Traffic Volume in Real-time Disaster Evacuation Guidance using Opportunistic Communications. In: Third International Conference on Intelligent Networking and Collaborative Systems (INCoS 2012), pp. 457–462 (2012)

QoS in Next Generation Mobile Networks: An Analytical Study

Javier Carmona-Murillo, José-Luis González-Sánchez, David Cortés-Polo, and Francisco-Javier Rodríguez-Pérez

GITACA Research Group, University of Extremadura, Spain

Abstract. Two of the major challenges for next-generation mobile systems are to achieve seamless mobility management in next generation wireless networks and to manage resources efficiently given the exponential growth that mobile data traffic has experienced over the last few years. To track host mobility, the IETF has made important efforts to develop mobility management protocols such as Mobile IPv6 and Proxy Mobile IPv6. These protocols establish a tunnel to connect the mobile node with its correspondent node. The tunneling method provided by MPLS can be profitably used to take advantage of MPLS traffic engineering capabilities in order to achieve faster re-routing when a mobile node changes its point of attachment to the network. Moreover, in order to deal with increasing mobile traffic demand, mobility management network architectures are being redesigned towards a more distributed operation. Given these scenarios, service disruption during handoffs continues to cause excessive packet loss that needs minimizing in order to support quality of service requirements for emerging applications. In this paper, a qualitative and quantitative analyses of the most representative host-based and network-based mobility management approaches is presented, including recent distributed mobility management approaches.

1 Introduction

The design of Next Generation Wireless Networks (NGWN) has two main goals. First of all, the possibility of maintaining connectivity while a user moves among heterogeneous networks. Secondly, the ability to provide a similar level of QoS (Quality of Service) while the node moves between these networks [1]. In order to achieve the first goal, the Internet Engineering Task Force (IETF) designed Mobile IPv6 [2] and Proxy Mobile IPv6 [3] to overcome the problems caused by handover in heterogeneous networks. Mobile IP (MIP) and Proxy Mobile IPv6 (PMIP) are widely accepted as the most appropriate protocols for addressing IP mobility management in future wireless mobile networks.

The second problem, related to the provisioning of enough network resources, has largely been studied in both wired and wireless environments. There are three general models used to provide network resources for quality of service (QoS) guarantees in the Internet: integrated services (IntServ), differentiated services (DiffServ) and MPLS (Multi-Protocol Label Switching) [4]. IntServ can provide quantitative QoS

C.X. Mavromoustakis et al., *Resource Management in Mobile Computing Environments*, Modeling and Optimization in Science and Technologies 3,
DOI: 10.1007/978-3-319-06704-9_2, © Springer International Publishing Switzerland 2014

guarantees to individual flows and DiffServ can provide qualitative QoS guarantees to multiple flows in an aggregate way. For its part, MPLS is a QoS technology with traffic engineering (TE) introduced to enhance the performance of the Internet's datagram model in terms of both management and delivery [5].

The integration of Mobile IP and Multiprotocol Label Switching has worked successfully due to the ability of MPLS to engineer efficient traffic tunnels, including constraint-based routing, survivability and recovery, thus avoiding congestion and enabling an efficient use of the available bandwidth. These features highlight the potential of MPLS for solving MIP's operational and architectural shortcomings such as: high handoff latency [5] and packet loss or high global signaling load and scalability issues. MPLS could also be viewed as a tunneling technology that overcomes the tunneling techniques proposed in Mobile IP standard. For this reason, it has been proposed to use both technologies together [7-10].

Moreover, these mobility management schemes developed for IP and cellular networks rely on a centralized mobility anchor entity, responsible for both the mobility signaling and user data forwarding. This traditional centralized approach is likely to have several issues or limitations which require costly network dimensioning and engineering to resolve. The IETF (Internet Engineering Task Force) has recently chartered the distributed mobility management (DMM) working group [11] with identifying limitations in the existing IP mobility support protocols and developing distributed mobility management protocols based on the existing IP mobility support protocols such as MIPv6 and PMIPv6.

This chapter presents qualitative and quantitative analyses on mobility management protocols and MPLS integration. In addition, the objective of this chapter is to provide a comprehensive comparison of the main alternatives and addresses the most important strong and weak points of each of these well-known mobility support protocols through numerical results. The rest of the chapter is organized as follows. In section 2, we present background knowledge about mobility management protocols. We have developed an analytical model used to derive the signaling cost function of registration update, link usage and packet loss. This is presented in section 3. In section 4 the numerical results are shown. Finally, concluding remarks are given in section 5.

2 Mobility Management in the Internet

Mobility management in the Internet is a key aspect of mobile communications and is the next step in the Internet evolution. It is practical now for a mobile node to roam between different access technologies and, in addition, it is reasonable to expect address continuity and session persistence across these handoffs. Anticipating these requirements, Mobile IPv6 protocol has been developed by the IETF. It is a host-based mobility management protocol requiring the participation of the host in all aspects of mobility management. An alternative approach is network-based mobility management protocols, where the host does not participate in any mobility related signaling. The main network-based protocol, recently developed by the IETF, is Proxy Mobile IPv6.

2.1 Host-Based Schemes

Until now, Mobile IP is the most representative mobile management scheme developed by the IETF on the way towards next generation mobile networks.

Mobile IPv6 allows nodes to remain reachable while moving around in IPv6 networks. Without specific support for mobility, packets destined to a mobile node would not be able to reach it while the mobile node is away from its home link. In order to continue communication in spite of its movement, a mobile node could change its IP address each time it moves to a new link, but the mobile node would then not be able to maintain transport and higher-layer connections when it changes location. Fig. 1 illustrates an overview of Mobile IPv6 and its basic terminology. Next, a brief description of the protocol is given.

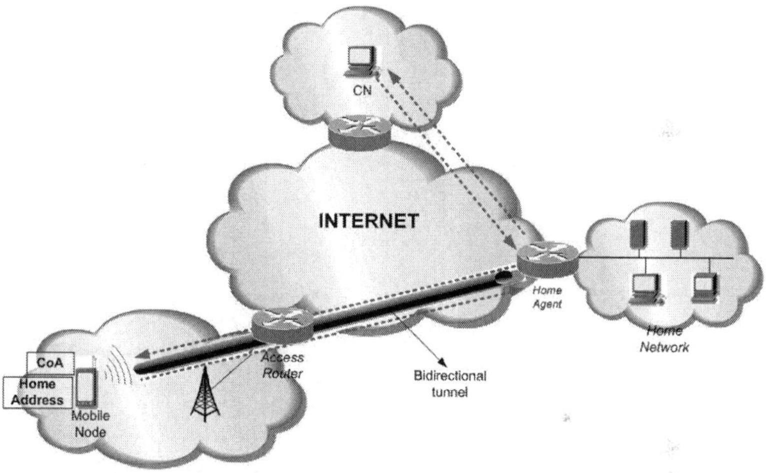

Fig. 1. Overview of Mobile IPv6

Mobile IPv6 basic operation is as follows. The Mobile Node (MN) establishes a connection with the Correspondent Node (CN). A Home Agent (HA) serves as the anchor node in the Home Network that tracks the network connection point (location) of a user as the user moves. Periodically, or whenever the user changes their point of attachment to the network, the user registers with the HA through Binding Update (BU) messages, informing of the user's current location and establishing a tunnel (IP-in-IP or GRE) between the HA and the MN located in a visited network. With this registration, the MN obtains a new address, called Care of Address (CoA), that belongs to the foreign network. The Home Agent is the critical part of the system since it is on the critical path of both signaling and data for mobile users.

The other alternatives based on Mobile IP, which integrate mobility and MPLS, are Mobile MPLS [7], FH Micro Mobile MPLS [8] and LinkWork Mobile MPLS [12]. All these schemes are briefly explained next.

Mobile MPLS was one of the first proposals to integrate Mobile IP and MPLS protocols. It aims to improve the scalability of the Mobile IP data forwarding process by removing the need for IP-in-IP tunneling from Home Agent (HA) to Foreign Agent (FA) using Label Switched Paths (LSPs).

FH Micro Mobile MPLS overcomes some limitations of Mobile MPLS. In this scheme the fast handoff mechanism anticipates the LSP procedure setup with an adjacent neighbor subnet that an MN is likely to visit. The main idea behind FH Micro Mobile MPLS is to set up an LSP before the MN moves into a new subnet to reduce service disruption. In this context, the authors consider active and passive LSPs. The active LSP is the one from the LERG (the root of the MPLS domain) to the current serving LER in the visited network. This LSP is used to transfer data. Passive LSPs are those from the LERG to the neighboring LER of the current foreign agent. These LSPs will not be used except when the MN moves to its own network. In this moment, the MN establishes its new active LSP and passive LSPs with neighboring subnets.

The last host-based proposal is LW Mobile MPLS, a proposal aimed to solve some problems detected in the previous alternatives. From our point of view, the need to setup a complete LSP after each movement increases the signaling over-head and reduces the overall performance of the network. In our proposal, we handle mobility efficiently by reducing the signaling overhead in an MPLS domain. This solution is based on the forwarding chain concept (set of forwarding paths). We introduce some special nodes in the MPLS, called Linkage Nodes (LN), which are responsible for the redirection of the LSP. This way, the LSP is composed of a set of forwarding paths that allow the signal to be localized and adapt in order to track host mobility. Fig. 2 shows the basic operation of this scheme.

Initially, when a mobile node moves to an adjacent network it disconnects from its previous LER/Access Router (AR) called PELER (Previous Egress LER) and it attaches to a new LER/AR called NELER, (New Egress LER) establishing a new LSP towards this router.

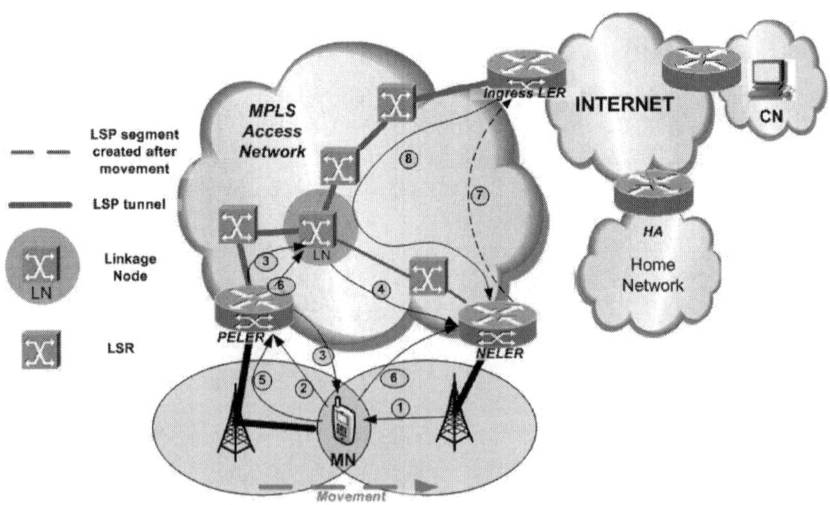

Fig. 2. LinkWork Mobile MPLS operation

When the MN moves to an adjacent network, it proceeds as follows. The MN enters an overlapped area of an adjacent subnet, it receives an L2 (layer 2) signal from the potential new base station (BS) (step 1). Next, the MN notifies the PELER of the possibility of a handoff by sending a HI (Handover Initiate) message which contains the new Base Station identifier. This information is going to be used to obtain the NELER IP address, thanks to a data structure that maintains a match between this identifier and each adjacent LER IP address (step 2). These 2 steps of the LW Mobile MPLS architecture are similar to those proposed in the FH-Micro Mobile MPLS scheme.

Once the PELER knows the subnet which the MN is going to move to, it sends a message upstream to the selected LN, notifying of a possible L3 handover, and beginning the setup of a new section of the LSP from LN to NELER (step 3) with the required QoS, using RSVP-TE. In this step, the PELER also informs the MN of the NELER IP address through a Neighbor Discovery message. At this moment, a new section of the LSP tunnel could be set up so data traffic can be forwarded towards the new location of the MN (step 4). When the signal strength received from the current base station falls below a certain threshold level, the MN notifies of the handoff to the PELER. Now, the PELER starts the mechanism responsible for minimizing packet loss (step 5).

Once the L2 handover is performed, the L3 (layer 3) handover is initiated by the MN with the NELER through MIPv6 registration process (step 6). The new LSP section from the LN to the new egress router will be used when the LN is aware of the movement. This happens when the PELER starts to return data packets back to the LN, which will be forwarded to the NELER through the new LSP section together with buffered packets according to the recovery mechanism. Finally, the NELER sends the Mobile IPv6 Binding Update message to ILER (step 7). The ILER will reply to the MN which is located in the new subnet.

2.2 Network-Based Schemes

As we have detailed in the previous section, host-based mobility management requires client functionality in the IPv6 stack of a mobile node. Exchange of signaling messages between the mobile node and the home agent enables the creation and maintenance of a binding between the mobile node's home address and its care-of address. Mobility in Mobile IPv6-based solutions requires the IP host to send IP mobility management signaling messages to the home agent, which is located in the network. This means that the protocol requires stack modification of the mobile node in order to support the mobility improvements. In addition, the requirement for the modification of mobile nodes may cause them to become increasingly complex. On the other hand, in a network-based mobility management approach, the serving network handles the mobility management on behalf of the mobile node; thus, the mobile node is not required to participate in any mobility-related signaling.

With these design goals, the IETF developed a network-based mobility management protocol which aims to cover:

- Support for unmodified Mobile Nodes: Unlike host-based mobility management protocols, the network-based protocol should not require any software modification for IP mobility support on the mobile nodes.
- Efficient use of wireless resources: The network-based protocol should avoid tunneling overhead over the wireless link, so it should minimize overhead with-in the radio access network.
- Reduction in handover-related signaling volume: Considering MIPv6, whenever an MN changes the subnets, various signaling messages are required. Therefore, in the network-based protocol the handover-related signaling should be performed as infrequently as possible.
- Support for IPv4 and IPv6: Although the initial design of the network-based protocol uses an IPv6 host, it is intended to work with IPv4 or a dual-stack host as well.

Compared to host-based mobility management approaches such as MIPv6 and its enhancements, a network-based mobility management approach such as Proxy Mobile IPv6 has several advantages.

From a deployment perspective, network-based mobility management does not re quire any modification of mobile nodes. This requirement can be considered one of the primary reasons Mobile IPv6 has not been widely deployed in practice.
From a performance perspective, due to the fact that wireless resources are very scarce, the efficient use of wireless resources can result in enhancement of network scalability. In host-based approaches such as MIPv6, the mobile node is required to participate in mobility related signaling. Thus, a lot of tunneled messages as well as mobility-related signaling messages are exchanged via the wireless links. Considering the explosively increase in the number of mobile subscribers, such a problem would cause serious performance degradation. On the contrary, in a network-based approach the serving network controls mobility management on behalf of the MN, so tunneling overhead as well as a significant number of mobility-related signaling message exchanges via wireless links can be reduced.

Another advantage is from a network service provider perspective. Network-based mobility management can enhance manageability and flexibility by enabling network service providers to control network traffic and provide differentiated services, among other things. In fact, some cellular systems such as IS-41 and Global System for Mobile Communications (GSM) can be considered network-controlled systems. Moreover, General Packet Radio Service (GPRS) has some resemblance to Proxy Mobile IPv6 in that they are both network-based mobility management protocols and have similar functionalities. Recently, the Third Generation Partnership Project's (3GPP) Evolved Packet System (EPS), commonly referred to as the 4G (Fourth Generation) Long Term Evolution (LTE) has adopted the Proxy Mobile IPv6.

PMIPv6, the main network-based protocol, shown in Fig. 3, is based on MIPv6 in the sense that it extends MIPv6 signaling and reuses many concepts such as HA functionality. The new principal functional entities of PMIPv6 are the mobile access gateway (MAG) and local mobility anchor (LMA). The MAG typically runs on the AR. Its main role is to detect the MN's movements and initiate mobility-related signaling with the MN'LMA on behalf of the MN. In addition, the MAG establishes a tunnel with the LMA to enable the MN to use an address from its home network prefix and emulates the MN's home network on the access network for each MN. On the other hand, the LMA is similar to the HA in MIPv6.

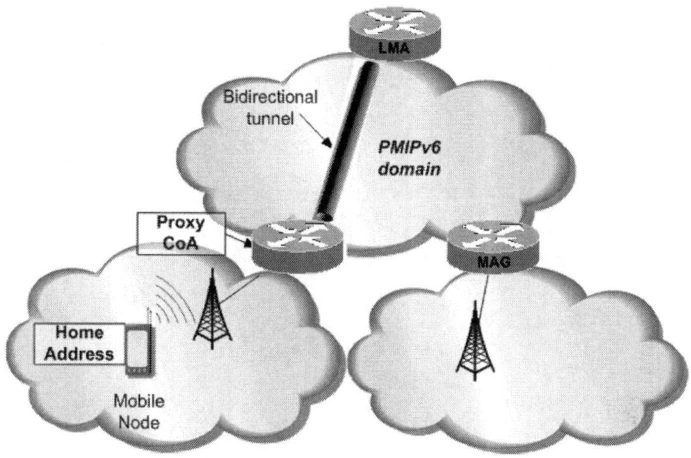

Fig. 3. Overview of Proxy Mobile IPv6

Its main role is to maintain access to the MN's address while it moves and to store information necessary to associate an MN with its serving MAG, enabling the relationship between the MAG and LMA to be maintained.

Other network-based proposals used in this analysis are FHPMIPv6 [13] and MPLS-PMIPv6 [14]. FHPMIPv6 is a fast handover extension for PMIPv6. The main idea behind this option is to establish a bi-directional tunnel between the PMAG and the NMAG so packets destined for the MN are forwarded from the PMAG to the NMAG through this tunnel.

MPLS-PMIPv6 is the first scheme which proposes MPLS as an alternative tunnel technology between a MAG and a LMA. Two kinds of labels are employed: a classical tunnel label and a Virtual Pipe label. The latter is introduced as a means to differentiate traffic with the same MAG-LMA end-points according to the operators of the various MNs served by the same MAG.

Fig. 4 shows the message flow of the operation in MIPv6 and PMIPv6, the most representative host-based and network-based mobility management protocols.

2.3 Distributed Mobility Management Approach

The mobility management proposals described in the previous section are based on a centralized mobility agent (Home Agent in Mobile IPv6 or LMA in Proxy Mobile IPv6) that allows a mobile node to remain reachable during its movement. Among other tasks, this anchor point ensures connectivity by forwarding packets destined to, or sent from, the mobile node.

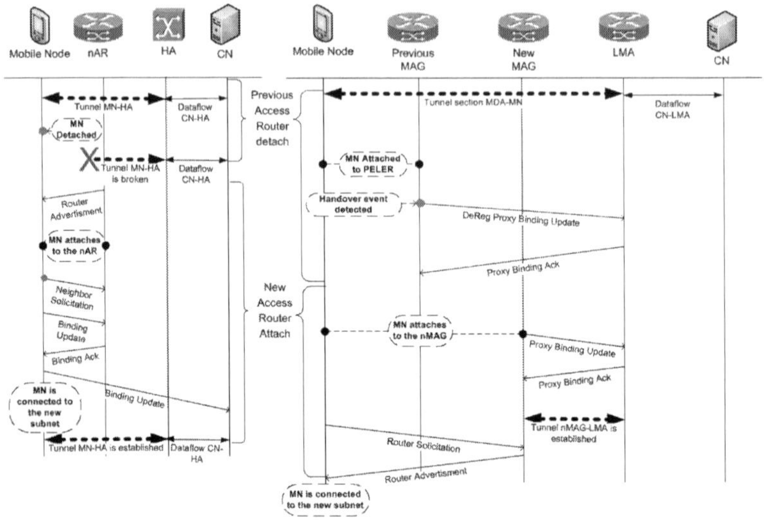

Fig. 4. Message flow in MIPv6 and PMIPv6

Nowadays, most of the deployed architectures have a small number of centralized anchors managing the traffic of millions of mobile users. This centralized approach brings several limitations such as non-optimal routing, scalability problems and reliability:

- Suboptimal routing: Since the (home) address used by an MN is anchored at the home link, traffic always traverses the central anchor, leading to paths that are, in general, longer than the direct one between the MN and its communication peer. This is exacerbated by the current trend in which content providers push their data to the edge of the network, as close as possible to the users, as for example by deploying content delivery networks (CDNs). With centralized mobility management approaches, user traffic will always need to go first to the home network and then to the actual content source, sometimes adding unnecessary delay and wasting operator resources.
- Scalability problems: Existing mobile networks have to be dimensioned to support all the traffic traversing the central anchors. This poses several scalability and network design problems, as central mobility anchors need to have enough processing and routing capabilities to be able to deal with all the user traffic simultaneously. Additionally, the entire operator's network needs to be dimensioned to be able to cope with all the user traffic.
- Reliability: Centralized solutions share the problem of being more prone to reliability problems, as the central entity is potentially a single point of failure.

Recent IP network usages such as multimedia content access and video streaming contribute to an exponential growth in bandwidth usage. The architectural limitation of centralized topologies require that data must first be routed to the HA or the LMA (centralized agents) which may be geographically far away from the mobile node, and then tunneled to the mobile node. Therefore, these limitations become clearer when the centralized mobility management needs to support mobile videos, which demand a large volume of data and often require quality of service (QoS) such as session continuity and low delay.

This motivates distributed mobility management solutions to efficiently handle ever increasing mobile traffic, the major portion of which carries video traffic [15]. Moreover, the IETF has recently created a working group called DMM (Distributed Mobility Management) that is identifying the limitation and defining the problem statements for achieving DMM with the existing IP mobility support protocols. A further description of the main DMM proposals can be found in [16] and [17].

3 Qualitative Analysis

In this section we investigate the qualitative model used to compare various existing well-known mobility support protocols, both host-based and network-based. This analysis is based on various evaluation criteria such as the cost functions of registration updates, total packet loss during a session, buffer size metrics and tunneling overhead. In order to evaluate the performance of these mobility protocols when an MPLS access network is introduced, some MPLS-based proposals are compared with non MPLS-based ones.

As we can see in the Table 1, we compare seven mobility protocols. However, not all of them are integrated with MPLS. Despite this, we consider them in our analysis due to their importance, given that Mobile IP and PMIPv6 were developed by IETF as the main Internet mobility solutions and 3GPP has used them to achieve mobility in the LTE (Long Term Evolution) evolved packet core [18].

Table 1. Protocols features

	Mobile IP-based protocols				Proxy Mobile IP-based protocols		
Protocol criteria	MIPv6	Mobile MPLS	FH Micro Mobile MPLS	LW Mobile MPLS	PMIPv6	FH PMIPv6	PMIPv6-MPLS
Required infrastructure	HA	HA-FA	LERG-FA	ILER-PELER / NELER	LMA MAG	LMA MAG	LMA MAG
Mobility Scope	Global	Global	Local	Local	Local	Local	Local
Tunneling protocols	IP-IP, GRE	MPLS	MPLS	MPLS	IP-IP, GRE	IP-IP, GRE	MPLS
MPLS integration	No	Yes	Yes	Yes	No	No	No

In order to simplify the analytical study, we suppose that every subnet is equidistant from the ingress LER in the MPLS domain, with a distance of δ (in terms of number of hops). In the same way, we do not consider the cost of the process that periodically updates the link (binding update) between the MN and the HA in order to update the cache, as is shown in [19]. We analyze the mobility behavior of the MN, keeping in mind a topology where a terminal could move to every neighbor network with the same probability. The parameters to be used are the following:

Parameters:
t_s Average connection time for a session;
t_r Average stay time at a visited network;
T_{ad} Time interval between Agent Advertisements messages;
N_h Average number of level 3 handover in a session ($N_h = t_s/t_r$);
N_g No. of remaining neighbors of the new LER not notified in a handover;
s_u Average size of a signaling message for record update;
s_l Average size of a message for LSP establishment;
$h_{x\text{-}y}$ Average number of hops between x and y in the wired network;
B_w Bandwidth of the wired link;
B_{wl} Bandwidth of the wireless link;
L_w Latency of the wired link (propagation delay);
L_{wl} Latency of the wireless link (propagation delay);
P_t Routing or label table lookup and processing delay;
λ_d Transmission ratio for a downlink packet;
T_{inter} Time between arrivals of consecutive data packets.

On the other hand, $t\,(s,h_{x\text{-}y})$ is the time spent for a packet with size s to be sent from x towards y across wired and wireless links. $t(s,h_{x\text{-}y})$ can be expressed in the form:

$$t(s,h_{x-y}) = c + h_{x-y} \cdot \left(\frac{s}{B_w} + L_w \right) + (h_{x-y} + 1) \cdot P_t$$

$$where \qquad c = \begin{cases} \dfrac{s}{B_{wl}} + L_{wl} & if \quad x = MN \\ 0 & if \quad x \neq MN \end{cases}$$

3.1 Signaling Cost

The total signaling cost of registration update for a session can be defined as C_u. This value depends on the traffic load when signaling messages are sent, i.e., this cost depends on the size of signaling messages and the number of hops in every level 3 handoff process during the time interval that communication of MN remains active. Therefore, the cost is defined by the messages size multiplied by the number of needed hops.

Each movement between neighboring subnets implies sending several signaling messages. In Mobile IP and Mobile MPLS cases, the registration update with the HA is needed, whereas in FH, LW and PMIP cases, the update is local with the root of the domain. Apart from signaling of the mobility management protocol, some proposals also add the cost of the LSP procedure set-up. This is the case of Mobile MPLS, FH-Micro Mobile MPLS, LW Mobile MPLS or PMIP-MPLS.

This way, we obtain the following values for the signaling cost when the registration update process occurs:

$$C_u(MobileIP) = 2 \cdot s_u \cdot h_{MN-HA} \cdot N_h$$
$$C_u(MobileMPLS) = 2 \cdot s_u \cdot h_{MN-HA} \cdot N_h + 2 \cdot s_l \cdot h_{FA-HA} \cdot N_h$$
$$C_u(FHMMM) = 2 \cdot s_u \cdot h_{MN-LERG} \cdot N_h + 2 \cdot s_u \cdot h_{FA-FA} \cdot N_h + 2 \cdot s_l \cdot h_{FA-LERG} \cdot N_g \cdot N_h$$
$$C_u(LW\ MobileMPLS) = 2 \cdot s_u \cdot h_{MN-ILER} \cdot N_h + 2 \cdot s_l \cdot h_{LN-NELER} \cdot N_h$$
$$C_u(PMIP) = 2 \cdot s_u \cdot h_{MAG-LMA} \cdot N_h + 2 \cdot s_u \cdot h_{nMAG-LMA} \cdot N_h$$
$$C_u(FHPMIP) = 2 \cdot s_u \cdot h_{nMAG-pMAG} \cdot N_h + 2 \cdot s_u \cdot h_{nMAG-LMA} \cdot N_h$$
$$C_u(PMIP-MPLS) = 2 \cdot s_u \cdot h_{pMAG-LMA} \cdot N_h + 2 \cdot s_l \cdot h_{nMAG-LMA} \cdot N_h$$

3.2 Packet Loss during a Session

Packet loss during a session (P_{loss}) is defined as the sum of lost packets per MN during all handoffs. Apart from FH, LW and FHPMIP, in the other schemes all in-flight packets will be lost during the handoff disruption time due to the lack of a buffering mechanism.

Our LW Mobile MPLS proposal also has a recovery mechanism that minimizes the packet loss. Its operation is explained briefly. When the MN informs PELER of an L2 handoff, this edge router does not send any more packets to the MN; instead it sends the packets back to the LN. When the first packet arrives back to the LN, it tags the next packet received and sends it and buffers all incoming packets from the PELER. Once the LN receives the tagged packet from the reverse path, it removes the tag, sends it through the new section of the LSP through the NELER. Once all incoming packets from PELER have been sent, buffered packets are forwarded to the NELER. This is how we avoid packet disorder and minimize packet loss. The main advantage of this alternative is that the packets are sent towards the new location of the MN in order so the MN task of reordering the information is significantly reduced.

Fig. 5 shows an example in which the path that a packet involved in the recovery mechanism follows ca be observed. d and d' are the distance between LN and the PELER and the distance between the LN and the NELER respectively.

FHPMIPv6 buffers data packets until the tunnel is established between the pMAG and the nMAG. Therefore, the value P_{loss} for each proposal is:

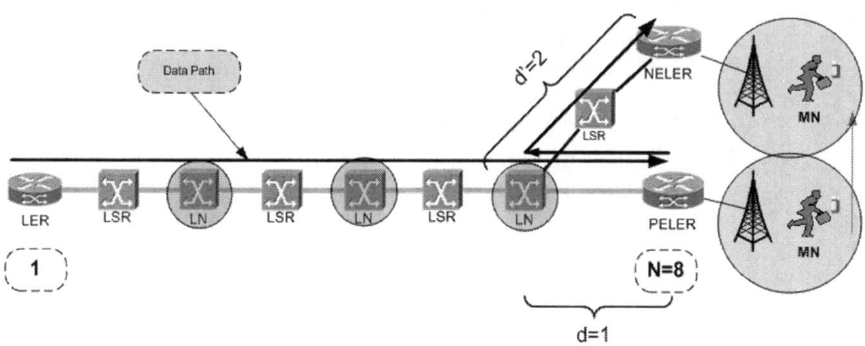

Fig. 5. Path followed by a packet involved in the recovery mechanism

$$P_{loss}(Mobile\,IP) = \left[\left(\frac{1}{2}T_{ad}\right) + T_c(MobileIP)\right] \cdot \lambda_d \cdot N_h$$

$$P_{loss}(Mobile\,MPLS) = \left[\left(\frac{1}{2}T_{ad}\right) + T_c(MobileMPLS)\right] \cdot \lambda_d \cdot N_h$$

$$P_{loss}(FH\ Micro\,Mobile\,MPLS) = t(s_u, h_{MN-FA}) \cdot \lambda_d \cdot N_h$$

$$P_{loss}(LW\ Mobile\,MPLS) = t(s_u, h_{MN-NELER}) \cdot \lambda_d \cdot N_h$$

$$P_{loss}(PMIP) = \left[\left(\frac{1}{2}T_{ad}\right) + T_c(PMIP)\right] \cdot \lambda_d \cdot N_h$$

$$P_{loss}(FHPMIP) = t(s_u, h_{MN-pMAG}) \cdot \lambda_d \cdot N_h$$

$$P_{loss}(PMIP-MPLS) = \left[\left(\frac{1}{2}T_{ad}\right) + T_c(PMIP-MPLS)\right] \cdot \lambda_d \cdot N_h$$

where T_c is the average time of the handover completion, which is defined as the sum of three terms: interruption time, establishment time and $T_{inter}/2$.

3.3 Buffer Size

The buffer for storing in-flight packets is located at the Linkage Node (LN) in the LW Mobile MPLS whereas in the FH-Micro Mobile MPLS the buffer is located in the LER/FA nodes. As for FHPMIP, the buffer can be found in the pMAG node. Therefore, the buffer size requirement (B_{size}) is listed as follows:

$$B_{size}(FH\ Micro\,Mobile\,MPLS) = \left(\frac{1}{2}T_{ad}\right) + t(s_u, h_{MN-FA} + h_{FA-FA}) \cdot \lambda_d$$

$$B_{size}(LW\ Mobile\,MPLS) = \left(\frac{1}{2}T_{ad}\right) + t(s_u, h_{LN-PELER} + h_{PELER-LN}) \cdot \lambda_d$$

$$B_{size}(FHPMIP) = \left(\frac{1}{2}T_{ad}\right) + t(s_u, h_{MN-pMAG} + h_{pMAG-nMAG}) \cdot \lambda_d$$

3.4 Tunneling Overhead

As we have seen in the previous sections, host-based and network-based mobility management protocols establish a tunnel to forward data packets. The IETF advises the use of IP-in-IP or GRE (Generic Routing Encapsulation) as tunneling methods. In this section we take a look at these technologies and compare them with MPLS tunnels.

IP-IP (IP in IP) is a protocol by which an IP datagram may be encapsulated (carried as payload) within an IP datagram, by adding a second IP header to each encapsulated datagram. However, IP-in-IP tunneling increases the overhead, because it needs an extra set of IP headers. Typically, a pure IP-over-IP tunnel configured with tunnel mode IP-IP has a 20-byte overhead, so if the normal packet size (MTU) on a network is 1500 bytes, a packet that is sent through a tunnel can only be 1480 bytes big.

GRE (Generic Routing Encapsulation) is another tunneling method that encapsulates any network layer packet. GRE requires the IP-in-IP encapsulation with the extra IP-IP header (20 bytes), but it also adds another 4 bytes of the GRE header to a packet, resulting in 24-byte overhead. After this increase the packet may need to be fragmented because it is larger than the outbound Maximum Transmission Unit (MTU).

On the other hand, an MPLS LSP tunnel has one label (4 bytes) or a stack of overhead labels (for example, when using Link Protection Fast reroute). MPLS adds four bytes to every datagram but, unlike GRE tunnel, MPLS does not change the IP header. Instead, the label stack is imposed on to the packet that takes the tunnel path.

The three approaches can also be compared in terms of the overhead they generate during data packet forwarding operations, i.e. when the MN communicates with the CN while remaining attached to the same foreign network access router. Table 2 shows this operational overhead.

From our analysis it emerged that MPLS can be profitably used to complement PMIPv6, as it enhances the tunneling paradigm with fast forwarding techniques and the potentially allows Traffic Engineering support. We showed that MPLS adds no extra overhead to MIPv6/PMIPv6; conversely it may even contribute to reductions in both handover delay and the operational overhead.

4 Numerical Results

In this section we focus on a quantitative analysis among the technologies presented in section 2. The parameter settings in our experiments are listed in Table 3. The settings of the distances $d_{x\text{-}y}$ values are represented by Fig. 5.

Fig. 6 presents the comparison of registration cost vs. resident time when parameters have their default settings.

Table 2. Operational overhead

Tunneling mechanism	Overhead
IP-IP	20 Bytes
GRE	24 Bytes
MPLS	4 Bytes

Table 3. Parameter settings

Parameter	Value
t_s	1000 sec.
t_r	5~50 sec (default 20)
t_{ad}	1 sec.
s_u	48 Bytes
s_l	28 Bytes
B_w	100 Mbps
B_{wl}	11 Mbps
L_w	1 msec.
L_{wl}	2 msec.
P_t	10^{-6} sec.
N_g	1
λ_d	64 Kbps

In this case, the Mobile MPLS scheme is the costliest alternative due to the need to establish a complete LSP tunnel from mobile node to HA separate from the specific Mobile IPv6 signaling. On the contrary, LW Mobile MPLS uses the resources in the MPLS access network efficiently as it reduces signaling to an area, and therefore doesn't overload links and nodes near ILER. This way, LW Mobile MPLS can significantly reduce the registration cost particularly when the MN hands off frequently (i.e. the resident time in each subnet is short). The introduction of LN nodes in the MPLS domain allows the signaling exchange to be reduced by the creation of a linkworked LSP that allows local registration.

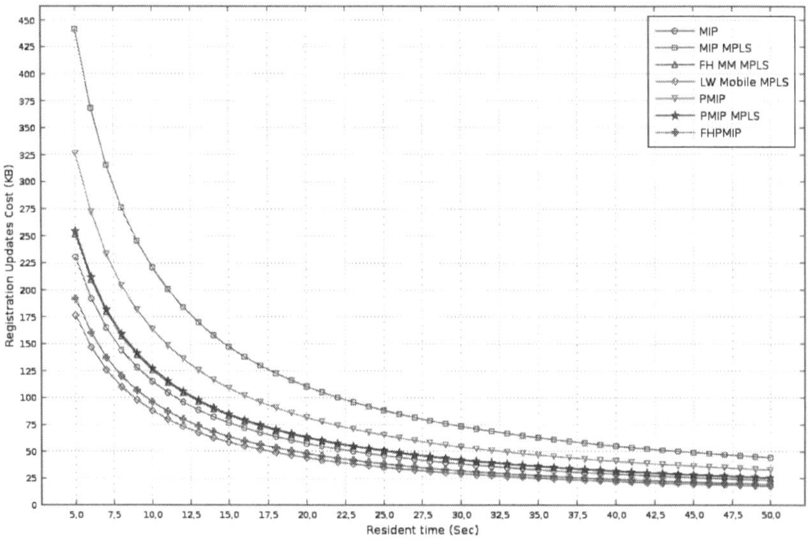

Fig. 6. Registration update cost

Fig. 7 shows the amount of lost packets during the whole connection session for different approaches. These results show the high difference between the proposals which have buffering mechanisms and those which do not. Both PMIP MPLS and Mobile MPLS have the largest amount of lost packets due to the higher establishment time (Tc) needed to setup an LSP between the HA / LMA and the new serving agent in the visited network.

In contrast, FH Micro Mobile MPLS, FH-PMIP and our proposal, LW Mobile MPLS, provide the best results thanks to the buffering and recovery mechanisms. Note the similar level of packet loss as all three proposals initiate the buffering mechanism at the same time. However, there are significant differences between them that explain the difference in the registration update cost performance. First of all, the LW Mobile MPLS approach performs the forwarding LSP chain in LN nodes, which are internal routers of the domain. In our opinion, the flexibility of the architecture can be improved by using a few nodes inside the domain as LNs which could also be easily adapted to the needs of the service provider. Secondly, the recovery mechanism proposed in our LW Mobile MPLS architecture is designed to deliver recovered packets in the correct order; this means that our proposal saves the upper transport layer from doing this task.

With respect to buffer size requirements, a buffer is needed to store in-flight packets during each handoff operation. As stated before, only LW Mobile MPLS, FHPMIP and FH Mobile MPLS do this. In this case, the LW proposal needs a smaller buffer than the others. This difference is based on the fact that the time in which the in-flight packets are stored is less than in the other proposals. Fig. 8 shows the buffer size vs. the bandwidth of the MPLS access network. In this graph we can observe that from 200 Mbps the size of the buffer remains rather stable, around 1,25~1,50 Kb.

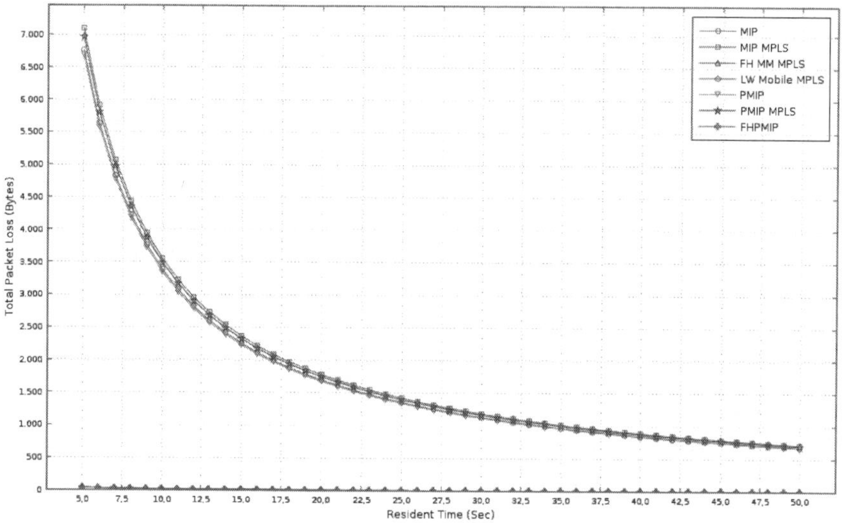

Fig. 7. Total packet loss during a session

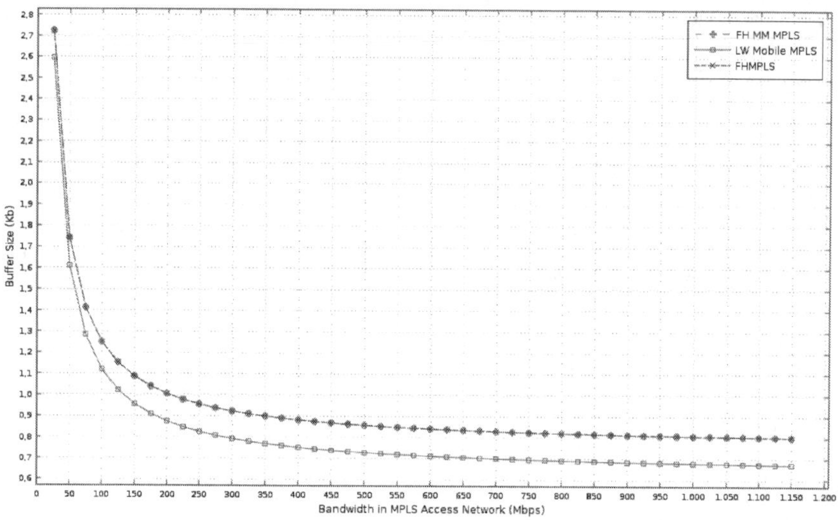

Fig. 8. Buffer size vs. bandwidth in MPLS access network

5 Conclusion

In this chapter, various well-known IP mobility support protocols have been evaluated and their performance has been measured when an MPLS domain is introduced into the access network. One of these schemes is called LinkWork Mobile MPLS, an architecture that offers efficient management through the use of special LSR that we call Linkage Nodes (LN). These nodes are responsible for rerouting the LSP tunnel to the LER that serves the mobile node in each handover. Also these nodes retrieve the packets in flight when a service interruption is provoked by a handover.

Through the analytical study and the simulations carried out we obtained the numerical results of seven protocols related to the links, the signaling costs, the packets lost during the movements in each session and the ideal buffer size needed to accomplish the objectives.

We highlight the small signaling cost of LW Mobile MPLS and also the great capacity to minimize the loss of packets compared to the alternatives. The analysis proves the need to use a buffer mechanism to store in-flight packets in order to achieve packet loss improvement.

Finally, from our study it emerged that MPLS can be profitably used to complement mobility protocols, as it enhances the tunneling paradigm with fast forwarding techniques and the possible support of Traffic Engineering. One of the main conclusions of this work is that MPLS adds no extra overhead and it may even contribute to reducing both handover delay and the overhead during data exchange.

References

[1] Langar, R., Bouabdallah, N., Boutaba, R.: A comprehensive analysis of mobility management in MPLS-based wireless access networks. IEEE/ACM Trans. Netw. 16(4), 918–931 (2008)

[2] Johnson, D., Perkins, C., Arkko, J.: Mobility Support in IPv6. IETF RFC 6275 (July 2011)

[3] Gundavelli, S., Leung, K., Devarapalli, V., Chowdhury, K., Patil, B.: Proxy Mobile IPv6- RFC 5213 (August. 2008)

[4] Rosen, E., Viswanathan, A., Callon, R.: Multiprotocol Label Switching Architecture. IETF RFC 3031 (January 2001)

[5] Taha, A.-E.M., Hassanein, H.S., Mouftah, H.T.: Extensions for Internet QoS paradigms to mobile IP: a survey. IEEE Communications Magazine 43(5), 132–139 (2005)

[6] Koodli, R.: Fast Handovers for Mobile IPv6. IETF RFC 4068 (July 2005)

[7] Ren, Z., Tham, C., Foo, C., Ko, C.: Integration of mobile IP and multi-protocol label switching. In: Proc. IEEE ICC, vol. 7, pp. 2123–2127 (2001)

[8] Langar, R., Tohme, S., Bouabdallah, N.: Mobility management support and performance analysis for wireless MPLS networks. International Journal of Network Management 16(4), 279–294 (2006)

[9] Sa-Ngiamsak, W., Krachodnok, P., Varakulsiripunth, R.: A Recovery Scheme for QoS Guaranteed Mobile IP Over MPLS Network. In: 2006 1st International Symposium on Wireless Pervasive Computing, January 16-18, pp. 1–5 (2006)

[10] Quintero, A., Del Frutos, E.: MPLS Based Architecture for Mobility and End-to-End QoS Support in Fourth Generation Mobile Networks. Journal of Computer Science 5(4), 255–262 (2009)

[11] IETF DMM working group charter page, `http://datatracker.ietf.org/wg/dmm/charter/` (accesed May 2013)

[12] Carmona-Murillo, J., González-Sánchez, J.L., Rodríguez-Pérez, F.J.: Performance evaluation of an architecture for localized IP mobility management. In: 2010 International Conference on Network and Service Management (CNSM), October 25-29, pp. 386–387 (2010)

[13] Yokota, H., Chowdhury, K., Koodli, R., Patil, B., Xia, F.: Fast handovers for Proxy Mobile IPv6. IETF Internet-Draft (May 2010) (expires November 2010)

[14] Garroppo, R.G., Giordano, S., Tavanti, L.: Network-Based Micro-Mobility in Wireless Mesh Networks: Is MPLS Convenient? In: Global Telecommunications Conference, GLOBECOM 2009, pp. 1–5. IEEE (December 2009)

[15] Zuniga, J.C., Bernardos, C.J., de la Oliva, A., Melia, T., Costa, R., Reznik, A.: Distributed mobility management: a standards landscape. IEEE Communications Magazine 51(3), 80–87 (2013)

[16] Shin, D.-H., Moses, D., Venkatachalam, M., Bagchi, S.: Distributed mobility management for efficient video delivery over all-ip mobile networks: Competing approaches. IEEE Network 27(2), 28–33 (2013)

[17] Lee, J.-H., Bonnin, J.-M., Lagrange, X.: Host-based distributed mobility management support protocol for ipv6 mobile networks. In: 2012 IEEE 8th International Conference on Wireless and Mobile Computing, Networking and Communications (WiMob), pp. 61–68 (2012)

[18] Ali, I., Casati, A., Chowdhury, K., Nishida, K., Parsons, E., Schmid, S., Vaidya, R.: Network-based mobility management in the evolved 3GPP core network. IEEE Communications Magazine 47(2), 58–66 (2009)

[19] Lo, S.-C., Lee, G., Chen, W.-T., Liu, J.-C.: Architecture for mobility and QoS support in all-IP wireless networks. IEEE Journal on Selected Areas in Communications 22(4), 691–705 (2004)

Performance Evaluation of Network Mobility Paradigms

Pedro Vale Pinheiro and Fernando Boavida

Faculdade de Ciencias e Tecnologia da Universidade de Coimbra, Coimbra, Portugal
{vapi,boavida}@uc.pt

Abstract. Despite the fact that Internet Protocol (IP) mobility – either for individual hosts or for entire networks – has been the subject of intense research for more than a decade, and that mobility solutions have been in existence for some time, the truth is that the behavior of these solutions under realistic conditions is largely unknown. In this respect, and specifically for the more demanding case of network mobility, several questions come to mind. Are the existing mobility paradigms adequate for generalized IP mobility? How will they behave in realistic scenarios and under considerable load? Are they scalable? This chapter addresses these questions by studying and comparing three existing network mobility paradigms: legacy, network-based and client-based. The study, performed through emulation in a scenario with real wireless links, actual mobility, and varying load, identifies the limitations of each paradigm, pointing to the very good potential of client-based network mobility solutions.

Keywords: network mobility, mobility models, route optimization.

1 Introduction

Moving entire networks is something that is envisioned as both interesting and necessary in the near future, opening up possibilities for enhanced connectivity, new services and exciting applications. Nevertheless, although several solutions for network mobility have been around for some time, there is lack of consensus on which one to adopt. Thus, technology is not mature enough in order to lead to stable, commercial deployment.

Up to now, the Internet Engineering Task Force (IETF) has considered three different network mobility paradigms, which, in the context of this text, we will name legacy, network-based and client-based. In the case of solutions developed according to the legacy paradigm (LG) there is little or no impact on end-nodes, in order to maximize compatibility with legacy networks and nodes. In the case of network-based solutions (NB), infrastructure systems such as mobile routers and/or mobility agents take up the management of most of the mobility-related tasks on behalf of the end-nodes. In client-based (CB) mobility solutions, end-nodes are aware of their mobility condition and perform most of the mobility functions, making mobility as transparent as possible to the network. This latter paradigm has hardly been explored

C.X. Mavromoustakis et al., *Resource Management in Mobile Computing Environments*, Modeling and Optimization in Science and Technologies 3,
DOI: 10.1007/978-3-319-06704-9_3, © Springer International Publishing Switzerland 2014

up to now, although some researchers claim that the system design would become much simpler, as no modifications to the network are needed.

Regardless the underlying paradigm, all network mobility solutions have one thing in common: they were tested – more often than not, using simulation – in relatively small-scale scenarios and with modest traffic loads. If this is enough to prove they can perform satisfactorily when the number of mobile nodes is small, it is not enough to allow us to determine their behavior in generalized mobility scenarios and under substantial load, as is expected in the (near) future Internet. Naturally, the authors of these solutions argue that it is unrealistic – and, thus, needless – to test their solutions in such scenarios because they do not yet exist. However, it is clear that if a solution is to be adopted in the future, it must scale.

The fact that explains the very limited testing of existing network mobility solutions is, nevertheless, very simple: lack of tools. Implementation and prototyping are not feasible but for small-scale scenarios. Existing simulators do not allow for true and effective large-scale scenarios. On the other hand, emulators require implementation-like development.

The central objective of this chapter is to provide data on the performance of the mentioned network mobility paradigms in realistic scenarios of medium-to-large scale, with true wireless links, actual mobility and varying load ranging from moderate to high values.

It should be noted that we do not aim at surveying mobility solutions in the Internet. An excellent survey of such solutions is available in [1]. Nevertheless, although reference [1] includes an informal and intuitive discussion of the pros and cons of various mobility approaches, it does not provide any data on the performance of existing mobility paradigms or solutions. The performance data provided in the current chapter is, thus, an important contribution, and will allow us to draw some conclusions on the potential of each paradigm for use in future Internet environments with generalized mobility.

Naturally, comparing paradigms can only be done through the comparison of representative implementations of each of them. Nevertheless, at the small scale used in most of the existing test-beds or simulations, the performance differences between solutions belonging to the same paradigm are minor and can only have very limited impact on the results.

The presented study results from a long-term research effort and would not have been possible without extensive, previous work. First of all, a proposal for a client-based network mobility solution and the associated specification had to be developed. The proposal was initially presented in [2], and was further detailed in [3]. Secondly, an emulation tool fit for large-scale scenarios was built [4]. This tool, named mobSim, was constructed and optimized for cluster operation so that it takes advantage of parallel processing capabilities and is horizontally scalable. mobSim allows for fine-grain emulation of mobility protocols, is highly flexible in terms of scenarios definition, and supports an implementation of one mobility solution belonging to each network mobility paradigm. Finally, the tool was used and stress-tested in two studies that compared the network mobility paradigms under somehow exaggerated and unrealistic situations: an extremely large-scale scenario with tens of thousands of mobile networks and up to sixteen levels of nested mobility [5], and a scenario with extremely high load [3]. The work presented in the current chapter clearly distinguishes itself from previous work, complementing and completing it by (as already mentioned) addressing realistic medium-to-large scale scenarios with moderate-to-high traffic loads. It thus provides an insight on how each of the network mobility paradigms will behave if and when implemented and deployed on the Internet.

Before proceeding to the presentation of the emulation study and respective results, a brief identification of related work is provided in section 2. This is followed by background information in section 3, which provides an overview of previous work on the client-based mobility solution and on the used emulation tool. The work in this section is mainly taken from references [3-5], and is provided for contextualization purposes only. In section 4 the emulated scenarios are detailed, followed by a presentation and discussion of the obtained results in section 5. The objective of this discussion is to highlight the advantages and drawbacks of each paradigm under study. Section 6 summarizes the key findings and identifies guidelines for further work.

2 Related Work

This section briefly describes the three mobility paradigms under consideration, and identifies the main network mobility solutions for each of the paradigms. More details on mobility paradigms can be found in [5]. As the purpose of the chapter is to assess and compare paradigms, we will not go into the detail of each particular proposal, as this can easily be found in the referenced literature. This section also includes some information on the potential and limitations of well-known simulator tools for simulation of network mobility, in order to clarify the rationale behind the development of the special purpose emulator used in the current study.

2.1 Legacy Paradigm

The main idea behind the legacy network mobility paradigm (LG) is to enable network mobility support without the need for changes in mobile network nodes (MNN) and correspondent nodes (CN). The Network Mobility (NEMO) Basic Support Protocol [6] is the reference solution for this paradigm.

In this case, when a packet whose the destination address corresponds to the mobile network prefix (MNP) reaches the home network, it is encapsulated by a home agent (HA) and sent to the care-of-address (CoA) of the mobile router (MR) through a tunnel (known as the MRHA tunnel). On reception of this encapsulated packet, the MR will decapsulate and deliver it to the mobile network node. In the reverse direction, packets from an MNN to a CN are encapsulated by the MR, sent over the MRHA tunnel to the HA, decapsulated at the HA and then normally routed to their final destination, i.e., the CN.

Despite the fact that the NEMO basic support solution is quite simple and totally compatible with legacy equipment, it has various limitations. These lead to several problems, of which the main ones are triangular routing, bottleneck in the home network and sub-optimality in nested mobile networks. These problems, addressed and analyzed in [7], [8] and [9], clearly derive from the lack of mechanisms for route optimization. This is also what distinguishes NEMO from network mobility solutions developed according to the network-based paradigm, described in the following sub-section.

2.2 Network-Based Paradigm

In the case of the network-based (NB) network mobility paradigm, most of the tasks inherent to network mobility – with emphasis on route optimization – are carried out by infrastructure elements, such as mobile routers, home agents and correspondent entities. The main idea behind this paradigm is that end nodes are as unaffected as possible by mobility, at the cost of higher infrastructure complexity.

Most of the existing proposals for network mobility are based on the NB paradigm class. As examples, one can refer the Optimized Route Cache (ORC) [10] [11] proposal, the Path Control Header (PCH) [12] [13] proposal, the Global HA to HA [14] proposal, and last but not least the Mobile IPv6 Router Optimization for Network Mobility (MIRON) [15] [16] [17].

Although differing in some of their details, the mentioned solutions are essentially similar in their main points. In general, a network element performs route optimization of all traffic to/from for local fixed nodes (LFN) and local mobile nodes (LMN). This network element can be a mobile router, a transit router, or even some sort of mobility agent, which, in this paper, we will simply call an MR. One important aspect to note is that, in this paradigm, MRs are responsible for optimizing the route of all packet flows that cross them and for maintaining the associated information. They also have to perform the return routability (RR) procedure every time the mobile network changes its position. This may require considerable resources from the MR, as the number of nodes and flows may be high.

For visiting mobile nodes (VMN), the MR must somehow provide a temporary address to the node and enable that address to be routable inside the moving network. Also, typically, MRs treat nested networks as VMNs.

2.3 Client-Based Paradigm

In the case of the client-based paradigm, end-nodes are involved in mobility management tasks, and this opens up a whole range of possibilities. This means, for instance, that end nodes can play an active part in route optimization, deciding when to optimize routes. In this case, mobile routers largely limit their role to plain routing.

The authors of the current chapter have proposed and developed a client-based network mobility solution named Optimized Mobility for Enhanced Networking (OMEN) [2]. In addition to detailed description of OMEN, reference [2] provides preliminary validation and assessment of the proposal, in a limited way and for small-scale scenarios.

It is interesting to note that although it explores an innovative paradigm, OMEN exclusively relies on existing protocols. For instance, as it will be seen in detail in section 1, OMEN mobile routers inform mobile nodes of their care-of-addresses. Instead of creating a new protocol for this purpose, OMEN uses the Neighbor Discovery (ND) protocol [29]. Thus, either as response to a router solicitation message or by its own initiative, an MR can send router advertisement messages that will be used by MNNs to learn their CoA. The CoA will be carried in a new option. As the definition of new options is already accounted for in RFC 4861, there is no need to change the protocol. On the other hand, route optimisation uses standard MIPv6 mechanisms.

An overview of OMEN will be provided in section 3, below, for contextualization purposes.

2.4 Simulation Tools

Most studies on host and network mobility were done by simulation. Nevertheless, none of the currently available simulation tools allow for reasonably large scenarios, and this clearly limits the applicability of the obtained simulation results. In this sub-section, a brief overview of the limitations of well-known simulation tools is provided.

The Network Simulator version 2 (NS-2) [18] is a simulation tool widely used by the networking research community. NS-2 has some native support for node mobility [19] and it also supports network mobility to a limited extent. For instance, MobiWan is an NS-2 extension for the study of mobility in wide-area IPv6 networks. This extension module was also used by some researchers as a basis to the development of NEMO Basic Support Protocol functionality [20]. Furthermore, although NS-2 has some capability for the simulation of large-scale scenarios [21], the system's hardware (namely its memory) limits the size of the scenarios. In addition, it should be mentioned that the NS-2 documentation states that erroneous simulation results may appear in scenarios involving high number of nodes and heavy traffic load.

The NS-3 simulator [22][23] is an evolution of NS-2, written in C++ and with optional support for a Python interface. The simulator aims at highly precise and reliable simulations for research environments. In addition it intends to solve some of the known limitations of NS-2, namely in what concerns debugging and memory management. By allowing the integration of the simulator with real system elements, such as kernel modules, interfaces, and programs, NS-3 seeks high fidelity and improved performance. Moreover, it is possible to connect several NS-3 instances running in different machines. Unfortunately, NS-3 has modest support for host and network mobility, mostly limited to the data link layer. There exists, nevertheless, some work on IP mobility simulation with NS-3 [24], although it does not pertain to network mobility.

OMNet++ [25] is another widely used simulator, developed in C++, comprising a framework and a set of libraries. In this case, extensibility is achieved through a modular architecture. OMNet++ has good documentation and there exist a considerable number of external contributions, resulting from the intense activity related to this tool. The simulator natively provides basic support for mobile IP. In addition, there is an externally developed mobility module, known as extensible MIPv6 (xMIPv6) [26][27], which, unfortunately, does not provide network mobility support. In 2003 there was a project whose objective was to explore parallel processing in order to increase the scalability of OMNet++ [28]. Unfortunately, little documentation concerning this project is available and, thus, it is unclear the extent to which it is officially supported.

The OPNET Modeler [31] is an object-oriented discrete event simulator with support for Mobile IPv6. Although support for RFC 3963 is not native, it was developed by some extension projects [32] [33]. OPNET's scalability is determined by the used hardware, having the ability to explore multi-core processors or multi-processor machines in order to speed up the simulations run-time. On the other hand, OPNET does not have horizontal scalability, that is, the possibility to simultaneously run on different machines, being limited by the hardware of the single platform on which it runs.

Summing up, all of the mentioned simulators pay significant attention to fidelity. Nevertheless, although most of them have some support for host mobility, network mobility support is not common and, when existing, is very limited. Moreover, scalability of these simulators is their weakest point, as none of them allow the simulation of large-scale network mobility scenarios.

3 Background

This section provides some background information for contextualization purposes. Namely, it provides an OMEN proposal overview and a mobSim emulation tool overview. Although detailed information pertaining to OMEN can be found in references [2] and [3], and detailed information concerning mobSim can be found in [4], it was decided to summarize it here, as it is essential to fully understand and assess the presented study.

3.1 OMEN Overview

Being a CB network mobility solution, OMEN mobile nodes are aware of their mobility and they take up an active role in the execution of route optimization procedures. In this way, MRs are freed from these tasks. To make this possible, after acquiring a care-of-address (CoA), an MR announces it to its network (more precisely, its inner network). This is done not by using a specially developed protocol but simply by using the standard Neighbor Discovery (ND) protocol (RFC 4861) [29], namely an optional field of the ND packet. When receiving this information, the mobile nodes in the inner network can use the announced CoA as if it were their own. The complete process is depicted in Fig. 1. Note that in this figure a simple notation is used just showing the main packet fields, such as source address (F, from), the destination address (T, to), the type of packet and relevant data.

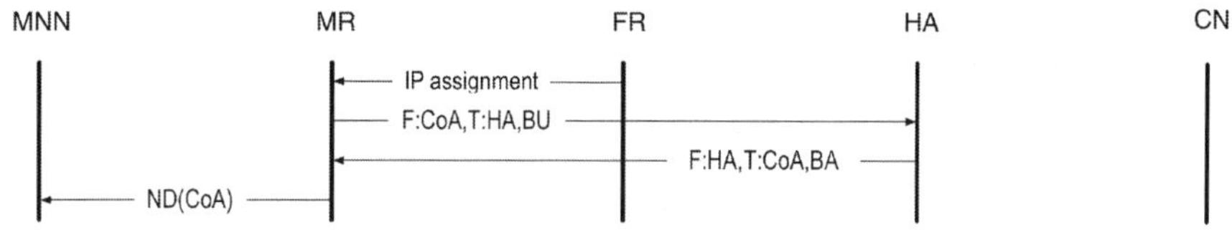

Fig. 1. OMEN route optimization process, phase 1: MR announces its CoA

In Fig. 1 one can see that after receiving the 'IP assignment' packet sent by a router of the foreign network (foreign router, FR), the MR sends a Binding Update message (BU) to its home agent and waits for the corresponding Binding Acknowledgement message (BA). When the BA packet is received, the MR announces its CoA to the MNNs on its inner network, using a Neighbor Discovery packet. After this, mobile nodes can perform route optimization without the intervention of their mobile router, as they have all the necessary information. Moreover, route optimization is simply performed by using the standard MIPv6 protocol.

In the case of the return routability procedure (RR), illustrated in Fig. 2, two essential fields of the MIPv6 protocol header enable the communication between MNNs and CNs. These are the Home Address Option and the Type 2 Routing Header (T2RH), defined in RFC 6275 [30].

As it can be seen in Fig. 2, when executing the RR procedure the MNN sends two packets to its CN: a Host Test Init packet (HoTI) is sent through the MRHA tunnel (this is identified by the double line arrow in the figure), i.e., using RFC 3963; and a Care-of Test Init packet (CoTI) is sent directly to the corresponding node. Both packets contain a token identifying the MNN. On reception of these packets, the CN answers each of them by sending back a Home Test message (HoT) through the MRHA tunnel, and a Care-of Test message (CoT) directly to the MNN. The HoT and CoT packets contain the reply to the token send in the HoTI and CoTI packets, respectively. When the return routability procedure is complete, the mobile network node initiates the binding update / binding acknowledgement procedure, as depicted in the last part of Fig. 2. This completes the route optimization process.

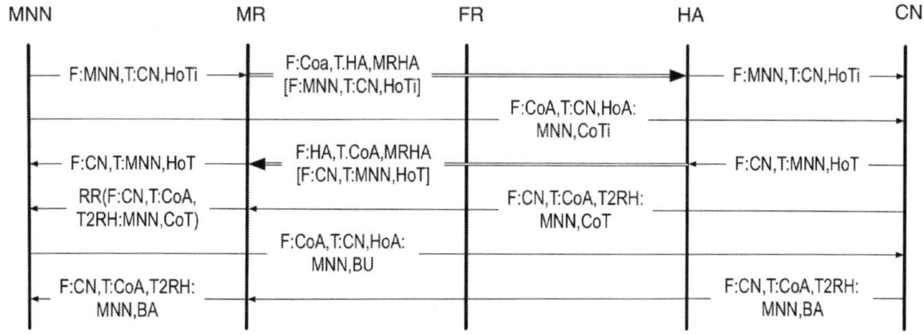

Fig. 2. OMEN route optimization process, phase 2: Return Routability procedure

After the completion of route optimization, communication between the MNN and the CN can be done directly. For this, the MNN uses the CoA address as source address. This is complemented by the T2RH or home address (HoA) fields so as to identify the next hop or the source address, respectively. Fig. 3 illustrates the direct communication between these two nodes.

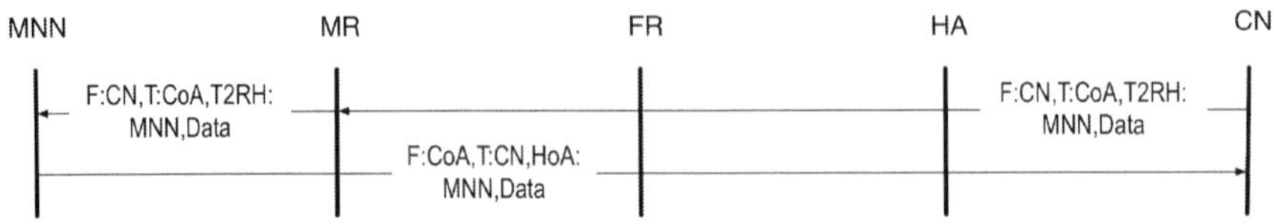

Fig. 3. OMEN route optimization process, phase 3: MNN and CN communication

Due to security issues, the process is slightly more complex in the case of VMNs, because it is necessary to confirm the visiting mobile node's identity. The VMN registration process is depicted in Fig. 4.

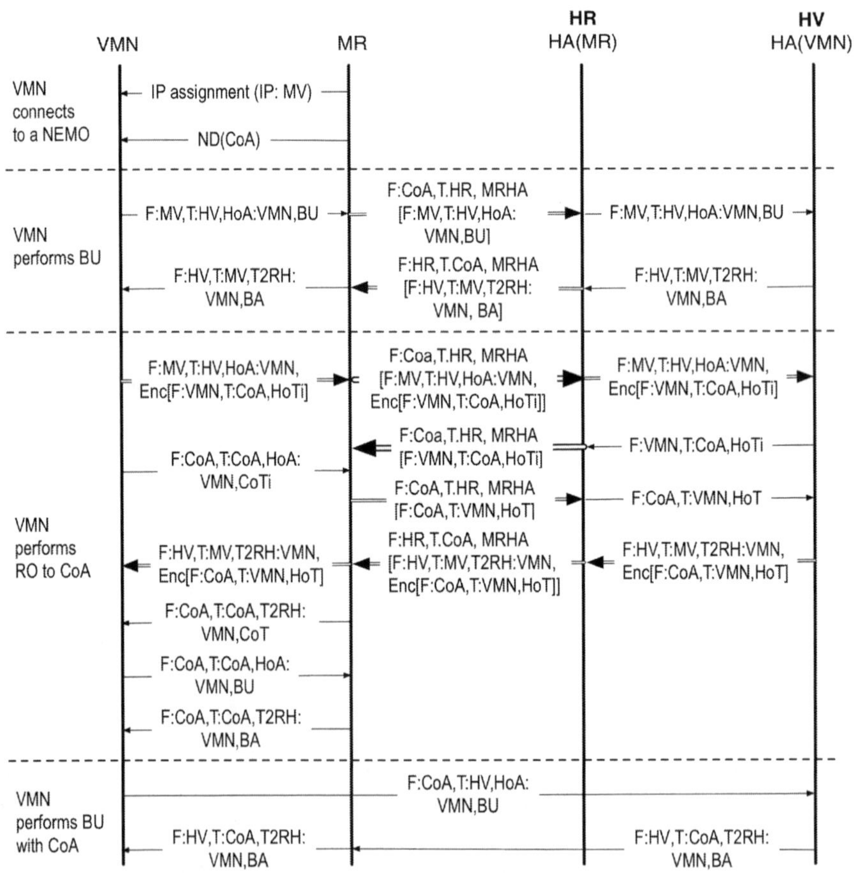

Fig. 4. OMEN: Registration of a visiting mobile node

When the VMN is assigned an IP address in the visited mobile network (the MV address, identified in the first part of Fig. 4, the Neighbor Discovery protocol is used to notify the VMN that it is inside a network with OMEN support and, thus, it should use the indicated care-of address. It is then necessary that the VMN registers its home address with the MR, so that the MR can correctly forward any packets destined to the VMN. To allow this, the VMN has to perform the BU/BA procedure with its HA, as depicted in the second part of Fig. 4.

Subsequently, the VMN executes the route optimisation procedure to the care-of address of the root-MR, using the MR's CoA, as can be seen the third section in Fig. 4. It should be noted that this route optimisation procedure to the root-MR is a necessary step, as it enables the root-MR to propagate the route of the registered addresses to sub-MRs. The RR procedure guarantees the identity of the VMN to the root-MR. When the return routability procedure is finished, the root-MR is sure that it is dealing with a legitimate VMN. The VMN must now perform a BU/BA procedure to the MR, which completes the route optimisation, as depicted in the final part of the third section of Fig. 4.

As last step, illustrated in the fourth section of Fig. 4, the visiting mobile node executes the binding update procedure to its home agent, using the CoA of its root-MR. After this, the VMN can safely communicate using the MR's CoA.

As can be seen from the previous explanation of procedures, OMEN solves several problems that exist in NEMO and in network-based mobility solutions. These are summarized in the following.

First, in NB solutions route optimisation is performed for all flows, whereas in OMEN this operation can be individually performed, on a per-flow and per-mobile-node basis. Additionally, in OMEN the end nodes decide which flows to optimize, which is clearly an advantage. In fact, end nodes are always in good position to know if a given flow is short-, medium-, or long-lived, as thus can decide on route optimization based on the nature of the flow.

Another distinguishing characteristic has to do with the issue of the load put on mobile routers by NB solutions. Although for small-scale networks this load may be reasonable, in medium-to-large scale scenarios with generalised mobility, such as the ones anticipated for the future Internet, the additional load may significantly affect the performance of mobile routers and the overall scalability of the solution.

By moving the decision and the tasks associated with route optimisation to end-systems, it is possible to keep mobile routers as light as possible, as they are only required to perform standard switching and routing.

As a final remark, it should be highlighted that the price to pay in the CB solutions in general (and the OMEN solution in particular) is very little. In the OMEN case, mobile nodes must be updated in order to support it, but this is a simple software update operation that only affects end systems. Conversely, there is high potential benefit, especially in scenarios with heavy traffic load and/or large dimension, as OMEN leads to lighter routers, requires no changes to the infrastructure, avoids unnecessary route optimisations and, last but not least, does not impose protocol changes.

3.2 mobSim Overview

As mentioned in section 2.4, existing simulators are not fit for large-scale network mobility scenarios. As this is clearly an obstacle to the study of network mobility proposals in such scenarios, the authors took the decision of developing a tool that would overcome this obstacle. Nevertheless, instead of developing another simulator, it was decided to construct an emulator tool as this would lead to better fidelity. The developed tool was named mobSim. Naturally, due to the necessarily large processing requirements of large-scale scenarios and the need for fidelity of the results, the following features were contemplated:

- horizontal scalability, taking advantage of parallel processing capabilities; for this reason, mobSim was constructed and optimised for cluster operation;
- allow for 'fine grain' simulation – mobSim level of detail is very close to an implementation, ranging from protocol header fields to the implementation of mobility mechanisms; the user can specify individual nodes, routers, networks and global scenarios;
- high flexibility in what concerns scenarios definition; this includes fixed and mobile networks, fixed and mobile nodes, no topology restrictions, and totally dynamic behaviour;
- allow the use of the exact same scenarios, parameters and conditions for the various solutions under consideration.

mobSim implements three mobility solutions, each one according to one of the network mobility paradigms under consideration in this chapter, namely the LG, NB and CB paradigms.

The LG paradigm implementation complies with the NEMO basic support specification (RFC 3963) [6]. In the case of the NB paradigm implementation, the actual approach is a mix of the most important features of the ORC and MIRON network mobility proposals. At first we had separate implementations for each of these, but the emulation results were very similar, confirming that the performance differences between solutions belonging to the same paradigm are very small. In the case of the CB paradigm implementation, the implemented solution is OMEN.

The mobSim tool implementation goes down to the level of protocol header fields, making it a very effective emulator. Table 1 presents the list of standard functionality implemented by mobSim.

mobSim was designed for running in clusters comprised of any number of nodes, thus achieving horizontal scalability. Its architecture, presented in Fig. 5, comprises one master node and a varying number of slave nodes.

The master node is organized into three modules, namely the controller module, the data analyzer module and the results database module.

The controller is the central module of the emulator. Its main tasks are the following: set up the emulation scenarios as per the user commands; build the code that will run in each slave node; send the code to each slave; initialize the simulations; get the results from the slaves and store them in the database; and shut down the slaves at the end of the emulation. Given the considerable number and the importance of tasks performed by this module, the controller does not perform any emulation itself, being dedicated to the overall management of the emulations.

Table 1. mobSim standard functionality

IPv6 Basic Support (RFC 2460)
Hop limit
Next header implementations
ICMPv6 (RFC4443)
ICMP echo, reply, unreachable, time exceeded
IPv6 encapsulation (RFC2473)
Mobility Header (RFC3775)
Type 2 routing header (routing type 2 for MIPv6 final hop HoA)
Home address option (specific for T2RH)
MIPv6 (RFC 4775)
Binding update
Binding acknowledgement (binding accept, reject)
Return routablity procedure
Home test init
Care-of test init
Home test
Care-of test
Nonce utilisation (RFC3775 section 5.2.2)
Binding refresh
Neighbour Discovery (RFC 4861)
Router advertisement
Router solicitation
Network Mobility (NEMO) Basic Support (RFC 3963)
Bidirectional tunnel (MRHA tunnel)
Binding update
Binding acknowledgement
Home agent implementation
Mobile router implementation

Slave nodes perform the actual emulations. Each slave can execute de code of up to a few hundred virtual devices, and runs IP-level code that could run in real routers and real end-nodes. In fact, although the devices are virtual, everything happens as if communication is taking place between real devices: real packets are actually assembled and sent between virtual devices. Routing tables are used and mobility mechanisms are implemented according to the respective RFCs.

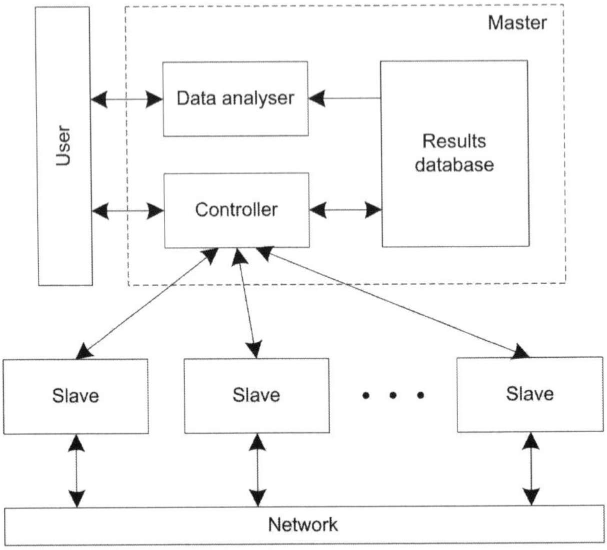

Fig. 5. mobSim architecture

Moreover, communication between the emulated devices is always performed via a real network, either wired or wireless, according to the specified emulation scenario. Which devices run in which slaves is determined by the controller during emulation set-up. When emulation finishes, the slaves send all the gathered data to the controller, which stores it in the database for subsequent, offline processing.

The data analyzer module is used for processing the emulation results, with the objective of extracting relevant information. Typical information includes packet round-trip times, handoff times, tunnel set-up times, return routability times, among other. This module also gives access to various logs, which can be used for more detailed analysis, including, for instance, routes taken by packets or tunneling actions. Logs can also be used for debugging purposes. Last but not least, it is also possible to access the raw emulation data.

As mentioned before, mobSim is extremely flexible, as it allows for any network topology, and imposes no restrictions on the number of networks, number of routers and nodes, or number of levels of nested mobility. In addition, parameters are independently configurable at node and/or network level, which makes it possible to define different values for different nodes and/or networks. Other configurable parameters include the time for data link layer handoff, the time needed for DHCP assignment of a new IP address, the BU and BA procedure times, line speed, the times for HoTI, CoTI, HoT, and CoT procedures and, last but not least, the time for encapsulation and decapsulation.

The controller automatically creates routing tables for each virtual device, according to the network topology defined by the user. The user can additionally define traffic characteristics, specify which nodes generate which traffic, and define the dynamic behaviour of the scenarios, more specifically the movement of nodes and networks.

All parameters, topologies and dynamic behaviour are kept unchanged for the various mobility solutions under study.

Two categories of devices are supported in mobSim: routing equipment and terminal equipment. Routing equipment includes top routers, routers, mobile routers and home agents. Terminal equipment may consist of local fixed nodes, local mobile nodes, visiting mobile nodes, and correspondent nodes. Correspondent nodes can be fixed or mobile (LFN, LMN or VMM). Although MIPv4 can also be configured, all devices use MIPv6 by default.

The mobSim emulator was specifically developed for use in large clusters. Consequently, special care was taken in order to optimize resources and minimize the CPU time, as these are normally expensive in these kinds of systems. This is the reason why emulations set-up is always done off-line, and the initialisation of the various slave nodes is done automatically by the master node (more specifically the controller module). The master node also controls the emulations execution, during which there is no user intervention, and subsequent shut down.

All configuration of virtual equipment is done by scripts that contain the appropriate commands. There are commands for the definition of IP addresses, gateway addresses, log server, debugging level, and location of code files, among other. Based on the scenario definition scripts, the controller module automatically generates the virtual equipment scripts.

4 Emulated Scenarios

Using mobSim, the base topology depicted in Fig. 6 was created, with the objective of comparing the performance of the three network mobility paradigms under a variety of conditions.

In this scenario, all routers were configured with the routes to every other router, so that communication between all networks was possible.

Several types of networks exist in this scenario. The mobile test network is composed of a home agent (the RA router), a mobile router (MRA), and a set of mobile nodes that, in total, can generate up to five hundred simultaneous packet flows (fl1 to fl500). Communication between RA and MRA is done through a wireless link.

There are four other mobile networks (the pairs RB/MRB through RE/MRE) that do not contain any nodes and are only used for creating nested networks (i.e., mobile networks within other mobile networks).

Network mobility scenarios without nesting are constructed by moving the MRA network to the FR1 network. Nested mobility scenarios are constructed by moving routers RB, RC, RD and/or RE to the FR2 network (e.g., RB moves to FR2, then RC moves to RB and then MRA moves to RC), in order to create up to 4 levels of nesting. Communication between any mobile router and its hosting network is always performed using a wireless link.

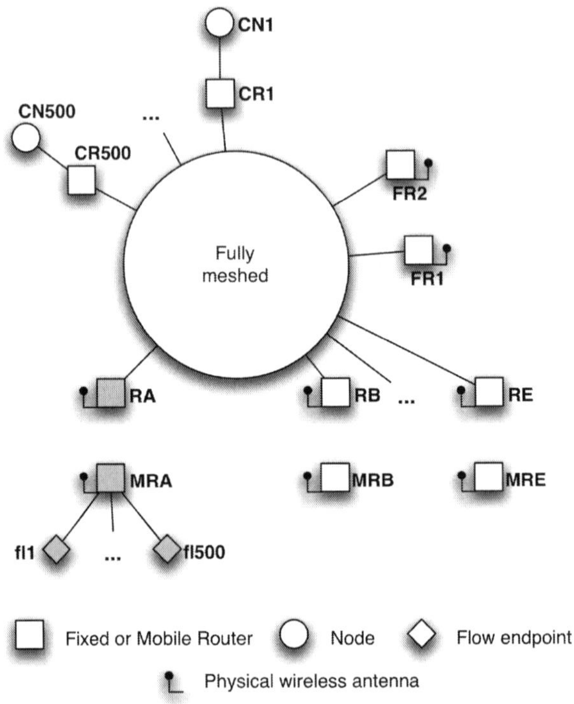

Fig. 6. Emulation scenario base topology

The corresponding networks are made up of a corresponding router (CR1 through CR500) and a corresponding node (CN1 through CR500). In order to avoid bottlenecks in the fixed part of the network, a corresponding node can only be the endpoint of a single flow. It should be noted that we intentionally isolated the test scenarios from mobility-extraneous factors, such as background traffic, use of different traffic models, or the imposition of bottlenecks in the fixed part of the network. In this way, the presented results exclusively derive from the architectural options of each of the network mobility paradigms under study.

The tests that were carried out cover all the combinations of the following parameter values, for each of the three network mobility paradigms under study:

- average packet inter-arrival times – 50 milliseconds (ms), 100 ms, 250 ms, and 500 ms; naturally, the smaller the average packet inter-arrival time the higher the network load;
- number of packet flows – 100, 200, 300, 400, and 500 flows;
- ratio of packet flows for which route optimization (RO) was performed – 2:10 (meaning 2 flows in 10 were optimized), 5:10 and 8:10;
- nesting level – no nesting and the mobile network was in a foreign network, 1 level of nesting (i.e., the mobile network was inside a non-nested mobile network), 2 levels of nesting, 3 levels of nesting, and 4 levels of nesting.

Thus, 300 different tests were made, for each network mobility paradigm. As each test was performed 3 times for each of the 3 mobility paradigms, a total of 2700 tests were performed.

In addition to the above, several delay parameters were used for all emulations. The chosen values are an approximation of actual values measured in a lab implementation, and were the following: DHCP delay – 300 ms; return routability delay – 200 ms; HoTi, CoTi, HoT or CoT messages processing delay – 100 ms; MAC-layer handoff delay – 500 ms; MR-HA tunnel setup delay – 10 ms; BU or BA messages processing delay – 10 ms.

5 Emulation Results

The objectives of the various tests were the determination of the response of each of the three paradigms to varying conditions. Specifically, we studied the responsiveness to traffic load variation, route optimization, and level of nesting. As previously mentioned, each test was run three times. The choice of the better result set was done using the Statgraphics tool (http://www.statgraphics.com/), which provides Tukey's Honestly Significant Difference (Tukey HSD) test, scatter diagrams, and analysis of variance (ANOVA). The following sub-sections present and discuss the results in detail.

5.1 Traffic Load Analysis

Load variation was achieved in two different ways: by changing the packets' mean inter-arrival time, and by changing the number of flows.

Fig. 7 shows the average round-trip time (RTT) of each of the mobility paradigms as a function of the packets' mean inter-arrival time.

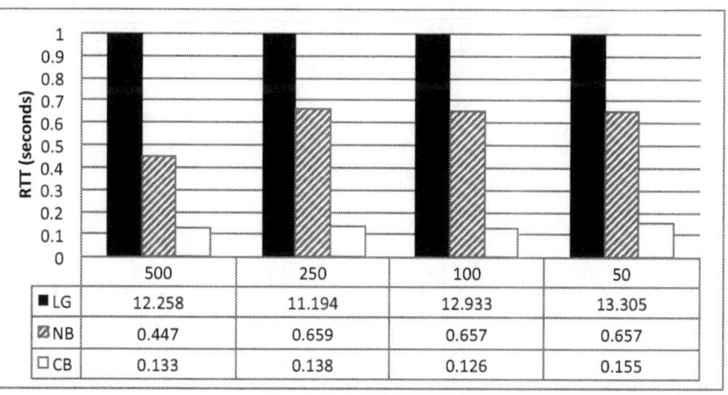

	500	250	100	50
■ LG	12.258	11.194	12.933	13.305
▨ NB	0.447	0.659	0.657	0.657
□ CB	0.133	0.138	0.126	0.155

Fig. 7. Average RTT as a function of the packets' mean inter-arrival time

Note that the average RTT for a given mean packet inter-arrival time is calculated using all emulations that were performed with this particular inter-arrival time, irrespectively of the values of the other parameters (number of packet flows, route optimization ratio, and level of nesting). The same is applicable to the remaining cases, for which, when we analyze a particular parameter value, we average the results of all emulations performed with that parameter value.

It should also be noted that for graphical intelligibility reasons, the columns pertaining to the legacy paradigm (LG) were cut at 1 second in all figures. The actual average RTT values can be found in the numerical part of the figure.

The first thing to note is that the worst performing paradigm is the LG paradigm, followed by the NB paradigm. The CB paradigm clearly outperforms the other two.

In the case of the LG paradigm, there is no route optimization and, thus, this leads to very poor RTT performance. Note that the extremely high RTT values result from the fact that several scenarios comprise various levels of nesting. This will be further addressed in sub-section 5.3.

By introducing route optimization, the NB paradigm significantly reduces the average RTT. As the load increases (i.e., as packet inter-arrival times decrease) NB's performance is slightly affected at first, but it quickly stabilizes due to the route optimization factor.

Nevertheless, it should be noted that the CB's average RTT is substantially better than the one for NB in all situations. The reason for this is that in the NB case all flows are optimized and almost all of the signaling load is put on mobile routers, which thus become critical bottleneck points. On the other hand, in the case of the CB paradigm, mobile routers simply perform routing and, thus, are quite unaffected by these load levels.

A direct view of the influence of the load level expressed as a function of the number of existing packet flows can be seen in Fig. 8.

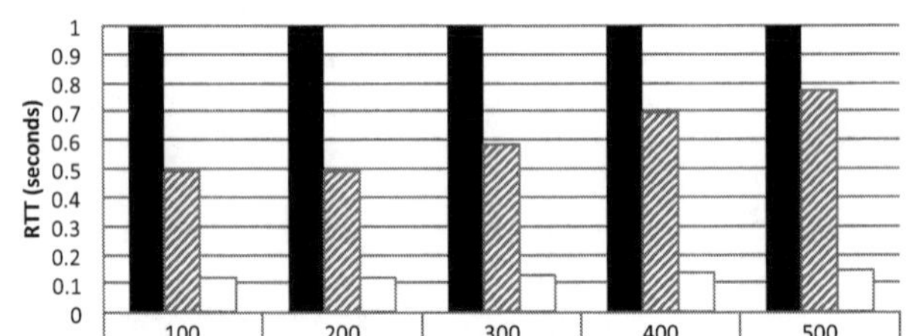

	100	200	300	400	500
■ LG	10.789	10.788	10.947	14.488	15.164
▨ NB	0.494	0.494	0.585	0.693	0.771
□ CB	0.123	0.124	0.128	0.139	0.146

Fig. 8. Average RTT as a function of the number of packet flows

The differences between the various paradigms are, again, quite obvious. Significantly, in this figure the performance degradation of the NB paradigm is clearly visible as the load increases, confirming the thesis that there is a bottleneck.

As there is no bottleneck in the fixed part of the network (all flows go to different corresponding networks and use separate paths), it is apparent that it lies in the mobile router. Moreover, using exactly the same topology, the CB paradigm performs much better and does not show significant performance degradation, thus confirming that relieving mobile routers from mobility management and putting it in client systems pays off.

As a final check of the influence of load on the performance of each of the three paradigms, Fig. 9 shows the total packet losses for the 100-flow, 300-flow and 500-flow cases.

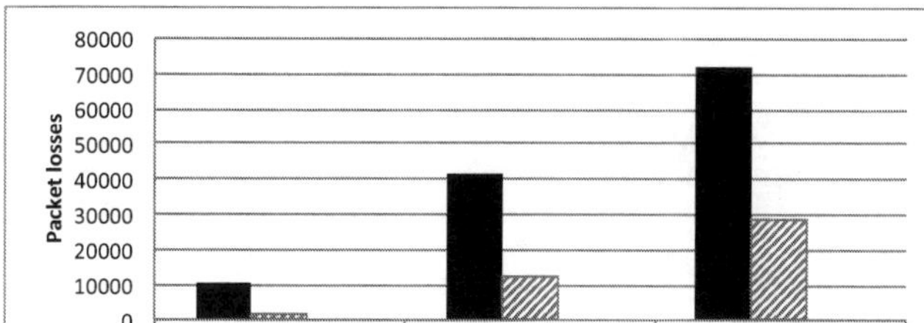

	100 flows	300 flows	500 flows
■ LG	10641	41173	71985
▨ NB	1560	12289	28773
□ CB	0	0	0

Fig. 9. Total packet losses as a function of the number of packet flows

A clear increase of the number of lost packets with the load increase is visible for the LG and NB paradigms. Interestingly, at these load levels the CB paradigm does not yet exhibit any packet losses.

We did perform a specific CB emulation test in order to determine the number of flows at which losses would occur and we arrived at the conclusion that they start at 900 flows.

5.2 Route Optimization Analysis

For the route optimization analysis, three different ratios of route optimization were used, in the case of the CB paradigm: 2:10 (i.e., route optimization was performed for 2 out of 10 flows), 5:10 and 8:10. Note that for the LG and NB paradigms, the route optimization ratio should have no influence, as there is no route optimization in LG, and all flows are optimized in NB.

The obtained emulation results concerning the average RTT as a function of the route optimization ratio are presented in Fig. 10.

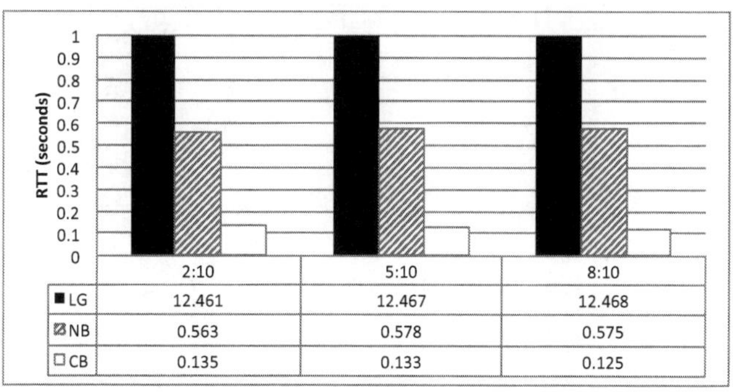

	2:10	5:10	8:10
■ LG	12.461	12.467	12.468
▨ NB	0.563	0.578	0.575
□ CB	0.135	0.133	0.125

Fig. 10. Average RTT as a function of the route optimization ratio

The results confirm that for the LG and NB cases there is no significant variation. Moreover, they also confirm that performing route optimization pays off, as the NB's RTT average values are significantly lower than the ones for LG.

In turn, the average RTT values for the client-based network mobility paradigm are significantly lower than the NB ones. In fact, as the route optimization ratio increases, CB's average RTT decreases, as more flows see the respective routes optimized. It should be noted, however, that the decrease is not very sharp because, in fact, RTT values are already at minimal levels in the case of the CB paradigm.

5.3 Level of Nesting Analysis

A final analysis addressed the behavior of the various network mobility paradigms under different levels of nesting.

In the emulated scenario, the reference mobile network, i.e., the one with the MRA mobile router (see Fig. 6), could move to a fixed network or to mobile networks inside other mobile networks, as explained in section 4, above. When MRA moves to a fixed network, it is said that there is 0 nesting. If it moves to a mobile network that is itself attached to a fixed network, then the level of nesting is 1. The maximum level of nesting considered in the study was 4.

The average RTT values as a function of the level of nesting are presented in Fig. 11.

The first thing to notice is that there is a sharp increase in the average round trip time with the level of nesting for the case of the LG paradigm. This, in fact, was expected and is a confirmation of the consistency of the emulation study, as in this paradigm there is no route optimization.

As the level of nesting increases, so increases the number of tunnels inside tunnels and, consequently, the number of networks that have to be traversed by the packets in order to reach the destination. The result is a dramatic increase in the round trip time with the level of nesting.

Another thing to notice is that in the case of the NB and CB paradigms, the level of nesting has a small, although quite perceptible, impact on the average round trip time. Both paradigms use route optimization and, thus, after route optimization is done, the RTT remains at low values. There is, nevertheless, a slight increase, which is higher in the case of the NB paradigm.

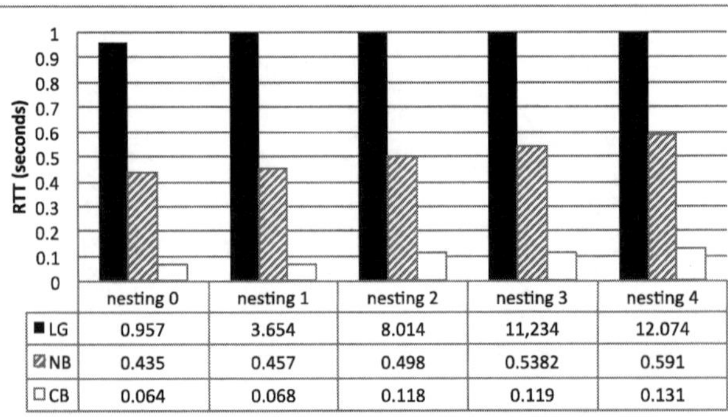

	nesting 0	nesting 1	nesting 2	nesting 3	nesting 4
■ LG	0.957	3.654	8.014	11,234	12.074
▨ NB	0.435	0.457	0.498	0.5382	0.591
□ CB	0.064	0.068	0.118	0.119	0.131

Fig. 11. Average RTT as a function of the level of nesting

The explanation for this higher increase in NB's average RTT lies in the amount of route optimization signaling and in the number of traversed wireless networks. In the NB case, all flows are route-optimized, which requires considerable signaling traffic. Naturally, for higher levels of nesting this traffic has to traverse a higher number of wireless networks, thus taking longer. In the case of the CB paradigm this increase is smaller, as not all routes need to be optimized.

As a final remark, it is again clear that the CB network mobility paradigm leads to much better results than the NB paradigm, thus confirming that, regardless the perspective, the client-based network mobility paradigm has significant advantages over the network-based network mobility paradigm and, consequently, over the legacy paradigm.

6 Conclusion

In this paper, an assessment of three network mobility paradigms has been presented: the legacy paradigm, used in the NEMO basic support protocol, for which it is assumed that network mobility must be possible even when legacy, mobility-unaware end-nodes are present; the network-based paradigm, according to which network mobility functions are mostly performed by mobile routers and other network elements, in the attempt of keeping end-nodes as unchanged as possible; and the client-based paradigm, in which end-nodes take up several mobility functions, relieving mobile routers and network elements from these tasks.

The main motivation for the presented study was to assess the potential and implications of each paradigm in realistic generalized mobility scenarios, such as the ones that are expected in the future Internet. A side motivation was to understand why the traditional network mobility paradigms – namely the legacy and the network-based paradigms – are failing to fulfill their promise of providing effective mobility.

In order to compare the mentioned paradigms, a realistic scenario with a varying number of traffic flows was built. This scenario was studied with the help of a specially built emulator, optimized for parallel operation. The emulator is extremely flexible, allowing the construction of any scenario, taking advantage of real or virtual machines, real networks (wired and/or wireless), and standardized mobility protocols. The emulator works at the IP level, sending and receiving real packets. Moreover, it implements three network mobility solutions, each one according to one of the network mobility paradigms.

It is important to notice that the objective of the paper was not to compare specific implementations, nor was it to obtain specific numeric values for specific output variables. The comparison should only provide information on the potential of each paradigm to support mobility in realistic scenarios and, ultimately, in an increasingly mobile Internet.

The analysis of the obtained results clearly confirms the known limitations of the legacy paradigm, which do not allow it to scale. Clearly, this paradigm was developed as an interim solution, in order to allow legacy nodes not to be left out of network mobility scenarios.

The network-based paradigm adopted the same base requirement as the legacy paradigm: avoid all changes to end-nodes. As a consequence, almost all mobility functions are performed by network elements, mostly by mobile routers, including route optimization functions. The emulation results concerning the network-based paradigm clearly prove that the overall performance of network elements is affected by this concentration of functionality, with consequences in terms of the resulting round trip time and overall packet losses.

In short, the results show that the legacy and network-based paradigms do not scale, thus confirming the intuitive analysis presented in [1].

The results also show that the client-based network mobility paradigm outperforms the other two paradigms and has very good potential. This was apparent even under relatively high traffic load. Relieving mobile routers from mobility tasks significantly improves the overall performance, as the burden is distributed by client systems instead of being concentrated in network systems. Making end-nodes mobility-aware thus results in considerable potential for improvement.

The study presented in this chapter opens up a large field for further research. Naturally, client-based network mobility solutions should be developed. In this chapter, a solution proposed by the authors was used, but many other client-based mobility solutions can and should be developed and studied. The study of their behavior under different network and mobility conditions should be carried out in the near future. Last but not least, implementations, more than simulations or even emulations, should be developed, analyzed and deployed.

Acknowledgments. This work was partly financed by projects PTDC/EIA-EIA/116173/2009 and CENTRO-07-ST24-FEDER-002003. The authors would like to thank Shivam Jain and André Carvalho for their contributions to and help with experimental evaluation, and Pedro Vieira Alberto, Luís Pinto and Pedro Almeida, of the Advanced Computing Lab of the University of Coimbra, for providing access to the Milipeia cluster, without which the work on the mobSim emulator and emulations would not have been possible.

References

1. Zhu, Z., Wakikawa, R., Zhang, L.: A Survey of Mobility Support in the Internet. RFC 6301, Internet Engineering Task Force (July 2011)
2. Pinheiro, P.V., Boavida, F.: OMEN – A New Paradigm for Optimal Network Mobility. In: Harju, J., Heijenk, G., Langendörfer, P., Siris, V.A. (eds.) WWIC 2008. LNCS, vol. 5031, pp. 115–126. Springer, Heidelberg (2008)
3. Pinheiro, P., Boavida, F.: Some Results on Network Mobility Stress Testing. In: Proceedings of BCFIC 2012 – 2nd Baltic Conference on Future Internet Communications, Vilnius, Lithuania, April 25-27 (2012)
4. Pinheiro, P.V., Boavida, F.: obSim – A Network Mobility Simulation Tool for Very Large-Scale Scenarios. In: Proceedings of NTMS 2011 - 4th IFIP International Conference on New Technologies, Mobility and Security, Paris, France, February 7-10 (2011)

5. Pinheiro, P.V., Jain, S., Boavida, F.: A Comparative Study of Network Mobility Paradigms. In: Masip-Bruin, X., Verchere, D., Tsaoussidis, V., Yannuzzi, M. (eds.) WWIC 2011. LNCS, vol. 6649, pp. 68–79. Springer, Heidelberg (2011)
6. Devarapalli, V., et al.: Network Mobility (NEMO) Basic Support Protocol. RFC 3963, Internet Engineering Task Force (January 2005)
7. Chan-Wah, N., et al.: Network Mobility Route Optimization Problem Statement. draft-ietf-nemo-ro-problem-statement-03, Internet Engineering Task Force (September 2006)
8. Chan-Wah, N., et al.: Network Mobility Route Optimization Solution Space Analysis. draft-ietf-nemo-ro-space-analysis-03, Internet Engineering Task Force (September 2006)
9. Carlos, J.: NEMO: Network Mobility in IPv6. Upgrade IV(2) (April 2005)
10. Wakikawa, R., et al.: Optimized Route Cache Protocol (ORC). draft-wakikawa-nemo-orc-01, work in progress, Internet Engineering Task Force (November 2004)
11. Wakikawa, R., et al.: ORC: Optimized Route Cache Management Protocol for Network Mobility. In: Proceedings of the 10th International Conference on Telecommunications, vol. 2, pp. 1194–1200 (February 2003)
12. Jongkeun, N., et al.: Route Optimization Scheme based on Path Control Header. draft-na-nemo-path-control-header-00, work in progress, Internet Engineering Task Force (April 2004)
13. Jongkeun, N.: Supporting Route Optimization in NEtwork MObility (NEMO). Technical Report INC2004-01, Seoul National University (December 2004)
14. Thubert, P., et al.: Global HA to HA protocol. draft-thubert-nemo-global-haha-01.txt, work in progress, Internet Engineering Task Force (October 2005)
15. Bernardos, C., et al.: MIRON: MIPv6 Route Optimization for NEMO. In: Proceedings of the ASWN 2004: 4th Workshop on Applications and Services in Wireless Networks, Boston University, Boston, Massachusetts, USA, August 8-11 (2004)
16. Bernardos, C., et al.: Mobile IPv6 Route Optimisation for Network Mobility (MIRON). draft-bernardos-nemo-miron-00, work in progress, Internet Engineering Task Force (July 2005)
17. Bernardos, C.: Route Optimisation for Mobile Networks in IPv6 Heterogeneous Environments. PhD thesis, Universidad Carlos III de Madrid, Spain (September 2006)
18. The Network Simulator – ns-2, `http://www.isi.edu/nsnam/ns/`
19. MobiWan: NS-2 extensions to study mobility in Wide-Area IPv6 Networks, `http://www.inrialpes.fr/planete/mobiwan/`
20. Shahriar, A.Z.M., Atiquzzaman, M.: Network Mobility in Satellite Networks, `http://citeseerx.ist.psu.edu/viewdoc/download?doi=10.1.1.125.149&rep=rep1&type=pdf`
21. Tmix: Internet Traffic Generation, Example, `http://www.isi.edu/nsnam/ns/doc/node575.html`
22. ns-3 discrete-event network simulator for Internet systems, `http://www.nsnam.org/`
23. ns-3 Wiki main page, `http://www.nsnam.org/wiki/index.php/Main_Page`
24. Mauchle, F., Frei, S., Rinkel, A.: Simulating mobile IPv6 with ns-3. In: Proceedings of SIMUTools 2010 - 3rd International ICST Conference on Simulation Tools and Techniques, Torremolinos, Malaga, Spain, March 15-19 (2010), doi:10.4108/ICST.SIMUTOOLS2010.8682
25. OMNet++, `http://omnetpp.org/`
26. An Accurate and Extensible Mobile IPv6 (xMIPv6) Simulation Model for OMNeT++, `http://www.kn.e-technik.tu-dortmund.de/content/view/232/lang,de/`

27. Yousaf, F.Z., Bauer, C., Wietfeld, C.: An accurate and extensible mobile IPv6 (xMIPV6) simulation model for OMNeT++. In: Proceedings of the 1st International Conference on Simulation Tools and Techniques for Communications, Networks and Systems & Workshops, pp. 1–8. ICST (Institute for Computer Sciences, Social-Informatics and Telecommunications Engineering), Marseille (2008)
28. Parallel Simulation with OMNeT++. CTIE, Monash University, Melbourne, Australia, http://ctieware.eng.monash.edu.au/twiki/bin/view/Simulation/ParallelSimulation
29. Narten, T., et al.: Neighbor Discovery for IP version 6 (IPv6). RFC 4861, Internet Engineering Task Force (September 2007)
30. Perkins, C., Johnson, D., Arkko, J.: Mobility Support in IPv6. RFC 6275, Internet Engineering Task Force (July 2011)
31. The OPNET Modeler, http://www.opnet.com/
32. Academic Research and Teaching with OPNET Software, http://www.ctr.kcl.ac.uk/opnet/opnet.html
33. OPNET Contributed Models, https://enterprise1.opnet.com/tsts/4dcgi/Models_SearchSubmit?QueryModels_what=FindAll&QueryRecordsPerPage=500

Scalable Routing Mechanisms for Mobile Ad Hoc Networks

David Palma and Marilia Curado

University of Coimbra, Palácio dos Grilos, Rua Ilha,
3000-214 Coimbra, Portugal
{palma,marilia}@dei.uc.pt

Abstract. Nowadays an increasing number of portable devices with wireless communication capabilities start to play an important role in our daily lives. People rely on being connected to computer networks where they can access services that empower both their professional and personal life experience, interconnecting computers, cell-phones, sensors and other common objects that start to offer these type of services. However, existing network infrastructures have not been designed for supporting such a large number of heterogeneous devices and seamless mobility is considered a great challenge. An existing alternative that departs from traditional network infrastructures focuses on the devices' ability to inter-connect themselves, creating wireless multi-hop networks. In fact, the Internet Engineering Task Force (IETF) Mobile Ad-hoc NETwork (MANET) working group, has defined two different routing protocols that explore the capabilities of wireless multi-hop networks by creating an ad-hoc network. In particular, the working group has defined a proactive routing protocol for dense and less dynamic networks with high load of traffic, as well as a reactive routing protocol aiming at tackling sparser networks with higher mobility. However, many other routing protocols and approaches have been proposed in the literature and, bearing in mind the main guidelines from the IETF working group and the challenges posed by mobility and the increasing number of devices, a thorough analysis of important routing protocols for Mobile Ad Hoc Networks will be presented. This analysis will consider the study of works relevant to the development of scalable routing approaches, capable of providing a solution for the current dissemination of wireless-capable devices and need to interconnect them. A complexity comparison of these protocols will also be presented, as well as performance evaluation of the main existing solutions for large-scale MANETs.

1 Introduction

Computer networking has long evolved since the first packet switching that existed in the early 60's. Not only has the number of connections between users and devices increased, but these connections have also diversified from copper cables, through optical fibre into the wireless medium. In particular, wireless technologies have registered a remarkable evolution in order to cope with the increasing portability of computers and other gadgets such as personal-digital-assistants, media players, cell-phones among others.

C.X. Mavromoustakis et al., *Resource Management in Mobile Computing Environments*, Modeling and Optimization in Science and Technologies 3,
DOI: 10.1007/978-3-319-06704-9_4,

Recent technological advances have promoted a massive dissemination of wireless capable devices with greater processing power, higher memory and autonomy, increasing the connectivity between users and different services and applications. As a result, in a near future, users are expected to own several hundreds of gadgets requiring wireless connections [1] amongst themselves and other users, motivating the development of networks capable of connecting them whilst supporting several applications' requirements, demanding a considerable amount of physical resources from the available infrastructures. Such demand of intra and inter networking capabilities will compel researchers and network providers to create alternative communication paradigms to the existing ones and deploy suitable infrastructures.

Despite the flexibility provided by new long-range wireless technologies, such as Worldwide Interoperability for Microwave Access (WiMAX) [2] and Long Term Evolution (LTE) [3], these networks are still expensive and do not scale with ease. For instance, in events where thousands of people are gathered, such as a football game or a concert, these networks are known to fail in delivering a good quality of experience when users try to share their emotions by sending emails, photos and other content. Moreover, in rural areas or disaster scenarios, the coverage provided by these approaches is usually limited or unavailable either by option from the operators or as result of existing damage on the infrastructures.

Another typical characteristic of the spreading wireless gadgets is their portability, creating new challenges related with mobility. This aspect is crucial for users who expect seamless connectivity regardless of their movement and action. However, different trajectories may reduce connectivity coverage or, on the other hand, increase the number of connections and consequently the number of packet collisions, resulting in the disruption of paths established by routing protocols.

Bearing in mind the necessity to handle the restrains of existing infrastructures, and the urge to provide alternatives where they are not available, the concept of Ad-hoc networks has been suggested. This has enabled the impromptu creation of wireless multi-hop networks, where each wireless node behaves as router. By using this approach, users are capable of maintaining their own network, being able to locally share their contents without requiring additional infrastructures. User mobility is of course an important requirement and thus Mobile Ad-hoc Networks (MANETs) must be able to handle the creation and destruction of new links between different users, a task usually delivered to a routing protocol.

While MANETs give users the freedom to create networks in the spur of the moment, without any particular restrictions, these networks may also suffer from scalability problems. In fact, the role of routing protocols may become extremely challenging when the number of connected nodes increases. This difficulty results from the non-existence of a well defined organisation and from interference phenomena intrinsic to wireless technologies.

Conventional routing used in wired and in infrastructure-based wireless networks could not be applied to these spontaneously created wireless networks, due to their dynamics. Distance-vector and link-state routing approaches have been used to establish routes in these networks, using techniques such as Multipoint Relay nodes, in order to optimize the forwarding of topology-related routing packets. However, these proactive routing schemes were not always suitable for networks where high mobility patterns were registered, motivating the creation of reactive on-demand routing alternatives. These protocols strive for typically having a reduced amount of control traffic, finding paths solely when required. Since on-demand routing suffered from an initial delay when retrieving paths and is prone to increased overhead in networks with high number of traffic flows, hybrid routing approaches were developed, trying to join the best of both proactive and reactive routing approaches. Other routing schemes take advantage of knowledge about nodes' positions. This class of routing protocols, geographical routing protocols [4], are characterized for having low overhead and memory requirements, however positioning information may not be available or it may be inaccurate in several scenarios, such as indoor scenarios or large and dense urban areas [5].

Wireless multi-hop networks have increasingly stood out for being available anywhere, without requiring any existing infra-structures, and for being self-organised, self-administrated and self-maintained. For this purpose, as previously mentioned, several works already exist on this topic. However, maintaining routing performance for large-scale networks is a critical issue [6]. Taking this problem into account, different works propose schemes involving techniques such as dynamic addressing, keeping network nodes organised in a well defined topology; geographic partitioning, in order to easily create stable clusters; and also typical clustering solutions, to simply reduce the total amount of routing traffic.

While some approaches aim at scalable routing using different approaches, they lack a thorough evaluation of the impact of different mobility models. In fact, regarding this aspect, most routing solutions disregard the dynamics of different mobility models, focusing only on one mobility pattern. Nevertheless, in order to appropriately evaluate the efficiency of an Ad-hoc network and the performance of routing protocols, these aspects have to be taken into account. Moreover, other works that study the impact of mobility fail to provide an extensive evaluation with existing mobility models [7].

A different perspective on wireless multi-hop routing has been provided with the definition of Delay-Tolerant Networks (DTNs). In these networks, routing protocols are designed to deliver traffic that is not delay sensitive, despite the sparse intermittently connected properties of such network. Conventional routing in wireless multi-hop networks is not suitable for highly dynamic scenarios, as it needs to establish an end-to-end path before starting the routing of data packets, which may not be possible at a given moment.

Even though most wireless networks are in fact intermittently connected due to interferences in the wireless medium, the mobility of nodes has also an important role in this aspect. Typical DTN solutions such as PRoPHET [8] are capable of operating with delay tolerant traffic when wireless connections are not reliable, but fails to perform well with completely unknown node mobility. Other approaches focus on more stable parameters, such as social interactions between nodes. For instance, the Friendship-based Routing (FBR) protocol [9] or the Social Aware Networking (SANE) scheme [10] take into account social interactions, both physical and virtual, in order to make a packet forward decision.

In this work, routing in multi-hop wireless networks is presented in Section 2, being also analysed existing proactive, reactive and hybrid routing approaches in Sections 3, 4 and 5 respectively. In these sections scalability issues of the protocols are considered and their complexity will be addressed in Section 6. A performance analysis methodology, which considers several routing parameters and compares them in two different scenarios, is presented in Section 7, followed by a through evaluation of three main routing protocols in Section 8. Finally, the conclusions and final thoughts on the current state of scalable routing in Mobile Ad-hoc Networks are presented in Section 9.

2 Presentation of Wireless Multi-hop Routing

In multi-hop wireless ad-hoc networks each wireless-capable device will behave as network router, creating the opportunity for other devices to use its resources in order to reach a distant node by issuing another hop. These networks, commonly referred as self-X networks, are expected to be spontaneously created, administrated and organised, relying on a routing protocol to maintain and acquire the paths between each node in the network.

Regarding the analysis of existing routing protocols for wireless multi-hop ad-hoc networks to be presented in this work, it will not consist on an exhaustive listing of existing routing protocols, but instead on a thorough analysis of works relevant to the development of insights towards scalable routing in MANETs. Other works provide a more generalized and broad list of routing protocols including position-based, multicast, multipath or even power-aware approaches [11,12,13]. However, these works do not focus specifically on the scalability of the routing concept and present protocols which provide only minor changes to other existing approaches with little added value. Some of these works are solely concerned with network awareness and dynamic routing, presenting extensions and new metrics to known routing protocols [14].

Currently there are several routing protocols that had a paramount importance in the development of ad-hoc networks which will be described and analysed. Nonetheless, this work focuses mainly in analysis of the protocols' ability to scale in large networks. A taxonomy of these routing protocols is depicted in Figure 1, showing the relationships between the different approaches to routing.

Despite all the provided mechanisms by each routing protocol, which will be presented, they are all subject to certain limitations and may fail in their purpose of scaling in large networks. Regarding the scalability of existing routing protocols their communication and storage complexity play an important role. Even though hierarchical solutions aim at being more scalable, this is not always true, since the complexity of these protocols is not necessarily better than flat solutions. Therefore, this work will also consider, for each class of routing protocols, a unique comparison table that highlights the techniques used by each routing scheme as well as their communication and storage complexities, in order to better understand their characteristics.

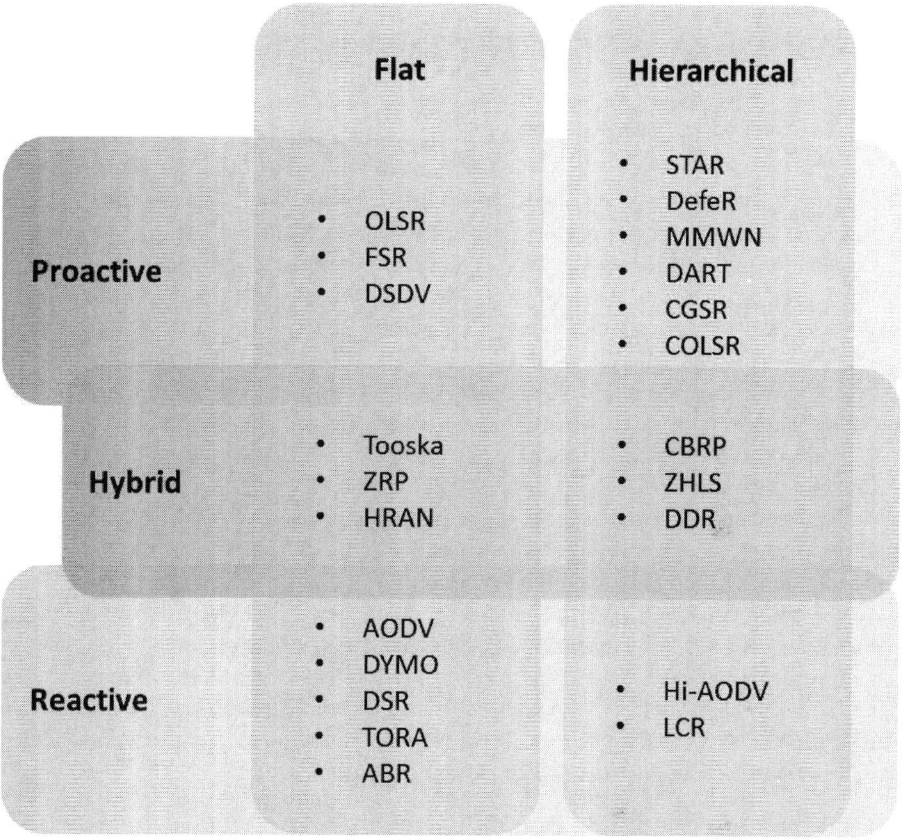

Fig. 1. Taxonomy of Wireless Multi-hop Routing Protocols

3 Proactive Routing Protocols

Following the inspiration provided by typical routing protocols used in wired networks, proactive routing protocols are based on the periodic exchange of routing messages in order to maintain updated routing tables. This paradigm allows a prompt retrieval of the next-hop to where data should be forwarded. However, this periodic update always occurs, even when there is no data to be transmitted, wasting resources without need.

3.1 Flat Proactive Routing Protocols

Flat Routing Protocols are characterized for not having any particular hierarchy to help in the organisation of the network. These are most the commonly found protocols and represent the foundation of MANET routing.

3.1.1 Highly Dynamic Destination-Sequenced Distance-Vector Routing Protocol

A routing protocol that results from a modification to the well known Distributed Bellman Ford algorithm [15], the Destination-Sequenced Distance-Vector (DSDV) protocol [16] is a routing solution where looping related issues are efficiently solved. It is a multi-hop pro-active protocol where each node stores a routing table with one entry to all possible destinations and the number of hops to each node [17]. In addition to this, not being dependent of any intermodal coordination mechanism allows the DSDV protocol to be robust solution for routing in MANETs. The protocol is also designed taking into account Medium Access Control (MAC) Layer details and sleeping nodes which should not be disturbed unless necessary, thus improving the total network lifetime.

The DSDV protocol periodically broadcasts update packets or whenever relevant information is available. These packets contain a new sequence number and information about the destination's address, the number of hops required to reach the destination and the sequence number of the previously received information regarding that destination. Routes containing the most recent sequence numbers are preferred when a path calculation is to be made.

A drawback from the standard DSDV implementation is observed when an existing path becomes invalid due to one or more broken links. When this occurs, the DSDV protocol assigns infinity to the path's metric and an odd sequence number (greater than the older one), which will be propagated through the network. However, while the link failure information is being propagated, some nodes will still drop several packets due to inaccurate information. This phenomenon is referred in literature as a stale route, requiring additional mechanisms to improve the response to a link failure. The Improved DSDV protocol [18] tackles this problem by maintaining a secondary routing table, which contains alternative routes to all the available destinations.

3.1.2 Optimized Link-State Routing Protocol

The Optimized Link-State Routing Protocol [19] is a variant of the typical link-state routing protocols which inherits the advantage of having routes immediately available while, at the same time, providing adequate optimisations for Ad-hoc Networks. The main mechanisms used by Optimized Link State Routing (OLSR) to improve its performance are the exchange of a reduced and non-synchronized amount of control packets used for link sensing and neighbourhood detection. This improvement consists of an efficient flooding technique based on the selection of Multipoint Relays (MPRs), minimizing the required bandwidth for protocol operations and avoiding the reception of redundant control messages [20]. An additional mechanism ensures that the required topology information is efficiently selected and diffused throughout the network.

Link Sensing

By periodically sending *HELLO* messages through the available wireless interfaces in which connectivity is confirmed (1-hop exchange between neighbour nodes), the OLSR protocol performs a link sensing operation. From this operation results a link set which contains the available information on "local" (1-hop) interfaces and on "remote" (2-hop) interfaces. This procedure may be replaced by link-layer information, if such feature is both available and sufficient to populate the link set, thus avoiding the exchange of *HELLO* messages.

Each link contained in the link set is described by a pair of interfaces, the local and the remote interfaces, and it has associated to itself the status of being either symmetric or asymmetric, depending on whether it can respectively send and receive data packets.

Neighbourhood Detection

The neighbourhood detection process consists in maintaining a set of neighbourhood tuples directly connected with the nodes' main address. The relationship between the OLSR main address and additional addresses is defined through the exchange of Multiple Interface Declaration (MID) messages.

There is a clear relationship between the neighbourhood set and the link set earlier described. In fact a node may only be considered "neighbour" of another *iif* there is a link between each other.

In addition to the neighbour set, there is a 2-hop neighbour set consisting of a set of nodes which have a symmetric link to a symmetric neighbour, being all this information gathered from the exchanged *HELLO* messages.

Still contained within the neighbourhood detection process, the population of both Multipoint Relay and Multipoint Relay Selector Sets is performed. MPRs are responsible for the existing flooding optimisation in OLSR as only them forward routing messages, avoiding a pure flooding approach where all nodes forward the protocol messages. Additionally they also avoid the transmission of duplicate messages by maintaining a Duplicate Set which records recently received messages as a "duplicate tuple" containing information about the originator address, message sequence number and a boolean indicating whether the message has been transmitted or not.

The selection of the MPR set is performed individually by each node which is responsible for selecting the most suitable nodes in its symmetric 1-hop set. This selection is performed in such a way that the node populating the set is able to reach all its strict 2-hop symmetric neighbours through the neighbours contained in the MPR set. Whenever changes occur in the 1-hop or strict 2-hop symmetric neighbours set, a complete recalculation of the MPR set is performed. Even though the MPR set does not have to be minimal, all strict 2-hop neighbours have to be reached through the selected MPR nodes and, in the worst case scenario, the MPR set may consist of the entire neighbourhood set, resulting in a typical link-state routing full flooding strategy.

The calculated MPR set may vary depending on the existing neighbourhood and on the nodes' willingness to act as MPR. This parameter is defined by the node depending on the available resources and other characteristics, being the values between WILL_NEVER and WILL_ALWAYS.

Finally, the MPR Selector set of a node *n* consists of all the addresses of nodes which have selected *n* as MPR.

Topology Discovery

By performing the already mentioned link sensing and neighbour detection procedures, each node is able to communicate with the directly connected neighbour nodes and it can participate in an optimised flooding mechanism. However, this information has to be disseminated through the entire network in order to allow the construction of routes to every node. This is done by MPR nodes which periodically send a Topology Control (*TC*) message with a set of links, known as Advertised Link Set, which contains the links to all the nodes in the MPR Selector set.

MPRs broadcast *TC* messages, flooding them to all the nodes in the network using other MPRs to efficiently improve the distribution of topology information, enabling greater scalability.

The generation of Topology Control messages is periodically performed by all MPR nodes in a time interval defined by the constant TC_INTERVAL, which can have several values such that, for a lower interval, a higher capacity of reaction to link failures is achieved. Related with failures, whenever a change to the MPR Selector set is detected, possibly due to a link failure, a new *TC* message should be sent earlier than the next interval generated message.

A common problem inherent to proactive protocols is the synchronization of control messages such as *HELLO* and *TC* messages. This increases the network overhead and may lead to losses due to collisions. In order to avoid this phenomenon, which typically arises with periodically sent messages, the OLSR protocol randomly defines a value named $jitter$ which should be between 0 and MAXJITTER. This value is used in the actual message interval by subtracting the $jitter$ value to it, thus varying the period in which messages are sent, and avoiding equal message transmission times which may synchronize.

OLSR v2

Currently still under development by the Internet Engineering Task Force (IETF) MANET working group, a new version of the OLSR protocol, the Optimized Link State Routing version 2 (OLSRv2) [21], proposes an update to the mechanisms of its predecessor. Even though the main algorithms are maintained, this new version offers a more modular and therefore flexible architecture, allowing, for instance, the addition of security extensions without compromising backwards and forwards compatibility [22]. Moreover, it also uses the Neighborhood Discovery Protocol (NHDP) [23] for the discovery of 1-hop and 2-hop neighbours, as well as discovering whether links are bidirectional, by sending *HELLO* messages similarly to the standard version of OLSR. The OLSRv2 protocol also implements the MPR Flooding process so that the link state information advertised by the protocol is efficiently propagated.

3.1.3 Fisheye State Routing Protocol

The proactive Fisheye State Routing (FSR) protocol [24] is inspired and takes its name from a well known technique proposed by Kleinrock et al. named Fisheye [25], originally used to reduce the size of samples required to represent graphical data. Similarly to a fish's eye, where the images are more detailed closer to the eye's focal point, a node using the FSR protocol has a better perception of its closer neighbourhood, updating information about more distant nodes with a lower periodicity.

As a link-state routing protocol, the FSR protocol maintains a topology map of the network at each node. However, instead of flooding a network change when it is detected, it proposes a different scheme for information dissemination [26].

In order to reduce routing control overhead, instead of sending routing updates at a fixed period, the FSR protocol uses different time intervals to exchange its routing information with nodes at different distances. Each node receives routing updates from further away nodes less frequently, maintaining a less accurate view of distant routes. However, whenever data is forwarded through the network, the precision of the used routes gradually improves as it gets closer to the desired destination.

3.2 Hierarchical Proactive Routing Protocols

The definition of specific hierarchies by different routing protocols has commonly been used, aiming at keeping the protocols more scalable. In contrast with typical flat routing protocols, hierarchical protocols usually exchange their routing information in different ways, according to a cluster or node hierarchy level.

The usage of hierarchies in conjunction with proactive routing approaches can be observed as a hierarchy of clusters, as an organised tree of addresses, or even as trees of paths forming a topology. Several schemes exist and all attempt to efficiently handle routing with the least overhead possible, as presented next.

3.2.1 Source-Tree Adaptive Routing Protocol

The Source-Tree Adaptive Routing (STAR) protocol [27,28] is a link-state protocol which has on average less overhead than on-demand routing protocols. Its bandwidth efficiency is accomplished by restraining the dissemination of link-state information only to the routers in the data path towards the desired destinations. STAR also creates paths that may not be optimal while avoiding loops, such that the total available bandwidth is increased. Moreover, STAR has specific mechanisms to know when update messages must be transmitted to detect new destinations, unreachable destinations, and loops.

Despite being able to scale, as each node only maintains a partial topology graph of the network, the STAR may suffer from large memory and processing overheads in scenarios where constant mobility may report different source trees, and routing paths are too big due to the network size.

3.2.2 Multimedia Support in Mobile Wireless Networks

The work entitled Multimedia support in Mobile Wireless Networks (MMWN) [29], the authors propose an architecture consisting of two main elements, corresponding to different node types, which can either be switches or endpoints. Both of these can be mobile, however only switch nodes can route packets and only endpoints can be sources or destinations for packets. This protocol also keeps a cluster hierarchy as a location management scheme, capable of obtaining the address of an endpoint. This information is kept as a dynamic distributed database, such that in each node there is a location manager node.

The proposed hierarchy allows the necessary amount of routing messages to be reduced, as only location managers are required to update their information and only then perform the location finding process [30]. However, this aspect is also negative on the overall performance of the protocol, as routing is strongly related with the hierarchy of the network, making the routing process complex and more vulnerable to disruptions when location managers change.

3.2.3 Cluster-Head Gateway Switch Routing

Using the mechanisms introduced by DSDV, another proactive hierarchical routing protocol is the Cluster-head Gateway Switch Routing (CGSR) protocol [31], which uses a routing approach where clusters are formed by electing a cluster head node, aiming to reduce the communication overhead, and thus making routing scalable and efficient. After the election of a cluster head, all nodes within its range will be considered as belonging to that cluster and all route updates should be done within its scope. All route discovery packets are forwarded through the cluster-head node.

One important task of this protocol is, essentially, the clusterhead election process. Authors argue that, when using distributed clustering algorithms, two possible choices are the lowest-Identifier (ID) algorithm and the highest-connectivity (degree) algorithm. The most important aspect to be taken into consideration when picking a clustering algorithm is stability. In order to avoid constant cluster head changes, which can harmfully impact the performance of other underlying protocols being used (such as DSDV), the algorithm chosen by the CGSR protocol is the Least Cluster Change (LCC) clustering algorithm [32]. This clustering algorithm is proposed as an improvement to existing algorithms, achieving enhanced stability.

Even though the proposed two-level cluster hierarchy may reduce the amount of flooding for dissemination of routing information, as only the cluster-heads are responsible for this task, the process of maintaining these clusters involves additional overheads, in particular the election of an appropriate cluster-head node. Moreover, this special node will always represent a bottleneck on each cluster, overloading it and possibly leading to a faster energy depletion and consequent cluster-head re-election.

3.2.4 Cluster-Based OLSR Extensions to Reduce Control Overhead in Mobile Ad Hoc Networks

The work entitled "Cluster-based OLSR (C-OLSR) extensions to reduce control overhead in mobile ad hoc networks" [33], proposes an extension to the OLSR protocol by introducing a cluster organised network. The authors propose a scheme where the existing clusters are considered as nodes themselves, using the MPR concept created by OLSR applied to clusters. This structure, in conjunction with the definition of Cluster *HELLO* (C-*HELLO*) and Cluster Topology Control (C-*TC*) messages, allows the maintenance of paths among the existing clusters while reducing the required amount of routing information, as only MPR Clusters generate C-*TC* messages.

Even though this paper uses the OLSR protocol for intra-cluster routing, proposing the mentioned C-*HELLO* and C-*TC* extensions to support a clustered network, the propagation of these new messages across clusters may have a negative impact. Moreover, the proposed mechanisms may suffer from mobility phenomena which, as in other approaches, require an additional overhead of updating the entire network structure.

3.2.5 Dynamic Address Routing for Scalable Ad-Hoc and Mesh Networks

Inspired on a previously work on a Dynamic Addressing paradigm, the authors propose Dynamic Address Routing (DART) for Scalable Ad-hoc and Mesh Networks [34], a proactive hierarchical approach that efficiently manages the organisation of nodes into zones for large scale networks. Address allocation and lookup are the main drawbacks of this proposal. However, the published work presents schemes to tackle these problems, showing how addresses can be allocated taking into account node positioning, by building a tree with l levels where l is the number of bits used in the routing address. A clear distinction is made between routing address and the identity of a node (a unique identification tag) since the routing address is dynamic and changes with node movement, contrasting with the node identifier which is always the same.

The three most important functionalities in DART are, first, the address allocation responsible for maintaining one routing address per network interface according to the movement and current position of a node; second, the routing which determines how to deliver packets from source to destination and, finally, the node lookup which consists in a distributed lookup table in charge of mapping identifiers to network addresses.

The DART proposal reveals to be an efficient solution for routing in large scale Adhoc networks. However, for small networks the Dynamic Address Heuristic has a strong overhead impact and in general it is difficult to implement, as the distributed lookup table is hard to manage.

Tree-Like Distance Vector

Inspired by the work presented in DART, the Tree-like Distance Vector (TLDV) routing protocol [35] uses a $2^b - ary$ tree locator and Distributed Hash Table (DHT), as opposed to DART's binary tree. The protocol also maintains at each node a routing and a neighbourhood table, being the routing table organised into $\lceil \log_{2^b} N \rceil$ rows with $(2^b - 1)$ entries each, in a network with N nodes. A major contribution from the TLDV protocol is not being restricted to the binary tree used by DART, exploiting a different space structure. However, the choice of parameter b needs careful consideration and strongly depends on the network's intrinsic properties. A trade-off between lower and higher values of b must be achieved between the size of the routing table, the amount of available locators and also route efficiency.

3.2.6 Deferred Routing Protocol

The Deferred Routing (DefeR) [36] approach consists on efficiently handling routing in clustered networks by defining a multiple view network hierarchy, achieved by aggregating clusters into different levels and by postponing routing decisions throughout traversed clusters until the final destination is reached [37]. Moreover, this routing protocol considers an enhanced mechanism that selects the most appropriate gateways by resorting to a link quality estimator [38].

This network organisation resembles the cartographic division of the world into continents, countries and cities, assigning identifiers with different granularities to each region. Another work with a similar approach, inspired by computational geometry techniques, is the Greedy Distributed Spanning Tree Routing protocol [39], which defines convex hull trees using nodes' absolute position information, which may not always be available, in order to optimize the routing process by sending packets to hulls which contain the desired destination's position.

In DefeR, a tree organisation of clusters is considered and, instead of traversing the entire tree looking for the desired cluster destination, the search can be optimized to a complexity of $O(\lfloor \log_2(n+1) \rfloor)$, for n clusters, without using any geographical position information [40]. Therefore, routes are established according to the cluster hierarchy, exploiting the different granularity levels of clusters within clusters. Moreover, the reliability of the links between clusters is taken into account, rather than minimizing the total hop count from source to destination.

One key advantage of using Deferred Routing is that, by keeping its optimised network hierarchy, it is able to limit not only the effects of micro but also macro-mobility, as clusters not involved in the mobility process of nodes are oblivious to changes in other clusters. Moreover, DefeR does not require additional routing messages for inter-cluster routing, being adaptable to any available link-state routing protocol with small changes to their own routing messages. The hierarchy employed by DefeR is based on a binary tree structure, motivated by the bisection that occurs in growing clusters and also by the base-2 logarithmic complexity of balanced binary-search-trees.

4 Reactive Routing Protocols

Proposed as an alternative to the expensive periodic update of proactive routing schemes, reactive protocols were introduced, performing route discoveries on-demand to avoid the waste of resources experienced with proactive solutions. This approach seems more suitable for mobile Ad-hoc networks where topology changes occur constantly. However, on-demand solutions suffer from an initial delay on retrieving a routing path which may not be acceptable, while at the same time the flooding of Route Request (RREQ) for route retrieval also adds an increased network overhead. In fact, several works aim solely at reducing the Broadcast Storm Problem (BSP), which was named after the broadcasting process typically involved in the dissemination of routing information of on-demand routing protocols [41].

4.1 Flat Reactive Routing Protocols

Similarly to proactive routing protocols, reactive approaches for wireless multi-hop routing can also be divided in two different categories where either flat or hierarchical network organisations are considered.

4.1.1 Ad Hoc On-Demand Distance Vector Routing Protocol

Designed for mobile wireless Ad-hoc networks, the Ad hoc On-demand Distance Vector (AODV) [42] routing protocol requires low memory and processing, while providing quick adaptation to dynamic link conditions. In addition to this, AODV has a low communication overhead and provides loop-free unicast routes in a reactive way, without having to maintain routes to destinations that are not currently in use.

Upon a request of a route to a destination, RREQ Packets are broadcasted throughout the network nodes until they reach their final destination or, alternatively, until an intermediate forwarding node, already containing an active/updated path to the destination, responds with a Route Reply (RREP) packet. Since each forwarding node keeps a reference of the source node triggering a RREQ, as well as the neighbour node likely to be used as next hop towards that destination, when processing a response message (a RREP), a node will route it back along the expected hops until it reaches the source node, making the new path available [43].

For efficiently managing the above described process, AODV uses sequence numbers to avoid loops and keep awareness of updated routes. Additionally, an assumption of bidirectional links is also present when a RREP is sent to its originating node. However, if unidirectional links exist, an alternative procedure needs to be used, in order to allow these packets to be correctly replied.

One important practice to be considered in AODV is the usage of an expanding ring search technique. This measure aims at preventing unnecessary network-wide dissemination of RREQ messages by controlling the extent to which these packets are broadcasted. This optimisation can be achieved by effectively setting some AODV specific parameters to the most appropriate level.

In order to keep accurate information about the active routes and avoid disruptive failures, each node monitors the link status of their next hops in these paths. The monitoring process is typically achieved by exchanging *HELLO* messages through the links, even though other mechanisms may be used. Upon the detection of a link break, a Route Error (RERR) message is used to notify the other nodes present in the path that a link loss occurred. Then, after receiving this message, the source node may decide to re-trigger a new RREQ, setting up a new route. Some extensions to AODV have already been proposed to address this specific point, for instance the Ad hoc On-demand Distance Vector Backup Routing (AODV-BR) Protocol [44].

The authors propose a modified version of the AODV protocol which not only uses the Expanded Ring Search (ERS) mechanism, but also a new approach named Hop Prediction, which improves the route search used by AODV. History records are maintained to each discovered route in order to optimise the ERS and reduce the overall routing overhead.

4.1.2 Dynamic Source Routing Protocol

An example of a completely reactive protocol with support of unidirectional links is the Dynamic Source Routing (DSR) [45] protocol proposed in the IETF MANET working group [46]. Aiming at scalability, in a network of at most two hundred nodes, the DSR protocol provides a soft-state approach where the two basic operations are Route Discovery and Route Maintenance, supporting asymmetric routes and assuming a typically small network diameter.

Designed for Internet Protocol version 4 (IPv4) addresses, provided by any mechanism such as Dynamic Host Configuration Protocol (DHCP) for dynamic assignment or static configurations, the DSR protocol is a loop-free protocol capable of quickly adapting to network topology changes. These adjustments of the topology only have an impact on the protocol when they affect paths currently active, being ignored by any other nodes. However, in order to avoid routing based only on flooding, topology changes related with mobility or other circumstances are not expected to happen so fast that the DSR protocol cannot adapt.

DSR uses explicit source routing, where an ordered list of nodes through which the discovery packet will pass, from source to destination, is used to allow multiple paths that enable the usage of load balancing mechanisms [47]. It also enhances the protocol robustness by tolerating path failures, choosing alternative ones immediately. Route caching is also an interesting feature that results from the forwarding and overhearing nodes' action of gathering information that can be used in the future, avoiding the Route Discovery process. When queried about a path, by performing a search in its local cache, a node can immediately retrieve the desired route and avoid further overheads of a Route Discovery process.

In the worst case scenario, when a complete Route Discovery has to be performed, the first node, the initiator, transmits a RREQ that will be broadcasted to all of the nodes until it reaches the destination node, the target. When this node is finally reached, it checks for a previous path cached to the initiator and sends a RREP. Otherwise it will start a new RREQ for the initiator, piggybacking the list retrieved by the first Route Discovery. Optionally, the target could simply reverse the path contained in the list given by the received RREQ, avoiding additional overhead but losing the asymmetric path support property.

Route maintenance in DSR states that each node is responsible for managing the flow over the link from that node to the next hop. This can be done either by using software or hardware acknowledgements, and a limited number of retransmissions. After the maximum number of retransmissions, a link is said broken and so the link is removed from Route Cache and a RERR is returned. If an alternative path exists in the initiator it shall be used, otherwise a new Route Discovery should be triggered.

4.1.3 Temporally-Ordered Routing Algorithm

Being a member of the link-reversal algorithms class, the multi-path and loop-free Temporally-Ordered Routing Algorithm (TORA) [48], is an on-demand source initiated routing protocol designed for multihop networks, which can also have destination initiated proactive routing for path optimisation and maintenance purposes.

Concerning routing, TORA routers only keep information about their one-hop neighbours and perform on-demand routing when retrieving a path to a destination. This operation performs best in networks with relatively sparse traffic patterns. At the same time, destination oriented mechanisms can also be triggered to maintain and monitor the path.

Summarizing TORA, it can be defined as four separate basic functions, namely creating routes, maintaining routes, erasing routes and optimising routes. For this, four different packet types are used: Query, Update, Clear and Optimisation [49]. TORA is an interesting protocol from the point of view that it does not use shortest paths to support its decisions and neither does it follow a link-state nor distance-vector algorithm.

4.1.4 Dynamic MANET On-Demand Routing Protocol

Much resembling with DSR and AODV, Dynamic MANET On-demand (DYMO) routing protocol [50] is a reactive loop-free routing protocol. Designed for networks with bidirectional links and capable of handling a wide range of mobility patterns, by dynamically determining routes in large scale networks, DYMO is best suited for sparse traffic scenarios. Having only to maintain minimal routing state information, it is a light-weight protocol applicable to devices with memory constraints.

The most relevant operations of the DYMO routing protocol are similar with DSR's Route Discovery and Route Maintenance. The former starts with the initiator node by sending a RREQ Packet to be broadcasted by all nodes until it reaches the desired target destination, which then replies with a RREP Packet through the best path, defined by a list that contains all the RREQ forwarding nodes. In order to reduce RREQ overhead, a forwarding node containing an active path to the destination may automatically respond with a RREP packet on behalf of the target node, avoiding further propagation of messages. An additional consideration is the usage of an adequate value for the HopLimit parameter which, to delimit the expanding ring of a RREQ, may be defined as described for the AODV protocol.

Complementing the above presented process, the Route Maintenance procedure is responsible for safeguarding the existing routes in use [51]. Route lifetime is extended by routers whenever a packet is correctly forwarded or a RERR packet is sent towards the packet source to indicate that the path contains an invalid or missing node. Additionally, by monitoring links over which traffic is flowing, any broken link detection should also immediately issue a RERR packet in order to swiftly notify DYMO nodes that certain routes are no longer available.

4.1.5 Associativity Based Routing

The Associativity Based Routing (ABR) [52] principle consists on the fact that after some migration process, where associativity ticks can be analysed, a certain stability time will exist, where a node will stay dormant within a cell before it moves again. The associativity tickets are analysed on the link layer level allowing to understand the degree of mobility of a node, where low associativity tickets are a synonym of a higher state of mobility and, on the other hand, high associativity ticks represent a stable state [53].

Route Discovery and Route Re-Construction are the two phases that compose the ABR protocol. During the Route Discovery phase a Query packet is broadcasted from source to destination, which then replies with a Reply message. The Query message is forwarded by every intermediate node that will keep the information of its upstream peer, removing it from the original packet and adding its own. If a duplicate Query is received by a node, it will be discarded. When a Reply message is sent back by the destination, nodes receiving this packet will set the path from source to destination as valid and active. Other nodes containing alternative paths will have them marked as invalid and will not relay packets to the destination, even if they hear the transmission.

Complementing the Route Discovery process, the Route Re-Construction phase handles possible failures caused by mobility or other situations by performing a partial route discovery, invalid route erasure, valid route update and, in the worst case scenario, new route discovery, which consists in the repetition of the entire processes described for the Route Discovery Phase.

4.2 Hierarchical Reactive Routing Protocols

The usage of Hierarchical Reactive Protocols is modest when compared with proactive or hybrid routing approaches. This is likely due to the fact that most well defined hierarchies require constant updates in order to be efficiently kept, going against the concept behind Reactive Routing, which only exchanges routing information when required. Nevertheless, some Hierarchical Reactive protocols do exist and are described in the following paragraphs.

4.2.1 Hierarchical AODV Routing Protocol

As the name indicates, the Hierarchical AODV (Hi-AODV) Routing Protocol [54,55] is a hierarchical version of the well known AODV routing protocol, using a tree based on cluster-heads for the creation of the concept of virtual nodes, which correspond to a typical cluster. The cluster-head is the only node responsible for handling control packets and managing the routing table of its own internal cluster. Having a tree composed of clusters seen as a virtual node allows Hi-AODV to reduce the number of control packets and avoid additional overhead.

In addition to the already mentioned challenges and overheads related to the maintenance of clusters and their cluster-heads (e.g. the cluster-head election process), again, it is clear that even though routing overheads can be reduced, the cluster-head will always have to be part of any routing path, leading to non-optimal paths, and additional interferences in the vicinities of cluster-heads.

4.2.2 Layered Cluster-Based Routing

The Layered Cluster-based Routing (LCR) protocol [56] is a hierarchical reactive protocol which exploits the main features of the Tiered Based Clustering Algorithm (TBCA) [57] also proposed by the same authors. This clustering scheme is organised into layered stages so that the number of nodes participating in the clustering process, at a given instant, is reduced. By the end of the clustering process a connected dominating set consisting of the elected Cluster-heads and Gateway nodes is formed.

Using an on-demand approach, the LCR protocol restricts its search space to the dominating set retrieved from TBCA. Whenever a new route is required to reach a destination, the initiating or source node broadcasts a RREQ and waits for a specific time interval before issuing a new request. This request is only propagated by dominating nodes, which maintain a table of previous requests (Table_request) in order to refrain a duplicate request, thus avoiding additional overhead. When the destination node receives a RREQ, similarly to the AODV protocol, it sends a RREP and initiates a route maintenance process which periodically exchanges *HELLO* messages between the nodes involved in the route, sending a RERR message if a route failure is detected.

Additional mechanisms used by LCR concern the sensing period of the source and dominating nodes. In fact, the source node's sensing wait time is set to a sensing period equal to Short InterFrame Space (SIFS), where the cluster-head's waiting time is equal to Point coordination InterFrame Space (PIFS) and the Gateway (GW)'s waiting time is equal to Distributed InterFrame Space (DIFS). These specific times are set in order to reduce the probability of collisions during the discovery phase.

Optimized Layered Cluster-Based Routing

An update to the LCR protocol was provided by its original authors [58], optimising the MAC-layer mechanisms to avoid collisions and defining a direction mechanism that reduces the number of dominating nodes involved in the routing process. This direction-based mechanism is free from any positioning techniques, such as Global Positioning System (GPS), using information included in the resulting layers from the clustering process and allowing dominating nodes to discard any RREQ when, for instance, it reaches higher layers than the layer where the destination is expected to be. In certain scenarios where this mechanism may not be available, the LCR protocol performs normally without any disadvantages.

5 Hybrid Routing Protocols

Recognising both the advantages and disadvantages of proactive and reactive routing protocols, hybrid routing protocols were proposed. The concept behind this new alternative is that the best of each approach (proactive and reactive) can be exploited together in the different tasks performed by a routing protocol.

5.1 Flat Hybrid Routing Protocols

Even though flat and hierarchical routing schemes can also be found in proactive and, even though less frequently, in reactive routing protocols, hierarchical routing approaches are usually more common in hybrid routing protocols which also consider flat network perspectives.

5.1.1 Zone Routing Protocol

Combining the advantages of the pro-active and reactive paradigms, the Zone Routing Protocol (ZRP) [59], proposes a zone based architecture where three embedded protocols, the Intra-zone Routing Protocol (IARP), the Inter-zone Routing Protocol (IERP) and the Bordercast Resolution Protocol (BRP), are responsible for maintaining the routing operation.

Assuming that a majority of the processed traffic occurs directly between neighbour nodes, the strategy of ZRP is to reduce the scope of proactive traffic into a zone centred on each node. The zones are defined as having a r radius expressed in hops, such that a zone includes nodes whose distance from a given node is at most r hops. Since zones overlap, ZRP is said to have a flat view of the network. This perspective results from the authors' statement that this approach can be used to detect optimal routes and to reduce network congestion.

The IARP is the protocol used within ZRP zones to proactively maintain routing tables up-to-date. In contrast, route discovery outside of a specific zone is made by the reactive IERP protocol. Using the information of IARP, an additional routing procedure is made by BRP, which consists in managing the packet delivery to the peripheral nodes in the border of a zone (bordercasting). The usage of this approach in conjunction with IERP allows a reactive route discovery to efficiently travel between zones [60].

The size of the zones used by ZRP can be managed by regulating the transmission power of devices (if such option is available). Additionally, mechanisms to efficiently and possibly dynamically choose r should be used, ensuring that a zone is big enough to provide a good connectivity between nodes, but not too big so that update traffic does not become excessive. However, such a dynamic process is complex and not easy to achieve [61]. Further works provide analytical models that determine the routing overhead incurred by the ZRP protocol and its variants. Some examples are the Independent Zone Routing Protocol (IZRP) [62], which proposes mechanisms for calculating the optimal zone radius of the node, being more efficient than the standard ZRP [63]. These mechanisms are known as min-searching and adaptive traffic estimation, and allow each node to have its own independent zone size. The Two-Zone Routing Protocol (TZRP) [64], that presents a zone-based architecture that decouples the (basic hybrid) protocol's ability to adapt to changing traffic patterns from the ability to adapt to different mobility models. And also the Fisheye Zone Routing Protocol (FZRP) [65], where the architecture defined by the ZRP uses Fisheye State Routing in its proactive operations.

5.1.2 Tooska Scheme and Mobility Aware Hybrid Routing

The Tooska Routing scheme [66] is a hybrid node-centric protocol which relies on AODV as its default routing protocol, switching to the Wireless Routing Protocol (WRP) [67] when appropriate. By selecting the nodes with more stable fixed neighbours, the core nodes, the protocol defines these intermediate nodes when data needs to be sent, through the analysis of the *HELLO* Message Counter (HMC) field stored by each node. Core nodes periodically update their routing tables by changing to the WRP protocol, informing all the remaining nodes of this change. In order to reduce the overhead introduced by the Tooska scheme, the number of core nodes is minimized by defining a minimum number of required stable neighbours.

Due to node mobility, the selection of core nodes can become inefficient in the Tooska scheme as it is proposed. Bearing this in mind, the Mobility Aware Hybrid Routing (MAHR) [68] defines an alternative selection method for core nodes, where the ratio of changing neighbour nodes is used. The routing process is similar to Tooska relying on the AODV protocol for route discovery, where the core nodes are responsible for the maintenance of routing tables by using the OLSR protocol.

5.1.3 Heat Routing for Ad-Hoc Networks

Parallel to the behaviour of heat trails in the physical world, wireless nodes using the Heat Routing for Ad-hoc Network (HRAN) protocol [69,70] emit a heat signal to be perceived by neighbour nodes. The amount of heat detected by each node depends on a gradient function such that nodes further away from the heat source register a lower level of heat when compared with 1-hop distant nodes.

The protocol's main mechanisms consist on the creation of a heat overlay, where each node proactively disseminates its topology information, with an amount of heat defined by a Time Aware Bloom Filter (TAB) which is a new type of Bloom Filter defined by the authors. The heat information is included in periodically sent *HELLO* messages, as the size of the used TAB never changes regardless of network size, creating a heat overlay or heat trails.

By using an on-demand approach, the second stage of the HRAN protocol consists on discovering a valid route from source to destination. This is achieved by issuing a predetermined number of Random Walk Request (RwREQ) queries, to be sent throughout the network. Upon receiving a RwREQ, a node checks if the received destination identifier is present in any of its registered heat trails. If a match is obtained, the random walk is terminated and a direct walk takes its place. This walk is started by sending a Follow Heat (FoHEAT) message which is only forwarded by nodes in the same heat trail, allowing the query to quickly reach the destination.

When a RwREQ reaches its intended destination node, it sends a Route Reply (RoREP) message to the source, using the discovered route, inverted. If after sending a RwREQ, a predefined time-out is reached and no RoREP is received by the source node, the protocol falls back to a typical reactive source routing protocol such as AODV. This mechanism is important as it allows the creation of heat tunnels which otherwise are only created after a route establishment, during the route maintenance process.

The final contribution of the HRAN protocol is the maintenance of routes which include the creation of heat tunnels, achieved by adding the destination's identifier in the proactively sent routing messages. This creates a "highway" for future route requests to this destination. Moreover, this process also ensures that failed routes are repaired by sending Route Repair (RoREPAIR) messages and it further aims at improving the found routing path. Since the first retrieved path may not be the shortest path due to the randomness of route discovery process, an additional message named as Route Improvement (RoIMP) is sent by the source within the heat tunnel, until it reaches the destination. In its turn, the destination sends back to the source a new RoREP, using an inverted more efficient path.

5.2 Hierarchical Hybrid Routing Protocols

Quite a few Hybrid Routing protocols for Ad-hoc networks can be found in the literature, however, despite the fact that many rely on clusters or well defined zones, not many implement a hierarchical routing scheme. The following protocols propose a hybrid routing scheme capable of retrieving inter-cluster information in a reactive approach, avoiding the necessity of restraining routing information in cluster-heads to reduce the overall overhead. However, on a downside, inter-cluster communication may be subject to route retrieval delay if no previous path has been maintained in cache.

5.2.1 Zone-Based Hierarchical Link-State Routing Protocol

The Zone-based Hierarchical Link-State (ZHLS) Routing Protocol [71], is characterized by dividing the network into non-overlapping zones where two different routing paradigms are used: proactive routing within the zones and reactive between different zones. This proposal alleviates single points of failure and bottlenecks by not being dependent on cluster-head nodes and, at the same time, by maintaining a scalable hierarchy based topology.

One important assumption, and a possible limitation from this protocol is that each node knows its own position (for instance, by using GPS) and consequently its zone ID which is directly mapped to the node position. With this approach, packets are forwarded by specifying in their header the zone ID and node ID of their destination.

The division of the network into a number of zones depends on factors such as node mobility, network density, transmission power and propagation characteristics. The geographic awareness is much more important in this partitioning process as it facilitates it when compared to radio propagation partitioning [72].

In addition to the limitation of requiring some positioning system, the ZHLS protocol requires that all nodes exchange inter-zone flooding information when only gateway nodes need this routing information for calculating the shortest path between different zones. Moreover, the ZHLS is susceptible to a route retrieval delay when establishing inter-zone paths, as reactive routing is used for this purpose.

In ZHLS, each node contains an intrazone and interzone routing table to manage routing between nodes from a same zone and from different zones respectively. The update of these tables is performed, by sending two types of Link State Packets (LSPs), node LSP and zone LSP for intrazone and interzone, in that order.

A proposal to enhance the routing, by ZHLS is given in [73], where the ZHLS Gateway Flooding (ZHLSGF) scheme is defined to reduce routing overheads and reduce routing tables' size. This modification is closely related with the nodes that act as a border between different zones, since they are responsible for calculating the shortest path between other gateway nodes, only sending interzone discovery packets between each other, thus avoiding unnecessary packet forwarding to other nodes within the zone.

5.2.2 Distributed Dynamic Routing

Another hierarchical hybrid routing protocol, the Distributed Dynamic Routing (DDR) algorithm [74], for mobile Ad-hoc networks, is a tree based routing protocol which consists of six different stages where an election of the preferred neighbour is made, followed by the forest construction which creates a suitable structure for the wireless network, allowing an improved resource utilisation. Afterwards intra and inter tree clustering is performed, followed by zone naming and partitioning. Zones are responsible for maintaining the protocol scalable and reducing the delay.

While DDR creates and maintains a dynamic logical structure of the wireless network, the Hybrid Ad Hoc Routing Protocol (HARP) [75] finds and maintains routing paths. The HARP protocol aims at discovering the most suitable end-to-end path from a source to a destination by using a proactive intra-zone routing approach and a reactive inter-zone scheme, by performing an on demand path discovery and by maintaining it while necessary.

Even though the DDR algorithm does not require any sort of cluster-head for cluster maintenance, the possibility of some nodes being chosen as preferred neighbours by other nodes may lead to the creation of bottlenecks, as they would be required to transmit an increased amount of both routing and data packets. It is important that the choice of preferred neighbours is balanced so that the overall performance of the protocol does not get compromised. Moreover maintaining the entire logical structure of the network may be somewhat heavy, depending on how dynamic nodes may be [76].

5.2.3 Cluster Based Routing Protocol

Aiming at a scalable, loop free routing protocol with support for asymmetric links, the Cluster Based Routing Protocol (CBRP) [77] proposes a variation of the "Min-Id" [78,79] for cluster formation, in which the purpose is to create a hierarchy consisting of overlapping 2-hop-diameter clusters where a node is elected as cluster head, responsible for maintaining cluster membership information. By exploiting the cluster architecture, flooding traffic used in the routing process is minimized.

As a 2-level hierarchy, this protocol can be scalable to a certain extent, however, the typical cluster formation and cluster-head election overhead still exists. Even though node mobility does not necessarily lead to inaccurate routing table calculations, as it would happen with a purely proactive approach, the inherent route retrieval propagation delay may lead to temporary loops.

In the Routing Process, RREQ packets are flooded from source to destination, but only cluster head nodes are used in this process. When these packets reach the target, a RREP is sent back to the initiator node [80]. Even though this process is triggered by an on-demand request, additionally, every node within a cluster zone periodically exchanges with its neighbours routing table information by using *HELLO* packets. This pro-active behaviour in conjunction with the reactive on-demand requests positions the CBRP within the hybrid family of routing protocols.

In addition to the Routing process, the CBRP also defines two other major components which are Cluster Formation and Adjacent Cluster Discovery. The Cluster Formation process consists on the usage of a variation of the "lowest ID" clustering algorithm where a set of rules for electing the cluster head are defined. Wrapping the whole protocol, the Adjacent Cluster Discovery process aims at discovering all bi-directionally linked adjacent nodes. The process is executed by broadcasting the summarised Cluster Adjacent Table of each cluster head as Cluster Adjacency Extension to the *HELLO* messages.

6 Routing Protocols' Complexity Analysis

Despite all the presented mechanisms, proposed by each routing protocol, they are all subject to certain limitations and may fail in their purpose of scaling in large networks. Regarding the scalability of existing routing protocols their communication and storage complexity play an important role. Even though hierarchical solutions aim at being more scalable, this is not always true, since the complexity of these protocols is not necessarily better than flat solutions. Therefore, for each class of routing protocols, considering the defined taxonomy, a comparison table highlighting the techniques used by each routing scheme as well as their complexity communication and storage complexity will be provided.

6.1 Flat Proactive Routing Protocols – Comparison

Proactive routing protocols stand out for always maintaining routes to all the available destinations. In flat organisations clustering is not typically used however other scalability mechanisms can be found. Table 1 shows these mechanisms for the presented routing protocols and analyses their complexity regarding storage and communication.

6.2 Hierarchical Proactive Routing Protocols – Comparison

Even though hierarchical proactive routing protocols present more scalability oriented features than flat ones, the communication and storage complexities are not necessarily better. Moreover, the mechanisms used for this purpose, presented in Table 2 can be quite complex and introduce additional overheads that are not accounted as routing overheads. Nevertheless, the analysed routing protocols require that all the presented aspects are available and therefore may not be as flexible as desired.

Table 1. Comparison of Flat Proactive Routing Protocols

Protocol	Cluster-based	Scalability Techniques	Communication	Storage
DSDV	no	n/a	$O(N^2)$	$O(N)$
FSR	no	fisheye updates	$O(N^2)$	$O(N)$
OLSR	no	MPR nodes	$O(M^2)$	$O(N)$

N : Total number of nodes
M : Total number MPR nodes

Table 2. Comparison of Hierarchical Proactive Routing Protocols

Protocol	Cluster-based	Scalability Techniques	Communication	Storage
CGSR	yes	Cluster-head	$O(C^2)$	$O(N)$
C-OLSR	yes	Cluster-MPRs	$O(C^2)$	$O(N)$
DART	yes (zones)	Dynamic Addresses	$O(\log_2 N)$	$O(N)$
DefeR	yes	Deferred Routing and Aggregated Networks Views	$O(C)$	$O(N)$
MMWN	yes	Location Managers	$O(N^2)$	$O(N)$
STAR	no	Partial Topology	$O(N)$	$O(D)$

N : Total number of nodes
D : Total number of destinations
C : Average number of nodes per cluster

6.3 Flat Reactive Routing Protocols – Comparison

Reactive routing protocols aim at being more lightweight than proactive ones by sending routing information only when necessary. However, this approach may result in expensive flooding of RREQ whenever a route is required. In addition to this limitation, which is more critical in scenarios with several traffic flows, these protocols also suffer from a route retrieval delay. The communication and storage complexity of reactive protocols is expected to be lower than a proactive routing protocol, as they only consider the necessary destinations. However, in a worst case scenario for reactive routing protocols, each node may be a source and destination node, resulting in a complexity similar to proactive protocols, as shown in Table 3.

Table 3. Comparison of Flat Reactive Routing Protocols

Protocol	Cluster-based	Scalability Techniques	Communication	Storage
ABR	no	Associativity Tickets	$O(N^2)$	$O(N)$
AODV	no	modified ERS	$O(N^2)$	$O(N)$
DSR	no	n/a	$O(N^2)$	$O(N)$
DYMO	no	HopLimit	$O(N^2)$	$O(N)$
TORA	no	Directed Acyclic Graph	$O(N^2)$	$O(N)$

N : Total number of nodes

6.4 Hierarchical Reactive Routing Protocols – Comparison

In the existing literature there are few Hierarchical Reactive Routing protocols since maintaining a hierarchy typically requires constant updates. Table 4 compares the performance of the two protocols which depend entirely on the robustness of the used clustering processes.

Table 4. Comparison of Hierarchical Reactive Routing Protocols

Protocol	Cluster-based	Scalability Techniques	Communication	Storage
Hi-AODV	yes	Cluster-heads as Virtual Nodes	$O(C^2)$	$O(N)$
LCR	yes	TBCA	$O(C^2)$	$O(N)$

N : Total number of nodes
C : Total number of dominating nodes or cluster-heads

6.5 Flat Hybrid Routing Protocols – Comparison

Table 5 presents a comparison of the main characteristics of the analysed hybrid routing protocols with a flat network organisation. As a direct consequence of employing both proactive and reactive routing protocols their complexity is similar to these protocols. However, these protocols also provide optimisations that may enhance the protocols performance in many situations. Moreover, the usage of zones by ZRP reveals an alternative to achieve a more scalable routing process.

Table 5. Comparison of Flat Hybrid Routing Protocols

Protocol	Cluster-based	Scalability Techniques	Communication	Storage
HRAN	no	Heat overlay	$O(N^2)$	$O(N)$
Tooska	no	HMC	$O(N^2)$	$O(N)$
ZRP	yes (zones)	Variable zone radius	$O(Z^2)$	$O(\frac{N}{Z})$

N : Total number of nodes
Z : Total number of zones or cluster-heads

6.6 Hierarchical Hybrid Routing Protocols – Comparison

Hierarchical Hybrid routing protocols provide most of the existing advantages in the previously analysed protocols. Their mechanisms and complexity are presented in Table 6, revealing that in a worst case scenario these protocols have similar complexities. The tree-based DSR protocol has a higher communication complexity as it constructs its own forest of connected zones, therefore being more complete than other protocols.

Table 6. Comparison of Hierarchical Hybrid Routing Protocols

Protocol	Cluster-based	Scalability Techniques	Communication	Storage
CBRP	yes	2-level Hierarchy	$O(\frac{N}{Z}^2)$	$O(N)$
DSR	yes (zones)	Preferred Neighbours	$O(N^2)$	$O(N)$
ZHLS	yes (zones)	No Cluster-heads	$O(\frac{N}{Z}^2)$	$O(N)$

N : Total number of nodes
Z : Total number of zones or cluster-heads

6.7 Summary and Considerations

In the existing literature several routing protocols have been created for Mobile Adhoc Networks. However not all of these protocols provide a significant new approach for routing, being many times small extensions of the most relevant routing schemes. Moreover, many works rely on complex or even unrealistic assumptions which are not suitable for dynamic networks such as MANETs.

The presented analysis of current routing protocols for MANETs highlighted the contributions provided by routing protocols separated into different routing classes, taking also into account improvements made to and provided by these protocols. Nevertheless, several issues still exist, motivating the creation of new routing mechanisms for increasingly larger autonomous networks.

The comparison of the analysed protocols showed that reactive routing protocols are not necessarily more scalable in worst case scenarios where many flows exist. Moreover, it also demonstrated that cluster-based alternatives are able to maintain a smaller communication complexity. Even though the most scalable approach is provided by the DART protocol, regarding the communication complexity, the mechanisms necessary for this scalability to be achieved involve themselves additional overhead which is neither accounted as communication nor storage complexity.

Moreover, in the presented taxonomy it becomes clear there are more proactive and reactive routing protocols, reflecting the IETF MANET working group decision to maintain only two main routing protocol approaches. In particular, there are several more proactive protocols rather than reactive ones, showing a trend in what is expected from Mobile Ad-hoc Networks. This shows that generally, the purpose of ad-hoc networks will be related with large-scale dense scenarios, where mobility is expected to be moderate.

7 Performance Analysis Methodology

The presented literature analysis shows that, despite the initial trend on Ad-hoc networks to follow reactive approaches due to mobility, a large number of works have moved towards proactive routing protocols, guaranteeing increased support for large-scale networks where mobility may still be present. Bearing this in mind, and for comparison purposes, the presented evaluation considers the popular OLSR protocol and two other alternatives C-OLSR and DefeR. These comprise, respectively, three different proactive routing approaches, employing flat un-clustered routing, flat clustered routing and hierarchical clustered routing. By analysing the three approaches it is easier to understand which technique is more suitable for large-scale networks. The OLSR protocol was chosen as a control subject, providing a basis for comparison due to its stability and popularity in MANETs, being a standard protocol – currently under improvement by the MANETs IETF working group in its second version [21] – and the other two for being evolutions of this protocol.

7.1 Objectives

In order to understand to what extent the existing Ad-hoc networks can scale, the presented performance evaluation will simultaneously assess, in different conditions, several routing aspects and metrics. The relevant parameters will be explained and varied, triggering different changes in performance of the already mentioned routing protocols. These changes are then analysed and conclusions about the impact of each change will be provided.

Such an analysis was not possible using a real testbed due to the involved network dimension and therefore a simulation environment was considered, allowing a significant amount of different repetitions and a proper validation of the presented analysis, as well as an accurate generalization of the scalability performance of the protocols.

7.2 Simulation Conditions

The provided results were obtained using the OPNET Modeler Wireless Simulator [81], where the considered wireless nodes follow the Institute of Electrical and Electronics Engineers (IEEE) 802.11g standard [82] at 2.4Ghz, and have a maximum range of 100 meters (Transmit Power of $3.7e^{-4}W$), which corresponds to the maximum obtainable range of common wireless cards [83,84], unless stated otherwise. Nonetheless, due to the accurate radio model implemented by default in the OPNET Simulator, asymmetric links or even unidirectional links may occur, as well as channel errors and multi-path interferences respectively. Moreover, the Consultative Committee on International Radio (CCIR) propagation model was used, configured to represent a small to medium city with a building coverage of 15.8 percent, as it is considered as an appropriate propagation model for MANETs [85]. The usage of this simulation environment strives for being more realistic when compared with other works, which use the outdated 802.11b with non-standard MAC layers and unlikely ranges (for instance, 250m). Each evaluated scenario has specific variations of several simulation parameters since they independently assess different characteristics. Simulation parameters not mentioned here or in the definition of the scenarios are defined with the values used by default in the OPNET Modeler Wireless Suite Simulator, version 16.0.A PL1.

All the different parameters varied in each of the defined scenarios were obtained after 30 runs per parameter, always using different seed values and the Linear-Congruential Random Number Generator Algorithm, for a total simulated time of 15 minutes (900 seconds per run), which allows routing protocols to be appropriately evaluated by guaranteeing enough mobility [86].

Taking into account the defined objectives of this evaluation and their statistical validity, all the presented results have a 95% confidence interval obtained from the central limit theorem, which states that regardless of a random variable's actual distribution, as the number of samples (i.e. runs) grows larger, the random variable has a distribution that approaches that of a Normal random variable of mean m, corresponding to the same mean as the random variable itself.

7.3 Evaluation Metrics

As previously mentioned, in order to provide a thorough evaluation of the chosen MANET routing protocols and their behaviour in large-scale networks, it is important to simultaneously assess the performance of different routing aspects, choosing appropriate comparison metrics. For this purpose the following items were considered in provided evaluation:

- Traffic Delivery Performance
  - Losses
  - End-to-end Delay.

- Routing Performance
  - Path Length
  - Routing Stability
  - Control Traffic Overhead.

Taking these different aspects into consideration, this performance assessment must involve the evaluation of a large scale network, measuring the stability and overhead of this concept, as well as its overall traffic delivery performance. Moreover, in order to allow a more exhaustive evaluation it is important to determine the protocol's ability to handle mobility phenomena, introducing dynamic scenarios with different mobility models.

The average percentage of losses and end-to-end delay reflect a protocol's competency to choose suitable paths and are taken into account in this evaluation in all the presented scenarios. The percentage of losses strongly influences the applicability of a routing protocol in different scenarios. However, in Mobile Ad-hoc Networks a high number of losses is expected due to its inherent nature, where nodes are intermittently connected and where interferences and collisions are frequent [87]. Moreover, the delay metric is also subject to these interferences, limiting the usage of real-time applications in some scenarios. Nonetheless, in an extreme outlook, where only MANETs may be available, the registered losses may not be significant and retransmission mechanisms can be used to successfully deliver the required data.

A different routing metric considers the path length (hop count), from source to destination, which typically is minimized by routing protocols in order to reduce the number of nodes that intervene in the data delivery process. By reducing the number of hops, protocols are expected to be more energy efficient. However, this is not always the best option, as bottlenecks may arise and collisions will not only originate more losses but also a faster energy depletion on nodes in "popular" paths. Regarding this aspect, the DefeR protocol follows a different approach from the other two, choosing paths that minimize the total number of cluster-hops, selecting the most suitable GW nodes according to their quality.

In addition to these metrics, it is also important to measure the required resources and, therefore, routing traffic overhead, as well as the stability of the existing routes. Regarding the latter aspect, mobility of nodes is responsible for most of the topology changes and it is the protocol's task to efficiently handle these changes.

The topology awareness of a routing protocol is a metric representative of a routing protocol's stability and knowledge about the network's structure, registering topology changes during the simulation. A topology change occurs whenever a new *TC* or a *TC* with a higher sequence number is received and also when a *TC* entry is deleted after expiry. Each topology change triggers a routing table recalculation, however, in order to reduce computational overhead, the routing table is only recalculated by default at most every 1 second, processing all the received topology changes between each recalculation. Such technique is compliant with the OLSR specification and used in existing implementations [88,89]. Moreover, all the analysed protocols use this improvement in order ensure a fair comparison between them.

The amount of processed topology changes in routing table calculation reflects a protocol's stability and will also be analysed, referred as Average Topology Changes per Routing Table calculation (AToCRT) and defined by equation 1.

$$AToCRT = \frac{\text{Number of Topology Changes}}{\text{Number of Routing Table Calculations}} \tag{1}$$

The number of routing table calculations possible in a T seconds simulation is defined in equation 2, with i being the simulation instant where n Topology Changes occur. Since the number of topology changes is influenced by the mobility of nodes, the different speeds used in an evaluation will be reflected in the AToCRT metric and also on the total number of routing table calculations. In particular, with higher speeds, an increased number of Topology Changes throughout the time will trigger a higher number of routing table calculations, with a maximum of 1 per second, as defined by $f(n)$.

$$\text{Routing Table Calcs} = \sum_{i=1}^{T} f(TopologyChanges_i),\ f(n) = \begin{cases} 0 \text{ if } n = 0 \\ 1 \text{ if } n > 0 \end{cases} \tag{2}$$

Topology Changes are propagated by *TC* messages sent by MPR nodes. These messages are forwarded to all the elected MPR nodes and represent most of the routing overhead, as they are the only forwarded messages sent throughout the network. The number of forwards per *TC* messages must then be analysed, in order to correctly assess the scalability of a protocol.

The overall routing overhead must also be considered taking into account the periodically sent and received routing traffic from both *HELLO* and *TC* messages. This will also reflect the protocols' ability to handle a large number of nodes.

Regarding the creation of clusters used by both the DefeR and C-OLSR protocols, a static definition of the areas comprised by each cluster was used and a mechanism for the nodes to automatically update their Cluster Identifier (CID) was implemented. However, this approach does not guarantee a constant density of nodes within each cluster. Such limitation impacts the performance of both protocols, since, in a worst case scenario, all the nodes might move into one single cluster. Nonetheless, in a realistic scenario clustering algorithms may not be able to guarantee constant density unless they introduce limitations of their own (such a single-hop cluster coverage) [90].

8 Simulation Results

The performance of a routing protocol can be assessed through several parameters. Typically, a protocol's competency to deliver data packets successfully, paired with the end-to-end delay of the chosen path, determines whether a protocol has a good performance or not. However, in Mobile Ad-hoc Networks there are several other metrics and characteristics that must be analysed. The presented simulation results consider the already introduced evaluation metrics that are indicative of the protocols' behaviour taking into account scalability and resilience to mobility.

8.1 Scalability Assessment

In order to assess how scalable the mechanisms of a protocol are, a set of results where the total number of nodes increases should be obtained. By increasing the number of traffic flows, it is also possible to understand how the protocol handles not only the size of the network, but also how it copes with a demanding network where several routes must be established.

Following an approach where a growing size network is used, the total number of node clusters is incremented presenting a scalability evaluation of the routing protocols. This evaluation depicts the behaviour of these protocols with both small and large-scale networks. It is a straightforward assessment which somewhat disregards the nature of MANETs, as it does not take into account the natural behaviour of moving people, being entirely random regarding both mobility and traffic flows.

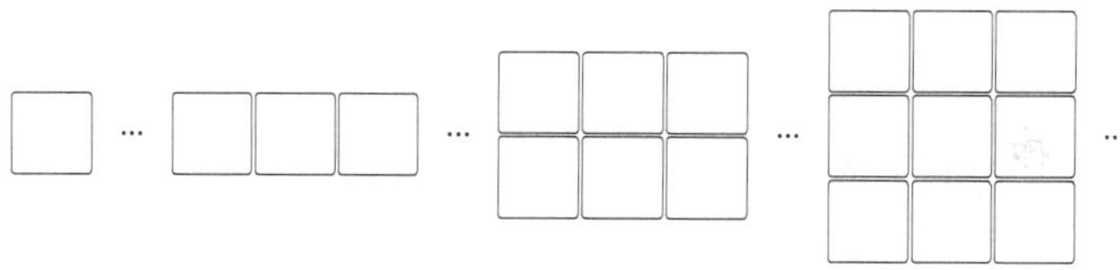

Fig. 2. Increasing Number of Clusters

A set of results from 1 cluster up to 10 clusters is provided, where each cluster has 49 nodes (which is the best number of nodes handled by OLSR [91]). The dimension of each cluster is of $500 \times 500m$, ensuring an initial constant density of the network. Figure 2 depicts the configuration of the network used in this scenario.

Regarding the smaller simulated networks with one cluster, both C-OLSR and DefeR behave exactly like OLSR, as both use it for intra-clustering and no inter-cluster operations are required. The provided results for smaller networks are important as other protocols designed for large-scale networks, such as Dynamic Address Routing, are known not to perform well in smaller networks [34].

In order to assess how the three protocols handle networks with a different number of traffic flows, the different size networks were also evaluated with 1, 4, 8 and 16 traffic flows. Each flow begins randomly after an interval between 50 and 250 seconds of simulation time, uniformly distributed, being concluded by the end of the simulation. The destination of each flow was randomly chosen, using a User Datagram Protocol (UDP) traffic type, with a constant bit rate of 8 packets of 4kbit per second, representative of typical interactive gaming, simple file transfers or information exchange [92].

In this scenario the DefeR protocol's ability to maintain a reduced overhead in scenarios where nodes are likely to move within nearby contexts is disregarded. All nodes randomly start their movement after an initial warm-up time, between 100 and 250 seconds (following an uniform distribution). The used mobility model is the Random Waypoint with a pause time of 60 seconds, without any distance or cluster restrictions, such that nodes are able to move freely across the entire network. The nodes' speed is uniform between 2 *and* $6km/h$, corresponding to pedestrians' walking speed [93].

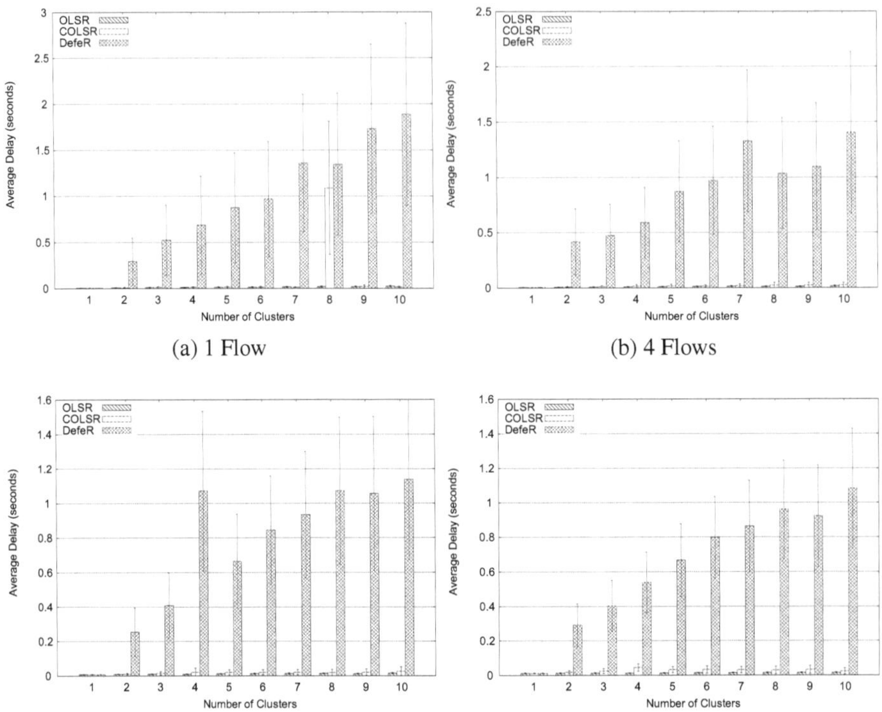

(a) 1 Flow (b) 4 Flows

(c) 8 Flows (d) 16 Flows

Fig. 3. Average Percentage of Losses

8.2 Obtained Results

Taking into account the discussed evaluation metrics, the obtained results in the defined scenario are presented next. Each metric is presented with the four different number of flows, side-by-side, in order to allow a better comparison.

8.2.1 Percentage of Losses

In any routing evaluation, the percentage of registered losses can be considered as an indicator of how a routing protocol performs. This is presented in Figure 3, where the obtained percentage of losses is clearly influenced by the number of clusters in the network. In a 1-cluster network, the three protocols have a similar performance, as all of them simply use the OLSR protocol for maintaining routing paths. While the growing number of flows varies only slightly the data traffic delivery performance, the increasing number of clusters has a higher impact, such that the C-OLSR protocol registers more than 80% of losses in networks with 4 or more clusters.

Regarding the overall percentage of losses, the DefeR protocol registers the best performance being able to constantly deliver more data packets than its competitors. However, the DefeR protocol still has a significant amount of losses in larger networks,

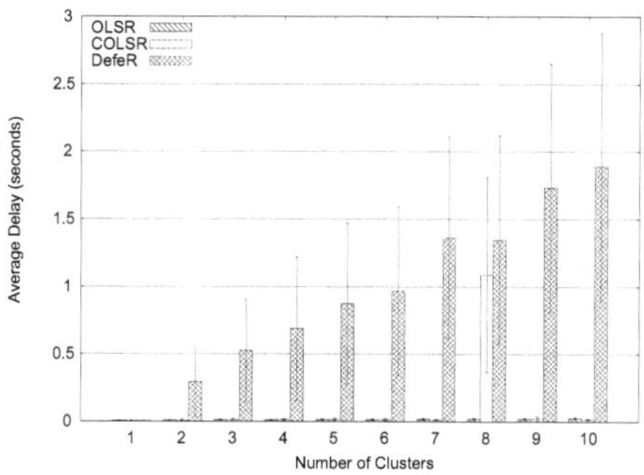

(a) 1 Flow

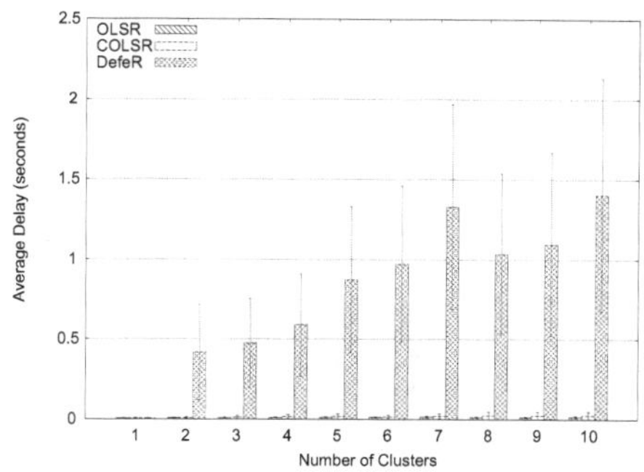

(b) 4 Flows

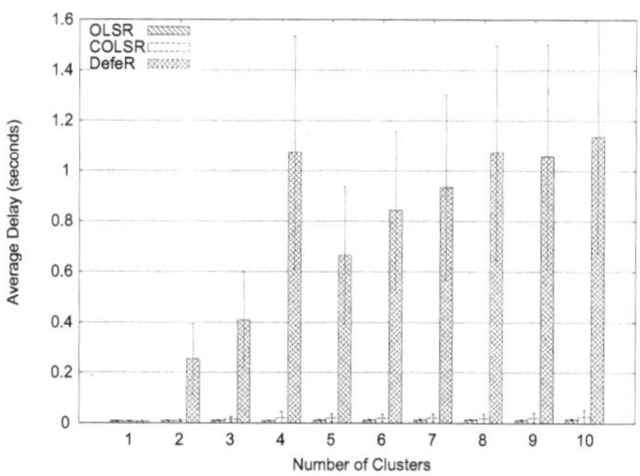

(c) 8 Flows

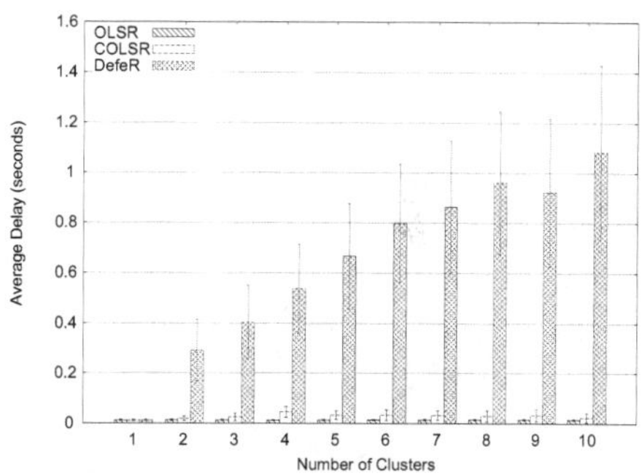

(d) 16 Flows

Fig. 4. Average End-to-end Delay

which is consistent with the performance of other protocols such as the DSDV or the DYMO routing protocols [94]. Though many losses are not desirable, this results from the intrinsic nature of MANETs. It is important to take into account that the proposed scenario is extremely demanding, where a path from source to destination may often not exist. Despite this fact, the proposed routing approach managed to perform two times better than the C-OLSR protocol in some network configurations.

8.2.2 End-to-End Delay

In realistic multi-hop wireless networks, as previously discussed, the constraint of an existing path between any two nodes cannot be guaranteed. As a result Delay-Tolerant Networks have been proposed [94], focusing on the delivery of data packets, regardless of the time interval it might take between source and destination. While the OLSR and C-OLSR protocols simply discard packets when a route is not found, the DefeR gateways are able to re-route packets if alternative paths exist. As a result of an improved traffic delivery, the DefeR protocol has a higher end-to-end delay, as seen in Figure 4. A similar delay is found in the C-OLSR protocol for an eight cluster scenario with solely 1 traffic flow, where this protocol has an abnormal improvement in traffic delivery (see Figure 3a).

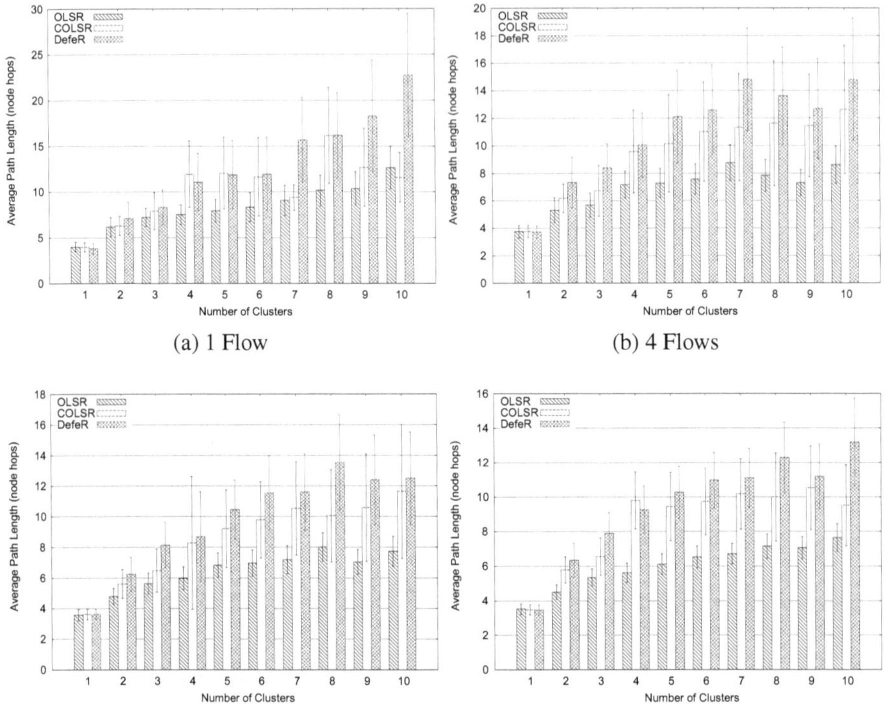

(a) 1 Flow (b) 4 Flows

(c) 8 Flows (d) 16 Flows

Fig. 5. Average Number of Hops

Considering the class of reactive routing protocols, the path discovery process is responsible for initial delays even higher than the ones registered by any of the three analysed protocols [70].

Even though the DefeR scheme is outperformed by the other two protocols, when delay is considered, its increased traffic delivery must not be disregarded as it helps to understand its origin. In fact, after a closer analysis of the obtained results, the high standard deviation reveals that the registered delay is only introduced by some flows, which are likely to be failed by the other protocols. This is the only reason for such a standard deviation, as the three protocols were equally simulated 30 times and only DefeR was this dynamic.

8.2.3 Path Length

The number of hops from source to destination is presented in Figure 5, where the OLSR protocol stands out for being able to achieve shorter routes. Regarding the cluster-based routing protocols, the DefeR protocol is able to keep up or even surpass the C-OLSR protocol's performance, while always delivering more data packets.

Once again, the increasing network size proportionally affects the metric results. However, while the average path length increases with the number of nodes, it decreases

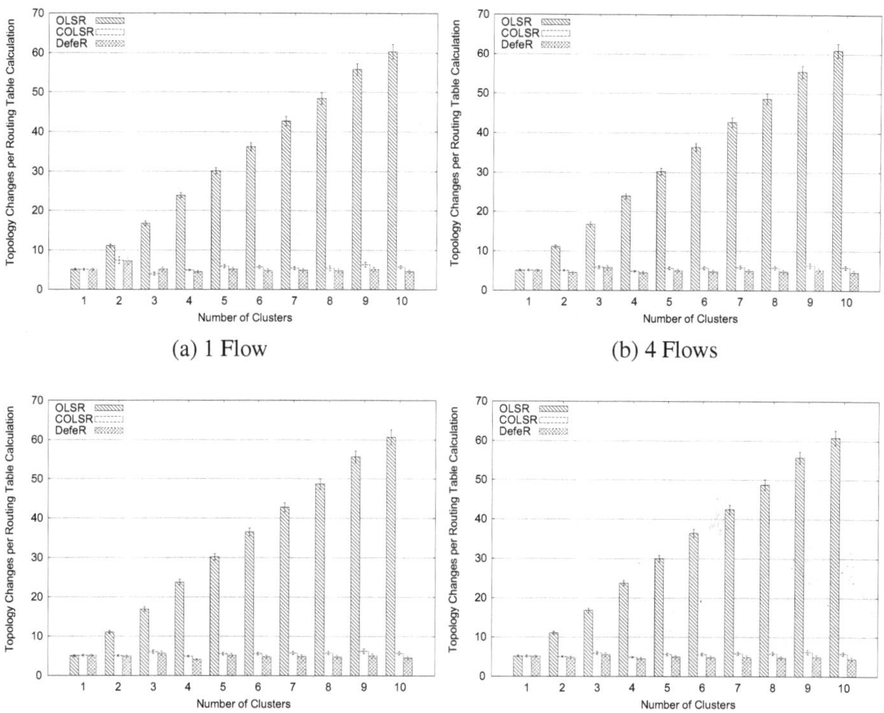

(a) 1 Flow

(b) 4 Flows

(c) 8 Flows

(d) 16 Flows

Fig. 6. Topology Changes per Routing Table Calculation (AToCRT)

with a higher number of traffic flows. A similar behaviour was found with the delay metric, as it is also influenced by the number of intervening nodes in the deliver of data packets.

8.2.4 Topology Changes Per Routing Table Calculation

In MANETs, topology changes are likely to occur very often, not only due to interferences but mainly due to the mobility of nodes. It is the routing protocol's responsibility to detect existing topology changes and reflect them when updating its routing table. However, too many topology changes have a strong impact on the overhead introduced by a routing protocol and may reveal that the protocol suffers from instability.

In Figure 6 the lack of scalability of the OLSR protocol becomes clear, resulting in a growing number of registered topology changes in networks with a higher number of nodes. On the other hand, the use of clusters by the DefeR and C-OLSR protocols allows them to achieve a more stable routing performance, keeping a fairly constant number of topology changes per routing table calculation. However, important topology changes cannot be disregarded by routing protocols. Regarding the overall routing performance of the C-OLSR protocol when compared with its unclustered version, even though it is more stable, it fails to achieve a similar traffic delivery, suggesting that its handling of topology changes does not have the same efficacy.

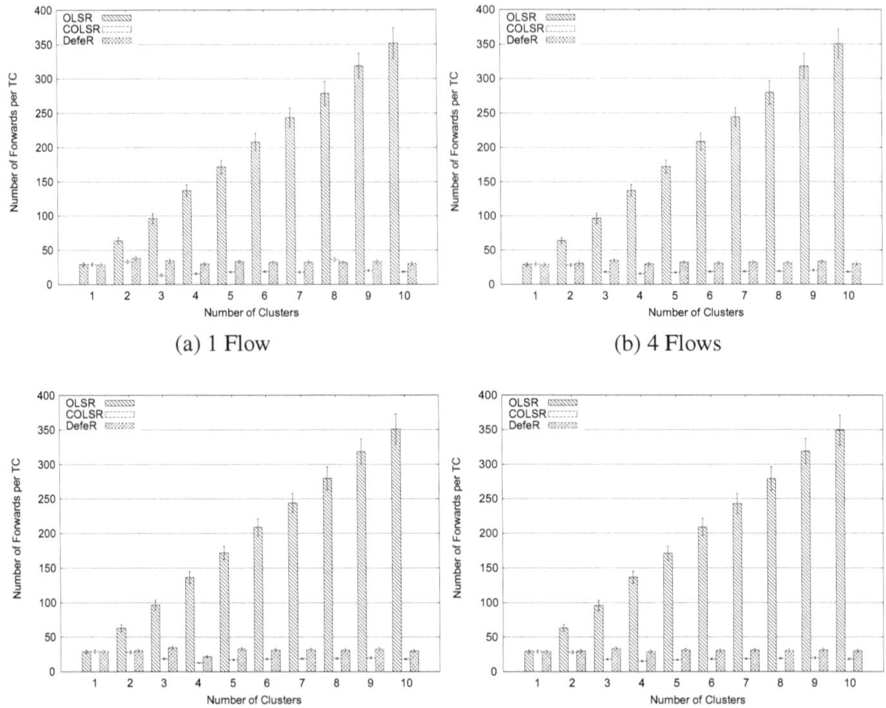

(a) 1 Flow (b) 4 Flows

(c) 8 Flows (d) 16 Flows

Fig. 7. Number of Forwards per *TC* message

8.2.5 Number of Forwards Per *TC* Message

Closely related with the detected number of topology changes, the ratio between sent and forwarded Topology Control messages is also a token of a protocol's ability to scale. The forwarding of *TC* messages deals with a large amount of overhead in the network and should be kept to a minimum. Due to containment of routing information within clusters, the DefeR and C-OLSR protocols require a rather small number of forwards in order to disseminate their routing information – though the C-OLSR protocol requires the smallest amount of forwards. However, once more, an excessively low number of updates may indicate that existing routes are not entirely valid.

Regarding the number of traffic flows, there is no obvious impact on this metric, as it only depends on the existing number of nodes and topology changes. The latter aspect is clearer in the OLSR protocol, as seen in Figure 7 which shows that it requires its *TC* messages to be forwarded to most of the nodes in the network.

8.2.6 Control Traffic Overhead

Since only purely proactive routing protocols are being considered in this evaluation, the number of traffic flows does not influence significantly the number of required routing messages. Figure 8 shows the total overhead of routing control traffic issued by each

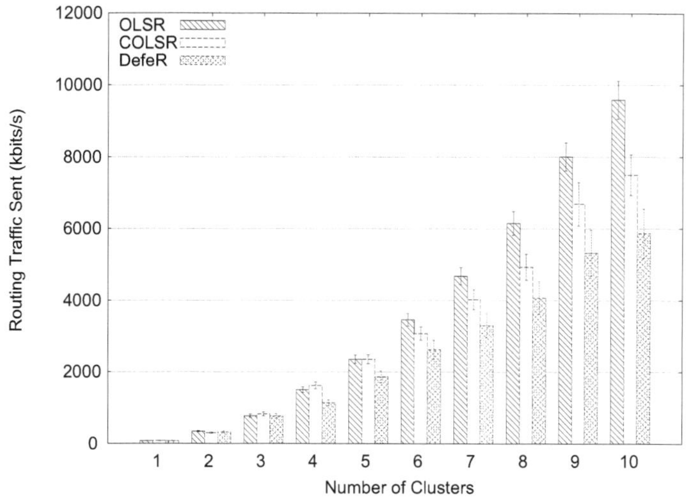

Fig. 8. Sent Routing Traffic Overhead

protocol in the scenario with 16 Flows. As the number of nodes increases, the amount of existing routing information also increases for any proactive protocol. However, the DefeR protocol increases its overhead slower than its competitors, since it requires less routing messages. Moreover, the performance of the proposed protocol can be further improved by using a clustering algorithm that provides a table with the mappings of each node to its CID, as they usually use such a table for cluster maintenance purposes.

The overhead felt by the sent routing messages is more clearly noticed by the received routing information in the entire network. While *HELLO* messages are only sent

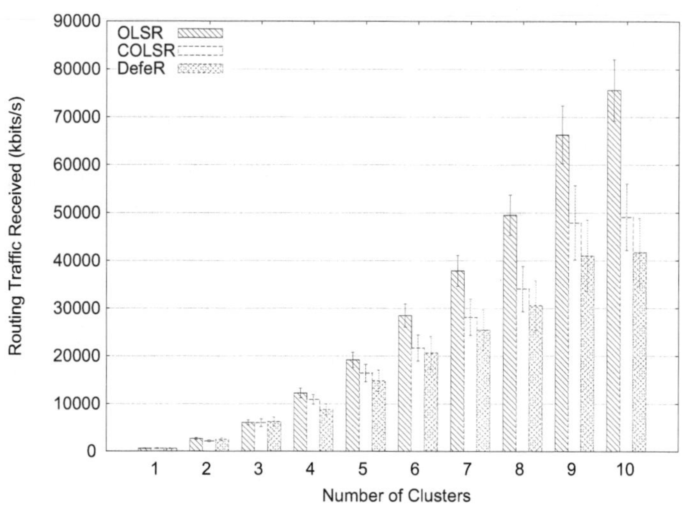

Fig. 9. Received Routing Traffic Overhead

locally, the previously analysed ratio between sent and forwarded *TC* messages determines how much more overhead is propagated through the network. Even though the C-OLSR protocol has a slightly lower ratio of forwarded *TC*s, when compared with DefeR, it has a higher received routing traffic overhead, as it sends more routing data per message. The received control traffic overhead for each protocol is presented in Figure 9.

8.3 Resilience to Mobility

The considered performance assessment must involve not only the evaluation of a large scale network, measuring the stability, overhead and overall traffic delivery performance, but also its ability to handle mobility phenomena, introducing dynamic scenarios with different mobility models.

Regarding this last aspect, even though many mobility models have been proposed in previous works, each one of them has unique characteristics, therefore not replacing one other.

In this evaluation, several mobility patterns will be taken into consideration. In order to do so, the BonnMotion tool [95] has been used to generate different node trajectories, later employed in conjunction with the OPNET Modeler Wireless Simulator. These trajectories were created assuming a plausible speed for a person walking [93], between 0.5 and 1.5 m/s and a pause time of 60 seconds, when applicable. The mobility generation disregarded the first 3600 seconds, solely using the follows 900 seconds of path randomization, avoiding the initial warm-up from the random number generations, thus achieving a more stable scenario. Moreover, the area of motion was of 1500 by 1500 meters, for a total number of 541 nodes. Higher speeds were not considered, as the sense of clusters would be faded away and the realm of vehicular Ad-hoc networks would be entered. Even though new mobility models already present similarities with human mobility, the used mobility patterns were chosen for the sake of comparison with existing works on this subject.

For illustration purposes, after being imported to the simulator, the resulting trajectories were then converted to image files and are depicted in figure 10, representing the Gauss-Markov (figure 10a), Manhattan (figure 10b), Nomadic Community (figure 10c), Random Direction (figure 10d), Random Waypoint (figure 10e) and Random Street (figure 10f) Mobility Models. These different mobility models are entirely random, but each one has its own specificities. By using them the intent is to demonstrate that the DefeR paradigm is suitable in the most diverse scenarios.

In order to evaluate the performance of the chosen proactive protocols, six scenarios incorporating different mobility models and an additional one with static nodes have been used. All these scenarios have the same area and number of nodes, using the trajectories defined by the BonnMotion tool, as previously detailed.

Another important aspect that motivates and influences wireless multi-hop networks is the establishment of data flows between nodes. In the defined scenarios, 24 traffic flows with different destinations were generated in each run. From these flows, 50% were randomly chosen throughout the network, while the remaining traffic destinations were set to nodes within the cluster of the source node. By using this approach, both

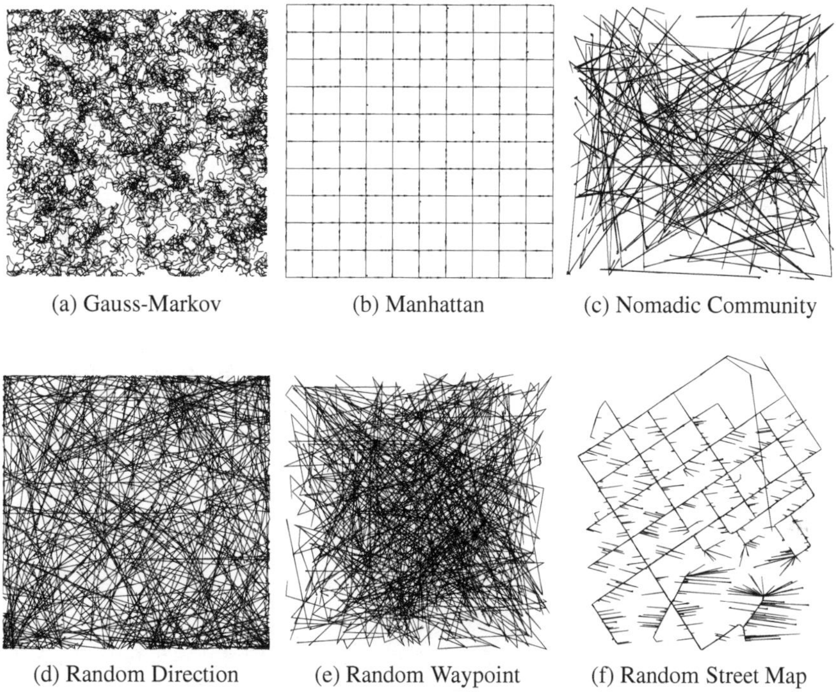

Fig. 10. Mobile Models' Trajectories

interactions within and outside clusters were assessed, providing a complete evaluation of the protocol's performance.

Each flow was defined with a constant bit rate of 8 packets of 4kbit per second (using UDP), representative of typical interactive gaming, simple file transfers or information exchange [92], which are all well suited applications for mobile Ad-hoc networks. The start time of each flow is randomly determined following a uniform distribution between 50 and 250 seconds of simulation time, being concluded by the end of the simulation.

8.3.1 Obtained Results

The purpose of this scenario is to clearly understand the impact of different mobility models on proactive routing. The following results show their efficiency and difficulty in dealing with several distinct patterns of mobility.

8.3.2 Percentage of Losses

Figure 11 illustrates the percentage of losses registered by the routing protocols in all the defined mobility variations. In these, the DefeR protocol stands out by dint of having almost less than half of the losses than the remaining protocols. Conversely, the C-OLSR protocol registers the worst performance, having always more lost packets than the remaining protocols.

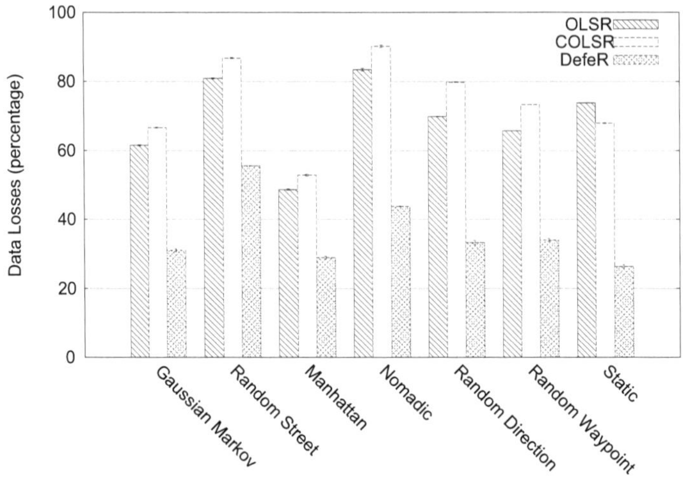

Fig. 11. Average Percentage of Losses

Regarding the Static scenario, the OLSR and C-OLSR protocols unexpectedly show worse delivery performance than in some mobile scenarios. This is a consequence of their inability to scale, as in the Static scenario more paths exist, whereas in the Manhattan scenario, for example, nodes are separated by the arrangement of the streets. However, the DefeR scheme is oblivious to the nodes' placement and has a similar performance in all the scenarios.

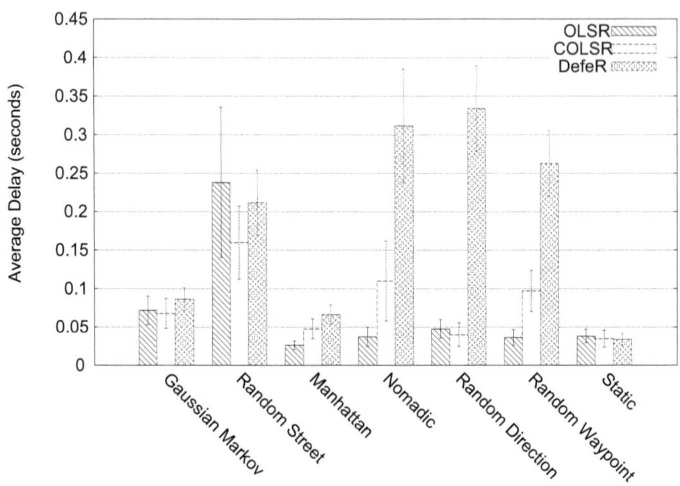

Fig. 12. Average End-to-end Delay

8.3.3 End-to-End Delay

The average end-to-end delay is presented in figure 12 for all the proposed mobility models. Being the static scenario the only exception, in the remaining scenarios the DefeR protocol presents a higher delay. This aspect may not be desirable for certain types of traffic, such as voice, which are not well suited for Ad-hoc networks. The explanation for the higher delay registered by the DefeR protocol repeats itself – as a consequence of the additional traffic delivery achieved, an increased load of traffic is forwarded instead of being dropped.

In fact, while the end-to-end delay is typically a result of a higher path length, the used metrics will show that this is not the case. Specifically, when analysing the Manhattan scenario, where the highest hop count of the all mobile scenarios is registered for DefeR (see Figure 13), it has at the same time the lowest delay of all the mobile scenarios. This confirms that the approach taken by DefeR, which sometimes uses longer but more stable paths, registers less losses and is efficient, not introducing any delay by itself. The higher delay times are not registered in the Manhattan model, as the nodes follow well defined trajectories, where the additional delay overhead in the other mobile scenarios is due only to the repairing of broken paths, allowing the increased performance in traffic delivery registered by DefeR.

The self-restoring property of the DefeR protocol may occur in demanding situations where, due to the mobility phenomena, instead of dropping packets while routing tables change, packets are held and re-forwarded to the appropriate route. Thus, as previously concluded, a higher total delay average is expectable. Moreover, when bottlenecks are avoided due to load-balancing, the re-routing process may also introduce a slight delay. However, as the DefeR scheme is able to reach more challenging destinations than its competitors, the additional delay overhead is justifiable and still suitable for many different applications.

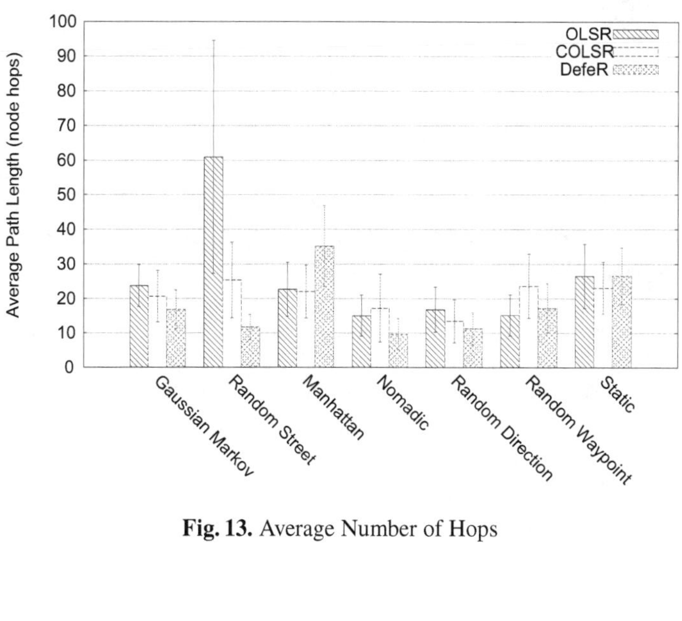

Fig. 13. Average Number of Hops

8.3.4 Path Length

Minimizing the path length is a typical target of routing protocols, with the purpose of reducing the network load and optimising packet delivery. However, due to network dynamics strongly influenced by node mobility, such a routing approach may reduce the protocols' traffic delivery as it disregards the stability of the chosen routes.

In most scenarios, the DefeR scheme is able to achieve a better path length than the remaining protocols while maintaining lower losses, as depicted in figure 13. Nevertheless, for the Manhattan, Random Waypoint and Static mobility models, the Deferred Routing Protocol has a slightly higher path length. This is a consequence of the scenarios' specificities and increased traffic performance of the DefeR, as it reaches more demanding destination nodes. The trade-off between path length and traffic efficiency, in order to achieve an increased traffic performance, should be therefore regarded as an important feature.

As a result of the randomly chosen destinations and of the wireless medium interactions, the confidence interval registered for the path length is higher than for other parameters. However, this interval is still similar to all the analysed routing protocols, validating the outcome of the parameter. The only observed exception worth of taking note occurs with the OLSR protocol in the Random Street Mobility Model. This mobility pattern is highly complex and it is clear that the OLSR protocol is not capable of dealing with the constant and close interactions between the moving nodes. In particular, the obtained standard deviation suggests that in certain occasions routing loops occur, drastically increasing the total number of hops.

8.3.5 Topology Changes Per Routing Table Calculation

When considering the scalability of a routing protocol, the stability of its routing tables is a key aspect on how it performs. The update of a routing table may be a costly procedure in terms of processing power and required energy, possibly leading to the creation and dissemination of additional routing messages, depleting the batteries of mobile devices faster than desirable.

Regarding this aspect, the OLSR protocol is clearly less scalable than the C-OLSR and DefeR protocols, which register a significantly smaller number of topology changes per routing table calculation, as shown in figure 14. In particular, the OLSR protocol has its worse performance in the static scenario. Such behaviour is a direct consequence of the wireless medium interactions of the nodes which are strongly connected in this scenario. In fact, in the mobile scenarios, where connectivity is often scarce, there is a clear reduction of the number of topology changes, suggesting once more that the OLSR protocol does not scale appropriately.

Considering the C-OLSR protocol, which benefits from the usage of clusters such as DefeR, it achieves a greater stability when compared with the standard OLSR. The number of topology changes per routing table calculation registered by this protocol is only slightly higher than the ones obtained from Deferred Routing. However, the overall performance of the C-OLSR protocol regarding traffic delivery suggests that its ability to timely register important topology changes is not appropriate, resulting in wrong or outdated routing paths. On the other hand, the DefeR awareness of the network is entirely different, detecting only the required amount of topology changes thus being more stable, leading to an increased traffic delivery performance, lower routing overhead and better energy efficiency.

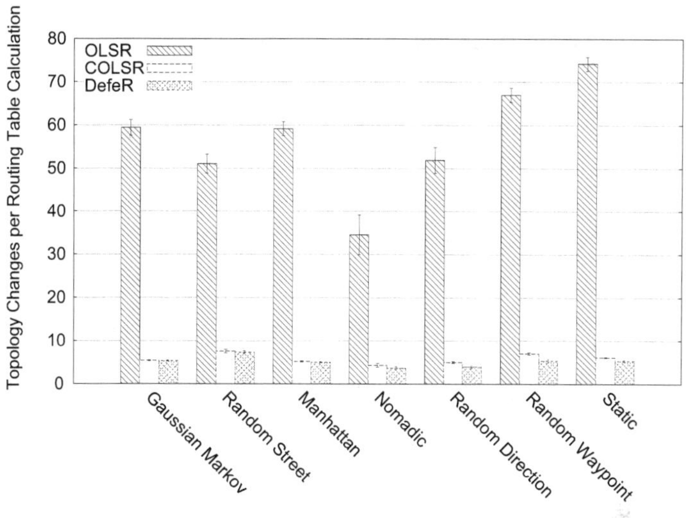

Fig. 14. Topology Changes per Routing Table Calculation (AToCRT)

8.3.6 Number of Forwards Per *TC* Message

The three considered routing protocols rely on Topology Control routing messages to propagate the required information. These messages are issued periodically and whenever a topology change is detected. Similarly to the previously analysed metric, the OLSR protocol is the worst performer, being at its lowest in the static scenario (Figure 15). The way that the OLSR routing protocol handles its routing information leads to an expensive propagation of its *TC* messages throughout the network.

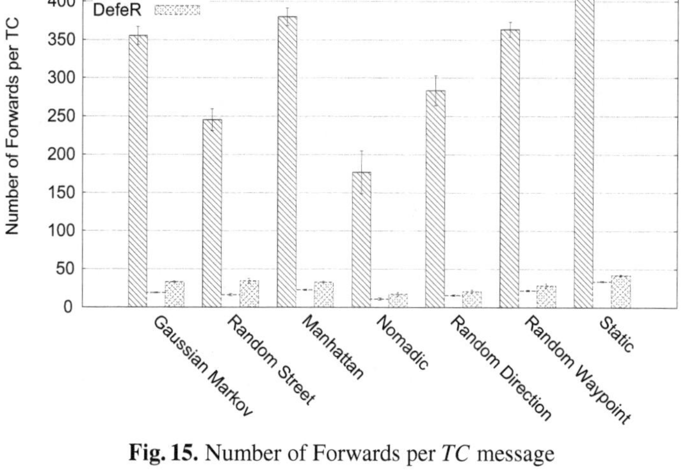

Fig. 15. Number of Forwards per *TC* message

On the other hand, the C-OLSR protocol requires less forwards per *TC* message than any of the other two protocols. Even though both C-OLSR and DefeR routing use the same clusters, it is clear that the usage of C-*HELLO* and C-*TC* messages by the C-OLSR protocol is able to reduce the ratio between forwarded and sent *TC* messages. However, the amount of information and validity contained in these messages, also needs to be considered, as the previously analysed metrics reveal.

8.3.7 Control Traffic Overhead

Figure 16 shows the total amount of routing traffic sent by each routing protocol using the different mobility models. The OLSR protocol once again stands out for having the worst performance. The lack of a well defined network structure, which can be more easily obtained by using clusters, originates an increased overhead. While in the Static scenario this protocol has a bad performance, it is in the Random Street model that more routing traffic is sent.

While the clustered version of the OLSR protocol is able to provide an improvement regarding sent routing traffic, as seen before, it is not capable of maintaining this improvement in terms of data traffic delivery. On the other hand, the proposed DefeR protocol not only outperforms the C-OLSR by having less overhead, but it also outperforms the OLSR protocol in traffic delivery, registering less losses.

Since the sent routing messages may be forwarded through several nodes, Figure 17 presents the control traffic overhead received throughout the network. These results confirm the superiority of Deferred Routing in the handling of different mobility models, being in accordance with the verified ratio between sent and forwarded *TC* messages. Moreover, these results are obtained without guaranteeing a uniform density of nodes within clusters, which would benefit the performance of the DefeR protocol even further, as presented in the following scenarios.

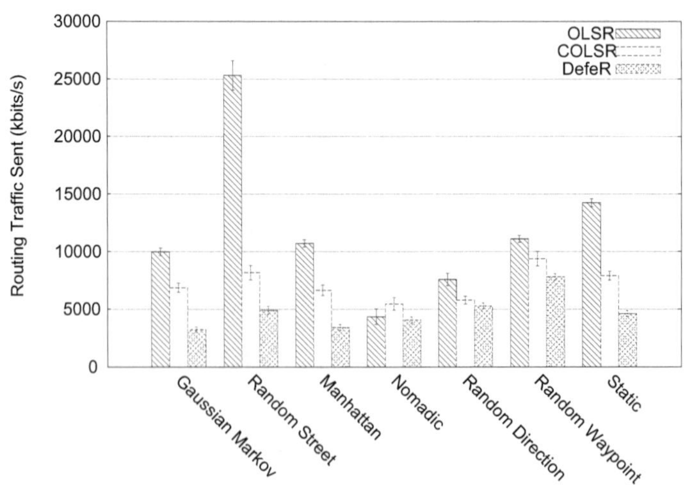

Fig. 16. Sent Routing Traffic Overhead

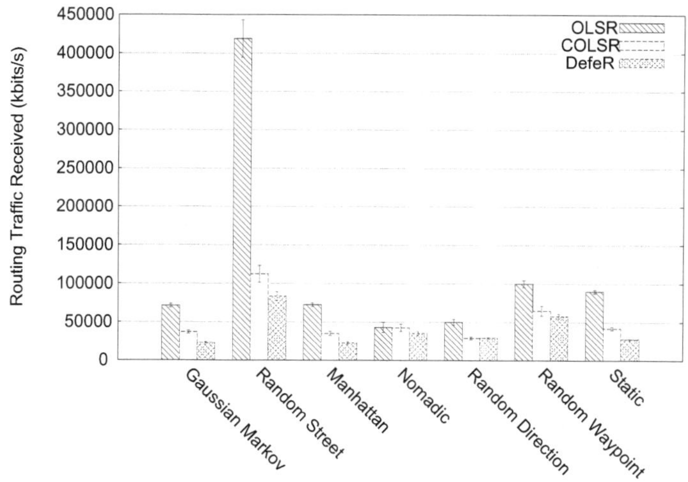

Fig. 17. Received Routing Traffic Overhead

8.4 Summary

The versatility of Mobile Ad-hoc Networks makes them suitable for a wide range of scenarios. Moreover, the dynamic nature of the wireless medium involves a large set of variables which influence the behaviour of these networks. Several parameters were considered for the assessment of three routing protocols, as well as two different scenarios with different characteristics. The protocols' scalability was tested by using a scenario with different size networks, while the effects of different mobile patterns were assessed in a scenario using seven different mobility models. The chosen routing protocols represent the proactive class of routing protocols as defined by the MANET working group, which aim at handling largely dense networks.

In the presented performance evaluation the DefeR scheme revealed that it is able to deliver more traffic than its competitors, even though it introduces some delay as a result of the path-repairing mechanism. Despite having a slightly higher delay, this protocol is still useful for many possible applications, being more stable regarding the number of required routing operations and having an overall smaller overhead, while the plain OLSR protocol showed difficulty in scaling.

The thorough evaluation obtained from all the defined scenarios and complete simulation environment, provided a good understanding of existing protocols' scalability. In particular, it revealed that existing routing protocols are already capable to handle large-scale networks, even though improvements are still desirable and questions about clustering techniques must still be addressed.

9 Conclusion

The usage of wireless multi-hop networks is undeniably important for a future world where thousands of wireless capable devices are expected to be connected. Despite the existing work on this topic, open issues such as routing scalability still exist. In this work, improved routing mechanisms as well as different scalability techniques for Mobile Ad-hoc Networks, have been described.

The IETF MANET working group presents two main routing classes of routing (proactive and reactive) however, only OLSR is aimed at dense networks and, despite using MPR nodes, there are still scalability issues. These and other approaches were analysed where the most relevant features and open issues were identified.

In order to thoroughly analyse the performance of the OLSR protocol and two other protocols aimed at scalability which consider this protocol, an extensive set of different scenarios is defined, where the protocols' performance is assessed regarding their scalability, stability and traffic delivery capabilities. The provided results were obtained from several simulations, taking into account the dynamic characteristics of the wireless link and different mobility patterns, which significantly influence MANETs.

The presented performance analysis provides a comprehensive and thorough evaluation that can be used to assess other routing protocols for MANETs. Several scenarios with different purposes are defined, scrutinising different aspects of the performance of a routing protocol, such as its scalability, regarding both the number of nodes and an increasing of flows, as well as its resilience to several distinct mobility patterns.

The importance of Mobile Ad-hoc Networks in a near future has been discussed throughout this work. From the obtained results, the performance of the routing protocols, in particular of the DefeR approach, motivate their usage in large-scale scenarios. However, the performance increase in scalability often results from the usage of clustered wireless networks, which allows routing information to be contained within limited contexts. Even though several routing protocols rely on this aspect, and considering that many clustering algorithms have already been proposed, there are still drawbacks from this approach. The modification of an existing clustering approach or even the definition of a new one should be addressed in a future work, taking advantage of the increasing availability of contextual information provided by sensors, databases, mobility and traffic patterns or even by the users themselves.

Acknowledgement. This work was supported by the Portuguese National Foundation for Science and Technology (FCT) through a PhD Scholarship (SFRH / BD / 43790 / 2008) and by the iCIS project (CENTRO-07-ST24-FEDER-002003), co-financed by QREN, in the scope of the Mais Centro Program and European Union's FEDER. The authors would also like to thank the OPNET University Program for the licenses provided for the OPNET Modeler Wireless Suite®.

References

1. Cimmino, A., Donadio, P.: Overall Requirements for Global Information Multimedia Communication Village 10th Strategic Workshop. Wireless Personal Communications 49, 311–319 (2009), `http://dx.doi.org/10.1007/s11277-009-9686-3`, doi:10.1007/s11277-009-9686-3

2. WiMAX Forum, IEEE Standard for Local and metropolitan area networks Part 16: Air Interface for Broadband Wireless Access Systems Amendment 3: Advanced Air Interface, IEEE Std 802.16m-2011(Amendment to IEEE Std 802.16-2009), pp. 1–1112 (2011), doi:10.1109/IEEESTD, 5765736
3. Shen, Z., Papasakellariou, A., Montojo, J., Gerstenberger, D., Xu, F.: Overview of 3GPP LTE-advanced carrier aggregation for 4G wireless communications. IEEE Communications Magazine 50, 122–130 (2012), doi:10.1109/MCOM.2012.6146491
4. Beigh, B.M., Peer, M.A.: Performance evaluation of geographical routing protocols: An empirical study. In: 2012 International Conference on Computer Communication and Informatics (ICCCI), pp. 1–6 (2012), doi:10.1109/ICCCI.2012.6158920
5. Piran, M.J., Murthy, G.R., Babu, G.P., Ahvar, E.: Total GPS-free Localization Protocol for Vehicular Ad Hoc and Sensor Networks (VASNET). In: 2011 Third International Conference on Computational Intelligence, Modelling and Simulation (CIMSiM), pp. 388–393 (2011), doi:10.1109/CIMSim.2011.77
6. Jemili, I., Chaabouni, N., Belghith, A., Mosbah, M.: An optimized layered cluster-based routing protocol for ad hoc networks. Wireless Advanced (WiAd), 156–163 (2011), doi:10.1109/WiAd.2011.5983304
7. Prasad, P.S., Agrawal, P., Sivalingam, K.M.: Effects of Mobility in Hierarchical Mobile Ad Hoc Networks. In: 6th IEEE Consumer Communications and Networking Conference, CCNC 2009, pp. 1–5 (2009), doi:10.1109/CCNC.2009.4784895
8. Lindgren, A., Doria, A., Davies, E., Grasic, S.: in Probabilistic Routing Protocol for Intermittently Connected Networks, PRoPHET, IRTF (2012), `http://tools.ietf.org/html/rfc6693.txt`
9. Bulut, E., Szymanski, B.K.: Friendship Based Routing in Delay Tolerant Mobile Social Networks. In: GLOBECOM 2010, 2010 IEEE Global Telecommunications Conference, pp. 1–5 (2010), doi:10.1109/GLOCOM.2010.5683082
10. Mei, A., Morabito, G., Santi, P., Stefa, J.: Social-aware stateless forwarding in pocket switched networks. In: INFOCOM, 2011 Proceedings IEEE, pp. 251–255 (2011), doi:10.1109/INFCOM.2011.5935076
11. Gupta, N., Gupta, R.: Routing protocols in Mobile Ad-Hoc Networks: An overview. In: 2010 International Conference on Emerging Trends in Robotics and Communication Technologies (INTERACT), pp. 173–177 (2010), doi:10.1109/INTERACT.2010.5706220
12. Alotaibi, E., Mukherjee, B.: A survey on routing algorithms for wireless Ad-Hoc and mesh networks. Computer Networks 56, 940–965 (2012), doi:10.1016/j.comnet.2011.10.011
13. Boukerche, A., Turgut, B., Aydin, N., Ahmad, M.Z., Bölöni, L., Turgut, D.: Routing protocols in ad hoc networks: A survey. Computer Networks 55, 3032–3080 (2011), doi:10.1016/j.comnet.2011.05.010
14. Paillassa, B., Yawut, C., Dhaou, R.: Network awareness and dynamic routing: The ad hoc network case. Computer Networks 55, 2315–2328 (2011), doi:10.1016/j.comnet.2011.03.010
15. Hutson, K.R., Schlosser, T.L., Shier, D.R.: On the Distributed Bellman-Ford Algorithm and the Looping Problem. INFORMS J. on Computing 19, 542–551 (2007), `http://dx.doi.org/10.1287/ijoc.1060.0195`, doi:10.1287/ijoc.1060.0195
16. Perkins, C., Bhagwat, P.: Highly Dynamic Destination-Sequenced Distance-Vector Routing (DSDV) for Mobile Computers. In: ACM SIGCOMM 1994 Conference on Communications Architectures, Protocols and Applications, pp. 234–244 (1994)
17. Spaho, E., Ikeda, M., Barolli, L., Xhafa, F., Biberaj, A., Iwashige, J.: Performance Comparison of DSDV and DYMO Protocols for Vehicular Ad Hoc Networks. In: 2012 IEEE 26th International Conference on Advanced Information Networking and Applications (AINA), pp. 629–634 (2012), doi:10.1109/AINA.2012.56

18. Lu, J., Zhang, B., Han, G., Wang, J., Dou, W.: A New Improvement on DSDV. In: 2011 7th International Conference on Wireless Communications, Networking and Mobile Computing (WiCOM), pp. 1–4 (2011), doi:10.1109/wicom.2011.6040278
19. Clausen, T., Jacquet, P.: Optimized Link State Routing Protocol (OLSR), vol. 3626 (2003)
20. Aponte, C.R., Bohacek, S.: OLSR and approximate distance routing: Loops, black holes, and path stretch. In: 2012 Fourth International Conference on Communication Systems and Networks (COMSNETS), pp. 1–10 (2012), doi:10.1109/COMSNETS.2012.6151347
21. Clausen, T., Dearlove, C., Jacquet, P.: The Optimized Link State Routing Protocol version 2, Internet Engineering Task Force, Internet-Draft (2012), `http://tools.ietf.org/id/draft-ietf-manet-olsrv2-15.txt`
22. Herberg, U., Clausen, T.: Delay Tolerant Networking with OLSRv2. In: 2011 IFIP 9th International Conference on Embedded and Ubiquitous Computing (EUC), pp. 144–149 (2011), doi:10.1109/EUC.2011.27
23. Herberg, U., Cole, R., Chakeres, I.: Mobile Ad Hoc Network (MANET) Neighborhood Discovery Protocol (NHDP), Internet Engineering Task Force, Internet-Draft (2012), `http://tools.ietf.org/id/draft-ietf-manet-nhdp-17.txt`
24. Pei, G., Gerla, M., Chen, T.W.: Fisheye state routing: a routing scheme for ad hoc wireless networks. In: 2000 IEEE International Conference on Communications, 6ICC 2000, vol. 1, pp. 70–74 (2000), doi:10.1109/ICC.2000.853066
25. Kleinrock, L., Stevens, K.: Fisheye: A Lenslike Computer Display Transformation (1971)
26. Sivakumar, N.R., Chelliah, C.: Simulation and Evaluation of the Performance on Probabilistic Broadcasting in FSR (Fisheye State Routing) Routing Protocol Based on Random Mobility Model in MANET. In: 2012 Fourth International Conference on Computational Intelligence, Communication Systems and Networks (CICSyN), pp. 371–376 (2012), doi:10.1109/CICSyN.2012.74
27. Garcia-Luna-Aceves, J.J., Spohn, M.: Source-tree routing in wireless networks. In: Proceedings of the Seventh International Conference on Network Protocols (ICNP 1999), pp. 273–282 (1999)
28. Garcia-Luna-Aceves, J.J., Roy, S.: On-demand loop-free routing with link vectors. IEEE Journal on Selected Areas in Communications 23, 533–546 (2005), doi:10.1109/JSAC.2004.842564
29. Kasera, K.K., Ramanathan, R.: A location management protocol for hierarchically organized multihop mobile wireless networks. In: 1997 IEEE 6th International Conference on Universal Personal Communications Record, Conference Record, vol. 1, pp. 158–162 (1997), doi:10.1109/ICUPC.1997.625502
30. Sehgal, S., Gondal, I.: Computational Intelligence in Medical Informatics. In: Karmakar, G., Dooley, L.S. (eds.) Information Science Reference, USA, vol. 1, pp. 185–206 (2007)
31. Ching-Chuan, C., Hsiao-Kuang, W., Winston, L., Mario, G.: Routing in clustered multihop mobile wireless networks with fading channel. In: Proceedings of IEEE SICON, pp. 197–211 (1997)
32. Suresh, A., Duraiswamy, K.: Scalable instant Way Point Routing protocol for MANET. In: 2010 International Conference on Computing Communication and Networking Technologies (ICCCNT), pp. 1–8 (2010), doi:10.1109/ICCCNT.2010.5591658
33. Ros, F.J., Ruiz, P.M.: Cluster-based OLSR extensions to reduce control overhead in mobile ad hoc networks. In: Proceedings of the 2007 International Conference on Wireless Communications and Mobile Computing, pp. 202–207. ACM, New York (2007), doi:http://doi.acm.org/10.1145/1280940.1280985
34. Eriksson, J., Faloutsos, M., Krishnamurthy, S.V.: DART: Dynamic Address RouTing for Scalable Ad Hoc and Mesh Networks. IEEE/ACM Transactions on Networking 15, 119–132 (2007), doi:10.1109/TNET.2006.890092

35. Wang, H., Mi, L., Xia, C., Lv, L., Chen, M.: TLDV: Tree-Like Locator Distance Vector routing protocol for MANETs. In: 2010 International Conference on Educational and Information Technology (ICEIT), vol. 2, pp. V2-291–V2-296 (2010), doi:10.1109/ICEIT.2010.5607575
36. Palma, D., Curado, M.: Scalable multi-hop routing in wireless networks. EURASIP Journal on Wireless Communications and Networking 2013, 86 (2013), http://jwcn.eurasipseriess.com/content/2013/1/86, doi:10.1186/1687-1499-2013-86
37. Palma, D., Curado, M.: Onto scalable Ad-hoc networks: Deferred Routing. In: Computer Communications, vol. 35, pp. 1574–1589. Elsevier Science Publishers B. V., Amsterdam (2012), http://dx.doi.org/10.1016/j.comcom.2012.04.026, doi:10.1016/j.comcom.2012.04.026
38. Palma, D., Araujo, H., Curado, M.: Link quality estimation in wireless multi-hop networks using Kernel based methods. In: Computer Networks, vol. 56, pp. 3629–3638. Elsevier North-Holland, Inc., New York (2012), http://dx.doi.org/10.1016/j.comnet.2012.07.012, doi:10.1016/j.comnet.2012.07.012
39. Leong, B., Liskov, B., Morris, R.: Geographic routing without planarization. In: Proceedings of the 3rd Conference on Networked Systems Design & Implementation, vol. 3, p. 25. USENIX Association, Berkeley (2006)
40. Palma, D., Curado, M.: DASH, Deferred aggregated routing for scalable ad-hoc networks. In: 2010 IFIP Wireless Days (WD), pp. 1–6 (2010), doi:10.1109/WD.2010.5657744
41. AlAamri, H., Abolhasan, M., Franklin, D., Lipman, J.: Optimised relay selection for route discovery in reactive routing. In: Ad Hoc Networks (2012), doi:10.1016/j.adhoc.2012.04.008
42. Perkins, C.E., Belding-Royer, E.M., Das, S.R.: Ad hoc On-Demand Distance Vector (AODV) Routing. In: Internet Engineering Task Force, Request For Comments 3561 (2003), http://www.ietf.org/rfc/rfc3561.txt
43. Gandhi, S., Chaubey, N., Tada, N., Trivedi, S.: Scenario-based performance comparison of reactive, proactive & Hybrid protocols in MANET. In: 2012 International Conference on Computer Communication and Informatics (ICCCI), pp. 1–5 (2012), doi:10.1109/ICCCI.2012.6158842
44. Lee, S.J., Gerla, M.: AODV-BR: backup routing in ad hoc networks. In: IEEE Wireless Communications and Networking Conference, WCNC 2000, vol. 3, pp. 1311–1316 (2000)
45. Johnson, D.B., Maltz, D.A., Hu, Y.C.: The Dynamic Source Routing Protocol (DSR) for Mobile Ad Hoc Networks for IPv4. Internet Engineering Task Force, MANET Working Group, Request For Comments 4728 (2007)
46. MANETWG (2012), http://www.ietf.org/html.charters/manet-charter.html
47. Rajkumar, G., Kasiram, R., Parthiban, D.: Optimizing throughput with reduction in power consumption and performance comparison of DSR and AODV routing protocols. In: 2012 International Conference on Computing, Electronics and Electrical Technologies (ICCEET), pp. 943–947 (2012), doi:10.1109/ICCEET.2012.6203855
48. Park, V., Corson, S.: Temporally-Ordered Routing Algorithm (TORA) Version 1 Functional Specification, Internet Engineering Task Force, MANET Working Group - draft-ietf-manet-tora-spec-04 (2001)
49. Lim, K.H., Datta, A.: An in-depth analysis of the effects of IMEP on TORA protocol. In: 2012 IEEE Wireless Communications and Networking Conference (WCNC), pp. 3051–3056 (2012), doi:10.1109/WCNC.2012.6214328
50. Chakeres, I., Perkins, C.: Dynamic MANET On-demand (DYMO) Routing, Internet Engineering Task Force, Internet-Draft (2012)

51. Kumar, S., Javaid, N., Yousuf, Z., Kumar, H., Khan, Z.A., Bibi, A.: DSDV, DYMO, OLSR: Link Duration and Path Stability. In: 2012 IEEE 11th International Conference on Trust, Security and Privacy in Computing and Communications (TrustCom), pp. 1862–1866 (2012), doi:10.1109/TrustCom.2012.133
52. Toh, C.K.: A novel distributed routing protocol to support ad-hoc mobile computing. In: Proceedings of the 1996 IEEE Fifteenth Annual International Phoenix Conference on Computers and Communications, pp. 480–486 (1996), doi:10.1109/PCCC.1996.493675
53. Preveze, B., Safak, A.: Comparative analysis of novel Long Life Routing Methods in Mobile Networks. In: 2010 IEEE 21st International Symposium on Personal Indoor and Mobile Radio Communications (PIMRC), pp. 1596–1601 (2010), doi:10.1109/PIMRC.2010.5671945
54. Ohta, T., Fujimoto, M., Oda, R., Kakuda, Y.: A class of hierarchical routing protocols based on autonomous clustering for large mobile ad hoc networks. Proc. IEICE Trans. Commun. E8 7-B, 2500–2510 (2004)
55. Oda, R., Ohta, T., Kakuda, Y.: An Efficient On-Demand Hierarchical Routing Protocol Based on Autonomous Clustering for Mobile Ad Hoc Networks. In: Eighth International Symposium on Autonomous Decentralized Systems, ISADS 2007, pp. 180–187 (2007), doi:10.1109/ISADS.2007.16
56. Jemili, I., Belghith, A., Mosbah, M.: A layered cluster based routing for an ad hoc environment. In: 2010 IEEE/ACS International Conference on Computer Systems and Applications (AICCSA), pp. 1–8 (2010), doi:10.1109/AICCSA.2010.5587016
57. Jemili, I., Belghith, A., Mosbah, M.: A Synchronous Tiered Based Clustering Algorithm for large-scale Ad hoc Networks. In: Mammeri, Z. (ed.) Wireless and Mobile Networking. IFIP AICT, vol. 284, pp. 41–55. Springer, Boston (2008)
58. Jemili, I., Chaabouni, N., Belghith, A., Mosbah, M.: An optimized layered cluster-based routing protocol for ad hoc networks. Wireless Advanced (WiAd), 156–163 (2011), doi:10.1109/WiAd.2011.5983304
59. Haas, Z.J., Pearlman, M.R., Samar, P.: The Zone Routing Protocol (ZRP) for Ad Hoc NetworksMarc R. Pearlman. Internet Engineering Task Force, MANET Working Group Internet Draft (draft-ietf-manet-zone-zrp-04) (2002)
60. Yelemou, T., Meseure, P., Poussard, A.M.: Improving ZRP performance by taking into account quality of links. In: 2012 IEEE Wireless Communications and Networking Conference (WCNC), pp. 2956–2960 (2012), doi:10.1109/WCNC.2012.6214310
61. Patel, B., Srivastava, S.: Performance analysis of zone routing protocols in Mobile Ad Hoc Networks. In: 2010 National Conference on Communications (NCC), pp. 1–5 (2010), doi:10.1109/NCC.2010.5430164
62. Samar, P., Pearlman, M.R., Haas, Z.J.: Independent zone routing: an adaptive hybrid routing framework for ad hoc wireless networks. IEEE/ACM Trans. Netw. 12, 595–608 (2004), doi:10.1109/TNET.2004.833153
63. Ravilla, D., Sumalatha, V., Reddy, P.: Performance comparisons of ZRP and IZRP routing protocols for ad hoc wireless networks. In: 2011 International Conference on Energy, Automation, and Signal (ICEAS), pp. 1–8 (2011), doi:10.1109/ICEAS.2011.6147133
64. Wang, L., Olariu, S.: A Two-Zone Hybrid Routing Protocol for Mobile Ad Hoc Networks. IEEE Trans. Parallel Distrib. Syst. 15, 1105–1116 (2004), doi:10.1109/TPDS.2004.73
65. Yang, C.C., Tseng, L.P.: Fisheye zone routing protocol: A multi-level zone routing protocol for mobile ad hoc networks. Comput. Commun. 30, 261–268 (2007), doi:10.1016/j.comcom.2006.08.014
66. Dargahi, T., Rahmani, A.M., Khademzadeh, A.: A hybrid of reactive and proactive approaches based on neighbors' conditions of nodes for mobile ad hoc networks. In: 4th International Conference on Information and Automation for Sustainability, ICIAFS 2008, pp. 425–430 (2008), doi:10.1109/ICIAFS.2008.4784004

67. Murthy, S., Garcia-Luna-Aceves, J.J.: An efficient routing protocol for wireless networks. Mob. Netw. Appl. 1, 183–197 (1996), doi:10.1007/BF01193336
68. Kang, M.S., Kum, D.W., Bae, J.S., Cho, Y.Z., Le, A.N.: Mobility Aware Hybrid Routing protocol for mobile ad hoc network. In: 2012 International Conference on Information Networking (ICOIN), pp. 410–414 (2012), doi:10.1109/ICOIN.2012.6164411
69. Trindade, J., Vazao, T.: HRAN: Heat routing protocol for Ad-Hoc Networks. In: 2011 The 10th IFIP Annual Mediterranean Ad Hoc Networking Workshop (Med-Hoc-Net), pp. 131–134 (2011), doi:10.1109/Med-Hoc-Net.2011.5970479
70. Trindade, J., Vazão, T.: HRAN: a scalable routing protocol for multihop wireless networks using bloom filters. In: Masip-Bruin, X., Verchere, D., Tsaoussidis, V., Yannuzzi, M. (eds.) WWIC 2011. LNCS, vol. 6649, pp. 434–445. Springer, Heidelberg (2011)
71. Joa-Ng, M., Lu, I.T.: A peer-to-peer zone-based two-level link state routing for mobile ad hoc networks. IEEE Journal on Selected Areas in Communications 17, 1415–1425 (1999)
72. Song, W.C., Lutfiyya, H.: Delivery-Guaranteed Geocast in MANETs by Using ZHLS. In: Fourth International Conference on Computer Sciences and Convergence Information Technology, ICCIT 2009, pp. 86–90 (2009), doi:10.1109/ICCIT.2009.194
73. Hamma, T., Katoh, T., Bista, B.B., Takata, T.: An Efficient ZHLS Routing Protocol for Mobile Ad Hoc Networks. In: DEXA 2006, pp. 66–70 (2006), doi:10.1109/DEXA.2006.24
74. Nikaein, N., Labiod, H., Bonnet, C.: DDR-distributed dynamic routing algorithm for mobile ad hoc networks. In: 2000 First Annual Workshop on Mobile and Ad Hoc Networking and Computing, MobiHOC, pp. 19–27 (2000), doi:10.1109/MOBHOC.2000.869209
75. Nikaein, N., Bonnet, C., Nikaein, N.: In: Proceedings of IST: International Symposium on Telecommunications (2001)
76. Kim, D.B., Lee, S.I.: A new hybrid routing algorithm: GHR (Group Hierarchical Routing). In: 2011 International Conference on ICT Convergence (ICTC), pp. 573–577 (2011), doi:10.1109/ICTC.2011.6082666
77. Jian, M., Li, J., Tay, Y.: Cluster Based Routing Protocol (CBRP). Internet Engineering Task Force, MANET Working Group, Internet Draft: draft-ietf-manet-cbrp-spec-01 (1999)
78. Gonzalez, T.F.: Tech. rep., Santa Barbara, CA, USA (1993)
79. Zhu, G., Jiang, X., Wu, C., He, Z.: A Cluster Head Selection Algorithms in Wireless Network Based on Maximal Weighted Independent Set. In: 2010 Proceedings of the 5th International Conference on Ubiquitous Information Technologies and Applications (CUTE), pp. 1–6 (2010), doi:10.1109/ICUT.2010.5678667
80. Parvathi, P.: Comparative analysis of CBRP, AODV, DSDV routing protocols in mobile Adhoc networks. In: 2012 International Conference on Computing, Communication and Applications (ICCCA), pp. 1–4 (2012), doi:10.1109/ICCCA.2012.6179145
81. OPNET Technologies, OPNET Modeler Wireless Suite Simulator (2012), http://www.opnet.com/
82. Ortiz, S.: IEEE 802.11n: The Road Ahead. IEEE Computer 42, 13–15 (2009), doi:10.1109/MC.2009.224
83. Anastasi, G., Borgia, E., Conti, M., Gregori, E.: Wi-Fi in Ad Hoc Mode: A Measurement Study. In: IEEE International Conference on Pervasive Computing and Communications, p. 145. IEEE Computer Society, Los Alamitos (2004), doi:ieeecomputersociety.org/10.1109/PERCOM.2004.1276853
84. Xing, B., Seada, K., Venkatasubramanian, N.: An Experimental Study on Wi-Fi Ad-Hoc Mode for Mobile Device-to-Device Video Delivery. In: IEEE INFOCOM Workshops 2009, pp. 1–6 (2009), doi:10.1109/INFCOMW.2009.5072111
85. Myers, W.: Comparison of Propagation Models, Richardson (1999), http://ieee802.org/16/tg2_orig/contrib/80216cc-99_13.pdf

86. Sagar, S., Javaid, N., Khan, Z., Saqib, J., Bibi, A., Bouk, S.: Analysis and Modeling Experiment Performance Parameters of Routing Protocols in MANETs and VANETs. In: 2012 IEEE 11th International Conference on Trust, Security and Privacy in Computing and Communications (TrustCom), pp. 1867–1871 (2012), doi:10.1109/TrustCom.2012.89
87. Cao, Y., Sun, Z.: Routing in Delay/Disruption Tolerant Networks: A Taxonomy, Survey and Challenges. In: IEEE Communications Surveys Tutorials, vol. PP, pp. 1–24 (2012), doi:10.1109/SURV.2012.042512.00053
88. Golubenco, T.: (2012), http://olsrdq.sourceforge.net
89. Tnnesen, A., Lopatic, T., Gredler, H., Petrovitsch, B., Kaplan, A., Tcke, S.O.: OLSRd - An Ad-hoc Wireless Mesh Routing Daemon (2012), http://www.olsr.org
90. Yang, H.S., Ritchie, L., Richa, A.W., Reisslein, M.: MANET Routing with Provably Low Complexity Through Constant Density Clustering and Route Request Broadcast. Wirel. Pers. Commun. 43, 605–621 (2007), doi:10.1007/s11277-007-9252-9
91. Palma, D., Curado, M.: Inside-Out OLSR Scalability Analysis. In: Ruiz, P.M., Garcia-Luna-Aceves, J.J. (eds.) ADHOC-NOW 2009. LNCS, vol. 5793, pp. 354–359. Springer, Heidelberg (2009)
92. ITU-T. Performance and quality of service requirements for international mobile telecommunications-2000 (imt-2000) access networks (2003)
93. 3GPP. Spatial channel model for Multiple Input Multiple Output (MIMO) simulations. Technical Specification Group Radio Access Network, 7 (2008)
94. Nordemann, F., Tonjes, R.: Transparent and autonomous store-carry-forward communication in Delay Tolerant Networks (DTNs). In: 2012 International Conference on Computing, Networking and Communications (ICNC), pp. 761–765 (2012), doi:10.1109/ICCNC.2012.6167525
95. Aschenbruck, N., Ernst, R., Gerhards-Padilla, E., Schwamborn, M.: BonnMotion: a mobility scenario generation and analysis tool. In: Proceedings of the 3rd International ICST Conference on Simulation Tools and Techniques, pp. 51:1–51:10. ICST (Institute for Computer Sciences, Social-Informatics and Telecommunications Engineering), Brussels (2010), http://dx.doi.org/10.4108/ICST.SIMUTOOLS2010.8684

Measurements on Modern Wireless Communication Technologies and Estimation of Human Exposure

Dimitrios Stratakis[1], Andreas Miaoudakis[1], Evangelos Pallis[1], Traianos Yioultsis[2], Thomas Xenos[2], George Mastorakis[3], and Constandinos X. Mavromoustakis[4]

[1] Technological Educational Institute of Crete, Department of Informatics Engineering, Estavromenos, Heraklion, 71500, Crete, Greece
{dstrat,miaoudakis}@ie.teicrete.gr, pallis@pasiphae.eu

[2] Aristotle University of Thessaloniki, Department of Electrical and Computer Engineering, Thessaloniki, Greece
{traianos,tdxenos}@auth.gr

[3] Technological Educational Institute of Crete, Department of Business Administration, Lakonia, Agios Nikolaos, 72100, Crete, Greece
gmastorakis@staff.teicrete.gr

[4] University of Nicosia, Department of Computer Science, 46 Makedonitissas Ave., 2414 Engomi, Nicosia, Cyprus
mavromoustakis.c@unic.ac.cy

Abstract. The evolution of wireless networking technologies in recent years leads to a faster, safer and more efficient knowledge transfer with the demand of growing up the volume of information handled. The ultimate aim of improving the living standards of citizens, requires a new foundation to evaluate and improve the existing to the present relevant measurement techniques and possibly the introduction of new ones. Also, new wireless information propagation models have to be proposed in the future and the old models have to be evaluated and modified for their improvement. This will be achieved by conducting reliable measurements and experimental tests. A key element of such measurements is the estimation of the present uncertainty. In this context, the purpose of this chapter is to study how the electromagnetic fields generated by modern wireless communication base stations can be measured and evaluated in terms of exposure.

Keywords: Channel power method, electromagnetic field measurements, electromagnetic field exposure, reference levels, spectrum analyser.

1 Introduction

Human exposure to non-ionizing electromagnetic fields (EMF) is of great concern in the last decade and a lot of work has been done in this area [1]-[3] leading to the adoption of exposure limits from International Organizations or individual countries [4]-[9]. The rapid growth of technology in mobile communication services has to

C.X. Mavromoustakis et al., *Resource Management in Mobile Computing Environments*, Modeling and Optimization in Science and Technologies 3,
DOI: 10.1007/978-3-319-06704-9_5, © Springer International Publishing Switzerland 2014

compromise between the transmitted power levels and the benefits of quality, speed and volume of data of the received signals. To fully implement the recent technology, various propagation models have been proposed [10]. Probably in the future more effective propagation models will be the state of the art, since 4G [11] and other technologies will be further applied [12].

These technologies are more consistent with the human exposure effects to non-ionizing electromagnetic fields, since very low power levels are transmitted. Unfortunately, as a result of lower transmitted power and since noise is more important, the noise reduction is the first step for accurate measurements. On the other hand, uncertainty is always present, even if highly expensive devises are used for the verification of propagation models or for exposure assessments. Thus, a thorough examination of measurement techniques and uncertainty effects on measurements concerning low level signals is of great importance.

Worldwide exposure guidelines and standards, regarding protection of people to non-ionizing electromagnetic fields, exist to control electromagnetic radiation levels, by introducing Basic Restrictions (BR) to EMF exposure. Such BR suggest maximum permissible levels for the current density, power density and Specific Absorption Rate (SAR) by human tissues (localized or whole body averaged). Evaluation of compliance to BR is a very difficult task requiring very special equipment that can be found only in specific laboratories. To assess in-situ EMF exposure compliance, Reference Levels (RL) regarding electric field strength (E), magnetic field strength (H), magnetic flux density (B) and equivalent plane wave power density (S) are used instead.

Most of the measurements for the evaluation of the EMF fields produced by modern base stations are being done using traditional Spectrum Analyzers (SA), while the measurement method and the particular settings of those instruments are of great importance to optimize accuracy and avoid common measurement errors. So far, several researchers have studied the characteristics of different types of signals in different frequency ranges, combined with the capabilities of the used instrumentation and proposed configuration instructions depending on the type of measured signals. For example, for DVB-T and DAB signals, related information can be found in [13] and [14] respectively, for GSM signals in [15], for UMTS signals in [16]-[18], for Wi-Fi signals in [19], for Bluetooth signals in [20], and for UWB signals in [21] and [22]. For various other signals employing OFDM (Orthogonal Frequency Division Multiplexing) or based on various versions of IEEE 802.11 protocol, information can be found in [23] and [24]. For some specific types of modern and rapidly growing communication systems, such as WiMAX signals, some information on measurement methods and SA settings can be found in [25]-[29].

In this chapter, we are dealing with two of the above mentioned wireless technologies, i.e.: Universal Mobile Telecommunications System (UMTS) and Worldwide Interoperability for Microwave Access (WiMAX). The first one is a third generation mobile cellular system widely spread nowadays, which uses wideband code division multiple access (W-CDMA). On the other hand, WiMAX technology supports point to multi-point broadband wireless access. WiMAX™ has been standardized by the IEEE 802.16 standard [12] and is promising to offer a data rate of 75Mbps at a distance of up to 50Km for Line of Sight (LoS) conditions and up to 15Km for mobile stations. The IEEE 802.16 standard is designed to operate in several frequency bands, covering the range from 2GHz to 66GHz and supports communication in non LoS conditions targeting to serve multimedia application to mobile users. WiMAX possible usages include the replacement of the low performance twisted pair based local loop for home Internet access and backbone connection for WiFi hot spots.

Since such signals usually have high peak-to-average ratios, the Channel Power or the Zero Span methods are considered to ensure accurate power measurements regardless of the characteristics of the individual signals [30]-[32].

In our work, for the generation of test UMTS and WiMAX signals, specialized software packages run in a laptop computer and a Vector Signal Generator (VSG) was used. Also, a horn type antenna was used to transmit these test bench signals and the received electric field was captured using a SA connected to a dipole type antenna in a certain distance (4 meters) from the transmitting antenna. Furthermore, a real UMTS signal produced by an operating UMTS base station at a distance of about 1600m from the measurements location, was measured for comparing purposes. In all the cases EMF exposure compliance evaluation is addressed.

The remaining of this chapter is organized as follows: measurement setup is presented in section 5.2. The measuring methodology and the procedure for measurement results evaluation are given in section 5.3. Finally, in section 5.4 conclusions from this work are presented.

2 Measurement Setup

2.1 Used Instrumentation

As mentioned in the introduction, test bench UMTS and WiMAX transmitters were created to provide reference UMTS and WiMAX signals. The measurement setup was implemented using the instrumentation of the Non-Ionizing Radiation Laboratory (NIRL) of the Department of Informatics Engineering at the Technological Educational Institute of Crete. More particularly, the experiment setup is described as follows.

a) For the test transmitters were used:
   - A vector signal generator (VSG) developed by Agilent Technologies (Agilent E4438C ESG),
   - A horn type Antenna developed by ETS-Lindgren (Model 3115: 1GHz-6GHz),
   - A suitable RF cable for the connection of the transmitting antenna with the VSG,
   - Specialized software packages developed by Agilent Technologies. Agilent N7600B Signal Studio for 3GPP WCDMA was used for the generation of the test UMTS signal [33] and Agilent N7613A Signal

Studio for 802.16-2004 WiMAX was used for the generation of the test WiMAX signal [34],

- A laptop computer connected via Ethernet to the VSG (Dell Inspiron 15R 7520 SE, running Microsoft Windows 8.1),

b) For the measurement instrumentation (for the real or the test signals) were used:

- A Spectrum analyzer (either the model ESA-E E4407B developed by Agilent Technologies, or the model FSH8 developed by Rohde & Schwarz can be used) for the capturing of the receiving signal,
- A precision conical dipole type antenna (PCD8250: 80MHz-3GHz) developed by Austrian Research Centers GmbH - ARC,
- An antenna rotator connected via USB cable to the controlling laptop computer (developed by Austrian Research Centers GmbH - ARC) was used for mounting and rotating the receiving antenna,
- Dedicated software developed by NIRL with the ability to control the Spectrum Analyzers to take measurements, capture and process the measurement data. Also, NIRL software is capable to control the ARC rotator,
- The same laptop computer as in case a), with a GPIB to USB interface (Agilent 82357A) for the connection to the Agilent E4407 SA, or for the connection via Ethernet to the R&S FSH8 SA,
- Suitable RF cable for the connection of the receiving antenna with the used spectrum analyzer,

Also for all the Ethernet connections a Switch was used (Linksys SD2008). A picture of the experiment setup is shown in Figure 1, whereas in Figure 2 the used measurement setup is depicted. The distance between transmitting and receiving antenna was 4 meters and the height of the antennas was 1.5 meters above the floor at a rooftop, where the measurements were taken.

Fig. 1. Picture of the instrumentation setup

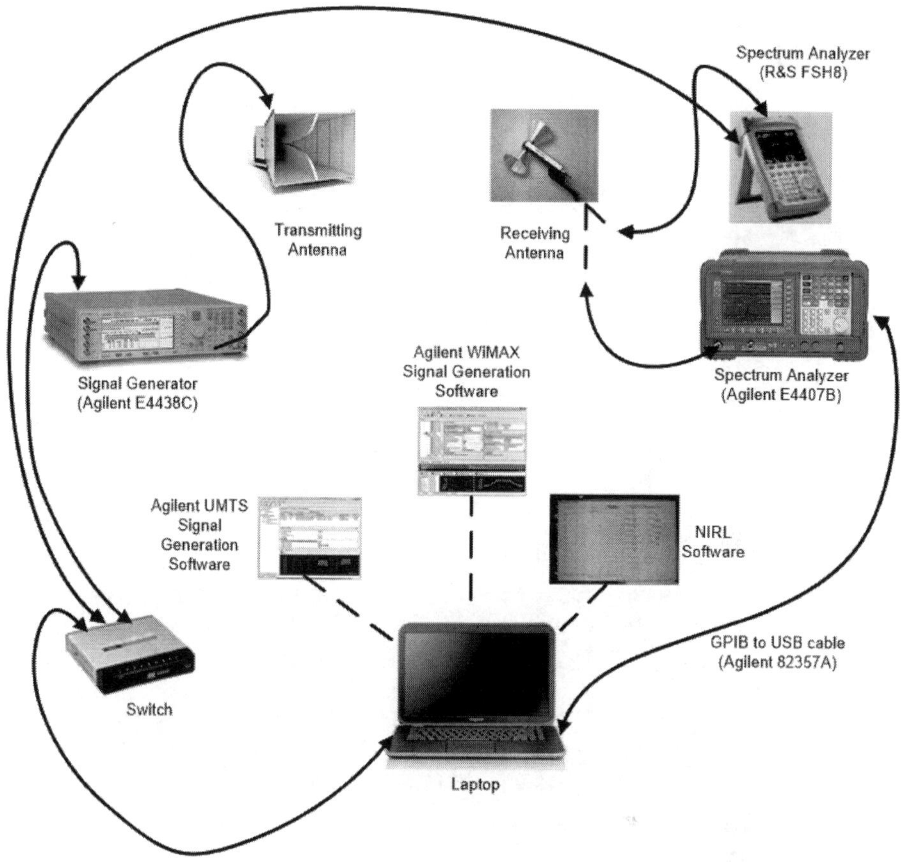

Fig. 2. Experimental setup

2.2 Generation of the Test UMTS and WiMAX Signals

For the generation of the bench UMTS signal, the Agilent N7600A Signal Studio for 3GPP W-CDMA was used and for the generation of the bench WiMAX signal, the Agilent N7613A Signal Studio for 802.16-2004 WIMAX were installed in the laptop computer. Both of these software packages use Standard Commands for Programmable Instruments (SCPI) over the General Purpose Interface Bus (GPIB) or Ethernet (TCP/IP) to communicate with the VSG generator.

In Figure 3, a screenshot of the N7600A Signal Studio is shown as used to produce the test UMTS signal carriers (two carriers were used with centralized at frequencies 2.13 and 2.14 GHz respectively). In Figure 4 the generation of the test UMTS signal via the VSG generator is shown and in Figure 5 the capturing of a real UMTS signal from E4407B SA is shown. Also in Figure 6 a screenshot of the N7613A Signal Studio is depicted as used to produce the test WiMAX signal carrier (one carrier was used in this case).

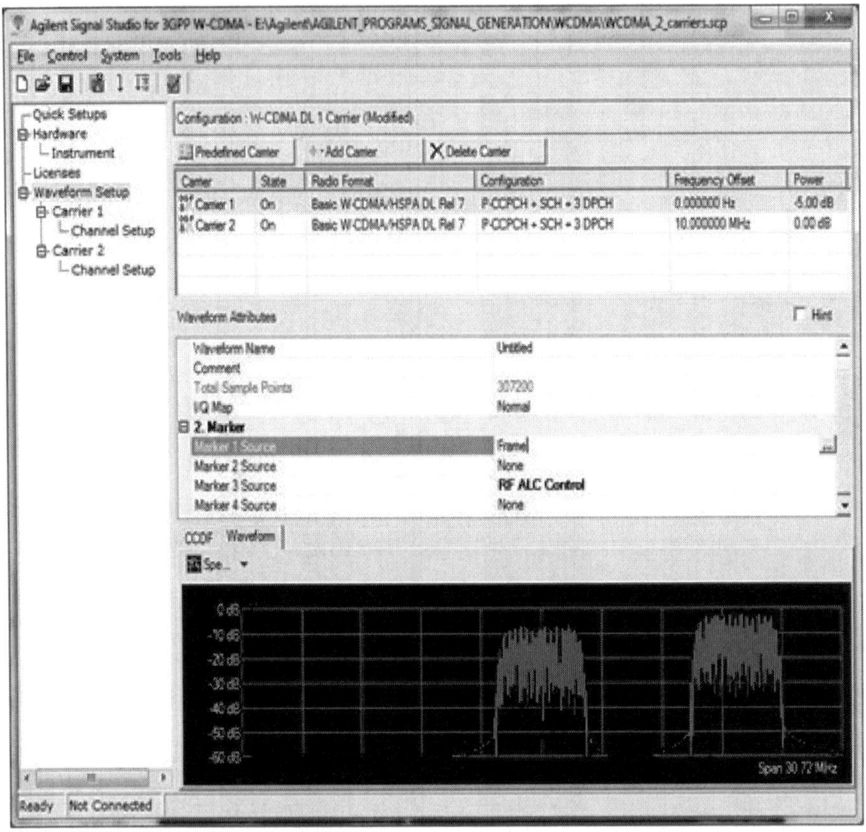

Fig. 3. Parameters and spectrum of the test UMTS signal

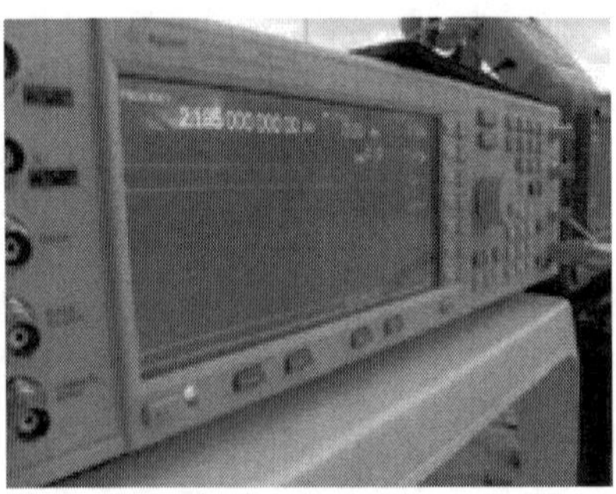

Fig. 4. Generation of the test UMTS signal

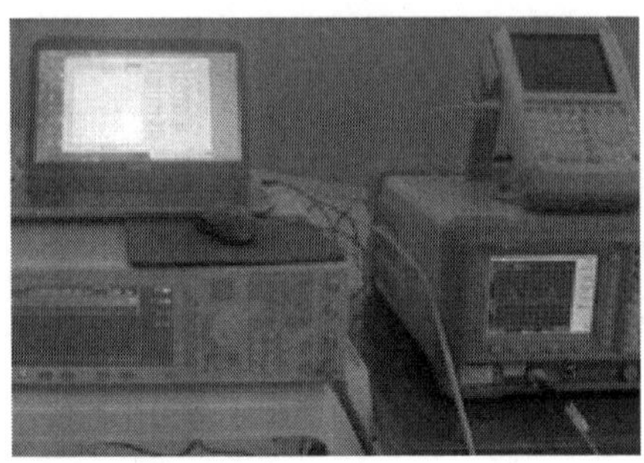

Fig. 5. Capturing of a real UMTS signal from Agilent E4407B Spectrum Analyzer

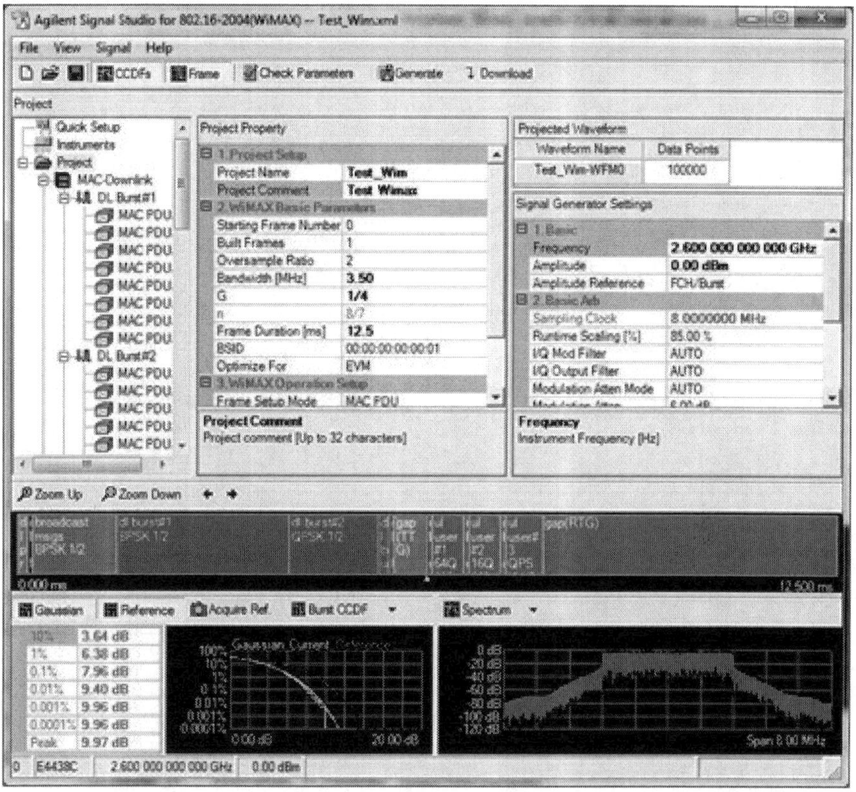

Fig. 6. Parameters, frame structure and spectrum of the test WiMAX signal

2.3 Transmission of the Test Signals

Towards transmitting the test signals, the VSG was connected via a suitable RF cable to a work type antenna (ETS-Lindgren Model 3115 with bandwidth 1GHz to 6GHz). The power output of the VSG was set the value of 0dBm allowing for a 10dB power amplifier back off (The maximum power setting of the used VSG is +10dBm). This was combined with the transmit antenna gain of about 8-9dBi and the connection cable losses of about 10-11dB at the measurement frequencies described in this chapter, resulting to an Equivalent Isotropic Radiation Power of 0-1dBm. The transmitted EIRP is lower than a real UMTS or WiMAX transmitter. However, as measurements were taken at short distances from the horn antenna, received signal strength was adequate for the experiments. In addition, the measurement distances were selected to lie in the far field region of the antenna radiation diagram. Far field conditions is a must for such type of measurements.

2.4 Reception of the Test Signals

At the received side, a precision conical dipole type antenna (PCD8250: 80MHz-3GHz) developed by Austrian Research Centers GmbH was incorporated. To achieve isotropic reception, an antenna rotator was used capable to place the antenna in three perpendicular to each other polarizations (named X, Y and Z polarizations). This way one measurement was performed for each polarization. Combining the three X, Y and Z measurement data using the Add3D [35] method isotropic behaviour of the receiving antenna is ensured.

Table 1. Basic settings of the Spectrum Analyser for the signal measurements

Spectrum Analyzer setting	Test UMTS signal	Test WiMAX signal	Real UMTS signal
Input Attenuation (dB)	0	0	0
Reference Level (dBm)	-50	-50	-60
Sweep Time (sec)	0.2	1.0056	0.2
Start Frequency (MHz)	2123,75	2595	2110
Stop Frequency (MHz)	2146,25	2605	2170
Resolution Bandwidth (kHz)	100	100	100
Video Bandwidth (MHz)	1	3	1
Average State	On	On	On
Number of Averages	10	100	100
Detector Type	RMS	RMS	RMS
Sweep points	401	401	401

The used antenna system (antenna with the rotator and the used cabling) was calibrated and the antenna factor data for the all the measurement bandwidths were available. The antenna receiving system could be connected to any SA (Agilent E4407B or R&S FSH8) to capture the received field.

To perform proper measurements, optimal setup values to the SA have to be selected. The basic settings that were used in the experiments described in this chapter, are presented in Table 1.

2.5 Measurement Procedure

During the measurements, the SA was controlled by the special purpose software developed by NIRL. The development platform of NIRL software was the Matlab® due to its easy connectivity and capabilities for remote programming measurement instruments using GPIB, USB or Ethernet connections and also due to its processing power. This software is an extension of the earlier used NIRL software [36]. This way measurement procedure repeatability is guaranteed eliminating possible personnel errors. The optimal setting for various types of measurements are stored in the software database and recalled to perform each measurement. Then NIRL measurement automation software passes these settings to the used SA using SCPI over GBIP or Ethernet.

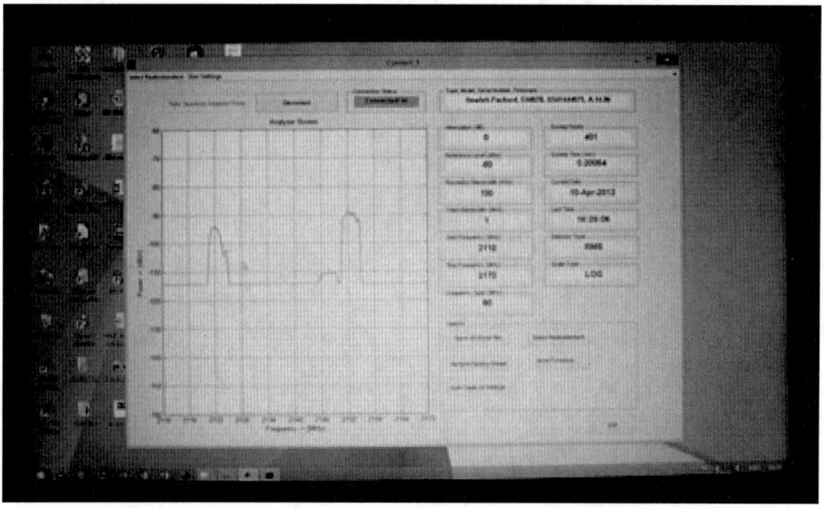

Fig. 7. Instance of NIRL software interface for controlling SA settings and saving measurement data during the capturing of a real UMTS signal data at the frequency range 2110-2170MHz

After the setting of the used SA, it performs the dedicated measurement and the NIRL automation software captures the measurement data. It then saves them to an Excel file along with the used SA settings, cable type, antenna type and the used antenna polarization. These settings are used along with measurement data for the evaluation of the electric field and the respective uncertainty. In Figure 7, a screenshot of NIRL specialized software during X polarization measurement for the real UMTS signal is shown. To perform electric field evaluation more other parameters have to be taken into account such as used Antenna Factor (AF), cable and connector losses, environmental parameters (temperature, humidity etc.) during the measurement period, SA calibration data and SA performance specifications. This is also done by NIRL software.

3 Processing of the Measurement Data

As mentioned before, to achieve isotropic behaviour of the receiver setup, measurements are being performed using three perpendicular to each other receiving antenna polarizations named X, Y and Z for every signal and frequency band of interest. The channel power method was used for the evaluation of the measurement results. This method as described in [32] and [37], computes the field strength out of measurement taking into account the resolution bandwidth of the receiver as well as the equivalent noise bandwidth. In Figure 8, the equivalent free space plane wave power densities S_x ,S_y and S_z for the real UMTS measurement as calculated from the measurement data, are plotted versus frequency (the respective Greek legislated limit is 6W/m^2). Of course and for the other type of measurements (test UMTS and WiMAX signals) described in this chapter, similar figures can be created from the NIRL software.

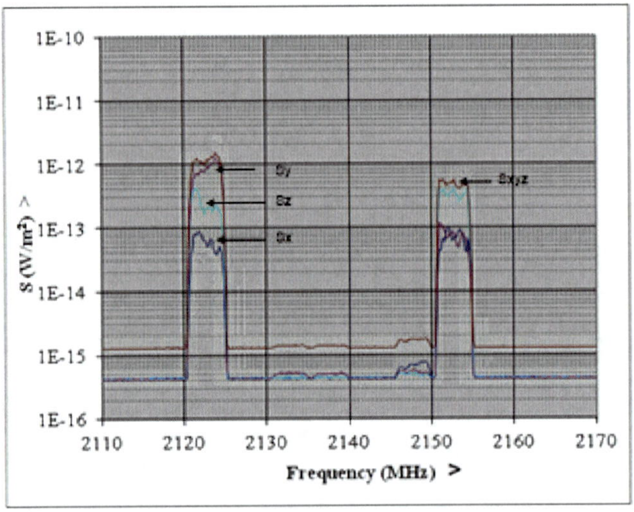

Fig. 8. Equivalent plane wave power densities for X, Y and Z polarization measurements (Sx, Sy and Sz), and the resulting total power density (Sxyz)

A metric for the EMF human exposure compliance assessment is the summation of exposure quotients (SEQ). The SEQ is taking into account that RLs are frequency depended, as SEQ is calculated by adding the square of the electric field strength value in every frequency of interest divided by the squared electric field strength limit at the same frequency. The SEQ value has to be below the unit to ensure EMF human exposure compliance.

For uncertainty estimation the approach proposed in [38] was used where several uncertainty contributors were taken into account (instrumentation performance data from either calibration data sheet or manufacturer specifications, environmental data etc). In case of spatial average measurements the approach proposed in [39] can be used. These approaches are compliant with the ISO Guide to uncertainty [40] and

Table 2. WiMAX signal measurement results

Parameter	Test UMTS signal	Test WiMAX signal	Real UMTS signal
Measurement bandwidth (MHz)	22.5	10	60
Frequency zone (MHz)	2123,75 - 2146,25	2995 - 2605	2110 - 2170
Total electric field at the place of the receiving antenna (V/m)	$2{,}06x10^{-3}$	0,12	$5{,}07x10^{-4}$
Expanded uncertainty of the total electric field (%)	24,11	23.53	19,74
Lowest electric field reference level for the frequency zone of the measurement (V/m)	47,25	47,25	47,25
Total plane wave equivalent power density at the place of the receiving antenna (W/m^2)	$1{,}13 \bullet 10^{-8}$	$3{,}91x10^{-5}$	$6{,}83 \bullet 10^{-10}$
Expanded uncertainty of the total plane wave equivalent power density at the place of the receiving antenna with confidence level 95.45%	48,22	47,06 %	39,47
Lowest reference level of the plane wave equivalent power density for the frequency zone of the measurement (W/m^2)	6,00	6,00	6,00
Summation of exposure quotients	$1{,}90x10^{-9}$	$6{,}60x10^{-6}$	$1{,}15x10^{-10}$
Expanded uncertainty of the Summation of exposure quotients with confidence level 95.45%	$\pm 5{,}11x10^{-11}$	$\pm 2{,}73x10^{-7}$	$\pm 3{,}34x10^{-12}$
Summation of exposure quotients with its expanded uncertainty (maximum value)	$1{,}95x10^{-9}$	$6{,}87x10^{-6}$	$1{,}19x10^{-10}$
Limits of exposure according to Greek legislation divided to the summation of exposure quotients with its expanded uncertainty	$5{,}12x10^{8}$	145.493,41	$8{,}43x10^{9}$

related documentation [41]-[43]. Table 2 presents the results of the EMF human exposure compliance estimation for the test bench UMTS and WiMAX signals and for the real UMTS signal as estimated by the described procedure. It must be mentioned that the presented results especially for the real UMTS base station, are time depended results since the total emitting power for such stations is depended on the number of subscribers served. For maximum exposure estimation in such cases the proposed methodologies in [44] and [45] have to be used.

4 Conclusions

The rapid growth of wireless technologies raised a worldwide concern about the effects of human exposure to the electromagnetic fields of such installations especially in rural areas where most of the new wireless infrastructures are installed. Worldwide and national authorities have adopted reference levels to the radiated EMF fields for the human protection. Among modern wireless technologies, the WiMAX standard offers to be a replacement to the twisted paired based home user last mile connection as well as to provide high data rate for mobile network access. Since WiMAX involve complex modulation schemes to produce a noise like wideband transmitted signal the evaluation of EMF human exposure compliance in such a case is also complex.

In this chapter the application of a methodology for EMF field human exposure compliance measurements is presented. For this purpose, test bench signals were generated using a vector signal generator with computer software. These signals were transmitted and measured using a spectrum analyser (narrowband measurement setup). Such software packages can be used for further study of the propagation characteristics of modern wireless communication signals in different environments. Issues related to measurement performance and measurement data processing were also addressed. The results of the measurements were presented accompanied to the measurement associated uncertainties. The measurements showed that for the test bench, the summation of exposure quotients (a metric for EMF human exposure compliance which should be below 1) is quite low. That was expected for the transmitting powers used, and also for the real UMTS station due to its distance from the location the measurements were taken. In real installations where equivalent isotropic radiation power can be very higher related to the presented setups, exposure limit violation may be found, depending on the distance from the transmitter, the surrounding topography, etc. The co-existing of other EMF radiation sources has to be taken into account as dictated by National and International exposure compliance standards. Furthermore the application of capable control software packages either for signal generation, or for signal evaluation showed that the simulations or the measurements could not be depended any more from the operators of the measurement system, avoiding this way any possible operator mistakes. Finally, several novel network architectures will be measured in the future based on the research work published by the authors in [46]-[50] in the field of Digital Video Broadcasting, as well as in [51]-[61] in the field of cognitive radio networks.

Acknowledgments. This work is dedicated to the memory of Professor Vassilios Zacharopoulos, who was the founder of Non-Ionizing Radiation Laboratory of the Informatics Engineering Department at Technological Educational Institute of Crete.

This project is implemented through the Operational Program "Education and Lifelong Learning", Action Archimedes III and is co-financed by the European Union (European Social Fund) and Greek national funds (National Strategic Reference Framework 2007 - 2013).

References

[1] Stewart, W.: Mobile phones and health. Independent Expert group on Mobile Phones, IEGMP (2000)

[2] Documents of health protection Agency, Series B: Radiation, Chemical and Environmental Hazards, "Power frequency electromagnetic fields, melatonin and the risk of breast cancer", Report of an independent Advisory Group on Non ionising radiation (2006)

[3] WHO. Electromagnetic Fields and Public Health - Radiofrequency electromagnetic fields effect. Promemoria (183) (1998)

[4] IEEE Standard for Safety Levels With Respect to Human Exposure to Radio Frequency Electromagnetic Fields, 3 kHz to 300 GHz. IEEE Std C95.1-2005 (2006)

[5] CENELEC, Human exposure to electromagnetic fields. High frequency (10 kHz to 300 GHz), ENV 50166-2, CENELEC (1995)

[6] Council Recommendation: 1999/519/EC of 12 July 1999 on the limitation of exposure of the general public to electromagnetic fields (0 Hz to 300 GHz). Official Journal L. 197 (1999)

[7] Ahlbom, A., et al.: Guidelines for limiting exposure to time-varying electric, magnetic, and electromagnetic fields (up to 300 GHz). International Commission on Non-Ionizing Radiation Protection. Health Phys. 74, 4 (1998)

[8] Greek Legislation: Law 3431, About Electronic Communications and Other Orders, vol. A, act. 13 (2006)

[9] Greek Legislation: Common Ministerial Decision-Protection measures for the exposure of the general public to all land based antenna stations, vol. B, act. 1105 (2000)

[10] Sarkar, T., Ji, Z., Kim, K., Medouri, A., Salazar-Palma, M.: A survey of various propagation models for mobile communications. IEEE Antennas and Propagation Magazine 45(3), 51–80 (2003)

[11] Evans, B.G., Baughan, K.: Visions of 4G. Electronics and Communication Engineering Journal (2002)

[12] Std 802.16e^{TM}-2005 and IEEE Std 802.16^{TM}-2004/Cor1-2005: IEEE Standard for Local and metropolitan area networks. Part 16: Air Interface for Fixed and Mobile Broadband Wireless Access Systems. Amendment 2: Physical and Medium Access. Control Layers for Combined Fixed and Mobile Operation in Licensed Bands and Corrigendum 1. IEEE, New York (2006)

[13] Angrisani, L., Capriglione, D., Ferrigno, L., Miele, G.: Power Measurements in DVB-T Systems: New Proposal for Enhancing Reliability and Repeatability. IEEE Trans. Instrum. Meas. 57(10) (2008)

[14] Bornkessel, C., Wuschek, M.: Exposure Measurements of Modern Digital Broadband Radio Services. In: German Microwave Conference, GeMiC 2006, Karlsruhe, Germany (2006)

[15] Olivier, C., Martens, L.: Optimal Settings for Narrow-Band Signal Measurements Used for Exposure Assessment Around GSM Base Stations. IEEE Trans. Instrum. Meas. 54(1) (2005)

[16] Eskerski, C., Braach, B.: UMTS measurements with the Selective Radiation Meter SRM-3000., Application Note AN_HF_1007, Narda Safety Test Solutions GmbH GmBH (2008)

[17] Fayos-Fernández, J., Victoria-González, F., Martínez-González, A.M., Morote-Marco, A., Sánchez-Hernández, D.: Effect of spectrum analyzer filtering on electromagnetic dosimetry assessment for UMTS base stations. IEEE Trans. Instrum. Meas. 57(6), 1154–1165 (2008)

[18] Olivier, C., Martens, L.: Optimal Settings for Frequency-Selective Measurements Used for the Exposure Assessment Around UMTS Base Stations. IEEE Trans. Instrum. Meas. 56(5), 1901–1909 (2007)
[19] Betta, G., Capriglione, D., Miele, G., Rossi, L.: Reliable Measurements of Wi-Fi™ Electromagnetic Pollution by Means of Traditional Spectrum Analyzers. In: Proceedings of IEEE International Instrumentation and Measurement Technology Conference, Victoria, Vancouver Island, Canada, pp. 206–211 (2008)
[20] Agilent Technologies, Bluetooth® Measurement Fundamentals, Agilent Technologies Inc. Manufacturing Part Number: 5988-3760EN, USA (2006)
[21] Agilent Application Note 1488, Ultra-Wideband Communication RF Measurements, Agilent Technologies, Inc., Manufacturing Part Number: 5989-0506EN, USA (2005)
[22] Takada, J., Ishigami, S., Nakada, J., Nakagawa, E., Uchino, M., Yasui, T.: Measurement Techniques of Emissions from Ultra Wideband Devices. IEICE Transactions on Fundamentals of Electronics, Communications and Computer Sciences E88-A(9), 2252–2263 (2005)
[23] Archambault, J., Surineni, S.: IEEE 802.11 Spectral Measurements Using Vector Signal Analyzers. RF Test and Measurement, pp. 38–49 (2004)
[24] Briggs, M., Martinez, J., Bare, D.: Power measurements of OFDM signals. In: International Symposium on Electromagnetic Compatibility, EMC 2004, vol. 2, pp. 485–488 (2004)
[25] Betta, G., et al.: The use of traditional spectrum analyzers to measure the electromagnetic pollution generated by WiMAX devices. In: Proceedings of XIX IMEKO World Congress, pp. 485–490 (2009)
[26] Huynh, D., Nelson, B.: Best Practices for Making Accurate WiMAX Channel Power Measurements. Agilent Technologies Inc. (2008)
[27] Joseph, W., et al.: Accurate assessment of electromagnetic exposure from WiMAX signals using a spectrum analyzer. IEEE Transactions on Instrumentation and Measurement 57(3), 518–521 (2008)
[28] Joseph, W., et al.: Accurate determination of the electromagnetic field due to WiMAX base station antennas. IEEE Transactions on Electromagnetic Compatibility 50(3), 730–735 (2008)
[29] Agilent Technologies, WiMAX Concepts and RF Measurements, Agilent Technologies Inc. Manufacturing Part Number: 5989-2027EN, USA (2005)
[30] Bertocco, M., Sona, A.: On the Measurement of Power via a Superheterodyne Spectrum Analyzer. IEEE Trans. Instrum. Meas. 55(5), 1494–1501 (2006)
[31] Rohde & Schwarz Application Note 1EF41, Measurement of Adjacent Channel Leakage Power on 3GPP W-CDMA Signals with the FSP, Rohde & Schwarz (2001)
[32] Agilent Technologies, Comparing Power Measurements on Digitally Modulated Signals, Agilent Technologies Inc., Manufacturing Part Number: 5968-2602E, USA (2000)
[33] N7600B Signal Studio for 3GPP W-CDMA. Agilent Technologies Inc. (2011)
[34] N7613A Signal Studio for 802.16-2004 (WiMAX™). Agilent Technologies Inc. (2008)
[35] Müllner, W., Neubauer, G., Haider, H.: Add3D, a new technique for precise power flux density measurements at mobile communications base stations. ARC Seibersdorfresearch GmbH (2000)
[36] Stratakis, D., et al.: Automation in electromagnetic field measurements. In: Proceedings of the Annual Conference on Telecommunications & Multimedia (2006)
[37] Agilent Application Note 1303, Spectrum Analyzer Measurements and Noise. Agilent Technologies Inc. Manufacturing Part Number: 5966-4008E, USA (2003)
[38] Stratakis, D., et al.: Overall Uncertainty Estimation in Multiple Narrow-Band in Situ Electromagnetic Field Measurements. IEEE Transactions on Instrumentation and Measurement 58(8), 2767–2779 (2009)

[39] Stratakis, D., et al.: On the spatial averaging of multiple narrowband electromagnetic field measurements: Methods and uncertainty estimation. IEEE Transactions on Instrumentation and Measurement 59(6), 1520–1536 (2010)

[40] Gum, I.S.O.: Guide to the expression of uncertainty in measurement, 1st edn. (1995)

[41] Taylor, B.N., Kuyatt, C.E.: NIST Technical Note 1297. Guidelines for evaluating and expressing the uncertainty of NIST measurement results (1994)

[42] UKAS: M3003, The Expression of Uncertainty and Confidence in Measurement, 2nd edn. UKAS (2007)

[43] EURACHEM/CITAC Guide CG4: Quantifying uncertainty in analytical measurement, 2nd edn. (2000)

[44] ECC Recommendation (02)04 (revised Bratislava 2003, Helsinki 2007), Measuring Non-Ionising Electromagnetic Radiation (9 kHz - 300 GHz), Edition 060207, Electronic Communications Committee, ECC (2007)

[45] CENELEC prEN 50492, Basic standard for the in-situ measurement of electromagnetic field strength related to human exposure in the vicinity of base stations, CENELEC (2006)

[46] Mastorakis, G., Kormentzas, G., Pallis, E.: A fusion IP/DVB networking environment for providing always-on connectivity and triple-play services to urban and rural areas. IEEE Network Magazine 21(2), 21–27 (2007)

[47] Mastorakis, G., Pallis, E., Kormentzas, G.: A DVB/IP QoS aware backhaul networking environment. Wireless Personal Communications 52(3), 637–649 (2010)

[48] Pallis, E., Mantakas, C., Mastorakis, G., Kourtis, A., Zacharopoulos, V.: Digital Switchover in UHF: the ATHENA concept for broadband access. European Transactions on Telecommunications 17(2), 175–182 (2006)

[49] Mastorakis, G., Pallis, E., Mantakas, C., Kormentzas, G., Skianis, C.: Exploiting digital switchover for broadband services access in rural areas. Journal of Communications 1(6), 45–50 (2006)

[50] Mastorakis, G., Markakis, E., Sideris, A., Pallis, E., Zacharopoulos, V.: Experimental Infrastructures for IP/DVB Convergence: an Actual Substantiation for Triple Play Services Provision at Remote Areas. In: IEEE 18th International Symposium on Personal, Indoor and Mobile Radio Communications, PIMRC 2007 (2007)

[51] Bourdena, A., Pallis, E., Kormentzas, G., Mastorakis, G.: A prototype cognitive radio architecture for TVWS exploitation under the real time secondary spectrum market policy. Paper published at the Special Issue on Cognitive Radio for LTE Advanced & Beyond. Physical Communication. Elsevier (2013), `http://dx.doi.org/10.1016/j.phycom.2013.11.003`

[52] Bourdena, A., Pallis, E., Kormentzas, G., Mastorakis, G.: Opportunistic TV White Spaces Exploitation through Efficient Radio Resource Management Algorithms in Cognitive Radio Networks. Paper Published at the Transactions on Emerging Telecommunications Technologies (ETT). Wiley (2013)

[53] Bourdena, A., Pallis, E., Kormentzas, G., Skianis, C., Mastorakis, G.: Real-Time TVWS Trading Based on a Centralised CR Network Architecture. Paper Presented at the IEEE Globecom2011, IEEE International Workshop on Recent Advances in Cognitive Communications and Networking, Texas, Houston, USA, December 05-09, pp. 964–969 (2011)

[54] Bourdena, A., Pallis, E., Kormentzas, G., Mastorakis, G.: A centralised broker-based CR network architecture for TVWS exploitation under the RTSSM policy. Paper Presented at the 2nd IEEE Workshop on Convergence among Heterogeneous Wireless Systems in Future Internet (CONWIRE 2012), IEEE ICC 2012, Ottawa, Canada, June 10-15, pp. 5685–5689 (2012)

[55] Bourdena, A., Mastorakis, G., Pallis, E., Karditsis, E., Kormentzas, G.: A radio resource management framework for TVWS exploitation under the RTSSM policy. Paper Presented at the International Conference on Telecommunication & Multimedia, IEEE TEMU 2012, Heraklion, Crete, Greece, July 30-August 1, pp. 1–6 (2012)

[56] Bourdena, A., Mastorakis, G., Pallis, E., Arvanitis, A., Kormentzas, G.: A Dynamic Spectrum Management Framework for Efficient TVWS Exploitation. Paper Presented at the 17th IEEE International Workshop on Computer-Aided Modeling Analysis and Design of Communication Links and Networks, IEEE CAMAD 2012, Barcelona, Spain, September 17-19, pp. 51–55 (2012)

[57] Bourdena, A., Pallis, E., Kormentzas, G., Mastorakis, G.: A radio resource management framework for TVWS exploitation under an auction-based approach. Paper presented at the 8th International Conference on Network and Service Management, IEEE CNSM 2012, The Mirage, Las Vegas, USA, October 22-26, pp. 204–208 (2012)

[58] Bourdena, A., Mastorakis, G., Pallis, E., Mavromoustakis, C.X., Kormentzas, G., Karditsis, E.: A Radio Resource Management Framework for Opportunistic TVWS Access. Paper Presented at the1st ACM Workshop on High Performance Mobile Opportunistic Systems, IEEE ACM HP-MOSys 2012, Paphos, Cyprus, October 21-25, pp. 33–38 (2012)

[59] Bourdena, A., Pallis, E., Kormentzas, G., Skianis, C., Mastorakis, G.: QoS provisioning and policy management in a broker-based CR network architecture. Paper Presented at the IEEE Globecom 2012, IEEE International Workshop on Recent Advances in Cognitive Communications and Networking, Anaheim, San Francisco, USA, December 3-7, pp. 1841–1846 (2012)

[60] Bourdena, A., Pallis, E., Kormentzas, G., Mastorakis, G.: Radio Resource Management Algorithms for Efficient QoS Provisioning over Cognitive Radio Networks. Paper Presented at the IEEE ICC2013, IEEE International Conference on Communications, Budapest, Hungary, June 9-13 (2013)

[61] Bourdena, A., Makris, P., Skoutas, D., Skianis, C., Kormentzas, G., Pallis, E., Mastorakis, G.: Joint Radio Resource Management in Cognitive Networks: TV White Spaces Exploitation Paradigm. In: Lagkas, T., Sarigiannidis, P., Louta, M., Chatzimisios, P. (eds.) Paper Published at the in the book Evolution of Cognitive Networks and Self-Adaptive Communication Systems (2013)

Radio Spectrum Issues and Cognitive Mobile Computing

Mahdi Pirmoradian[1,2], Olayinka Adigun[1], and Christos Politis[1]

[1] WMN Research Group, Faculty of SEC, Kingston University London, UK
[2] Islamic Azad University Islamshahr Branch, Tehran, Iran
m.pirmoradian@iiau.ac.ir, {o.adigun,c.politis}@kingston.ac.uk

Abstract. The rapid proliferation of mobile wireless devices, astronomical data traffic transmission and development of multifarious technologies has resulted in huge demand for usable radio spectrum bands. Allocation of new spectrum bands and maximizing the usage of currently allocated radio spectrum bands have become of vital importance in emerging mobile computing networks. This chapter highlights the importance and issues of radio spectrum in mobile computing networks and also presents cognitive functionalities as an appropriate solution to cope with the scarcity of usable radio spectrum in emerging cognitive mobile computing networks.

1 Introduction

Today, large portions of the radio spectrum are assigned to the authorized users, and this static spectrum allocation mechanism causes radio spectrum bands to be underutilized at various times and locations. Beside the tremendous growth of wireless devices and increased volumes of multimedia data transmission, there is a need for more spectrum bandwidth to be made available for emerging wireless technologies. Exploiting the usage of underutilized spectrum bands is one of the ways in tackling the problem of spectrum band shortage. This process takes time and needs close cooperation between governmental and technical bodies to efficiently implement and enforce new spectrum allocation policies.

Experimental measurements have shown that current radio spectrum bands are utilized both spatially and temporally at different times and locations. Recent initiatives have also led to the refarming of parts of the Ultra High Frequency (UHF) and Very High Frequency (VHF) bands following the switch-over from analogue TV broadcast to digital technology. This refarmed spectrum bands are expected to be opened to new radio technologies. Future mobile computing data traffic will be expected to achieve the required high data rates and desired quality of service by using a combination of advanced technologies such as; advanced antennas, improved modulation techniques, dynamic network topologies and the use of additional spectrum band.

Cognitive Radio (CR) is a key technology to overcome the shortage of useful spectrum bands available for wireless communications by exploring the Dynamic Spectrum Access (DSA) technique. The DSA techniques have the potentials to

C.X. Mavromoustakis et al., *Resource Management in Mobile Computing Environments*, Modeling and Optimization in Science and Technologies 3,
DOI: 10.1007/978-3-319-06704-9_6,

dramatically increase spectrum utilization in cognitive networking environments. The DSA approach can be implemented on both the licensed and license-exempt platforms. The use of licensed spectrum bands for cognitive networking has been the main challenge in this area; however, there are different ongoing researches on radio spectrum regulation approaches such as Licensed Spectrum for Exclusive Usage, Licensed Spectrum for Shared usage, Licensed Exempt Spectrum and Open Spectrum [1]. Next generation mobile computing networks should be designed to incorporate cognitive functionalities such ad self-spectrum management, self-organization and self-optimization. The rest of the chapter is organized as follows: section two presents spectrum requirements and polices, section three discusses spectrum handoff and mobility techniques, section four highlights cognitive functionalities in mobile computing networks, section five presents the critical challenges and issues in cognitive mobile computing and the chapter is summarized in section six.

2 Spectrum Requirements and Polices

2.1 Spectrum Requirement and Solution

As the number of wireless connections and expected data-rates for wireless networks increase, spectrum demand and spectrum congestion will become critical challenges in the forthcoming all-encompassing wireless world. Increased throughput, higher quality of service requirement, seamless mobility, diversity of wireless devices based on multiple wireless standards are in more demand than ever before. To cope with these critical problems, sufficient spectrum management mechanisms and flexible spectrum usage techniques are being considered by scientists and engineers to exploit currently underutilized spectrum bands while protecting the licensed transmission against harmful interference. The concept of Cognitive Radio (CR) technology, which was coined by Mitola in 1999 [2], is a promising technology to deal with the spectrum scarcity problems envisaged in the future. The cognitive radio device is an intelligent radio, which can perceive its radio environment and adapts its radio transceiver's parameters based on internal states and radio environment knowledge. This multi-standard equipment is a prime tool that allows interoperability between communication systems of different standards, which normally will not operate together (see figure 1).

The predominant feature of CR is improving spectrum efficiency using the dynamic spectrum access mechanism without harming licensed transmission with interference. The main functionalities of DSA are radio spectrum awareness, spectrum analysis, spectrum decision and radio parameter adaptation. The DSA concept allows CR devices (secondary users) to access the unused portion of licensed and license-exempt radio frequency bands (spectrum holes) at any time and any location either vertically (spectrum sharing between heterogeneous networks) or horizontally (spectrum sharing between licensed-exempt users).

a)

b)

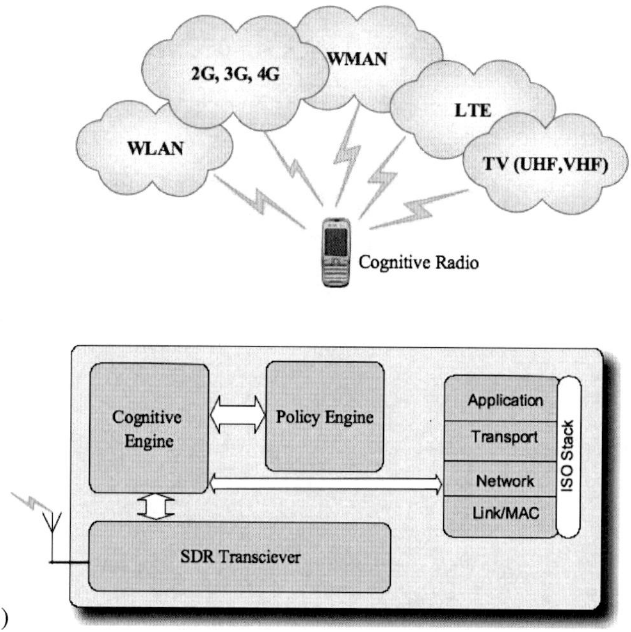

Fig. 1. a) Connection of cognitive radio to technologies b) the architecture of a cognitive radio

2.2 Spectrum Demand Forecasting

With the exponentially increasing numbers of wireless equipment and increased consumers demand for high data-rates and multimedia applications, mobile data traffic will continue to grow at an alarming rate. Wireless network operators such as Nokia Siemens Networks and Cisco have forecasted that the volume of data transmitted over mobile networks will increase by 13-fold between 2012 and 2015 [3]. Operators use several methods to augment capacity, such as increasing the number of cell sites and deploying more efficient technologies. However, these techniques only cannot meet and satisfy the consumers' demands. An effective way to cope with the high data-rates traffic demands with strict QoS requirement is to allocate more spectrum bandwidth for wireless communication services. How much spectrum and the dates of the spectrum allocation are prominent decisive factors in emerging mobile communications technology. Forecasting the use of spectrum bands is a crucial challenge, which can be estimated by different techniques such as historical growth of the number of land/mobile systems or monitoring new technologies and noting their spectrum requirements. Figure 2 shows the spectrum demand forecasting from 2010 to 2020 [5]. It can be seen that "Partially Increased Spectrum" produces a 50% increase of spectrum relative to the present by 2020, whereas "Fully Increased Spectrum" produces an approximate 100% increase of

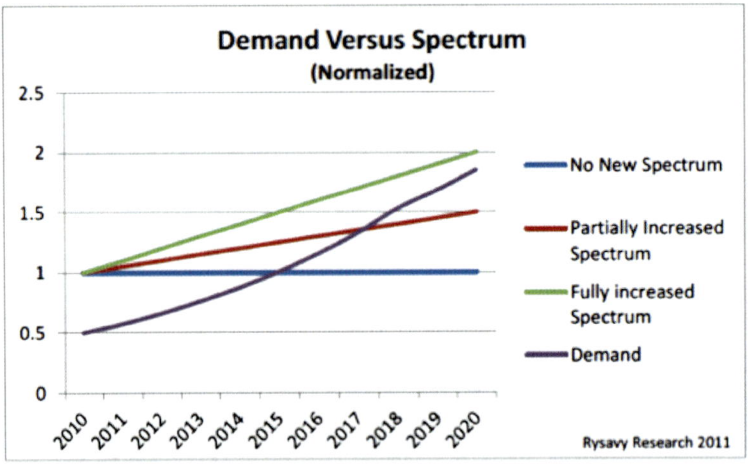

Fig. 2. Normalized spectrum demand from 2010 to 2020 (Y axis: demanded capacity, from [5])

spectrum by 2020. In [6], authors studied the estimation of radio spectrum requirement for wireless broadband services from 2010 to 2030. The forecasting algorithm defined is based on ITU-R M.1768-1.

2.3 Spectrum Sharing Approaches

Spectrum sharing could be implemented either in a cooperative or a coexistence manner. In the former, devices with different technologies cooperate to share spectrum while avoiding harmful interference to each other. To do this, a common protocol needs to be defined and distributed among users. In coexistence mode, devices are allowed access to spectrum bands without producing harmful interference. The devices must be able to sense targeted bands to avoid interference (e.g. CR). With regards to the regulatory policies, sharing techniques are categorized into four methods namely [7]: (I) sharing among equal primary users, (II) sharing among primary and secondary users, (III) sharing among equal secondary users and (IV) sharing among equal regional infrastructure.

Sharing among equal primary users; spectrum is shared among users without priority. In this case, users are allowed access to the shared bands either in a cooperative or a coexistence manner.

Sharing among primary and secondary users; this can be implemented as either cooperative sharing or coexistence sharing. In the cooperative scenario, a secondary user (SU) communicates with the primary users (PU) to know the availability of licensed channels at the demand time. Basically, this cooperation guarantees QoS for both the primary and secondary transmission. In coexistence sharing, two sharing mechanisms namely; Opportunistic Spectrum Access (OSA) and Underlay Spectrum

Access (USA) are proposed. In OSA, secondary users are allowed access to either temporary spectrum holes[1] or spatial spectrum holes.[2]

In the USA approach, the secondary users are allowed access to licensed channels simultaneously under low transmission power and interference level constraints.

Sharing among equal secondary users; secondary users can be in licensed or license-exempt spectrum. In both cases, users are prevented from causing harmful interference to each other.

Sharing among equal regional infrastructure; in this form a regional service provider coordinates and shares spectrum bands among users, thereby seeking a cost effective way to exploit spectrum resources.

2.4 Policies on Spectrum Utilization

There are several ongoing techniques in improving the utilization of current spectrum bands and these are explained below.

- **Light Licensing**

In this utilization technique, all stations must be registered in a defined database spectrum map. Light Licensing is a novel concept in the US in the 3,650 to 3,700 MHz range [4], it allows systems to share spectrum on a co-primary basis, whereby geographical location and signal sensing mechanisms are employed to ensure that no interference is caused among the involved systems. The IEEE 802.11y Working Group is one standard body that has focused on such concepts.

- **Spectrum fragmentation**

Spectrum fragmentation is another reason why new techniques are needed to exploit spectrum gaps and white spaces among the different spectrum assignments. This approach is based on the joint simultaneous use of different frequency bands in the frequency range of about 400 MHz up to about 6 GHz including the license-exempt frequency bands for WLAN applications.

- **Spectrum Harmonisation**

Harmonisation means defining technical conditions including spectrum, band plan and technology at a global and regional level. This helps to ensure efficient spectrum usage, seamless services over wide areas including roaming, system co-existence and global circulation of user equipment across borders.

---

[1] *Temporary spectrum holes*; where primary and secondary users are located in the same coverage area. SUs are allowed to exploit temporary unoccupied licensed frequency bands during specific time instances. A strong dynamic spectrum access mechanism is needed to avoid disruption to licensed transmission.

[2] *Spatial spectrum holes*; also called regional spectrum access, the secondary transmitter is located far from the licensed coverage area. The secondary transmitter is equipped with a geographical positioning system (GPS) system and is supported by an updated regional radio spectrum database.

- **Spectrum Liberalisation**

Essentially, liberalisation of spectrum means the removal of technology and regulatory restrictions on policies to enable new access technologies to be deployed within the same band or bands as existing and legacy technologies, for example, UMTS or HSPA could be deployed in spectrum bands where traditionally GSM, CDMA or TDMA have been used.

2.5 Spectrum Release Plans

As explained earlier, spectrum bands are predominantly authorized and largely lightly used by the assigned owners. However, there is still a significant increase in spectrum demand from stakeholders including broadcasters, business radio and mobile telecommunications companies, who use wireless communications to deliver their services (e.g. voice calls, video streaming, social networking, TV broadcast). Several investigations have been focused on releasing various spectrum bands for communications. The Ministry of Defence (MoD) in the UK identified two spectrum bands with priority to be released by March 2015 (2310-2390 MHz and 3400-3600MHz) [8]. These bands together comprise of 160MHz of spectrum (around one third of 500MHz spectrum required by 2020[3]). Moreover, Ofcom (the British regulator and competition authority) informed that some portions of UHF/VHF frequency bands (especially channels 61 and 62) would be released after the digital switchover.

3 Spectrum Handoff and Mobility

3.1 Spectrum Handoff Techniques

Spectrum handoff is an initiated process in mobile terminals. Handoff can be categorized as hard *handoff* and *soft handoff* in wireless communications. In the former, a mobile user can connect to only one base station at any time instant, so the target base station is selected considering the measured link quality. In hard handoff, an interruption will be caused to the transmission process, which might degrade the quality of data transmission. In a soft handoff mechanism, a mobile user can connect to a number of base stations during the handoff process. The best base station needs to be targeted among available base stations using accurate link quality measurement techniques. Soft handoff is complex and expensive to implement but has better energy efficiency and is a reliable technique compared with hard handoff [9]. Table 1 shows a brief description of applications and comparison of the aforementioned spectrum handoff approaches.

---

[3] The Spending Review announces that at least 500MHz of public sector spectrum below 5GHz will be released over the next ten years for new mobile communication technologies, including mobile broadband.

Table 1. Comparison of different spectrum handoff techniques

Specifications / Mode	Application	Connection	Power Consumption	Delay	Complexity	Effective Interference
Hard Handoff	TDMA, IS-54, PDC, GSM	Single	High	High	High	Low
Soft Handoff	CDMA	Multi	Low	Low	High	High

3.2 Spectrum Handoff and Mobility in Cognitive Networks

In cognitive radio environment, spectrum handoff occurs when a primary user appears/reappears in its own frequency band, which is being used by a secondary user. In this case, the secondary user must change its operating frequency to another unoccupied frequency band through a fast, efficient and reliable algorithm. Soft handoff mode is seen to be a strong candidate according to the architecture (software based) of the radio nodes and cognitive capability of the nodes. In cognitive radio networks, handoff schemes can be implemented in two different strategies; *reactive handoff* and *proactive handoff*. In the former, a secondary user changes its current operating frequency after detecting a link failure in the operating channel, while in the proactive technique, a cognitive user observes radio channels and predicts future activities in the current and available frequency bands, hence, an appropriate channel will be selected and channel switching is performed before link failure occurs. It should be noted that degradation in the quality of transmission and fast channel switching are prominent features of the reactive approach.

As a result of this, a robust and flexible multi-layer spectrum mobility management scheme is required to accomplish spectrum handoff in an efficient and reliable manner. It is necessary to learn and adapt new parameters and ensure that the transition stage from the old parameters to new parameters is carried out rapidly and successfully during mobility task. The nature of spectrum mobility in cognitive radio networks can be divided into the following categories:

Spectrum mobility in time domain; where a CR device adapts its operating frequency bands to newly available unoccupied spectrum band over various time slots.

Spectrum mobility in space domain; where a CR device changes its operating frequency bands based on its operational geographical region, i.e., when it moves from one place to another, the operating frequency changes accordingly. Thus, the prominent challenge is the continuous allocation of spectrum in order to avoid degradation in the QoS of the PUs and SUs transmission [10], [11]. Various spectrum handoff and mobility issues in different cognitive network architecture and spectrum mobility techniques are studied in [12].

4 Cognitive Functionalities in Mobile Computing Networks

The volume of data transmitted over mobile networks will be significantly increased by about 800% over the next four years [4], also 7 trillion wireless devices will be serving 7 billion people by 2017, implying 1000 wireless devices per person [13]. This tremendous growth of wireless devices and its associated massive data processing requirement resulting from applications like multimedia services and video transmission will require advanced and intelligent devices as well as more spectrum bandwidth. Cognitive Radio technology is a well known solution to enhance better spectrum utilization and improved energy efficiency, the prominent specifications of cognitive mobile computing devices are multiprocessing, multitasking and reconfigurability. These advanced and intelligent devices require more power, radio spectrum, advanced signal processing and diverse modulation schemes in order to meet the desired QoS in networks. Cognitive devices can perform complex computational tasks in order to satisfy future users' demands. These context aware users place cognitive functionalities (i.e. observation, analysis, decision making and adaptability) on top the software defined radio firmware to make mobile networks highly smart with features like self-organization, self-management, self-optimization and self-healing.

Figure 3 shows a typical infrastructure free (Ad Hoc) cognitive mobile computing network. The cognitive users within the coverage area are allowed to sense, learn from their radio environment and adapt their transceiver parameters using the cooperative central decision strategy. This multi standard network collaborates with the spectrum coordinator and spectrum database centre in order to have access to appropriate spectrum band for data transmission within the cognitive mobile network.

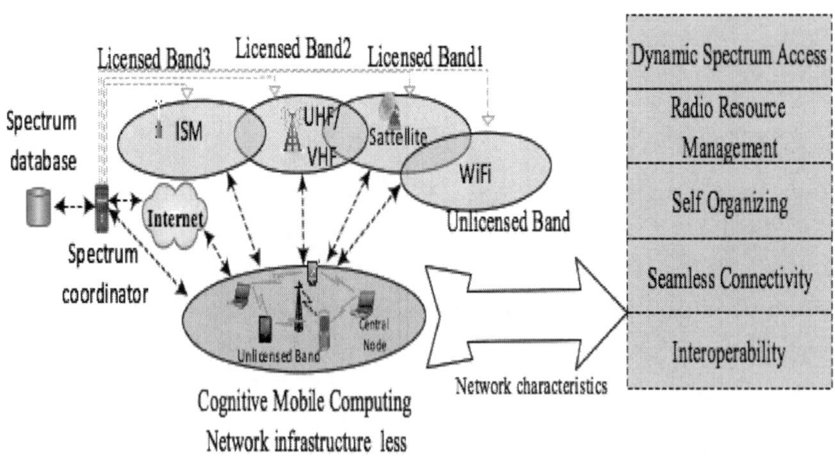

Fig. 3. A typical cognitive mobile computing network and its characteristics

5 Critical Challenges and Issues in Cognitive Mobile Computing

Many challenges and research areas have been identified in the field of cognitive mobile computing. The key challenges are: Quality of connectivity, low bandwidth, hierarchical mobility and location awareness (i.e. handoff, network selection, mobile and station migration), power consumption, security and privacy (i.e. data encryption, authentication and authorization), mobile routing and efficient route selection, and resource reservation (appropriate channel reservation and power allocation). Some open challenges are categorized and discussed in the following:

5.1 Cooperative Decision

The cooperative strategy is defined to be a network of multiple dynamic users that share information to accomplish a common objective. In order to implement the computation of complex tasks, cooperative computing is an efficient strategy in mobile computing networks. In cooperative computing, information from multiple unlicensed users is exchanged among each other to make appropriate spectrum decisions. The decision-making process in a cooperative computing network can be either centralized or distributed. By using cooperative exchange of information, the accuracy of radio resources allocation is significantly improved in a heterogeneous network. However, the main problem is the additional computational overhead when compared with non-cooperative computing networks.

5.2 Seamless Connectivity

Mobility is a critical problem in cognitive mobile networks. As mobile users migrate to new locations, there should be good considerations on seamless high quality connectivity, efficient routing adaptation, location awareness and real-time optimization. These are top challenges that need to be addressed. Changing channel frequency and migrating to newly unoccupied channel (licensed or license-exempt) without causing disruption to the data transmission is a paramount need in cognitive mobile computing.

5.3 Interoperability

Interoperability is an indispensable characteristic of cognitive devices in the next generation multi-vendor, multi-technology and multi-service environment in order to achieve seamless connectivity and guaranteed quality of service. Interoperability is necessary at various levels namely; device to device, device to network, network to network and even service to service. There are several standardization bodies working on interoperability within telecommunication technology such as; International Telecommunication Union-Telecommunication sector (ITU-T), European Telecommunication Standard Institute (ETSI), Committee T1 Telecommunications

(Committee T1), Institute of Electrical and Electronic Engineering (IEEE), Internet Engineering Task Force (IETF), Network Management Forum (NMF) and the 3rd Generation Partnership Project (3GPP).

5.4 Security and Privacy

Security and privacy is an intrinsic requirement of a cognitive mobile computing network. As the number of mobile computing devices such as smart-phones, tablets and other mobile devices grow, the need for security is vital in sending and retrieving information among users. A cognitive mobile device can intelligently connect to a diverse range of available wireless technologies in its radio environment, therefore secure data transmission with embodied authentication should be considered. In essence, intelligent collaborative security algorithms based on internal radio parameters and radio environment knowledge are required to achieve a secured multi technology environment.

5.5 Energy Optimization

Many efforts have been focused on energy optimization in mobile networks while meeting the Quality of Service (QoS) requirement of various applications at minimal cost [14], [15], [16]. Advanced low energy consumption integrated circuits have been a key consideration to meet the power consumption criteria in mobile computing networks. Several techniques have been proposed in the field of energy efficient networks such as; the exploitation and the control of power management capabilities (i.e., sleeping and rate adaptation) inside components of network equipment [17], [18], [19]. In this respect, at the physical layer level, advanced programmable hardware is the main goal in academic and industrial efforts. Also, at the link layer, efficient transmission and error detection techniques will effectively decreases retransmission of data packets and eliminate the energy consumption associated with retransmission. Transmit power control schemes are the most targeted mechanism to optimize energy consumption subject to the interference constraint and desired QoS in the network.

5.6 Self-organization

Self-organized networks are seen as one of the promising network types for an operator to augment the quality of data transmission as well as reduce operator's expenditure. Cognitive mobile computing devices and the network have capabilities of self-configuration, self-optimization and self-healing. The self-configuration starts either in pre-operational stage or joining and moving users. Self-optimization task works in operational state, which starts when all radio parameters are measured and collected from the environment. In the self-optimization techniques, the configuration parameters are auto-tuned to make the most appropriate use of the network resources. Self-healing is the feature of automatic failure detection and automatic adjustment of parameters in order to solve various failures' classes in the cognitive mobile computing network.

5.7 Radio Resource Management (RRM)

Efficient and powerful radio resource management strategies based on accurately perceived and estimated radio parameters, achieved using spectrum sensing, interference avoidance schemes, efficient power allocation techniques and appropriate channel selection schemes should be used in cognitive mobile computing networks. Dynamic and real-time RRM algorithms should be executed to optimize the use of radio resources and meet the desired QoS in the network. The increased flexibility of algorithms enables cognitive computing networks to operate with diverse technologies. Dynamic RRM strategies can be implemented in centralized or distributed architectures. A dynamic radio resource management for multiple access points in WLAN is proposed in [20]. IEEE has also suggested a distributed RRM architecture under the 802.16h standardization activities. 802.16h proposes a coexistence protocol to enable all related functions such as detection of the neighborhood topology, registration to the defined database and negotiation for sharing radio spectrum. A common dynamic radio resource management algorithm can be performed over available networks (licensed and licensed exempt) and cognitive mobile computing network. This strategy significantly improves network performance and the QoS of the end-to-end users.

5.8 Potential Health Hazard

Cognitive computing devices have complex computing and application capabilities without a predefined connection to a network. These multi-standard mobile devices are allowed access to licensed and licensed-exempt spectrum bands, which may interfere with sensitive medical equipment. Therefore, comprehensive health/spectrum regulatory policies need to be implemented for cognitive mobile users, medical equipment and information technology.

6 Summary

With the development of mobile computing technologies, there still exist a vacuum of technologies and applications that are needed to overcome the challenges of emerging mobile networks. Cognitive devices and networks are good candidates to tackle the challenges of present day mobile computing networks and provide intelligent solutions that will enhance the next generation of mobile networks. The capabilities of cognitive devices (i.e. observation, learning and adaptation) have the capability to turn current mobile networks to a dynamic, reconfigurable, interoperable and multi-standard network.

References

[1] Marques, P., Rodrigez, J., Delaere, S., Delahaye, P., Lecroart, B., Gundlach, M., Triantafyllopoulou, D., Moessner, K., Noguet, D.: Spectrum Sharing in the EU and the Path Towards Standardization. In: Future Networks & MobileSummit (2013)

[2] Mitola III, J.: Cognitive radio: An integrated agent architecture for software defined radio, KTH Royal Institute of Technology (2000)

[3] Cisco, Cisco Visual Networking Index: Global Mobile Data Traffic Forecast Update, 2012–2017 (2013)

[4] Wireless Operations in the 3650-3700 MHz Band, Federal Communications Commission, Report 2007 (2007)

[5] Wireless Technology & Spectrum Working Group, Optimising use of available mobile radio spectrum, Wireless Technology & Spectrum Working Group, Digial Communications Knowledge transfer Network, London (2009)

[6] Real wireless, Study on the future UK spectrum demand for terrestrial mobile broadband applications (June 2013)

[7] Peha, J.M.: Sharing Spectrum Through Spectrum Policy Reform and Cognitive Radio. Proceedings of the IEEE 97(4), 708–719 (2009)

[8] Enabling UK growth –Releasing public spectrum Making 500 MHz of spectrum available by 2020. Department for culture media and sport (2011)

[9] Chung, Y.-U., Lee, H., Park, S.: Performance Comparison of Handoff Modes in Terms of Energy. In: Kim, T.-h., Lee, Y.-h., Kang, B.-H., Ślęzak, D. (eds.) FGIT 2010. LNCS, vol. 6485, pp. 154–163. Springer, Heidelberg (2010)

[10] Akyildiz, I.F., Lee, W.-Y., Vuran, M.C., Mohant, S.: A Survey on Spectrum Management in Cognitive Radio Network. IEEE Communications Magazine (April 2008)

[11] Akyildiz, I.F., Lee, W.-Y., Vuran, M.C., Mohanty, S.: NeXt generation/dynamic spectrum access/cognitive radio wireless networks: A survey. Computer Networks 50, 2127–2159 (2006)

[12] Christian, I., Moh, S., Chung, I., Lee, J.: Spectrum Mobility in Cognitive Radio Networks. IEEE Communications Magazine (June 2012)

[13] David, K., Dixit, S., Jefferies, N.: 2020 Vision, IEEE Vehicular Magazine, 22–29 (September 2010)

[14] Gür, G., Alagöz, F.: Green wireless communications via cognitive dimension: An overview. IEEE Network 25(2), 50–56 (2011)

[15] Nedevschi, S., Popa, L., Iannaccone, G., Ratnasamy, S., Wetherall, D.: Reducing network energy consumption via sleeping and rate-adaptation. In: NSDI 2008: Proceedings of the 5th USENIX Symposium on Networked Systems Design and Implementation, Berkeley, pp. 323–336 (2008)

[16] Grace, D., Chen, J., Jiang, T., Mitchell, P.D.: Using Cognitive Radio to Deliver Green' Communications. In: 4th International Conference on Crowncom (2009)

[17] Oto, M.C., Akan, O.B.: Energy-Efficient Packet Size Optimization for Cognitive Radio Sensor Networks. IEEE Transaction on Wireless Communications 11(4), 1544–1553 (2012)

[18] Lee, W.-Y., Akyildiz, I.F.: Optimal Spectrum Sensing Framework for Cognitive Radio Networks. IEEE Transaction on Wireless Communications 7(10), 3845–3857 (2008)

[19] Yi, L., Shengli, X., Yan, Z., Rong, Y., Victor, C.M.L.: Energy-Efficient Spectrum Discovery for Cognitive Radio Green Networks. Mobile Network Application 17, 64–74 (2012)

[20] Hui, L., Kin, K.L., Ming, Y.: A dynamic radio resource management technique for multiple APs in WLAN. IEEE Transaction on Wireless Communications (September 2005)

Resource Request Mapping Techniques for OFDMA Networks

Adelina Basholli and Thomas Lagkas

Computer Science Department, The University of Sheffield International Faculty, CITY College, Thessaloniki, Greece
{abasholli,tlagkas}@city.academic.gr

Abstract. Wireless broadband networks are designed to provide high quality services to multiple mobile users simultaneously. The IEEE 802.16e Mobile WiMAX standard uses Orthogonally Frequency Division Multiple Access (OFDMA) schema for frame structure. OFDMA defines rectangular resource allocation of time slots and frequency carriers, separating in this way the channel into multiple subcarriers. This structure is used for arranging the incoming user requests. Aimed to increase the bandwidth utilization, while arranging the incoming user requests into an effective way, we have considered and evaluated Bin-Packing algorithms. Subsequently in the following sections are presented analysis and design of various Bin packing algorithms developed in our simulator. Moreover, a combined algorithm is proposed, named Guillotine First Fit Algorithm and a new version of the Shelf First Fit Algorithm. Simulation's results concerning performance of implemented algorithms in different input values are gathered, analyzed and compared. The results are encouraging and provide indications regarding usage of the proposed algorithms in practice.

1 Introduction to Wireless Communication and Bin Packing Algorithms

The development of mobile communication technologies is influenced by furious rate improvements in information technology and industry. These transformations enabled greater usage of mobile communication facilities [1]. Another indication of mobile technologies expansion includes the improvements of mobile communication services from generation to generation [2-4]. Developed services comprise of: data transfer, international communication - roaming, multimedia services and voice over IP services, e-commerce, global positioning system services, and many other related services [5-8]. These facts lead to perceptions that mobile communication technologies are integral part of human's everyday life. Considering new services provided by different generations of mobile communication technologies [9], it is worth to mention that changes are required in both radio access part of the networks and in the core part, too [10]. Concerning the radio access part, various multiplexing techniques have been developed, such as: Time Division Multiple Access (TDMA), Frequency Division Multiple Access (FDMA), Code Division Multiple Access

C.X. Mavromoustakis et al., *Resource Management in Mobile Computing Environments*,
Modeling and Optimization in Science and Technologies 3,
DOI: 10.1007/978-3-319-06704-9_7, © Springer International Publishing Switzerland 2014

(CDMA), Wide Code Division Multiple Access (WCDMA), or Orthogonally Frequency Division Multiple Access (OFDMA).

Regardless of the generation of mobile communication technologies, the prior aim is the user satisfaction with the quality of offered services and the ability to be always ready to support different user requirements. For these reasons, large capacity of bandwidth is required. This implies higher data rates, support of more services and fulfillment of more user requirements in an efficient way. In order to accomplish these needs, various multiplexing techniques may help. One of them is OFDMA that is used in fourth generation (4G) technologies [11]. OFDMA is known as a technique used in 4th generation standard- WiMAX [12], and it defines a rectangular resource space (time by frequency) in which multiplexed user requests are arranged. This technique is widely used in new wireless networks of IEEE standard, digital television, audio broadcasting and 4G mobile communications [13].

Bin packing algorithms present a method which may be used to increase utilization of bandwidth and to organize incoming OFDMA multiplexed user requests in an efficient way. Two dimensions are needed, corresponding to time and frequency. These dimensions are used in algorithms as input values for width and height of one rectangle which needs to be arranged into a specific shelf/bin. Different bin-packing algorithms exist with different scaling factors, which imply the usage of different properties for arrangement. We can mention: Shelf algorithms, Guillotine-able algorithms, Skyline algorithms, Maximal Rectangle algorithms, etc. Besides the algorithm to use, the objective is to arrange user requirements in a most possible efficient way, in order to provide high bandwidth utilization.

The focus here is to arrange different requests, which are multiplexed using OFDMA schema. User requests are arranged in different time slots and frequency sub-carriers. These parameters are used as input dimensions for Bin Packing algorithms. Therefore, our first aim is to present a new version of Shelf First Fit Algorithm and a combined-new algorithm which we have named it Guillotine First Fit Algorithm. These algorithms are part of Bin Packing algorithms. Our second aim is to simulate and compare gathered results based on performance of these algorithms and their applicability on industrial world.

The structure of the chapter is as follows: Section 2 presents related research and proposed solution for bandwidth utilization; Section 3 discusses concatenation of user request dimensions in OFDMA and Bin packing algorithms, while section 4 presents a thorough view of the developed simulator. Next, the implemented algorithms are analyzed, especially Guillotine First Fit and Shelf First Fit. Section 6 presents the simulation results and critically discusses usage of each algorithm in industry. Finally, the last section provides the overall conclusions.

2 Related Research and Proposed Solutions for Bandwidth Utilization

A frequent problem that telecommunication industry confronts is the usage of a method which will increase bandwidth utilization while organizing incoming frames in an efficient way. Towards this problem, different schemas have been suggested. All these schemas propose usage of a specific algorithm, in some cases proposed an algorithm and in others developed one. However, compared to them, our application comprise of five developed and implemented algorithms each with a different working scenario, simulated with the aim to address the bandwidth optimization problem.

In [14] a simple heuristic algorithm is proposed for the two-dimensional rectangular mapping. This schema may be used for downlink bursts in IEEE 802.16e Mobile WiMAX, ensuring strict QoS requirements. The algorithm is called eOCSA (enhanced-One Column Striping with non-increasing Area first mapping) and it presents an approach for handling the rectangle mapping problem. It first maps the resources for each subscriber into a downlink burst in a rectangular type. There is also a prior version (OCSA) which maps the resource allocation blocks from bottom to top and right to left [15]. The eOCSA aim is to maximize the throughput through sorting the resource requests in a descending order (largest first). Then, they are mapped from bottom to top and right to left to allow the space for the variable.

Aimed to minimize the waste of utilized bandwidth, a group of researches in [16] proposed a method that dynamically adjusts the downlink-to-uplink ratio. Dynamic Ratio Determination (DRD) monitors the mapping operation of both downlink and uplink sub-frames, and while considering load balance it receives feedback from both processes and proceeds to the appropriate selection of the forthcoming downlink-to-uplink ratio.

The idea how to arrange two-dimensional blocks which consists of time and frequency of multiple-users in OFDMA networks is analyzed in [17]. The authors in this paper prove that the problem of resource allocation in IEEE 802.16 in NP-complete. They propose a low complexity heuristic algorithm to solve the length/width variable of the two-dimensional packing problem. The Weighted Less Flexibility First (WLFF) algorithm provides a set of guidelines to choose one allocation schema for each iteration. Gathered simulation results showed that the performance of WLFF is comparable to the recursive searching methods, while the complexity is much lower.

Similarly, aimed to maximize the available radio resources in WiMAX, the Greedy Scheduling Algorithm (GSA) is proposed [18]. This algorithm is capable of transforming 2D WiMAX OFDMA link packing problem into 1D searching problem.

3 Affiliation of User Request Dimensions in OFDMA and Bin Packing Algorithms

Bin Packing algorithms present methods which are used while observing packing problems. In these cases, a resource (one or more dimensional) is given and an amount of items which need to be packed there is defined. Two-dimensional Bin packing problems deal with allocation of a set of *n* rectangles, each having width and height dimensions, into a predefined number of bins without overlapping [19-22].

Hence, the idea is to use the minimum number of bins with the same capacity, in order to accommodate the incoming user requests [23].

In this type of problems, a number of elements with different parameters have to be packed into a finite number of bins [24- 26]. The objective is to minimize the number of used bins while arranging incoming rectangles. The affiliation with Orthogonally Frequency Division Multiple Access technique can be found in usage of Bin Packing algorithms to arrange user requests which are multiplexed using the same technique. Similarly to OFDMA, two dimensions are needed, which correspond to rectangle dimensions used in Bin Packing algorithms. In this way multiple user requests multiplexed in time and frequency diversity are represented as rectangles which have two dimensions, width and height corresponding to time and frequency.

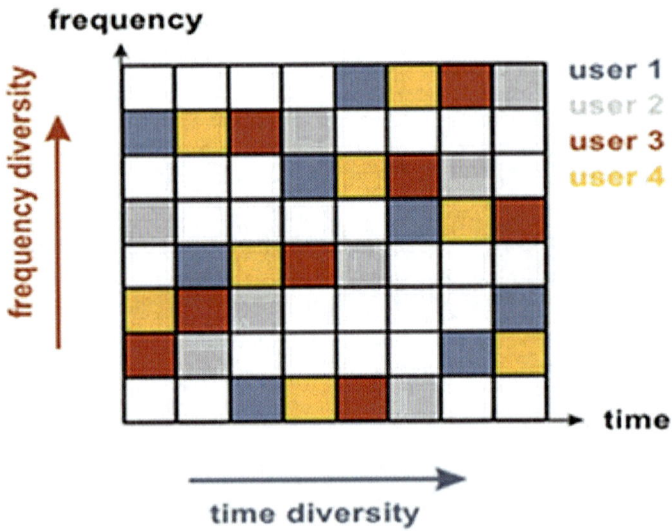

Fig. 1. OFDMA multiplexing schema [27]

Orthogonally Frequency Division Multiple Access presents a multiplexing technique which enables simultaneous transmission and reception of multiple user requests even within a single channel on the so-called subchannels [28, 29]. In order for multiple user requests not to interfere or overlap, the time and frequency dimensions are used (Fig.1.). These dimensions are used in algorithms as input values for width and height of one rectangle which needs to be placed into a specific shelf/bin. Bin packing algorithms may consider different scaling factors which imply the usage of different properties for arrangement. In [30] the following most common algorithms are described:

- Shelf algorithms

  Present a case were a shelf is defined to be a subrectangle of the bin with width *Wb* and height *Hs*, and then as packing proceeds the rectangles are placed into shelves, bottom-up, and from left to right. In this category, different versions with the same base logic are included. Some of those algorithms are: Shelf Next Fit (SHELF-NF), Shelf First Fit (SHELF-FF), Shelf Best Width Fit (SHELF-BWF), Shelf Best Height Fit (SHELF-BHF), Shelf Best Area Fit (SHELF-BAF), etc. However, these algorithms seem to waste a lot of space.

- Guillotine-able algorithms

  Present a procedure of placing a rectangle to a corner of a free rectangle of the bin, after which the remaining L-shaped free space is split again into two disjoint rectangles. As in the first case, here different subcategories of this method are also defined, such as: Guillotine Best Area Fit (GUILLOTINE-BAF), Guillotine Best Short Side Fit (GUILLOTINE-BSSF), Guillotine Worst Fit Rules, The Rectangle Merge Improvement (-RM), etc. These algorithms provide improvement compared to Shelf Algorithms, but the split line boundaries still cause problems with the practical performance.

- Skyline algorithms

  These algorithms propose a similar solution with Shelf Algorithms. However, the specific approach completes packing a lot faster than the ones using the Maximal Rectangles data structure.

- Maximal Rectangles algorithms

  These algorithms store a list of free rectangles that represent the free area of the bin and they perform an operation that essentially corresponds to picking both split axes at the same time.

Regardless the used algorithm, the primary aim is to map user requirements, which from algorithms' point of view are rectangles.

Before presenting more thorough analysis corresponding to the implementation of the chosen Bin Packing algorithms, a brief overview of the simulator features is presented in the next section.

4 Proposed Simulator of Network Request Mapping Techniques

The generic aim of our simulator is evaluation and optimization of resource sharing in OFDMA networks (like WiMAX or LTE). In this way we developed a program that has implemented different Bin Packing algorithms. The focus here is to arrange different requests that come from subscribers into a compact schema that will allow effective use of bandwidth. These requests are arranged in different time slots and frequency sub-carriers. Targeted groups for this application are internet or telephony provider companies that operate in 4G platform.

The user request mapping simulator, is developed using Java programming language and practicing Software Engineering methodologies. This application implements five Bin Packing algorithms. In order to simulate the considered algorithms, one should provide the input values to the application. The following are considered to be the input values:

- Dimensions of the bin
- Number of bins
- Dimensions of user requests

Provision of user request dimensions is implemented using three different methods, as presented in Fig. 2.

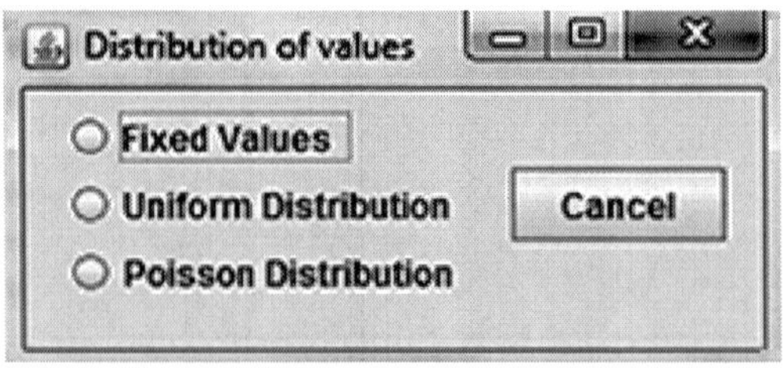

Fig. 2. Distribution of values- GUI

Specifically, they are set according to:

- Fixed value distribution
- Uniform distribution
- Poisson distribution

Considering the first method, two dimensions should be provided. These are: the user request dimensions of width and height. Final results, for all methods, include the number of arranged user requests into predefined frame dimensions, the number of idle slots, and their combination of values saved in a spreadsheet. Fixed value distribution, shown in Fig.3 a), is used while simulating our application only for testing reasons. The reason consists of: facts that in telecommunication applications there are rare or no cases where user requests have same sizes.

In this application, the random number generator is used to set the user request dimensions and a maximum number is pre-defined. This value is used as a maximum range value for generating user request dimensions. Hence, if the simulator user provides number 9, user request dimensions will have values ranging from 0 to 8. However, while simulating, we always add 1 in order to avoid user requests of null dimensions. The corresponding user interface while choosing Uniform Distribution is shown in Fig.3 b).

The third method of inputting user request dimensions is based on the Poisson value generator (Fig.3 c).). This method requires two mean values; one for the width of the user request and the other for the height. In this way, the generated user requests have dimensions which range close to the provided mean value.

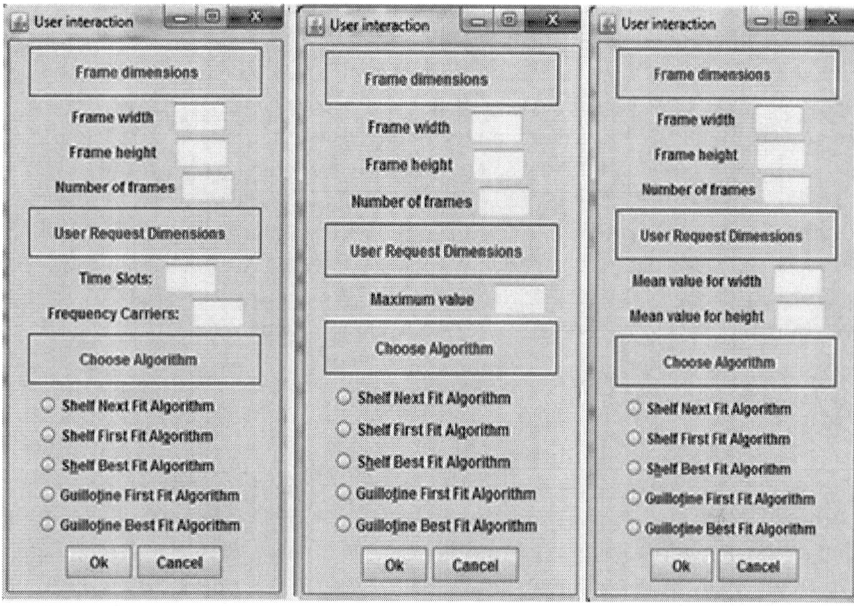

a). b). c).

Fig. 3. Distribution value types- GUI: a). Fixed values; b). Uniform Distribution; and c). Poisson distribution

Regardless of the used input method for user request dimensions, in the end of the simulation, the gathered results will be presented as a total number of arranged user requests and total number of idle slots. In order to view the combination of idle slots, we have also created the output part in a spreadsheet. This is part of the application, too.

The next section presents the detailed analysis and design steps of the implemented algorithms, starting with Shelf Algorithms and continuing with Guillotine-able Algorithms.

5 Implemented Bin Packing Algorithms

Developed simulator consists of Bin packing algorithms that are applicable to our practical aim. Considering our objective- to provide an effective bandwidth utilization solution to a dynamic system (telecommunication system which has unpredictable user requests), we chose Shelf and Guillotine algorithms. In order to develop the application, which consists of five Bin Packing algorithms, we have used Software Engineering methodologies. While working with the simulator, terms like:

- Rectangle, present the user request. Implying that the user request time and frequency dimensions are equivalent with rectangle's width and height dimensions.
- Shelf and bin, present the frames that are going to be used in order to arrange user requests. Similarly shelf's dimensions correspond to frame dimensions.

These terms (rectangles, bin, or shelf) are used just for adoption with terms of Bin-packing algorithms.

Shelf algorithms, as the simplest methods of packing rectangles, are defined by two dimensions. Specifically, a shelf is considered a subrectangle of the bin with width (*Wb*) and height (*Hs*). The packing area is arranged with placement of rectangles into shelves from bottom-up, and from left to right. Hence, when packing a rectangle (*w, h*) onto the shelf (*Wb, Hs*) we have to choose whether to rotate it or not.

The orientation procedure, presented in Fig. 4, is applied to all Shelf Algorithms. This procedure considers the suitable orientation in which the rectangle will fit into the allocated area and arrange rectangles so as to save the predefined space. The orientation procedure consists of decision making, such as placing rectangles as follows:

- Upright: *R= (min (w, h), max (w, h));* or
- Sideways: *R= (max (w, h), min (w, h)).*

However, first of all we should have shelf dimensions. As it is presented in Fig. 4, we have initialized the Width to zero (*Wt*=0). It corresponds to the total used Width, which at the beginning of arrangement is zero. After we have the rectangle with its dimensions, we need to determine whether this is the first rectangle on a new shelf:

- If so, then we store it sideways in order to minimize the height of the new shelf.

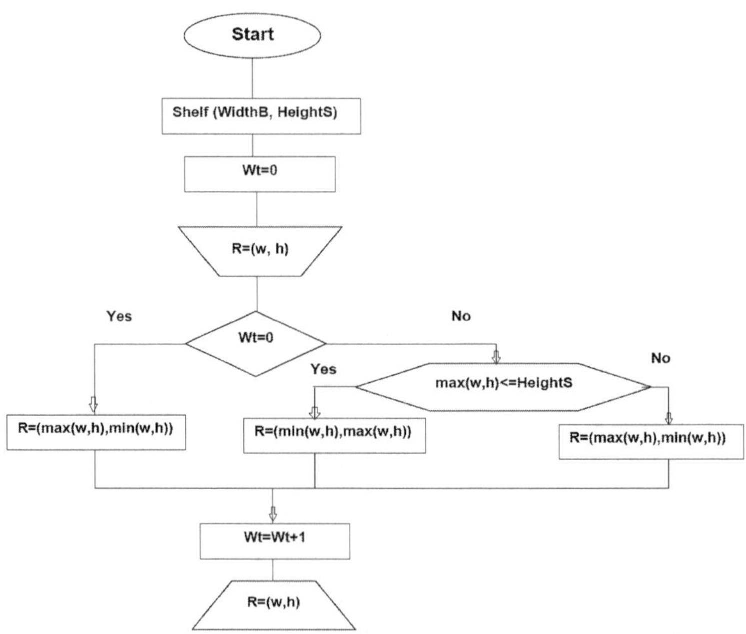

Fig. 4. Orientation process in Shelf Algorithms

– If not, then we need to see if the rectangle fits upright. For this reason in Fig. 4 we have the condition: *max (w, h) <= HeightS*. This condition checks if the maximum of rectangles' dimensions is lower than the shelf's height. If this condition is fulfilled, then the rectangle is stored upright, if not sideways.

The Shelf Next Fit algorithm is one of the Shelf Algorithms that helps solve the packing problem. It first determines the proper orientation of the rectangle, and then it tries to fit it in the current open shelf. If the rectangle does not fit there, the algorithm opens a new shelf, if there is room for that; and if not it terminates execution. A closed shelf will not be opened again.

Shelf Best Width Fit Algorithm tries to minimize the remaining capacity or width of the shelf space. This algorithm is very similar to Shelf First Fit Algorithm which we will present in more detail in the next subsection.

5.1 Shelf First Fit Algorithm

Shelf First Fit Algorithm is an extension of Shelf Next Fit Algorithm. Therefore, based on the Shelf Next Fit Algorithm work methodology, we can conclude that it is wasteful to leave behind free spaces in old shelves while opening new ones [20, 22, 30, 31]. Shelf First Fit tries to look into those left areas, which in telecommunication point of view are idle slots, and fit the incoming rectangle/user request there. What differentiates our approach from the proposed working methodology of Shelf First Fit Algorithm proposed in [30], is that we first try to fit the incoming user request into the current open shelf and if it does not fit there, we search into the left areas starting from the lowest indexes. As shown in Fig. 5, this working methodology promotes a new version of Shelf First Fit algorithm. By all means of algorithm's performance, the execution time compared to Shelf Next Fit algorithm is longer.

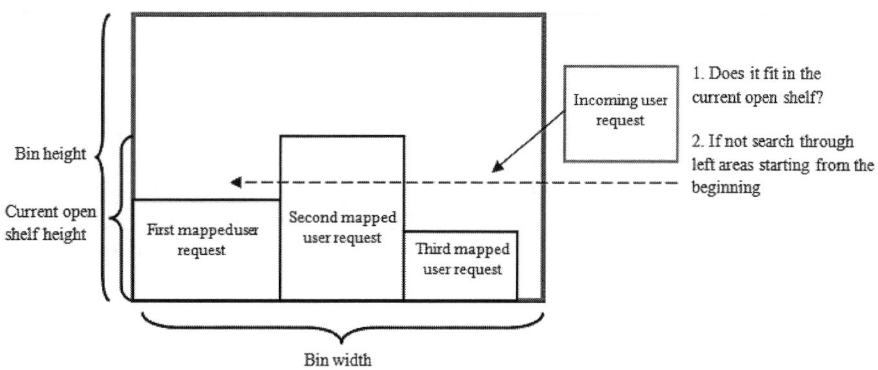

Fig. 5. Shelf First Fit Algorithm's operating logic

5.2 Guillotine First Fit Algorithm

Guillotine Algorithms present our second proposed pack of algorithms for better bandwidth utilization. These algorithms are based on the operation of guillotine split placement. The objective here is to place an incoming rectangle in the corner of a free-picked rectangle of the bin. Then the guillotine split procedure is applied and the L-shaped free space is composed of two disjoint free rectangles which can further sub-divided. In this way we need to maintain a list of the composed rectangles which can be used to arrange other incoming rectangles.

Split choices used in Guillotine algorithms and operating logic are presented in Fig.6. The bin dimensions are represented by W and H (indicating Width and Height), while w and h present rectangle dimensions. Two composed rectangles from split procedure are presented as F and F'. The composed rectangles will have the remaining dimensions of the respective bin, as shown in Fig.6.

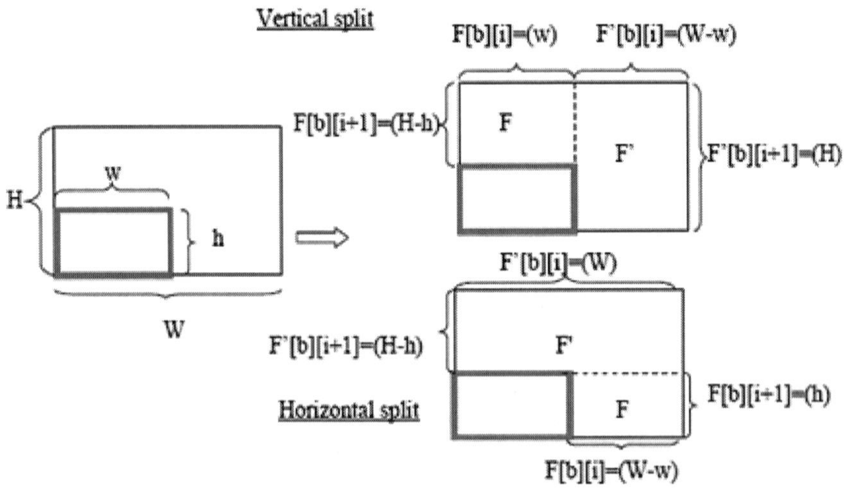

Fig. 6. Split choices used in Guillotine Algorithms

Split procedure of Guillotine algorithms implies the possibility to choose between two possible directions. On behalf of this, we have combined two methods. The Shorter/Longer Axis Split Rule [30] is used to choose the direction of the split procedure. This method includes two possible prerequisites, that for the horizontal split ($w<h$) and for the vertical one ($w>h$). Therefore, while observing rectangle's dimensions we choose which one of the split directions to apply. Finally, we will have an arranged rectangle, and another two free and disjoint rectangles with corresponding dimensions as presented in Fig.6.

Guillotine Algorithm's operating logic combined with First Fit method provide a new combined algorithm which is used to arrange the incoming user requests (Fig. 7.). Guillotine First Fit Algorithm takes into consideration all free spaces through split rectangles and places the incoming rectangle into one of them, making the choice based on the numbering (as the name implies-first fit).

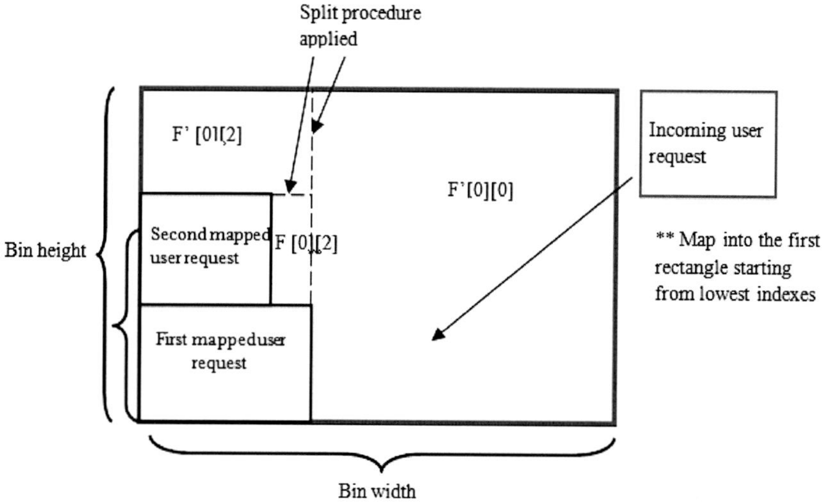

Fig. 7. Guillotine First Fit Algorithm's operating logic

Detailed design milestones of proposed algorithms can be found in Appendix A. Workflow diagrams are presented based on these algorithms' operating logic. It is worth to mention that the implementation procedure is completely based on those diagrams.

6 Analysis of Simulation Results

In this section we present and discuss the results gathered via our simulation application. The corresponding results are summarized based on the distribution mode of user requests. Whereas simulation user should at the beginning choose the method of distribution of the values for the corresponding user requests.

Consecutive simulations showed that while using Fixed Value Distribution, all Shelf Algorithms perform similarly. This due to the same user request dimensions. It is reminded that the Fixed Value Distribution uses once entered values for time slots and frequency carriers. In this way, while simulating using the same input dimensions for all algorithms we have gathered results shown in Table 1.

Table 1 presents the input dimensions of the shelf, the number of shelves, and the values of time slots and frequency carriers. These parameters are provided by the user at the beginning of the execution. In the subsequent columns the total number of mapped slots (arranged user requests) and the total number of idle slots (left areas) are presented. Total values of the slots are calculated by summing the multiplied user request dimensions.

Table 1. Fixed Value Distribution

	Shelf dimensions	Number of Shelves	Time slots	Frequency carriers
Fixed Value Distribution	(27, 31)	30	12	15
	Total Mapped Slots	Total Idle Slots		
Shelf Next Fit	21600	3510		
Shelf First Fit	21600	3510		
Shelf Best Width Fit	21600	3510		
Guillotine First Fit	13500	11610		
Guillotine Best Fit	5400	19710		

The gathered results are promising for Shelf Algorithms. By observing the performance of the Guillotine Algorithms, we can notice high number of idle slots. This, due to low number of used frames (30 in this case scenario) for arrangement of user requests. While increasing the number of frames, the performance of Guillotine algorithms is enhanced, as presented in next scenarios. However, as mentioned previously, Fixed Value Distribution is used in our application for testing reasons.

Focused to highlight results of our proposed algorithms, in this case Shelf First Fit new version and combined Guillotine First Fit algorithm, we tried to provide a comparison of their performance. Therefore, while changing the number of used frames we got following graphs, Fig.8 and Fig.9. We provided the same dimensions for both scenarios, but the number of used frames in the first case is 70 frames, while in the second case is 150. Hence, we used the Uniform Distribution or Random number generator, with bin dimensions (34, 36) and a maximum value of 22 (automatically generated used request dimensions will range from 1 to 22).

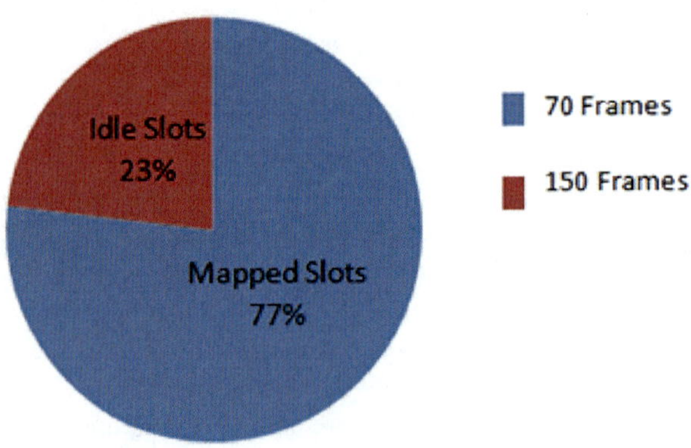

Fig. 8. Uniform Distribution – Shelf First Fit Algorithm

Simulation results of Shelf First Fit Algorithm while using the Uniform Distribution are presented in Fig.8. As one can notice by doubling the number of used frames for user request arrangement, also the number of mapped and idle slots is doubled. From industrial application's point of view this is seen as positive performance. Taking into consideration telecommunication applications where the number of user requests most of the time is unforeseen, so does the number of frames that the application should use.

Same results are gathered for Guillotine First Fit algorithm (Fig.9). However, we can summarize performance of proposed Bin packing algorithms as follows:

- The performance of Shelf First Fit algorithm is highlighted in most of the case scenarios
- Guillotine First Fit algorithm performs better for higher number of frames, implying the possibility to arrange more user requests

Nevertheless considering others conclusions presented in [30], and our experience, the memory consumption and execution time make Guillotine algorithms more resourceful and useful for industrial applications.

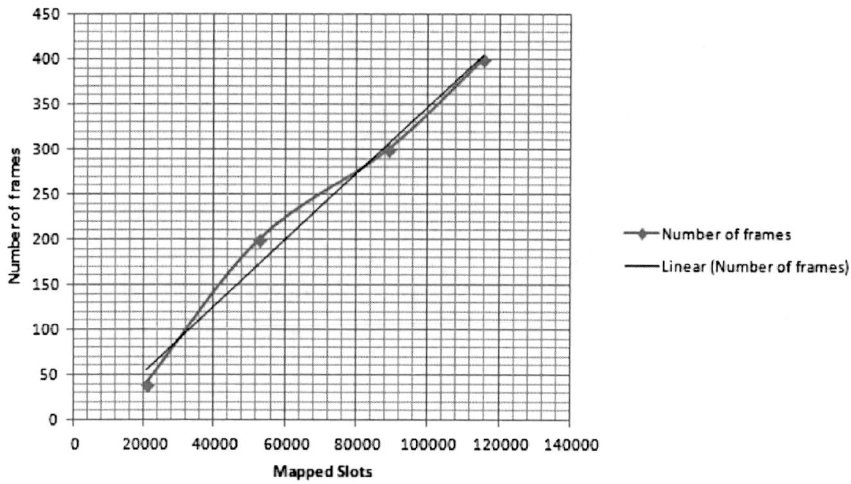

Fig. 9. Uniform Distribution – Guillotine First Fit Algorithm

The third method of distribution of values implemented in our application implies Poisson distribution. This method generates user request dimensions based on manually entered mean values for the width and the height. For the simulations that are based on the Poisson distribution we used 200 frames, with user request dimensions having mean values of the width 24 and the height 27.

Overall results as total number of mapped slots and idle slots for each algorithm, while using Poisson distribution are presented in Fig.10. Similarly to the previous scenarios, the Shelf First Fit algorithm managed to arrange the highest number of user requests while generating the lowest number of idle slots. On the other hand, the Shelf

Best Width Fit algorithm is very close to the best performance, too. It is reminded that the Shelf Best Width Fit uses the methodology of arranging the incoming user request into the left position where the remaining capacity is the lowest. Therefore, as we can notice from the corresponding graphs, the number of idle slots is lower compared to the Shelf Next Fit algorithm.

Guillotine First Fit algorithm performance compared to Shelf First Fit algorithm is considered proximate in the context of mapped user requests. Even though the total number of mapped and idle slots differs, the memory usage and simulation time in the case of Shelf First Fit algorithm are higher. This implies a tradeoff between delay, memory usage and efficiency.

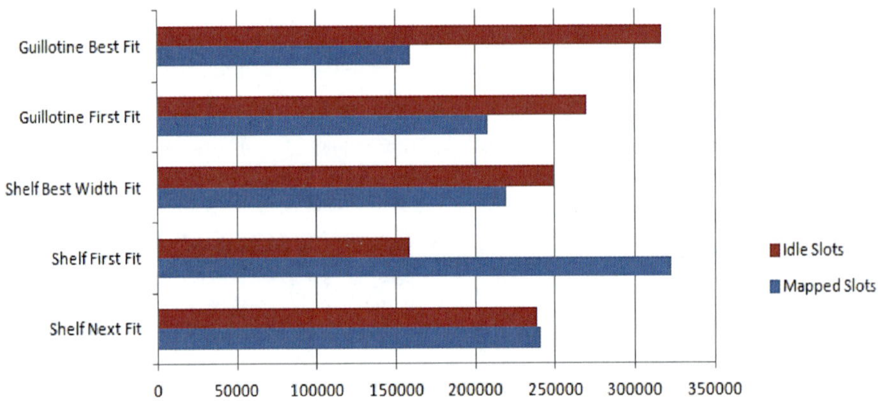

	Shelf Next Fit	Shelf First Fit	Shelf Best Width Fit	Guillotine First Fit	Guillotine Best Fit
Idle Slots	238726	158698	249265	269991	316803
Mapped Slots	241213	322395	219728	208409	159769

Fig. 10. Gathered results while using Poisson Distribution

7 Conclusions

Towards lowering the resource allocation problem in telecommunication areas, we propose different Bin-Packing algorithms which are designed for use in OFDMA networks.

The corresponding resource allocation techniques can be incorporated into small wireless broadband systems which use OFDMA as multiplexing scheme. Incoming parameters are provided as multiplexed values (time slots and frequency carriers) of user request, which should be arranged into the system's resource space depending on the available bandwidth. Hence, as a result at the end of execution the user requests are mapped to specific time-frequency slots, in a manner exploiting most of the free space left within the frame. The corresponding left areas are considered as idle slots, which vary depending on the specific case.

The implemented Bin-Packing algorithms exhibit different features and functionalities, which enable to compare and distinguish them for specific case scenarios where they can provide better bandwidth utilization.

The analysis of the results reveals that our new version of the Shelf First Fit algorithm was the most resourceful algorithm. However, the high memory requirements and execution time can be compensated by using the proposed Guillotine First Fit algorithm which also exhibits high performance. In the future, we plan to enhance the algorithms for considering QoS requirements as well.

References

1. Bannister, J., Mather, P., Coope, S.: Convergence Technologies for 3G Networks IP, UMTS, EGPRS and ATM. In: Background to Convergence 1rt end. PO19 8SQ, England. John Wiley & Sons Ltd., The Atrium (2004) ISBN 0-470-86091-X
2. Romero, J.P., Sallent, O., Augusti, R., Diaz-Guerra, M.A.: Radio resource management strategies in UMTS. In: The Mobile Communications Sector 1st edn. PO19 8SQ, England. John Wiley & Sons Ltd., The Atrium (2005) ISBN-10: 0470022779
3. Junaid, M.A., Farooq, A., Shah, A.: Evolution and Development towards 4th Generation (4G) Mobile Communication Systems. Journal of American Science 6, 1545–1003 (2010) ISSN: 1545-1003
4. Tanaka, K.: The Mobile Communications Environment in the European Union: Systems, Regulations, and Consequences in the Past, Present, and Future. In: IEEE 2001 Conference on the History of Telecommunications, St. John's Newfoundland Canada, pp. 25–27 (July 2001)
5. Marjanovic, D., Müller, C., Schönenberger, T.: New Opportunities for B2B and B2C Services Offered by New Mobile Communication Technologies. Presentation Lectures: Department of Informatics - Communication Systems Group. University of Zurich
6. Tellabs: 4G, the What, Why and When. The worldwide adoption and growth of wireless are the fastest technological achievements in history. In: White paper, Innovative solutions at your fingertips. 74.1948E Rev. C 2/12 (2012)
7. Borcoci, E.: WiMAX Technologies: Architectures, Protocols, Resource Management and Applications. In: CTRQ Conference, Bucharest, Romania, June 29-July 5 (2008)
8. So-In, C., Jain, R., Tamimi, A.-K.: Scheduling in IEEE 802.16e Mobile WiMAX Networks: Key Issues and a Survey 27(2) (February 2009)
9. Hwang, J.S., Consulta, R.R., Youn, H.J.: Opportunities & Threats for implementation of 4G cellular networks. PTC 2007 Proceedings (2012), Paper published in online library at `http://www.scribd.com/doc/6852728/4G`
10. Rumney, M.: IMT-Advanced: 4G Wireless Takes Shape in an Olympic Year. Agilent Measurement Journal, 5989-9793EN (2008)
11. Omerovic, S.: WiMax Overview. Faculty of Electrical Engineering. Dissertation, University of Ljubljana, Slovenia
12. Toh, C.K.: 4G LTE Technologies: System Concepts. In: Technology White Paper, VERSION 1.0 (November 16, 2011)
13. Ballal, B.R., Chadha, A., Satam, N.: Orthogonal Frequency Division Multiplexing and its Applications. International Journal of Science and Research (IJSR) 2(1) (January 2013) ISSN: 2319-7064

14. So-In, C., Jain, R., Al-Tamimi, A.: eOCSA: An Algorithm for Burst Mapping with Strict QoS requirements in IEEE 802.16e Mobile WiMAX Network. In: 2nd Inter. Conf. on Comp. and Automation Engineering, ICCAE 2010 (2010)
15. So-In, C., Jain, R., Al-Tamimi, A.: OCSA: An algorithm for Burst Mapping in IEEE 802.16e Mobile WiMAX Networks. In: 15th Asia-Pacific Conference on Comm, APCC 2009 (2009)
16. Sarigiannidis, A., Nicopolitidis, P., Papadimitriou, G., Sarigiannidis, P., Louta, M.: Using Learning Automata for Adaptively Adjusting the Downlink-to-Uplink Ratio in IEEE 802.16e Wireless Networks. In: ISCC 2011 Proceedings of the 2011 IEEE Symposium on Computers and Communications, pp. 353–358 (2011), doi:10.1109/ISCC.2011.5983863 ISBN: 978-1-4577-0680-6
17. Wang, T., Feng, H., Hu, B.: Two-Dimensional Resource Allocation for OFDMA System. In: IEEE Communications Society Subject Matter Experts for Publication in the ICC 2008 Workshop Proceedings (2008)
18. Zubowa, A., Murb, D.C., Costab, X.P., Favaroc, P.: Greedy Scheduling Algorithm (GSA)-Design and Evaluation of an Efficient and Flexible WiMAX OFDMA Scheduling Solution. In: Elsevier Editorial System (tm) for Computer Networks, August 10 (2009)
19. Lodi, A., Martello, S., Vigo, D.: Recent advances on two-dimensional Bin packing problems 123(1-3), 379–396 (2002) PII: S0166-218X(01)00347-X
20. Zhang, G.: A new version of on-line variable-sized Bin packing 27(3), 193–197 (1997) AMS classification: 90B35, 9OC27
21. Bin Packing. Presentation, The Icon Programming Language: The University of Arizona, `http://www.cs.arizona.edu/icon/oddsends/bpack/bpack.htm`
22. Dosa, G., Sgall, J.: First Fit packing: A tight analysis. In: Leibniz International Proceedings in Informatics Schloss Dagstuhl, pp. 1–15 (1998), doi:10.4230/LIPIcs.xxx.yyy.p
23. Baldi, M.M., Crainic, T.G., Perboli, G., Tadei, R.: Worst – Case Analysis for New Online Bin Packing Problems, CIRRELT (February 2013)
24. Kühn, P.J.: Designing a Frequency Selective Scheduler for WiMAX using Genetic Algorithms. Dissertation, University of Stuttgart, Institute of Communication Networks and Computer Engineering (2008)
25. Davis, T.: Bin Packing (2006), `http://www.geometer.org/mathcircles`
26. Lodi, A.: Algorithms for Two-Dimensional Bin Packing and Assignment Problems. Doctorate thesis, Universita Degle Studi Di Bologna (1999)
27. History of OFDMA and how it Works (2005), `http://www.webopedia.com/DidYouKnow/Computer_Science/2005/OFDMA.asp`
28. Shazzad, K.M.: Orthogonal frequency-division multiplexing, `http://www.mikrobiokosmos2009.org/7-orthogonal-frequency-division-multiple-access-ofdma.html`
29. Ahmadi, S.: Introduction to mobile WiMAX Radio Access Technology: PHY and MAC Architecture. Presentation: Wireless Standards and Technology Intel Corporation (2006)
30. Jylänki, J.: A Thousand Ways to Pack the Bin – A Practical Approach to Two-Dimensional Rectangle Bin Packing. Dissertation (2010), `http://clb.demon.fi/files/RectangleBinPack.pdf`
31. Bin packing. Institute of Computer Science, Faculty of Mathematics and Natural Sciences II, part of the Lecture notes HYPERLINK, ch. 10, `http://www2.informatik.hu-berlin.de/alkox/lehre/lvws1011/coalg/bin_packing.pdf`

Appendix A

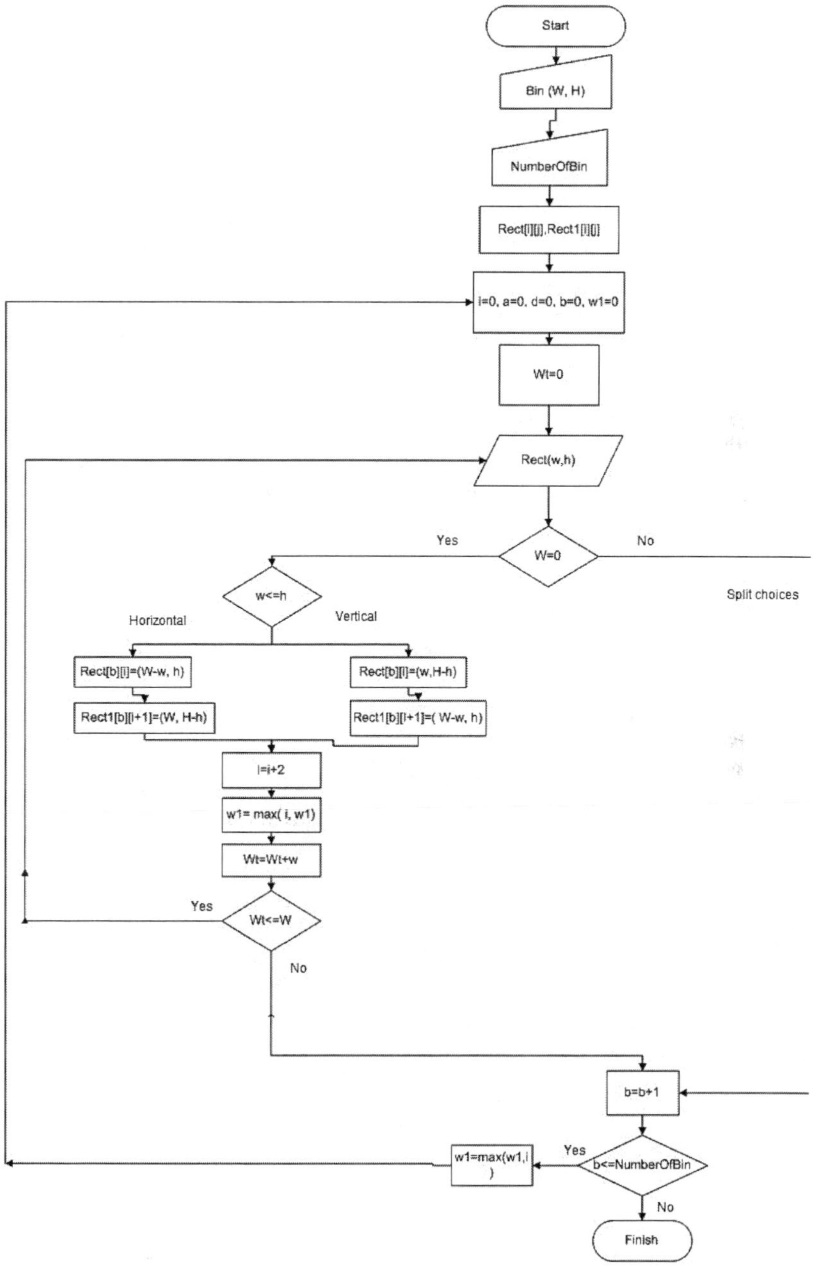

Fig. 11. Workflow diagram of Guillotine First Fit Algorithm- Part 1

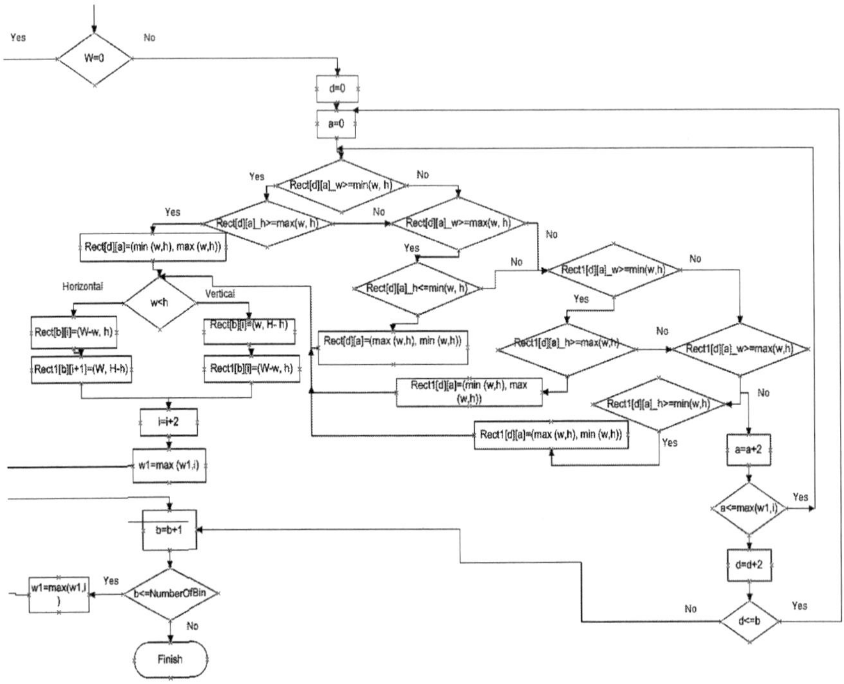

Fig. 12. Workflow diagram of Guillotine First Fit Algorithm- Part 2

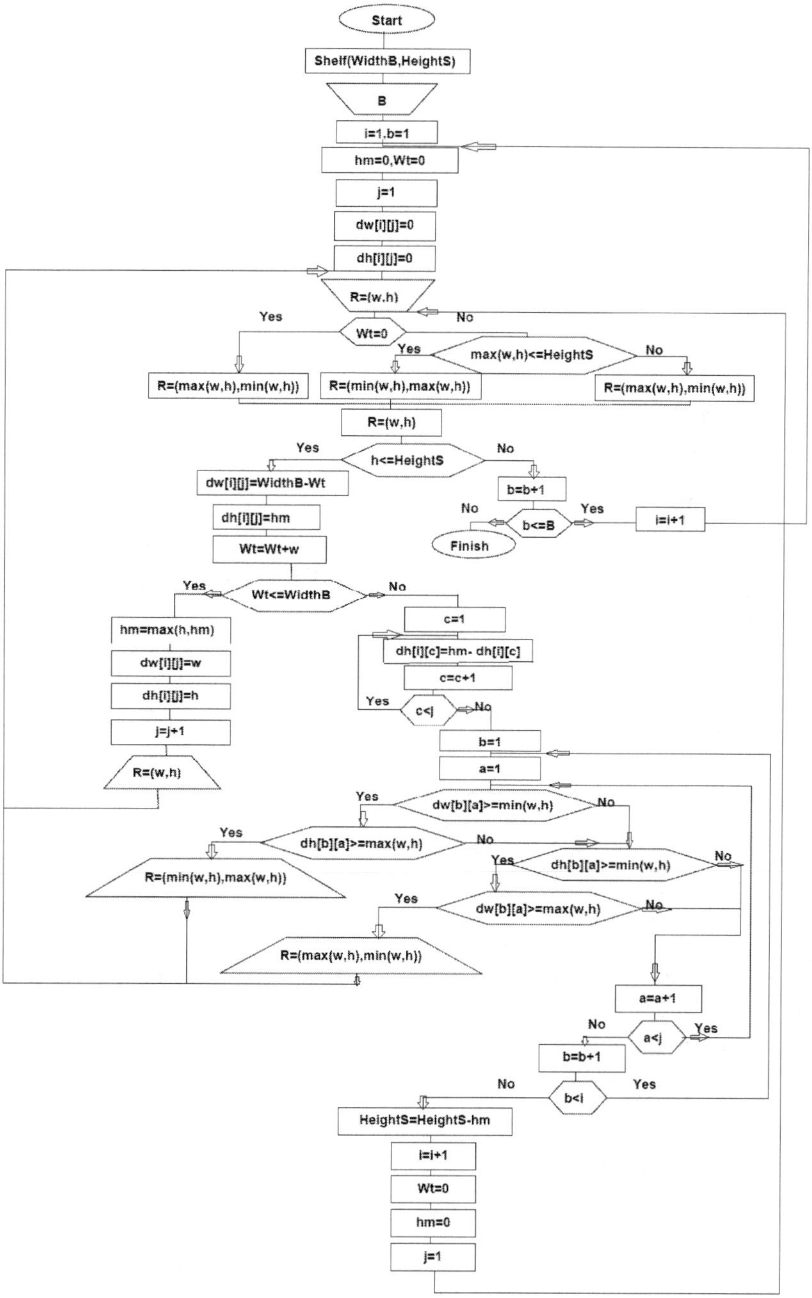

Fig. 13. Workflow diagram of Shelf First Fit Algorithm

Part II

Peer-to-Peer Systems for Mobile Computing

Generic IP Network Traffic Management from Measurement through Analyses to Simulation

Seferin Mirtchev[1], Constandinos X. Mavromoustakis[2], Rossitza Goleva[3,*], Kiril Kassev[4], and George Mastorakis[5]

[1,3,4] Technical University of Sofia, Kl. Ohridski blvd. 8, Sofia, 1756, Bulgaria
{stm,rig,kmk}@tu-sofia.bg
[2] University of Nicosia, 46 Makedonitissas Avenue, 1700 Nicosia, Cyprus
mavromoustakis.c@unic.ac.cy
[5] Technological Educational Institute of Crete, Estavromenos, Heraklion, 71500, Crete, Greece
gmastorakis@ieee.org

Abstract. The aim of this chapter is to present different approaches to network traffic management applicable to the IP, transport and application layers in IP, 3G, WiMAX and 4G technologies. The proposed technology for analysis is flexible enough to different types of traffic in opportunistic networks. We start with traffic measurements and obtain accurate data for detail network simulations and precise analysis. Then, we highlight the self-similar nature of the incoming traffic at network nodes. In our next analysis, we look at mapping the measured data with the Polya arrival process by Pareto and gamma distributed inter-arrival times. Polya, Pareto and gamma distributions have the capability to change shape and scale in a way to simulate different types of observed traffic. A proper analytical description of the end-recipient traffic flows and point process of self-similarity inputs are applied for a better user behavior specification. During an end-to-end simulation, more complex queuing models with priorities are proposed. The behavior of the system at its bounds is shown. We map the data from measurements and simulations with the application layer requirements, cross-layer Quality of Service (QoS) and Quality of Experience (QoE) parameters. This is done by traffic fractality analyses, codec-dependent resource reallocation and Fibonacci backward difference traffic moments analyses. All of them demonstrate special moments in the breakdown of the shaping effect. Finally, we express views on open-research issues for offering optimization in the Internet traffic analyses.

Highlights

- Polya Arrival Process, gamma and Pareto distributions.
- Description and evaluation of Polya/D/1 model.
- Numerical results of different peakedness of the traffic input flows.
- Priority queuing and waiting time limits.
- Distributed QoS and QoE management.
- Applicability in Internet of Things and opportunistic networks.

* Corresponding author.

C.X. Mavromoustakis et al., *Resource Management in Mobile Computing Environments*, Modeling and Optimization in Science and Technologies 3,
DOI: 10.1007/978-3-319-06704-9_8,

Keywords: Self-similar traffic, traffic measurements, traffic simulations, Polya arrivals, Pareto and gamma distribution, point process, end-to-end queuing, scheduling, packet level QoS, application level QoE.

1 Introduction

Network design and management in the Internet of Things and especially in distributed mobile computing environments is a challenging subject. The availability of free resources for files, music, movies, and pictures through virtual storages is considered natural by the audience. This resource seems to be unlimited but the truth is that it is finite by nature. A smart way of resource allocation is crucial for the service perception and overall Quality of Experience [1-4]. These features require a new approach to network design and configuration as well as special attention to the Quality of Service.

Traffic engineering based on the works of Kleinrock has to be carefully adapted due to the mobility of the objects, additional traffic shaping points, possible resource sharing, priorities, and disruptions in the network structure [5-10]. The simplest way to start adaption is to study carefully the traffic in the network. Furthermore, it is important to look for a traffic model that is flexible to traffic diversity. Taking into account new network features, we demonstrate that the classical traffic engineering formulae are carefully revised. Pareto, Polya and gamma distributions are proposed after behavior analyses of queuing systems. The conclusion comes from detailed study of the end-user behavior, traffic measurements, and end-device features. We also highlight new approaches to network design and analysis based on mapping between traffic and configuration parameters and end-to-end traffic shaping. The proposed steps for traffic engineering and resource management take into account the dynamics of the distributed mobile computing environment. We show that it can be flexible enough to support resource reallocation almost in real-time [11-12].

2 Traffic Measurements

Network traffic analysis used to be always a continuous process. Nowadays cloud computing and the Internet of Things as well as technologies like Disruptive Tolerant Networks and smart grids make this task much more challenging. Traffic analysis that is based on traffic measurements is becoming a crucial task for network configuration, management and design. This is essential in all cases when ad hoc network design is not applicable. The nature of the traffic in different domains [13-14] changes depending on the service profile and it can be represented in the system analyses by different traffic distributions. Multimedia services generate packet streams with different inter-arrival times and packet lengths within a single interface. By precise distribution function, mean value, variance, skewness and kurtosis we are capable to simulate exactly the behavior of the system and investigate its performance in unusual working conditions. In addition, the approach allows forecasting of domain behavior when the service profile is changed. Further mapping to the configuration parameters is evident.

In order to obtain data for traffic analyses we perform measurement in a 3G network [15]. The services observed are Voice over IP (VoIP), video over IP, and ftp. The measurement is not related to the specific resource allocation procedures at Radio Network Controller (RNC) (Fig. 1) [13, 16]. Every measurement session continues one minute. Data is collected on the connection between Node B and Session Border Gateway Controller (SBC) via RNC, Serving GPRS Support Node (SGSN), Gateway GPRS Support Node (GGSN) and the backbone IP network. The data is captured at the Gi interface in the forward and backward direction.

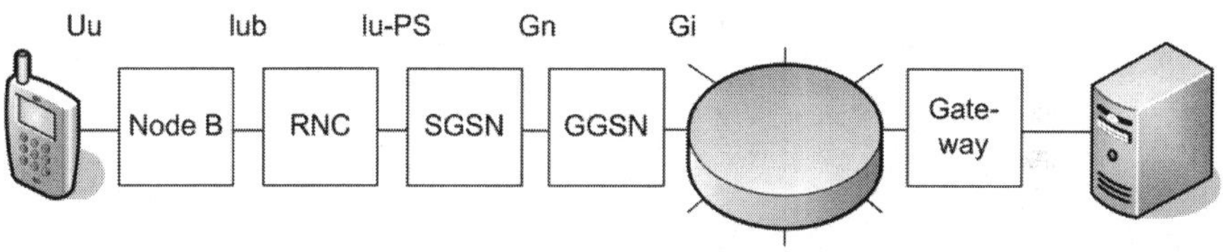

Fig. 1. The observed network architecture

The data presented is influenced by resource allocation, metering, marking, aggregating, shaping, and policing procedures in the network. The calling-party and called-party are UMTS users. GSM 6.10, G.711 A-law, and G.729 codecs are applied for VoIP. Video calls use H.263, H.263+ (1998), H.264 (low and high end) codecs. The packet-level traffic characteristics [17-18] of the observed data are presented in Table 1. The statistical analyses are performed by Wireshark, Excel, and Crystal Ball tools [14]. The results show deterministic, multimodal, log-normal, exponential, gamma nature of the inter-arrival times (Table 1) as well as a difference between the mean value, the variance and the skewness. The long-range dependence between the burst duration and number of packets in the burst leads to a non-trivial distribution of the number of packets versus the inter-arrival time (Fig. 2). The packets from the burst are observed in the time interval of approximately one millisecond and have the same size and deterministic distribution (first peak on the figure). The second peak corresponds to inter-arrival times between short bursts in variable rate voice and video streams. The third and fourth peaks correspond to multiple bursts and different session lengths.

The outbound and inbound traffic as well as uplink and downlink traffic characteristics are different. The packets at the receiving end are policed and shaped by QoS procedures. The shaping effect depends on the priority of the packets. Real-time applications also depend on packet fragmentation. More details are presented in [13].

Table 1. Traffic source specification in 3G network at Gi interface in forward direction

Experiment	Packet size (bytes)	Transport protocol	Inter-arrival time distribution	Packet size distribution
VoIP (GSM 06.01)	100	UDP	Exponential (mean 12 ms)	Deterministic
VoIP (A-law)	230	UDP	Gamma (mean 20 ms)	Deterministic
VoIP (G.729)	80	UDP	Log-normal (mean 12 ms)	Deterministic
Video over IP (H263)	230	UDP	Exponential (mean 8 ms)	Deterministic
Video over IP (H264)	230	UDP	Gamma closed to exponential (mean 8 ms)	Deterministic
FTP	40 or 1500	TCP	Gamma (mean 17 ms)	Multi-modal
TV	208	UDP	Almost deterministic	Deterministic

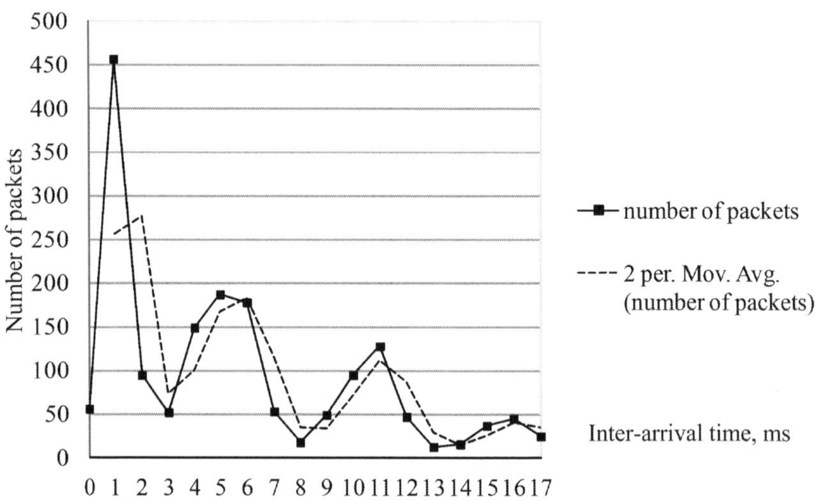

Fig. 2. Number of packets versus packet inter-arrival times at Gi interface for all services

In a fixed IP network, we performed measurements on four traffic types, namely, HTTP transfer, point-to-point (P2P), TV, and VoIP flows. The derived traffic type characteristics are summarized in Table 2. We collected HTTP measurements in a period of 24 hours for a worldwide set of queried locations on a personal computer. Queries are performed during peak network hours. The data provided below is the most typical set. The browser applied is Google Chrome. Point-to-Point (P2P) traffic is observed for over 500000 packets in Skype and μtorrent sessions of variable duration. For the TV traffic type, we observed a random channel at iptv.bg for 10 minutes, while for VoIP the traffic generated by Cisco IP Communicator between mobile and IP network is monitored.

We notice from the results in Table 2 that the packet intensity, i.e. the mean number of packets per second, and the session duration vary considerably depending on the type of application. Applications with typically large content transfer such as P2P and TV generate much more packets and have considerably longer sessions than lightweight applications such as HTTP or VoIP. Although this is a somewhat expected result, its identification is of extremely important for QoS provisioning. The network should be able to provide a fair treatment of traffic classes with distinct characteristics. Moreover, applications with longer sessions are more vulnerable to fluctuations in the QoS over the different domains of a heterogeneous network.

Table 2. Traffic source observed data in fixed IP network

Parameter	HTTP	P2P	TV	VoIP
Packets intensity, num/sec	50	600	400	60
Mean packet size, bytes	500	800	1000	100
Session duration and mean inter-arrival time	10 sec	1 hour	1 hour	360 sec
Distribution of session duration and inter-arrival time	Exponential	Exponential	Exponential	Exponential
Distribution of packets within single burst	Deterministic	Deterministic	Deterministic	Depends on application
Priority of packets	Low	Low	Medium	High

The measurement data presented are specific at a given reference point in the network. The distributed nature of today's cooperative communications requires coordination of different nodes. Therefore, an appropriate mechanism that takes into account the nature of the traffic and traverses each node in a region is necessary. Dynamic regulation of the data flows is the only solution that compromises between the growing needs in communication network resources and the limitations of the user demands. The data and control message volume grows exponentially [19], whereas the associated traffic dynamics in the channels become complicated and chaotic [20-21]. There is a great need to explore these traffic dynamics through models that primarily investigate the behavior of the traffic through time. This has also made the need for a multi-scale modeling apparent, where different scales carry essential information.

By simple observation of the IP interface in different scales, we explore the self-similar nature of the traffic (Fig. 3). The aggregated traffic of packets within 1, 10 and 100 seconds points to similar behavior. Therefore, the probabilistic mapping of the object characteristics in different time scales is similar [22].

These self-similar features lead to accurate definitions of a traffic statistical behavior. In most cases, we may use a probabilistic way to model these features as stochastic processes. Traffic self-similarity is a phenomenon where through a repetition time-window the stochastic properties re-occur and therefore, can be approximately forecasted [23]. The first foundations of the self-similar processes are identified in [24-25] by modeling the statistical behavior of the system. A self-similar phenomenon looks the same or behaves in the same way when viewed at different degrees of magnification or when the scales change on a given dimension [25]. It is considered bursty over all time scales [1, 26]. The measured traffic traces correspond to the stochastic self-similarity (fractality) features of different window frames through time. The fractal self-similarity makes it possible to take into consideration the stochastic nature of many network devices and events that together influence the network traffic and measure the aggregated traffic at the destination within a specified time interval.

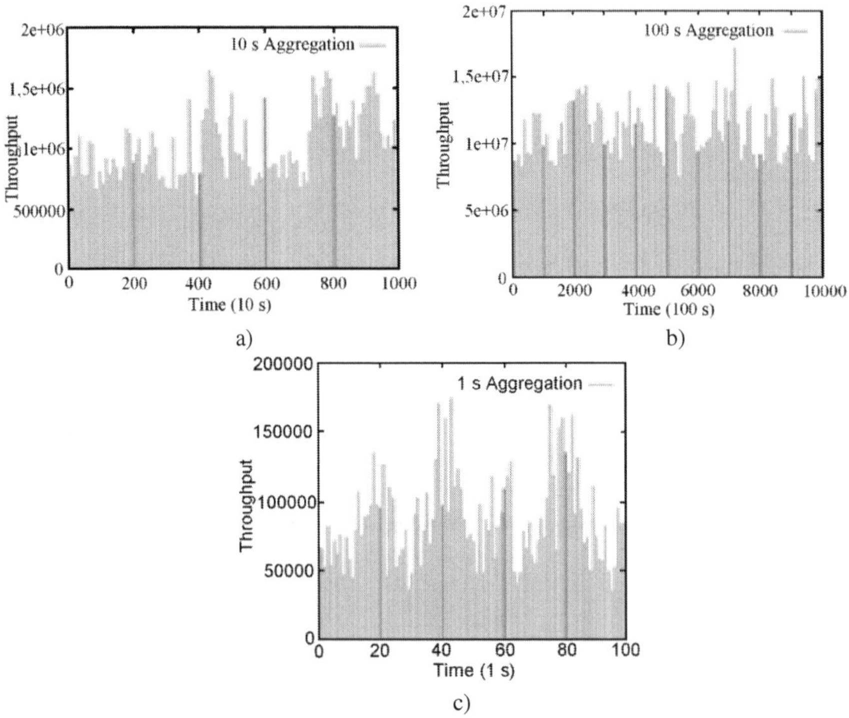

Fig. 3. Self-similarity phenomenon observed with regard to the throughput with different scaling

For years, networks have been planned according to the mathematical foundation given by Erlang's and Engset's loss formulae. This is one of the reasons why mobile operators face traffic congestions after introducing packet switched services. Most of the operators double the capacity of data channels. There is a theoretical foundation for packet switched networks planning given by Kleinrock in [5] that is not customized for technologies like 3G, 4G, WiMAX, ad-hoc and sensor networks. The COST 290 group [6] has done serious traffic and QoS analyses of 3G networks based on the multimodal and exponential traffic distributions. Krendel proposes Pareto distribution for traffic behavior intensity in multimedia environment [7]. Details regarding the behaviour of the queue can also be seen in [27]. Ricciato proposes a new approach to QoS and traffic analyses based on the traffic measurements in 3G networks [15]. He investigated the nature and behavior of TCP flows. The monitoring framework is shown at large and fine scale measurements as well as on the user plane and control plane. Our recent observations show that the amount of UDP traffic in the network is growing and its impact on the behavior of the interfaces is significant [16].

Further studies shown in [18], and [27-28] explore analytical solutions having features similar to the ones of the data observed. The self-similarity of the traffic at the user interface, edge or core parts of the network allows us to apply similar analytical solutions at any segment or scale of the connection as well as end-to-end [29-32]. A special study of the Long Range Dependence (LRD) using Markov Modulated Poisson Process (MMPP) that allows flexible tuning and evaluates the Internet traffic source performance is shown in [33]. In [34] a complicated packet flow analysis by Hidden Markov Model is demonstrated. Interesting and complex statistical analyses of Skype and P2PTV traffic are done in [35-38]. Peer-to-peer traffic statistical analyses are shown in [39], worm traffic in [40], email traffic in [41], traffic of popular games in [42], as well as a traffic classification in [43].

3 Polya Arrival Process

On the basis on the results from the measurements we looked for discrete and continuous time distributions that can be mapped to the inter-arrival times and number of packets in the systems during a given time interval. We propose gamma and Pareto distributions for the inter-arrival times simulation. This makes the arrival process (number of packets in the system) Polya distributed or negative binomial (a special case is called Pascal distribution). All three distributions can change the shape and the scale of random sequences [27-28], [44].

The complex traffic functions measured above can be described by Polya distribution as a negative binomial distributed number of arrivals in a fixed time interval and gamma distribution as inter-arrival times. This is exactly the behavior of the IP interface. The Polya arrival process is a peaked process defined by two parameters – the mean value and the variance [45]. The Polya distribution changes depending on its parameters from pure Poisson through geometric to distribution with very big peakedness. It is actually capable of covering all cases of observed distributions in a discrete time scale. We present a study of the influence of the offered traffic peakedness on the blocking probability, the mean number of the packets in the system and the mean waiting time. Similar but more complex approaches can be found in [5], [33], [35], [11-12, 46-53].

The simulation model of Polya/D/1 queue is configured as a long-tailed queuing system. We calculate the state probabilities, the probability that the queue exceeds a definite number of packets, the average delay and the waiting time distribution. We show that the variance of the input stream changes significantly the characteristics of the waiting system. The aim is to find appropriate parameterized systems for Internet access, central processing units, ad hoc IP networks, delay tolerant networks, core networks, and cellular systems. We consider that the network analysis requires a technique that can represent any kind of traffic, peaked or smooth, within the same model [54]. The Polya arrival process is a pure birth process with an average arrival rate λ [8]. The probability $P_i(t)$ of i arrivals in an interval with a duration of t seconds is given by

$$P_o(t) = (1 + \beta\lambda t)^{-\frac{1}{\beta}}$$

$$P_i(t) = \left(\frac{\lambda t}{1+\beta\lambda t}\right)^i \frac{(1+\beta)\ldots[1+(i-1)\beta]}{i!} P_o(t)\,. \tag{1}$$

The mean value (the average number of arrivals in an interval of length t) is

$$M(t) = \textstyle\sum_{i=1}^{\infty} i\, P_i(t) = \lambda t\,. \tag{2}$$

The variance of the number of arrivals in an interval of length t is

$$V(t) = \sum_{i=0}^{\infty} [i - M(t)]^2 P_i(t) = \lambda t(1 + \beta\lambda t). \tag{3}$$

The peakedness of the Polya input flow is

$$z(t) = \frac{V(t)}{M(t)} = 1 + \beta\lambda t > 1. \tag{4}$$

When β = 0, M(t) = V(t) = λt i.e. it is a regular Poisson process. When β=1 the Polya distribution is a geometric distribution. The system has unlimited waiting places and τ is the constant service time. This queuing system is a non-Markovian model or renewal process (Fig. 4). We analyze it using the theory of the embedded Markov chain. The finite state-transition probability diagram of the embedded Markov chain is shown on Fig. 5. We show only transitions in state n (0 < n < ∞).

We present an algorithm for calculating the state probability from which the performance measures such as the time congestion probability, the average number of packets and the average delay are calculated. The offered and carried traffic is equal to the probability of a busy server $A = A_o = 1 - P_o$.

Let τ be the length of an observed interval that falls at random in time. The packet in service at time t should have left the system at time t + τ because of the constant service time. Then, we have the relation

$$P_n = P_o Q_n(\tau) + \sum_{i=1}^{n+1} P_i Q_{n-i+1}(\tau) \quad n = 0,1,2,\ldots,\infty, \tag{5}$$

where Pn is the steady state probability that n packets exist in the system, Qn(τ) is the probability that n packets arrive during the service time, and τ is the service time.

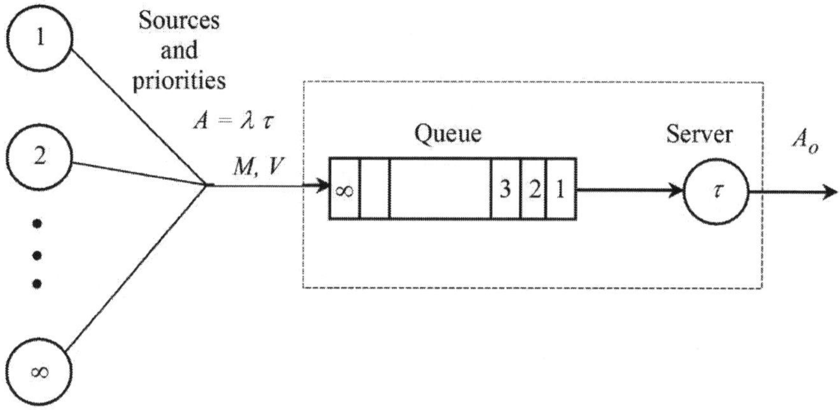

Fig. 4. Generalized queuing model with peaked input flow

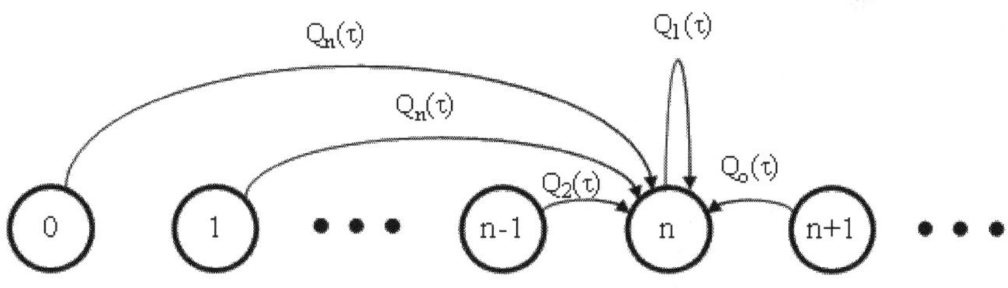

Fig. 5. How to reach state *n (0<n<∞)* at the end of an observed time interval of length τ

Because of the Polya arrival process, the probability that n packets arrive in the service time is

$$Q_o(\tau) = (1+\beta\lambda\tau)^{-\frac{1}{\beta}}$$

$$Q_n(t) = \left(\frac{\lambda\tau}{1+\beta\lambda\tau}\right)^n \frac{1(1+\beta)\ \ldots\ [1+(n-1)\beta]}{n!} Q_o(\tau). \tag{6}$$

In this way, we can calculate the steady state probability of the queuing system. Fig. 6 a) illustrates the probability that the queue exceeds k-packets in a single delay system with peakedness z as a function of the buffer size. The peaked input flow increases the probability that the queue exceeds k packets vastly. Fig. 6 b) presents the normalized mean system time (W/τ) as a function of the offered traffic intensity with different peakedness of the Polya input flow.

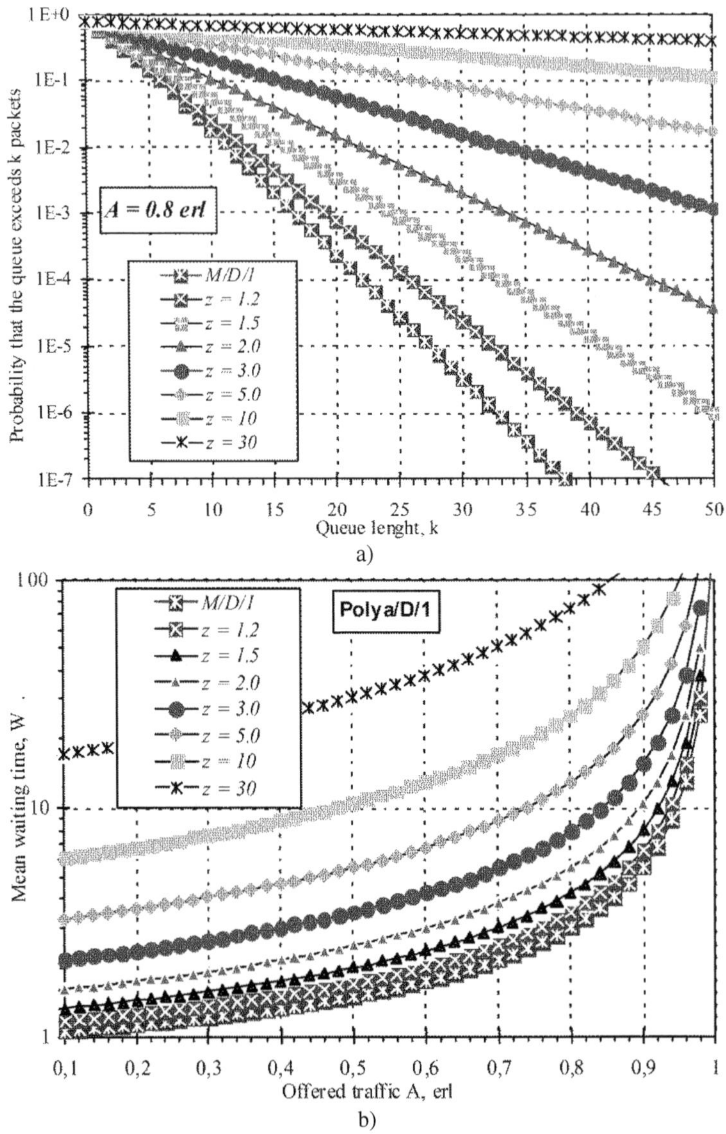

Fig. 6. a) Polya/D/1 time congestion probability with different peakedness; b) Polya/D/1 normalized mean system time with different peakedness

The influence of the peakedness on the mean system time is significant. The similarity of Polya distribution in all cases observed is obvious and can be obtained directly from the graphs in Fig. 6. VoIP traffic queues are short (up to 10 packets, with low variance and therefore low z). Data traffic queues are long (can be more than 50 packets, with high variance and high peakedness z). The estimated data packet delay can be many times higher than the delay of a VoIP packet. Video traffic queues will suffer from lack of queue resources. IPTV non-real-time queues are long (can be more than 50 packets, with low variance in case of an additional jitter buffer).

By applying different mean values and variances (with indirect influence on the peakedness using different β) we can model all types of traffic as a composition of Polya arrivals. The inter-arrival times can be modeled in a similar way by gamma and Pareto distributions. It is possible to map Polya/D/1/k queuing system to different type of traffic in the opportunistic environment and predict the behavior of the system in lack of resources. The peaked input flow contributes to the raising the congestion and waiting time. This is of great importance to the real-time traffic and end-to-end performance measures.

4 Point Process of the Self-similar Inputs

In the previous section, we showed a simplified approach to traffic source modeling by Polya/gamma/Pareto distributions. In this section, we prove the applicability of the Fractal Renewal Process (FRP) in complex ON/OFF traffic source superposition. We also demonstrate how the destination end node can be characterized.

The self-similar structure of the network traffic on the magnified level (i.e. the aggregated traffic generated by all active hosts) provides a basis for a new understanding of the traffic dynamics and the traffic structure generated by different hosts. Self-similarity of LAN traffic leads to structural models that can be reduced to ON/OFF sources. The ON and/or OFF periods follow Heavy Tails Distribution (HTD) with infinite variance [55]. According to the superposition theorem, ON/OFF processes define the direct connection between the self-similarity characteristics on the magnified level and the heavy tails phenomenon observed in the aggregated traffic flow and the traffic structure that is typical of the separate pairs of source-destination [24] (Fig. 7). We explore the Fractal Renewal Process concept (FRP) [55-58] by applying the fractality principles in deterministic wireless LANs [30].

In 3G networks, we observe less than 1% packets in the flows that are beyond 3σ interval where σ is the standard deviation of the inter-arrival times (Fig. 2). They form the long-tail of the distribution function. We suppose that the Fractal Renewal Process is also applicable to the case of 3G traffic when the variance of the inter-arrival times is significant. This is also directly related to the kurtosis and skewness of the measured flows.

The fractality principles have independent time intervals between stochastic points, by definition. The network traffic model is often based on the idea of the theory of random point processes (streams). The process models possessing this property have infinite power [24], [30] when the self-similarity property for the whole time interval or for frequency range is demonstrated. Fig. 7 (b) shows ON and OFF durations of this point process for which the PDF tail of time moments between arrivals decays through the power law:

$$w(t) = \begin{cases} kt^{-(\gamma+1)} \forall A < t < B \\ 0 \; otherwise \end{cases}. \tag{7}$$

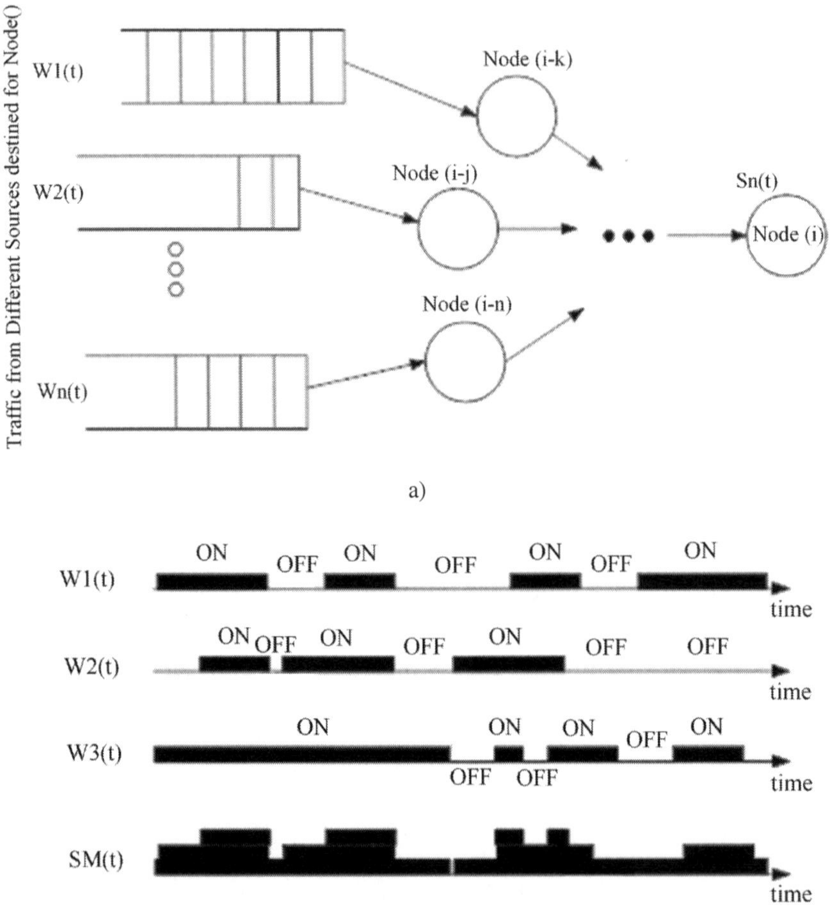

Fig. 7. a) Different sources super positioning the traffic onto destination, and b). M sources with $W_1(t)$, $W_2(t)$ and $W_3(t)$ and their sum $S_3(t) = W_1(t) + W_2(t) + W_3(t)$ for M=3.

Here A and B are threshold parameters [16], [21], [56-57], γ is a fractal parameter ($0 < \gamma < 2$) and k is a normalizing constant defined by the normalization requirement $\int_0^\infty w(t)dt = 1$. For $0 < \gamma < 1$ the FRP process is fully fractal. Therefore, the power spectrum density, the coincidence rate, the index of dispersion for counts and even the PDF of time moments between arrivals demonstrate the traffic ranges with regard to the time set for these values in the range between A and B. The related exponents are fully defined as the correlation of the second-order moment function of random traffic intensity.

The second-order moment function of the point process random intensity by definition equals to:

$$G_n(\tau) = \lim_{\Delta t \to 0} \frac{M(\Delta N_t \Delta N_{t+\tau})}{\Delta N_t^2}. \tag{8}$$

Here ΔN_t characterizes the appearance of at least one point inside the infinitesimal interval $(t - \Delta\tau, t)$. In addition, t is the time interval between the events of point's appearance. The process N_i describes the number of events between the initial and the current time comprising t moments in the analysis. The point process dN_t describes the covariance, i.e. Reoccurring Rate (RR) [59]. When having a multisource to destination transmissions, the end-recipient node can be characterized by the point process random traffic intensity as in (9).

Fig. 7 (b) shows the aggregated ON/OFF periods that define the activity periods of the traffic. The ON/OFF activity periods are considered to be a function of the peak traffic of the end-to-end connection. The peak aggregated traffic can be seen through the superposition of the aggregation of the traffic on a destination node. Suppose an M node exists and can handle the superpositioned aggregated traffic with independent and homogeneously distributed ON/OFF sources. Let each source j transmits its sequence of packet series with its contribution order $\{W^{(j)}(t), t \geq 0\}$. The superposition of packets at time moment t will be marked as $S_m(t) = \sum_{j=1}^{M} W^{(j)}(t)$. Having rescaled the time for factor T, the aggregated packet process over the interval [0,Tt] is:

$$W_M^*(Tt) = \int_0^{Tt} \left[\sum_{j=1}^{M} W^{(j)}(u)du\right]. \tag{9}$$

The equation shows the statistical behavior of the stochastic process $\{W_M^*(Tt), t \geq 0\}$ for large M when T depends on ON and OFF period distributions. ON and OFF distributions can be chosen so that as $M \to \infty$ and $T \to \infty$ the process behavior is reduced in an appropriate manner. It would be equivalent to the process $\{\sigma_{\lim} B_H(t), t \geq 0\}$, where $\sigma_{\lim}$ is constant and B_H is fractional Brownian motion [60]. The applicability of the gamma distribution for ON and OFF intervals is obvious. The gamma distribution is a composite type of distribution, i.e. its parameter is a gamma function that has the same distribution itself. This interesting feature allows mapping between packet inter-arrival times and ON/OFF time intervals. Similar results are also obtained by Pareto distribution. Furthermore, the number of packets during ON interval is Polya distributed in this case [17, 28, 44].

The scheme for applying the above stochastic approximations can be implemented with the aim to increase the wireless device lifetime. The conservation of the power will be part of the offered QoS to the end-users [29-32]. We map the exhibited nature of the traffic depicted in [22] onto a model-based approach to enable greater time assigned to each node for the sleep time duration. The monofractality properties [30] are taken into consideration for modeling the energy schedules and for greater associativity with the self-similar behavior [29, 32]. The window of the traffic duration is t [30] and is expressed by

$$t_w(s,d) = \{\lim_{t \to k \cdot t} F_{n(t)} \in t(s,d) : R_N(t) \approx R_N(t - k\tau) \forall k < 2\}, \tag{10}$$

where $t_w(s, d)$ is the time window measure for the multipath pair source-destination model with limit $k.\tau$ for the determined window size. In addition k should be less than 2 in order to satisfy the monofractality property of the repetition index of the incoming traffic. A calculated example for message delivery can be found in [32]. The mono-fractal scheme [29-30] also takes into consideration the caching mechanism proposed. The requested packets are 'cached' in an intermediate node or nodes in order to prevent any loss.

5 Complex Queuing Models

Having considered the traffic source behavior, the source and the destination end-points modeling and superposition, we propose the use of priorities in network nodes. The observed distributions are applied in complex traffic models with parallel and cascaded queues, a non-preemptive servicing order, ON/OFF sources, and waiting time limits. The models are used for end-to-end dynamic QoS and traffic management [18, 61-63].

The simulation is performed for $\sum G_i / \sum D_i/1/ \sum k_i / p_i$ traffic systems with flexible limits of waiting time, loss and priority (i is the priority level, Fig. 4). The results of the simulation of two cascaded leaky buckets at a single IP interface are presented in Table 3. The simulation model is flexible to the dynamic traffic changes and priorities. In addition to the models, which are already known, we add different limits for waiting times in the queues based on the traffic source behavior. The accuracy of the model derived is not changed end-to-end due to the fact that the interfaces at both ends are the same and have the same characteristics. The results are obtained for Priority Queuing scheduling in DiffServ networks and demonstrate better traffic shaping of real-time traffic. In Table 3 we observe the regulation influence of the waiting time limit on the real-time services like VoIP with short queues. It is also visible that the same waiting time limits are useless for non-real-time services like email and ftp because the TCP algorithm is flexible enough to manage error corrections and sliding window changes.

The delay and loss limits as well as priorities are used to adjust the router interfaces behavior. It can be done per aggregated service and aggregated packet level [63-64]. Queue behavior is complex due to the priorities and limits on waiting times and places. Many parameters have been derived from the model including probability of packet loss due to the lack of place in the queue, probability packet to be dropped due to the exceeding of the waiting time limit, probability to wait for different types of traffic, observations on the distribution of packets intervals, queue lengths, delay, delay jitter. The statistical accuracy of the derived results is proven by the batch mean method for output results analysis with Student distribution and confidence probability 0.95.

Table 3. Numerical results on utilization and loss probabilities

Parameter	Value					
VoIP time limit, s	0.002	0.001	0.0001	0.0001	0.0001	0.00004
VoIP queue limit, packets	10	5	3	3	2	7
FTP time limit, s	0.07	0.06	0.04	0.04	0.03	0.0004
FTP queue limit, packets	50	25	5	5	5	24
Email time limit, s	0.09	0.08	0.06	0.06	0.04	0.0004
Email queue limit, packets	50	25	5	5	5	24
Queue size, packets	110	55	13	13	12	55
Prob. of loss due to place limit for VoIP	0	0	0	0.0001	0.0056	0
Prob. of loss due to place limit for FTP	0	0	0.0007	0.0309	0.0978	0.1804
Prob. of loss due to place limit for email	0	0	0	0.0017	0.1389	0.5485
Prob. of loss due to time limit for VoIP	0	0	0	0	0	0.0006
Prob. of loss due to time limit for FTP and email	0	0	0	0	0	0
Utilization	0.4732	0.458	0.468	0.6132	0.752	0.8218
Utilization for VoIP	0.0453	0.044	0.0447	0.0681	0.0997	0.1389
Utilization for FTP	0.423	0.41	0.4185	0.523	0.6150	0.648
Utilization for email	0.0045	0.005	0.0046	0.0221	0.0373	0.0349

Under almost the same utilization factor, the fraction of the queue per service is changed. The scenario limits are shown in the first six lines in Table 3. It can be seen in the last column of the table that a tight limit to the waiting time will keep the VoIP queue short with low probability of waiting time loss. The total load on the interface is significant but the occupation of the interface by VoIP packets is very low in comparison to other cases. Similar results are obtained from simulation of an end-to-end connection. The adjustments of the waiting time and waiting place limits in NSIS (Next Steps in Signaling) and WRED (Weighted Random Early Detection) protocols are considered equal along the connection with the exception of access routers. The comparison to the traffic measured in forward and backward directions end-to-end also proves the applicability of the model.

The proposed solution for traffic shaping is demonstrated against the queue priority and traffic heterogeneity nature [62]. It is less complex than the heavy tail distribution analyses proposed in [65-67]. Such an application to technologies like 4G or Disruption-Tolerant Networks (DTN) requires a dynamic and flexible dimensioning approach [9]. The connection is usually not performed on a single path. Multipathing and multihoming in an aggressive environment is the only way to transmit traffic bundles in DTN reliably. Relay traffic in 4G should be taken into account [10] because it occupies the interfaces additionally. Further mapping of the proposed complex queuing model to Polya/gamma/Pareto distributed arrivals and fractal renewal process is going to be analyzed.

6 Traffic Fractality Influence over Application Level

In the previous section, we showed a linear approach to the end-to-end connection management. Here we try to apply a stochastic approach known as fractality influence over the service availability. In [32, 68-69] an efficient energy conservation scheme is shown. The traffic fractality influence (using a variable length repetition/correlation window) can be estimated using the Backward Traffic Difference (BTD) scheme [69]. The model designed takes into consideration the multi-fractality principle [24]. It guarantees the end-to-end availability of the resources requested while significantly reducing the energy consumption and maintaining the requested scheduled transfers. The innovative aspect of this work is that each node uses different assignment(s) of sleepwake schedule estimation. It is based on the traffic mono-fractality difference through time. The one-level Backward Difference of the traffic is evaluated by estimating the difference of the traffic while the Node(i) is set in the sleep-state for a period:

$$\nabla C_{N_i(1)} = T_2(\tau) - T_1(\tau - 1)$$

$$\nabla C_{N_i(2)} = T_3(\tau - 1) - T_2(\tau - 2)$$

$$\dots$$

$$\nabla C_{N_i(n+1)} = T_n(\tau - (n-1)) - T_{n-1}(\tau - n)\,, \tag{11}$$

Here $\nabla C_{N_i(1)}$ denotes the first moment traffic/capacity difference that is destined for Node(i) and cached onto Node (i-1) for time τ. Also $T_2(\tau) - T_1(\tau - 1)$ is the estimated traffic difference while packets are being cached onto (i-1) node for recoverability. The first equation depicts the BTD estimation for one-level comparisons. This means that the moments are only estimated for one-level ($T_2(\tau) - T_1(\tau - 1)$). The traffic difference is estimated so that the next sleep-time duration can be directly affected according to:

$$\delta\big(C(T)\big) = C_{total} - C_1, \forall C_{total} > C_1, T \in \{\tau - 1, \tau\}\,, \tag{12}$$

where the traffic that is destined for Node(i), urges the node to remain active for

$$\frac{\delta(C(T))}{C_{total}} . T_{prev} > 0\,. \tag{13}$$

The difference estimates the self-similarity among the different moments of the traffic. The sleep-time duration is assigned according to the scheme in a dissimilar form to enhance node prolonged hibernation (if needed). It avoids mutation that will result in network partitioning and resource sharing losses. The aggregated traffic onto each node (buffered traffic), is considered in the evaluation of the cumulative amount of traffic arrivals. When a node admits traffic, the traffic flow t_f can be modeled as a stochastic process [60] and denoted in a cumulative arrival form as $A_{t_f} = \{A_{t_f}(T)\}_{T \in N}$. Here, $A_{t_f}(T)$ represents the cumulative amount of traffic arrivals in the time space [0..T]. Then, the

$$A_{t_f}(s,T) = A_{t_f}(T) - A_{t_f}(s)\,, \tag{14}$$

denotes the amount of traffic arriving in time interval (s, t]. Hence, the next sleep-time duration for Node (i) can be evaluated as:

$$L_i(n+1) = \frac{\delta(C(T)|A_{t_f}(s,T))}{C_{total}}.T_{prev}, \forall\delta(C(T)) > 0\,. \tag{15}$$

If $\delta(C(T)) < 0$ than:
$\delta(C(T)) = C_{total} - C_1, \forall C_{total} < C_1, T \in \{\tau - 1, \tau\}$, and
$\frac{\delta(C(T))}{C_{total}}.T_{prev} < 0, \forall T_{prev} > T_{prev}(\tau - 1),\ C_{N_i} < 0$. The total active time increases gradually according to the following estimation:

$$T_{sleep} = T(\tau - t_1) - (-C_{N_i}) = T(\tau - t_1) + T_{C_{N_i}}\,. \tag{16}$$

where $T_{C_{N_i}}$ is the estimated duration of the capacity difference for $C_{N_i} < 0$. The sleep-time duration decreases according to Equations (15) and (16). Considering the above estimations the traffic flow can be expressed [69] as:

$$A_{t_f}(T) = m_{t_f}(T) + \hat{Z}_{\,t_f}(T)\,, \tag{17}$$

where $m_{t_f}(T)$ is its mean arrival rate and $\hat{Z}_{\,t_f}(T) = \sqrt{a_{t_f} m_{t_f}(T)}\hat{\hat{Z}}._{\,t_f}(T).a_{t_f}$.

The coefficient a_{t_f} is the variance coefficient of $A_{t_f}(T)$. In addition, $\hat{\hat{Z}}._{\,t_f}(T)$ is the smoothed mean [69] and with $E\left(\hat{\hat{Z}}._{\,t_f}(T)\right) = 0$ satisfies the following variance and covariance functions:

$$v_{t_f} = a_{t_f} m_{t_f}.T^{2H_{t_f}}$$

$$\sigma_{t_f}(s,T) = \frac{1}{2} a_{t_f} m_{t_f}.\left(T^{2H_{t_f}} + s^{2H_{t_f}} - (T-s)^{2H_{t_f}}\right), \tag{18}$$

Here, $H_{t_f} \in [\frac{1}{2}, 1]$ is the Hurst parameter, indicating the degree of self-similarity in the Fractality process. Estimations in (19) can only be valid if the capacity of the Node (i-1) can host the aggregated traffic destined for Node (i) satisfying the $\sup_{s \le T}\{\sum_{t_f=1}^{N} A_{t_f}(s,T) - C_{t_f}(T)\}$, for traffic flow t_f at time T. Also, $C_{t_f}(T)$ represents the service capacity of the Node(i-1) for this time duration.

Taking into consideration the above stochastic estimations, the Energy Efficiency EE_{t_f} can be defined as a measure of the capacity of the Node(i) over the total power consumed by the node, as:

$$EE_{t_f}(T) = \frac{C_{t_f}(T)}{TotalPower} \tag{19}$$

This equation can be used as the primary metric for the lifespan extensibility of the wireless node in the system. Indicative real-time results hosting the traffic-based scheme onto MICA2dot motes sensor nodes are shown in Fig. 8. The queuing policy is considered to be as in [70-72] and the classification scheme as in [72] for extracting the comparative response evaluations.

Dynamic capacity management algorithms are shown in [73-75] and verified at the application level. We propose proper admission control policy for QoE support. The admission decision policy is codec-dependent and relies on the concept of user-perceived voice quality. The well-known bufferless fluid-flow method is applied and new simple exact formulas for Call Admission Control performance evaluation are derived [74]. An analysis of an IEEE 802.16-based network, serving voice traffic with the carrier grade QoS restrictions can be found in [73]. The burstiness of voice traffic with Voice Activity Detection is taken into account. The proposed VAD and codecs dependent bandwidth reservation is further enhanced by the adjustment of VoIP packets payload [75]. This will enhance the use of the WiMAX cell in its full capacity. Cases where more than one traffic moments may be considered in combination with the traffic and energy-aware measurements are presented in the next section.

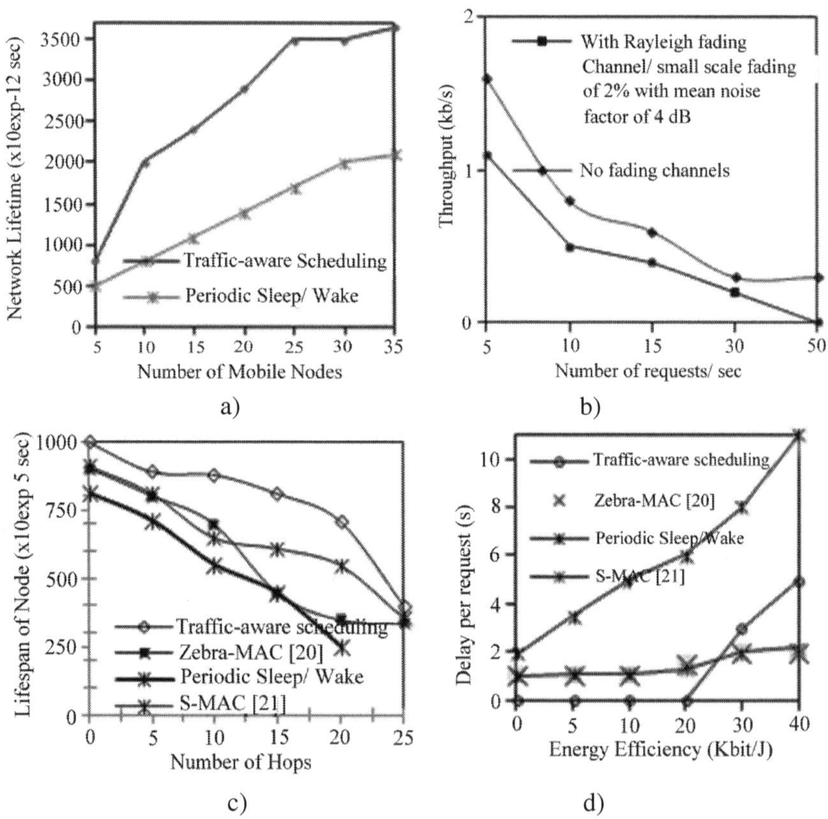

Fig. 8. a) Network lifetime with the number of mobile nodes; b). Throughput response of the system hosting the proposed scheme with the number of requests for certain fading measures; c) Lifespan of each node with the number of hops in a 50 node evaluation sensor network (MIC2dot/xbot); d). Energy efficiency with the traffic-delay impact using real-time sensor network (MIC2dot/xbot).

7 Fibonacci Backward Difference Traffic Moments for Nodes' Lifetime Extensibility

Let $C(t)$ be the capacity of the traffic that is destined for the Node i in the time slot (duration) t, and $C_{N_i(t)}$ is the traffic volume capacity that is cached onto *Node (i-1)* for time t. Then, according to the one-level Backward Difference of the traffic, the difference in the capacity measure can be estimated as the difference of the traffic while the *Node(i)* is set in the sleep-state – and admits traffic - for a period, as follows (Fig. 9):

$$F_{N_i(3)} \begin{bmatrix} T(\nabla C_{N_i(1)}) = T_2(\tau) - T_1(\tau - 1) \\ T\nabla C_{N_i(2)} = T_3(\tau - 1)) - T_2(\tau - 2) \end{bmatrix}$$

$$\vdots$$

$$F_{N_i(i+n+3)} \left| T\nabla C_{N_i(n+1)} = T_n(\tau - (n-1)) - T_2(\tau - (n-2)) \right. \tag{20}$$

where $\nabla C_{N_i(1)}$denotes the first moment traffic/capacity difference that is destined for Node(i) and cached onto Node (i-1) for time τ, $T_2(\tau) - T_1(\tau - 1)$ is the estimated traffic difference while packets are being cached/buffered onto (i-1) hop for recoverability [29-32, 66].

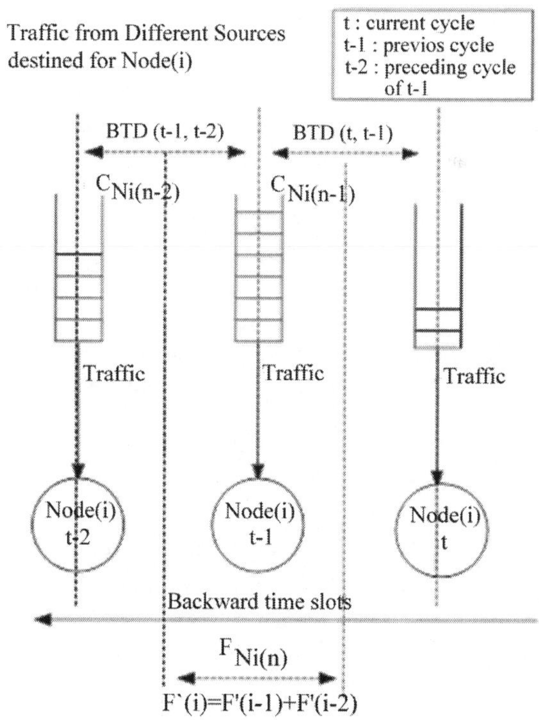

Fig. 9. Two-level traffic moments for Node(i) and the association between the T_1, T_2, T_3, for obtaining the $F_{N_i(n)}$.

Equation (20) depicts the Backward Difference Traffic (BDT) estimation for one-level traffic comparisons and denotes that the moments are only estimated for one-level $(T_2(\tau) - T_1(\tau - 1))$. The associated Fibonacci Backward Difference Traffic moments can be evaluated by using the aggregated traffic as in section 6 in terms of $\nabla C_{N_i(i-1)}, \nabla C_{N_i(i-2)}$ for the $\nabla C_{N_i(i)}$ where:

$$F_{N_i(n)} = T\left(\nabla C_{N_i(i-1)}\right) + T\left(\nabla C_{N_i(n-2)}\right), \forall N_i(t) < t_s \tag{21}$$

where $t_s$is the upper allowed end-to-end delay for the *s* stream for adequately reaching the destination, and $N_i(t)$ is the time that the specific packet chunks are cached onto the intermediate node. $F_{N_i(n)}$ is the Fibonacci Backward Difference Traffic moments for the two previous BDT moments $\nabla C_{N_i(n-1)}$ and $\nabla C_{N_i(n-2)}$. Equation 21 is valid only if the F-BDT is greater than the next pair BDT evaluation as follows:

$$F_{N_i(n)} > F_{N_i(n+1)} \tag{22}$$

Let t be the duration function of the active period of the Node i, then the sleep-time duration can be measured using the following expression:

$$T_s(i)|^n = \frac{t\left(C_{N_i(\tau-1)}\right) + t(F_{N_i(\tau)})}{1+\Delta T_s}, \tag{23}$$

where $T_s(i)|^n$ is the estimated sleep-time of the node during the next cycle, $t(C_{N_i(\tau-1)})$ is the time that the $C_{N_i(\tau-1)}$ capacity of the traffic needs to be processed, and $t(F_{N_i(\tau)})$ is the F-BDT if Eq. 23 is satisfied. The ΔT_s is measured according to the formula:

$$\Delta T_s = \frac{a_i(\tau)}{t_s} + T(\nabla C_{N_i(n)}) \tag{24}$$

where $\nabla C_{N_i(n)}$ is the normalized BDT, $a_i(\tau)$ is the incoming traffic for node i during the (current) time τ and $t_s$is the upper allowed end-to-end delay for the *s* stream for adequately reaching the destination. In the case where Eq. 23 is not satisfied, the sleep-time duration of the node is measured according to the following:

$$T_s(i)|^n = \frac{t(C_{N_i(\tau-1)})}{1+\Delta T_s} = \tag{25}$$

The above measures take place according to the algorithmic steps in Table 4.

The above algorithm is valid when incoming traffic reaches a node, so that the quantitative measures of Table 4 take place. The associated mechanisms ensure that when the buffered traffic that traverses a node reaches a certain limit, the sleep time for the node can be adjusted accordingly in order to overcome the saturation state during the assignment of the sleep-time duration. Finally, the activity period of nodes can be bounded within certain limitation ranges in order to enable them to conserve energy, thus allowing extension of lifetime.

Table 4. Basic steps of the proposed BDT scheme

for Node(i) that there is $C(t)$ *>0 {*
  while ($C_{N_i(t)}$ *>0) { //cached Traffic measurement*
    Evaluate ($T(\nabla C_{N_i(1)})$ *);*
if (Activity_Period for node(i)>0 && $C_{N_i(t)} \forall t(N_i) > 0$ *)*
//Measure Sleep-time duration
Evaluate $\nabla C_{N_i(1)}$ *and* $\nabla C_{N_i(2)}$*;*
if ($N_i(t) < t_s$ *)*
    estimate the $F_{N_i(n)}$ such that $F_{N_i(n)} > F_{N_i(n+1)}$
    sleep for $T_s(i)\big|^n = \frac{t(C_{N_i(\tau-1)}) + t(F_{N_i(\tau)})}{1+\Delta T_s}$; //during the next cycle
  else
    sleep for $T_s(i)\big|^n = \frac{t(C_{N_i(\tau-1)})}{1+\Delta T_s}$;//during the next cycle
  } //while
}//for

8 Conclusion and Future Work

The optimization of the Internet connections based on the Internet traffic analysis is a difficult and non-trivial task. The nature of the self-similar traffic and specific distributions seen after different services and service configurations require flexible approach to QoS, QoE and network configuration [2]. Network devices should be smart enough to keep the information about connection segments, technology requirements, Service Level Agreement requirements and to optimize the configuration data using QoS negotiation and QoS guarantee protocols. This is especially valid when the technologies along the connections change and require relays, application of additional scheduling policies, and dynamic QoS or QoE coordination [3-4]. New streams in the traffic analysis open research issues include novel QoS algorithms and their applicability that will take into account the dynamic nature of the traffic in the network and can guarantee the end-to-end QoS. The protocol should support fine-scale and course-scale network segments with or without aggregated traffic. The short-term goal is to map Polya/gamma/Pareto distributions and traffic generator parameters with pdf, mean, variance, skewness and kurtosis of the measured data. After defining the system bounds and taking into account the self-similarity nature, we aim at estimating the parameters of the user level SLA. Further research also encompasses mutli-fractal estimations where different moments of the traffic variations will be considered and modeled using self-similar, N-window fractality measurements. In addition, currents trends in traffic analysis include the energy-traffic profiling and tagging of wireless sensor networks according to the QoE parameters of the tagged users. End-to-end QoS and QoE modeling will be proposed based on distributed traffic shaping phenomena.

Acknowledgments. Our thanks to ICT COST TMA IC0703 Traffic Monitoring and Analyses project and ICT COST Action IC1303: Algorithms, Architectures and Platforms for Enhanced Living Environments (AAPELE) funded by the European Commission.

References

[1] Gibbens, R.J.: Traffic Characterization and Effective Bandwidths for Broadband Network Traces. In: Stochastic Networks, Theory and Applications, pp. 169–179. Oxford Science Pub. (1996)

[2] Adami, D., Callegari, C., Giordano, S., et al.: Design and Performance Evaluation of Service Overlay Networks Topologies. JNW 6(4), 556–566 (2011)

[3] Matijasevic, M., Skorin-Kapov, L.: Design and Evaluation of a Multi-user Virtual Audio Chat. In: Future Generation Computer Systems (2003), doi:10.1016/S0167-739X(02)00149-8

[4] Plissoneau, L., Biersack, E.: A Longitudinal View of HTTP Video Streaming Performance. In: MMSys 2012, ACM (2012)

[5] Kleinrock, L.: Queueing Systems, vol. I and II. John Wiley and Sons (1976)

[6] Koucheryavy, Y., Giambene, G., Staehle, D., Barcelo-Arroyo, F., Braun, T., et al. (eds.): Traffic and QoS Management in Wireless Multimedia Networks: COST 290 Final Report. LNEE, vol. 31. Springer, Heidelberg (2009)

[7] Krendel, A.: Network Planning Aspects for 3G/4G Mobile Systems. Thesis, Tampere University (2005) ISBN 952-15-1436-1

[8] Korn, G., Korn, T.: Mathematical Handbook for Scientists and Engineering. McGraw Hill (1968)

[9] 3GPP, TS 23.107 V6.4.0 - Quality of Service (QoS) Concept and Architecture (2006), `http://www.3gpp.org/ftp/Specs/archive/23_series/23.107/23107-640.zip`

[10] Dimitrova, D.C., van den Berg, H., Heijenk, G., Litjens, R.: Flow Level Performance Comparison of Packet Scheduling Schemes for UMTS EUL. In: Harju, J., Heijenk, G., Langendörfer, P., Siris, V.A. (eds.) WWIC 2008. LNCS, vol. 5031, pp. 27–40. Springer, Heidelberg (2008)

[11] Sleurs, K., Li, D., Lil, E., et al.: How Different Queuing Systems Affect the Discrete Representation of a Packet Stream. In: Clobecom 2008, New Orleans, USA (2008)

[12] Lee, I., Fapojuwo, A.: Modeling Wireless TCP Connection Arrival Process. In: IEEE Clobecom, San Francisco, USA (2006)

[13] Goleva, R., Mirtchev, S., Atamian, D., et al.: Experimental Analysis of QoS Provisioning for Video Traffic in Heterogeneous Networks. In: Joint ERCIM eMobility and MobiSense Workshop (with WWIC 2012) (June 2012)

[14] Goleva, R., Atamian, D., Mirtchev, S., et al.: Traffic Sources Measurement and Analysis in UMTS. In: Proceedings of HP-MOSys 2012, Paphos, Cyprus (2012)

[15] Ricciato, F.: Traffic Monitoring and Analyses for the Optimization of a 3G Network. IEEE Wireless Communications, 42–49 (December 2006)

[16] Goleva, R., Mirtchev, S., Atamian, D., et al.: Traffic Measurements and Flow Analyses in 3G Network. In: ICEST 2011, Nic, Serbia, pp. 95–98 (2011)

[17] Mirtchev, S., Goleva, R.: Evaluation of the Influence of the Input Flow Peakedness on the Polya/D/1 Queueing System. In: Telecom 2010, Sofia, pp. 110–117 (2010)

[18] Goleva, R.I., Goleva, M., Atamian, D., et al.: Quality of Service System Approximation in IP Networks. Serdica Journal of Computing 2, 101–112 (2008) ISSN 1312-6555

[19] Hilbert, M., López, P.: The World's Technological Capacity to Store, Communicate, and Compute Information. Science 332(6025), 60–65 (2011)

[20] Willinger, W., Taqqu, M.S., Leland, E., et al.: Self-Similarity in High-Speed Packet Traffic: Analysis and Modeling of Ethernet Traffic Measurements. Statistical Science 10, 67–85 (1995)

[21] Willinger, W., Taqqu, M.S., Sherman, R., et al.: Self-Similarity through High Variability: Statistical Analysis of Ethernet LAN Traffic at the Source Level. IEEE/ACM Transactions Networking 5, 71–86 (1997)

[22] Deane, J.H.B., Smythe, C., Jefferies, D.J.: Self-similarity in a Deterministic Model of Data Transfer. International Journal of Electronics 80(5), 677–691 (1996)

[23] Leland, W.E., Taqqu, M.S., Willinger, W., et al.: On the Self-Similar Nature of Ethernet Traffic (ext. ver.). IEEE/ACM Trans. Networking 2(1), 1–15 (1994)

[24] Sheluhin, O., Smolskiy, S., Osin, A.: Self-Similar Processes in Telecommunications (2007) ISBN: 978-0-470-01486-8

[25] Kolmogorov, A.: Logical Basis for Information Theory and Probability Theory. IEEE Transactions on Information Theory 14(5), 662–664 (1968)

[26] Callegati, F., Lakshman, S.T.: On Optical Burst Switching and Self-Similar Traffic. IEEE Communications Letters 4(3), 98–100 (2000)

[27] Mirtchev, S., Goleva, R.: Evaluation of Pareto/D/1/k Queue by Simulation. In: International Book Series "Information Science&Computing" No.1, Supplement to the International Journal Information Technologies & Knowledge, vol. 2, pp. 45–52 (2008)

[28] Mirtchev, S., Goleva, R.: Discrete Time Single Server Queueing Model whit a Multimodal Packet Size Distribution. In: Atanasova, T. (ed.) Proceedings of a Conjoint Seminar "Modeling and Control of Information Processes", Sofia, Bulgaria, pp. 83–101 (2009) ISBN: 978-954-9332-55-1

[29] Mavromoustakis, C.X., Karatza, H.D.: Quality of Service Measures of Mobile Ad Hoc Wireless Network using Energy Consumption Mitigation with Asynchronous Inactivity Periods. Simulation: Transactions of the Society for Modeling and Simulation International 83, 107–122 (2008)

[30] Mavromoustakis, C.X., Zerfiridis, K.G.: On the Diversity Properties of Wireless Mobility with the User-centered Temporal Capacity Awareness for EC in Wireless Devices. In: Proceedings of the Sixth IEEE International Conference on Wireless and Mobile Communications, ICWMC 2010, Valencia, Spain, pp. 367–372 (2010)

[31] Mavromoustakis, C.X., Dimitriou, C.D., Mastorakis, G.: Using Real-Time Backward Traffic Difference Estimation for Energy Conservation in Wireless Devices. In: Proceedings of the Second International Conference on Ambient Computing, Applications, Services and Technologies, AMBIENT, Barcelona, Spain, September 23-28 (2012)

[32] Mavromoustakis, C.X., Karatza, H.: End-to-end Layered Asynchronous Scheduling Scheme for Energy Aware QoS Provision in Asymmetrical Wireless Devices. In: IEEE International EDOC Conference-Enterprise Computing Conference 2008, AQuSerM: Advances in Quality of Service Management Workshop, Munchen, Germany (2008)

[33] Muscariello, L., Mellia, M., Meo, M., et al.: Markov Models of Internet Traffic and a New Hierarchical MMPP Model. Computer Communications Journal 28(16), 1835–1851 (2005)

[34] Dainotti, A., Pescape, A., Salvo, R., et al.: Internet Traffic Modeling by Means of Hidden Markov Models. Computer Networks (Elsevier) 52(14), 2645–2662 (2008)

[35] Markovich, N., Krieger, U.: Statistical Analysis and Modeling of Skype VoIP Flows. Computer Communications (COMCOM) 31(suppl. 1), S11–S21 (2010)

[36] Eittenberger, P., Krieger, U., Markovich, N.: Measurement and Analysis of Live-Streamed P2PTV Traffic. In: Czachyrski, T. (ed.) Performance Modelling and Evaluation of Heterogeneous Networks HET-NETs, pp. 195–212 (2010)

[37] Xi, B., Chen, H., Cleveland, W., et al.: Statistical Analysis and Modeling of Internet VoIP Traffic for Network Engineering. Electron. J. Statist. 4, 58–116 (2010)

[38] Molnar, S.: Fractal Analysis and Modeling of VoIP Traffic. Networks (2004)

[39] Dainotti, A., Pescape, A., Ventre, G.: A Packet-level Model of Starcraft Traffic. In: Proc. of IEEE Hot-P2P, pp. 33–42 (2005)

[40] Dainotti, A., Pescape, A., Ventre, G.: Worm Traffic Analysis and Characterization. In: IEEE International Conference on Communications, ICC 2007 (2007)

[41] Ohri, R., Chlebus, E.: Measurement Based E-mail Traffic Characterization. In: Proc. of Int. Symp. on Performance Evaluation of Computer and Telecommunication Systems (2005)

[42] Feng, W., Chang, F., Feng, W., et al.: A Traffic Characterization of Popular On-line Games. IEEE/ACM Transactions on Networking 13(3), 488–500 (2005)

[43] Tammaro, D., Valenti, S., Rossi, D., et al.: Exploiting Packet Sampling Measurements for Traffic Characterization and Classification. International Journal of Network Management (2012)

[44] Mirtchev, S., Goleva, R., Alexiev, V.: Evaluation of Single Server Queueing System with Polya Arrival Process and Constant Service Time. In: Proceedings of the International Conference on Information Technologies (InfoTech 2010), Bulgaria, pp. 203–212 (2010)

[45] Ramos, H., Almorza, D., Garcia-Ramos, J.: On Characterizing the Polya Distribution. ESAIM: Probability and Statistics 6, 105–112 (2002)

[46] Arya, V., Duffield, N., Veitch, D.: Temporal Delay Tomography. IEEE INFOCOM (2008)

[47] Foh, C., Zukerman, M., Tantra, J.: A Markovian Framework for Performance Evaluation of IEEE 802.11. IEEE Transactions on Wireless Communications 6(4) (2007)

[48] Abdrabou, A., Zhuang, W.: Service Time Approximation in IEEE 802.11 Single-Hop Ad Hoc Networks. IEEE Transactions on Wireless Communications 7(1) (2008)

[49] Klemm, A., Lindemann, C., Lohmann, M.: Modeling IP Traffic Using the Batch Markovian Arrival Process. Performance Evaluation Journal 54(2), 149–173 (2003)

[50] Dainotti, A., Pescape, A., Ventre, G.: A Packet-level Characterization of Network Traffic. In: Proc. of IEEE CAMAD, pp. 38–45 (2006)

[51] Ibe, O.: Markov Processes for Stochastic Modeling. Academic Press (2008) ISBN-10: 0123744512

[52] Baccelli, F., Machiraju, S., Veitch, D., et al.: The Role of PASTA in Network Measurement. IEEE/ACM Transactions on Networking 17(4) (2009)

[53] Mandjes, M., Zuraniewski, P.: Tail Asymptotics of the M/G/inf Model. Stochastic Models 27(1), 77–93 (2011)

[54] Mandjes, M., Zuraniewski, P.: M/G/inf Transience, and Its Applications to Overload Detection. Performance Evaluation 68(6), 507–527 (2011)

[55] Delbrouck, L.E.N.: A Unified Approximate Evaluation of Congestion Functions for Smooth and Peaky Traffics. IEEE Trans. on Commun. 29(2), 85–91 (1981)

[56] Lowen, S.B.: Fractal Stochastic Processes. PhD thesis, Columbia University (1992)

[57] Lowen, S.B., Teich, M.C.: Fractal Renewal Processes Generate 1=f noise. Physical Review E 47, 992–1001 (1993)

[58] Berger, J.M., Mandelbrot, B.B.: A New Model for the Clustering of Errors on Telephone Circuits. IBM Journal of Research Development 7, 224–236 (1963)

[59] Ryu, B.K., Lowen, S.B.: Point Process Approaches to the Modeling and Analysis of Self-Similar Traffic: Part I–Model construction. In: Proceedings of IEEE INFOCOM (1996)

[60] Ryu, B.K.: Fractal Network Traffic: from Understanding to Implications. PhD thesis, Columbia University (1996)

[61] Jin, X., Min, G., Wang, L.: A Comprehensive Analytical Model for Weighted Fair Queuing under Multi-Class Self-Similar Traffic. In: ICC 2009, pp. 1–5 (2009)

[62] Goleva, R., Mirtchev, S.: Traffic Modeling in Disruption-Tolerant Networks. In: Annual Seminar of the PTT College, Modeling and Control of Information Processes, CTP, Sofia, pp. 6–20 (2010) ISSN: 1314-2771

[63] Goleva, R., Mirtchev, S.: End-to-End Queue Dimensioning in IP Network. In: ICEST 2010, Ohrid, Macedonia, pp. 289–292 (2010)

[64] Jin, X., Min, G.: Modeling Priority Queuing Systems with Multi-Class Self-Similar Network Traffic. In: ICC 2007, pp. 550–555 (2007)

[65] Markovitch, N., Krieger, U.: The Estimation of Heavy-tailed Probability Density Functions, Their Mixtures and Quantiles. Elsevier, Computer Networks 40, 459–474 (2002)

[66] Hohn, N., Papagiannaki, K., Veitch, D.: Capturing Router Congestion and Delay. IEEE/ACM Transactions on Networking 17(3) (2009)

[67] Loiseau, P., Gonçalves, P., Dewaele, G., et al.: Investigating Self-Similarity and Heavy-Tailed Distributions on a Large-Scale Experimental Facility. IEEE/ACM Transactions on Networking 18(4) (2010)

[68] Barabási, A.L., Albert, R.: Emergence of Scaling in Random Networks. Science 286(5439), 509–512 (1999)

[69] Charalambous, M.C., Mavromoustakis, C.X., Yassein, M.B.: A Resource Intensive Traffic-Aware Scheme for Cluster-based Energy Conservation in Wireless Devices. In: Proceedings of IEEE HPCC 2012 and WNM 2012, Liverpool, UK, June 25-27 (2012)

[70] Rincón, D., Minerva, F., Sallent, S., et al.: Towards an Improved Characterization of Fractal Network Traffic. In: Third International Working Conference, Performance Modeling and Evaluation of Heterogeneous Networks, West Yorkshire, UK (2005)

[71] Jin, X., Min, G.: Modelling and Analysis of Priority Queueing Systems with Multi-Class Self-Similar Network Traffic: A Novel and Efficient Queue-Decomposition Approach. IEEE Transactions on Communications 57(5), 1444–1452 (2009)

[72] Askar, S., Zervas, G., Hunter, D.K., et al.: Evaluation of Classified Cloning Scheme with Self-similar Traffic. In: CEEC, pp. 23–28 (2011)

[73] Tsankov, B., Koleva, P., Kassev, K.: Scheduling Algorithms for Carrier Grade Voice over IEEE 802.16 Systems. In: Proceedings of MELECON, pp. 126–131 (2008)

[74] Kassev, K., Mihov, Y., Kalaydzhieva, A., et al.: Call Admission Control Dimensioning for VoIP Traffic over Wireless Access Networks: From Network to Application-specific Perspective. IARIA International Journal on Advances in Networks and Services 3(3&4), 333–345 (2010)

[75] Mihov, Y., Kassev, K., Tsankov, B.: WiMAX Cell Capacity Analysis for Voice Traffic under Consideration of Packet Duration. Elektrotechnica & Elektronica (E+E) Magazine 45(3-4), 15–20 (2010)

Ambient Assisted Living Tools for a Sustanaible Aging Society

Existing Solutions and Future Challenges

Andrés L. Bleda[1], Rafael Maestre[1], Antonio J. Jara[2], and Antonio Gomez Skarmeta[2]

[1] Dept. of Electronics and Home Automation Tech. Centre of Furniture and Wood (CETEM)
[2] Dept. Information and Communications Engineering University of Murcia, Spain (UM)
{al.bleda,r.maestre}@cetem.es, {jara,skarmeta}@um.es

Abstract. In the last decade there has been a huge increase in Ambient Assisted Living (AAL) technologies that try to address the continuous growing aging population. These demographic changes have led to new challenges for the society, which has to provide assistance to old people while promoting other aspects such as their autonomy and independence at home, as well as supporting their common daily life activities. The R&D community has been well-aware of these changes and over the last few years has come up with new AAL approaches. This chapter discusses the changes that the new demographic scenario implies and the major forces that shaped it. We will analyze the needs and limitations that the elderly have to face in their everyday lives, and why new AAL technologies are a viable solution. The major scientific results will be described along with their derived commercial products. Similarly, a complete analysis of the companies' effort, progress and products on this area will be presented. Finally, the future research paths will be discussed.

1 Introduction

Due to increasing life expectancy and decrease of fertility rates, the proportion of people over 60 years is increasing faster than any other age group in almost all countries. The population ageing can be considered a success of public health policies and socioeconomic development, but also a challenge to society, which must provide health services and try to improve old people functional capabilities, security and social participation.

According to WHO (World Health Organization) [1], the world's population of people 60 years old and over is forecast to reach 2 billion by 2050. There is a real need to prolong independent living for older people at their own homes, because it clearly increases their quality of life, and reduces costs for families and the Social Security System.

These demographic changes require the development of new functionalities and the integration of new technologies at home environments. Decades ago, home automation technologies started to emerge with the goal of making people's life easier and increasing energy savings. It began with simple features related to the automation

C.X. Mavromoustakis et al., *Resource Management in Mobile Computing Environments*, Modeling and Optimization in Science and Technologies 3,
DOI: 10.1007/978-3-319-06704-9_9, © Springer International Publishing Switzerland 2014

of basic functionalities such as lighting control with motion sensors and detectors. The complexity of these systems has been increasing. The integration of smart functionalities and AmI (Ambient Intelligence) at home results in Ambient Assisted Living (AAL) environments that provide care, support and assistance to the elderly.

Actually, the same systems with some minor adjustment can also target a different population segment such as the disabled, who may need in some cases a special level of care and assistance that family members cannot directly provide, or even the impaired person wants to live independently.

This chapter structure contains a first section describing demographic changes that have led the emergence of AAL environments highlighting the importance of the elderly and its growth rates in the world population and more specifically in three of the most significant economies in the world such as Europe, the United States and Japan. After presenting this study of population profiles, the related work section starts with an exhaustive research about the necessities of the elderly. Once these necessities have been identified it is easier to establish the goals AAL systems should aim for and what kind of assistance and care they should provide. Then, a complete identification of existing devices for AAL support is detailed, making a distinction between different specific hardware systems such as robotics, mobile devices, wearable or environmental sensors and smart homes. After the description of existing hardware, a complete characterization of the most common software techniques in AAL is included (user location, context modelling, activity recognition, anomaly detection or scheduling). Finally, the Use Cases section contains examples of implementation and full existing AAL systems in Europe, United States and Japan. After the existing techniques and systems have already been described, the fourth section identifies the applications of Opportunistic Wireless Access Networks (WAN) in AAL environments and how this kind of networks can be used to improve Ambient Intelligent systems. To conclude, the future challenges section gives details of the immediate future work that needs to be done to ensure security and privacy of the great quantity of information managed in AAL systems.

2 Societal Framework in AAL

Ambient Assisted Living (AAL), as the name suggests, refers to intelligent systems of assistance for a better, healthier and safer life focused in living environments. According to the European Commission, its goal is to prolong the time people can live in a decent way in their own home by increasing their autonomy and self-confidence, supporting everyday activities, and monitoring and caring for the elderly or ill person, in order to enhance security and save resources.

2.1 Population Trends

Obviously, the emergence and consolidation of the AAL concept came out due to demographic population changes and the arising of new necessities which need to be addressed by new technological systems and products. These changes are because of the fact the world is undergoing a gradual population aging process over the last decades. This is particularly significant in more developed countries where, in recent years, it has become more intense, basically due to the increased longevity of older people and emphasized by a low birth rate.

The report of the European Commission about "Active ageing and solidarity between generations" [2] compiles a few world statistics which provide some valuable information about population ageing. For instance, the United Nations estimated the world's population at 6,895 million in 2010, and during the period from 1990 to 2010 it grew, on average, by 1,3% per annum. Also in this period the share of those aged 65 or more in the total population increased by 1.4 percentage points.

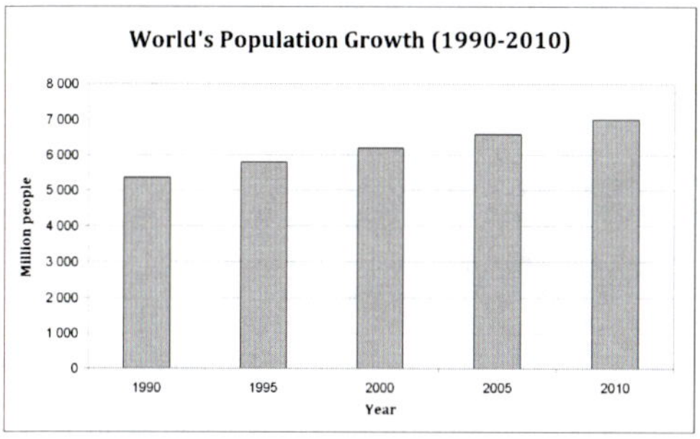

Fig. 1a. World's population and elderly growth (1990-2010)

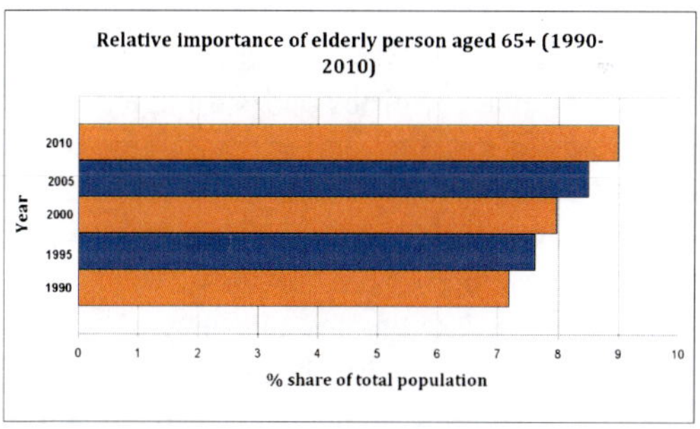

Fig. 1b. Relative importance of elderly (1990-2010)

The projections for the coming years show the increasing relevance of the elderly in the world population. The United Nations projections suggest that there will be a population growth until 2050, on average, of 0.8 percentage points, reaching an amount of about 9,300 million people. But what is more significant is that the elderly percentage of the total population will grow continuously and it will be close to 17 % by 2050. This means an increase going from 7% in 2010 to 17% in 2050.

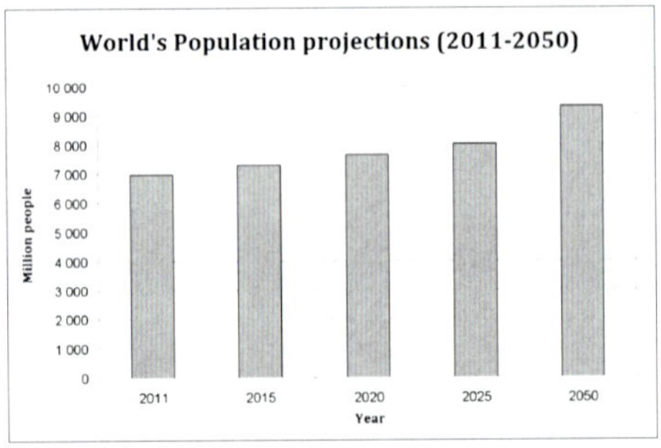

Fig. 2a. World's population and elderly projections (2011-2050)

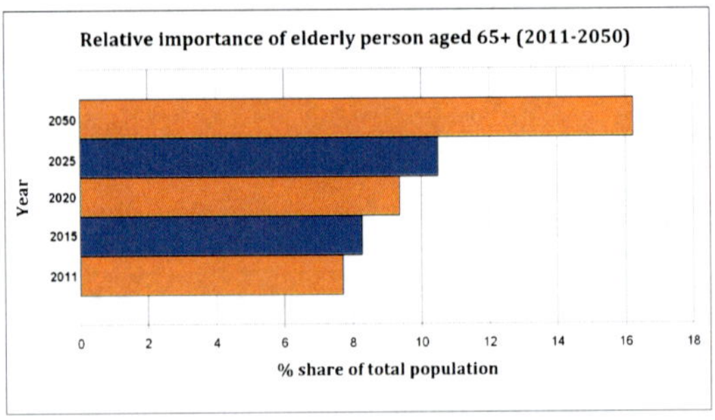

Fig. 2b. Relative importance of elderly (2011-2050)

These demographical changes in the world population are caused by a main factor, the difference between the number of births and the number of deaths. High fertility rates experienced in the postwar decades started to decrease over the years while life expectancy increased.

To emphasize the importance of demographical changes, a population comparison has been made between the three main regions which are working in AAL developments (European Union, United States and Japan).

According to the United Nations [3], the structure of the population in the European Union is expected to become dramatically older. There will be less than two people in age of work for every older person aged 65+ in Europe by 2060. Nowadays this quantity is almost four to one. In 2010 the elderly was the 17,4% share of total population, but projections estimate that by 2050 they will be 29%.

In the United States the aging of population is not estimated to grow as fast as in the EU-27, but the percentage of the elderly in the total population will also increase significantly. According to the US Department of Health and Human Services [4], in 2010 people over 65 represented 13.1% of the population, but it is expected to grow to be 21.2 % in 2050. However, the most impressive data information about the elderly comes from Japan, which has the largest increase in elderly over 65 years between 1990 and 2010, reaching an amount of 22.7% share of total population, and projections estimate that in 2050 there will be a 35.1% share.

Table 1. World's elderly population projections (2011-2050)

	Projected relative importance of elderly persons (% share of total population)	
	Aged 65+	
	2010	**2050**
World	7,6%	17,4%
EU-27	17,4%	29%
EEUU	13,1%	21,2%
Japan	22,7%	35,1%

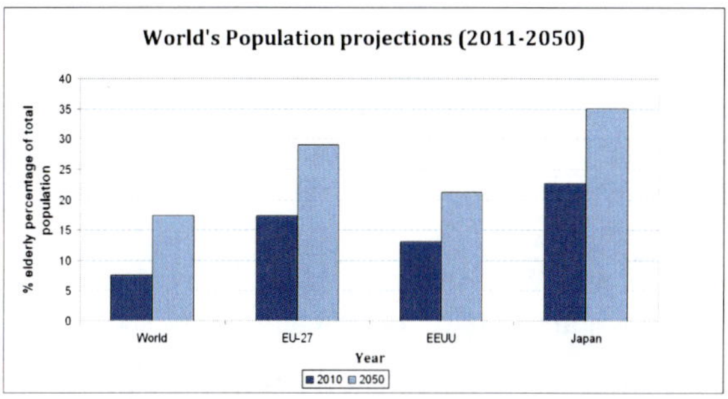

Fig. 3. World's elderly population projections (2011-2050)

Clearly, this new scenario, where population over 65 starts to have a great weight in the society, creates new opportunities for companies and researchers to intensify their efforts and develop new solutions with the objective of covering the necessities of this growing population sector.

But not all elderly are in the same conditions, neither need identical assistance. For instance, in the United States about 29.3% of older person lived alone in 2010, and other 4.1% lived in institutional settings such as nursing homes. Also approximately 2.4% of the elderly lived in senior housing with at least one supportive service available to their residents [5]. The percentage of elderly living alone for future years rises in the entire world, particularly in Europe, according to the European Commission report [6]. It predicts that persons older than 65 suffering from at least one disability in activities of daily living will more than double until 2060; dependent older people receiving care in institutions would almost triple; those receiving formal care at home would more than double; and those receiving informal or no care would almost double. Most of the care that those elderly require is provided by family or friends and, in many cases, they do not receive any payment.

Despite their limitations and sometimes even disabilities, many old people want to remain independent and active for as long as possible, aging at home, instead of moving to institutions or nursing homes. There is a close relationship between aging and disability, and a proper use of information and communication technologies can help improve the quality of life of aging people. Maintaining independence with aging is an important goal to most people, disabled or not, and it depends on personal conditions (health and capacities), but also the physical and social environment of the person. The physical environment can be enhanced to include technology that makes life easier, more enjoyable and comfortable improving the possibilities for perception, communication, information processing, mobility and health maintenance.

Young disabled people can also significantly benefit from these AAL systems that have been initially designed for the elderly. In some cases they require minimum adaptation, but can make their daily life much easier and safe. There are the following disability types:

- Physical: can be defined as the reduction or absence of motor functions or physical members (lack of a hand, leg, foot, etc.), which prevents their daily normal development.
- Sensory: corresponds to the visually impaired, deaf and those who can present problems in communication and language. There are scales to mark different degrees of sensory impairment.
- Psychic: presents "adaptive behaviour disorders, predictably permanent".
- Intellectual or mental: is characterized by intellectual functioning below average, which coexists with limitations in two or more of the following skill areas: communication, self-care, home living, social skills, community use, self-direction, health and safety, functional classroom content, leisure and work.

According to the WHO [7], about 15% of the world's population has some form of disability; these people need more health care than other population sector and have greater unmet needs. Therefore, it is obvious that the development of AAL systems will improve their quality of life, make their daily life easier and help their caregivers to do their jobs in an efficient and more stress-free manner.

From the above, we can draw the following conclusions:

1. The sector of elderly population in need of care and assistance is growing continuously.
2. Older people would like to maintain their independence.
3. Elderly and disabled people are growing in number and need more healthcare than other population sector, therefore the governments' public expenditure per person on these people keeps growing exponentially much faster than the working population.
4. Family and friends are usually responsible for the care of elderly and disabled people.
5. People spend more time at home as they age.

Also, it has to be noted that AAL functionalities could potentially be extended to other population sector (not necessarily elderly or disabled), where these kind of systems help to improve the user comfort at home.

This scenario paves the way for the emergence of AAL solutions which goal is to simplify and improve the quality of life by providing support in situations that need assistance, focused on home environments and based in the use of new ICT solutions. Even worldwide organizations, as WHO, guides and supports Member States to increase the awareness about elderly and disabled issues, and promotes its inclusion in national and international health policies and programmes. For instance, the European Union launched “The Ambient Assisted Living Joint Programme” (AAL JP) [8] which is a funding activity running from 2008 to 2013, with the concrete aim of enhancing the quality of life of older people and strengthening the industrial base in Europe through the use of ICT. Similarly to the European Union, other governments are supporting the development of AAL projects, because they are sensitive to the importance of these population sectors and cost savings in the long run.

2.2 Elderly and Disabled Needs

The population aged over 65 is increasing dramatically and most of these people experience some limitations and needs, which poses a new challenge for the society. Providing adequate care and assistance to this population in an economically viable way requires new solutions. First, we will discuss the main limitations that the elderly have to face:

1. **Physical limitations**: many adults experience physical limitations as they age. According to [9] a physical limitation refers to having difficulty performing any of eight different physical activities. It is important because of its relationship with the ability to live independently and to the overall quality of life [10]. Obviously it emerges as a restriction in a person’s range of motion, strength, endurance or balance, and these limitations are more common in older people. For instance adults aged over 80 have three or more physical limitations compared to people between 50-59 years old.
2. **Perceptual limitations**: most people experience perceptual changes as they age. These perceptual changes are mainly related to vision and hearing.
   - Vision impairment occurs in more than 13% of the elderly population and is one of the leading causes of morbidity in adults over the age of 65. This limitation increases with age, from 1% in 65-69 range, to 17% in the over-80 population [11].
   - Hearing loss is the most common sensory deficit in the elderly, affecting 1 in 3 people older than 60 and half of those older than 85 [12].
3. **Cognitive limitations**: cognitive functions establish processes by which the person receives, stores and uses the perceived real information and relate it with itself info. Attention, orientation, perception, fixation and memory, among others, can change with age, and this affects to fundamental processes, functions and psychological capacities of our life.
4. **Alzheimer’s disease**: is suffered by a noteworthy population sector. Actually, the number of people living with Alzheimer in 2010 was about 4.7 million patients and, what is more remarkable, this quantity is predicted to climb to 13.8 million people by 2050 [13].

Once the main elderly challenges have been explained, it is really important to identify ways AAL systems can help satisfy elderly needs, and it is essential to highlight that many older individuals consider the capacity to carry out activities of daily living (i.e., functional independence) to be of greater concern than prevention of disease [14]. So, the following fields have been identified where AAL systems have to focus:

1. **Daily Activities**:
   - Activities of Daily Living (ADL): ADL are basic routine tasks, such as grooming, dressing, eating, continence or being able to transfer oneself from bed to armchair and back. There are people who need assistance to perform these activities [15].
   - Instrumented Activities of Daily Living (IADL): these tasks are more complex and require some physical conditions, organization skills and sound judgement (using telephone, shopping, preparing meals, managing medications, maintaining the home, being able to drive or use public transportation). Elderly people who adequately perform these activities are considered to be able to live safely and independently [15].
   - Enhanced Activities of Daily Living (EADL): EADL include participation in social activities, learning new skills and engaging hobbies [16].
2. **Memory Functions**: obviously the age is an important factor to consider in memory functions. Memory loss and damages in cognitive functions (as language, perception, reasoning or calculation) are common in the elderly.
3. **Health Monitoring**: older people usually have more health disorders than the rest of the population and governments spend a lot of money on assistance for them.
4. **Decrease the caregiver's workload**: not all caregivers are care professionals. Familiars and friends work hard in taking care of senior people and, in many cases, they do not receive any economic compensation for it. In any case, any help that decreases the caregiver's workload will be welcome.

Current AAL developments are focused in making easier elderly daily life activities, helping to properly cover their needs, preventing for possible health problems and decreasing caregiver's efforts.

3 AAL Tools and Devices

The evolution of computer systems has generated incredible achievements since their inception in the 1960s. In the last few years a new computing paradigm has emerged, the so-called "Ambient Intelligence" (AmI), which refers to electronic environments characterized by human-centric computer interaction and that are sensitive and responsive to the presence of people. These systems use embedded technologies where networked devices are seamlessly integrated in the environment and work in concert to support people in carrying out their everyday life activities in an easy, natural way. AmI systems are capable of recognizing the users and their situational context (what they are doing); then anticipate and respond to the user needs at every particular moment. The AAL concept combines Assisted Living with Ambient Intelligence in one complete system and with the clear goal of improving the user life in his daily environments. Figure 4 [17] shows a historical description of computer systems until Ambient Intelligence systems.

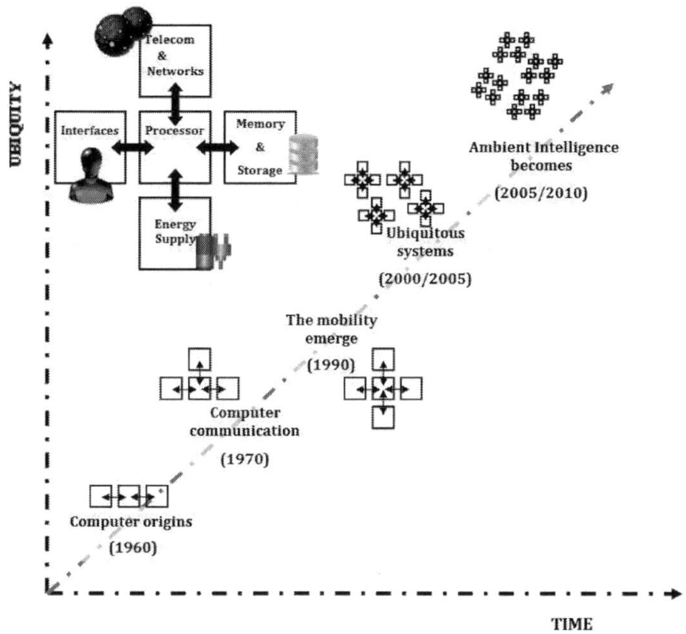

Fig. 4. Historical description of Ambient Intelligence Systems [18]

As described in section 2, the main users of AAL systems are elderly and disabled people who need care and assistance in their daily life activities. But what kind of existing devices and tools could be focused in elderly care and assistance?

There are many existing tools very useful for AAL purposes. The next subsections highlight some of the most representative.

3.1 Mobile Devices

Mobile devices are distinguishable for its portability characteristic. For its proper use, these devices have to be intentionally taken by the user, as opposed to wearable devices that are usually integrated in wearable clothes, in a seamless way, as an integral part of it.

There are many mobile medical devices designed for specific illnesses and focused in monitoring different functions. For instance, an IoT-based device for diabetes therapy management [19], an armband designed for monitoring patients with chronic pulmonary disease [20], or a system that captures mechanical activity and bio-impedance of the skin called HealthWear [21].

Also the importance of smart phones has greatly grown in recent years and they can be easily used for detecting user activity and mobility. Today, most smart phones integrate different sensors such as gyroscopes, accelerometers, proximity sensors or Global Positioning System (GPS), and the obtained valuable information from these sensors is very useful to detect user activity for AAL goals.

3.2 Robotics

Robotics with assistive goals are focused in helping seniors in their daily life activities (ADL, IADL and EADL), but also they are helpful in monitoring [22] or providing interfaces to technology [23].

- **ADL**: robotics focused on helping in ADL, assist on tasks related to grooming, feeding, dressing or basic mobility. For example, the DU-O-MI robot [24] is a nursing robot system for the elderly and the disabled that assists users in moving independently in indoors. The RIBA Robot [25] helps user to move from a bed or armchair including getting up, the Care-O-bot is a robot suitable for catching typical household objects and change their placement [26], or Dusty robot [27] which retrieves dropped objects from the floor.
- **IADL**: robotics focused in IADL assist on tasks related to telephone use, shopping, preparing meals, managing medications or housekeeping. For example, if the user's arms have some damage, there are robots as PerMMA [28] that is able to help in grooming, meal preparation, mobility or shopping tasks. Also uBot5 [29] can assist in those mentioned activities and check vitals, or Pearl robot [30] helps on giving medication.
- **EADL**: robotics focused in EADL assist on tasks related to social activities, hobbies or new skills. For example, companion robots (Care-O-bot [26] or Paro robot [31]) help to improve the user well-being playing the role of a mate.

The use of robotics is influenced by a strong cultural component, and this sector is more relevant in Asian countries, as Japan and Korea, where there are many research institutes and companies working on robots.

3.3 Wearable Sensors

Wearable sensors concept refers to sensors that are an integral part of our everyday clothing and assist final users in different everyday situations, even in mobility situations.

One of the origins of smart clothes was the Smart Shirt [32] that allows monitoring different vital parameters as heart rate, ECG or respiration. Nowadays there are many commercial examples of these products (LifeShirt by Vivometrics [33], Hexoskin [34] or Equivital [35] among others), but also medical devices are integrated in other clothing types. For example a belt [36] that monitors the intensity of movements, smart shoes or smart baby bodies [37].

But not only wearable sensors are integrated in clothing, also in other complements. For example, watches that measure the user activity or insomnia (Actiwatch [21]), monitor blood oxygen [38], measurement of glucose [39]. Rings with saturation sensors [40] and other smart jewellery items as developed in [41].

So wearable sensors are an easy way to develop AAL functionalities in elderly at home because the users do not have to remember and intentionally grab any specific device to obtain their vital signs measures, they only have to continue wearing their clothes or complements.

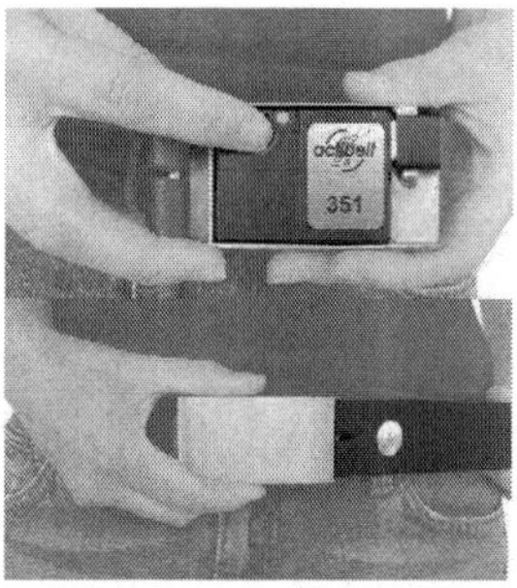

Fig. 5. Actibelt [36]

3.4 Environmental Sensors

Elderly people spend most of their time at home, and this is a suitable environment where to deploy sensors and obtain information from final users. As with wearable sensors, this time the user is not forced to intentionally grab a specific device. In this case, sensor devices are integrated in common home objects that cover its functionality without the user awareness, in many cases due to a really good ubiquitous integration of devices in common home items. Evidently, concepts such as ubiquitous or pervasive integration of devices become more important in these device types, where it is really necessary that the user does not perceive the existence of any anomalous device in their familiar environment, avoiding any kind of rejection by the user.

Basically we distinguish between two different of sensor-based environments. On the one hand, there are environments that integrate sensors for obtaining in-formation about the final user behaviour in their own home, and from this context information identify possible warning situations [42]. In this way, there are previous works with many different sensors, but with the same goal. For example: monitoring seat occupancy patterns and functional status with PIR (Presence Infrared Sensor) characterization [43], detection of movement in bed using load cell sensors [44], or the use of cameras to sense the environment and detect user behaviour [45].

On the other hand, some sensor-based environments include enhanced functionalities and can even monitor vital signs [46]. There are common home objects that can check the user pulse, respiration or body temperature like the Smart Pillow [47], or the Smart Sofa [48] that is able to identify individuals sitting on it.

Deploying sensor systems in home environments and wearable sensors are ways to assist and make elderly life easier. These are really autonomous systems, and with a proper seamless integration of the system, the user will not perceive the existence of it.

3.5 Smart Homes

Smart Home generic concept arises from the evolution of Home Automation, which main aim is to centralize and automate home tasks and control its devices in order to improve final user comfort. But new functionalities have been added over the years. An example related to energy efficiency is described in [49-50], where common home devices have been interconnected in order to gather energy consumption data and provide to the user a control tool. Another area of focus is eHealth monitoring, like the home-based health care systems linked to hospitals presented in [51-52]. Finally, other assistance services are focused on providing general improvements to the life people at home (e.g. activity/mobility monitoring [53], reminder technologies [47], fall detection [53] or social issues [54]).

4 AAL Software Techniques

There are many computational techniques related to AAL goals which allow the systems to identify and infer some useful information such as where the user is, what kind of activity he is doing, and detect any anomalous situation or remind the user some special task. The following subsections show a description of the most frequent techniques used in AAL applications.

4.1 Context Modelling

AAL systems are designed to facilitate the user's daily life at home. These systems react to the current information about the users' activities and status, as well as their surroundings. This information is called context, and consequently those systems are referred to as context-sensitive. The information is represented by a context model and is extracted by processing sensor data from multiple sources [55]. An AAL system needs a context model, and a pattern is necessary to define and store the context into a format that can be processed by a computer.

The system architectures that manage context information must be open, therefore independent of the applications that use or produce this information, so that information can be shared among different applications. It should also be dynamic in the sense that it permits the addition of new devices on the fly and, therefore, the addition of new information to the context representation [56]. An example of this kind of architectures is SOCAM [57], a decentralized architecture in which information producers and consumers are decoupled.

Context modelling systems represent information using different sources of data such as situation modelling languages [58], key/value-based models [59] or markup-based models [60], among others. But for AAL applications, ontology models are very interesting.

Developing flexible and useful context ontologies is a challenging task. There are several approaches to model context; Strang and Linnhoff-Popien [61] make a summary of these approaches, which are based on the data structures used to represent and exchange this contextual information. Ontologies represent concept descriptions and their relationships. They are the most commonly used form of representing context information. The ontologies can be used even if there is no exact match between the available information and the one required by the system

[57,62-64]. Moreover, they enable the use of developed technologies for the Semantic Web. According to Strang and Linnhoff-Popien, ontologies are one of the most expressive models and fulfil most of their requirements, which are distributed composition, partial validation and quality of information. There are many proposals for generic ontologies, for instance [64-67].

4.2 Activity Recognition

One of the most important information in AAL systems is reliable recognition of the monitored person's behaviour during their ADLs. The activity recognition can be done at different levels, for example identify a single movement, various actions that result in a complete activity, or even a group of activities. For this purpose, it is possible to highlight two different main data sources; wearable and ambient sensors that generate data samples, and vision-based system like cameras and thermographic devices that produce video data.

- Sensor-based data information sources, such as accelerometers and gyro-scopes, provide data and time series that can be useful to detect physical activities (walking or running). There are some techniques to post-process the data and time series and identify the user activity. One of them is the unsupervised model "motif discovery" which automatically detects activity patterns [68]. Sensor networks can also be used to recognize more complex activities. There are some algorithms based in mixture models [69], neural networks [70], decision trees [71], or popular graphical models [72] which infer activities based on sensor variables, these types of algorithms include Hidden Markov Model [72], Dynamic Bayesian network [73], or conditional random fields [74] among others.
- Vision techniques are based in the process of labelling image sequences with action labels [75]. In general terms, vision-based techniques are more computationally complex than previous ones; however, they provide detailed context information. There are two main Vision-based categories [76]: single layered (it is used for recognizing simple actions) and hierarchical (it is used for more complex activities). Although vision techniques provide detailed information, it is also important to consider privacy concerns of final users.

4.3 User Location

GPS is the typical outdoor navigation system, but indoor environments require the use of different techniques for localizing the user. Indoor location is very useful in AAL systems to track and monitor user behaviour at home, and in the last decades some advances and researches in this field have been developed.

There are many indoor location techniques: RFID-based systems like Land-mark [77] or SpotOn [78] (they use RFID readers to determine user location in indoor environments). Pressure sensors integrated on floors [79] or infrared motion sensors. Ultrasonic systems based on the measure of "time of flight" (Cricket system: it links the time of flight with a specific indoor location [80]) or WiFi-based systems such as Radar [81].

4.4 Anomaly Detection

Anomaly detection refers to the problem of finding patterns in data that do not conform to expected behaviour [82]. This concept is very broad and is used in a wide variety of applications such as intrusion detection for cyber-security, fault detection systems, military goals, and fraud detection for credit cards or health care.

Related to AAL, anomaly detection is focused on detecting anomalous user behaviour that do not match with daily life usual patterns. There are techniques based on temporal relations, rule-based or similarity-based comparisons [83-85]. This kind of detection is also used to identify dangerous situations for the final user, who usually has the same behaviour everyday pattern.

4.5 Scheduling

Other important component in AAL is automatic planning focused on people with memory problems who need to be reminded about carrying out their ADLs. Some forms of memory impairment are much more frequently in the elderly and scheduling techniques offer the capacity to identify such situations and prioritize among different types of reminders. Mainly, there are two plan configuration models, when designers provide plan details to the system [86-87] and when the system learns and adapts the schedule based on real observations [88].

It is important to consider that there are some risks when an external system reminds some tasks; for instance if the system issues too many prompts, it could result into final user rejection. So it is necessary to reduce the number of prompts as much as possible, avoiding incorrect reminders and prioritising them. This kind of problem can be mitigated with planning, inference or reinforcement learning techniques [87, 89].

5 Full AAL Developments and Systems

Although there are some full solutions offered by important multinational companies (such as Bosch [90], Philips [91], Intel [92] or Honeywell [93]), in order to provide an overview of the developed work around the world, this section focuses in three of the most representative economies (Europe, United States and Japan) and their related work within AAL. There are many differences in the way they set out a plan to support AAL and, depending on the specific zone, they are more specialized in one subject than others.

5.1 Europe

Due to the huge increase of the proportion of elderly in population, the European Commission realised that they should take the necessary measures to support initiatives related to elderly care. This would result in future savings in the budget that governments use for that purpose. The AAL Joint Programme (AALJP) was founded as an activity that started in 2008, with 23 countries working together with the goal to develop a programme to improve the quality of life of older adults through the application of ICT.

Since 2008 there has been six different "Calls" focused in different concrete aspects related to AAL [94]:

1. Call I: "ICT-based solutions for Prevention and Management of Chronic Conditions of Elderly People".
2. Call II: "ICT-based solutions for Advancement of Social Interaction of Elderly People".
3. Call III: "ICT-based Solutions for Advancement of Older Persons' Independence and Participation in the "Self-Serve Society".
4. Call IV: "ICT-based solutions for advancement of older persons' mobility".
5. Call V: "ICT-based solutions for (self-) management of daily life activities of older adults at home".
6. Call VI: "ICT-based solutions for supporting occupation in life of older adults".

There are more than a hundred projects completed or under development in the different calls with a final total average available funding budget of 55 Millions € per call.

Some of the developed or ongoing projects associated with different fields are:

- **Mobile devices:** the "AMICA" [95] project uses a dedicated mobile device to send physiological signals and emulate the medical consultation at home. The "CAPMOUSE" [96] project is focused on identifying the end user and find out their needs. The "BANK4ELDER" [97] project standardises mobile bank interfaces targeting the elderly, whereas the "ASSISTANT" [98] project enables older users to confidently and safely use public transport. The "EMOSION" project [99] combines existing mobile platforms, an IP-connected server platform and home security sensor network.
- **Robotics**: some examples of the use of robotic technologies are the "DOMEO" [100] project, aimed helping elderly to stay longer and safer at home and also help caregivers in their daily work, and the "ALIAS" [101] project whose robot interacts with elderly users, monitors and provides cognitive assistance and promotes social inclusion.

Fig. 6. DOMEO [102] and ALIAS [103] robots

- **Wearable sensors** related projects have been also supported by the AALJP. For instance the "ECAALYX" [104] project uses conductive textile material to develop a comfortable garment for monitoring purposes. The "PAMAP" [105] system which is based on a network sensor which are worn by the subjects to extract parameters of their physical activity. Also the "SOFTCARE" [106] project is based on a bracelet containing a 3D accelerometer, and the "eSTOCKINGS" [107] project is based in a new generation of smart compression stockings with integrated ICT. The HEALTH@HOME [108] project uses wearable sensors to acquire patients' physio-pathological cardiovascular and respiratory parameters and sends them to a remote server.

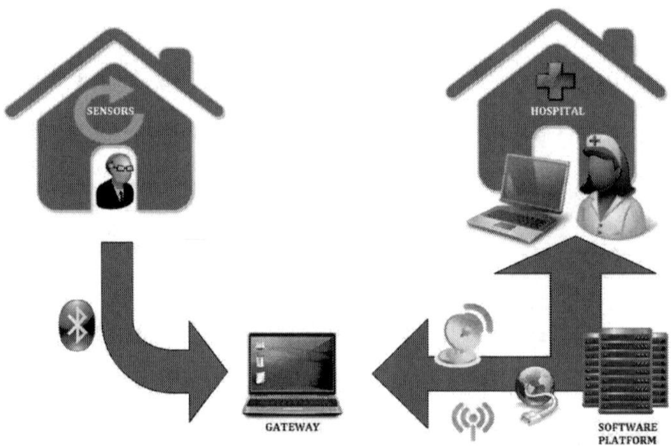

Fig. 7. Health@Home architecture [108]

- **Environmental sensors**: for instance the "A2E2" [109] project stimulates beneficial levels of exercising in the elderly, or the "BEDMOND" [110] project that developed an assistant system based on behavioural patterns. These projects are based in the deployment of sensors in home environments.
- **Smart Homes**: the "HOPE" [111] project is a solution installed at the senior people's homes, and provides services for lifelong care management and health support, self-monitoring and decision making. To enhance the social interaction of elderly people, projects like "HOMEdotOLD" [112] provide a TV-based platform, or "CARE@HOME" [113] that, apart from social interaction, also supports health wellness and well-being of the elderly at home.

Apart from projects related to the AALJP, there are other developments made in Europe. For instance the IMMED [114] project which monitors dementia user tasks using a wearable camera, other e-textiles projects such as MagIC [115] or WEALTHY [116], or other internet of thing-based devices with eHealth functionalities such as the diabetes therapy management device [117] or the drug identification and interaction checker [118].

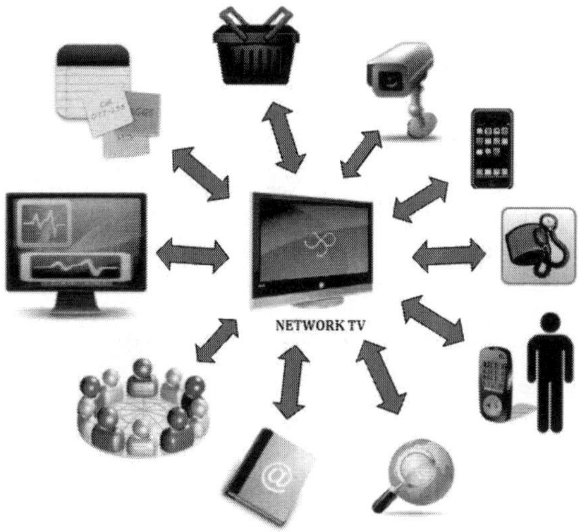

Fig. 8. Care@Home functionalities [113]

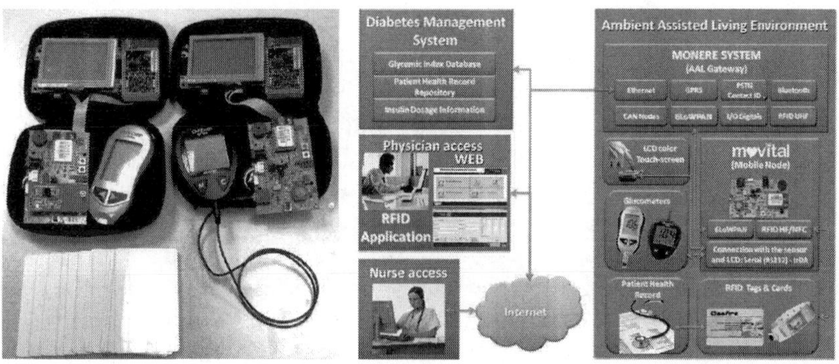

Fig. 9. a) Diabetes therapy management device. b) Architecture diagram.

5.2 United States

A large number of AAL developments has been made in the United States, although they have many different goals and are not focused on the same specific objective. For example the CASAS [119] project at Washington monitors some user ADL tasks to verify that user completes the activities in a normal way. For that purpose, the system uses the simultaneous frequent-periodic activity pattern mining to recognise activities and this results in a non-invasive assistive environment system for dementia users. Other projects were focused on recognizing social activities [120] or estimates walking times [121] to directly link with ADL.

The "Aging in place" project [52] in Missouri, which provides a long-term care for the elderly, supports e-health functionalities and allows residents to remain in their apartments. It has two main parts, an integrated sensor network in home environments (bed sensors, chair sensors, temperature or motion sensors), and the Sinclair Home Care system which provides services such as vital sign checkups or medication assistance.

The "Aware Home" [122,123] at the Georgia Institute of Technology in Atlanta, which has been focused in supporting daily routines and maintaining a familiar vigilance throughout the development of computational perception techniques with a variety of sensing devices and the design of interfaces to facilitate the Human-Computer interactions.

The Elite Care Technologies Company (ECT), in Oregon, has developed an integrated monitoring and electronic documentation system, which emulates the knowledge that family caregivers would have of their care. It is focused in resident environments but it also could be transferred to home environments. The system consists of simple components. Residents and staff members carry a badge that emits radio-frequency and infrared signals picked up by sensors stationed around the residence. Load cells installed under each bed continuously monitor every resident's weight and analysis of this data provides trends in weight, time spent in bed, rising and sleeping time, and restlessness while in bed. All sensors are integrated with an alert system so that staff can be informed when any unusual event takes place [124].

The Massachusetts Institute of Technology (MIT) also has developed multidisciplinary projects associated with the design of the home and its related technologies, products, and how services should evolve to better meet the opportunities and challenges of the future. The House_n project [125] was one of these projects led by researchers at the MIT Department of Architecture.

Pittsburg and Carnegie Mellon Universities have developed a personal robotic assistant (Pearl robot [126]) that could be used to prompt people with failing memories to take medicine, to provide remote telepresence for professional caregivers or assist with difficult tasks.

Falls are very prevalent among the elderly; it is the first cause of death for people above 79 years old. Also in the United States some projects related to fall detection have been developed. For instance a smart passive floor-vibration is presented in [127], completely passive and unobtrusive to the user which detects a vibration pattern on the floor when a human falls.

5.3 Japan

Another world economy where there are important efforts in Ambient Assisted Living is Japan. As previously outlined, cultural features affect the way different countries direct their work. In Japan, the use of robotics with this aim is more relevant than in other countries, furthermore the existence of many research institutes and companies working in robotics in Japan benefits the developments in this area.

For instance, the RIBA robot [128] is able to transfer the user from one place to another using human-type arms, apart from handling heavy objects. It has a motion control unit by touching to interact with the user and the caregiver.

The Mamoru Robot [129] is an attendant robot that closely observes a person's daily life, keeping track of everyday activities such as taking medicine, and then notices if the person has completed those activities that day or not.

The Paro Therapeutic robot [130] developed by AIST, which reduces user and caregivers stress by stimulating the interaction between them and improving the socialization of users.

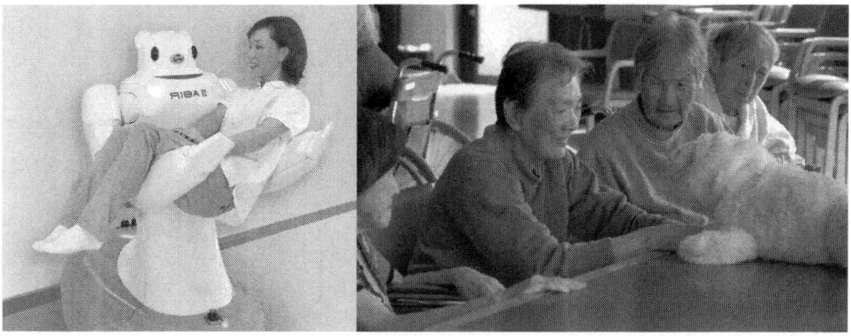

Fig. 10. Riba [129] and Paro [130] (AIST, Japan) robot

Apart from robotics developments, other AAL projects have been developed in Japan. An example is a monitor watching TV for diagnosing health condition [133]; not only physical condition but also mental. It uses image processing algorithms to reconstruct monochrome images and diagnoses the subject's health. An-other example is a sleep monitoring mattress [134], which monitors heart rate, breathing or body movement. Similarly, [135] presents an air-filled mat, which obtains information about perturbations in the air pressure and extracts information about heart and respiratory rates.

Also it is important to mention the SELF [45] system, based on observation of daily human behaviour in their home environment and able to monitor body movement, breathing or oxygen in the blood, among others. Finally, the airbag fall-protection system for the elderly [136] reduces the impact force of falls.

6 Opportunistic Wireless Access Networks in AAL Environments

Opportunistic networks (Oppnets) are a particular case of delay tolerant networks which have been a huge increase in recent years, thanks to the popularity of wireless technologies and the miniaturization of devices. The hard working conditions of these kinds of wireless networks make them very attractive for the study and development of protocols to increase the reliability and quality of information.

In Oppnets, mobile nodes are able to communicate with each other even if a connecting route between them never existed. It stands out for its dynamically routing characteristic, the messages between the origin and destiny can be routed through any network node that can opportunistically be used as next hop. These characteristics can be used as a solution in absence of an end-to-end path between the sender and the receiver (hoping messages between different intermediate nodes) or setting up a new node inside a previously constituted network.

In AAL environments this routing method can be used for different goals. For instance, one of the most important functionalities of an AAL system is to be able to locate and predict the user movements at home, so it is necessary to consider their spatial and temporal patterns of movement that it will be directly related to the repeated sequences of jumps in an Oppnet [137].

Other Oppnet functionalities, such as medical monitoring, can also be applied for AAL goals. Remote medical monitoring using wireless communications and body sensors in a non-intrusive way have increased their popularity in the last few years. A use example is described in [138], where several medical care applications are identified for the use of Oppnets in Bluetooth networks, this concept can be transferred to AAL environments and caregivers support.

One of the key concepts in AAL is privacy and security. One of the main important things is to maintain the user privacy in their own home environment and avoid possible intruders accessing their personal information. In [139] an investigation about three different helper privacy approaches in Oppnets has been done (automatic trust negotiation, Semantic Web technologies and automatic enforcement of privacy policies). Also it is necessary to consider the difficulty to protect data privacy because of the fact that there is not any shared secret key in a multi-cast from the controller because of a simple capture of even a single device could lead to the failure of the whole system.

Opportunistic networks can have a practical use in AAL systems because of the nature of these environments. The mobility of the user in their home and the add of new nodes to the network in "opportunistic" occasions, as a medical, caregiver or familiar visit, could be one of the many chances that Oppnets can have more potential in AAL applications.

7 Future Challenges

Nowadays the existing hardware and technological devices allow us to obtain a large variety and quantity of information that can be used to address AAL goals. Although the current technological solutions provide an excellent source of information, and will always be in continuous growth including new improvements, the great challenges in AAL solutions are the **standardization in the management of this enormous quantity of information and security aspects to maintain users' privacy.**

As described in subsection 3.3.1, the management of information in AAL systems requires a context model pattern to store context information that can be computer-processed. Ontology models are an interesting solution to store and establish a suitable structure of the obtained information. Different groups are working on setting ontology standards that are involved in different fields of in-formation such as Smart Cities, Smart Grids and generally IoT [140, 141].

Due to the management of a great quantity of sensitive personal data and information in AAL solutions, security and privacy is also other of the main challenges in this field. It is critical to maintain and preserve user privacy preventing unauthorized people from having access to this personal information. More specifically, the R&D community is working to enhance some aspects:

- Unauthorized access protection (identification and authorization are required not only for people but also for devices and services).
- Users' identification, particularly in multi-user contexts such as home.
- Protection of privacy, with techniques related to anonymity, pseudonimity, unobservability and unlinkability.
- Techniques to ensure the identity when handling a device.
- Creating easy reliable and secure personal identification.

For instance, to protect security and privacy of information in a data transmission between sensor, home station and a hypothetical help centre, authentication and encryption mechanisms are applied and nowadays technical and standardization problems have to be solved. Safety data information storage is also necessary in any situation when health information about patient has to be shared with doctors. This case requires a data backup mechanism and an electronic patient record, but is still under discussion.

In what relates to functionality and areas of applications, most existing AAL systems focus on providing support and assistance to the elderly. They try to replace human care in detecting hazardous situations (e.g. falls) that usually need immediate attention. More recent systems carry out health monitoring for people with medical needs. The next natural step would be to integrate both type of systems into one that is capable of simultaneously monitoring behaviour and health variables.

One of the major future challenges that remains to be addressed is to develop intelligent systems not only capable of detecting problems when they have already arisen, but when they are in the dormant or initial phases of development. Disease prevention is an area that can really contribute to increase our healthy life expectancy in an economically viable way. This will not only improve the quality of life of old people, but it will also reduce the medical and assistance costs. It has always been a big truth that "prevention is better than cure".

In the first stages of AAL, the target has been the elderly. Similarly, disease prevention will likely be applied to old people first, since they have a higher probability of developing serious diseases. However, AAL could also be applied to the general public with a more generic goal that is the promotion of better and healthier life styles. This will also have a big positive impact on the wellness of people in the short term, as well as their health in the medium and long terms.

Nowadays the health systems of developed countries spend huge amounts of money on the treatment of diseases that were originated by wrong life styles. Very simple things like eating healthy food, regular exercising and walking, and having enough sleep can really benefit our health. Sometimes people do not do it because of laziness, ignorance or bad habits. The future AAL systems will be able to know what people are doing, and motivate people to change and to improve. Some initial work that uses the power of social network has started, but it is not yet applied to the field of AAL. These future AAL systems will use hardware that combines home-integrated sensors, as well as wearable and mobile devices. They will use software in the cloud for analysing the measured data, but that also exploit the tools of social networks and information sharing with doctors.

Without any doubt technologies like AAL paint a better, happier and longer life for us all. We just need to research and implement towards what we think is possible.

References

[1] World Health Organization, http://www.who.int/topics/ageing/en/index.html (last access: March 05, 2013)

[2] European Commission Report: Active ageing and solidarity between generations: A statistical portrait of the European Union 2012, 2012 Edition (2012)

[3] Department of Economic and Social Affairs of the United Nations, http://esa.un.org/wpp/unpp/panel_population.htm (last access May 02, 2013)

[4] U.S Department of Health and Human Services, http://www.hhs.gov/ (last access May 02, 2013)

[5] U.S. Department of Health and Human Services: A Profile of Older Americans (2011)

[6] European Commission, Long Term Care for the elderly: Provisions and providers in 33 European countries (2012)

[7] World Health Organization, http://www.who.int/mediacentre/factsheets/fs352/en/ (last access May 02, 2013)

[8] Ambient Assisted Living Joint Programme (AAL JP), http://www.aal-europe.eu/ (last access May 02, 2013)

[9] Holmes, J., Powell-Griner, E., et al.: Aging Differently: Physical Limitations Among Adults Aged. NCHS Data Brief, No. 20 (July 2009)

[10] Kaplan, G.A.: Maintenance of functioning in the elderly. Annals of Epidemiology 2(6), 823–834 (1992)

[11] Chen, Y.A., Thomas, M.: Vision screening in the elderly: current literature and recommendations. University of Toronto Medical Journal 87(3), 166–169 (2010)

[12] Centers for Disease Control and Prevention (CDC), http://www.cdc.gov/healthliteracy/developmaterials/audiences/olderadults/understanding-challenges.html (last access: May 03, 2013)

[13] CBS NEWS, http://www.cbsnews.com/8301-204_162-57568046/alzheimers-rates-expected-to-triple-by-2050-because-of-aging-baby-boomers/ (last access: May 03, 2013)

[14] Paterson, D., Govindasamy, D., Vidmar, M., et al.: Longitudinal study of determinants of dependence in an elderly population. J. Am. Geriatr. Soc. 52, 1632–1638 (2004)

[15] Huang, Y., et al.: Physical fitness, physical activity, and functional limitation in adults aged 40 and older. Medicine and Science in Sports and Exercise 30(9), 1430 (1998)

[16] Rogers, W.A., et al.: Functional limitations to daily living tasks in the aged: A focus group analysis. Human Factors: The Journal of the Human Factors and Ergonomics Society 40(1), 111–125 (1998)

[17] J.: B Waldner. adapté de "Nano-informatique et intelligence ambiante". Hermes Science Publishing (2007)

[18] Wikipedia (Evolution computer 1960-2010 image), http://en.wikipedia.org/wiki/File:Evolution_computer_1960-2010-fr.jpg (last access: September 23, 2013)

[19] Jara, A.J., Zamora, M.A., Skarmeta, A.F.G.: An internet of things-based personal device for diabetes therapy management in ambient assisted living (AAL). Pers. Ubiquit. Comput. 15, 431–440 (2011)

[20] Pitta, F., Troosters, T., Probst, V.S., et al.: Quantifying Physical Activity in Daily Life with Questionnaires and Motion Sensors in COPD. European Respiratory Journal 27(5), 1040–1055 (2006)

[21] Leonhardt, S.: Personal Healthcare Devices. In: Mukherjee, S., et al. (eds.) AmIware: Hardware Technology Drivers of Ambient Intelligence, pp. 349–370. Springer, Dordrecht (2006)

[22] Harmo, P., Taipalus, T., Knuuttila, J., et al.: Needs and solutions—Home automation and service robots for the elderly and disabled. In: Proc. Int. Robot. Syst. Int. Conf., pp. 3201–3206 (2005)

[23] Raytron, "Chapit" (2011)

[24] Park, H.K., Hong, H.S., Kwon, H.J., et al.: A Nursing Robot System for The Elderly and The Disabled. In: Proc. of the International Workshop on Human-Friendly Welfare Robotic Systems, Daejeon, Korea (2002)

[25] Mukai, T., Hirano, S., Nakashima, H., et al.: Development of a nursing-care assistant robot RIBA that can lift a human in its arms. In: Proc. Intell. Robot. Syst. Int. Conf., pp. 5996–6001 (2010)

[26] Graf, B., Reiser, U., Hagele, M., et al.: Robotic home assistant care-o-bot. In: Proc. Adv. Robot. Social Impacts Workshop 2009, pp. 139–144 (2009)

[27] Xu, Z., Deyle, T., Kemp, C.: 1000 trials: An empirically validated end effector that robustly grasps objects from the floor. In: Proc. Robot. Autom. Int. Conf. 2009, pp. 2160–2167 (2009)

[28] Brose, S., Weber, D., Salatin, B., et al.: The role of assistive robotics in the lives of persons with disability. Amer. J. Phys. Med. Rehabil. 89(6), 509–521 (2010)

[29] Kuindersma, S., Hannigan, E., Ruiken, D., et al.: Dexterous mobility with the uBot-5 mobile manipulator. In: Proc. Int. Conf. Adv. Robot., pp. 1–7 (June 2009)

[30] Pollack, M.E., et al.: Pearl: A mobile robotic assistant for the elderly. In: AAAI Workshop on Automation as Eldercare, pp. 85–91 (2002)

[31] Wada, K., Ikeda, Y., Inoue, K., et al.: Development and preliminary evaluation of a caregiver's manual for robot therapy using the therapeutic seal robot Paro. In: Proc. IEEE ROMAN, pp. 533–538 (September 2010)

[32] Georgia Institute of Technology, From Research to Market: Smart Shirt Moves. Georgia Institute of Technology, GA, USA (2004)

[33] Lifeshirt by Vivometrics, http://vivonoetics.com/products/sensors/lifeshirt/ (last access: May 8, 2013)

[34] Hexoskin, http://www.hexoskin.com/en (last access: May 08, 2013)

[35] Equivital, http://www.equivital.co.uk/ (last access: May 08, 2013)

[36] Actibelt (Image published with permission of Martin Daumer, SLCMSR-The Human Motion Institute & Trium), http://www.actibelt.com/ (last access: May 08, 2013)

[37] Linti, C., et al.: Sensory baby vest for the monitoring of infants. In: International Workshop on Wearable and Implantable Body Sensor Networks, BSN 2006, pp. 3–137. IEEE (2006)

[38] Wahr, J., Tremper, K.: Non-Invasive Oxygen Monitoring Techniques. Critical Care Clinics 11(1), 199–217 (1995)

[39] Tamada, J.A., Lesho, M., Tierney, M.J.: Keeping Watch on Glucose: New Monitors Help Fight the Long-Term Complications of Diabetes. IEEE Spectrum 39(4), 52–57 (2002)

[40] Asada, H.H., et al.: Mobile monitoring with wearable photoplethysmographic biosensors. IEEE Engineering in Medicine and Biology Magazine 22(3), 28–40 (2003)

[41] Starner, T., Auxier, J., Ashbrook, D., et al.: The Gesture Pendant: A Self-Illuminating, Wearable, Infrared Computer Vision System for Home Automation Control and Medical Monitoring. In: Proceedings of the IEEE International Symposium on Wearable Computing, pp. 87–94. IEEE Press (2000)

[42] Tapia, E.M., Intille, S.S., Larson, K.: Activity recognition in the home using simple and ubiquitous sensors. In: Ferscha, A., Mattern, F. (eds.) PERVASIVE 2004. LNCS, vol. 3001, pp. 158–175. Springer, Heidelberg (2004)

[43] Kaushik, A.R., Celler, B.G.: Characterization of PIR detector for monitoring occupancy patterns and functional health status of elderly people living alone at home. Technol. Health Care 15(4), 273–288 (2007)

[44] Adami, A., Pavel, M., Hayes, T., et al.: Detection of movement in bed using unobtrusive load cell sensors. IEEE Trans. Inf. Technol. Biomed. 14(2), 481–490 (2010)

[45] Nishida, Y., Hori, T., Suehiro, T., et al.: Sensorized environment for self-communication based on observation of daily human behavior. In: Proc. Int. Robot. Syst. Int. Conf., pp. 1364–1372 (2000)

[46] Ferguson, G.T.: Have Your Objects Call My Objects. Harvard Business Review 80(6), 138–143 (2003)

[47] Park, S.H., Won, S.H., et al.: Smart Home - Digitally Engineered Domestic Life. Personal Ubiquitous Computing 7(3+4), 189–196 (2003)

[48] Legon, J.: 'Smart Sofa' Aimed at Couch Potatoes. In: CNN (September 23, 2003)

[49] Capone, A., Barros, M., Hrasnica, H., et al.: A New Architecture for Reduction of Energy Consumption of Home Appliances. In: TOWARDS eENVIRONMENT, European Conference of the Czech Presidency of the Council of the EU (2009)

[50] eDIANA (Embedded Systems for Energy Efficient Buildings), `http://s15723044.onlinehome-server.info/artemise/` (last access: May 09, 2013)

[51] Yamazaki, T.: Beyond the smart home. In: International Conference on Hybrid Information Technology, ICHIT 2006, pp. 350–355. IEEE (2006)

[52] Dishman, E.: Inventing wellness systems for aging in place. Computer 37(5), 34–41 (2004)

[53] Chan, M., Campo, E., Estève, D.: Assessment of activity of elderly people using a home monitoring system. International Journal of Rehabilitation Research 28(1), 69–76 (2005)

[54] CO-LIVING project, `http://www.project-coliving.eu/` (last access: May 09, 2013)

[55] Van Kasteren, T., Englebienne, G., et al.: Recognizing activities in multiple contexts using transfer learning. In: AAAI AI in Eldercare Symposium (2008)

[56] Euzenat, J., et al.: Dynamic context management for pervasive applications. Knowledge Engineering Review 23(1), 21–50 (2008)

[57] Gu, T., Hung, K.P., Da Qing, Z.: A service-oriented middleware for building context-aware services. Journal of Network and Computer Applications 28(1), 1–18 (2005)

[58] Brdiczka, O., Crowley, J.L., Reignier, P.: Learning situation models for providing context-aware services. In: Stephanidis, C. (ed.) Universal Access in HCI (Part II), HCII 2007. LNCS, vol. 4555, pp. 23–32. Springer, Heidelberg (2007)

[59] Henricksen, K., Indulska, J., Rakotonirainy, A.: Modeling context information in pervasive computing systems. In: Mattern, F., Naghshineh, M. (eds.) PERVASIVE 2002. LNCS, vol. 2414, pp. 167–180. Springer, Heidelberg (2002)

[60] Salber, D., Dey, A.K., Abowd, G.D.: The context toolkit: aiding the development of context-enabled applications. In: Proceedings of the SIGCHI Conference on Human Factors in Computing Systems: the CHI is the Limit. ACM (1999)
[61] Strang, T., Linnhoff-Popien, C.: A context modeling survey. In: Workshop Proceedings (2004)
[62] Nieto, I., Botía, J.A., Gómez-Skarmeta, A.F.: Information and hybrid architecture model of the OCP contextual information management system. Journal of Universal Computer Science 12(3), 357–366 (2006)
[63] Chen, H., et al.: Intelligent agents meet the semantic web in smart spaces. IEEE Internet Computing 8(6), 69–79 (2004)
[64] Flury, T., Privat, G., Ramparany, F.: OWL-based location ontology for context-aware services. In: Proceedings of the Artificial Intelligence in Mobile Systems (AIMS 2004), pp. 52–57 (2004)
[65] Sadeh, N., Gandon, F.: Using semantic web services for context-aware mobile applications (2004)
[66] Fu, G., Jones, C.B., Abdelmoty, A.I.: Ontology-based spatial query expansion in information retrieval. In: Meersman, R. (ed.) CoopIS/DOA/ODBASE 2005. LNCS, vol. 3761, pp. 1466–1482. Springer, Heidelberg (2005)
[67] Ngoc, K.A.P., Lee, Y.-K., Lee, S.-Y.: OWL-based user preference and behavior routine ontology for ubiquitous system. In: Meersman, R. (ed.) CoopIS/DOA/ODBASE 2005. LNCS, vol. 3761, pp. 1615–1622. Springer, Heidelberg (2005)
[68] Tanaka, Y., Iwamoto, K., Uehara, K.: Discovery of time-series motif from multi-dimensional data based on MDL principle. Machine Learning 58(2-3), 269–300 (2005)
[69] Barger, T.S., Brown, D.E., Alwan, M.: Health-status monitoring through analysis of behavioral patterns. IEEE Transactions on Systems, Man and Cybernetics, Part A: Systems and Humans 35(1), 22–27 (2005)
[70] Mozery, M.C., et al.: The neural network house: An overview (1999)
[71] Maurer, U., et al.: Activity recognition and monitoring using multiple sensors on different body positions. In: International Workshop on Wearable and Implantable Body Sensor Networks, BSN 2006. IEEE (2006)
[72] Liao, L.: Location-based activity recognition. Diss. University of Washington (2006)
[73] Van Kasteren, T., Krose, B.: Bayesian activity recognition in residence for elders, pp. 209–212 (2007)
[74] Zhao, L., et al.: Motif discovery and feature selection for crf-based activity recognition. In: 2010 20th International Conference on Pattern Recognition (ICPR). IEEE (2010)
[75] Poppe, R.: A survey on vision-based human action recognition. Image and Vision Computing 28(6), 976–990 (2010)
[76] Aggarwal, J.K., Ryoo, M.S.: Human activity analysis: A review. ACM Computing Surveys (CSUR) 43(3), 16 (2011)
[77] Ni, L.M., et al.: LANDMARC: indoor location sensing using active RFID. Wireless Networks 10(6), 701–710 (2004)
[78] Hightower, J., et al.: Design and calibration of the spoton ad-hoc location sensing system (August 2001) (unpublished)
[79] Orr, R.J., Abowd, G.D.: The smart floor: a mechanism for natural user identification and tracking. In: CHI 2000 Extended Abstracts on Human Factors in Computing Systems. ACM (2000)
[80] Priyantha, N.B., Chakraborty, A., et al.: The cricket location-support system. In: Proceedings of the 6th Annual International Conference on Mobile Computing and Networking. ACM (2000)

[81] Bahl, P., Padmanabhan, V.N.: RADAR: An in-building RF-based user location and tracking system. In: Proceedings of the IEEE Nineteenth Annual Joint Conference of the IEEE Computer and Communications Societies INFOCOM 2000, vol. 2, IEEE (2000)
[82] Chandola, V., Banerjee, A., Kumar, V.: Anomaly detection: A survey. ACM Computing Surveys (CSUR) 41(3), 15 (2009)
[83] Perry, M., et al.: Multimodal and ubiquitous computing systems: Supporting independent-living older users. IEEE Transactions on Information Technology in Biomedicine 8(3), 258–270 (2004)
[84] Franco, C., et al.: Behavioral telemonitoring of the elderly at home: detection of nycthemeral rhythms drifts from location data. In: 2010 IEEE 24th International Conference on Advanced Information Networking and Applications Workshops (WAINA). IEEE (2010)
[85] Aztiria, A., Augusto, J.C., Basagoiti, R., Izaguirre, A.: Accurate Temporal Relationships in Sequences of User Behaviours in Intelligent Environments. In: Augusto, J.C., Corchado, J.M., Novais, P., Analide, C. (eds.) ISAmI 2010. AISC, vol. 72, pp. 19–27. Springer, Heidelberg (2010)
[86] Mihailidis, A., et al.: The use of an intelligent prompting system for people with dementia. Interactions 14(4), 34–37 (2007)
[87] Pollack, M.E., et al.: Autominder: An intelligent cognitive orthotic system for people with memory impairment. Robotics and Autonomous Systems 44(3), 273–282 (2003)
[88] Chu, Y., et al.: When did you start doing that thing that you do? Interactive activity recognition and prompting. In: AAAI 2011 Workshop on Artificial Intelligence and Smarter Living: The Conquest of Complexity (2011)
[89] Kamar, E., Horvitz, E.: Jogger: models for context-sensitive reminding. In: Proceedings of International Conference on Autonomous Agents and Multiagent Systems, AAMAS 2011 (2011)
[90] Bosch healthcare solutions, `http://www.bosch-telehealth.com/en/us/startseite/home.html` (last access: May 30, 2013)
[91] Philips Healthcare, `http://www.philips.com/about/company/businesses/healthcare highlights` (last access: May 30, 2013)
[92] Intel Healthcare, `http://www.intel.com/content/www/us/en/healthcare-it/healthcare-overview.html` (last access: May 30, 2013)
[93] Honey well healthcare, `http://honeywell.com/Products-Services/Pages/healthcare-medical.aspx` (last access: May 30, 2013)
[94] Ambient Assisted Living Joint Programme, `http://www.aal-europe.eu/` (last access: May 27, 2013)
[95] The "AMICA" project, `http://www.aal-europe.eu/projects/amica/` (last access: 27/05/2013)
[96] The "CAPMOUSE" project, `http://www.aal-europe.eu/projects/capmouse/` (last access: May 27, 2013)
[97] The "BANK4ELDER" project, `http://www.aal-europe.eu/projects/bank4older/` (last access: May 27, 2013)
[98] The "ASSISTANT" project, `http://www.aal-europe.eu/projects/assistant/` (last access: May 27, 2013)
[99] The "E-MOSION" project, `http://www.aal-europe.eu/projects/e-mosion/` (last access: May 27, 2013)
[100] The "DOMEO" project, `http://www.aal-europe.eu/projects/domeo/` (last access: May 27, 2013)

[101] The "ALIAS" project, http://www.aal-europe.eu/projects/alias/ (last access: May 27, 2013)

[102] DOMEO robot image (Image published with permission of ROBOSOFT), http://www.gerontechnologie.net/nouveau-prototype-pour-kompai-le-robot-domestique-de-teleassistance/311996 (last access: September 23, 2013)

[103] The "ALIAS" project (Image published with permission of TUM-Technische Universität München, MMK-Institute for Human-Machine Communications), http://www.aal-europe.eu/projects/alias/ (last access: May 27, 2013)

[104] The "eCAALYX" project, http://www.aal-europe.eu/projects/ecaalyx/ (last access: May 27, 2013)

[105] The "PAMAP" project, http://www.aal-europe.eu/projects/pamap/ (last access: May 27, 2013)

[106] The "SOFTCARE" project, http://www.aal-europe.eu/projects/softcare/ (last access: May 27, 2013)

[107] The "E-STOCKINGS" project, http://www.aal-europe.eu/projects/e-stocking/ (last access: May 27, 2013)

[108] The "HEALTH@HOME" project, http://www.aal-europe.eu/projects/healthhome/ (last access: May 27, 2013)

[109] The "A2E2" project, http://www.aal-europe.eu/projects/a2e2/ (last access: May 27, 2013)

[110] The "BEDMOND" project, http://www.aal-europe.eu/projects/304/ (last access: May 27, 2013)

[111] The "HOPE" project, http://www.aal-europe.eu/projects/hope/ (last access: May 27, 2013)

[112] The "HOMEdotOLD" project, http://www.aal-europe.eu/projects/homedothold/ (last access: May 27, 2013)

[113] The "CARE@HOME" project, http://www.aal-europe.eu/projects/bank-4-elder/ (last access: May 27, 2013)

[114] Mégret, R., et al.: The IMMED project: wearable video monitoring of people with age dementia. In: Proceedings of the International Conference on Multimedia, pp. 1299–1302. ACM (2010)

[115] Di Rienzo, M., et al.: MagIC system: A new textile-based wearable device for biological signal monitoring. Applicability in daily life and clinical setting. In: 27th Annual International Conference of the Engineering in Medicine and Biology Society, IEEE-EMBS 2005, pp. 7167–7169. IEEE (2005)

[116] Paradiso, R., Loriga, G., Taccini, N.: A wearable health care system based on knitted integrated sensors. IEEE Transactions on Information Technology in Biomedicine 9(3), 337–344 (2005)

[117] Jara, A.J., Zamora, M.A., Skarmeta, A.F.: An internet of things—based personal device for diabetes therapy management in ambient assisted living (AAL). Personal and Ubiquitous Computing 15(4), 431–440 (2011)

[118] Jara, A.J., Zamora, M.A., Skarmeta, A.F.: Drug identification and interaction checker based on IoT to minimize adverse drug reactions and improve drug compliance. Personal and Ubiquitous Computing, 1–13 (2012)

[119] Rashidi, P., Cook, D.J.: Keeping the resident in the loop: Adapting the smart home to the user. IEEE Transactions on Systems, Man and Cybernetics, Part A: Systems and Humans 39(5), 949–959 (2009)

[120] Chen, D., Jie, Y., Howard, D.W.: Towards automatic analysis of social interaction patterns in a nursing home environment from video. In: Proceedings of the 6th ACM SIGMM International Workshop on Multimedia Information Retrieval. ACM (2004)

[121] Hayes, T.L., et al.: Unobtrusive assessment of activity patterns associated with mild cognitive impairment. Alzheimer's and Dementia 4(6), 395–405 (2008)
[122] Abowd, G.D., et al.: The aware home: A living laboratory for technologies for successful aging. In: Proceedings of the AAAI 2002 Workshop "Automation as Caregiver" (2002)
[123] Kientz, J.A., et al.: The georgia tech aware home. In: CHI 2008 Extended Abstracts on Human Factors in Computing Systems. ACM (2008)
[124] Elite Care Company, http://www.elitecare.com/ (last access: May 28, 2013)
[125] MIT, House_n, http://architecture.mit.edu/house_n/ (last access: May 30, 2013)
[126] Pollack, M.E., et al.: Pearl: A mobile robotic assistant for the elderly. In: AAAI Workshop on Automation as Eldercare, vol. 2002 (2002)
[127] Alwan, M., et al.: A smart and passive floor-vibration based fall detector for elderly. In: Information and Communication Technologies, ICTTA 2006, 2nd edn., vol. 1. IEEE (2006)
[128] Mukai, T., et al.: Whole-body contact manipulation using tactile information for the nursing-care assistant robot RIBA. In: 2011 IEEE/RSJ International Conference on Intelligent Robots and Systems (IROS). IEEE (2011)
[129] Watching Recall Support Robot Mamoru: http://www.jsk.t.u-tokyo.ac.jp/research/irt/mamoru.html (last access: May 30, 2013)
[130] The Paro Therapeutic Robot, http://www.parorobots.com/index.asp (last access: May 30, 2013)
[131] Riba robot image (Image provided by RIKEN-TRI Collaboration Center for Human-Interactive Robot Research), http://mentalfloss.com/article/30898/10-robots-very-specific-tasks (last access: September 23, 2013)
[132] Paro robot image Image published with permission of AIST, Japan, http://web.cecs.pdx.edu/~gerry/class/ME370/notes/cases/Case_study_Paro_robot.html (last access: September 23, 2013)
[133] Nambu, M., et al.: An algorithm for the automatic detection of health conditions. IEEE Engineering in Medicine and Biology Magazine 24(4), 38–42 (2005)
[134] Andoh, H., et al.: Home health monitoring system in the sleep. In: SICE 2003 Annual Conference, vol. 2. IEEE (2003)
[135] Masuda, Y., et al.: An unconstrained monitoring system for home rehabilitation. IEEE Engineering in Medicine and Biology Magazine 24(4), 43–47 (2005)
[136] Shi, G., et al.: Mobile human airbag system for fall protection using MEMS sensors and embedded SVM classifier. IEEE Sensors Journal 9(5), 495–503 (2009)
[137] Karamshuk, D., Boldrini, C., Conti, M., Passarella, A.: Human mobility models for opportunistic networks. IEEE Communications Magazine 49(12), 157–165 (2011)
[138] Cho, D.-K., et al.: Opportunistic medical monitoring using bluetooth P2P networks. In: 2008 International Symposium on World of Wireless, Mobile and Multimedia Networks, WoWMoM 2008. IEEE (2008)
[139] Lilien, L., et al.: Opportunistic networks: the concept and research challenges in privacy and security. In: Proc. of the WSPWN, pp. 134–147 (2006)
[140] Gyrard, A.: A machine-to-machine architecture to merge semantic sensor measurements. In: Proceedings of the 22nd International Conference on World Wide Web Companion, International World Wide Web Conferences Steering Committee (2013)
[141] Pfisterer, D., et al.: SPITFIRE: toward a semantic web of things. IEEE Communications Magazine 49(11), 40–48 (2011)

Adaptive Heuristic-Based P2P Network Connectivity and Configuration for Resource Availability

Andreas Andreou[1], Constandinos X. Mavromoustakis[1], George Mastorakis[2], Athina Bourdena[3], and Evangelos Pallis[3]

[1] University of Nicosia, 46 Makedonitissas Avenue, 1700 Nicosia, Cyprus
[2] Technological Educational Institute of Crete, Department of Business Administration, Lakonia, Agios Nikolaos, 72100, Crete, Greece
[3] Technological Educational Institute of Crete, Department of Informatics Engineering, Estavromenos, Heraklion, 71500, Crete, Greece
mavromoustakis.c@unic.ac.cy, andeou.and@st.unic.ac.cy, gmastorakis@staff.teicrete.gr, {bourdena,pallis}@pasiphae.eu

Abstract. Many schemes have been examined in the recent past in order to define a methodology for efficient resource sharing in combination with the temporal characteristics of dynamically varying topologies whereas others create a ground for a combination of temporal and spatial techniques. This chapter provides a brief introduction to, and motivation for, topology formation for P2P network connectivity and configuration from a network resource availability viewpoint. Efficient topology formation is probably the crucial parameter for enabling end-to-end reliability and data integrity between peers. The research background is presented using different approaches in the field of topology formation with respect to the parameters that affect the reliability issues of each node. Self-awareness is presented as well as the choice of mechanisms for efficient and reliable communication in a dynamically changing peer-to-peer system. Thereafter the recent adaptive heuristic-based P2P network connectivity approach for topology configuration for resource availability is presented along with some conducted results extracted from the research bibliography. Finally the chapter ends with the conclusion based on the general discussion of the covered issues in the chapter and the open ended research issues that come along with the discussed ones.

Keywords: Cloud resources, Scheduling management, APIs, Virtual Data Centres (VDC), Resource allocation, Cloud management, Virtual Engine Migration (VEM), Virtual-to-Virtual (V2V) resource migration, High Resource Availability, Integration and composition.

1 Introduction

Modern peer-to-peer systems are complex systems composed of many heterogeneous nodes, links and users. Peer-to-peer networks operate through hybrid environments where network resources like node remaining capacity and or energy or channel's link

Modeling and Optimization in Science and Technologies 3,
DOI: 10.1007/978-3-319-06704-9_10,

quality are factors which significantly degrade the end users perception in experiencing the so called Quality of Service (QoS). Additionally some spatial constraints like the application data (e.g., the spatial and temporal location of data) and the user behaviour (e.g., user request pattern) add degradation streams to the end users' offered service. Changes that occur over time constraints can be disastrous and cause severe service interruption.

In this chapter the adaptive topology for connectivity establishment is discussed and through previous researches is thoroughly explored. Basically the chapter examines three dynamic factors that trigger network adaptation with regard to the crucial factors for maintaining network connectivity. Namely the chapter addresses the dynamic factors of expending the network resources and the parameterized metrics that are associated with these factors and contribute into saving crucial resources concerning the application data and dynamic user behaviours. Furthermore the self-organization capabilities of peer-to-peer networks are examined with the auto configuration in dynamically changing topologies. Adaptive techniques are illustrated by describing representative network schemes that adapt to the existing dynamic factors in such peer-to-peer environments. For each scheme, this chapter describes the methodology that was used to adapt into these frequent changes. Finally the chapter ends with a comprehensive analysis of the existing heuristic-based adaptive techniques and the extracted result in the existing researches by using this adaptivity, for forming peer-to-peer network accuracy. Interesting open research issues are then discussed in the conclusion section.

1.1 Background

The data availability is extremely difficult to be achieved in dynamically changing topologies and the underlying systems that are supporting them. Peer-to-peer networks can become very complex particularly in terms of interconnectivity as discussed in [1, 13, 16]. Transmission characteristics of such systems can provide beneficial feedback to the reliability diversities in order to reduce significantly the packet loss rate (and increase the Successful Delivery Rate) and the associated likelyhood, the packet redundancy and untargeted replication [2, 3] and corruption of packets. Additionally the user behavior can significantly affect the parameters that characterize the reliability. An additional dynamic factor that is strictly associated with users' behavior is the application data, which refers to data generated by the application in a network such as files in a peer-to-peer network, and sampling triggered data that is collected in a sensor peer-to-peer network environment. Users' application data may dynamically change in terms of availability (which changes in time as expressed in [2, 4, 5, 12, and 16]) and in terms of amount. Thus both problems should be faced in a dynamic way, which means a crossed and close collaboration between the dynamic factors such as network topology and the dynamic and static factors which are included in the network resources. Activities of network users like sudden user behaviors should be aware of the network topology in a step of time, in order to anticipate the reliable service provision to end users. Users' behaviors may change due to the connectivity and due to their locations, as well as the service requirements and data request patterns. Availability of resources should also be aware of all the factors discussed above in order to offer QoS to end users and prevent data loss.

Related research work in the field of information diffusion with targeted and untargeted end-recipients has already been explored in recent literature. In dynamically changing topologies the minimum end-to-end response time is required otherwise the packets are dropped and the system is not efficient. Dense systems in terms of the number of users solve the problem of network partitioning in MP2P systems but according to the studies in [2, 3] the problems of overpopulation and the information redundancy during the diffusion process are degrading the performance of the system. The problem of overhearing which is addressed in [7] causes to the network to generate a significant amount of overhead data packets which in turn create bandwidth impairment. Overhearing and the side effects of the generated messages are exchanged and are further degrading the communication among peers.

Mobile Peer-to-Peer systems with their passive retransmissions can lead to another significant impairment factor of the offered bandwidth, the so-called broadcast storm problem [4]. In the broadcast storm problem nodes have significantly degraded performance and almost impaired reliability while exchanging resources. Authors in [5] propose an epidemic-like scheme which based in local interactions of the nodes according to their state. In this way nodes manage the replicated databases in a robust way in order to face possible losses and failures during the communication process. Epidemic-like algorithms are very beneficial particularly in dynamically changing Mobile Peer-to-Peer (MP2P) devices, as these devices experience high data losses due to the intermittent connectivity and their motion. A flooded with redundant information network [7] causes collisions and further reduction in performance due to additional traffic caused by queries that are generated in the system. In addition, some other approaches that were proposed, like data replicas in [8] based on the increment of accessible data. In [8] authors consider the accessibility without taking into consideration the delay caused by multiples queries. Significant improvements of the basic flooding approaches for topology formation using the "passive adverts" in a geographic region have also been studied in [9] recently. In [10] a communication scheme is being developed which allows users to enjoy high data rates whereas users may be moving with different speeds. Authors in [11] associate the capacity of each wireless device with each one of the users' motion pattern in order to provide improvement in the performance of the system. An issue that should be considered is the regional capacity measurements with respect to the variations in the resource exchange performance and efficiency.

Managing the issues above and below the transmission layer for the topology formation concept should precisely be considered. One novel approach for enabling dynamic topology formation is addressed in [12] where a reliable dissemination clustered landcast technique is used in order to overcome possible access problems when a node is no longer available or failed. The proposed scheme in [12] is utilized as location-based broadcasting methodology which taken into consideration the created clusters to enable efficient intercluster resource sharing.

This chapter closely examines network topologies and the adaptive network formations. It examines the factors that trigger the behavior of the network with respect to the network adaptation. These factors can be the available through time network resources and user behaviors (requests and sharing capability). Furthermore it introduces four basic network functions and illustrates typical adaptive techniques heuristic and non-heuristic. For each scheme, this chapter describes the algorithms used to adapt, and their subsequent benefit. This chapter ends with a comprehensive analysis of existing adaptive techniques. The rest of the chapter is organized as follows. Section 2 discusses the self-organization in dynamically changing topologies followed by the section 3 where the topological issues are discussed. Section 3.1 then discusses the topology formation using self-awareness and depicting the self-awareness through the epidemic nature. Thereafter the self-managing networks in hop-by-hop node clustering notation is presented along with the heuristic-based topology formation for resource availability using dynamically changing factors in the peer-to-peer network. The topology formation is combined with the novel heuristic approach for network adaptive formation and is presented along with results conducted by different researches. Section 4 provides the final conclusions about the existing and the discussed techniques, mentioning some future references based on the existing innovative techniques. Finally the section 4 concludes with a discussion on the basic issues of the chapter.

2 Self Organization in Dynamically Changing Topologies

Peer-to-peer networks are subject to dynamic changes. These changes are characterized by many factors like available resources referring to both node and link resources and to factors that are dynamically changing through time. Factors like the available link resources such as bandwidth may change in time based on the temporal traffic load. In addition the availability of resources cannot be guaranteed as sudden network partitioning and intermittent connectivity problems occur.

Different adaptive mechanisms are implemented within four basic network functions: hop-by-hop connectivity in the end-to-end path; routing the requested data and re-route in order to be more efficient ([20] chapter 5); efficient scheduling data transmission; and adjusting transmission rate [14]. There are many Peer-to-Peer middlewares that take into consideration these new challenges of the shared environments. In the following section some topological issues are being explored using different dimensions and perspectives for offering resource exchange reliability.

3 Topological Issues

3.1 Topology Formation Using Self-awareness and the Epidemic Nature

Epidemic information dissemination or so-called bio inspired methods can be applied in communication networks and according to the study cases [2-8] can be very beneficial. Different mathematical models were explored in the recent past like the Susceptible, Infected, Recover state models targeting the maximization in the information dissemination process. The goal in epidemic diffusion is rather to 'infect' the maximum possible nodes by diffusing (sending/forwarding) targeted information to them. This issue was in many researches strongly associated with the network topology formation. Eugster et al. [15] identify four key requirements on algorithms for epidemic network formation as follows:

- Membership – how nodes perceive neighboring nodes, or crucial requirements for the forwarding/dissemination activity.
- Network parameterized self awareness – awareness of connectivity between nodes in order to form the network topology.
- Buffering policy– denotes which information to drop when the storage buffer of a node is full.
- Cross-layer Design – as in [34-36] it refers to protocol design which has as a point of interaction different layers with different associated functionalities.

Autonomic systems which basically the peer-to-peer philosophy lays-on, is a self-aware system which uses some policies. The autonomous peer-to-peer systems should be able to analyze the changes to ensure reliability. The main advantage of having the epidemic adoption into any system -and particularly to dynamically changing topology systems like the P2P-is that these schemes are triggering on any system a positive feedback, mirroring the changes and according to some metrics, form the topology that is of need [39]. This epidemic adoption to the frequent network changes reflects any detection or reconfiguration in the case of failures as it proactively chooses the next hops when they are needed. The passive forwarding capability of a node enables the system to have additional features in the response and the behavior of the system in terms of:

- Scalability: The large number of nodes within a Peer-to-Peer network may degrade or play a positive role in the performance in terms of latency in the sharing process, reliability and the "dense recoverability" of data.
- Dynamic network formation: Network nodes are self-tuned due to the applied heuristic epidemic-based technique as depicted in [4, 25-29], and basically have a self-tuned way in informing the network topology before entering into a "crucial state".
- Self-state-knowledge–Denotes that the system may be able to adjust the behavior to different conditions (self-awareness [25-26]) and this may be applied via reinforcement learning methodologies.

In figure 1 the initial network formation is presented according to general rules of the peers. These rules are only based on the connectivity and on the self-awareness of each node (i.e. like remaining capacity or remaining resources available for the given node [39]).

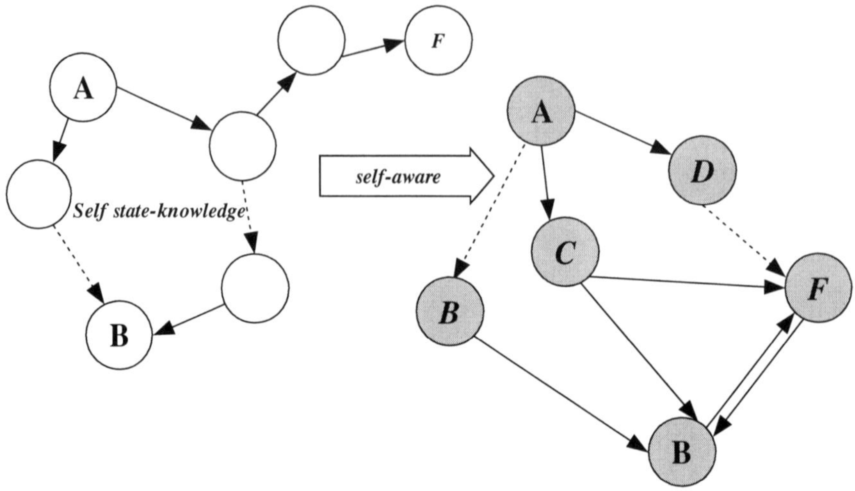

Fig. 1. The partial network formation configured in a peer-to-peer system according to node self-awareness

Each node basically is aware of the self-resources, and according to the availability, each node measures the connectivity.

In order to have this self awareness the node should make an evaluation based on the resources that are needed to cope in terms of service rate system, and thus the resources that will be needed to enable connectivity between peers (of i.e. two partitioned networks). This is usually done using a Node Duty Cycle Management which is a mechanism for self awareness that controls the transmissions of a node.

3.2 Creating the Prospects for a Relay Path or Hop-by-Hop Connectivity

Relay path or in an individual basis hop-by-hop concept is a principle of controlling the flow of data in a network where different paths can be created and at the same time an end-to-end connection is established and maintained. In this way the assembled file chunks of data are passively forwarded from node to node in a store-and-forward manner. Dynamic relay path-oriented changes in a system may cause intermittent connectivity to occur. Therefore, constructing hop-by-hop connectivity is of primary importance. Factors that directly can affect the topology and the connectivity concern basically the 1-hop communication of a certain node to others. A node obtains information regarding its neighboring nodes like capacity limitations or links' states to a neighboring node, and/or distance or delay latency to a neighboring node. Based on the acquired information, nodes evaluate their neighboring nodes (as in figure 2 where the peer 1 needs to send data to peer 4 via peer 3) according to the quality of the connectivity among them. According to [16-18] constructing hop-by-hop connectivity which adapts to changes is characterized by three dynamic factors: network resources; application data; and user behaviors. The following section introduces a representative scheme that adapt to those changes. Since in a hop-by-hop communication only a node is concerned about the next possible hop and recursively the next hop is concerned about the next one, the mechanism allows data to be forwarded even if the path between source and destination is not permanently connected during end-to-end communication.

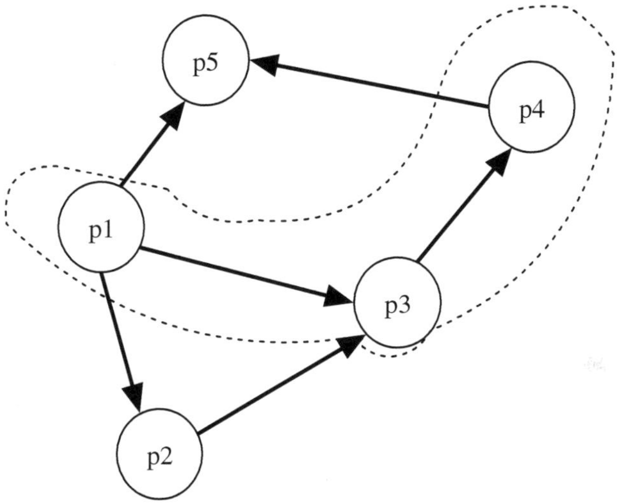

Fig. 2. Establishing a peer to peer connectivity according to node self-awareness

3.2.1 Self-Managing Networks in Hop-by-Hop Node Clustering

Self-Managing Networks should enable the dynamic adaptation to sudden changes, particularly in terms of network resources. Constructing the hop-by-hop connectivity is a scheme that establishes a connected topology for a P2P network like the well known Gianduia (Gia) in [16]. In [16] the network configuration enables a hop-by-hop connectivity for P2P file sharing by adapting to dynamic node resources such as computational power for both source and destination, available bandwidth and available storage capacity. This scheme a-priori enables the dynamic adaptation of the existing network resources into the adjustment of the critical resources of the nodes that are forming this P2P network. More specifically each node gathers the required data from its neighboring nodes. This data can be the look-up table of their available resources. Each wireless node evaluates each one of its neighboring nodes using the obtained information. Based on its preference (e.g., a node may prefer neighboring nodes with redundant resources or selfish resources [22, 23]), the node establishes the direct or indirect links (directly or via other nodes) with the neighbors and updates the topology formation table (or the routing table containing the associated links). Each node also sets the upper limitation for the number of links that a node can maintain, and this is evaluated and determined based on the available resources of the node and its neighbors. In such a way, nodes can adopt the available resources and adjust the capabilities according to the state of the node to neighboring nodes' links conditions.

More available links imply that there is more traffic and higher workload. However Gia [16] is evaluated in this way and according to the scheme it was found that it improves the efficiency and scalability of a P2P network by allowing nodes with more resources to establish and maintain more links. This adaptivity also can be done in adaptive network topology formation as shown in figure 3. Figure 3 shows the network formation using a cluster-based scheme where the topology is formed using a cluster based scheme. A node decides to become a head node, based on its self awareness (self available resources).

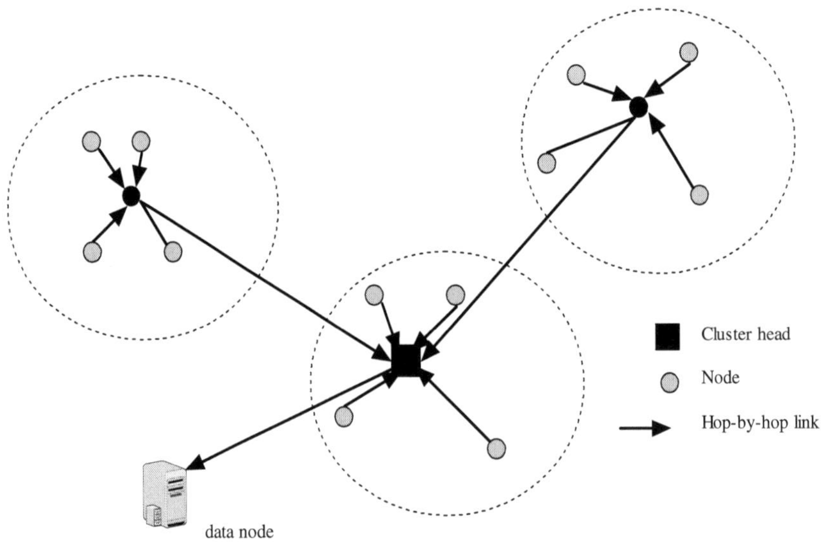

Fig. 3. Clustering nodes (Peer/Node clustering) in order to establish peer to peer connectivity

3.2.2 Heuristic Topology Formation for Resource Availability

Autonomous resource sharing scheme has essentially as a vital part the reliability issues and the resource availability at any time in the network when requested. Many schemes have been already presented like "availability" schemes where nodes make an explicit request for a resource, and the whole network is flooded with a query, as well as resource discovery algorithms [8]. However, this approach may not be valid when the link failure probability is taken into consideration. Researches in [10-11] have shown significant improvements in terms of the diffusion methodology using flooding whereas they make use of the location oriented characteristics. In [12] authors depict that there is a need to associate the regional characteristics in terms of capacity measures and the diversities in the regional capacity in a mobility framework scenarios. This as authors' initial investigation will affect the throughput of the system and the response of each one of the wireless nodes. It was recently found that a combined mathematical model with a heuristic approach can solve partially or in total many resource availability problems along with the network topology formation support.

Dynamic –in a heuristic way-network formation should enable nodes to be self-tuned and use the heuristic approach to form the network topology before entering into a “crucial state”. Thus such a scheme provides a combination of proactiveness and inactiveness approaches. Such a scheme for dynamically changing topologies is the so-called Hybrid Mobile Infostation System (HyMIS) in organized cluster-based landscapes. This hosts the methodology for targeting high reliability by using the epidemic heuristic selection of a forwarding node in order to enable resource sharing.

A host may have a unique interconnection and should have a redundant topology [21]. Capacity and topology parameters play a major role in the final state topology formation, since these parameters can affect the final state of the network and thus the topology and connectivity. All above issues should be expressed in a reliable manner. Reliability can be ensured through various ways that have already been described in the recent literature. However the new notation of using a modeled approach of the epidemic mimetic scheme provides a promising way to enrich reliability and ensure the end to end connectivity in P2P systems, as shown in the next subsection.

3.2.3 The Novel Heuristic Approach for Network Adaptive Formation

A recently propose scheme enables the reliability through self-awareness which measures the topology of the network, and then it reserves available resources for usage. The scheme makes use of the well-known epidemic approach [22-26] for fast searching and finding the available resources in a P2P network. The scheme utilizes the Hybrid Mobile Infostation System (HyMIS) which basically is of fixed parameterized resources and uses the epidemic based topology according to some states to maintain and enhance the reliability of file/ resource sharing process among dynamically changing topology nodes. There are however many storage constraints where in order to maintain this stability the epidemic backup node selection is adopted, merging the advantages of epidemic file dissemination through static or non static Infostations [23-24].

Heuristic cooperation schemes between Infostations and ordinary peers in organized topological landscapes have also been taken into account in order to enable greater reliability. Some research results presented in [22], show that the average throughput per source-destination pair, can be stabilized when the number of nodes that are moving within a vicinity is increasing through time. The new concept in [22] is to assemble the data packets into smaller pieces of each source node and diffuse them to as many nodes as possible. Taking into account all the above reasons, authors in [24] introduced the idea of Infostation [23] which was adopted but in a dynamically changing topology framework. Infostation comprises of a hybridized version of the static Infostation concept, which structures a layered (clustered) tree-based scheme using a purely mobile Infostation concept. Figure 4 shows the basic model for the mobile Infostation in four-regional, geographical cluster-based areas. There is a concept adopted of the Primary Infostation (PI) which is temporarily set by a node [24]. PI acts as a messager (message ferry) to other mobile devices that are in the same geographical area. Requested packets are forwarded by the PI’s to users based on their demands and channel’s requirements. A model adopted in [24, 25] evaluate the capacity limitations and considers whether a node can be the PI, otherwise another node (candidate PI (cPI)), becomes a PI. By using this configuration to play the role of broadband hotspots (like static PIs) this itself comprises a way to improve reliability in a centralized form. Infostations [23] consist of high bit rate connectivity and can be seen as independent access ports to the Internet or specified Intranet, like a static Gateway or a centralized access server.

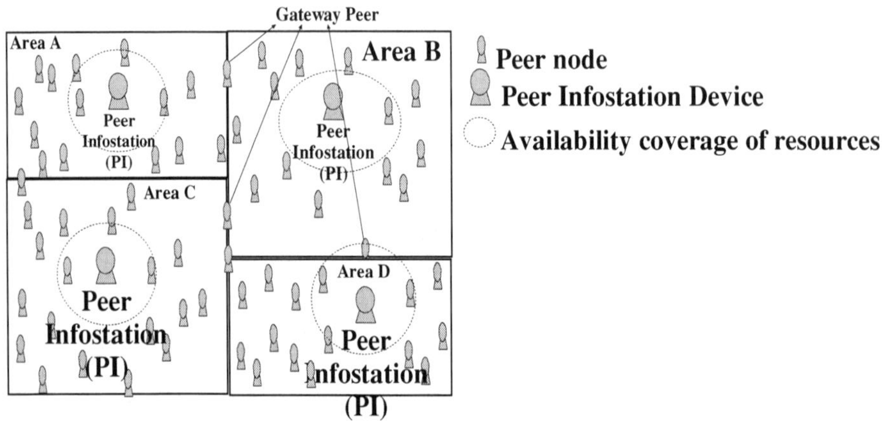

Fig. 4. Infostation concept in four-regional cluster-based areas

In addition with the above scenarios the topology can be formed also for mobile peer-to-peer nodes. As proposed in [24] the set Primary Infostation (PI) which is purely mobile [25], utilizes the Hybrid Mobile Infostation System (HyMIS) architecture. HyMIS architecture is thoroughly discussed in [25] where all the different scenarios are explored. Figure 5 shows the coverage area of each device in the Landscape, and the cPI and PI principle with the tree formation described earlier.

In [25] the storage node scenario is used. PI node plays this role by being the control storage node [26] of the clustered area. Moreover the tree formation shown in Figure 5, enables the local view of any device that shares resources with any other device in the landscape. PI is selected[1] and located in every landscape covering a certain predetermined area. A cPI mobile peer is chosen to be PI only if it has high residual energy and capacity, and moves with low velocity (or stall). This means that for a certain time distance T (until it remains as the PI of the landscape) user $U^1_{q(i)}$ will match all selection criteria until other cPI will substitute the current PI node. In Figure 2, user $U^1_{i,j,k}$ downloads simultaneously from users MI_i , MI_j, MI_k . Within the coverage of PI, some cPIs are approaching to become a PI based on the criteria previously explained.

[1] Selection is based on the available resources criteria and -if on move-the trajectory of each device.

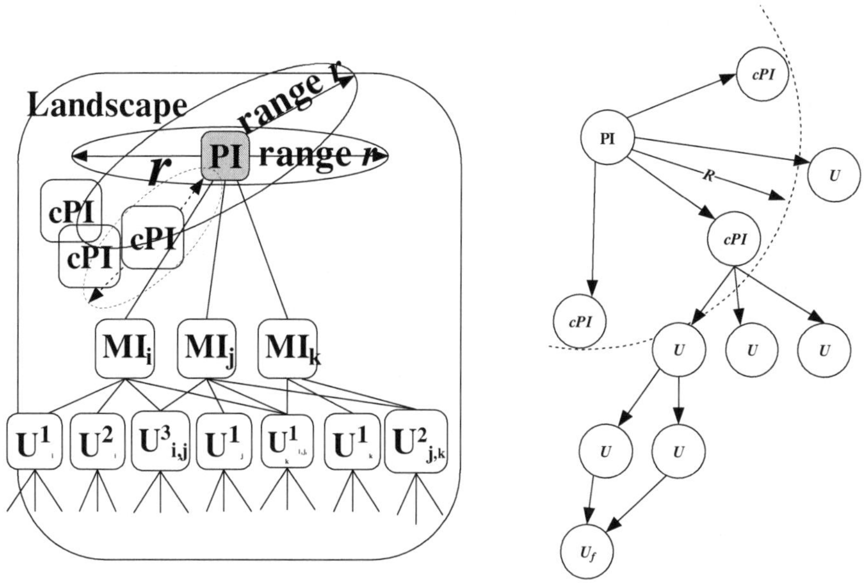

(i) (ii)

Fig. 5. (i) The hybrid peer-to-peer mobile Infostation based architecture, with cPIs and PI in a formed cluster, and (ii) the corresponding PI-cPI tree based configuration

3.2.4 Resource Availability Using the Object Sharing Metamodeling of the Epidemic Replica

In [26], [27], [28], and [29-31] there is an evidence that the utilized schemes with disseminating proactively the requested data can significantly improve the performance of the system. Considering that by nature each epidemic has duration of time t, and it is applied to different species or organisms, in this work a similar mimetic mechanism is used. In [25] an isolated system comprising of a fixed number of mobile nodes is assumed, confined in 4 predefined (clustered) geographic regions (landscapes). Researches in [30-34] considered that there is a tradeoff between reliable coverage and data rate. Roughly speaking, the limited connectivity coverage that MP2P systems offer [32], [33] [34], results a significant delay in downloading a message or file [24] (group of packets). At any time step each wireless device can be transferred to a nearby landscape and also can change the state into "die". Each mobile host m_k has a predetermined capacity M(t) at time t. At any time in the network each m_k has a state. The epidemic nature consists of three discrete time states: the susceptible state S(t) represents the number of hosts in the system which are "susceptible" and has a duration $\sigma(t)$ as shown in [35] and [36], the infected state I(t) represents the number of "infected" hosts, and R(t) represents the "recovered" hosts. The duration of S(t) is evaluated by $S(m_{M(t)},\sigma(t))$, of I(t) is evaluated as

$I(m_{M(t)}, \imath(t))$ where $\imath(t)$ is the duration of the infection of $m_{M(t)}$, and the duration of R(t) is $R(m_{M(t)}, \rho(t))$. It is also assumed that the recover time duration $\rho(t) < \imath(t)$ and $\rho(t) << \sigma(t)$.

A host is in susceptible state *S(t)* if the device does not share any information with any other host. In turn a host is in infected state *I(t)* if a file(s) sharing occurs. Finally a host is in "recovered" state *R(t)* if any shared file(s) are no longer pending file sharing processes. The average delay or the duration of the infected state can be modeled and evaluated by associating the $I(m_{M(t)}, \imath(t))$ with buffer's capability. At this point we introduce the buffer capability metric which characterizes each node's buffer capabilities with respect to capacity, neighboring recoverability and channel's access time from a source *i* to a destination *j*. The buffer's capability metric is evaluated as:

$$B_{i->j} = \overline{B}_{i->j} \cdot \xi \cdot e^{-R_C} - \overline{B}_{i->j} \cdot \xi \cdot e^{-(R_C - R_{C(k)})} \tag{1.1}$$

$$B_{i->j} = \overline{B}_{i->j} \cdot \xi \cdot e^{-R_C} (1 - e^{R_{C(k)}}) \tag{1.2}$$

where ξ (in the range of 0.7<ξ<1 as [25] depicts for high SDR) is a constant representing the recoverability of node j, $R_{C(k)}$ is the mean residual capacity of *j*'s neighbors, and $\overline{B}_{i->j}$ is the average buffer capability for all nodes. Equation (1.2) above speculates the exponential nature of capability of node's buffer associated with the recoverability and the average buffer capability for all nodes.

In the model of SIR, it is considered that with a certain probability $P_{M(t)}$ the node will be infected. This assumes that $P_{M(t)}$ is the likelihood of a node to have specified requested packet(s) or file(s) from other node(s). In turn the probability of a node that has a requested packet, to be infected equals to: $P_{\inf ected} = \frac{1}{e^{e^C}}$, where C is a fixed parameter according to the infection/recover model. The duration for a disease to reach every node that has a proper requested packet for a source node, might follow the *non-recover* model (if more than one node continuously demand a requested packet(s)) of *f* targets [15]. Thus considering the number of nodes is N, then the number of rounds that the system needs to reach and infect M nodes N>>M is $R = \frac{\log(M)}{\log(\log(M))} + O(1)$. Therefore for *f* targets that have to be reached, the number of rounds that the system needs to reach and infect M nodes is $R = \log_{f+1}(M) + \frac{1}{f}\log(M) + O(1)$. The logarithmic steps are significantly quicker to reach and infect for sharing purposes any resources among nodes. If now the equation

$P_{\inf ected} = \frac{1}{e^{e^C}}$, with C fixed parameter, is associated with buffer's capability metric, then C could be expressed as a function of the remaining capacity. Thus C could be expressed as $C = \frac{\bar{c}_i \cdot \tau}{\mathrm{T}_{precise}}$, where T is the precise time that a resource requires to be shared, and τ is the estimated time that the resource will be in sharing process[2]. For reliable transmission the $\tau << \mathrm{T}_{precise}$ must be satisfied.

Therefore the Markov chain model of an infectious disease with susceptible, infected, and recovered states can be used, as shown in Figure 6, to infect and recover each node at a time *t*. Markov chain model evaluates the storage requirements and characteristics of each node and it is used to specify and determine the file sharing termination criteria. In Figure 6 SIR model were modeled for evaluating the transition likelihood. Thus if a resource is available the node is infected with $P_{M(t)}$ at time *t*.

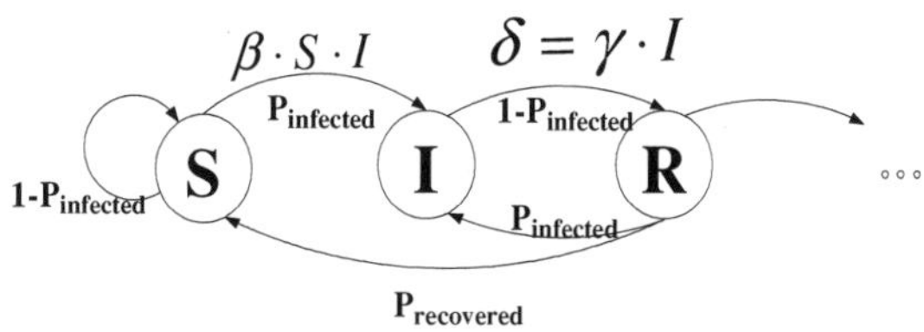

Fig. 6. Markov chain model of an infectious disease with susceptible, infected, and recovered states

Suppose there are *k* hosts in the system, then a host is sharing a resource with *β(k–1)* other hosts per unit time. Thus *S (k 1)* nodes do not have yet the disease. Therefore, the transition rate from state S to state I becomes:

$$Filesharing = (nu\_\inf ected) \cdot (dld\_Rate) \cdot (nu\_NOT\_share) \tag{2.1}$$

$$Filesharing = I[\beta(k-1)] \cdot \left[\frac{S}{k-1}\right] \tag{2.2}$$

where β is the contact rate for *k* hosts. Then the downloaded (no longer pending) rate is:

$$\delta = \gamma \cdot I \tag{2.3}$$

where γ is the download rate and I is the number of infected devices.

The $\beta \cdot S \cdot I$ is called π coefficient which indicates the enforcement degree of the diffusion process. π has the dimension of $\left[\frac{1}{Time}\right]$. Previous examinations of the

---

[2] Sharing process denotes infection.

behavior of small scale systems [17] showed that relatively small populations could be faced with a stochastic model. If a file is no longer requested for sharing as shown in Figure 6, the node becomes either infected or recovered with the certain likelihood.

Taking into account that π depends on the number of S(t) and I(t) and the probability of transmitting the information, we can derive S(t) as follows:

$$\begin{aligned} \frac{dS}{dt} &= -\beta \cdot S \cdot I = -\pi \\ \frac{dI}{dt} &= \beta \cdot S \cdot I = \beta(N-I) \cdot I = \beta NI - \beta I^2 \end{aligned} \tag{3.1}$$

By solving the first order differential equation the outcome is:

$$I(t) = \frac{N}{1 + e^{-\beta \cdot N \cdot t}(N-1)} \tag{3.2}$$

According to the definition of spreading ratio equation (3.2) becomes:

$$\begin{aligned} I'(t) &= \frac{I(t)}{N} = \frac{N}{N \cdot (1 + e^{-\beta \cdot N \cdot t}(N-1))} \\ I'(t) &= \frac{1}{1 + e^{-\beta \cdot N \cdot t}(N-1)} \end{aligned} \tag{3.3}$$

Equation 3.3 is referred as the cumulative distribution function.

An issue is when the locations will be updated. This issue can be measured as follows:

$$L(t) = L(t-1) + S_t \cdot \vec{d} \tag{4}$$

where L(t) is the new location L(t-1) the previous location at step time (t-1), S_t is the speed of each device and $\vec{d}$ is the directed unit vector [15]. Additionally the distance from a node to the closest PI can be measured as in [24] where [(Node's position - center of PI's communication area) - (radius of area)]. The average delay experienced by all the peers in downloading a multi-part file is as follows:

$$\bar{d}^{(m)} \approx \frac{\tau_0}{m} \log_2 n \tag{5}$$

where m is the number of identical sized chunks that the file is divided, n is the number of peers and τ_0 is the amount of time taken to download the whole file, if downloaded from a single peer.

Two different user-based cases where the communication might be disturbed: (i) when source user's communication fails and (ii) user's destination communication fails. In case (i) the source users might move to a point that no communication coverage exists and as a result connection failure will occur, and the prospective resource for download will be lost. Considering a file[3] download from a node A to a node B it is helpful to evaluate both cases. In (i) the user's device (mobile node) chooses in epidemic form (infection) which of user's device neighboring nodes will be MI. This classification is based on candidate's node residual energy, capacity and signal transmission power [23]. PI only communicates with MI and not with pure users. Node "A" then copies packets to the chosen "infected" MI for time t, $t > \varphi_T$, where φ_T is the time estimated for complete download. MI in turn copies these packets to PI of the landscape to which is the only user which communicates directly [15, 23]. This file lies in PI buffer for time t and then is deleted. This mechanism occurs recursively for the forthcoming chosen MIs, in the case where MI's communication fails. Finally when downloading is completed, file is being removed from PI and "infected" MI even if time t, $t > \varphi_T$, has not yet ended. Then every infected node recovers. If any change in PI occurs, then PI copies all requested packets to the new PI and to (one) best chosen cPI.

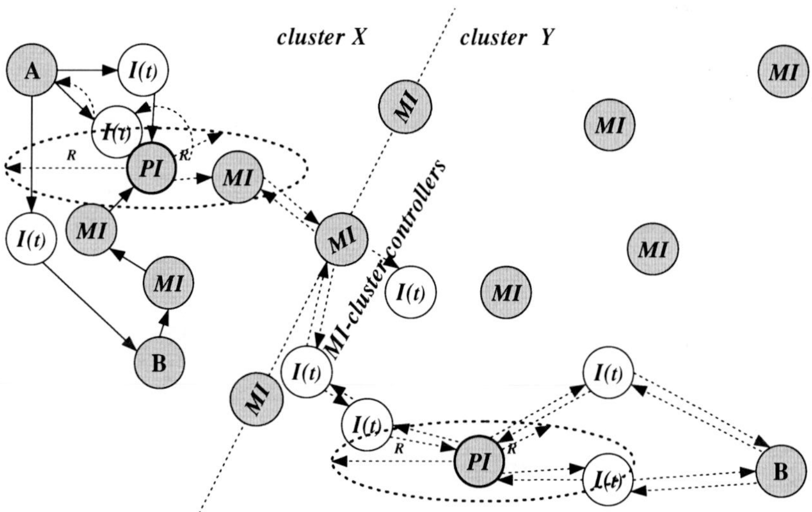

Fig. 7. Scenarios for two different user-based cases, where communication is dynamically corrupted. Directed "infections" of different dynamically changing topology users (MIs) are shown in the case for where user's destination communication fails, and user changes dynamically a cluster.

In the case of (ii) where the destination node "B" is to be moved influencing the communication between A and B, a similar mechanism is activated along with (i) as shown in Figure 7.

Some indicative results extracted are shown in Figures 8-12. In Figure 8 the HyMIS configuration enables significantly higher SDR compared with HIS. Figure 8

[3] Stream(s) of packets.

depicts that the deployment of a fully dynamic scenario in terms of capacity and resource availability, enables end-users to experience high packet delivery rates.

Due to the chained and inherent nature of peer-to-peer network architectures, a major problem in such networks is the performance optimization, and to guarantee the availability of resources, so as to ensure that all users receive their requested QoS, while the network capacity is reasonably high. Multi-hop capacity analysis requires the consideration of metrics like the CPU cycles that are needed for this transmission, and delay constraints. Static and dynamic scenarios described in [24, 25, 26], with respect to node's capacity limitations/constrains, in the presence of real time traffic and the arisen reliability issues were examined while file sharing process is in progress. The heuristic method [26] attempts to enhance file sharing reliability on demand. To demonstrate the heuristic methodology discussed exhaustive discrete time simulations were performed under several different conditions. Some of the results are shown in figures 8-12.

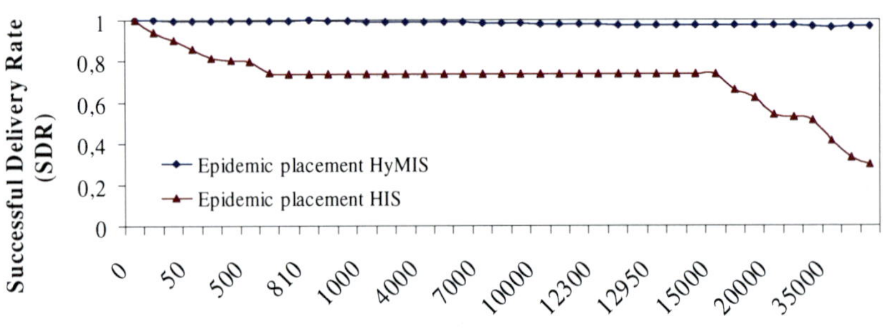

Fig. 8. The ratio of successful packet delivery comparing two different schemes [ref. 4, fig 12]

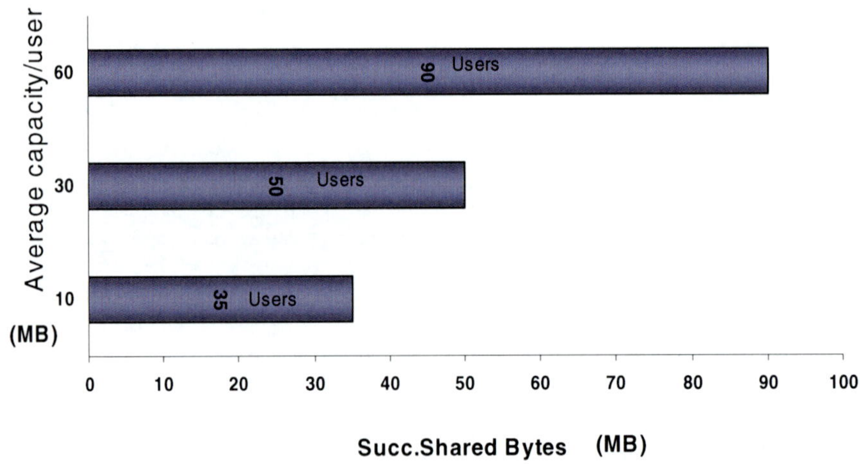

Fig. 9. Impact on population size [ref. 4, fig 18]

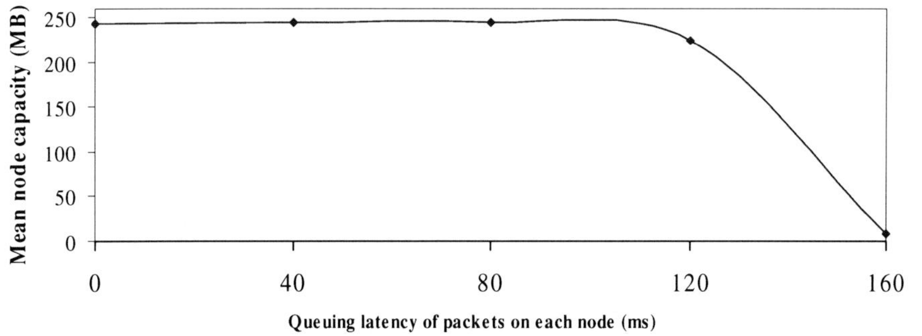

Fig. 10. Mean node capacity with the mean queuing latency of packets on each node with topology formation on a peer-to-peer basis [ref. 4, fig 10]

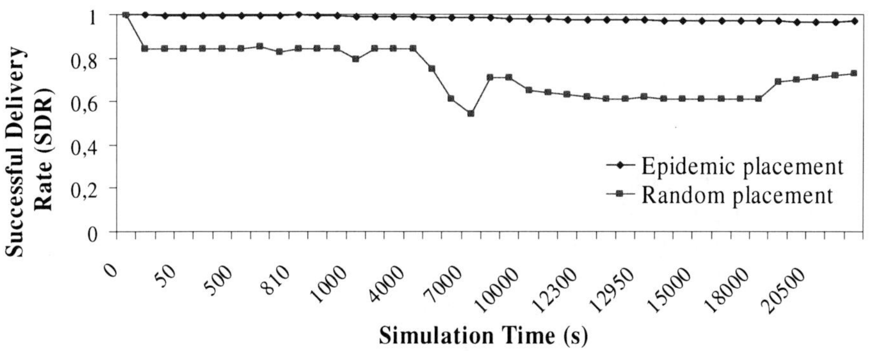

Fig. 11. The ratio of successful packet delivery [ref. 24, fig 5]

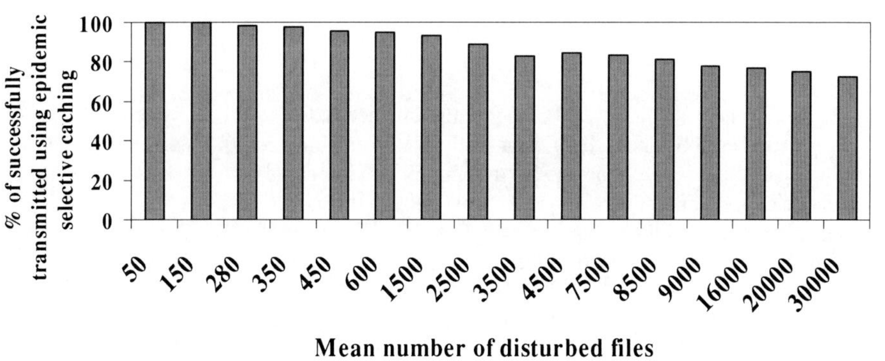

Fig. 12. Mean number of disturbed files with the number of successfully transmitted using epidemic selective caching [ref. 24, fig 10]

4 Conclusions

This chapter has provided a brief overview for the heuristic-based scheme that targets the efficient topology formation and the adaptive P2P network connectivity and configuration, from a network resource availability viewpoint. Efficient topology formation is probably the crucial parameter for enabling end-to-end reliability and integrity between peer devices. Network topology formation should have mechanisms so that when applied by nodes can dynamically alter each node functionalities and/or behavior, so that topology can then fit and be in accordance with the changing needs of its users. Recent researches in the field of topology formation have shown that there are many crucial parameters like the capacity measurements and the related parameters that affect the reliability offered by the Peer-to-Peer systems. Self-awareness is again one basic parameter of the node in order to take measures for the reactive behavior of the system. The choice of mechanisms for efficient and reliable communication in a dynamically changing peer-to-peer system, is influenced by many factors including, network topology formation and the related parameters for information diffusion, mode of transmission and the various storage constraints of mobile users.

Future directions are expected to be focused in P2P network's connectivity with respect to internet traffic. Different patterns are expected to characterize the self-similarity of the traffic dimensions. An interesting open research issue is the severe examination of the anti-entropy nodes, or a history based framework for each device in a region, and the estimated correlation in the epidemic dissemination process. Under certain circumstances and node characteristics, there may be locality to the request patterns of the peer/ nodes, as well as a correlation function with node's contribution in the diffusion process. This comprises an open research issue for dynamically changing topologies of peer-to-peer networks.

References

1. Xiao, L., Yunhao, L., Ni, L.M.: Improving unstructured peer-to-peer systems by adaptive connection establishment. IEEE Transactions on Computers 54(9), 1091–1103 (2005)
2. Sharman Networks, Worldwide distribution of licensed works using peer-to-peer technology, `http://www.kazaa.com` (retrieved October 2012)
3. Ganesan, D., Krishnamachari, B., Woo, A., Culler, D., Estrin, D., Wicker, S.: Complex behavior at scale: An experimental study of low-power wireless sensor networks. Technical Report UCLA/CSD-TR 02-0013, UCLA (2002)
4. Mavromoustakis, C.X., Karatza, H.D.: Under storage constraints of epidemic backup node selection using HyMIS architecture for data replication in mobile peer to peer networks. Journal of Systems and Software, Elsevier 81(1), 100–112 (2008)
5. Demers, A., Greene, D., Hauser, C., Irish, W., Larson, J.: Epidemic algorithms for replicated database maintenance. In: Proceedings of the Sixth Annual ACM Symposium on Principles of Distributed Computing, pp. 1–12. ACM Press (1987)

6. Kulik, J., Heinzelman, W., Balakrishnan, H.: Negotiation-based protocols for disseminating information in wireless sensor networks. Wireless Networks 8(2-3), 169–185 (2002)
7. Hong, X., Xu, K., Gerla, M.: Scalable routing protocols for mobile ad hoc networks. IEEE Network Magazine 16(4), 11–21 (2002)
8. Hara, T.: Effective Replica Allocation in Ad Hoc Networks for Improving Data Accessibility. In: Proceedings of the IEEE INFOCOM 2001 Conference, pp. 1568–1576 (2001)
9. Jivodar, B., Tchakarov, N., Vaidya, H.: Efficient Content Location in Wireless Ad Hoc Networks. In: IEEE International Conference on Mobile Data Management (MDM 2004), pp. 74–82 (2004)
10. Gavrilovich, C.D., Ware, G., Freindenrich, L.: Broadband Communication of the Highways of Tomorrow. IEEE Communications Magazine 39(4), 146–154 (2001)
11. Grossglauser, M., Tse, D.: Mobility Increases the Capacity of Ad Hoc Wireless Networks. In: Proceedings of IEEE InfoCom 2001, pp. 312–319 (2001)
12. Chungshen, X., Bo, X., Chien-Chung, S.: A novel layered graph model for topology formation and routing in dynamic spectrum access networks. In: First IEEE International Symposium on New Frontiers in Dynamic Spectrum Access Networks, DySPAN 2005, November 8-11, pp. 308–317 (2005)
13. Younis, O., Fahmy, S.: Distributed clustering in ad-hoc sensor networks: a hybrid, energy efficient approach. In: Proceedings of IEEE INFOCOM, Hong Kong, China, March 7-11, vol. 1, pp. 633–640 (2004)
14. Moore, D., Hebeler, J.: Peer-to-Peer: Building Secure, Scalable, and Manageable Networks. MCGraw Hill Book Company (2002) ISBN: 9780072192841
15. Eugster, P.T., Guerraoui, R., Kermarrec, A.-M.: Epidemic information dissemination in distributed systems. IEEE Computer 37(5), 60–67 (2004)
16. Chawathe, Y., Ratnasamy, S., Breslau, L.: Making Gnutella-like P2P systems scalable. In: Proceedings of ACM SIGCOMM, Karlsruhe, Germany, August 25-29, pp. 407–418 (2003)
17. Skraba, P., Aghajan, H., Bahai, A.: Distributed passive routing decisions in mobile adhoc networks. In: IEEE 60th Vehicular Technology Conference, VTC 2004, September 26-29, vol. 4, pp. 2814–2818 (2004)
18. Hara, T., Murakami, N., Nishio, S.: Replica allocation for correlated data items in ad hoc sensor networks. ACM SIGMOD 33(1), 38–43 (2004); Special Issue: Special section on sensor network technology & sensor data management (Part II) table of contents, ISSN:0163-5808
19. Cardei, M., Cardei, I., Du, D.-Z., Helmy, A.: Efficient Resource Discovery in Wireless AdHoc Networks: Contacts Do Help. In: Resource Management in Wireless Networking. Network Theory and Applications, vol. 16. Springer, US (2005) ISSN:1568-1696
20. Setton, E., Girod, B. (eds.): Peer-to-Peer Video Streaming, 1st edn. Springer (2007) ISBN-10: 0387741143
21. Wai-Sing, L. (ed.): Peer-to-Peer Computing Building Supercomputers with Web Technologies, 1st edn. Springer (2006) ISBN-10: 1846283817, ISBN-13: 978-1846283819
22. Garetto, M., Giaccone, P., Leonardi, E.: On the Capacity of Ad Hoc Wireless Networks Under General Node Mobility. In: 26th IEEE International Conference on Computer Communications, INFOCOM 2007, May 6-12, pp. 357–365. IEEE (2007)
23. Iacono, A.L., Rose, C.: Mine, mine, mine: Information theory, Infostation networks, and resource sharing. In: Wireless Communications and Networking Conference, WCNC IEEE, September 23-28, vol. 3, pp. 1541–1546 (2000)

24. Mavromoustakis, C., Karatza, H.: Segmented File Sharing with Recursive Epidemic Placement Policy for Reliability in Mobile Peer-to-Peer Devices. In: Proceedings of the 13th Annual Meeting of the IEEE International Symposium on Modeling, Analysis, and Simulation of Computer and Telecommunication Systems (MASCOTS), Georgia Tech, Atlanta, Georgia, September 26-29, pp. 371–380 (2005)
25. Mavromoustakis, C., Karatza, H.: On the Performance Analysis of Recursive Data Replication Scheme for File Sharing in Mobile Peer-to-Peer Devices Using the HyMIS Scheme. In: Proceedings of the 5th International Workshop on Performance Modeling, Evaluation, and Optimization of Parallel and Distributed Systems (PMEO-PDS 2006) in Conjunction with IEEE International Parallel & Distributed Processing Symposium (IPDPS), Rhodes Island, Greece, April 25-29, p. 8 (2006)
26. Mavromoustakis, C., Karatza, H.: Reliable File Sharing Scheme for Mobile Peer-to-Peer Users Using Epidemic Selective Caching. In: Proceedings of IEEE International Conference on Pervasive Services (ICPS), Santorini, Greece, pp. 169–177 (July 2005)
27. Mavromoustakis, C.X., Dimitriou, C.D.: Using Social Interactions for Opportunistic Resource Sharing using Mobility-enabled contact-oriented Replication. In: International Conference on Collaboration Technologies and Systems (CTS 2012), In Cooperation with ACM, IEEE, Internet of Things, Machine to Machine and Smart Services Applications (IoT 2012), The Westin Westminster Hotel, Denver, Colorado, USA, May 21-25 (2012)
28. Mavromoustakis, C., Karatza, H.: Performance evaluation of opportunistic resource sharing scheme using socially-oriented outsourcing in wireless devices. The Computer Journal (2013) (to appear)
29. Mavromoustakis, C., Karatza, H.: Embedded socio-oriented model for end-to-end reliable stream schedules by using collaborative outsourcing in MP2P systems. The Computer Journal 54(4), 19 (2011)
30. Newman, J.: The Movements of North Pacific Blue Whales During the Feeding Season off Southern California and their Southern Fall Migration, Santa Fe Institute 01-12-073, vol. 15(4), pp. 1246–1257 (2006) (published online: August 26, 2006)
31. Mavromoustakis, C.X.: Synchronized Cooperative Schedules for collaborative resource availability using population-based algorithm. Simulation Practice and Theory Journal, Elsevier 19(2), 762–776 (2011)
32. Mavromoustakis, C., Karatza, H.: Dispersed information diffusion with level and schema-based coordination in mobile peer to peer networks. In: Cluster Computing. Computer Communications & Networks, vol. 10(1), pp. 33–45. Springer, Netherlands (2007)
33. Kawadia, V., Kumar, P.R.: A cautionary perspective on cross-layer design. IEEE Wireless Communications 12(1), 3–11 (2005)
34. Mavromoustakis, C., Karatza, H.: Community oriented neighbouring feedback for efficient end-to-end delay-sensitive MP2P streaming via temporal relay-based replication. Journal of Algorithms & Computational Technology 5(2), 177–198 (2011)
35. Dimitriou, C., Mavromoustakis, C.X., Mastorakis, G., Pallis, E.: On the performance response of delay-bounded energy-aware bandwidth allocation scheme in wireless networks. In: Proceedings of the IEEE/ICC 2013 International Workshop on Immersive & Interactive Multimedia Communications over the Future Internet, Organized in Conjunction with IEEE International Communications Conference (ICC 2013), Budapest, Hungary, June 9-13, pp. 641–646 (2013)
36. Bourdena, A., Mastorakis, G., Pallis, E., Mavromoustakis, C., Kormentzas, G., Karditsis, E.: A Radio Resource Management Framework for Opportunistic TVWS Access. In: Proceedings of the 15th ACM International Conference on Modeling, Analysis and Simulation of Wireless and Mobile Systems (MSWiM 2012), Paphos, Cyprus, October 21-25, pp. 33–38 (2012)

Resource and Power Management in Mobile Computing Systems

Evaluation of Ubiquity Support in Mobile Computing Systems

Bruno Sousa, Kostas Pentikousis, and Marilia Curado

University of Coimbra, EICT, University of Coimbra, Palácio dos Grilos, Rua Ilha, 3000-214 Coimbra, Portugal

Abstract. Ubiquitous computing (UbiComp) systems rely on heterogeneous technologies and diverse protocols to enable the Anywhere, Anything, Anytime paradigm. Besides networking, the typical definition of a UbiComp system includes also soft- ware components and user interaction. The design of a UbiComp system is not an easy task, as architectural choices need to be made regarding technologies, applications and protocols to be employed. In this chapter we present a methodological framework for evaluating the support of ubiquity in mobile computing systems and protocols. This framework can be employed in different network architectures to design and evaluate protocols for future mobile networks, without incurring the overhead of prototyping, simulation or ethnographic studies. We illustrate the importance of protocol selection in UbiComp systems by presenting a case that com- pares ubiquity support in Mobile IPv6 (MIPv6) and Host Identity Protocol (HIP) protocols. The case study shows that without recurring to experts in the field (e.g. UbiComp systems experts) we can demonstrate that HIP supports ubiquity better than MIPv6.

1 Introduction

Ubiquitous Computing (UbiComp) systems are systems that aim to fulfill the Mark Weiser vision, in which computing systems would become embedded in everyday artifacts, and would be used to support daily activities[1]. Computing platforms have evolved in the last decades at a tremendous rate [2]. New technologies appear to maximize the realization of the Anywhere, Anything, Anytime paradigm. Besides technologies, software has evolved to run smoothly on different environments and user contexts. For instance, applications can run on a smartphone or on a laptop. Not only applications evolved, protocols have been updated to accommodate new functionalities that come closer to meeting the requirements of UbiComp systems.

UbiComp systems are characterized by a set of technical aspects, which include among others, accessibility, stability, adoptability, invisibility, and extensions, which enhance technical characteristics [3]. Moreover, UbiComp systems can be de-composed on three distinct components: (H)ardware that includes technologies and physical aspects of the computing platforms, (S)oftware that includes applications,

C.X. Mavromoustakis et al., *Resource Management in Mobile Computing Environments*, Modeling and Optimization in Science and Technologies 3,
DOI: 10.1007/978-3-319-06704-9_11,

protocols and operating systems and what we will refer to as the (U)ser component, which addresses all aspects related to the interfaces with the system and the user.

With respect to mobility management, MIPv6 [4] in the foundational Internet mobility management protocol. MIPv6 has several performance extensions to enable seamless support of heterogeneous technologies. Indeed, the Multiple Care of Address (MCoA) and Flow Bindings specifications [5] enhance MIPv6 to support multihoming, and thus are suitable for the heterogeneous and overlapping technologies of UbiComp systems [6]. Other extensions, such as Proxy Mobile IPv6 (PMIPv6) [7] or Network Mobility (NEMO) [8] enable a network-oriented mobility management approach, as mobility is managed by elements in the network- side. Fast Mobile IPv6 (FMIPv6) follows the user-oriented approach, similarly to MIPv6, but enhances the core mobility management protocol to be a pro-active protocol. Moreover, FMIPv6 can be combined with cross layer mechanisms, such as IEEE 802.21 [9] [10] standard, to enable make-before-break handover approaches. In addition, this standard enables handovers between heterogeneous and overlap- ping technologies.

The Host Identity Protocol (HIP) [11] is a protocol with different fundaments than MIPv6 and related protocols. HIP focuses on the identification of a node, which does not change across attachments to new mobile networks. This way, HIP performs a locator/identifier split approach, which has enhanced multihoming support, mobility (i.e. identity of node does not change) and security (e.g., HIP has protection against main-in-the-middle attacks) [6].

When designing UbiComp systems, diverse approaches can be followed: First, experimentation based on working systems or prototypes to test new proposals. Notwithstanding the experimentation overhead (i.e. human and physical resources) evaluating the different architectural choices is time consuming and practically infeasible [12]. Rapid-prototyping techniques reduce time and required resources, but cannot be applied, in a satisfactory form, to all the UbiComp scenarios [13]. Second, experimentation based on simulation allows to test scalability aspects, to a greater extent (e.g. high number of users or devices) but has associated implementation overheads, as well [14]. Finally, analytical models are a valuable approach, as they facilitate the comparison between different proposals, and do not have implementation or monetary costs [15] [16], but are typically limited in capturing the full range of sophistication that UbiComp systems exhibit.

We note that the choice of protocols for UbiComp systems must not be taken lightly, as choices for each component must be optimized. After all, UbiComp systems rely on extensive communication with their environment, thus protocols are a foundational block in their operation. If one pursues an experimentation-based design process, endless choices would have to be evaluated. In this chapter we review and enhance the Ubiquity Evaluation Framework (UEF) [15], by completing the description of its specification. Ubiquity Evaluation Framework (UEF) is an evaluation tool that enables the characterization of a protocol regarding its ubiquity support. With UEF it is possible to compare different protocols without requiring experts on the field, working systems or even prototypes. UEF specifies an analytical evaluation model that accommodates protocol specificities and functionalities. In particular, UEF assesses UbiComp aspects and the degree of mobility supported. While the former includes technical characteristics and extensions of UbiComp systems the latter focuses on the functionalities of protocols regarding mobility management support.

The remaining of the chapter includes a comprehensive review of the state of the art in Section 2, followed by the detailed specification of UEF and a study case in Section 3. The evaluation methodology is defined in Section 4. We present and discuss our ubiquity evaluation results in Section 5. The chapter concludes with Section 6.

2 State of the Art

This section reviews the state of the art on Ubiquitous Computing (UbiComp) Systems and considers as well, mobility management protocols that can be deployed in UbiComp systems to manage IP mobility.

2.1 Ubiquitous Computing Systems

Ubiquitous computing (UbiComp) was first introduced by Mark Weiser to spell out a vision for the future of computing technologies. According to Weiser, computing technologies would become embedded in everyday artifacts, and would be used to support daily activities, which could include both professional or leisure activities [17,18]. For such, the computing technologies would be embedded in such a way that the user would not notice their presence (also referred as invisible computing or calm technology, as referred by Mark Weiser). Indeed, the user would not need to change his/her style of life (i.e. sitting on a desk to interact with computer), as computing technologies would possess the capability to adapt their operations to provide the best user experience [18], or even to anticipate user needs [2]. Take the example of a refrigerator that orders automatically missing items, without requiring user intervention. UbiComp can be used interchangeably with terms such as pervasive computing [1].

In fact, UbiComp systems aim to be a system instantiation of the Ubiquitous Computing vision of Mark Weiser. A UbiComp system needs to include heterogeneous devices with different capabilities and functionalities to allow a many-to-many usage model, where an user uses multiple devices and several users may use the same device (e.g., wall displays, public computers). Nowadays, not only users inter- act with such systems, but the different devices of such systems are able to communicate with each other. Such kind of interactions is formally known as smart spaces or urban spaces [19] [20] in which different systems adapt their behaviour according to the actions of users.

The design of a UbiComp system is not an easy task, as architectural choices need to be made regarding technologies, applications, protocols, user interfaces to be employed, in order to enable the Anywhere, Anything, Anytime paradigm. Several concerns arise, such as security, privacy, scalability, trustiness, which need to be addressed from early conception steps [18] [21]. Moreover, not only technological aspects should drive the design of UbiComp systems. Users need to understand how the system works and to what extent security is supported, as their personal data may be involved. One such everyday example pertains to credit card transactions [22].

An efficient UbiComp system design needs to consider meticulously three distinct components: (**H**)ardware, (**S**)oftware and (**U**)ser [23,24,25]. The hardware component includes all the technologies and computing platform aspects of a UbiComp system. The software component includes all the aspects related to applications, protocols, middleware, operating systems that enable services in UbiComp systems [26]. The user component addresses all the interactions with the user to specify intelligent interfaces that do not restrict user activities (for instance, force the user to use a mousse to interact with the system), that may anticipate user behaviour [19], or simply characterize user satisfaction [20].

2.2 UbiComp Evaluation Techniques

An effective UbiComp system design, besides addressing all three components mentioned above, ought to evaluate the different architectural aspects of hardware, software and user components, in order to guarantee that the different components integrate with each other to form the UbiComp system. Different evaluation methodologies can be followed, as summarized in Table 1.

Prototyping is an evaluation methodology that provides the proof-of-concept to demonstrate the feasibility of a UbiComp system [12]. Realistic and confident results can be achieved, as all the UbiComp system components are assessed. In addition ethnographic studies (i.e. user surveys, interviews) can be conducted, for in-stance to determine user satisfaction. Nonetheless, prototyping cannot be employed in early phases of UbiComp system design and involve a log of human resources, experts and users, if user feedback is gathered, as happens in [19]. Also, prototyping requires investments both in terms of money and time.

UCAN is a Ubiquitous Computing Application Development and Evaluation Process Model [23] that allows the evaluation of ubiquitous applications (e.g., for radio frequency identification applications). The evaluation includes different stages and methods. The original idea can be evaluated by using interviews, while the prototype (pilot), tailored to specific applications, is assessed through user acceptance methods. The evaluation of UbiComp systems can benefit from rapid-prototyping techniques that alleviate monetary costs and do incur in the development time of a full prototype [13]. Nonetheless, not all the devices or UbiComp systems components can be prototyped in a timely fashion using rapid-prototyping. Moreover, if such development relies on existent technologies evaluations may be biased.

Real-world deployment of UbiComp systems, corresponds to the final step of UbiComp systems design. As such, evaluations performed on the UbiComp systems should be performed to assess the functionality of the system and user satisfaction

Table 1. UbiComp systems evaluation

Evaluation Approach	Advantages	Issues
Prototypes	Realistic and confident results; Proof-of-concept demonstrations; Include all the components.	Requires UbiComp experts; Time consuming; Cannot be employed in all the phases.
Rapid Prototyping	Proof-of-concept demonstrations; Includes all the components.	Requires UbiComp experts; Cannot be employed in all design phases; Evaluation results may be biased;
Deployment	Realistic and confident results; Proof-of-concept; includes all the components.	Requires UbiComp experts and users; Time consuming; Cannot be employed in all design phases.
Simulation	Allows to test scalability;	Requires UbiComp experts; Not fully representative; Results can be biased.
Analytical Models	Abstracted from technological aspects; Allows to isolate performance of specific components.	May not include all the components; Results can be subjective.
Ethnographic study	User feedback is important to tailor system to user needs	Requires UbiComp experts; Requires prototypes or real-world deployments to be effective.
Context based	Generic	Requires UbiComp experts; Requires prototypes or real-world deployments to be effective; No objective results.

[27,28]. Users can contribute to the system conception by presenting their expectations regarding services and evaluation considers how such expectations have been fulfilled [20]. Moreover, it can be important to assess the best configuration options of a running system.

Alternatively, simulation which is an important and widely used evaluation tool, as can be used, thus avoiding all overheads associated with actual prototyping. Moreover, simulation also has the advantage of assessing scalability aspects to a greater extent (e.g. one can explore system operation and performance with higher orders of magnitude concerning the number of devices, simulated user engagement, traffic patterns and so on). Notwithstanding, simulation results must be evaluated for consistency and correctness as certain assumptions (both obvious and hidden) may lead to inaccurate results. In other words, simulation setups strive to represent the real-world environment. Otherwise, one runs the risk of basing system design decisions on results that are not fully representative, since not all the components are addressed [14].

Analytical evaluation, on the other hand, is an interesting approach, as the analytical models facilitate the comparison between different proposals, and do not have implementation or monetary costs. A model driven development is pointed as an evaluation model that allows us to make architectural choices without having to attain full-scale technological details [16]. Moreover, such kind of evaluation automates the process of selecting optimized architectural choices. The UEF employed in this chapter follows this approach to enable an objective comparison between protocols regarding ubiquity support.

Other types of evaluation methodologies only address a specific component of UbiComp systems. For instance, approaches considering context, only address the user component, as the focus relies on the user and system interactions [29,30]. Ontonym [31] is a framework that models context based on ontologies. For instance, people are modeled by using classes with different attributes such as Name and ReligiousName. The evaluation considers three aspects: design principles, (e.g., extensibility and documentation); content (e.g., clarity and consistency) and purpose (on which domain evaluation is performed). Despite using established standards, Ontonym focuses on the context representation problem, therefore not providing objective and comparable metrics to evaluate UbiComp systems.

Ethnographic studies are commonly combined with other evaluation techniques, such as prototyping and deployment approaches. For instance, UbiComp systems can be evaluated in terms of quality, assessing the supported technical characteristics [3,32]. The assignment for each capability/item usually relies on interviews with experts in the field (e.g., with ubiquitous computing experience), giving a classification in the range of *{1 7}*. These solutions require the involvement of experts, limiting a generalized utilization of this type of methodology and may require prototypes or real-world deployments to be effective.

2.3 Mobility Management Protocols

Wireless networks in UbiComp environments are moving to all-IP architectures in line with the Anywhere, Anything, Anytime paradigm. In this section we review the fundamentals of the most salient Internet mobility management protocols standardized by the Internet Engineering Task Force (IETF), namely MIPv6 [33], the core protocol for IPv6 mobility management and HIP [34,35,11].

2.3.1 Mobile IPv6

MIPv6 [7,4] is to a large degree the archetypical mobility management protocol for IPv6 networks. Mobile nodes, through the assistance of the home agent, can maintain established communications with correspondent nodes, while moving across networks. To maintain established communications, MIPv6 specifies methods for signaling and tunnel management to assure that packets towards the mobile node are delivered no matter its location. For such, the home agent, after the mobile node has registered its location (i.e. new IPv6 address), forwards packets that were sent to the home address of the mobile node (i.e. address that the mobile node has when it is at the home network).

Several drawbacks have been identified in the initial specification of MIPv6 [6]. First, it does not support multihoming. If a node has multiple addresses when it has several interfaces, MIPv6 only supports the registration of a single address. Second, MIPv6 requires modifications on end-nodes to support mobility, which somehow can be a restriction on UbiComp systems, as each device must include support for this protocol. Third, MIPv6 has associated signaling overheads [36]. Fourth, MIPv6 is a reactive protocol, which performs mobility management only when movement is detected.

2.3.2 Mobile IPv6 Extensions

MCoA [37] and Flow Bindings [38,5] enhance MIPv6 to enable multihoming support. MCoA enables the registration of multiple addresses that a multihoming mobile node can have if connected to multiple networks simultaneously. Moreover, the Flow Bindings specification, on top of MCoA, enables the definition of policies for the applications flows. For instance, Voice over IP (VoIP) flows can be mapped to the address of an interface with lower delay, while data flows can be mapped to addresses associated with higher throughput. This constitutes an important aspect, as user preferences can be implemented easily by employing flow policies. Concerns with Quality of Service (QoS) in MIPv6 have also been studied in [39,40].

NEMO [41,8] is an extension to MIPv6 that enables network mobility. Instead of managing mobility of a single-node, a group of nodes (in the same network, formally known as a mobile network) is managed in terms of mobility. Besides the home agent, correspondent and mobiles nodes, NEMO introduces the mobile router, which is responsible to manage mobility in a transparent form to the nodes inside the group. In this perspective, NEMO does not require modifications to end- nodes to assure session survivability. Nonetheless, it assumes that is the mobile network that moves and not a single node on the network. With deployment concerns, PMIPv6 [7,42,43] extends MIPv6 to avoids any modification on each end-node. In fact, mobility is managed by the network, through a local mobility anchor (LMA) and the mobile access gateway (MAG). LMA corresponds to the point on which mobile end-nodes connect, while MAG manages mobility (performs all the signaling on behalf of end-nodes). As with the NEMO protocol, PMIPv6 has advantages in terms of deployment in UbiComp systems, as the support for mobility management is on the network side. Off-the-shelf devices do not observe session disruption when moving between wireless networks, due to PMIPv6 transparent assistance.

Hierarchical Mobile IPv6 (HMIPv6) [44,45,46] is an extension to MIPv6 that organizes the network in hierarchical levels to reduce the signaling overhead. MIPv6 considers a global area, where the home agent and mobile nodes manage mobility.

This has the disadvantage of introducing too much overhead in the home agent and on the paths between mobile nodes and home agent. This way, HMIPv6 limits the signaling to a more restricted domain, managed by the Mobility Anchor Point (MAP). It is worth to point that HMIPv6 shares the MIPv6 deployment issues, as modifications are required in all nodes.

FMIPv6 [45,47] transforms MIPv6 into a pro-active protocol to reduce packet loss and handover delay. With FMIPv6, handovers are prepared beforehand, as signaling is triggered in advance, and not as a reaction to a mobility event. Despite the improved performance in terms of handover delay, FMIPv6 requires cross-layer mechanisms to be efficient, such as IEEE 802.21 [9]. All the advantages of the protocol rely on the assumption that handovers are known in advance. For such, Layer 2 information is needed (e.g., a new wireless network is detected). Moreover, FMIPv6 has the advantage of being easily combined with other protocols to improve handover efficiency, and there are extensions to PMIPv6 to improve handover latency and packet loss [48].

2.3.3 Host Identity Protocol

The Host Identity Protocol (HIP) [35,34,11] introduces a new host identity namespace and a new host identity layer between the network and the transport layers. In addition, HIP decouples identifiers (used by transport layer protocols) from locators (used for routing purposes). In short, the transport layer sockets and the IP security associations are bound to host identifiers, which in the end are tied to IP addresses. HIP is one of the first protocols to support the locator-split approach, which is considered a key design choice for future networks [6]. Moreover, HIP specification includes cryptographic mechanisms to improve security. For instance, instead of including the addresses on each packet, a host identity tag is employed to encode the host identifier, which is based on public/private key mechanisms.

HIP, due to its intrinsic nature of being an identification protocol, manages mobility in a distinct fashion from MIPv6. Indeed, two approaches can be followed: *LOCATOR* parameter and *RendezVous* service [34]. Using the *LOCATOR* parameter approach, a HIP host can notify a correspondent peer about alternate locations (e.g., addresses) through which it is reachable. With the HIP *RendezVous* service, each HIP host publishes its host identifier with a *RendezVous* server. The *RendezVous* server maintains the mapping between the host identifiers and the current locators. Hosts aiming to communicate with a HIP node, need to contact the *RendezVous* server to get the updated location of the HIP node.

2.4 Mobility Management Evaluation

The performance of IP-based mobility management protocols has attracted significant attention from both academia and industry and is still a very active research topic. This section discusses proposals addressing mobility management protocols and their performance metrics support, which include Packet Delivery Cost (PDC), handover latency, location update cost, signaling cost, multiple interfaces, simultaneous mobility support and energy efficiency. Through the evaluation of such metrics, protocol performance comparison is possible and an intelligent choice for UbiComp systems can be performed.

UbiComp systems need to include also protocols that enable energy efficiency, the so called green computing. In wireless networks, devices and protocols need to accommodate mechanisms that minimize energy consumption without sacrificing performance (e.g., bandwidth) [49]. Paging support metric is associated with being energy efficient, as it allows mobile nodes to have periods when each radio interface can be shutdown to improve battery life and reduce signaling overhead [50,51]. Nonetheless, several evaluation studies evaluate paging support by assessing the power consumption cost, the paging delay cost in a technology-dependent way, or rely on application session rates [43,52], which does not allow to isolate the energy-efficiency of a protocol. For instance, to simply determine if the functionality of the protocol is disrupted when a radio interface is in sleep mode.

The Packet Delivery Cost (PDC) metric determines the cost (e.g., processing or transmission) of the different packet delivery mechanisms (e.g., tunnel, direct forwarding) [53], and is commonly associated to MIPv6, NEMO and PMIPv6 protocols [54,55]. For instance, this metric quantifies the overhead that occurs when delivering packets, as extra information on packets or signaling packets needs to be exchanged to allow delivery.

The handover delay metric assesses how efficient is a mobility management protocol regarding the way mobility is managed. Low handover delay indicates that all the procedures to manage mobility, such as movement detection, address configuration, security operations and location registration are performed in a timely fashion, that is reducing the impact in applications. Indeed, this metric is considered in almost all studies devoted to mobility management performance evaluation [56,57,58,59]. Different approaches are followed to measure this metric. For instance simulation and analytical models focus on scalability and comparison between different protocols (MIPv6 vs PMIPv6), while experimentation in testbeds allows to characterize the performance of protocols in a working scenario, but normally resorting to a small number of devices.

The signaling cost is a compound metric that combines the packet delivery cost with handover cost, commonly designated as location update cost [53,60]. This metric quantifies all signaling performed, and not only what is necessary during handovers. The location update cost is determined according to the network model (e.g., number of hops, number of domains, wired and wireless links), message rate and respective message length. The difference between existing proposals resides on the fact that some of them include the functions of each involved entity (e.g., home agent, correspondent node), while others only include the mobile node or only the cost of specific operations (e.g., tunneling) [61].

In mobile computing systems, no assumptions should be taken regarding nodes that communicate with mobile nodes. Indeed, all nodes can be mobile, as it is common nowadays. Just imagine two users in different cities having a phone conversation while they are going to work. As such, the support of simultaneous mobility metrics cannot be neglected, as happens often in evaluations, by assuming fixed correspondent nodes [53,61]. Evaluations should at least consider the probability of simultaneous movement [62].

3 Framework

This section is devoted to the framework to assess the ubiquity support of a protocol. We revisit our Ubiquity Evaluation Framework (UEF) [15], explaining in de- tail each of the evaluation steps and clarifying related concepts. Indeed, UEF aims to work as the foundation to compare protocols feasible to be deployed on UbiComp systems. As such, UEF can be employed by network architects that aim to design/evaluate/ compare protocols for mobile networks. UEF considers all the aspects of UbiComp systems (Accessibility, Authentication, etc) and the particular purpose of the protocol, namely mobility management. The following subsections provide the detailed specification of UEF.

3.1 Objectives and General Aspects

UEF [15] was proposed to provide an objective evaluation methodology that can be used at any stage of the development of a UbiComp system, without requiring experts on the field (i.e. not using interview methods). UEF establishes the base for protocol comparison regarding ubiquity support, without requiring any prototype or working system to assess the ubiquity support.

Definition 1. Ubiquity is the ability to support secure and optimized mobility to enable access to services according to the Anywhere, Anything, Anytime paradigm and with acceptable quality levels.

UEF considers ubiquity as stated in Definition 1. This definition combines aspects of UbiComp systems with the functionalities of a protocol. UbiComp aspects are divided into two major groups: technical features that include mandatory features of UbiComp systems, such as security; and supported extensions, which can enhance the performance of UbiComp systems. For instance, security mechanisms can include authentication as well as authorization schemes. Protocol functionality includes the degree of mobility supported by a protocol for a UbiComp system. Such degree of mobility support assesses how efficient the protocol mobility management procedures are. For instance, if the protocol introduces signaling overhead, if includes procedures to detect movement of devices in a systematic form, to enable the Anywhere, Anything, Anytime paradigm in UbiComp systems [63]. The considerations to formulate UEF are summarized in Table 2.

Table 2. UEF Consideration

Type	Description	Units	Values	Examples
$\Delta(t)$	time variables	milliseconds (ms)		Handover, processing delay, idle intervals
$S(x)$	size variables	bytes		size of data structures
$C(x)$	capacity Variables	byte/s	constant	
$P(x)$	ratio		[0, 1]	procedure finalization and preparation rates

3.2 General Formulation

Ubiquity, as per Definition 1, combines lC-technical capabilities and lU -extensions of UbiComp systems, according to the Ψ-degree of mobility supported by a protocol. Equation 3 formulates Ubiquity - UMH , where w_{lC} and w_{lU} are the weights for technical characteristics and extensions aspects, respectively. Weights assignment satisfy the following constraint: $w_{lC} + w_{lU} \leq 1$. A simple rule to assign weights can be based on the number of technical or extension items and the overall number of items (technical + extensions), as Equation 1 and Equation 2 show for technical and extensions items respectively.

$$w_{lC} = \frac{nLC}{nLC + nLU} \tag{1}$$

$$w_{lU} = \frac{nLU}{nLC + nLU} \tag{2}$$

$$U_{MH} = (w_{lC} \,.\, lC + w_{lU} \,.\, lU).\Psi \tag{3}$$

Technical capabilities and extensions are determined for each component of a UbiComp system, namely user (U), software (S) and hardware or computing platform (H), as shown in Table 3 and Table 4, respectively. MIPv6 and HIP, as proto- cols enabling mobility management, fall in the software component (S). Thus, their capabilities and extensions are evaluated by considering only the capabilities and extensions of the software component. A Table entry marked with "√" means that the respective capability is supported; "0" means that it is not supported; and " " means that the capability is not applicable.

Table 3. Technical Capabilities of UbiComp Systems in UEF for (S)-Software, (U)- User and (H) - Hardware components (in a total of 39)

Technical Capability	(H)ardware	(U)ser	(S)oftware	MIPv6	HIP
Accessibility	√	–	√	√	√
Accuracy	√	√	√	√	√
Adaptability	√	–	√	√	√
Adjustability	√	√	√	√	√
Adoptability	–	√	–	–	–
Analyzability	√	√	√	√	√
Compatibility	√	√	√	√	√
Configurability	√	–	√	√	√
Connectivity	√	–	√	√	√
Credibility	–	√	–	–	–
Customizability	√	–	√	√	√
Decomposability	√	–	√	0	0
Downloadable	–	–	√	√	√
Embeddedness	√	–	√	√	√
Effectiveness	√	–	√	√	√
Efficiency	√	–	√	√	√
Extensibility	√	–	√	√	√
Integrability	√	–	√	√	√
Interoperability	√	–	√	√	√
Interpretability	–		–	–	–
Invisibility	√	–	–	–	–
Learnability	–	–	√	0	0
Maintainability	√	–	√	√	√
Mobility	√		√	√	√
Portability	√	–	√	0	0
Predictability	√	–	√	√	√
Proactiveness	–	–	√	0	0
Reconfigurability	√	–	√	0	0
Reliability	√	–	√	√	√
Reusability	√	–	√	√	√
Scalability	√	–	√	√	√
Security	√	–	√	√	√
Sensibility	√	√	√	√	√
Shareability	√	–	√	0	0
Stability	√	–	√	√	√
Testability	–	–	√	√	√
Understandability	–	√	–	–	–
Usability	–	√	–	–	–
Wearability	√	–	–	–	–
Total	**30**	**11**	**32**	**26**	**26**

Table 4. Extensions of UbiComp systems in UEF for (S) - Software, (U) - User and (H) - Hardware components (in a total of 22)

Extensions	(H)ardware	(U)ser	(S)oftware	MIPv6	HIP
Authentication	–	–	√	–	√
Authorization	–	–	√	–	√
Automation	√	–	√	√	√
Autonomy	√	–	√	√	√
Context Reusability	–	–	√	–	–
Durability	√	–	–	–	–
Entity Tracking	–	–	√	–	√
Identity Tracking	–	–	√	–	√
Inferred Context	–	–	√	–	–
Location Tracking	–	–	√	–	√
Negotiation	–	–	√	–	–
Response Time	√	–	√	√	√
Seamlessness	√	–	√	√	–
Self-Control	–	√	–	–	–
Service Coverage	√	–	√	√	
Standardization	√	–	√	√	√
Trust	–	√	–	–	–
User Context	–	–	√	–	–
User preference	–	–	√	–	
User profile	–	–	√	–	–
User Satisfaction	–	√	–	–	–
Utility	–	√	–	–	–
Total:	**7**	**4**	**17**	**6**	**9**

Capabilities and extensions are evaluated using a Boolean scale (0-not supported and 1-fully/partially supported). Moreover, to avoid ambiguity in the evaluation, UEF employs the meaning of each capability/extension according to standard dictionaries [64,65,66,67,68]. Finally, UEF considers non-overlapping capabilities and extensions as opposed to [3,32] that evaluates an item twice, namely as a capability and as an extension, which increases complexity in the evaluation process. Each capability/extension is determined according to Equation 4, as specified by Kwon and Kim [3], where *n* is the number of supported capacities/extensions, C_i is the value of the capacity (0 or 1) with *MaxScale* = 1, and n_ξ is the number of capacities/extensions that apply to the component.

$$C_{\xi} = \frac{\sum_i^n = 1\, C_i}{n_\xi . MaxScale} \quad with\ \xi \in \{U, S, H\} \tag{4}$$

This simplistic evaluation of technical capabilities and extensions in UEF promotes comparison between different proposals for UbiComp systems. Besides including all the components of such systems, UEF only requires a general knowledge of a specific solution. For instance, between the choice of WiFi or Bluetooth technologies, the general knowledge includes coverage, transmission power, data transfer rater characteristics, among others.

UEF assesses mobility support through the degree of mobility Ψ , which is a compound metric of performance and cost aspects of the mobility management process, as per Equation 5. Costs represent processes that introduce overhead, such as signaling. Earlier approaches [53,61] consider cost aspects only. Performance aspects include the level of energy efficiency $E f$ and handover procedure preparation rate $\lambda prep$. Cost aspects include handover $H c$ and signaling Sc costs, as well as the handover procedure finalization rate $\lambda fina$. The term $N \cdot maxS$ corresponds to the number of cost aspects and the maximum cost value, respectively, with $maxS = 1$, and $N = 3$. βm is used to distinguish performance and cost aspects, as employed in additive von Neumann Morgenstern utility functions [69]

$$\Psi = N.maxS + \beta_m(E_f + \lambda_{prep}) - (1 - \beta_m).(H_c + S_c + \lambda_{fina}) \quad (5)$$

In IP mobility management evaluation, UEF includes metrics for energy efficiency and the procedure preparation rate, an improvement over previous work that only evaluates mobility management performance by assessing costs. UEF explores the end-host mobility approach, when all procedures are triggered by the mobile node, and includes support for simultaneous mobility events. In the latter case, the correspondent node plays a dual role as it is also a mobile node. The procedure rates include λ_{prep} -procedure preparation rate before the handover and λ_{fina} -procedure finalization rate after handover. Considering a total of n_{proc} procedures and $n_{proc} = n_{prep} + n_{fina}$, Equation 6 and Equation 7 formulate λ_{prep} and λ_{fina}, respectively.

$$\lambda_{prep} = \frac{n_{prep}}{n_{proc}} \quad (6)$$

$$\lambda_{fina} = \frac{n_{fina}}{n_{proc}} \quad (7)$$

Hc, the handover cost, quantifies cost in terms of handover delay, d, measuring the sum of procedure delays in the n_e entities. Handover delay is determined according to Equation 8, with n_{je} procedures executed at entity e with Δt_{proc} processing time.

$$d = \sum_{e=1}^{n_e} \sum_{j=0}^{n_{je}} \Delta t_{proc\ j,e} \quad (8)$$

The handover cost, as per Equation 9, includes the cost of procedures invoked only after handover. A sigmoid function normalizes delay values that have increased granularity by a factor of d_g (=1000 by default). Handover cost could consider other metrics, such as handover delay at Layer 2 [70], but this would tie UEF to a specific radio access technology and prevent the framework from apportioning the performance of the assessed protocol. Moreover, this does not restrict the evaluation on a specific phase of UbiComp systems.

$$Hc = \frac{1}{1 + e^{-\frac{\sqrt{d+1}}{d_g}}} \tag{9}$$

The signaling cost, as per Equation 10, determines the procedure overhead of the protocol, for all the procedures that are employed for signaling [53]. Signaling cost is considered for the mechanisms of a protocol, which correspond to a set/group of procedures *Gp*. For instance, in MIPv6, registration of addresses includes several messages such as Binding Update (BU), Binding Acknowledgment (BA) and Binding Refresh Request (BRR).

$$Sc = \left[1 + e^{-\sqrt{\frac{\sum Cp}{max(Cp)}}}\right]^{-1} \quad \forall_p \in Gp \tag{10}$$

The relation between the sum of all procedures $\sum Cp$ and the maximum cost *max(Cp)* of all the procedures is the base for the signaling cost formulation. In UEF, the cost of a procedure *Cp* is formulated according to the message size, the message transmission frequency or the number of transmissions, and the processing cost ϕ of each entity. Most approaches rely only on the message size [53]. Equation 11 determines the cost of a procedure invoked *nI* times, with message size *Li* and transmitted *nTx* times or at a frequency *Qi*.

$$Cp = \sum_{n=0}^{nI}\left[\sum_{t=1}^{nTx}\sum_{i=1}^{nM}\left(L_{n,t,i} \times Q_{n,t,i} \times \sum_{e \in \{X\}} \phi_{n,t,i,e}\right)\right] \tag{11}$$

For the number *nTx* and frequency *Qi* of transmissions, we make the following assumptions:

- *Qi* =1, if *nTx* > 1, i.e. when there are retransmissions;
- *nTx* = 0, if *Qi* >1, for instance, messages that do not require any reliability but are sent frequently (e.g., router advertisements).

Equation 11, by including the *nTx* number of transmissions or *Qi* frequency, aims to have a broader applicability, since protocols can perform signaling based on the number of transmissions or simply by sending messages after a certain interval (e.g., heartbeat messages of Stream Control Transport Protocol (SCTP) [71]).

$P\phi_e$-processing cost of an entity e is the relation between the Pc_e-operation cost of a procedure and Nif_e-number of interfaces of entity, as depicted in Equation 12.

$$P\phi_e = Nif_e \times Pc_e \tag{12}$$

UEF considers multihomed nodes and does not rely on upper-layer parameters (e.g., session rate) to determine the processing cost. Instead, *Pc* corresponds to the relation between the processing delay *pDelay* and the operation complexity, as per

Equation 13. Complexity is modeled by the number of operations *nOper*, and the size of data structures *sizeData*. Whilst the size of the data structures can be dynamic, UEF only considers the size of a single record, for simplicity. When procedures do not involve data structures, *sizeData* = 0. *sizeData* differs from message length, since it accounts for the size of data structures necessary to perform the operations in procedures (e.g., record in a routing table).

$$Pc = [nOper \times (1 + sizeData)] \times pDelay \tag{13}$$

Energy efficiency, *Ef*, considers the rates of reducing the active area $\lambda_{rdActArea}$ and the paging cost λ_{rdPagC}, as per Equation 14. Paging cost includes all the signaling mechanisms to enable paging. *N* is the number of cost aspects and *maxS* the maximum value of these costs, with *N* = 1 and *maxS* = 1. Power saving mechanisms, at the physical layer, are not included in order to meet the technology independence requirement, as well as the possibility to perform evaluation isolated, this is with- out requiring a specific technology to evaluate a protocol.

$$Ef = N \times maxS + \beta_e \times \lambda_{rdPagC} - (1 - \beta_e) \times \lambda_{rdActArea} \tag{14}$$

The rate of active area reduction, $\lambda_{rdActArea}$, is the relation between the domain *dArea* and the paging *pArea*, as formulated in Equation 15.

$$\lambda_{rdActArea} = \frac{dArea - pArea}{dArea} \tag{15}$$

The paging area is determined by considering the node that initiates paging till the endpoint (e.g., mobile node). Additionally, the area can consider the radius cover- age (in meters), or simply the number of hops between the paging initiator and the endpoint [43]. The domain area is limited by the prefix management entity, for instance an IPv6 router, and the endpoints. Values close to 1 indicate that the paging area is too small, with reduced costs, but with few optimizations. The paging cost, *PagC* is given in Equation 16, where *L* represents the message size, transmitted *nTx* times. Each entity e participating in the paging group *Ga* has *Nif* interfaces in idle state during Δt interval, and for each paging message the processing cost is *Pc*. The paging group *Ga* includes all entities involved in paging signaling.

$$PagC = \sum_{t=1}^{nTx} L_t \times \sum_{\forall_e \in Ga} (Nif_{e,t} \times Pc_{e,t} \times \Delta t_{e,t}) \tag{16}$$

The processing cost, *Pc*, is determined according to Equation 13 with *sizeData = 0*. The ratio of paging cost reduction, λ_{rdPagC},, depicted in Equation 17, is the relation between paging cost at effective idle intervals, Δt_idle, and theoretical intervals, Δt_Tidle, during which the a mobile end-node could remain in idle state (e.g., no data transfer and no mobility management signaling exchanges).

$$\lambda_{rdPagC} = \frac{PagC_{\Delta t\_idle}}{PagC_{\Delta t\_Tidle}} \tag{17}$$

The following subsection formulates ubiquity support using UEF for MIPv6 and HIP protocols.

3.3 UEF Study Case: Ubiquity in MIPv6 and HIP

In order to clarify the usability of UEF, this section specifies MIPv6 [33] and HIP [11] ubiquity support. These protocols have been chosen as MIPv6 is the main management mobility protocol for IPv6 networks, and HIP is a protocol that supports Locator/Identifier split paradigm, an important aspect in multihoming support and future Internet architectures [6]. HIP supports mobility management with the RendezVous extension. The ubiquity support is determined according to its technical capabilities and extensions.

The analysis, herein performed is based solely on MIPv6 [33] and HIP [11] specifications. No expert on mobility management or in UbiComp systems was inquired. Technical capabilities of MIPv6 and HIP are similar as presented in Table 3, $lC_{MIP} = lC_{HIP} = {}^{26}/_{32} = 0.81$. Extensions are different between MIPv6 and HIP, as summarized in Table 4 where $lU_{MIP} = {}^{6}/_{17} = 0.35$, and $lU_{HIP} = {}^{9}/_{17} = 0.53$. According to UEF, HIP seems to be better suited for UbiComp systems regarding the technical and extensions capabilities. Nonetheless, the mobility management functionality cannot be ignored. Therefore, the assessment on how the degree of mobility is supported in MIPv6 and HIP needs to be determined. The following paragraphs illustrate the study case for MIPv6 and HIP to assess degree of mobility support and consequently, ubiquity.

The formulation of the degree of mobility includes diverse procedures:

- **Registration** (RG) - register new location information;
- **Security** (AA) - protect and secure identity of mobile node;
- **Address Configuration** (AD) - configure addresses in new networks;
- **Movement Detection** (MD) - detect availability of new networks.

The different procedures employ specific messages to convey signaling for the operations being supported, as summarized in Table 5, for MIPv6 and HIP protocols.

Each procedure is specified according to Equation 4. Mobility management in MIPv6 includes the Mobile Node (MN), Home Agent (HA) and Correspondent Node (CN) entities, as illustrated in Figure 1. In HIP, the HIP Initiator (HI), the RendezVous Server (RVS) and the HIP Responder (HR) manage mobility, as depicted in Figure 2. As some procedures rely on IPv6 mechanisms, *E1* denotes the MN or the HI, while *E2* stands for the HA or the RVS. Finally the *E3* represents CN or HR entities.

Table 5. Messages and procedures of mobility management

Procedure	MIPv6	HIP
Registration	BU, BA, BRR	I1, R1, I2, R2
Security	HoTI, HoT, CoTI, CoT	Included in Registration
Address Configuration	Unsolicited NS, unsolicited NA, RS, RA, DAD mechanism	
Movement Detection	solicited NS, solicited NA, RS, RA, DAD mechanism	
Tunnelling	Header information in IPv6 packets	

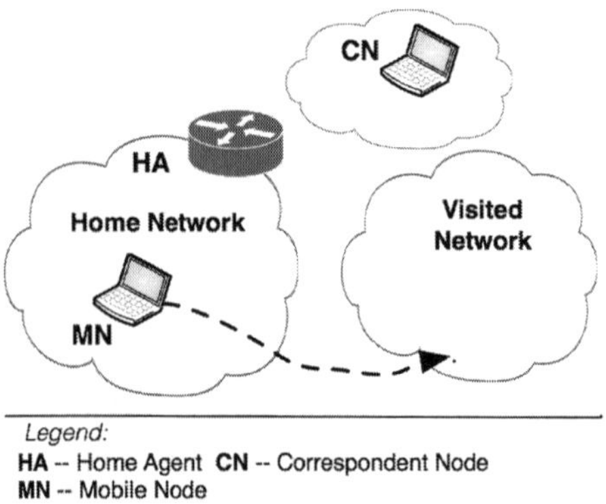

Fig. 1. MIPv6 Mobility scenario

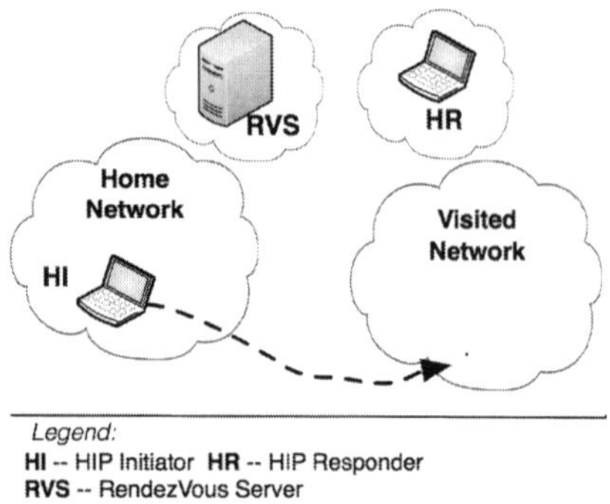

Fig. 2. HIP Mobility scenario

3.3.1 Registration

Mobile IPv6 (MIPv6) registration is based on binding messages. MN sends BUs to the HA and CN when new addresses are available to trigger the registration process. Binding Acknowledgment is transmitted to acknowledge the reception of a BU and the status of treatment initiated by the BU. Moreover, binding can be refreshed using BRR, or the CN can inform the MN about errors using Binding Error (BE) message. The cost is determined in the same fashion as HIP, Equation 18, but with Binding Update, Binding Acknowledgment, Binding Error and Binding Refresh Request messages.

HIP registration is performed in three steps (*s1*, *s2*, *s3*) according to the handover phase. In the s1 step, HI and HR register with RVS using *I1*, *R1*, *I2* and *R2* messages. The cost of this step is determined by Equation 18. Such messages are exchanged between HI and RVS nodes and between HI and HR nodes, to perform registration in RendezVous Server and HIP Responder nodes. The base exchange (step *s2*) corresponds to a four-way handshake between HIP Initiator and HIP Responder and only involves the RendezVous Server to forward I1 messages.

$$\begin{aligned} C_{RG-HIP_{s1,r2}} = & \sum_{t=1}^{nTx} L_{I1,t} \cdot \sum_{e \in \{HI,RVS,HR\}} (Nif_{e,t} \cdot Pc_{e,t}) \\ & + L_{R1,t} \cdot \sum_{e \in \{HI,RVS,HR\}} (Nif_e \cdot Pc_e) \\ & + L_{I2} \cdot \sum_{e \in \{HI,RVS,HR\}} (Nif_e \cdot Pc_e) \\ & + L_{R2} \cdot \sum_{e \in \{HI,RVS,HR\}} (Nif_e \cdot Pc_e) \end{aligned} \tag{18}$$

After the handover (step *s3*), HI needs to update the locator information on *dest* nodes, which include RVS and HR. For such purpose, it employs the update message with locator information, and issues an *echo_request*, which status of update is reported in the *echo_response* message. The registration cost of the update is determined according to Equation 19.

$$\begin{aligned} C_{RG-HIP_{s3}} = & \sum_{t=1}^{nTx} L_{UPD(locator),t} \cdot \sum_{e \in \{HI,dest\}} (Nif_{e,t} \cdot Pc_{e,t}) \\ & + L_{UPD(echo\_request)} \cdot \sum_{e \in \{dest,HI\}} (Nif_e \cdot Pc_e) \\ & + L_{UPD(echo\_resp)} \cdot \sum_{e \in \{HI,dest\}} (Nif_e \cdot Pc_e) \end{aligned} \tag{19}$$

3.3.2 Security

MIPv6 can rely on external mechanisms, such as IP Security [72], to enable higher levels of security. Nevertheless, this study focuses on the return routability, since it is an internal procedure of MIPv6 that allows the verification of addresses when the MN is at visited networks. Equation 20 formulates the cost of this procedure relying on the Home of Test Init (HoTI), Care of Test Init (CoTI) and respective reply messages, namely Home of Test (HoT) and Care of Test (CoT).

$$\begin{aligned} C_{AA-MIP} = & \sum_{t=1}^{nTx} L_{HoTi,t} \cdot \sum_{e \in \{MN,HA,CN\}} (Nif_{e,t} \cdot Pc_{e,t}) \\ & + \sum_{t=1}^{nTx} L_{CoTi,t} \cdot (Nif_{MN,t} \cdot Pc_{MN,t} + Nif_{CN,t} \cdot Pc_{CN,t}) \\ & + L_{HoT} \cdot \sum_{e \in \{MN,HA,CN\}} (Nif_e \cdot Pc_e) \\ & + L_{CoT} \cdot (Nif_{MN} \cdot Pc_{MN} + Nif_{CN} \cdot Pc_{CN}) \end{aligned} \tag{20}$$

Integrity protection and encryption is performed in HIP by employing the Encapsulating Security Payload (ESP). The registration cost already includes the security cost C_{AA-HIP}, as ESP security association is part of the base exchange. Regarding security, HIP establishes a distinction comparatively with MIPv6, as security is included in the registration process.

3.3.3 Address Configuration

Address configuration in MIPv6 and HIP nodes relies on IPv6 schemes that include Router Solicitation (RS), Router Advertisement (RA) and the messages in the Duplicate Address Detection (DAD) mechanism. Neighbour Solicitation (NS) messages are sent to multicast addresses with the reply of Neighbour Acknowledgment (NA) messages. In addition, IPv6 routers (at home and foreign networks, Rtr_h and Rtr_f, respectively) advertise prefixes via RA frequently, while Router Solicitation messages are retransmitted on error events. Equation 21 defines the cost of address configuration.

$$\begin{aligned} C_{AD} = & \sum_{t=1}^{nTx} L_{NS,t} \cdot \sum_{e \in \{E1,E2,E3\}} \left(Nif_{e,t} \cdot Pc_{e,t}\right) + L_{NA} \cdot \sum_{e \in \{E1,E2,E3\}} \left(Nif_e \cdot Pc_e\right) \\ & + \sum_{t=1}^{nTx} L_{RS,t} \cdot \sum_{e \in \{E1,E2,E3\}} \left(Nif_{e,t} \cdot Pc_{e.t}\right) \\ & + Q_{RAhome} \cdot L_{RAhome} \cdot \sum_{e \in \{E1,Rtrh,E2,E3\}} \left(Nif_e \cdot Pc_e\right) \\ & + Q_{RAforeign} \cdot L_{RAforeign} \cdot \sum_{e \in \{E1,Rtrf,E2,E3\}} \left(Nif_e \cdot Pc_e\right) \end{aligned} \tag{21}$$

The DAD mechanism is employed to assure the uniqueness of a configured address, since, on IPv6 networks there can be configured based on the advertised prefixes in RA messages.

3.3.4 Movement Detection

Movement detection also relies in IPv6 schemes, namely the Neigbour Unreachability Detection (NUD) mechanism. NUD uses solicited Neighbour Solicitation and Neighbour Acknowledgment messages and the respective cost is formulated according to Equation 22.

$$\begin{aligned} C_{MD} = & \sum_{t=1}^{nTx} L_{NS,t} \cdot \sum_{e \in \{E1,E2,E3\}} \left(Nif_{e,t} \cdot Pc_{e,t}\right) \\ & + L_{NA} \cdot \sum_{e \in \{E1,E2,E3\}} \left(Nif_e \cdot Pc_e\right) \end{aligned} \tag{22}$$

The NUD mechanism enables a node to determine the reachability of a router. For instance, when a router does not respond to Neighbour Solicitation, it means that the node is moving to another network.

3.3.5 Tunneling

Finally, MIPv6 includes the tunnel cost, since packets can be forwarded to MNs at visited networks via tunnels. The cost of tunnel establishment is determined in an application independent fashion, as tunneling relies on IPv6 encapsulation mechanisms. The tunnel establishment cost, as per Equation 23, considers only the size of message headers and respective processing cost in MN, HA and CN.

$$C_{TU} = \sum_{t=1}^{nTx} HdrT_{MN,t} \cdot (Nif_{MN,t} \cdot Pc_{MN,t}) + \sum_{t=1}^{nTx} HdrT_{HA,t} \cdot (Nif_{HA,t} \cdot Pc_{HA,t}) + \sum_{t=1}^{nTx} HdrT_{CN,t} \cdot (Nif_{CN,t} \cdot Pc_{CN,t}) \tag{23}$$

4 Evaluation

This section details the methodology used when applying UEF in an evaluation study. After deriving all the formulation for MIPv6 and HIP protocols, an evaluation is performed to assess the impact that the number of handovers and communication nodes can have on such protocols, and consequently on the performance of UbiComp systems employing these protocols.

4.1 Objectives

The goals for this evaluation are two-fold:

1. Determine the ubiquity support of MIPv6 and HIP, assessing which protocol is best suited for UbiComp systems. Consider the same underlying technologies and applications.
2. Determine the impact that different configurations, such as number of handovers and number of correspondent nodes have on ubiquity.

The second evaluation goal is associated with scalability, in terms of the velocities achieved and the number of simultaneous communications that can be supported.

4.2 Evaluation Scenario

Figure 3 illustrates the evaluation scenario used for this case study. The nodes (e.g., MN, HI, CN, HR) are configured with three interfaces, a common configuration in mobile terminals, if considering the example of a laptop with WiFi, Bluetooth and 3G capable interfaces. Moreover, Mobile Node/HIP Initiator can communicate simultaneously with several correspondent nodes or HIP responders, which can be located in different networks.

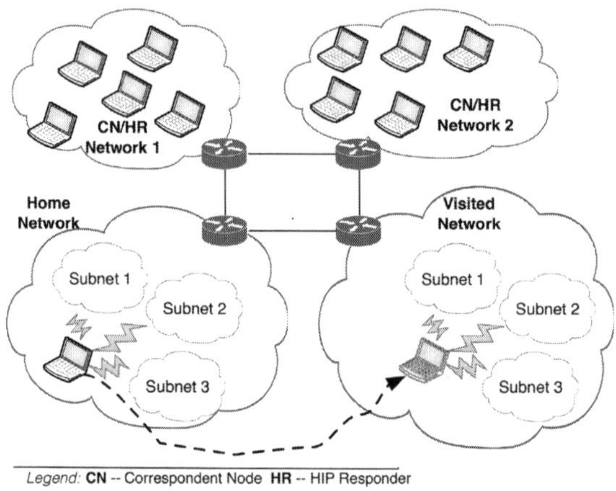

Fig. 3. Evaluation Scenario

4.3 Evaluation Parameters

UEF is applied in a study where the choice of a protocol to manage mobility in UbiComp system is considered at the design phase. MIPv6 and HIP protocols are assumed to be operating with the maximum message size (e.g., with all options filled). In addition, only the mandatory messages are considered; optional messages, such as HIP - NOTIFY, are not included.

The nodes with three interfaces, *nif* = {3} can communicate with *ncns* = {1, 5, 10} other nodes. In addition, nodes move with different speeds, thus having to handle a number of handovers *nho* = {10, 50, 100, 200}. All sessions last for 300s. Assuming we are in the design phase of a UbiComp system, values of processing delay cannot be measured (as no prototype is available). Thus for this study case, we take all processing times to follow normal and exponential distributions, with different means $x = \{1s, 10s\}$ and rate $\lambda = 1$. Different distributions are used to accommodate different modeling mechanisms for processing times.

The analytical evaluation has been performed using the R framework [73] and considering that both MIPv6 and HIP do not include energy efficiency mechanisms *Ef* = 0, as no paging schemes are incorporated. We stress that such kind of evaluation enables ubiquity support determination, promoting the choice of most performant protocols regarding ubiquity support.

We use ubiquity weights $w_{IC} = 0.65$ and $w_{IC} = 0.35$ according to the number of items in technical and extensions categories, as summarized in Table 3 and Table 4. The degree of mobility weight is equal to $w_m = 0.5$, as no energy efficiency mechanisms are considered and thus the degree of mobility relies mainly on the cost. Values higher than 0.5 tend to neglect the impact of cost in mobility support.

5 Results

The results reported in this section are based on 100 runs to improve statistical significance, and are reported considering the number of handovers, for MIPv6 and HIP protocols. Ubiquity, as a compound metric is discussed in first place. The remaining metrics, such as the degree of mobility supported, handover cost and signaling overhead are also discussed.

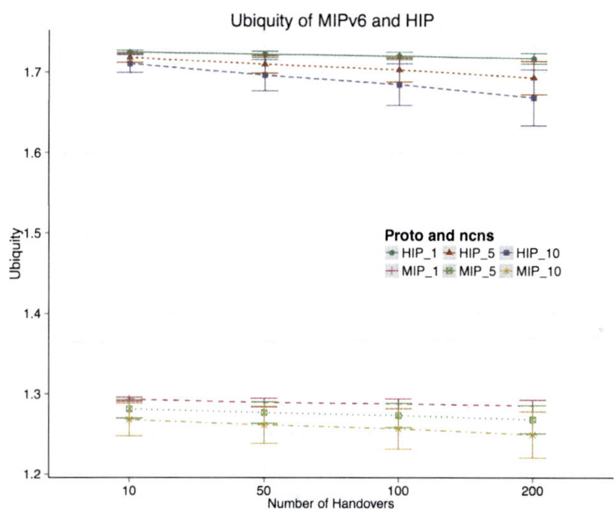

Fig. 4. Ubiquity Support in MIPv6 and HIP protocols

UEF can assess ubiquity taking into consideration protocol functionalities in different conditions that UbiComp systems can face. Under mobile scenarios, the number of handovers impacts the performance of mobility management protocols, as more signaling is required.

Figure 4 depicts the ubiquity support of MIPv6 and HIP protocols for different number of handovers. HIP supports ubiquity to a greater extent, as values rely ≈ 1.7 when compared to MIPv6 that has values ≈ 1.3. Ubiquity results consider technical capabilities and extensions, as well as the degree of mobility supported (recall Equation 3). As previously determined, the technical capabilities of MIPv6 and HIP are equal, $\mathrm{IC}_{\mathrm{MIP}} = \mathrm{IC}_{\mathrm{HIP}} = {}^{26}/_{32} = 0.81$, but the number of supported extensions is different, with HIP as a protocol more prone to be employed in UbiComp systems, $\mathrm{IU}_{\mathrm{MIP}} = {}^{6}/_{17} = 0.35$, and $\mathrm{IU}_{\mathrm{HIP}} = {}^{9}/_{17} = 0.53$.

The number of handovers and the number of correspondent nodes have an impact in the ubiquity support as explained bellow. Figure 5 depicts the degree of mobility for both protocols. With higher number of handovers, all procedures required to handle mobility are triggered often, introducing degradation in the performance, as signaling overhead increases. With the increased number of correspondent nodes, ubiquity support and the degree of mobility support are lower since updates need to be forwarded to more nodes.

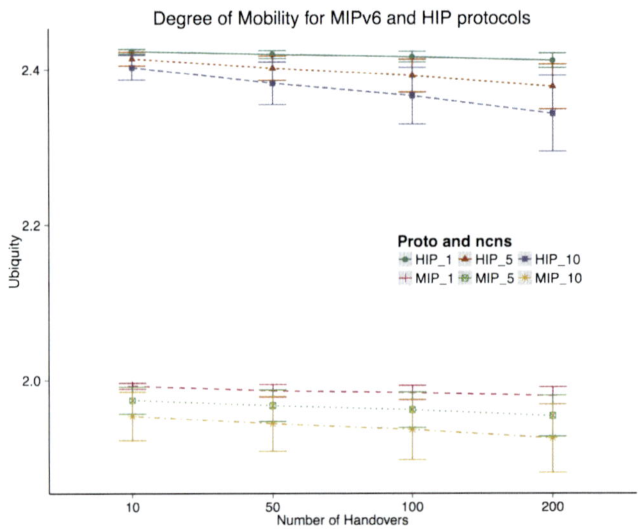

Fig. 5. Degree of Mobility support in MIPv6 and HIP protocols

The handover cost, depicted in Figure 6, is determined according to Equation 8 and accounts the processing delay in both protocols. As per the evaluation methodology, this time has not been measured, but was assumed to follow an exponential distribution, which means that the processing delay is the same in both protocols. Nonetheless, it can be observed that the handover cost increases linearly with the number of handovers and with the number of correspondent nodes.

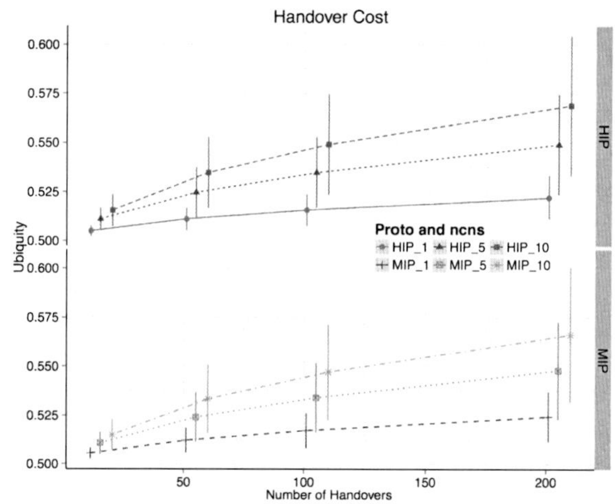

Fig. 6. Handover Cost of MIPv6 and HIP protocols

The signaling cost, depicted in Figure 7, is determined according to Equation 11 and accounts for the overhead of a protocol. HIP is more impacted with the number of handovers, since signaling cost of HIP increases linearly. For instance, with 200 handovers it is above 0.60 while MIPv6 is $\approx$ 0.58. In contrast, MIPv6 is more impacted with the number of correspondent nodes than with the number of handovers. For instance, signaling cost does not increase as in HIP with the number of handovers.

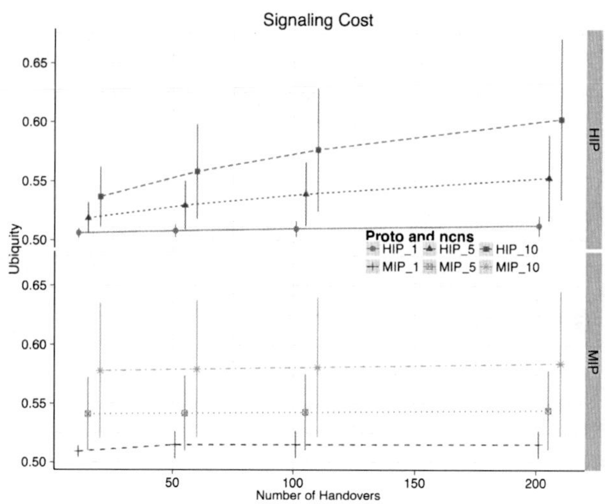

Fig. 7. Signaling Cost of MIPv6 and HIP protocols

The difference between MIPv6 and HIP relies on the registration and security procedures, as movement detection and address configuration share the same DAD and NUD mechanisms. First, the four-way handshake nature of HIP leads to more message exchange. Finally the messages exchanged have higher sizes, when compared to MIPv6. HIP signaling overhead results UEF are inline with similar approaches dedicated to the evaluation of HIP and MIPv6 protocols [74].

Such results allow to determine that HIP supports ubiquity to a greater extent in comparison to MIPv6. For a UbiComp system, HIP is preferable in comparison to MIPv6, mainly due to the locator/identifier split paradigm. Moreover, such results are inline with related evaluations, that point HIP as a protocol with stronger security mechanisms [75].

6 Conclusions

This chapter presented a framework that assesses ubiquity support in protocols which can be employed in UbiComp systems. We showed that Ubiquity Evaluation Framework can be used in any phase of UbiComp systems development and presented a study case, using Mobile IPv6 (MIPv6) and Host Identity Protocol (HIP) as example of protocols that can be evaluated in this context. Our results demonstrate the capacity of UEF in assessing ubiquity support. Indeed, the analytical study was based on the generic and technical characteristics of each protocol, which can be easily compiled from the respective specifications. Of course, the framework can take advantage of input from experts on UbiComp systems and mobility management fields to enhance the accuracy of results, but this is not a hard requirement for UEF. A key design advantage of UEF is that it is not tied to a specific protocol, technology or scenario.

The results of the use case evaluation with UEF highlights HIP as providing an better/extended ubiquity support. Notwithstanding, such enhanced ubiquity support is associated with deployment issues in existing IP networks, as it implements a locator/identifier split paradigm and changes are required in end-nodes and on the network side.

Acknowledgments. This work has been supported by CMU- PT/RNQ/0015/2009, TRONE A Trustworthy and Resilient Operations in a Network Environment and iCIS project (CENTRO- 07-ST24-FEDER-002003), co-financed by QREN, in the scope of the Mais Centro Program.

References

[1] Krumm, J.: Ubiquitous Advertising: The Killer Application for the 21st Century. Pervasive Computing 10, 66–73 (2011)

[2] Symonds, J., Khosrow-Pour, M.: Ubiquitous and Pervasive Computing: Concepts, Methodologies, Tools, and Applications. Information Science Reference-Imprint of IGI Publishing, Chicago (2009)

[3] Kwon, O., Kim, J.: A Multi-layered Assessment Model for Evaluating the Level of Ubiquitous Computing Services. In: Ubiquitous Intelligence and Computing, NA, pp. 1059–1068 (2006)

[4] Ali, I., et al.: Network-based Mobility Management in the Evolved 3GPP Core Network. Communications Magazine 47(2), 58–66 (2009)

[5] De La Oliva, A., Bernardos, C., Calderon, M., Melia, T., Zuniga, J.: IP flow mobility: smart traffic offload for future wireless networks. Communications Magazine 49(10), 124–132 (2011)

[6] Sousa, B., Pentikousis, K., Curado, M.: Multihoming Management for Future Networks. Mobile Networks and Applications (MONET) 16, 505–517 (2011)

[7] Kong, K.-S., Lee, W., Han, Y.-H., Shin, M.-K., You, H.: Mobility Management for All-IP Mobile Networks: Mobile IPv6 vs. Proxy Mobile IPv6. Wireless Communications 15(2), 36–45 (2008)

[8] Kuntz, R.: Deploying Reliable IPv6 Temporary Networks Thanks to NEMO Basic Support and Multiple Care-of Addresses Registration. In: SAINT, p. 46 (2007)

[9] IEEE, IEEE Standard for Local and Metropolitan Area Networks - Part 21: Media Independent Handover Services (January 2009)

[10] Sargento, S., et al.: Mobility Management Architecture for WiMAX Networks. In: WiMAX Networks, pp. 251–287. Springer (2010)

[11] Gurtov, A.: Host Identity Protocol (HIP): Towards the Secure Mobile Internet. Wiley (2008)

[12] Connelly, K., et al.: Evaluating Pervasive and Ubiquitous Systems. Pervasive Computing 7(3), 85–88 (2008)

[13] Fitton, D., et al.: Rapid prototyping and user-centered design of interactive display-based systems. Pervasive Cpmputing 4(4), 58–66 (2005)

[14] O'Neill, E., Lewis, D., Conlan, O.: A simulation-based approach to highly iterative prototyping of ubiquitous computing systems. In: SIMUTOOLS, vol. 56, pp. 1–10 (2009)

[15] Sousa, B., Pentikousis, K., Curado, M.: UEF: Ubiquity Evaluation Framework. In: Masip-Bruin, X., Verchere, D., Tsaoussidis, V., Yannuzzi, M. (eds.) WWIC 2011. LNCS, vol. 6649, pp. 92–103. Springer, Heidelberg (2011)

[16] Giner, P., Torres, V., Pelechano, V.: isbd2007-06: Building ubiquitous business process following an mdd approach. Latin America Transactions 6(4), 347–354 (2008)

[17] Weiser, M.: The Computer for the 21st Century. Scientific American 265(3) (1991)

[18] Krumm, J.: Ubiquitous Computing Fundamentals. Chapman & Hall/CRC (2009)

[19] Gilman, E., Davidyuk, O., Su, X., Riekki, J.: Towards interactive smart spaces. Journal of Ambient Intelligence and Smart Environments 5(1), 5–22 (2013)

[20] Kukka, H., et al.: This is not classified: everyday information seeking and encountering in smart urban spaces. Personal and Ubiquitous Computing 17(1), 1617–4909 (2013)

[21] Beckwith, R.: Designing for Ubiquity: The Perception of Privacy. Pervasive Computing 2(2), 40–46 (2003)

[22] Hong, J.I.: Minimizing security risks in Ubicomp systems. Computer 38(12), 118–119 (2005)

[23] Resatsch, F.: Ubiquitous Computing Developing and Evaluating Near Field Communication Applications. Gabler (2010)

[24] Kumar, S.: Challenges for Ubiquitous Computing. In: ICNS 2009, pp. 526–535 (2009)

[25] Mostefaoui, S.K., Mostefaoui, S.K., Maamar, Z., Giaglis, G.M.: Advances in Ubiquitous Computing: Future Paradigms and Directions. IGI Publishing (2008)

[26] Raychoudhury, V., Cao, J., Kumar, M., Zhang, D.: Middleware for pervasive computing: A survey. Pervasive and Mobile Computing 9, 177–200 (2013)

[27] Fox, A., Davies, N., de Lara, E., Spasojevic, M., Griswold, W.: Real-World Ubicomp Deployments: Lessons Learned. Pervasive Computing 5(3), 21–23 (2006)

[28] Storz, O., et al.: Public Ubiquitous Computing Systems: Lessons from the e-Campus Display Deployments. Pervasive Computing 5(3), 40–47 (2006)

[29] Rehman, K., Stajano, F., Coulouris, G.: An Architecture for Interactive Context-Aware Applications. Pervasive Computing 6(1), 73–80 (2007)

[30] Marquardt, N., Greenberg, S.: Informing the Design of Proxemic Interactions. Pervasive Computing 11(2), 14–23 (2012)

[31] Stevenson, G., Knox, S., Dobson, S., Nixon, P.: Ontonym: A Collection of Upper Ontologies for Developing Pervasive Systems. In: Proceedings of the 1st Workshop on Context, Information and Ontologies, pp. 1–8 (2009)

[32] Scholtz, J., Consolvo, S.: Toward a framework for evaluating ubiquitous computing applications. Pervasive Computing 3(2), 82–88 (2004)

[33] Johnson, D., Perkins, C., Arkko, J.: Mobility Support in IPv6, ETF RFC: 6275 (July 2011)

[34] Gurtov, A., Komu, M., Moskowitz, R.: Host Identity Protocol. The Internet Protocol Journal 12(1) (March 2009)

[35] Moskowitz, R., Nikander, P.: Host Identity Protocol (HIP) Architecture, IETF RFC: 4423 (May 2006)

[36] Langar, R., Bouabdallah, N., Boutaba, R., Sericola, B.: Proposal and analysis of adaptive mobility management in IP-based mobile networks. Trans. Wireless. Comm. 8(7), 3608–3619 (2009)

[37] Wakikawa, R. (ed.): RFC 5648 - Multiple Care-of Addresses Registration, IETF RFC:5648 (October 2009)

[38] Tsirtsis, G., Soliman, H., Montavont, N., Giaretta, G., Kuladinith, K.: Flow Bindings in Mobile

[39] Alnasouri, E., Mitschele-Thiel, A., Diab, A.: Handling Mobility Management and QoS Aspects in All-IP Networks. Mob. Netw. Appl. 16(1), 83–95 (2011)

[40] Bouabdallah, R.N., Boutaba, R., Sericola, B.: Proposal and analysis of adaptive mobility management in ip-based mobile networks. IEEE Transactions on Wireless Communications 8(7), 3608–3619 (2009)

[41] Devarapalli, V., Wakikawa, R., Petrescu, A., Thubert, P.: Network Mobility (NEMO) Basic Support Protocol, IETF RFC: 3963 (January 2005)

[42] Gundavelli, S., Leung, K., Devarapalli, V., Chowdhury, K., Patil, B.: Proxy Mobile IPv6, IETF RFC: 5213 (August 2008)

[43] Lee, J.-H., Chung, T.-M., Pack, S., Gundavelli, S.: Shall we apply paging technologies to proxy mobile IPv6? In: Proceedings of the 3rd International Workshop on Mobility in the Evolving Internet Architecture, pp. 37–42 (2008)

[44] Soliman, H., Castelluccia, C., ElMalki, K., Bellier, L.: Hierarchical Mobile IPv6 (HMIPv6) Mobility Management, IETF RFC: 5380 (October 2008)

[45] Al-Surmi, I., Othman, M., Ali, B.M.: Review: Mobility management for IP-based next generation mobile networks: Review, challenge and perspective. J. Netw. Comput. Appl. 35(1), 295–315 (2012)

[46] Upadhyay, P.C., Tiwari, S.: Distributed and Fixed Mobility Management Strategy for IP-based Mobile Networks

[47] Koodli, R.: Mobile IPv6 Fast Handovers, IETF RFC: 5568 (July 2009)

[48] Yokota, H., Chowdhury, K., Koodli, R., Xia, B.P., Frank: Fast Handovers for Proxy Mobile IPv6, IETF RFC: 5949 (September 2010)

[49] Newaz, S.H.S., Akbar, R.U., Jun, K.C., Gyu, M.L., Crespi, N.: Efficient IP Mobility Management for Green Optical and Wireless Converged Access Networks. In: Vehicular Technology Conference (VTC Fall), pp. 1–5 (2012)

[50] Taehwan, C., Laeyoung, K., Jeongeun, N., Jooseok, S.: Combinatorial mobile IP: a new efficient mobility management using minimized paging and local registration in mobile IP environments. Wirel. Netw. 10(3), 311–321 (2004)

[51] Haitao, T., et al.: Paging Issues and Methods for Multiaccess. In: Communications and Networking in China, Chinacom, pp. 769–776 (2007)

[52] Hung, T.D., Yoshikuni, O.: A comparison of different paging mechanisms for mobile IP. Wirel. Netw. 13(3), 379–395 (2007)

[53] Qi, W., Abu-Rgheff, M.A.: Signalling analysis of cost-efficient mobility support by integrating mobile IP and SIP in all IP wireless networks. International Journal of Communication Systems 2, 225–247 (2006)

[54] Sang-Jo, Y., Seung-Joon, C., Dong, S.: Analysis of Fast Handover Mechanisms for Hierarchical Mobile IPv6 Network Mobility. Wireless Personal Communications 48(2), 215–238 (2009)

[55] Jong-Hyouk, L., Ernst, T., Tai-Myoung, C.: Cost analysis of IP mobility management protocols for consumer mobile devices. IEEE Transactions on Consumer Electronics 56(2), 1010–1017 (2010)

[56] Ki-Sik, K., Wonjun, L., Youn-Hee, H., Myung-Ki, S., Heung, R.Y.: Mobility management for all-IP mobile networks: mobile IPv6 vs. proxy mobile IPv6. Wireless Communications 15(2), 36–45 (2008)

[57] Jong-Hyouk, L., Tai-Myoung, C., Gundavelli, S.: A comparative signaling cost analysis of Hierarchical Mobile IPv6 and Proxy Mobile IPv6. In: Personal, Indoor and Mobile Radio Communications, pp. 1–6 (2008)

[58] Ilkwon, C., Koji, O., Tae, W.K., Choong, S.H.: Performance analysis of IP mobility with multiple care-of addresses in heterogeneous wireless networks. Wireless Networks, 1–12 (2013)

[59] Nie, G., Qing, X.H.: Comparison and Handover Performance Evaluation of the Macro-mobility Protocol. Journal of Networks 8(1) (2013)

[60] Lin-Li, M.: Research on Mobile IPV6 Technology and Handover Performance Optimization. In: Yang, G. (ed.) Proceedings of the ICCEAE2012. CCIS, vol. 181, pp. 941–946. Springer, Heidelberg (2013)

[61] Makaya, C., Pierre, S.: An Analytical Framework for Performance Evaluation of IPv6-Based mobility Management Protocols. Transactions on Wireless Communications 7(3), 972–983 (2008)

[62] Daniel Wong, K., Dutta, A., Schulzrinne, H., Young, K.: Simultaneous mobility: analytical framework, theorems and solutions. Wireless Communications and Mobile Computing 7(5), 623–642 (2007)

[63] Sarigiannidis, M.D.L., Zournatzis, P., Ws, S.K., Panagiotis, G., Demetropoulos, I.: Towards realization of the ABC vision: A comparative survey of Access Network Selection. In: ISCC, pp. 472–477 (2011)

[64] IEEE, IEEE Standard Computer Dictionary: A Compilation of IEEE Standard Computer Glossaries 610 (1990)

[65] European Union, InterArchive Terminology for Europe, IATE (2013), `http://iate.europa.eu/`

[66] International Electrotechnical Commission, International Electrotechnical Commission Glossary, `http://std.iec.ch/glossary`

[67] International Electrotechnical Commission, Electropedia: The World's Online Electrotechnical Vocabulary, http://www.electropedia.org/
[68] Oxford University Press, Oxford Dictionaries (2013), http://oxforddictionaries.com/
[69] von Neumann, J., Morgenstern, O., Rubinstein, A.: Theory of Games and Economic Behavior (Commemorative Edition). Princeton University Press (2007)
[70] Liu, Y., Li, M., Yang, B., Qian, D., Wu, W.: Handover for Seamless Stream Media in Mobile IPv6 Network. In: Boavida, F., Monteiro, E., Mascolo, S., Koucheryavy, Y. (eds.) WWIC 2007. LNCS, vol. 4517, pp. 55–66. Springer, Heidelberg (2007)
[71] Shaojian, F., Atiquzzaman, M.: SCTP: State of the Art in Research, Products, and Technical Challenges. IEEE Communications Magazine 42(4), 64–76 (2004)
[72] Kent, S., Atkinson, R.: Security Architecture for the Internet Protocol, IETF RFC: 2401 (November 1998)
[73] R Development Core Team, R: A Language and Environment for Statistical Computing (2010), http://www.r-project.org

Peer-to-Peer Constellations in an IMS-Enabled Network Architecture Based on Interactive Broadcasting

Evangelos Markakis[1], Charalabos Skianis[1], Evangelos Pallis[2], George Mastorakis[3], Constandinos X. Mavromoustakis[4], and Antonis Antonas[2]

[1] University of the Aegean, Department of Information and Communication Systems Engineering, Karlovassi, 83200, Samos, Greece
{emarkakis,cskianis}@aegean.gr

[2] Technological Educational Institute of Crete, Department of Informatics Engineering, Estavromenos, Heraklion, 71500, Crete, Greece
pallis@pasiphae.eu

[3] Technological Educational Institute of Crete, Department of Business Administration, Lakonia, Agios Nikolaos, 72100, Crete, Greece
gmastorakis@staff.teicrete.gr

[4] University of Nicosia, Department of Computer Science, 46 Makedonitissas Ave., 2414 Engomi, Nicosia, Cyprus
mavromoustakis.c@unic.ac.cy

Abstract. This chapter elaborates on emerging advances and solutions to introduce the convergence of Peer-to-Peer (P2P) mechanisms into interactive broadcasting systems, enhanced with IP Multimedia Subsystem capabilities. The proposed solutions enable for the proper users' access on P2P-based services from dispersed locations, as well as the deployment of a novel users' ecosystem to gear-up new businesses and increase revenue prospects. The P2P-based network architecture is validated through several experimental tests, performed on an actual prototype providing real transmission and reception of data. The experimental results justify the effectiveness of the solution and indentify open issues for further investigation and experimentation.

Keywords: Peer-to-Peer networks, interactive broadcasting, IP multimedia subsystem, experimental prototype.

1 Introduction

A large number of applications, services and user-created multimedia content has been emerged through the recent adoption of sophisticated networking infrastructures. Such content include interactive broadcasting services, live streaming, video and audio on demand, as well as Peer-to-Peer (P2P) services and applications. One of the attributes of these services include the active contribution of users during the content development and distribution process, by respecting several Quality of Service (QoS) constraints. Towards this direction, new challenges arise to address issues related with the efficient exploitation and management of the resources in emerging networking

C.X. Mavromoustakis et al., *Resource Management in Mobile Computing Environments*, Modeling and Optimization in Science and Technologies 3,
DOI: 10.1007/978-3-319-06704-9_12,

architectures. More specifically, dynamic network configuration schemes and resource management techniques have to be investigated and adopted over best-effort networks, when the broadband access capability is an open issue for users, located in dispersed locations. In this context, this chapter proposes promising technological solutions based on the convergence of Internet-based systems, interactive broadcasting and emerging networks for the deployment of novel prototypes, according to three individual systems and mechanisms:

- Regenerative interactive broadcasting systems for terrestrial transmission (DVB-T) [1], [2] of multimedia services and deployment of wireless network infrastructures for broadband access in dispersed locations.
- IP Multimedia Subsystem (IMS) [3] for services management, addressing issues related with security and heterogeneity of access technologies.
- P2P mechanisms to foster self-organization and increase resources exploitation.

Recent research approaches in the field of interactive broadcasting highlight that regenerative DVB-T configurations enable for the proper deployment of wireless networking infrastructures, providing multiple services to users located in isolated areas. Based on such approaches, interactive broadcasting in regenerative configurations acts as a middle-mile network to make available the core-backbone infrastructures inside the entire coverage area. This approach enables for the proper development of high speed access networks in rural locations, as well as the effective distribution of user-created content to the entire network. Even though this approach can be exploited to bridge the digital divide among the rural and urban areas users for broadband network access, the limited resources availability of the interactive broadcasting infrastructure can lead to reduced deployment scalability. In addition, such research approaches cannot provide guaranteed QoS in a per-user basis.

On the other hand, P2P technology is an emerging paradigm to efficiently exploit the network resources and enable the proper content delivery through best-effort networks. This approach employs resources-monitoring algorithms [4] used among multiple peers. In addition, "content bottleneck" issues are addressed through the adoption of coding techniques [5] and the problem of the aggregate load is overcome using load-minimization mechanisms [6], [7]. Moreover, P2P applications for live streaming [8] use mechanisms [9] for small setup times. Nevertheless, the diverse nature of the peers in the network, their active behavior, as well as their location [10] have not yet been considered. In a general context, even though flexibility and scalability have lead to a broad adoption of Peer-to-Peer based networks and applications, several open issues still exist regarding the central management and security. In addition, emerging networking infrastructures are based on the convergence of IP-based systems, such as the IMS. IMS provides solutions for the access networks heterogeneity, the security challenges, as well as the mobility management.

In this context, this chapter elaborates on emerging advances and solutions to introduce the convergence of P2P mechanisms into interactive broadcasting systems, enhanced with IMS capabilities. The proposed solutions enable for the proper users' access on P2P-based services from dispersed locations, as well as the deployment of a novel users' ecosystem to gear-up new businesses and increase revenue prospects. More specifically, Section 2 and Section 3 elaborate on related research approaches and present the proposed network architecture based on the convergence of IMS, interactive broadcasting and P2P mechanisms. Section 4 is dedicated to present performance evaluation results to validate the proposed approach and Section 5 finally concludes this chapter, by highlighting fields for further investigation.

2 Related Work

The network architecture [11], [12], [13],[14] depicted in Figure 1 is based on a converged environment, consisting of a main transmission point (i.e. regenerative DVB-T) and several Cell Main Nodes (i.e. CMNs) located inside the broadcasting area.

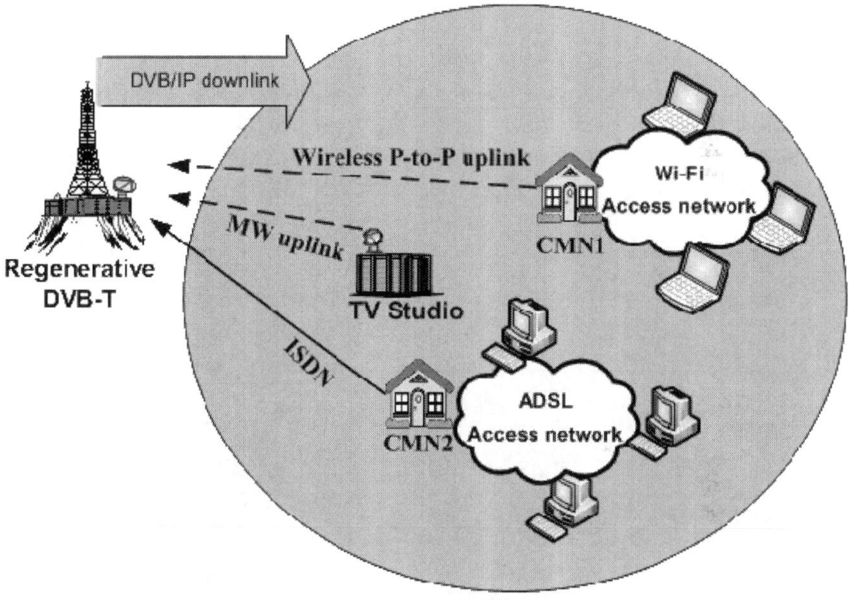

Fig. 1. A DVB/IP network architecture

The users are connected with each CMN of this infrastructure through several wireless or wired communication links. In this configuration, all users' IP traffic is collected from the CMNs, which are in charge to forward it to the regenerative DVB-T point through several uplinks. The IP traffic that is received by the regenerative DVB-T point is multiplexed in a single stream together with digital television programmes to create the DVB/IP "bouquet". However, following this architecture and by considering the limited resources of the network, scalability issues become a major challenge, especially when access to user-generated services under the Client/Server model comes to the foreground. A very promising solution that relaxes the DVB-T utilisation and provides a better exploitation of its scarce resources is the realisation of network overlays that utilise P2P technology [15] see Figure 2.

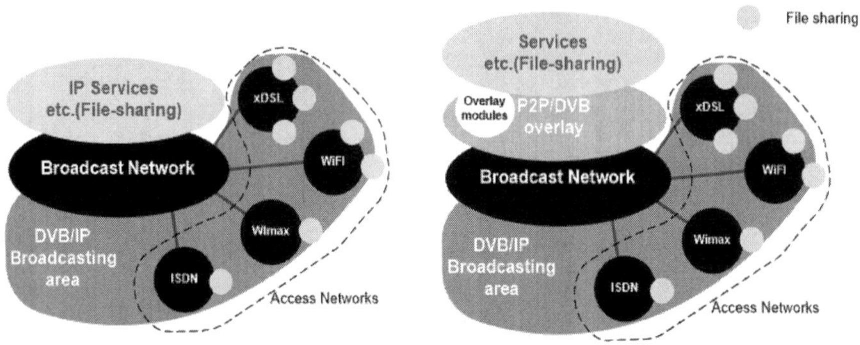

Fig. 2. Layered structured of iDVB-T vs P2P/iDVB-T

Such network overlay may enable end-users to fully exploit all network segments, including the uplink channels in the access network, by taking advantage of the resources being available at their neighbours' level. In such a case, the configuration of the previously described iDVB-T architecture with P2P-capabilities is presented in Figure 3.

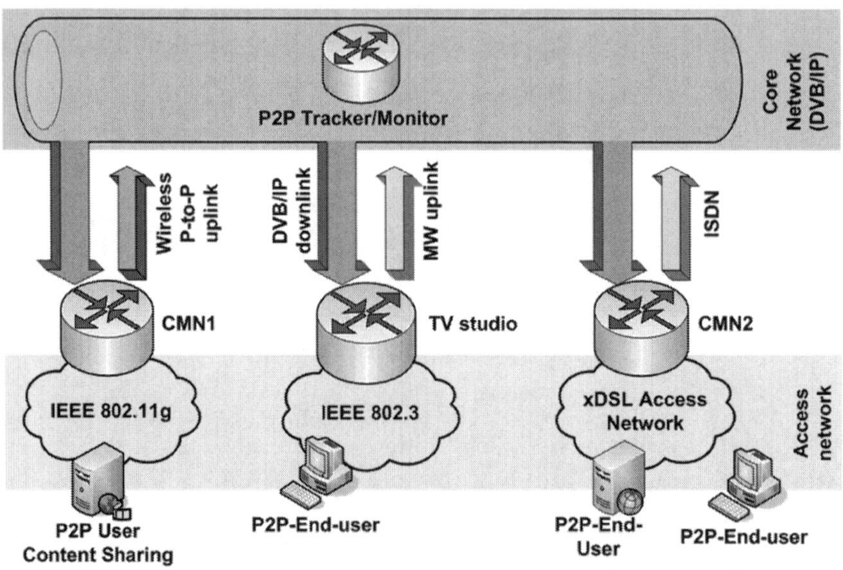

Fig. 3. Overall configuration of the P2P-capable iDVB-T network

Peers are interconnected via the P2P-enabled CMNs, which take the responsibility of forwarding data traffic and P2P signalling information to the regenerative DVB-T module. In turn, peers receive indexing information from a P2P tracker/monitor module located in the central broadcasting point, along with the data concerning the location of services, the Round Trip Time [16] (RTT) and the "location" of other end users (peers). Authors in [17] experimentally verified that in such network, the service "locality" information, along with network distance knowledge in the P2P overlay network, can significantly improve the overall system performance and provide better resource exploitation, especially in the limited-resources DVB-T link.

In addition, authors in [18] provide a literature review on the latest developments in the convergence of networks. The network convergence elaborates on the increasing tendency of users to have access to any service, anytime, anywhere and from any device. Authors in other related works are referred to the QoE (Quality of Experience), which is an indicator of the degree of user satisfaction from the experience in the field of networks convergence. Also, other quality characteristics are mentioned, such as the Grade of Service (GoS) and the Quality of Resilience (QoR). A great emphasis is also placed on the convergence between fixed and wireless networks, as well as between heterogeneous wireless networks.

The research approaches in [19] show what network topologies are chosen to broadcast multimedia content on a large-scale peer-to-peer network. In addition, an analysis of advantages and disadvantages in each case is performed. The main problems to be tackled are the Dynamic Uptime and the Limited dynamic peer bandwidth. Furthermore, the need for convergence in heterogeneous networks becomes greater. It must be also mentioned here that in wireless networks, the system becomes unstable. The problem of instability of wireless networks in [20] remains and it prevails the need to provide services to mobile devices. The main topic of [21] is the fairness on resources sharing within peer-to-peer networks and live streaming, as well as the fact that some powerful peers, such as universities with large upload rate do not provide to the network the available resources, affecting normal users with home DSL connections and consequently, affecting the overall system performance. Authors in [22] propose a Peer-to-Peer based network topology, which minimizes transmission delays and shows no problems of playback delays and playout lags, while streaming content.

Research work in [23] deals with the increase of data flows in the network and the ISP backbone induced by IPTV applications that have been recently developed. It studies in depth the data growth caused by a famous application, the PPLive. Also, authors in [24] study the options of IPTV transmission compared between Peer-to-Peer topology and IP Multicast. According to this study, the transmission by connecting to a large number of users on a server is the preferred option and the IP Multicast is always more efficient, in both mobile and fixed networks. However in cases when the channels broadcasted are monitored by a low number of users on a Peer-to-Peer network, the performance is acceptable and comparable to the method of IP Multicast. In addition, authors in [25] test other applications than PPLive, such as SopCast and TVAnts. This approach highlights that applications such as TVAnts and PPLive are more friendly towards the network since they prefer to distribute data among peers that are within the same autonomous system. Authors, in [26] discuss the topic of the security of IPTV services over a Peer-to-Peer network. They propose the use of the protocol SIP and AES encryption to hide transmission paths. This is done by creating an overlay named SIPTVMON, which apart from security issues will play an important role in load-balancing and providing stability of the system. SIPTVMON will also provide opportunities for convergence of heterogeneous networks and Internet applications, Finally, the research work in [27] takes into account the important role of each user for the normal operation of an IPTV system over Peer-to-Peer networks. The classification proposed is based on some technical features. This will help in placing stable nodes in the upper levels of the path, while nodes with less stability are placed lower. The simulation of classification showed a reduction of the impact abnormal exits on the overall system.

3 IMS-Based iDVB-T Network Architecture

According to the network architecture presented above, this section of the chapter elaborates on an overlay interoperability solution to enable the proper convergence of P2P mechanisms with an IMS-enabled interactive broadcasting network (see Figure 4).

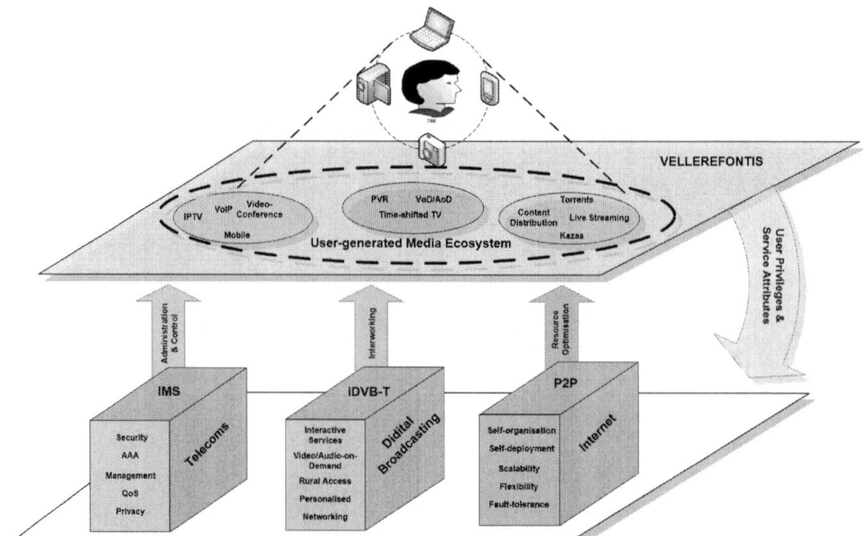

Fig. 4. Overlay interoperability solution based on P2P mechanisms

In addition, a hybrid convergence configuration is depicted in Figure 5, considering and unifying IMS, P2P and iDVB-T technologies. This configuration enables content developers to support the provision of applications under QoS restrictions to end users, who are located in dispersed locations. Is can also be exploited by users for the development and distribution of their own content through P2P overlays.

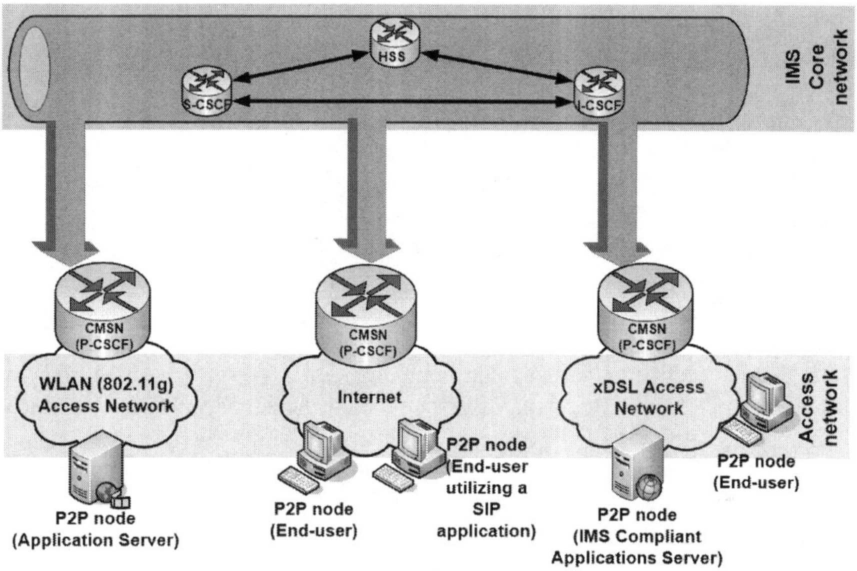

Fig. 5. Overall configuration of an IMS-based DVB/IP infrastructure

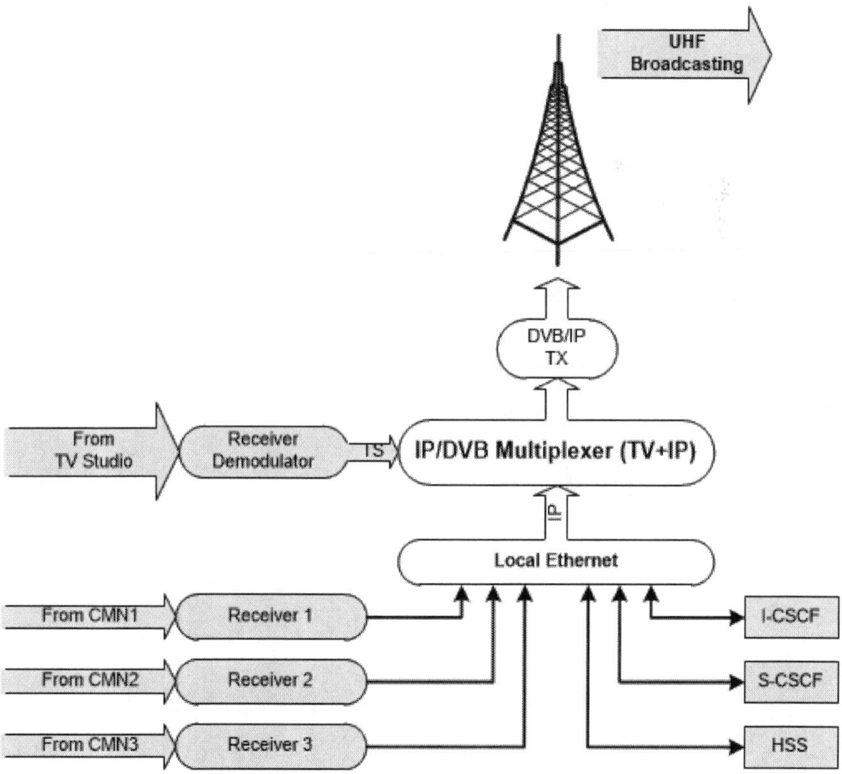

Fig. 6. Regenerative IMS-based DVB-T configuration

The configuration above includes a core network that is able to transfer traffic stemming from the peers, as well as the signalling information from the IMS through the dedicated uplink connections. This is performed through Proxy Call Session Control Function (P-CSCF). For the case of a registration, the Interrogating Call Session Control Function (I-CSCF) is in charge to receive IMS signalling, trying to find the Serving Call Session Control Function (S-CSCF) (see Figure 5). The I-CSCF in then in charge to route the SIP request to the assigned S-CSCF. The core network of this configuration and the P2P module are also responsible to change the priority scheme, towards increasing the connectivity of the peers that are connected in the related CMN. In addition, Figure 6 depicts a view regarding the IMS-based core network. The multiplexing device in this network is in charge to receive P2P-based traffic, as well as the IMS signalling data together with digital television programs. It then multiplexes them in a single stream that is broadcasted through the entire coverage area.

The configuration of CMN based on IMS mechanisms is presented in Figure 7. This CMN exploits an aDSL-based access network to enable for its proper communication with the users located inside its coverage area.

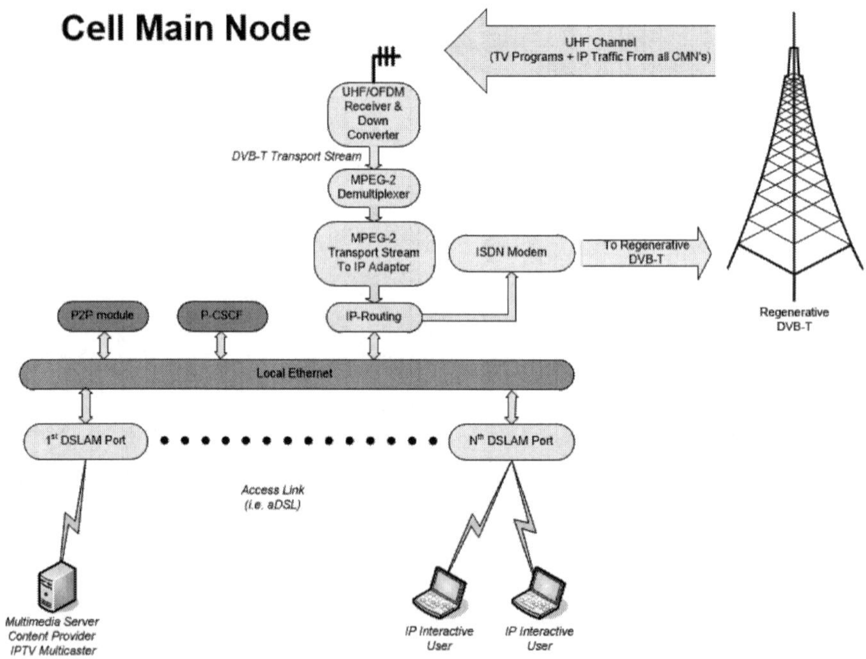

Fig. 7. Overall configuration of an IMS aware CMN

4 Performance Evaluation Analysis, Experimental Results and Discussion

Based on the above-mentioned approaches, this section is dedicated to present the development of an actual prototype, serving as an experimental test-bed to perform evaluation tests. This test-bed is presented in Figure 8 including the following modules:

- A DVB-T transmission module, operating in the range of 622-630MHz, under an 8K operation mode, 1/32 guard interval, 7/8 code rate, 16QAM modulation scheme and the multi-protocol encapsulation scheme.
- A P2P/IMS module to provide data regarding the location of the peers, as well as the completion time.
- A CMN (i.e. CMN1 in Figure 8) to support file sharing applications to inter-connected peers.
- 4 CMNs that are located in a dispersed location. Each CMN serves five peers through the access network of this network architecture.

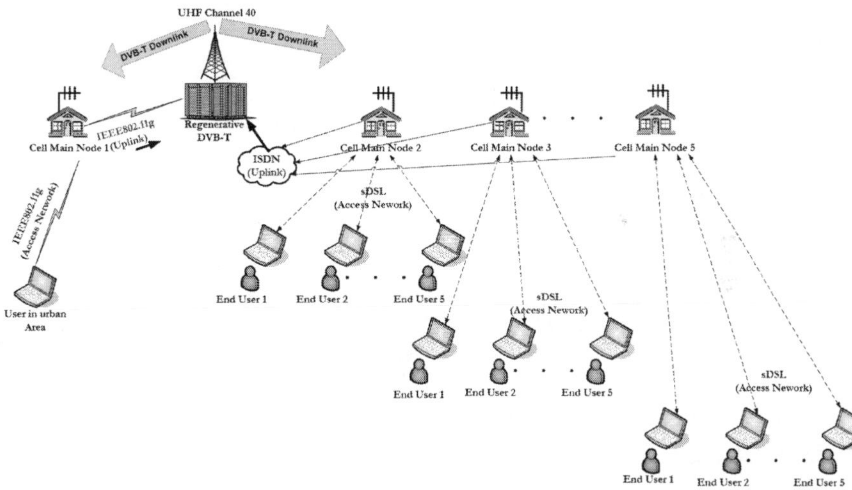

Fig. 8. Test-bed deployment

According to this test-bed, several experimental tests were conducted, towards evaluating the performance of the proposed system. The peers located in dispersed locations request a multimedia file that is hosted in the peer's terminal connected to CMN1. When five peers were concurrently requesting the same multimedia file, the results indicated that the DVB/IP downlink was underutilized (12%) and the utilization of each uplink in the access network of the peers was almost 90%. The next set of experiments validated the overall network architecture by increasing the number of the peers. In this case, ten peers were concurrently requesting the same multimedia file. The evaluation results indicated that DVB/IP downlink was underutilized (25%) and the utilization of each uplink in the access network was 98%. In addition, when five peers were concurrently requesting the same file, the losses [28] in the DVB-T network were about 1%, as shown in Figure 9. If the number of the peers is increasing, then the losses are also increased.

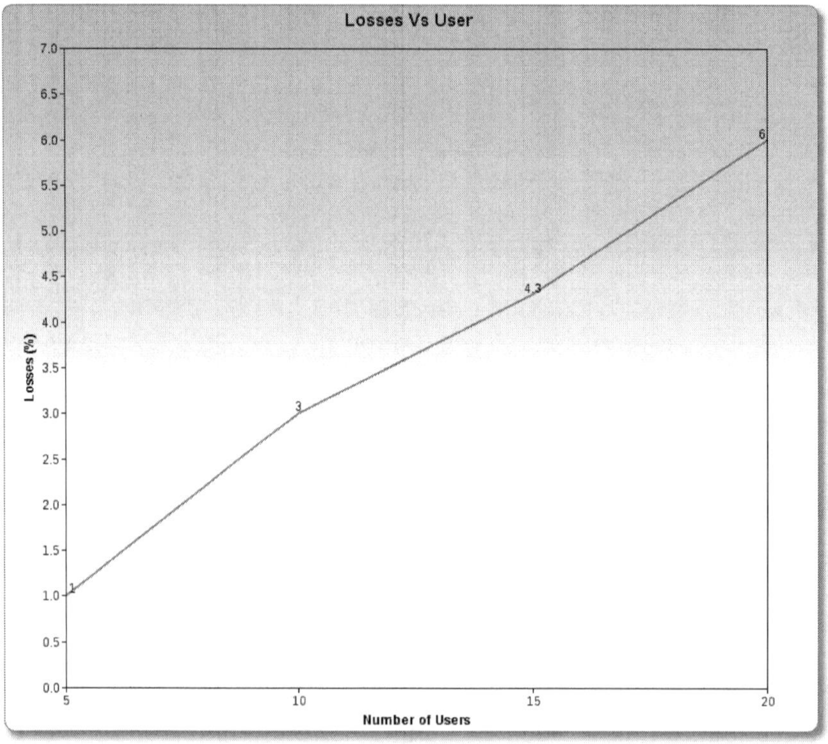

Fig. 9. Losses Versus Users

5 Conclusions and Further Research

The evaluation of the proposed converged network architecture was conducted in a real environment to state the strong and weak points of this approach. The solution presented is based on broadcasting networks that are the most common type of access network, as well as the most rapidly growing category of networks with successful examples (e.g. LTE, Wimax). The proposed network was successfully validated, in terms of network performance under actual conditions, towards verifying its validity. Peer-to-peer solutions need not only to provide compelling services to content providers and consumers, but also need to address the concerns of network service providers. The approach in this chapter is using the Broadcast P2P overlay, in order to improve the network performance and free more resources from the "expensive" DVB-T channel providing the ability for new services' to be deployed. This chapter indicated several open issues for future work. The content/service aware functionality has to be investigated to increase the system performance, in terms of guaranteed QoS provision and maximum possible recourse exploitation.

Acknowledgments. This project is implemented through the Operational Program "Education and Lifelong Learning", Action Archimedes III and is co-financed by the European Union (European Social Fund) and Greek national funds (National Strategic Reference Framework 2007 - 2013).

References

1. ETS 300 802, Digital Video Broadcasting (DVB); Network-independent protocols for DVB interactive services, ETSI (1997)
2. Mastorakis, G., Markakis, E., Sideris, A., Pallis, E., Zacharopoulos, V.: Experimental Infrastructures for IP/DVB Convergence: an Actual Substantiation for Triple Play Services Provision at Remote Areas. In: IEEE 18th International Symposium on Personal, Indoor and Mobile Radio Communications, PIMRC 2007 (2007)
3. Technical Specification Group Services and System Aspects, IP Multimedia Subsystem (IMS), Stage 2, TS 23.228, 3rd Generation Partnership Project (2006)
4. Christakidis, A., Fiedler, J., Efthymiopoulos, N., Koutsopoulos, K., Markakis, E., Garvey, S., Denazis, S., Tombros, S., Dempsey, S., Pallis, E.: Integrating P2P with Next Generation Networks. In: Future Internet Services and Service Architectures, pp. 15–73. River Publishers (2011)
5. Gkantsidis, C., Rodriguez, P.: Network coding for large scale content distribution. Infocom (2005)
6. `http://www.wcl.ece.upatras.gr/index.php?sid=&lang=el&id=51`
7. Kangasharju, J., Ross, K.W., Turner, D.A.: Optimizing File Availability in Peer-to-Peer Content Distribution. In: Infocom (2007)
8. Castro, M., Druschel, P., Kermarrec, A.-M., Nandi, A., Rowstron, A., Singh, A.: SplitStream: High-Bandwidth Multicast in Cooperative Environments. In: SOSP 2003 (2003)
9. Rowstron, A., Druschel, P.: Pastry: Scalable, Distributed Object Location and Routing for Large-scale Peer-topeer Systems. In: Guerraoui, R. (ed.) Middleware 2001. LNCS, vol. 2218, p. 329. Springer, Heidelberg (2001)
10. Magharei, N., Rejaie, R., Guo, Y.: Mesh or Multiple-Tree: A Comparative Study of Live P2P Streaming Approaches. In: INFOCOM, Anchorage, Alaska (2007)
11. Mastorakis, G., Kormentzas, G., Pallis, E.: A fusion IP/DVB networking environment for providing always-on connectivity and triple-play services to urban and rural areas. IEEE Network Magazine 21(2), 21–27 (2007)
12. Mastorakis, G., Pallis, E., Kormentzas, G.: A DVB/IP QoS aware backhaul networking environment. Wireless Personal Communications 52(3), 637–649 (2010)
13. Pallis, E., Mantakas, C., Mastorakis, G., Kourtis, A., Zacharopoulos, V.: Digital Switchover in UHF: the ATHENA concept for broadband access. European Transactions on Telecommunications 17(2), 175–182 (2006)
14. Mastorakis, G., Pallis, E., Mantakas, C., Kormentzas, G., Skianis, C.: Exploiting digital switchover for broadband services access in rural areas. Journal of Communications 1(6), 45–50 (2006)

15. Markakis, E., Pallis, E., Skianis, C., Zacharopoulos, V.: Optimised network resource exploitation in interactive broadcasting environments via P2P constellations. In: 2012 IEEE International Conference on Computing Networking and Communications (ICNC), pp. 417–420 (2012)
16. Almes, G., Kalidindi, S., Zekauskas, M.: A Round-trip Delay Metric for IPPM. RFC 2681, IETF (1999)
17. Markakis, E., Pallis, E., Skianis, C., Zacharopoulos, V.: Exploiting Peer-to-Peer Technology for Network and Resource Management in Interactive Broadcasting Environments. In: Proc. IEEE GLOBECOM (2010)
18. Rost, P., Boutaba, R., Doppler, K., Gumaste, A.: Recent Advances in Network Convergence. Computer Networks 55(7), 1455–1458 (2011)
19. Yiu, W.P., Jin, X., Chan, S.H.: Challenges and approaches in large-scale p2p media streaming. IEEE Multimedia 14(2), 50–59 (2007)
20. Hareesh, K., Manjaiah, D.H.: Comparisons of various approaches of content delivery in peer-to-peer video on demand system. World Journal of Science and Technology 2(5) (2012)
21. Wu, D., Liang, Y., He, J., Hei, X.: Balancing performance and fairness in p2p live video systems. IEEE (2013)
22. Mykoniati, E., Landa, R., Spirou, S., Clegg, R., Latif, L., Griffin, D., Rio, M.: Scalable peer-to-peer streaming for live entertainment content. IEEE Communications Magazine 46(12), 40–46 (2008)
23. Hei, X., Liang, C., Liang, J., Liu, Y., Ross, K.W.: A measurement study of a large-scale P2P IPTV system. IEEE Transactions on Multimedia 9(8), 1672–1687 (2007)
24. Bikfalvi, A., García-Reinoso, J., Vidal, I., Valera, F., Azcorra, A.: P2P vs. IP multicast: Comparing approaches to IPTV streaming based on TV channel popularity. Computer Networks 55(6), 1310–1325 (2011)
25. Ciullo, D., Garcia, M.A., Horvath, A., Leonardi, E., Mellia, M., Rossi, D., Veglia, P.: Network awareness of P2P live streaming applications: a measurement study. IEEE Transactions on Multimedia 12(1), 54–63 (2010)
26. Wang, C.H., Chu, Y.H., Wei, T.T.: SIPTVMON: a secure multicast overlay network for load-balancing and stable IPTV service using SIP. In: 2011 IEEE Conference on Computer Communications Workshops (INFOCOM WORKSHOPS), pp. 97–102 (2011)
27. Ullah, I., Bonnet, G., Doyen, G., Gaiti, D.: Toward user-classified P2P IPTV systems: A persona-based approach. In: 2013 IFIP/IEEE International Symposium on Integrated Network Management (IM 2013), pp. 1187–1190 (2013)
28. Miras, D., Sadagic, A., Teitelbaum, B., Leigh, J., El Zarki, M., Liu, H.: A Survey of Network QoS Needs of Advanced Internet Applications. Internet2 QoS Working Group (2002)

Part III

Resource Management in Mobile Cognitive Radio Networks

Cross-Layer Based Resource Management Frameworks for Mobile Cognitive Radio Networks

Vasileios Karyotis, Evangelos Anifantis, and Symeon Papavassiliou

School of Electrical and Computer Engineering, National Technical University of Athens (NTUA), Iroon Polytechniou 9, Zografou, 15780, Athens, Greece
{vassilis,vangelis}@netmode.ntua.gr, papavass@mail.ntua.gr

Abstract. Resource management has been one of the most important aspects of Mobile Cognitive Radio Networks (MCRNs) since their inception. It engulfs fundamental network operations such as spectrum management, bandwidth allocation, channel access, data routing, transmission power assignment and energy management. All such mechanisms are necessary for achieving the delay/throughput requirements desired by modern network services and users. Compared to traditional wireless mobile networks, the MCRN environment poses additional challenges and complicated requirements to the above functions and resource management in general, which eventually lead to an urgent need for reconsideration of the employed wireless protocol stack. The objective in MCRNs would be to apply more efficient protocol hierarchies and achieve better management of the common available resources for both secondary and primary users, at the minimum possible cost and by leaving the minimum possible resource footprint, as required for all modern communications systems. The cross-layer design approach has been proven extremely useful towards this direction, especially for network types such as sensor, mesh and cognitive radios. It allowed exploiting the most appropriate network mechanisms, while bypassing monolithic and outdated requirements when possible. In this chapter, we focus exactly on this aspect of protocol architectures and network management, and present state-of-the-art cross-layer based resource management approaches for MCRNs. Initially, we present three important resource management frameworks, and then we provide directions for developing generic and adaptive resource management protocol architectures for MCRNs, based on cross-layer and component-based design principles emerging from the previously presented approaches. This is the first attempt to highlight the suitability of combining cross-layer with component-based network design, and this work will also sketch qualitatively the potential benefits of such approach. Finally, some practical considerations for the proposed architecture, along with directions for future research conclude the chapter.

1 Introduction

The proliferation of wireless devices and networks was followed by a surge of mobile services that can be nowadays easily accessed anywhere and anytime (Anything-Anywhere-Anytime paradigm, which is also referred to as pervasive or ubiquitous computing [1, 2]). Mobile computing was initially available only for professionals, e.g. the military, rescue teams, doctors, engineers, etc. However, nowadays it is commercially

C.X. Mavromoustakis et al., *Resource Management in Mobile Computing Environments*, Modeling and Optimization in Science and Technologies 3,
DOI: 10.1007/978-3-319-06704-9_13,

accessible even to the novice user for diverse purposes of varying importance [3], e.g. mobile applications, cloud services, etc.

On the other hand, the emergence of commercial mobile applications and the rapidly increasing rate of users and services, call for reliable and efficient management of resources within such stringent environments, along with novel communications paradigms that will be able to seamlessly support the future and further increasing volumes of users and traffic. Among others, mobile cognitive radios have been conceptually designed to facilitate such ubiquitous network availability, while providing flexible and efficient access to large quantities of distributed resources, avoiding disturbing the entrenched users [4].

Wireless mobile networks are broadly classified into cellular and multihop [5, 6]. The first consist of well controlled architectures and certified devices that strictly abide with the protocols and specifications of wireless operators. Wireless multihop networks on the other hand, such as ad hoc, sensor, vehicular, mesh, delay-tolerant, etc., are distributed systems [6, 7], sensitive to environmental conditions or user patterns, and inherently flexible in cases of, e.g. crisis management, formation, tactical scenarios, etc. Fig. 1 depicts typical examples of centralized and multihop network paradigms for the cognitive radio environment. Relevant and similar, in principle, examples of network topologies can be found for other wireless networking paradigms as well.

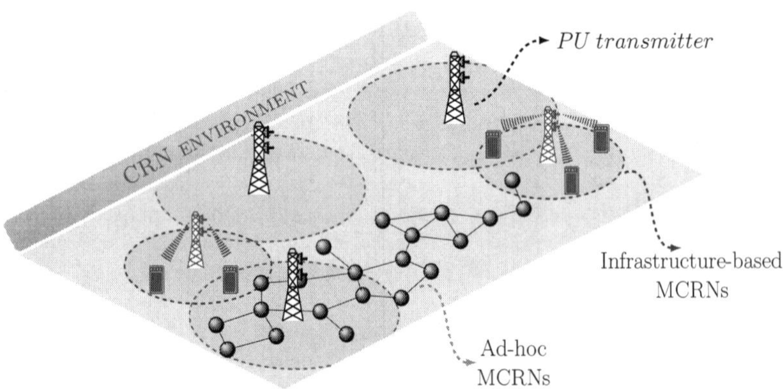

Fig. 1. Multihop and infrastructure based Mobile Cognitive Radio Networks (MCRNs) with primary user (PU) infrastructure. Secondary users (SUs) can employ both cellular or multihop based topologies (as shown in this figure), or even hybrid ones (not depicted in this figure explicitly) of both cellular and multihop topologies.

As services and applications become more sophisticated and demanding, wireless mobile networks will be required to become more self-organized, capable of reconfiguring their structures and mechanisms instantaneously, thus adapting to dynamic operational conditions. Within such broader operational environment for future networks, it is instigated that future wireless devices will consist of both software and hardware reconfigurable modules, offering greater flexibility and design degrees of freedom, in a manner similar to the cognitive radio networking paradigm [8]. However, it is also expected that the 'wireless Internet' will require significantly more elements than simply putting together the currently available wireless and Internet knowledge (wireless Internet $>$ wireless + Internet [9]). Through a potentially reconfigurable and adaptive network and resource management approach, all parameters associated with such wireless and potentially decentralized networks will be treated as variables at all available layers of the protocol stack.

As the technology aligns more with this direction, nowadays, the wireless landscape is composed of numerous applications, a plethora of wireless standards (e.g. DVB-T, GSM, 3G/4G, WiFi, WiMax, LTE, etc.) and billions of devices that have to work in unison [10]. At the same time, wireless networks are inherently characterized by scarce resources, i.e. limited usable transmission spectrum, energy-constrained transmission, mobility constraints, etc. In this overcrowded and stringent wireless scenery, the shortage of usable spectrum becomes one of the prominent impediments of future wireless networks, originating not only from physical transmission constraints (since not all wavelengths can be commercially used), but also from human regulation policies. Mobile Cognitive Radio Networks (MCRNs) have recently emerged as one of the promising technologies that can alleviate the aforementioned gridlock and enable the development of flexible and self-organized networks. However, such new-generation networks further pose new challenges that are currently still under investigation, especially regarding their advanced resource management capabilities and their operation.

In this chapter, following the motivation set by the previous discussion, we will focus on the subject of resource management that are particularly suitable for MCRNs and their implications for wireless protocol architectures in general. Our aim is to identify, classify and systematically reveal those trends and fundamental principles that can be exploited in future designs and developments in resource management for MCRNs and possibly in other types of wireless networks as well. Especially, we focus on frameworks that adopt and extend the cross-layer design approach and allow for holistic management of resources in diverse environments and missions. At the same time, we advocate the use of the component-based approach combined with cross-layer design in order to redefine and eventually optimize mechanisms of the wireless protocol stack in novel ways. MCRN concepts and features will have a key role in this evolution and this chapter aspires to become a stepping stone for boosting and sustaining the corresponding effort towards this direction. For this reason, one of the main objectives of this work will be to contribute towards the potential standardization of cross-layer resource management within a MCRN framework.

The rest of this chapter is organized as follows. In Section 2 we provide some background on previous approaches regarding the interplay between cross-layer design, cognitive radios and wireless protocol stacks. Section 3 presents an overview of the cognitive radio paradigm from a wireless mobile computing perspective, while Section 4 discusses resource management within the framework of cross-layer and mobile computing in general, and MCRNs in particular. More specifically, subection 4.1 presents an early cross-layer resource management framework leading to the back-pressure technique, while subection 4.2 describes a broader utility maximization formulation and architecture framework. Subection 4.3 provides an overview of a potential function based cross-layer framework (Markov Random Fields). Section 5 identifies and presents the features of cross-layered resource management approaches in a component-based network architecture proposed for MCRNs (CCRM framework) and provides an overview of open issues for future study. Finally, Section 6 summarizes the contributions of the chapter and some elements on the potential evolution of future MCRNs.

2 Related Work

Irrespective of the type of a communication or computer network, its resources have to be carefully used in order to ensure correct and efficient transferring of information. Traditionally, resource management has been based on the definition of a strict protocol stack, which serves the purpose of properly separating the involved mechanisms and properly coordinating access to the resources of the network by its entities [11]. For most of the current commercial networks this protocol stack is based on the TCP/IP protocol suite [12], whereas some fewer networks apply the OSI protocol stack [13]. However, nowadays, even for the more recent wireless multihop networks, the employed protocol stacks are mainly based on the TCP/IP paradigm.

In order to overcome the emerging challenges, mainly in the wireless setting, various 'violations' of the traditional protocol stacks have been proposed. Such designed violations involve various types of interactions between mechanisms and parameters of different layers and are cumulatively denoted as cross-layer design (explained in more detail in subsection 4). The effort for obtaining protocols relying on interactions across traditional layers has increased significantly, mainly due to the obtained benefits for the network and/or users, as described extensively in [14–18] and references therein.

To an extent, cross-layer design has emerged as one approach to re-establish radio network architectures and resource management by breaking the original barriers of the traditional protocol stack and exploiting input/signaling/control information across protocol layers, especially for the lowest protocol substrates [14, 15, 17]. However, such approaches have been shown to yield significant complications and raise numerous concerns, especially with respect to the involved scales of operation, as well as implementation issues [18]. More specifically, several works have focused on formalizing cross-layer design in cellular, e.g. [14, 15], and multihop, e.g. [16, 17], networks, respectively. Various others have addressed emerging complications and dangers, which could eventually risk the consistency of the whole protocol stack [18]. Most cross-layer approaches adopt a performance improvement perspective, where the cross-layer interactions aim at improving some metrics related to information transferring, e.g. throughput. This often comes at the cost of modular design and especially implementation efficiency, synchronization, signaling, etc.

Various resource management approaches have been suggested in the past, attempting to address the special characteristics of different types of networks [19]. A survey of resource management approaches for heterogeneous wireless networks was presented in [20], while several requirements and limitations for radio resource management specifically have been presented in [21]. Resource management addressing capabilities such as dynamic channel assignment, dynamic power control, and load sharing were presented in [22], while in [23] the key aspects of the evolution which impact radio resource management for the mobile broadband wireless networks were examined, along with areas that need to be addressed for providing varying quality of service requirements.

In this chapter, we will focus on cross-layer resource management. We will take a more cautionary perspective by developing guidelines and principles for cross-layer architectures. The target will be to allow decoupling the acquiring of cross-layer information and decision-making from the actual protocol operations and thus ensure modular (component-based) protocol design and distributed decision-making, while maintaining the objectives and performance of the traditional wireless stack. Furthermore, we highlight these principles, using MCRNs as an application domain. On one hand, MCRNs create a more complicated environment with additional challenges that need to be addressed. On the other hand, the MCRN networking paradigm allows the instigated cross-layer functionality that will be described in later parts of the chapter in a seamless and flexible manner.

Cross-layer design was considered in Cognitive Radio Networks (CRNs) and MCRNs since their inception. As will be explained in the next chapter in more detail, cross-layer interactions are allowed in CRNs and MCRNs inherently towards developing efficient resource management. In this direction, [24] highlights the need for "true cross-layering" concept in CRN solutions, which should be characterized by direct interaction and bidirectional flow of information between protocol layers. Additionally, with respect to individual protocol mechanisms, [25] presents a cross-layer ARQ-based optimization for CRNs, which builds on the MAC ARQ mechanism, while [26] presents a cross-layer approach for CRNs that jointly considers routing and dynamic spectrum access. Furthermore, [27] presents a cross-layer approach for CRNs which builds on the beamforming mechanism, while [28] proposes a cross-layer call admission control mechanism especially for WiMAX based CRNs.

Apart from the above approaches that focus on a specific mechanism of the considered CRN architecture and develop cross-layer interactions with other layers, more holistic cross-layer frameworks have been proposed, which are closer in concept with the architectural features presented in this chapter. Such frameworks attempt to address cross-layer protocol design with respect to various mechanisms of the protocol stack concurrently. Among others, notable works include [29] that suggests cross-layer frameworks for optimizing throughput in CRNs, [30] and [31] that both propose distributed cross-layer optimization frameworks for CRNs, and [32] which suggests a cross-layer mechanism for QoS provisioning in CRNs.

In the following, we first present a brief overview of the wireless networking setting of MCRNs, which constitutes the application domain of interest, and then touch the cross-layer based resource management topic from a more educated point of view.

3 Mobile Cognitive Radio Networks

Wireless mobile communications have progressively evolved over the recent years to become one of the dominant networking technologies, covering probably all aspects of human activities. The emerging requirements posed by user trends and the intrinsic characteristics of the wireless environment have paved the way towards their structural evolution, eventually transforming the wireless nodes from plain passive protocol executors to rational autonomous agents. This in turn, stirred the interest for more agile wireless communications, such as cognitive radios and mobile communications.

3.1 From Wireless to Mobile Cognitive Radio Networking

Considering the aforementioned available spectrum shortage and the requirement for wireless devices to support mobility in a seamless and efficient manner, several new wireless technologies have emerged. In order to alleviate the emerging spectrum occupancy gridlock caused by the tremendous increase of wireless services and traditional static regulations, the Dynamic Spectrum Access (DSA) concept has been introduced [4,33,34], addressing latest findings that some of the allocated spectrum bands are frequently under-utilized [35,36]. DSA strategies can be broadly classified in three distinct models, namely *dynamic exclusive use*, *open sharing* and *hierarchical access*, which are described in detail in [33] along with their variations. DSA has the full potential to create a breakthrough in the wireless landscape. However there are political, social, economic, and technological factors all together along with the interplay between DSA players (i.e. policy maker, academia, industry and end-users) that are expected to shape the future regarding the commercial success of DSA [37]. With respect to wireless communications and in response to DSA requirements, the emergence of cognitive radios in the early 2000s [38] has created an impetus in today's mobile communications field for improving the traditionally monolithic spectrum management policies, and developing cheaper, more efficient and flexible mobile communications architectures.

The cognitive radio paradigm aspires to alleviate holistically various impairments of wireless communications by relying on theoretical concepts from network information theory and learning, as well as exploiting latest advances in the field of wireless mobile communications, e.g. the availability of Software Defined Radios (SDRs). SDR describes cumulatively an approach where components that have been typically implemented in hardware (e.g. mixers, filters, amplifiers, modulators/demodulators, detectors, etc.) are instead implemented by means of software [39,40]. However, cognitive radios should not be considered as a substitute term for SDRs, since they are expected to incorporate additional components and capabilities, enforce policies, exploit SDR reconfigurability, provide incentive and coexistence mechanisms, and further implement cognition in their operation, as explained in more detail in subsection 3.2.

In the above sense, Cognitive Radio Networks (CRNs) have come to bridge the gap between the theoretical benefits of DSA concept and its commercial deployment, while addressing the need for self-reconfigurable, flexible and efficient wireless networks. At a glance, cognitive radios aspire to implement DSA concepts in order to address the scarceness of available spectrum and evolve the conventional methodologies providing a vast untapped potential for wireless mobile networking, especially regarding the management of the scarce available resources. More important is the fact that their concepts can be further exploited to holistically improve wireless networking architectures and devices, as it is already the case with standards such as LTE-12, and as will be highlighted in subsection 5.

3.2 The Anatomy of Mobile Cognitive Radio Networks

The cognitive radio paradigm [38] has been initially introduced as a novel approach for improving the utilization of the radio electromagnetic spectrum [41] [42]. By its conception cognitive radios were designed as an intelligent system (as shown in Fig. 2) that becomes aware of its environment and uses learning and adaptation on the received input in order to ensure two objectives. The first is, as already mentioned, spectrum utilization and the second is to achieve highly reliable and efficient communications whenever and wherever needed. Especially the latter can be very tough in tactical and rescue scenarios, and in cases where permanent infrastructure is difficult or expensive to deploy. Cognitive radios aspire to provide the means for achieving high quality data transfers, in very stringent, mobile and dynamic environments.

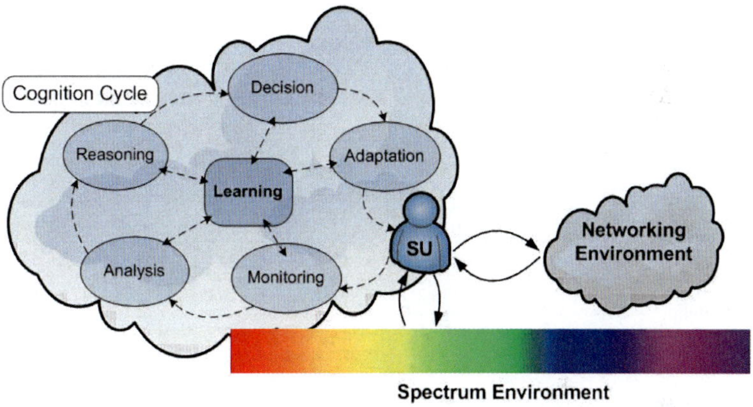

Fig. 2. Cognition cycle and cognitive radio secondary user (SU) functionalities. The cognitive cycle starts by sensing the environment and raising ambient information, which is then exploited for decision-making and/or adaptations. Monitoring throughout ensures proper and flexible operation in a timely manner.

Leaving behind the earliest years of academic research [43], cognitive radio technology day-by-day is finding its way into emerging standards and protocols, e.g. IEEE 802.22, IEEE 802.16h, IEEE 802.11af, IEEE P1900 standard series, etc. At the same time the cognitive radio vendor ecosystem becomes larger, while the industrial interest increases towards inserting cognitive radios into more widespread and mobile technologies, such as upcoming and future LTE versions, i.e. LTE Release 12 and beyond [44]. Even though centralized cognitive radios seem to be a feasible reality and lately more paradigms are employing such features [45], multihop cognitive radios need to be better studied in order to exploit their full potential. Unlike the centralized cognitive radios that resemble the infrastructure-based networks, e.g. IEEE 802.22 [46], the multihop point-to-point architecture can create wide-area self-organized cognitive radio backhaul networks, where nodes operate in a peer fashion acting as relays for their neighbors (Fig. 1). The main objective is to exploit spatial reuse for increasing the capacity of the network as in traditional multihop networks, while facilitating ubiquitous and flexible connectivity and increased coverage through spectrum scavenging.

At first glance, multihop CRNs appear similar to traditional multi-channel ad hoc networks. However, a set of emerging characteristic features of the first uniquely distinguish them from the latter. These are:

- dynamic availability of spectrum (in spatial and temporal terms),
- wider and dynamic ranges of used frequencies,
- heterogeneity of radio frequencies (varying propagation characteristics across employing frequency bands), and finally,
- dynamically changing topology due to primary user (PU) activity accompanied by incomplete network information.

At the same time, all the aforementioned characteristics create a more stringent landscape in which MCRNs are required to operate and eventually perform in a more efficient manner.

CRNs improve spectrum utilization by making it possible for a non-served SU to access a spectrum hole, namely a radio spectrum chunk currently unoccupied by its nominal PU, at the right location and time, conditional on the PU status. In addition, by inception, CRNs require multiple technologies/mechanisms to collaboratively produce their outcome [41]. All these functionalities that need to be supported can be described by six key features that stand out in the typical definition of CRNs and characterize uniquely their operation with respect to DSA and SDR approaches:

- *awareness*,
- *intelligence*,
- *learning*,
- *adaptivity*,
- *reliability*, and
- *efficiency*.

These features (depicted in Fig. 2) are of course not unique among networking paradigms. However, cognitive radios in general constitute the first networking regime to exhibit them cumulatively. In fact, cognitive radio is the first paradigm that actually requires most of these features as basic functionalities in order to ensure seamless operation. For example, spectrum sensing is the fundamental operation of CRNs, without which any notion of spectrum scavenging would fail in practice.

All the above features can be found in various capacities in mechanisms residing across the protocol stack, from the physical up to higher protocol layers. This distribution of functionalities to multiple protocol stack components unveils in an emphatic fashion, more than in any other type of network, the interdependencies among different communication layers that call for novel architectures espousing in cross-layer design. For example, the selection of optimal relay node(s) depends on all three lowest layers. The process starts by the routing function, but also depends on the availability of the optimal spectrum bands (signaling and data transmission). Once the medium is accessed, the physical layer should determine the required power and rate given the propagation characteristics of the selected spectrum bands. In that sense, the operations of each protocol layer can drastically influence the conditions of other layers and consequently the optimal provided solutions. Other similar issues, such as interference mitigation, coordinated spectrum sharing among secondary users (SUs), topology awareness, minimization of reconfiguration overhead [47], etc., currently remain open in the research literature of MCRNs [48, 49].

The key resource within the framework of MCRNs is the white spaces of the wireless spectrum, i.e. available transmission frequency channels that can be employed by the SUs of the MCRN. However, as cognitive radios address not only the shortage of spectrum, but also aim at providing robust, efficient and mainly improved communications anywhere and at any time, several other resources of their volatile architecture require efficient and prompt management, as described in Table 1. The transmission power of SUs and the knowledge acquired from the neighborhood regarding radio parameters are very important resources, as well as traffic patterns, user trends, etc. Depending on the employed cognitive radio paradigm (interweaved, overlayed, etc., [50]), transmission power will be assigned in a different manner and this in turn will affect the energy resources of cognitive radio devices, which especially in the case of mobile SUs can be critical for their existence. Furthermore, CR-enabled nodes have to exploit/manage the arsenal of the flexible and reconfigurable radio functionalities offered by the recent SDR advancements and select among different design options for the employed radio interface. Last but not least, in order to support seamless and efficient cognitive radio operations, diverse protocol mechanisms and parameters should be efficiently managed towards holistically addressing the aforementioned CRN challenges, in manners similar to those explained later in Section 5. Thus, resource management requires significantly more in-depth study, especially for MCRNs, in order to be able to bear the flexibility and efficiency levels desired in modern and future wireless communications.

Taking this into account and with respect to Table 1, it becomes evident that resource management in MCRNs necessarily needs to jointly take into account multiple layers of

Table 1. Classification of resource types emerging in CRNs

Classes of Resources	Associated Operations/Parameters
White Spaces	Optimal spectrum sharing; Bandwidth availability and network capacity; Network connectivity; etc.
Energy	Transmission power; Topology control; Discovery of energy-efficient routing paths; etc.
Neighborhood Awareness	Connection to external resources (e.g. spectrum databases); Self-constructing knowledge methodologies; Coalition between SUs; Traffic patterns; Interference levels; etc.
Radio Parameters	Selection of coding/modulation schemes; Carrier aggregation; Transmission rate; Spectrum sensing techniques; etc.
Dynamic protocol functionality	Selection among cross-layered solutions; Component-based architecture; Protocol parameters distributed across layers; etc.

the traditional network protocol stack, in order to provide the most efficient and feasible solutions to the emerging problems. New forms of cross-layer aware/enabled resource management mechanisms/architectures should be developed and analyzed, especially considering the peculiarities of MCRNs, in order to enable the desired levels of efficient and flexible wireless networking in the near future. This will be the topic of the next section, where the relation of MCRNs and cross-layer analysis/design will be presented and thoroughly analyzed.

4 Cross-Layer Based Resource Management

As already mentioned in the previous sections, MCRNs need to make intelligent decisions across the employed protocol stack in order to efficiently manage and exploit the available resources. Fig. 3(b) depicts the generic form of a wireless protocol architecture (which is also employed by current MCRNs), and compares it with the traditional OSI protocol stack (Fig. 3(a)), as proposed in the late 1970s for computer networking. By comparing subfigures (a) and (b) in Fig. 3, it becomes apparent that the typical wireless protocol stack aims at reducing the overhead imposed by various higher protocol layers like those in the OSI model, while retaining the basic functionality and modularity of the involved operations into a more compact, yet well defined protocol hierarchy. This is the main reason that the protocol layers in the wireless stack are fewer, which however, allows the potential for higher layer protocols to access directly protocols in lower layers, signifying the potential for cross-layer design as well.

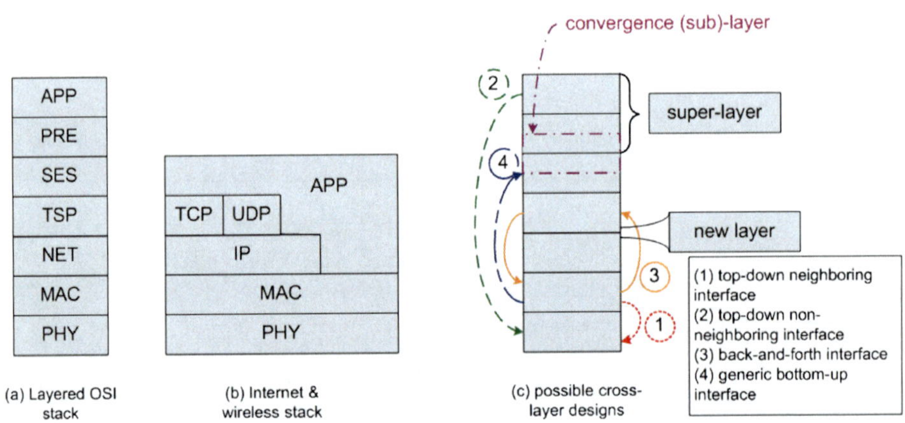

Fig. 3. Comparison of layered and cross-layered protocol stacks. A non-exhaustive set of examples of cross-layer interactions are shown in the third subfigure. (PHY = physical, MAC = medium access control, NET = network, TSP = transport, SES = session, PRE = presentation, APP=application.)

For both OSI and most of the times for the wireless stack cases shown in Fig. 3, according to the framework of the traditional layered protocol stack paradigm, each layer, e.g. routing mechanisms at the network, abstracts the complexity of the layer(s) below and provides a service to the layer immediately above [12]. Different layers iterate on different subsets of decision variables among all of network parameters, and using layer-local information they attempt to achieve a global (network-wide) objective (global optimum).

However, as already described in the previous section, the MCRN environment is characterized by more stringent operational conditions than traditional mobile communications environments such as cellular or multihop topologies, e.g. lack of licensed spectrum and high time-variability. Such additional challenges cannot be always addressed to the desired degree and/or at a tolerable cost by the traditional, strictly layered protocol architectures described before. The main reason is that resource management and most of the emerging tradeoffs, as addressed by recent designed protocols and mechanisms in the research literature, involve parameters that belong to different and typically non-neighboring layers with respect to the traditional protocol stacks depicted in Figs. 3(a), (b). For example, throughput optimality can be achieved if one permits using arbitrary paths in a multihop network, which might come at the expense of higher end-to-end delay [51, 52]. Such throughput-delay tradeoff demonstrates the interdependence of the protocol mechanisms (medium access-routing) addressing these parameters, which typically lie at different layers (MAC - Network, respectively). Using information from other layers in the decision-making process would violate the layering principles of the protocol stack.

In addition, layered architectures such as the one of Fig. 3(a) create significant overhead as the information of the top layer is encapsulated in frames of the lower ones and header overhead is appended in each intermediate layer. Thus, towards addressing some of these and other relevant concerns in future wireless networks and especially MCRNs, cross-layer design seems to be a very promising approach for dealing with the emerging complications of protocol overhead and signaling. Cross-layer may be considered as a reinvented approach enabling the modification of the traditional wireless protocol stack in an efficient and flexible manner for MCRN devices, and in general for reconsidering the wireless protocol stacks of the future.

The term cross-layer describes cumulatively a family of approaches that violate the traditional protocol layer module segregation, in order to exploit information available in one or more layers for making improvements, optimizations and adaptations in other layers [53]. Typical examples of possible cross-layer forms are shown in Fig. 3(c). Essentially, any protocol design based on the slightest violation of a strictly layered reference architecture, i.e. any violation of the principle that forbids direct communication of nonadjacent layers, can be characterized as cross-layer design with respect to the particular reference protocol stack. Such violation may have the form of information flowing upwards/downwards in the protocol stack to/from non-neighboring layers (1, 2, 3, 4 in Fig. 3(c)), or the form of a layer/ sublayer modification/ insertion/ compression. Another option is to create a new sublayer/superlayer or a convergence layer, as will be explained next. The depicted arrows in Fig. 3(c) represent information exchanges between layers that violate the modular approach of the traditional stack. In this chapter, we will focus on spatial (MRF in subsection 4.3) and temporal (DBRA in subsection 4.1) cross-layer approaches, that leverage topological or temporal elements of the system, in order to reconfigure the functions of the network and achieve efficient resource management. For the three approaches presented in this chapter, we introduce the basic concepts and then provide some fundamental elements of the relevant analysis that will be useful in better understanding their principles.

Regarding the examples presented in Fig. 3(c), and especially the creation of a sublayer/superlayer or the design of a new layer to substitute two or more previously neighboring ones (convergence layer), very characteristic cases of these can be found in Delay Tolerant Networking (DTN) [54–57]. DTN implementations were envisioned and designed as overlay architectural approaches to tie together a wide range of challenging networks with diverse intrinsic characteristics and different protocol stacks, hence, forming an "internetwork of challenged internets" [54]. Within their implementation structure [54,55], convergence layers play a significant role since in this manner adaptations and augmentations can be implemented allowing oftentimes cross-layer optimization, e.g. [56]. An indicative example, aiming at energy-efficient and low complexity DTN solutions, would be the creation of a convergence layer for the Bundle Protocol (BP) [57] of DTN so that it could be implemented directly over the Data Link Layer (DLL). In this case, a new layer (convergence layer) would be introduced to connect the BP layer that typically resides above the network and transport layers, directly to the DLL (MAC layer), while discarding (bypassing) completely the two intermediate ones.

Cross-layer design can be achieved by one of three possible ways, namely, direct communication signaling between layers, a shared database across the layers and completely new abstractions [17]. Taking an architectural shortcut can lead to a performance gain, and typically, this is the main incentive behind cross-layer design. These architectural 'violations' seem to be better accommodated in wireless networks, where the medium allows richer modalities of communication and cross-layer than wired networks. It can be also observed from current trends, that cross-layer designs are typically of short-term nature. For instance, it is easier for a network to employ cross-layer for a specific period, e.g. towards increasing throughput, than adopt a completely new protocol stack for attaining a similar objective. Longer-term benefits are usually based on more strictly layered architectural schemes, where the incurred added investment will be better exploited. With respect to such obtained benefits, it is explained for instance in [18] that cross-layer design is possible to lead to a throughput improvement by at most a constant factor, which even though not radical, it will be a very important benefit nevertheless, especially in absence of alternative approaches.

On the other hand, cross-layer approaches come at a cost, which affects their adoption for commercial purposes and poses challenges to the network designers. Once the modular layering is broken, the luxury of safe design is lost, and even a slight change of signaling or the location of a mechanism in the protocol stack can create loops, which as it is well-known from control theory, might put at stake the stability of the whole protocol architecture. In addition, when layered design is broken, some interactions are not easily foreseen, as in the modular scenario. Besides stability, this can affect the overall robustness of the network. Another disadvantage of this approach is that cross-layer design cannot lead to the seamless proliferation of technology to massive commercialization, at least not at the scales that layered architectures do so. Mostly, this is due to the fact that cross-layer approaches are not been currently under standardization processes. It is for this reason, that one of the main objectives of this work is to contribute towards this direction within a mobile cognitive radio framework. Last but not least, timescale separation in cross-layer design based protocol stacks is no longer possible within the layers involved in the cross-layer. Especially for that, synchronization considerations regarding the communication of the involved mechanisms might emerge, and as will be explained in the following sections, great caution is required.

However, towards an efficient and holistic management of the available resources in complex MCRNs, it seems that cross-layering deserves further investigation, due to the inherent protocol interplay in MCRNs, described in subsection 3.2. The potential benefit obtained and the reduced implementation cycle (compared to developing a new stack completely from the beginning) or even the need to optimize by incorporating cross-layer adaptations cannot be ignored and in fact in many cases, it emerges as the most suitable approach. Indeed, several cross-layer optimization opportunities present themselves when holistically considering the protocol stack, and the following three subsections are devoted to the presentation and analysis of relevant approaches.

As explained in [18], even though layered architectures have attained considerable success for wired networks, and having substantial impact in network design in general, it is not at all obvious that layered architectures, as the ones depicted in Fig. 3(a) - Fig. 3(b), are a priori appropriate for wireless networks. In addition, it is argued that compared to their wired counterparts and due to the intrinsic properties of the wireless medium, the emerging challenges of wireless (mobile) networks, and especially of MCRNs, encourage the potentials to dispel existing notions and concepts and start a clean-slate approach in designing future wireless networks. The following parts of this chapter are devoted exactly to this purpose.

Towards this direction, in the next subsections we focus on the most important cross-layer approaches that emerged previously or lately for wireless networks in general, and we show how they can be less or more suitable for exploiting their cross-layer principles through distributed decision-making and decentralized operation in MCRNs. The presentation follows an evolutionary approach scaling along the thread of research evolution of utility based network design. It will be described how the initial need for utility based cross-layer decision-making was eventually developed into a broader architectural framework and into the potential for obtaining holistic protocol architectures that exploit cross-layer utility functions, thus accommodating efficient and feasible resource management mechanisms in MCRNs.

4.1 Cross-Layer Design in Wireless Networks and the Emergence of Back-Pressure and Utility Functions

As explained before, in wireless networks in general, and especially in MCRNs, the different layers interact in a nontrivial manner in order to achieve the best possible resource utilization. The first coherent and broad enough approach that enabled capturing these cross-layer interaction for arbitrary wireless networks, but also in a fashion appealing to MCRNs as well, is a multi-commodity data flow formulation, which will be explained in detail in this subsection. Such model is capable of capturing cross-layer interactions in the traditional wireless protocol stack, from the physical to the transport layer. In subsection 5.2, we will explain how this model can be potentially combined with other holistic cross-layer approaches for obtaining more intelligent and flexible MCRN resource management architectures.

The basic formulation considers an arbitrary wireless network from a pure data flow perspective across neighboring protocol layers. Such formulation accommodates MCRN data transfer in a straightforward manner. The main resources considered are the allocated transmission rates, protocol layer queue lengths (which determine packet delay) and the corresponding power/energy cost. Additional network resources can be considered in a straightforward manner. Regarding data transmission, the network throughput (sum-rate) function can be used for cumulatively studying network performance. Assuming a time-slotted system, $\mu(t) = [\mu_{ab}(t)]$ represents the matrix of transmission rates offered over each link (a,b) during slot t (in units of bits/slot). The bit rates of individual transmission links can be represented by a function $\mu_{ab}(t) = C_{ab}(t) = C_{ab}(I(t),S(t))$ for each link (a,b). Then the link transmission rates across the network are determined by a link transmission rate function $\mathbf{C}(I(t),S(t))$:

$$\mu(t) = \mathbf{C}(I(t),S(t)) \tag{1}$$

where $\mathbf{C} = [C_{ab}]$ is the matrix of $C_{ab}(t)$, $S(t)$ represents the network topology state during slot t (network parameters related to topology, such as channel conditions, interference, mobility and node locations), while $I(t)$ represents a link control action taken during slot t (in MCRNs resource allocation options include power, bandwidth, channel access, etc.). Thus, $S(t)$ describes the uncontrollable factors of the network, i.e. information that can be collected and exploited in decision making (observable part), and $I(t)$ the controllable actions, namely resources of the problem to be managed and assigned to users/devices.

At every time slot, the network controller (a concept described in more detail in subsections 5 and 5.1 in terms of cross-layer and component-based design) observes the current topology state $S(t)$ and chooses a resource allocation $I(t)$, according to some resource management policy that takes into account environmental and resource related factors. The resource management policy can be centralized or distributed, depending on the MCRN type, even though distributed implementations are always desired. Various examples of different network types and their corresponding link rate functions $\mathbf{C}(I(t),S(t))$ can be found in [58].

All information entering and carried in the considered network is associated with a commodity. Commodities represent specific types of data destined for a specific sink node, from specific sources, priority classes, etc., and combinations thereof. Thus, the notion of commodity generalizes that of data flow in the network of interest. Fig. 4 depicts the protocol architecture and data flow between layers of such a network along with the modeling parameters involved in the corresponding cross-layer resource management. $A_i^{(c)}(t)$ is the amount of commodity data c that exogenously arrives to source node i during slot t[1]. $L_i^{(c)}(t)$ is the backlog of commodity c stored in the transport layer queue at node i. Similarly to $A_i^{(c)}(t)$, $R_i^{(c)}(t)$ is the amount of commodity c data allowed to enter the network layer from the transport layer for node i. $U_i^{(c)}(t)$ represents the

[1] Whether the data arrives progressively in the duration of slot t, or at bulk at one time instant has no impact on the analysis.

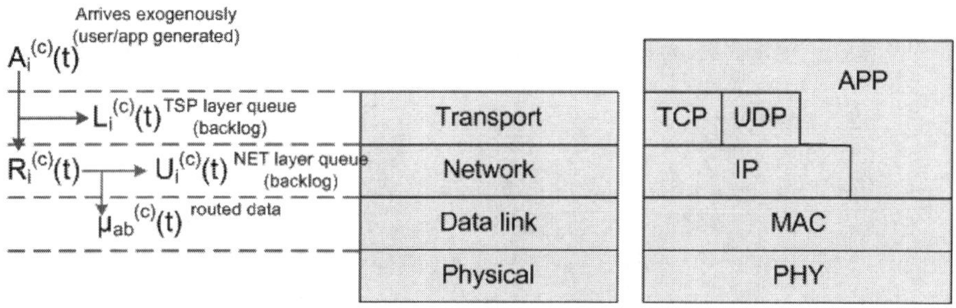

Fig. 4. Multi-commodity flow cross-layer formulation and layer interaction. Each class (c) of data corresponds to a different end-to-end data flow (commodity) between users. The figure depicts data flow between neighboring layers for a single data class.

unfinished work (backlog) of commodity c data in the network layer queue of node i, while $\mu_{ab}^{(c)}(t)$ is the routing decision variable that denotes the amount of commodity c data transmitted over link (a,b) and it is decided by the employed routing protocol.

Given routing constraints defined over the routing decisions $\mu_{ab}^{(c)}(t)$, the dynamics of the queue backlogs (transport-network), link constraints imposed by topology, and routing paths in the form of the set of available links for commodity data c, the latter denoted by $\mathcal{L}_c$ (set of permitted transmission links), a network control algorithm jointly makes decisions about routing, scheduling and resource allocation in reaction to current topology state and queue backlog information. Such constraints can be expressed in the following form:

$$\sum_{c\in\mathcal{K}} \mu_{ab}^{(c)}(t) \leq \mu_{ab}(t) \tag{2}$$

$$\mu_{ab}^{(c)}(t) = 0, (a,b) \notin \mathcal{L}_c \tag{3}$$

$$U_i^{(c)}(t+1) \leq \max\left[U_i^{(c)}(t) - \sum_b \mu_{ib}^{(c)}(t), 0\right] + R_i^{(c)}(t) + \sum_a \mu_{ai}^{(c)}(t) \tag{4}$$

$$L_i^{(c)}(t+1) \leq \min\left[L_i^{(c)}(t) - R_i^{(c)}(t) + A_i^{(c)}(t), L_i^{\max}\right] \tag{5}$$

where inequality (2) denotes that the rate is constrained by the capacity of a link, equality (3) denotes the fact that the rate of a non-existent link is zero and inequality constraints (4) and (5) describe the data flow dynamics between layers, as shown in Fig. 4. Parameter $\mathcal{K}$ is the set of employed commodities. $L_i^{\max}$ is the transport layer storage capacity, which can be finite or infinite, i.e. $0 \leq L_i^{\max} \leq \infty$.

The above set of constraints defines a cross-layer resource management function. The cumulative effect of this resource management cross-layer function can be analyzed and studied through the capacity region of the network, which is the set of all end-to-end traffic load matrices that can be supported under the appropriate selection of the network control policy. More formally, the network capacity (stability) region is the closure of the set of all arrival rate matrices $[\lambda_i^{(c)}]$ that can be stably supported by the network (considering all possible strategies for choosing the control variables). The arrival rate matrices are defined by the arrival process, i.e. parameters $\lambda_i^{(c)}$ are the time average rates of the arrival process $A(t) = \{A_i^{(c)}(t)\}_i^c$. It should be segregated from the capacity region of a specific policy, which is the collection of all traffic load matrices that are sustainable by the specific policy. Essentially, the *network capacity* region is the union of the individual *policy capacity regions*, taken over all possible control policies. As provided in [58], the existence of the capacity region is subject to a fixed rate matrix $[G_{ab}]$ (G_{ab} is a fixed rate at which data can be transferred over link (a,b)) together with multi-commodity flow variables $f_{ab}^{(c)}$ that satisfy typical flow conservation, routing and link capacity constraints. These types of constraints are typical in network flow problems and can be found in [59]. Intuitively, they express the conservation of (data) flow in the network.

One of the important outcomes of the above formulation is the possibility to exploit general algorithms for stabilizing such generic networks without requiring the knowledge of arrival rates or topology state probabilities. In fact, several such algorithms exist, such as the Borst Algorithm [60], the Max μ_i/r_i Algorithm [61] and the MWM Algorithm [62], described cumulatively in [58], all of which are based on the basic observation that bounding the transmission rate of a node does not ensure stability, but bounding the queue length does. This is a very strong result, which has been extensively used in the literature of throughput and delay analysis in wireless multihop networks. Thus, in general it will be of more benefit for the network to empty the most crowded queue, which is a very simple policy and very appealing for MCRN environment, where simple and distributed processes are required. The only case that the queue length in each user will become unimportant is when the network is completely saturated, since all queues will be constantly busy and full. In these cases, it is the transmission rate that matters most, as each packet will experience the same waiting time and the performance metrics will depend on the transmission rate of each queue, rather than queue lengths.

With respect to the aforementioned algorithms, the Dynamic (max-weight) Backpressure and Resource Allocation (DBRA) algorithm efficiently and provenly stabilizes the network whenever the vector of arrival rates lies within the capacity region [63]. This inherently cross-layer algorithm is appropriate for generic wireless networks, but also rather suitable for MCRNs, due to the efficient and simple operation it yields. DBRA determines the optimal commodity that maximizes the differential backlog $U_a^{(c)}(t) - U_b^{(c)}(t)$ over all links (a,b) and defines the optimal weight, respectively:

$$c_{ab}^*(t) = arg\,max_{c|(a,b)\in\mathcal{L}_c} \left[U_a^{(c)}(t) - U_b^{(c)}(t)\right] \tag{6}$$

$$W_{ab}^*(t) = \max\left[U_a^{(c_{ab}^*(t))}(t) - U_b^{(c_{ab}^*(t))}(t), 0\right] \tag{7}$$

The control action can be easily selected as the solution to the optimization problem

$$\max \sum_{ab} W_{ab}^*(t) C_{ab}(I(t), S(t)) \tag{8}$$

$$s.t. \qquad I(t) \in \mathcal{I}_{S(t)}$$

where $\mathcal{I}_{S(t)}$ denotes the set of control actions that can be supported under state $S(t)$. For each link (a,b) such that $W_{ab}^*(t) > 0$, data commodity $c_{ab}^*(t)$ will obtain a transmission rate of $\mu_{ab}(t) = C_{ab}(I(t), S(t))$.

The computation of weights demands only the backlog sizes of neighboring nodes, which is a very reasonable and low-overhead signaling exchange. It could be obtained either explicitly via direct short one-hop messages, or piggybacked in routing information exchanged in the network. At every timeslot, the network controller observes only its own and neighboring queue backlogs and the topology state variable. The paths are chosen dynamically at each timestep according to the back-pressure between neighboring nodes. The resulting algorithm assigns higher transmission rates to links with larger differential backlog and zero transmission rates to links with negative differential backlog. In principle, the obtained controller is centralized, because it requires knowledge of the whole network state. For some cases, one of which is described in [64], the above problem can be solved in a distributed manner with only local state information on each outgoing channel. The latter will be more suitable for distributed (multihop) MCRNs, while the centralized solution is applicable to centralized MCRNs.

Most of the properties and especially the stability of networks under the DBRA algorithm can be proved using the *Lyapunov function*, which is a non-negative function defined on the aggregate congestion of all queues in the network (sum of quadratic queue lengths). If a network has L queues and $\mathbf{U}(t) = [U_1(t)...U_L(t)]$ is the vector of backlog in each queue process, then the quadratic Lyapunov function can be defined as $L(\mathbf{U}) = \sum_{i=1}^{L} U_i^2$. The corresponding Lyapunov drift

$$\Delta(\mathbf{U}(t) = \mathbb{E}\{L(\mathbf{U}(t+1)) - L(\mathbf{U}(t))|\mathbf{U}(t)\} \tag{9}$$

is negative whenever the queue state is outside of a bounded region, thus creating a tendency to draw it back to the bounded region. The main drawback of this approach is that the underlying optimization problem is not always easily amenable to a distributed solution, as this would require full knowledge of $S(t)$ and complete coordination of network nodes. Thus, such approach is more suitable for infrastructure-based rather than purely ad hoc MCRN types.

The Enhanced Dynamic Backpressure Routing Algorithm (EDR) originally presented in [64] and also described shortly in [58] can be used to sacrifice some rate optimality, i.e. yield lower transmission rates (reduced capacity region) for less delay in the end-to-end paths, by incorporating a shortest path bias into the weights of the DBRA algorithm, thus revealing an inherent optimality-efficiency tradeoff. In any case, the EDR variation stabilizes the network, whenever the DBRA algorithm does so. Like DBRA, the EDR algorithm is also a centralized approach. We also note that by considering the dual problem of (8), where dual variables correspond to queue backlogs rather than network prices, the back-pressure algorithm can be shown to be a dynamic implementation of a subgradient search algorithm [58, 65]. More details on the optimization-related nature of DBRA can be found in [58] and references therein.

Perhaps the greatest value of DBRA is revealed when operating outside but close to the capacity region. In this case, the cross-layer optimization can be performed on the basis of a utility function, along with flow control necessary for maintaining stability (since the operational point will be out of the capacity-stability region of the network), and employed to tradeoff stability for fairness and delay performance (stochastic network optimization). In this case, the flow control is necessary for limiting the amount of admitted data. The model remains the same as in Fig. 4, where now a utility $g_n^{(c)}(r)$ is defined, which is non-decreasing and concave, and denotes the satisfaction of user n from rate r. In the previously presented cases, the assumption $A_n^{(c)}(t) = R_n^{(c)}(t)$ (no transport buffering allowed and any excess transport layer flow is discarded) was possible, while now outside the capacity region the additional constraint $R_n^{(c)}(t) \leq L_n^{(c)}(t) + A_n^{(c)}(t)$ for all (n,c), and time slots t is required. Different choices of the utility function, e.g. logarithmic or sigmoidal, lead to different properties, e.g. fairness [66], power scheduling [67], etc.

To address the new optimization obtained by adding the above constraint, a cross-layer algorithm (CLC1) was proposed by suggesting a pricing approach [58]. CLC1 employed the DBRA for routing and scheduling, while flow control was based on the pricing rule: $\max \sum_{c=1}^{K} \left[V g_n^{(c)}(r_n^{(c)}) - r_n^{(c)} U_n^{(c)}(t) \right]$, where $V > 0$ is a chosen constant that affects the performance of the algorithm. Given that, the allocation of resources at link layer and transmissions can be determined. Such approach can be readily employed in centralized MCRN users, since the implementation complexity and signaling overhead required are limited. In case of linear utilities, the result is a maximization of weighted sum of throughput and the resulting strategies are of bang-bang type, i.e. yielding the maximum or the minimum in each case [68]. Proportional fairness[2] may be achieved by maximizing the sum of utilities of the form $g_n^{(c)}(r) = \log(1 + \beta r_n^{(c)})$ for some constant $\beta > 0$ over any convex set of input rates.

Dynamic back-pressure can be employed in cases of unsaturated networks, where no infinite backlog assumption is applicable and the input rates can be arbitrary. The employed utilities can be linear or nonlinear and the problem can be solved by using auxiliary variables (denoted as virtual cost queues). In these scenarios, the main objective is to efficiently utilize/manage the available energy resources. A modified DBRA algorithm can be obtained (denoted by CLC2b [58]), based on the definition of a penalty/cost $\mathbf{x}(t) = [x_1(t), \ldots, x_{M_x}(t)]$ (incurred by the network control decisions at time slot t) and a reward $\mathbf{y}(t) = [y_1(t), \ldots, y_{M_y}(t)]$ (earned at time slot t) vectors respectively, and the utility function to be minimized $f(\mathbf{x}) - g(\mathbf{y})$ as a function of the penalty/reward vectors. The penalty could be a function of transmission powers and the reward can be related to transmitted data, but other types of penalties/rewards can be employed as well, depending on the objectives of the analyzed problem. The involved utility functions are assumed to be non-negative, continuous, bounded, concave and entry-wise decreasing. Such utilities may be employed in centralized MCRNs, while for multihop topologies, the generalized CLC (GCLC) algorithm described in [58], where the reward depends only on flow control variables, can be used and solved in a distributed manner. The typically addressed relevant resource management problems include maximizing throughput with average power constraints, minimizing power expenditure in multihop MCRNs, and minimizing energy scheduling. Additional problems include proportional

[2] Proportional fair is a compromise-based scheduling algorithm. It is based upon maintaining a balance between two competing interests: trying to maximize total throughput while at the same time allowing all users at least a minimal level of service. This is done by assigning each data flow a data rate or a scheduling priority (depending on the implementation) that is inversely proportional to its anticipated resource consumption.

fairness, quality of service optimizations and other objectives that can be expressed via proper utility functions and pricing mechanisms.

From the above discussions, it becomes evident that the potentials for utility based formulations with pricing and decomposition mechanisms developed for centralized multi-commodity problems to be solved by distributed algorithms (even in a sub-optimal form) is a promising direction, especially appealing to the nature and requirements posed in MCRNs. In the following subsection, this approach will become more concrete and yield more powerful cross-layer formulations for cross-layer design in wireless networks and MCRNs more specifically.

4.2 Network Utility Maximization: Utilities and Decomposition

Wireless protocol architectures and TCP/IP based protocol stacks (Fig. 3(b)) have been traditionally obtained in an *ad hoc* fashion and in general, they lead to piecemeal design approaches. The DBRA framework described above and various centralized or distributed variations have set the basis for developing more flexible and powerful cross-layer protocol stacks for wireless networks, and partially address some of the challenges posed by resource management in more stringent environments, such as MCRNs as well. In this direction it was made possible to cast the cross-layer inter-dependence of wireless protocols as global optimization problems, in the form of generalized Network Utility Maximization (NUM) problems [69]. NUM is a more systematic approach than the multi-commodity cross-layer formulation, capable of yielding broader-scope solutions and covering diverse types of applications, design objectives and network types. *Network utility* denotes the sum of utility functions of network nodes [70, 71], and the corresponding problems that refer to network utility, model the behavior of networks with respect to cumulative (social) emerging trends, whether these correspond to communications parameters (throughput, delay, etc.), or application oriented operations (trust, recommendations, etc.). In the following we describe a broad framework for network optimization, which has been shown to accommodate in a straightforward manner resource management/control in wireless networks, and exhibits a lot of promise for MCRNs as well.

NUM is capable of extending cross-layer design across the protocol stack via the utility function design. The overall communication network functionality is modeled by a generalized NUM problem. The key idea is that each layer of the protocol stack will correspond to a decomposed subproblem of the general NUM optimization problem, while the interfaces among layers will be quantified as functions of the optimization variables coordinating the subproblems. There can be many alternative decompositions of the generalized maximization problem, all of which correspond to a choice of different layering architectures.

From the above, it becomes evident that there are two main aspects of NUM that make it a valuable approach. The first one is considering the network as a distributed solution to some global optimization problem (network as an optimizer). The second one is the concept of *layering as decomposition*, where obtaining an optimal solution to a generalized NUM in a modular and distributed way suitable for low cost wireless devices, e.g. MCRNs, can be achieved through the theory of nonlinear optimization decomposition.

One of the advantages of NUM compared to other existing cross-layer frameworks is that there are many different ways to decompose a given problem. Such decompositions correspond to different layering schemes and are characterized by different degrees of efficiency, robustness, messaging complexity and implementation feasibility. NUM theory allows a systematic exploration of the space of alternative decompositions and determining which are better than others with respect to specified criteria. It also offers a complete and flexible framework and a level ground for fair comparison among the variety of cross-layer designs. Furthermore, it provides a unified perspective and proper decision-making framework that takes into account the fundamental limits and the impacts of cross-layering on network performance and robustness.

The basic steps of the framework is to first formulate a specific NUM problem, then obtain/devise a modularized and distributed solution following a particular decomposition, and finally explore the range of all possible decompositions that provide a choice of layered stacks, along with implementation complexity, involved time scales of operation, etc. The basic NUM problem belongs to the category of monotropic programming [65] and it is quantified as:

$$\begin{aligned} \max \quad & \sum_s U_s(x_s, P_{e,s}) + \sum_j V_j(w_j) \\ s.t. \quad & \mathbf{R}\mathbf{x} \leq \mathbf{c}(\mathbf{w}, \mathbf{P}_e), \\ & \mathbf{x} \in \mathcal{C}_1(\mathbf{P}_e), x \in \mathcal{C}_2(\mathbf{F}) \text{ or } \in \Pi(\mathbf{w}), \\ & \mathbf{R} \in \mathcal{R}, \mathbf{F} \in \mathcal{F}, \mathbf{w} \in \mathcal{X} \end{aligned} \tag{10}$$

where x_s denotes the rate of source s, $P_{e,s}$ the desired decoding error probability of node s, and w_j the physical layer resource at network element j. Utility functions U_s and V_j may be any nonlinear, monotropic functions. $\mathbf{R}$ is the routing matrix, $\mathbf{x}$ the source rate vector, $\mathbf{c}$ the logical link capacities (functions of physical layer resources $\mathbf{w}$ and the desired decoding error probabilities $\mathbf{P}_e$). The rates may be also constrained by hop-by-hop error control mechanisms $\mathcal{C}_1(\mathbf{P}_e)$, such as ARQ. $\mathcal{C}_2(\mathbf{F})$ represents medium access constraints, where $\mathbf{F}$ is the contention matrix, or more generally the scheduling constraint set Π. Thus, the above two constraints capture the fact of scheduled or random access based medium access, applicable in all types of distributed networks and MCRNs as well. The optimization variables are $\mathbf{x}, \mathbf{w}, \mathbf{P}_e, \mathbf{R}, \mathbf{F}$ and describe resources to be assigned. $\mathcal{R}$, $\mathcal{F}$, $\mathcal{X}$ are the corresponding sets of routing policies, contention schemes and physical layer resources respectively.

While utility models lead to objective functions, the constraint set described above (and in fact the more general ones as well), incorporate two types of constraints. The collection of physical, technological, and economic restrictions in the communication infrastructure, and secondly, the set of per-user, hard, inelastic QoS constraints that cannot be violated at the equilibrium. Having available a generalized NUM formulation, it is desired, especially in MCRNs, to modularize the solution method by exploiting decomposition approaches, so that each decomposed subproblem will control only a subset of variables and observe only a subset of constant parameters (distributed computation). This corresponds to the limited control and observation capability desired for each layer in the protocol stack, but it is also suitable for the resource-constrained users of MCRNs.

The basic strategy for resource management according to NUM employing decomposition is to divide the original large (and centralized) optimization problem into smaller subproblems, coordinated by a master problem, via information exchange corresponding to signaling. Typical examples are primal and dual decompositions, where the original or its Lagrange dual problem are decomposed [72]. Primal decomposition corresponds to the master problem directly giving each subproblem an amount of resources it can use, while its role is to properly allocate available resources. In dual decomposition, the master problem sets the prices for each resource and each subproblem decides the amount of resources to use given the specific prices. The master problem in these cases needs to obtain the best pricing strategy. Both approaches are capable to provide a mathematical language and solution methodology for the management of available approaches. This approach is rather appealing for MCRNs and especially the distributed ones, where each SU needs to autonomously take decisions regarding resources, while respecting the rules of a centralized master entity. Apart from the frequently used dual-based distributed algorithm, various other types of decompositions may be employed, each of which provides possibly an alternative network architecture, as mentioned before. A more targeted treatise, specifically focused on methods for decomposing generalized NUM problems in order to obtain distributed and modularized solutions is provided in [72], where some specific decomposition methods more appealing to MCRNs (cellular or distributed) can be found.

Given the previous discussion and from a network architecture point of view, two broad classes of decompositions are possible, namely horizontal and vertical. Roughly speaking the first regards the protocols of each layer individually, while vertical decomposition corresponds to alternative modularized control over multiple functional modules or layers.

The horizontal decomposition potentials of NUM have contributed significantly towards the geographical decomposition of problems, namely allowing distributed computation by network elements for various network protocols, such as TCP, BGP, IEEE 802.11DCF and physical layer algorithms, supporting the systematic attempt to put many of these on a mathematical foundation [69]. In fact, NUM framework is one of the first such systematic and generic enough approaches that enable reverse-engineering of network protocols. A thorough summary of the NUM based frameworks developed for reverse-engineering the TCP congestion control function [73] and random access MAC [74] is provided in [69] and [72]. Horizontal decompositions allow multiple distributed agents, like the ones in ad hoc MCRNs, to solve otherwise centralized problems, as mentioned above, and thus allow the development of flexible resource management frameworks with minimum signaling for such networks.

On the other hand, in vertical decomposition, the general NUM problem itself is modularized and becomes distributed across the protocol stack of a single user, via dual decomposition or some primal penalty function. This approach allows for cross-layer designs and it can be exploited in MCRN mechanism design. The individual modules can range from adaptive routing and distributed matching to information-theoretic source coding and signal processing, coupled with explicit message passing of selected layering prices. Some of the obtained solutions include the following:

- joint congestion control and adaptive coding or power control,
- joint congestion and contention control,
- joint routing and power control,
- joint congestion routing and scheduling,
- joint routing scheduling and power control,
- joint routing resource allocation and source coding,
- TCP/IP interactions and HTTP/TCP interactions, and
- joint congestion control and routing.

Further details and a more thorough overview of these vertical decompositions may be found in [69]. These already established algorithms in various wireless settings may constitute a basis for the design of cross-layered MCRN mechanisms, centralized or distributed, either by straightforward adaptations when possible, or by properly modifying the original formulations and following the corresponding methodologies for obtaining the desired solutions.

In both horizontal and vertical decompositions, the decomposition approach can be segregated into three major categories, i.e. primal/dual decompositions, pricing for decoupling objective function and alternative methods. In order to decompose coupled constraints, the original problem is split into subproblems which are then coordinated by a master problem by means of some kind of signaling, often at the benefit of not requiring centralized solution of the master problem. Primal and dual decompositions are applicable in such cases, and typically the primal one is realized through adaptive slicing, while the dual via pricing feedback. It should be noted that a given decomposition may lead to more than one distributed algorithm, depending on the choice of update method, ordering of variable updates and the time scales involved, as explained in the following regarding hierarchical problem decompositions.

In general, basic NUM allows a number of features for developing distributed algorithmic solutions for end-to-end mechanisms, which are suitable for networks such as MCRNs. The optimization problem formulation is characterized by:

1. separable objective function,
2. additivity of constraint functions,
3. interchangeability of utility summation index,
4. zero duality gap.

The first and second allow for obtaining easier decomposition methods, while the third allows for asynchronous updates, which are rather desired in a distributed environment as the one in MCRNs. The fourth ensures optimality of the obtained solutions for the general problem. Even though this is not of interest for practical purposes due to the involved overhead and computational burden imposed, it can be exploited for assessing the effectiveness of more practical approximation solutions obtained for real implementations.

In the following, we provide some additional guidelines for selecting the appropriate decomposition method in MCRNs. Due to the diversity of emerging problems, e.g. centralized or distributed topologies, different degrees of mobility and inter-dependence, etc., no piecemeal decomposition methodologies exist, and all factors have to be considered before making a proper selection. The following guidelines allow identifying the special conditions/requirements of each emerging problem in each analyzed

MCRN setting and could aid in starting with the right decomposition technique for each scenario.

A *primal decomposition* is appropriate when the problem has a *coupling variable* such that, if it is fixed to some value, the rest of the optimization problem decouples into several subproblems. On the contrary, *dual decomposition* is appropriate when the problem has a *coupling constraint*, which when relaxed, the optimization problem decouples into several subproblems. Thus, primal-dual decomposition is dictated by the existence of a coupling variable or constraint respectively.

Violation of the 'separability of objective function' feature provided above, yields problems with coupled utilities and potentially coupled constraints. However, by introducing auxiliary variables and additional equality constraints transfers the coupling from the objective function to constraint coupling. Such auxiliary variables are denoted as prices and can be updated using a subgradient method [65]. In addition, violation of the 'zero duality' property mentioned above, leads to alternative decomposition methods in both horizontal and vertical directions.

Relevant alternative methodologies span a very broad range of research literature. One of the most characteristic ones is the application of primal-dual decomposition in a recursive manner, leading to a hierarchy of decompositions, where smaller and smaller subproblems are obtained. In these cases, convergence and stability are guaranteed if the lower level master problem is solved on a faster time scale than the higher level master problem, so that each iteration of a master problem occurs given that all problems at lower levels[3] have already converged.

In all decompositions, there exists an implicit underlying choice of constraint representation, while several such representations may be obtained for each NUM problem. Each representation of a particular NUM may lead to different decomposability structures, different decomposition approaches, and the associated different distributed algorithms. In addition, each such algorithm represents a different way to distribute the desired control over the network architecture (vertical decomposition), or over the network topology (horizontal decomposition). Also, the above scene of layered protocol architectures reveals an inherent interconnection between neighboring layers and implicitly a virtual interconnection between distant layers. This once more motivates the potentials of cross-layer design in wireless networks and especially for designing efficient and flexible resource management mechanisms, as desired especially in MCRN environments.

As mentioned already, the NUM framework extends the utility concept to a broader framework for network architectures and control and as such, it allows for significantly more flexible designs. Especially the latter, could constitute a broad extension of layered protocol architecture approaches, into flexible and powerful mathematical frameworks dictating the controls that need to be taken in order to obtain the most desired operations by the developed infrastructures. However, several important issues remain and currently prevent it from massive implementation in current and short-term future network devices. One of the most important ones is the required computational and signaling burden they impose. Especially in MCRNs, where devices are resource-constrained

[3] Here level denotes decomposition hierarchy, rather than protocol hierarchy as in the rest of this chapter.

and the time-scales of decision-making are faster than in traditional wireless networks, NUM approaches do not seem to scale properly in their current form. Significant computational resources are required for determining the right decompositions, while signaling overhead increases substantially as the network becomes larger, especially for distributed topologies. In order to address such considerations, techniques that further build on the utility framework have emerged, which on the contrary, focus on retaining the signaling overhead and the required complexity low. To do so, they sacrifice some degree of optimality as will be described in more detail in the following subsection.

4.3 Markov Random Field Optimization

An alternative resource management framework to NUM that emerged lately in the context of CRNs [75], and exhibits promising outcomes for performing well especially in networks requiring distributed operation and management is based on Markov Random Fields (MRFs). This approach exploits variations of utility functions called potential functions in order to reduce the computational requirements and allow less overhead while sacrificing some degree of optimality of the obtained solutions.

Discrete MRFs [76,77] are undirected probabilistic graphical models that have been of fundamental importance to many computer science problems, such as computer vision, image processing, swarming intelligence, machine learning, etc. MRF optimization focuses on global optimal solutions based on spatially restricted information and it is mainly expressed in the sense of energy minimization problem due to the close relationship between MRFs and Gibbs Random Fields (GRFs), as described by the Hammersley-Clifford theorem [76,77]. The Hammersley-Clifford theorem asserts that a GRF distribution (given by equation (12)) with energy function expressed in terms of clique potentials leads to an MRF conditional probability (given by equation (11)) and vice-versa. Typically, MRF optimization serves as a framework for optimal decision-making and distributed optimization in class of problems characterized by large space of possible configurations, and thus it has attracted the attention of the respective fields for over 20 years now. This makes MRFs rather suitable for MCRNs, and especially for multihop types, where optimal coordinated resource management is desired by means of spatial interactions and subject to spatio-temporal resource limitations.

An MRF can be defined as a strictly positive probability measure on a set S, $|S| = n$ of spatially dependent random variables X_s (also called sites $s \in S$) that cumulatively determine the global system state. The set of possible states of each site (phase space) is denoted by Λ, namely X_s takes a value $x_s \in \Lambda$. The family of random variables $X = \{X_s, s \in S\}$ describes the system state at a specific time and a configuration $\omega = \{(x_1, \ldots, x_s, \ldots, x_n) : x_s \in \Lambda, s \in S\}$ corresponds to one of all possible states of the system. The finite product space $\Omega = \Lambda^n$ denotes the configuration space, i.e. the set of all possible configurations. Examples of such random variables could be a collection of pixels in the context of image analysis [78] or a set of SUs with strong spatial dependencies in a MCRN context [75], where the possible states of each random variable would represent the assigned resources for each SU. Those sites for which their states influence the state of a particular site s, are called neighbors of s, hence, defining a corresponding neighborhood system. The key property of MRFs is that the state of each site is only dependent on the states of its neighbors, thus leading to inherently distributed approaches and low computational cost by exploiting knock-on effect, i.e. local decisions give rise to long-range adaptations towards global optimization. This is expressed by:

$$\mathbb{P}(X_s = x_s \mid X_r = x_r, r \neq s) = \mathbb{P}(X_s = x_s \mid X_r = x_r, r \in \mathcal{G}_s) \tag{11}$$

where a neighborhood system on S is defined as a family $\mathcal{G} = \{\mathcal{G}_s\}_{s \in S}$ of subsets $\mathcal{G}_s \subset S$, such that for every $s \in S$, $s \notin \mathcal{G}_s$ and $r \in \mathcal{G}_s$ if and only if $s \in \mathcal{G}_r$. $\mathcal{G}_s$ denotes the neighborhood of s. In this fashion, MRFs can be considered as a generalization of the Markov property in space. Similarly, a random field X is called a Gibbs Random Field (GRF) if it satisfies:

$$\mathbb{P}(X = x) = \frac{1}{Z} e^{-\frac{U(x)}{T}} \tag{12}$$

where $Z := \sum_{x \in \Lambda^n} e^{-\frac{U(x)}{T}}$ is called the partition function and $T = T(n)$ is called the temperature of the system. $U(x)$ is called the MRF system energy and represents an 'energy' metric of configuration x.

MRF optimization describes the task of finding minimal energy configurations, i.e. assigning values at each random variable (e.g. channel resources at each SU) to minimize the global system energy[4]. In fact, an MRF distribution is characterized by its energy function U that can be decomposed into a family of functions V, called potentials, which in turn are associated with a single site or interactions of neighboring sites and can play a role analogous to the utility functions in NUM. Considering pairwise[5] nearest-neighbor potentials (namely, the potential functions associated with a single site or a pair of neighboring sites [77, 79]), MRF optimization aims at selecting states, x_s, for each site s such that the system objective function is optimized, i.e. minimizing the MRF energy:

$$U = \sum_{\forall s} V_s(x_s) + \sum_{\forall neighbors \{s_1, s_2\}} V_{s_1 s_2}(x_{s_1}, x_{s_1}). \tag{13}$$

As an optimization framework, MRF theory can be straightforwardly used for resource management in MCRNs, both centralized and distributed. Especially for distributed MCRNs, MRF framework can be used to represent their special structure as multihop wireless networks, i.e. multiple nodes interact with their peers in a finite communication range, while their local decisions can influence the overall network performance (describing a knock-on effect, similarly in spirit to the MRF definition (11)). Compared to previously presented approaches that were inherently centralized and distributed solution were based on heuristics, MRF theory can seamlessly support distributed network optimization in resource management and other functions in multihop MCRNs. In this context, the MRF sites represent the wireless CR-aware nodes, the states of each site stand for their assigned resource and the MRF potentials are appropriately defined so as to reflect resource management objectives. For example, based

[4] The MRF system energy refers to the value of potential functions employed and not the battery energy of the terminals used for transmission and computation.

[5] Although pairwise nearest-neighbor potentials are commonly used to capture the interaction of communication pairs, higher order potential functions can be also utilized in a broader sense depending on the nature of the underlying problem.

on local only information and MRF optimization, the spectrum sharing problem among SUs can be efficiently addressed by appropriately assigning radio resources to each SU such that the overall spatial reuse of radio resources is optimized without causing interference to higher priority users [75]. In a slightly different approach, the dynamic selection of the most suitable routing protocol for the prevailing network conditions can be achieved through an MRF maximum a posteriori (MAP) optimization problem [7], where nodes make autonomous decisions by coherently combining their observations with prior beliefs.

The diversity of available resources in MCRNs, as well as their impact throughout the traditional protocol stack, can be facilitated by the construction of MRF potentials based on general cross-layer utility functions. In this manner, an MRF-based optimization framework is capable of contributing to network performance improvements, where cross-layer objectives are accomplished by an efficient and holistic resource management representation and assignment. It should be noticed that MRF optimization framework allows for practical non-convex utility functions (without restrictive assumptions), since the convergence to global optimal can be achieved through stochastic relaxation approaches, such as Gibbs sampling [78], by which each node interacts with neighbors based on an asynchronous message-passing system. This is a major difference compared to the DBRA and NUM approaches, since probabilistic rather than deterministic optimization is employed. However, based on such stochastic relaxation methods it theoretically takes an infinite time to reach the global optimal solution, thus sacrificing some degree of optimality in practical scenarios (where finite computation time is required) for the favor of implementation simplicity. On the other hand, alternative approaches for addressing discrete MRF-based optimizations are exploiting powerful dual decomposition techniques. In such cases, the corresponding MRF energy minimization problem is first decomposed into a set of appropriately chosen sub-problems and then their solutions are combined in a principled way [80]. In this manner, the MRF energy minimization problem is transformed into a constrained linear integer problem and auxiliary variables are added towards decoupling the objective function. Using dual decomposition, as presented in Section 4.2, the initial optimization problem is decomposed into a master problem and smaller MRF subproblems that are trivial to be solved. As analyzed in [80], the type and the size of the chosen subproblems, e.g. tree-structured MRFs, loopy subproblems, etc., can affect significantly the quality of the obtained solution as well as the convergence speed.

From an implementation perspective, the network designer should prefer MRF optimization framework for system configurations characterized by small size local neighborhoods such that the overall network will not become overburden, e.g. by the messages during Gibbs sampling solution approach. Two additional implementation issues can raise some concerns for applicability in real networking environments and especially in distributed MCRNs. First, if the convergence time to the optimal solution is too long such that the network conditions have changed appreciably in the meantime, then the optimality of solution cannot be guaranteed. Secondly, the appropriate tuning of potential function parameters poses challenges for an efficient design in real-time systems. However, compared to the signaling overhead and computational requirements of DBRA and NUM, MRF decision-making and implementation remain more efficient for distributed cases of MCRNs. It should be also noted that the availability of external resources and infrastructure based systems, such as connection to spectrum databases, can significantly differentiate the MRF formulation and implicitly determine the most beneficial methodology for MRF optimization.

All the above facets regarding decomposition, local decision-making, computational and signaling requirements and simplicity versus optimality will be jointly considered for the three approaches, namely back-pressure, NUM and MRF based, in the rest of this chapter. This joint consideration will be developed into a generic network architecture, which in turn can lead to cross-layer and component based resource management frameworks for MCRNs.

5 Joint Cross-Layer and Component-Based Resource Management (CCRM) for MCRNs

In the previous subsections three different, but related, cross-layer optimization frameworks were presented in the context of resource management, mostly relying on mathematical advances in network analysis (utility design). Even though these approaches have been applied in multiple cases in wireless networks and have demonstrated their promising potential for systematic resource management in various types of wireless topologies, the heterogeneity of the proposed optimization solutions in conjunction with the absence of formal cross-layer driven standardization efforts contribute to a complicated state that without doubt discourages commercialization. Towards this direction, the overall network architecture needs to be reconsidered holistically, in order to capture the emerging trends of cross-layer design and offer a flexible implementation platform for adopting the diverse existing optimization techniques in a unifying manner.

Network architecture determines functionality management rather than just resource allocation. In this context, it is desired to maintain the basic functionalities of existing protocols, while providing augmentations and additive features in a holistic and coordinated manner through the dynamic incorporation of different optimization methodologies. For these reasons, and based on the results of the component-based approach developed in software engineering [81], hereinafter, we discuss the design benefits of a component-based network architecture, where the principles of cross-layer design are offered in a modularized approach. Component-based design emphasizes the separation of concerns in respect of the wide-ranging functionality available throughout a given system. It is a reuse-based approach to defining, implementing and composing loosely coupled independent components into systems. This practice aims to bring about an equally wide-ranging degree of benefits in both the short-term and the long-term for the system itself. In this manner, it is possible to facilitate the implementation of diverse optimization techniques as different modules and circumvent complicated interactions between multiple layers, thus, reducing the required signaling among layers and retaining the simplicity of the traditional protocol layers, by conveying the complexity of each cross-layer optimization framework in distinct modules.

The objectives of this holistic architecture should combine the fundamental features of MCRNs, cross-layer analysis and utility (potential function) optimization. The following list briefly summarizes these objectives and in the following we highlight the directions and design choices that need to be made in order to fulfill them.

- The protocol stack should be responsive to variations in the underlying network conditions so that the optimal operating point is always maintained.
- The benefits of DBRA, NUM or MRF approaches should be exploited on a per case basis, which could mean instantaneous exploitation of each approach depending on the operational status of the network.
- As in the case of NUM, vertical and horizontal interdependence associated with network terminals should ensure that the underlying protocol mechanisms operate in appropriate time scales with each other and respond promptly to environmental/operational variations, while ensuring system stability and scalability.
- Spatial dependencies and limitations should be possible to be modeled and addressed holistically in the design and control of the network, and especially mitigate potential excessive interference emerging due to environmental and traffic conditions.
- Above all, the architecture should allow the achieved decomposition to eventually lead to network synthesis of different and possibly diverse mechanisms, when this leads to more efficient implementations, higher performance benefits at the same or at least slightly increased cost.

In the following, we first describe the proposed joint Cross-layer and Component-based Resource Management (CCRM) architecture and then the suitability and potential fit of each of the aforementioned cross-layer approaches within this framework.

5.1 The CCRM General Architecture

Taking the above into account, the envisioned network architecture is depicted in Fig. 5, where the protocol stack of each user is shown separately. For each protocol stack, a control module spanning across the protocol layer mechanisms is depicted and corresponds to the incorporated cross-layer logic, e.g. DBRA, NUM or MRF and possibly combinations thereof. Signaling is exchanged between the cross-layer module and each protocol layer, while protocol mechanisms of the same user communicate between well defined interfaces (as in the traditional protocol stack). Signaling between protocol layers and the control module is local (within the node) and thus bears no additional overhead, while signaling between mechanisms of the same layer at different users is considered as external overhead (delocalized). As long as this overhead is restricted between neighboring nodes, it is acceptable for distributed operation, otherwise the imposed cost is even higher. Protocol mechanisms of a user communicate with peer protocol mechanisms lying in other users by utilizing the services provided by the underlying protocol layers. Thus, Fig. 5 demonstrates the corresponding potentials for developing control mechanisms within the CCRM network architecture with respect to the cross-layer paradigms presented before.

One of the characteristic elements of the architecture depicted is the horizontal and vertical interdependence between protocol mechanisms. The vertical interdependence refers to layer interdependence in order to fulfil the data transferring operation, while the horizontal refers to peer protocol mechanisms that collaboratively achieve a broader distributed computation or social benefit, e.g. distributed routing and spectrum-aware

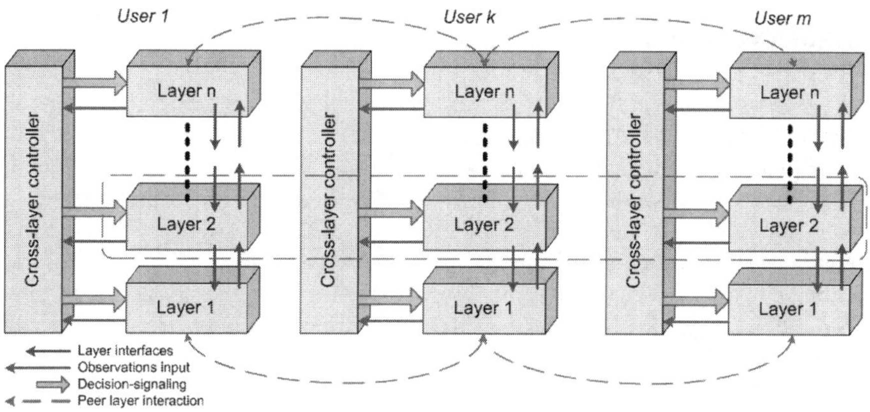

Fig. 5. Vertical and horizontal interdependence of CCRM network architecture. Horizontal involves communication between agents lying at the same layer at different users, while vertical interdependence is achieved not only between neighboring layers, but across the whole protocol stack via the cross-layer controller.

agents in MCRNs. This inherent two-tier interdependence of the protocol stack motivates the use of methodologies such as the NUM framework, where decomposition techniques can be exploited for assigning the different functionalities across the layers and users, while also ensuring convergence and synchronization.

The vertical cross-layer module receives in each node all observations regarding the network and environmental state (both user internal and external information available), user behavior, traffic conditions, etc., and based on the employed cross-layer framework it makes proper decisions for exploiting in the most suitable manner available resources at various layers. Feedback signaling (internal) is returned to protocol mechanisms in order to make the proper resource assignments and allocations. It should be also noted that in the envisioned architecture, the signaling information collected at the cross-layer decision module is of local nature, already available at each user, or at most obtained through simple message exchange (possibly piggybacked in exchanged data frames at no significant additional cost) in the immediate (one-hop) neighborhood of each user (external information).

The above architecture is rather suitable for implementing efficient resource management in multihop MCRNs, since it accommodates their distributed nature in collecting local information from the network and environment, assess the current state within a user's neighborhood and then make intelligent/informed decisions regarding the allocation and manipulation of available resources in the best possible manner, essentially implementing a form of the cognitive cycle operation presented in Fig. 2. The architecture of Fig. 5 seamlessly enables the cognitive operation of MCRNs, and in addition, it adopts and extends a novel cross-layer design approach, based on modular operation and control. The cross-layer decision-making can be implemented according to various frameworks, while the information collected should provide as many indications as possible, e.g. SINR values, interference, noise, etc. The resulting mechanisms should be able to respond in a timely and efficient manner to all variations, as shown in the cognitive cycle of Fig. 2.

The component-based design approach employed in the CCRM architecture, allows for great flexibility in terms of horizontal and vertical decision-making in the system. Through the part of the architecture that is common to all users (traditional protocol mechanisms, e.g. routing, medium access, coding), MCRN users are capable of maintaining their connectivity and communication, thus fulfilling the fundamental role of network formation. At the same time, the decision-making part of each user (cross-layer controller module) is decomposed from the rest of the protocol stack and thus potentially, apart from making different decisions, it can even vary across different users, according to the resources of each device, e.g. CPU, memory, energy, transceiver capabilities.

These diverse implementations of the cross-layer controller in each user may be applying completely different means for allocating resources. However, in any case, each different cross-layer controller makes decisions compatible with the rest controllers in the network, e.g. ensuring that capacity is fairly distributed if required, or that congestion is not concentrated in a single network part, or more importantly, that interference does not become excessive for the network, etc. This is implicitly ensured via peer signaling communication provided by protocol mechanisms of the common protocol layer stack.

The latter alone is a very significant challenge that requires considerable effort to achieve. Furthermore, it has major implications on the cumulative signaling of the architecture, often posing limitations to the achieved control over the resources and environmental conditions. Especially in ad hoc (multihop) MCRNs, all these processes do not have the luxury of centralized solutions, and thus, as will be explained in more detail in the following, great caution along with thorough analysis and testing is required, in order to achieve the seamless operations of the depicted architecture with the existing available frameworks.

5.2 Applicable Frameworks for the CCRM Architecture

With respect to the CCRM architecture described in the previous subsection, it became apparent that the selection of framework for designing a cross-layer controller is crucial for the operation and performance of the system. A first approach would be to employ a single framework, e.g. develop a controller according to NUM principles. However, the real value of CCRM is that a combination of frameworks, may potentially and under specific circumstances, yield hybrid cross-layer controllers. For instance, the latter could lead to designs where a back-pressure max weight algorithm is employed for scheduling and routing and an MRF decision module for channel allocation in MCRNs, or a back-pressure framework is used by one user and an MRF based controller by another user for managing the same types of resources by different users, e.g. both approaches used for channel allocation purposes.

The multi-commodity cross-layer approach presented previously can successfully capture the cross-layer data flow across the vertical stack, from the transport to the physical layer. However, applying it directly in the CCRM architecture as a sole methodology for the cross-layer controller at each user is not straightforward in the context of

CCRM, particularly for multihop MCRNs. The main reason is that the framework does not currently allow for efficient ways to describe the horizontal interdependence of decision-making and it does not provide methods for decomposition of protocol functionalities, as the NUM and MRF frameworks do. The multi-commodity formulation takes for granted a protocol layer separation (namely that congestion control takes place at the transport layer, routing at the network, etc.) and accounts for the data flowing through the boundaries of each layer. The latter does not allow for determining such boundaries dynamically, as desired in environments such as MCRNs. NUM and MRF frameworks currently cover better these aspects of dynamic operation decomposition. Thus, additional work would be required to ensure that the back-pressure approach can be holistically applied in the CCRM architecture.

Dynamic operation decomposition can be better accommodated by the NUM or MRF frameworks. More specifically, NUM fundamentally allows collecting environmental and network state input into cross-layer modules and then obtaining distributed solutions corresponding to decomposed protocol mechanisms and proper resource allocations, through formal decomposition methods. These also provide system stability and scalability guarantees (obtained efficiently and in a feasible manner mainly for centralized MCRNs) increasing the robustness of the infrastructures. NUM also allows dynamic redefinition of the protocol stack, as explained before, and thus allows both horizontal and vertical flexibility of the architecture.

Similarly, the MRF framework enables most of the aforementioned benefits of NUM, at the added feature of simpler and less complex implementation through stochastic relaxation methods. Of course, in order to achieve this, it sacrifices some degree of optimality of the obtained network adaptations, as explained before. However, the significant benefit of this approach is that it ensures distributed and feasible collaboration of the cross-layer network controllers across network users (mainly in multihop MCRNs) in a guaranteed manner and through strictly local (one-hop) signaling information, with formal convergence guarantees in most cases.

There are many alternatives to decompose the cross-layer function of each approach (NUM, MRF) into distributed control and assignment of networking functions across layers. In addition, the time-scales of protocol interactions within the stack of a user or across same layers of multiple users vary considerably. Some preliminary discussion regarding decomposition and timing scales was provided previously, and especially with respect to NUM. However, the response timing scales within the CCRM architecture will need to be considered across the whole architecture for different or even hybrid implementations of DBRA/NUM/MRF frameworks. Furthermore, such considerations should take into account the alternative decomposition solutions emerging in each case. Since various aspects of their operation, e.g. convergence, implementation, etc., differ inherently, the architecture should ensure that even though the employed control framework can be switched according to the component-based methodology, the employed decision-making timescales of the developed mechanisms will remain consistent, ensuring the stability and scalability of the whole system.

The CCRM architecture allows addressing the spatial properties of the formed network via formal methods, since both NUM and especially MRF inherently incorporate the spatial dependencies in their cross-layer formulation. An evolved form of DBRA would be applicable too. Thus, this can be further exploited in the decomposition of protocol mechanisms and signaling decision-making, so as to either overcome spatial limitations, e.g. at network borders, or exploit spatial relations that potentially allow the decomposition to be implemented in a more efficient manner and possibly with less cost (e.g. less message passing, etc.). A general trend emerging according to the characteristics of each approach presented, dictates that NUM appears more suitable for centralized, while MRF more suitable for multihop CCRM architectures. An evolved DBRA approach with protocol decomposition features would be applicable for both network types and hybrid designs would favor one or another network type, depending on the specific features employed and the design objectives to be satisfied.

On a final note, the proposed component-based CCRM framework allows properly addressing the benefit versus cost of collaboration tradeoff, which is especially important for the MCRN paradigm. The dynamic protocol and parameter adaptation, enabled by decomposition, allows for balancing such tradeoff towards the desired direction according to the operational objectives and via the provided mathematical tools (primal/dual decomposition, stochastic relaxation, etc.). Various such examples are already available in the literature, e.g. those provided in [69], [75]. In addition, this tradeoff leads to a broader tradeoff between performance and architecture, where on the one hand higher performance can be achieved by centralized approaches such as NUM based paradigms, while simplicity is favored by relaxation-based methods, e.g. Gibbs sampling in MRF paradigms. Given this tradeoff between performance and implementation, the tension between the two opposing objectives can relate to realizing short-term versus longer-term gains respectively, and their corresponding operational time-scales imposed by environmental or application factors.

5.3 The Road Ahead

From the previous discussions, it has become evident that considerable work has been already accumulated in the analysis and development of cross-layer protocol architectures for MCRNs and other types of wireless networks. However, significant additional steps need to be taken in order to ensure that such schemes operate as seamlessly and efficiently as intended to. Especially combining different features of the available approaches, e.g. the simplicity of MRFs with the stability properties of the back-pressure technique, is a direction currently under-explored and possibly even under-rated. Thus, several aspects of the described frameworks should be further investigated and extended to more systematic methodologies, which will be possible to be exploited in standard network analysis and design.

NUM offers a mathematical component-oriented solution to network coordination and resource allocation. This modular approach of cross-layer provided by NUM should be transformed into a broader framework that defines more explicitly resource management and control. The component-based approach developed in software engineering combined with cross-layer DBRA, NUM or MRF methodologies could provide the desired platform and theoretical tools for developing more holistic and computationally tractable techniques to required network adaptations, while also ensuring the desired performance with as little cost as possible. However, while a number of cross-layer design proposals already exist, and many of them are suitable for application in MCRNs, it is not clear which are the most suitable for specific types of emerging problems in MCRNs. The following, is a non-exhaustive list of related aspects that require additional study, and we believe that the field is mature enough to tackle them successfully in the near future. More in-depth analysis of these aspects will reveal the importance of each approach within diverse problem and goal sets. Other broader and more ambitious open issues are discussed by the end of this subsection.

- Investigation of the possibility to use concurrently DBRA/NUM/MRF based controllers at different users of the same network, e.g. some users employ NUM-based controllers, while others employ MRF-based or DBRA-based. Signaling issues will require thorough investigation in this case.
- Network optimization versus reconfiguration cost (feasibility-complexity issues). Should the controller perform reconfiguration or is it better to perform complete optimization from first grounds?
- Systematic and automated approach to switch between available cross-layer implementations, i.e. back-pressure, NUM or MRF based. Uniformly switch between presented controllers concurrently by all nodes based on the escalating network conditions.
- A systematic treatment on the variety of vertical decompositions, which will contribute to a rigorous understanding of the architectural choices for allocating functionalities to control modules.
- Explore if enumeration and comparison of alternative decompositions can be done systematically and in an automated fashion.
- Stability of a cross-layer architecture (harmonization and coordination among parts operating at different timescales).
- Study of control loop conflicts for parameters controlled by multiple mechanisms spanning over two or more layers.
- Investigation of the long-term architectural value (benefit) of the proposed protocol stack modifications in a holistic rather than fragmented fashion.
- Real experimentation of available cross-layer resource management approaches in diverse use cases by exploiting existing cognitive radio platforms.

In order to properly assess the magnitude of each of the above open issues regarding the CCRM architecture, and in general the cross-layer resource management, a thorough cost-benefit analysis is required in terms of implementation complexity versus the performance improvement yielded by each of them. Then, a more thorough quantitative analysis of each aspect should be performed, and progressively incorporated into the rest of the architecture and corresponding implementations.

Perhaps the major issue that requires attention and cannot be answered in a definite manner at the moment, is how different and diverse cross-layer design approaches can coexist within an even broader network analysis and control framework that allows adapting networks to anything a designer or user can think of, dynamically and by consuming the least of resources. Furthermore, the question of whether a truly generic framework suitable for MCRNs exists, a framework that could optimally identify network conditions and respond to them effectively, can drive ahead the corresponding research for many more years.

6 Conclusions

In this chapter, we focused on the area of mobile cognitive radio networks (MCRNs), and especially on relevant resource management methodologies. We explained the anticipated role of dynamic and self-organizing devices employed in MCRNs and presented the features they should bear in order to achieve efficient resource management. In MCRNs, cross-layer analysis emerged as a natural approach for tackling the diversity of involved parameters and mechanisms both for resource management and network optimization. Following the presentation of relevant state-of-the-art approaches, such as the back-pressure, NUM and MRF based, we introduced a cross-layer and component-based resource management architecture (CCRM) for mobile cognitive radios that extends the cross-layer paradigm and utility optimization principles, and aims at a systematic and potentially automated network reconfiguration approach. The CCRM architecture seems promising in potentially achieving efficient resource management and dynamic network optimization. We believe there is still much road to be covered regarding optimal network reconfiguration and resource management and this chapter aspired to provide a cornerstone towards this direction.

On a closing note, we iterate that the proposed approach for resource management in MCRNs is a bold attempt to exploit mathematical advances in network analysis in the last decade in the fields of optimization and discrete mathematics, as well as achieved advances in DSA approaches. Several problems pose considerable challenges, some of which are constrained by the current state-of-the-art available in the corresponding mathematics and device technology fields. Among others, such problems include the interplay between cross-layer control frameworks (DBRA/NUM/MRF) and DSA limitations regarding interference mitigation. The progress achieved in the near future will indicate the degree to which the problems mentioned in this chapter will be tackled. In any case, the effort in this chapter aspires to offer valuable insight and it is expected to contribute to even more fruitful results in the future.

Acknowledgment. This research was co-financed by the European Union (European Social Fund) and Hellenic national funds through the Operational Program 'Education and Lifelong Learning' (NSRF 2007-2013).
E. Anifantis was partially funded by the Foundation for Education and European Culture (IPEP).

References

1. Hansmann, U., Merk, L., Nicklous, M.S., Stober, T.: Pervasive Computing: The Mobile World, 2nd edn. Springer Professional Computing (August 2003)
2. Greenfield, A.: Everywhere: The Dawning Age of Ubiquitous Computing. New Riders, USA (2006)
3. Satyanarayanan, M.: Mobile Computing: The Next Decade. In: Proc. 1st ACM Workshop on Mobile Cloud Computing & Services: Social Networks and Beyond, MCS (June 2010)
4. Akyildiz, I.F., Lee, W.-Y., Vuran, M.C., Mohanty, S.: NeXt Generation / Dynamic Spectrum Access / Cognitive Radio Wireless Networks: A Survey. Computer Networks Journal (Elsevier) 50, 2127–2159 (2006)

5. Peterson, L.L., Davie, B.S.: Computer Networks: A Systems Approach, 5th edn. Morgan Kaufmann, USA (2012)
6. Akyildiz, I.F., Wang, Z.: A Survey on Wireless Mesh Networks. IEEE Communications Mag. 43(9), S23–S30 (2005)
7. Doyle, L.E., Kokaram, A.C., Doyle, S.J., Forde, T.K.: Ad Hoc Hetworking, Markov Random Fields, and Decision Making. IEEE Signal Processing Mag., 63–73 (September 2006)
8. Pujolle, G.: The Architecture of the Future Wireless Internet. In: Cho, K., Jacquet, P. (eds.) AINTEC 2005. LNCS, vol. 3837, pp. 157–167. Springer, Heidelberg (2005)
9. Tekinay, S., Goodman, D.J.: Wireless Internet > Wireless + Internet. In: Furht, B., Ilyas, M. (eds.) Wireless Internet Handbook, ch. 2. CRC Press (2003)
10. Molisch, A.F.: Wireless Communications. Wiley-IEEE Press (2005)
11. Tanenbaum, A.S.: Computer Networks, 4th edn. Prentice-Hall (2002)
12. Comer, D.: Internetworking With TCP/IP: Principles Protocols, and Architecture, 6th edn., vol. 1. Addison-Wesley (2013)
13. Zimmermann, H.: OSI Reference Model-The ISO Model of Architecture for Open Systems Interconnection. IEEE Trans. Communications 28(4), 425–432 (1980)
14. Raisinghani, V.T., Iyer, S.: Cross-layer Design Optimizations in Wireless Protocol Stacks. Elsevier Computer Communications 27(8), 720–724 (2004)
15. Lin, X., Shroff, N.B., Srikant, R.: A Tutorial on Cross-Layer Optimization in Wireless Networks. IEEE Journal on Selected Areas on Communications 24(8), 1452–1463 (2006)
16. Foukalas, F., Gazis, V., Alonistioti, N.: Cross-layer Design Proposals for Wireless Mobile Networks: A Survey and Taxonomy. IEEE Communications Surveys & Tutorials 10(1), 70–85 (2008)
17. Srivastava, V., Motani, M.: Cross-Layer Design: A Survey and the Road Ahead. IEEE Communications Magazine 43(12), 112–119 (2005)
18. Kawadia, V., Kumar, P.R.: A Cautionary Perspective on Cross Layer Design. IEEE Wireless Communication Magazine 12(1), 3–11 (2005)
19. Cardei, M., Cardei, I., Du, D.-Z.: Resource Management in Wireless Networking. Springer Network Theory and Applications, vol. 16 (2005)
20. Gavrilovska, L.M., Atanasovski, V.M.: Resource Management in Wireless Heterogeneous Networks (WHNs). In: 9th Int'l Conference on Telecommunication in Modern Satellite, Cable, and Broadcasting Services (TELSIKS), pp. 97–106 (October 2009)
21. Zander, J.: Radio Resource Management in Future Wireless Networks-Requirements and Limitations. IEEE Communications Mag. 35, 30–36 (1997)
22. Hills, A., Friday, B.: Radio Resource Management in Wireless LANs. IEEE Communications Mag. 42(12), S9–S14 (2004)
23. Periyalwar, S., Hashem, B., Senarath, G., Au, K., Matyas, R.: Future Mobile Broadband Wireless Networks: A Radio Resource Management Perspective. Wireless Communications and Mobile Computing 3(7), 803–816 (2003)
24. Cesana, M., Cuomo, F., Ekici, E.: Routing in Cognitive Radio Networks: Challenges and Solutions. Ad Hoc Networks 9(3), 228–248 (2011)
25. de Baynast, A., MŁhnen, P., Petrova, M.: ARQ-based Cross-layer Optimization for Wireless Multicarrier Transmission on Cognitive Radio Networks. Elsevier Computer Networks 52(4), 778–794 (2008)
26. Ding, L., Melodia, T., Batalama, S., Matyjas, J., Medley, M.: Cross-Layer Routing and Dynamic Spectrum Allocation in Cognitive Radio Ad Hoc Networks. IEEE Trans. Veh. Technol. 59(4), 1969–1979 (2010)
27. Liu, J., Chen, W., Cao, Z., Zhang, Y.J.A.: Cooperative Beamforming for Cognitive Radio Networks: A Cross-Layer Design. IEEE Trans. Commun. 60(5), 1420–1431 (2012)
28. Lu, J., Ma, M.: A Cross-Layer Cognitive Radio-Based Framework and CAC Scheme in WiMAX Networks. Springer Wireless Personal Communications 71(1), 255–273 (2013)

29. Ma, M., Tsang, D.H.K.: Cross-Layer Throughput Optimization in Cognitive Radio Networks with SINR Constraints. International Journal of Digital Multimedia Broadcasting 2010 (April 2010)
30. Shi, Y., Hou, Y.T., Zhou, H., Midkiff, S.F.: Distributed Cross-Layer Optimization for Cognitive Radio Networks. IEEE Trans. Vehicular Tech. 59(8), 4058–4069 (2010)
31. Guo, W., Huang, X.: Distributed Cross-layer Optimization for Wireless Regional Area Network-based Cognitive Radio Networks. Wiley Wirel. Commun. Mob. Comput (2011)
32. Mihov, Y.Y.: Cross-Layer QoS Provisioning in Cognitive Radio Networks. IEEE Commun. Lett. 16(5), 678–681 (2012)
33. Zhao, G., Sadler, B.M.: A Survey of Dynamic Spectrum Access. IEEE Signal Processing Mag. 23(3), 79–89 (2007)
34. Ren, P., Wang, Y., Du, Q., Xu, J.: A Survey on Dynamic Spectrum Access Protocols for Distributed Cognitive Wireless Networks. EURASIP Journal on Wireless Communications and Networking 2012, 60 (2012)
35. Zhang, Y., Zheng, J., Chen, H.-H. (eds.): Cognitive Radio Networks: Architectures, Protocols and Standards. CRC Press (Taylor and Francis Group) (May 2010)
36. Axell, E., Leus, G., Larsson, E.G., Poor, H.V.: Spectrum Sensing for Cognitive Radio: State-of-the-Art and Recent Advances. IEEE Signal Processing Mag. 29(3), 101–116 (2012)
37. Song, M., Xin, C., Zhao, Y., Cheng, X.: Dynamic Spectrum Access: From Cognitive Radio to Network Radio. IEEE Wireless Communications Mag. 19(1), 23–29 (2012)
38. Mitola, J., Maguire Jr., G.Q.: Cognitive Radio: Making Software Radios More Personal. IEEE Pers. Commun. 6(4), 13–18 (1999)
39. Dillinger, M., Madani, K., Alonistioti, N. (eds.): Software Defined Radio: Architectures, Systems and Functions. John Wiley and Sons, New York (2003)
40. Staple, G., Werbach, K.: The End of Spectrum Scarcity. IEEE Spectrum 41(3), 48–52 (2004)
41. Haykin, S.: Cognitive Radio: Brain-Empowered Wireless Communications. IEEE Journal Sel. Areas in Commun (JSAC) 23(2), 201–220 (2005)
42. Pawelczak, P., Nolan, K., Doyle, L., Oh, S.W., Cabric, D.: Cognitive Radio: Ten Years of Experimentation and Development. IEEE Communications Magazine 49(3), 90–100 (2011)
43. Wang, B., Liu, K.: Advances in Cognitive Radio Networks: A Survey. IEEE Journal of Selected Topics in Signal Processing 5(1), 5–23 (2011)
44. Akyildiz, I.F., Gutierrez-Estevez, D.M., Chavarria Reyes, E.: The Evolution to 4G Cellular Systems: LTE-Advanced. Physical Communication 3(4), 217–244 (2010)
45. Venkataraman, H., Muntean, G.-M. (eds.): Cognitive Radio and its Application for Next Generation Cellular and Wireless Networks. LNEE, vol. 116. Springer, Heidelberg (2012)
46. Cordeiro, C., Challapali, K., Birru, D., Shankar, S., Res, N., Manor, B.: IEEE 802.22: The First Worldwide Wireless Standard based on Cognitive Radios. In: Proc. 1st IEEE Int. Symp. New Frontiers in Dynamic Spectrum Access Netw (DySPAN), pp. 328–337 (November 2005)
47. Arkoulis, S., Anifantis, E., Karyotis, V., Papavassiliou, S., Mitrou, N.: On the Optimal, Fair and Channel-aware Cognitive Radio Network Reconfiguration. Elsevier Computer Networks 57(8), 1739–1757 (2013)
48. Javed, F., Mahmood, A., Shafi, I., Ali, S.: Open Research Areas in Cognitive Radios. In: 4th Int'l Conference on Computational Intelligence, Modelling and Simulation (CIMSiM), pp. 375–380 (September 2012)
49. Sengupta, S., Subbalakshmi, K.P.: Open Research Issues in Multi-Hop Cognitive Radio Networks. IEEE Commun. Mag. 51(4), 168–176 (2013)
50. Goldsmith, A., Ali Jafary, S., Maric, I., Srinivasay, S.: Breaking Spectrum Gridlock with Cognitive Radios: An Information Theoretic Perspective. Proc. of IEEE 97(5), 894–914 (2009)

51. Grossglauser, M., Tse, D.: Mobility Increases the Capacity of Ad Hoc Wireless Networks. IEEE/ACM Trans. Networking 10(4), 477–486 (2002)
52. Neely, M.J., Modiano, E.: Capacity and Delay Tradeoffs for Ad-Hoc Mobile Networks. IEEE Trans. Information Theory 51(6), 1917–1937 (2005)
53. Shakkottai, S., Rappaport, T.S., Karlsson, P.C.: Cross-Layer Design for Wireless Networks. IEEE Commun. Mag. 41(10), 74–80 (2003)
54. Fall, K.: A Delay-Tolerant Network Architecture for Challenged Internets. In: Proc. 2003 Conference on Applications, Technologies, Architectures, and Protocols for Computer Communications, pp. 27–34. ACM (August 2003)
55. Cerf, V., et al.: Delay Tolerant Network Architecture. IETF RFC 4838, Internet Society (April 2007)
56. Demmer, M., Ott, J., Perreault, S.: Delay Tolerant Networking TCP Convergence Layer Protocol, `http://tools.ietf.org/pdf/draft-irtf-dtnrg-tcp-clayer-07.pdf` (accessed October 2013)
57. Scott, K., Burleigh, S.: Bundle Protocol Specification. DTN Research Group, RFC 5050 (November 2007)
58. Georgiadis, L., Neely, M.J., Tassiulas, L.: Resource Allocation and Cross-Layer Control in Wireless Networks. Foundations and Trends in Networking 1(1), 1–144 (2006)
59. Cormen, T.H., Leiserson, C.E., Rivest, R.L., Stein, C.: Introduction to Algorithms, 3rd edn. The MIT Press, Boston (2009)
60. Borst, S.: User-level Performance of Channel-aware Scheduling Algorithms in Wireless Data Networks. IEEE/ACM Trans. Networking 13(3), 636–647 (2005)
61. Tse, D.N.C.: Optimal Power Allocation over Parallel Broadcast Channels. In: Proc. Int'l Symposium on Information Theory (June 1997)
62. Tassiulas, L., Ephremides, A.: Dynamic Server Allocation to Parallel Queues with Randomly Varying Connectivity. IEEE Trans. Information Theory 39(2), 466–478 (1993)
63. Tassiulas, L., Ephremides, A.: Stability Properties of Constrained Queueing Systems and Scheduling Policies for Maximum Throughput in Multihop Radio Networks. IEEE Trans. Automatic Control 37(12), 1936–1949 (1992)
64. Neely, M.J., Modiano, E., Rohrs, C.E.: Dynamic Power Allocation and Routing for Time Varying Wireless Networks. IEEE Journal on Selected Areas in Communications (JSAC), Special Issue on Wireless Ad-hoc Networks 23(1), 89–103 (2005)
65. Rockafellar, R.T.: Convex Analysis. Princeton Univ. Press, Princeton (1970)
66. Eryilmaz, A., Srikant, R.: Fair Resource Allocation in Wireless Networks using Queue-Length-Based Scheduling and Congestion Control. In: Proc. IEEE INFOCOM (March 2005)
67. Lee, J.W., Mazumdar, R.R., Shroff, N.B.: Opportunistic Power Scheduling for Dynamic Multiserver Wireless Systems. IEEE Trans. Wireless Communications 5(6), 1506–1515 (2006)
68. Kirk, D.E.: Optimal Control Theory: An Introduction. Dover Publications, USA (1998)
69. Chiang, M., Low, S.H., Calderbank, A.R., Doyle, J.C.: Layering as Optimization Decomposition: A Mathematical Theory of Network Architectures. Proc. IEEE 95(1), 255–312 (2007)
70. Kelly, F.P., Maulloo, A., Tan, D.: Rate Control for Communication Networks: Shadow Prices, Proportional Fairness and Stability. J. Operations Res. Soc. 49(3), 237–252 (1998)
71. Saraydar, C., Mandayam, N., Goodman, D.: Pricing and Power Control in a Multicell Wireless Data Network. IEEE J. Sel. Areas Commun. (JSAC) 19(10), 1883–1892 (2001)
72. Palomar, D.P., Chiang, M.: A Tutorial on Decomposition Methods for Network Utility Maximization. IEEE Journal Sel. Areas Commun. (JSAC) 24(8), 1439–1451 (2006)
73. Low, S.H.: A Duality Model of TCP and Queue Management Algorithms. IEEE/ACM Trans. Netw. 11(4), 525–536 (2003)
74. Lee, J.W., Tang, A., Huang, J., Chiang, M., Calderbank, A.R.: Reverse-engineering MAC: A Game Theoretic Model. IEEE J. Sel. Areas Comm. (JSAC) 25(6), 1135–1147 (2007)

75. Anifantis, E., Karyotis, V., Papavassiliou, S.: A Markov Random Field Framework for Channel Assignment in Cognitive Radio Networks. In: Proc. 7th IEEE Int'l Conf. on Pervasive Computing and Commun (PerCom) Workshop on Pervasive Wireless Netw (PWN) (March 2012)
76. Kindermann, R., Snell, J.L.: Markov Random Fields and their Applications. American Mathematical Society, Prov (1980)
77. Winkler, G.: Image Analysis, Random Fields and Markov Chain Monte Carlo Methods: A Mathematical Introduction, 2nd edn. Springer (2003)
78. Geman, S., Geman, D.: Stochastic Relaxation, Gibbs Distributions, and the Bayesian Restoration of Images. IEEE Trans. Pattern Anal. Mach. Intell. 6(6), 721–741 (1984)
79. Tan, X., Xi, W., Baras, J.: Decentralized Coordination of Autonomous Swarms Using Parallel Gibbs Sampling. Automatica 46(12), 2068–2076 (2010)
80. Komodakis, N., Paragios, N., Tziritas, G.: MRF Energy Minimization and Beyond via Dual Decomposition. IEEE Trans. on Pattern Analysis and Machine Intelligence 33(3), 531–552 (2011)
81. Håkansson, J., Carlson, J., Monot, A., Pettersson, P., Slutej, D.: Component-Based Design and Analysis of Embedded Systems with UPPAAL PORT. In: Cha, S(S.), Choi, J.-Y., Kim, M., Lee, I., Viswanathan, M. (eds.) ATVA 2008. LNCS, vol. 5311, pp. 252–257. Springer, Heidelberg (2008)

Energy-Efficient Routing in Cognitive Radio Networks

George Mastorakis[1], Constandinos X. Mavromoustakis[2], Athina Bourdena[3], Evangelos Pallis[3], Giorgio Sismanidis[4], Dimitrios Stratakis[3], and Stelios Papadakis[1]

[1] Technological Educational Institute of Crete, Department of Business Administration, Lakonia, Agios Nikolaos, 72100, Crete, Greece
{gmastorakis,spap}@staff.teicrete.gr
[2] University of Nicosia, Department of Computer Science, 46 Makedonitissas Ave., 2414 Engomi, Nicosia, Cyprus
mavromoustakis.c@unic.ac.cy
[3] Technological Educational Institute of Crete, Department of Informatics Engineering, Estavromenos, Heraklion, 71500, Crete, Greece
{bourdena,pallis}@pasiphae.eu, dstrat@ie.teicrete.gr
[4] Politecnico di Milano, School of engineering of computing systems, Piazza Leonardo da Vinci 32, 20133, Milano, Italy
giorgio.sismanidis@mail.polimi.it

Abstract. This chapter proposes a novel routing protocol enriched with a capacity-aware scheme that enables energy conservation and efficient data flow coordination among communication nodes with heterogeneous spectrum availability in distributed cognitive radio networks. Efficient routing protocol operation, as a matter of maximum energy conservation, maximum-possible routing paths establishments and minimum delays is obtained by utilizing both a signaling mechanism and an energy efficient scheme that were implemented based on a simulation scenario. This simulation scenario includes a number of secondary communication nodes, operating over television white spaces (TVWS) under the "spectrum of commons" regulation regime. The validity of the proposed energy efficient routing protocol is verified, by conducting experimental simulations and obtaining performance evaluation results. Simulation results validated routing protocol efficiency for minimizing energy consumption, maximizing resources exchange between secondary communication nodes and minimizing routing delays.

Keywords: Cognitive Radio Networks, Routing Protocols, TV White Spaces, Energy Conservation.

1 Introduction

Cognitive Radio (CR) technology [1] provides an emerging communication paradigm that exploits efficiently radio spectrum resources, enabling for the deployment of future sophisticated wireless networks. CR networks are comprised of communication nodes, capable of adapting their technical characteristics, based on interactions with

C.X. Mavromoustakis et al., *Resource Management in Mobile Computing Environments*, Modeling and Optimization in Science and Technologies 3,
DOI: 10.1007/978-3-319-06704-9_14, © Springer International Publishing Switzerland 2014

the surrounding spectral environment. They can sense a wide radio spectrum range, dynamically identify locally unused/unexploited frequencies and efficiently access them. This capability opens up the possibility of designing new dynamic radio spectrum access strategies/policies with the purpose of opportunistically reusing under-utilized frequencies at local level, such as "television white spaces" (TVWS) [2]. TVWS comprise VHF/UHF radio spectrum portions that are either resulted by switchover process from analogue to digital terrestrial television, or are completely under-utilized due to frequency planning principles ("Interleaved Spectrum") [3]. Therefore, introduction of CR networks in TVWS represents a disruption to the current "command-and-control" paradigm of TV/UHF spectrum management, thus the exploitation of CR technology is highly intertwined with the regulation models that would eventually be adopted [4], [5]. "Spectrum of Commons" regime is a well-suited regulation model, adopted in distributed cognitive radio networks, where resource allocation is presided locally by communication nodes, instead of using a centralized entity, such as a spectrum manager [6].

The flexibility in radio spectrum access phase by CR networks caused new challenges along with increased complexity in the design of communication protocols at different layers. More specifically, the design and adoption of efficient routing schemes, is a vital process for such an emerging networking paradigm. Ad-hoc CR networks are characterized by completely self-configuring architectures [6], where routing is challenging and different from routing in a conventional wireless network. A key difference is that spectrum availability in an ad-hoc CR network highly depends on primary communication nodes presence, thus, it is difficult to exploit a Common Control Channel (CCC), in order to establish and maintain a fixed routing path between secondary communication nodes.

Energy conservation figures an important aspect for the high performance deployment in ad-hoc CR networks. On one hand, the Energy Conservation scheme has to be reactive so that the Energy levels of wireless nodes will be tuned according to the estimated parameters (i.e. capacity, traffic [7] of the nodes). On the other hand, an Energy-efficient scheme has to take into consideration the bounded end-to-end delays of the transmissions. As the network lifetime is closely related to the transmission characteristics [8] of a source node to a destination node and the underlying routing protocol used [9], a mechanism that combines the temporal traffic-aware behavior of the node [10] and the efficient routing scheme in an end-to-end path has to be investigated.

In this context, this chapter elaborates on the design, development and experimental evaluation of an energy efficient routing protocol for ad-hoc CR network architectures, enabling for the effective communication of secondary communication nodes that operate under the "Spectrum of Commons" regime. More specifically, a signaling mechanism combined with an energy efficient scheme is proposed, based on the Backward Traffic Difference estimation [7, 8, 10] in contrast to the end-to-end bounded delay of the transmission. Based on the underlying routing scheme and the volume of traffic that each node receives/transmits, the proposed scheme aims at minimizing the Energy consumption by applying asynchronous, non-periodic Sleep-time assignment slot to the secondary wireless nodes. Following this introductory section, Section 2 elaborates on related work and research motivation, while section 3 presents the design and development of a novel green aware routing protocol, enabling for the energy efficient data transition, across secondary communication nodes with different TVWS availability. Section 4 elaborates on performance evaluation analysis of the proposed research approach, discussing experimental results and Section 5 concludes this chapter by highlighting fields for future research.

2 Related Work and Research Motivation

Conventional routing algorithms exploited in wireless ad-hoc networks, enable for the optimization of network performance metrics, such as end to end delay, switching delay and backoff delay. A rich literature on conventional routing protocols is available based on network-wide broadcast messages, without using any local hops information, towards improving the choice of optimum routing paths. Such approaches are not suited for wireless CR networks, since there is no support for concurrently considering radio spectrum availability of secondary communication nodes, as well as the effect on other primary nodes that share spectrum resources. In a general context, several research approaches have been recently proposed in [11], [12], [13], [14], towards addressing routing issues in CR networking environments.

Furthermore, a routing protocol is proposed in [15], exploited to combine geographical routing and radio spectrum assignment, towards avoiding regions with high presence of primary communication nodes. It also determines optimum routing path channel combinations that reduce delays in the network. A spectrum aware data adaptive routing algorithm is proposed in [16], where the end to end route selection depends on the amount of data to be transferred. Furthermore, the proposed routing protocol in [17] builds a forwarding mesh based on a set of available routes to the destination and opportunistically adapts during the forwarding process, according to the dynamic radio spectrum conditions. Moreover, a joint approach of on-demand routing and spectrum band selection is proposed in [18] for CR networking environments and a delay based metric is used to evaluate the quality of alternative routes. Most of the previous schemes are based on on-demand routing protocols and discover paths between source and destination communication nodes.

On the other hand, the routing mechanism has to be strictly associated with the Energy-efficiency when the CR networking architecture hosts wireless nodes requesting spectrum, via which the traffic will be transferred. Therefore, the routing mechanism in collaboration with an Energy efficient scheme should guarantee the end-to-end availability of requested resources whereas it should be able to significantly reduce the Energy Consumption. In addition, the mechanism should be able to maintain the requested scheduled transfers and the entire end-to-end connectivity. Many recent measurement studies [19] have convincingly demonstrated the impact of Traffic on the End-to-End connectivity [20] and thus showed the impact on the Sleep-time duration and the Energy Consumption. Measures extracted in real-time using realistic traffic [19-20] have shown that the impact of the responsiveness of the routing scheme in regards to the end-to-end transmission reliability is significant. Real-Time communication networks and multimedia systems, exhibit noticeable burstiness over a number of time scales [21-22]. Based on the stochastic traffic modeling, the traffic in most of the cases can be expressed in time exhibiting fractal-like characteristics. The problem of hosting a scheme, where in collaboration with the routing mechanism used, takes into account the traffic characteristics in order to conserve energy has not yet explored. The scheme will be able to tune the wireless interfaces of the nodes to the Sleep or Active state according to the incoming Traffic and a model which considers the next Sleep-time duration. Notwithstanding, many Sleep-time scheduling strategies were introduced that model the node transition between ON and OFF states. Existing scheduling strategies for wireless nodes could be classified into three categories: the coordinated sleeping [23], where nodes adjust their sleeping schedules, the random sleeping [24], where there is no certain adjustment mechanism between the nodes in the sleeping schedule with all the pros and cons [10], and on-demand adaptive mechanisms [25], where nodes enter into Sleep-state depending on the environment requirements whereas an out-band signaling is used to notify a specific node to go to sleep in an on-demand manner.

Although there are many schemes developed addressing different Energy Conservation methodologies, the combination of a traffic-aware scheduling scheme with the routing protocol supported by the CR networking architecture, has not yet been explored. The latter poses a fertile ground for the development of new approaches with the association of different parameters of the communication mechanisms, in order to reduce the Energy Consumption. Such schemes are classified into active or passive mechanisms. Active techniques conserve energy by performing energy conscious operations, such as transmission scheduling and energy-aware routing. Mavromoustakis et al. in [19] consider the association of Energy Conservation problem with different parameterized aspects of the traffic (like traffic prioritization) and enable a mechanism that tunes the interfaces scheduler to sprawl in the sleep state according to the activity of the traffic of a certain node in the end to-end path in real-time.

The main target of the proposed scheme and research approach of this chapter, is to exploit the incoming Traffic pattern in order to minimize Energy Consumption of secondary communication nodes. The scheme takes into consideration the repetition pattern of the Traffic, as well as the delay limitation (bounded delay) of each transmission. The proposed scheme then estimates the Backward Traffic Difference for extracting the time duration for which the nodes are allowed to Sleep (overcoming the network partitioning problems and considering the delay limitations of the transmission). This mechanism, in order to enable further recoverability and availability of the requested resources, uses promiscuous caching [10] in an opportunistic manner, to cache the packets destined for the node with turned-off interfaces (sleep state) onto intermediate nodes. This methodology enables, through the Backward Traffic Difference estimation, the next Sleep-time duration of the recipient node to be adjusted according to the activity duration and the volume of the traffic in collaboration with the routing mechanism.

3 Energy Efficient Routing Scheme Using Backward Traffic Difference Estimation

The transmission of secondary communication nodes in an ad-hoc CR network is based on radio spectrum opportunity, where routing has to take into account the availability of spectrum in specific geographical locations at local level. Spectrum awareness, route quality and route maintenance issues have to be investigated for different routing schemes, in order to enable for efficient data transfer across regions with heterogeneous radio spectrum availability, even when the network connectivity is intermittent or when an end to end path is temporarily unavailable. Figure 1 illustrates a simulation scenario, where primary nodes operate over specific channels in three geographical areas (i.e. Area A, B and C in Figure 1). Secondary nodes (i.e. 43 nodes were defined in simulation scenario) opportunistically operate, by utilizing remaining available channels in each geographical area (i.e. TVWS in Figure 1). It has to be noted here that a CCC does not exist between secondary nodes, which are located in neighboring geographical areas (i.e. Area A, B and C in Figure 1). In this case, secondary communication nodes, which are positioned in locations with higher TVWS availability (e.g. locations outside areas A, B and C) operate as intermediate relay nodes, switching between alternative channels. Therefore, such relay nodes enable for ad-hoc connections among secondary nodes, located inside areas A, B and C.

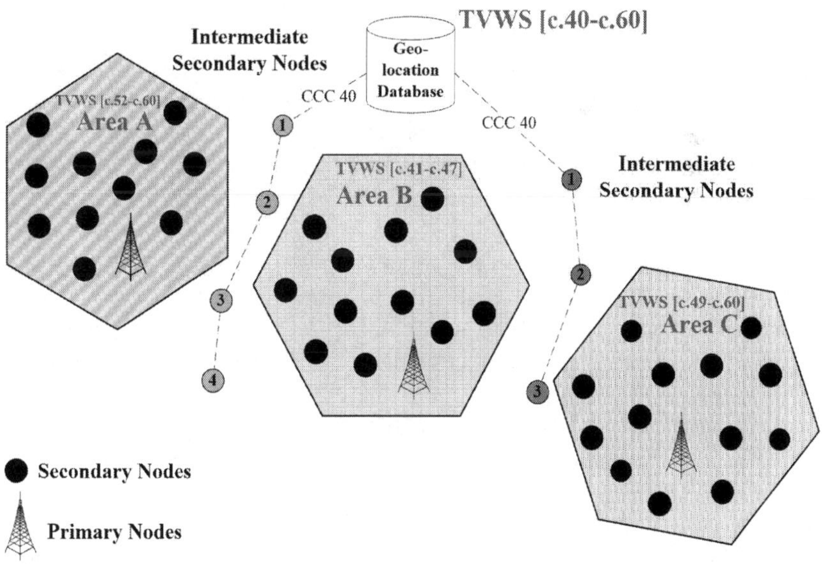

Fig. 1. Secondary communication nodes operating over heterogeneous TVWS

Taking into account this simulation scenario, spectrum awareness has to be investigated, regarding routing in such an ad-hoc CR network, where secondary nodes are prohibited to operate on spectrum bands occupied by primary nodes. The main target of routing in this CR networking environment is to provide optimal, high throughput data transfer by efficiently selecting the best routing paths among secondary nodes. In this framework, a novel routing protocol has to be adopted, in order to enable routing path discovery capabilities, considering TVWS heterogeneity of different geographical areas. Route quality issues have also to be investigated since the actual topology of such multi-hop CR networks is highly influenced by primary nodes behavior and classical ways of measuring/assessing the quality of end-to-end routes (nominal bandwidth, throughput, delay, energy efficiency and fairness) should be coupled with novel measures on path stability. In a general context, routing in an ad-hoc CR network over TVWS constitutes a rather important but yet unexplored problem, especially when a multi-hop network architecture is considered. Therefore, a novel routing protocol is vital to be designed and developed, in order to overcome the above mentioned challenges, towards establishing and maintaining optimum routing paths, among communication nodes with different radio spectrum availability.

3.1 Signalling Routing Mechanism

Secondary nodes located outside geographical areas A, B and C in the above mentioned scenario, are able to operate over all available channels (i.e. c.40-c.60) and act as intermediate nodes, connected with a Geo-location database that includes information regarding TVWS availability. They are also enhanced with routing mechanisms capabilities, enabling to determine routing paths between secondary nodes with different radio spectrum availability in such areas. Towards enabling for an optimum data transfer, among secondary communication nodes, a novel routing protocol was designed, developed and evaluated, by conducting experimental simulations. This routing protocol is based on the exchange of AODV-style messages [26] between secondary nodes, including two major steps in the route discovery process (i.e. route discovery and route reply step). Figure 2 presents the detailed signaling mechanism of the proposed routing protocol for handling both RREQ and RREP messages. A source node initiates a flow (i.e. New Flow in Figure 2), transmitting a RREQ message to an intermediate node located in a neighboring location. This intermediate node determines if it is possible to accommodate the incoming flow based on information stemming from the Geo-location database. In case that it is possible to accommodate it, performance metrics are evaluated and RREQ message is forwarded to the next hop. When destination node receives this message, it is informed of TVWS availability along the routing path from the source node. Destination node replies by sending a RREP message to the source node that includes relevant information, concerning channel allocation. Such data/information is mainly exploited to enable secondary nodes setting their channel of operation along the routing path. When source node receives RREP message, routing path has been established and useful data transmission is initiated.

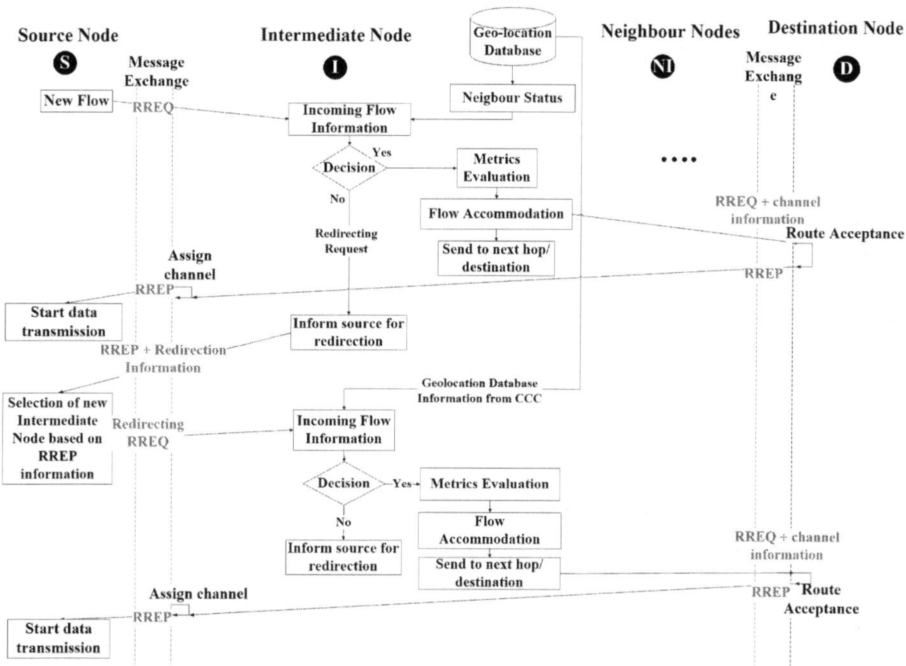

Fig. 2. Message exchange process of the proposed routing protocol

In the case when the intermediate node is not capable to accommodate the incoming flow (i.e. new flow in Figure 2), a redirection mechanism (redirection process in Figure 2) is in charge of informing the source node, about the neighboring node, which could possibly act as an alternative intermediate node. Towards further optimizing the proposed routing protocol, an assigning mechanism was designed and adopted to alleviate service load of intermediate nodes. This process is adapted to each intermediate node, which is further able to determine if a neighbor node performs better during the process of routing paths establishment. More specifically, when a source node initializes a new flow, by sending a RREQ, the intermediate node is informed regarding the status of neighboring nodes from the geo-location database through the CCC. Then, the intermediate node evaluates the new flow (i.e. evaluation of performance metrics) and encapsulates the evaluation results in a message that it is forwarded to all neighboring nodes. Once neighboring nodes receive a redirecting request, they check its validity with the corresponding flow, ensuring that they are not the source/destination nodes or next-hop nodes of that flow. Then the neighboring nodes initiate a process, in order to evaluate the flow and they send to the intermediate node the result of the evaluation through a redirecting reply message. Once the intermediate node receives the redirecting reply from several of its neighboring nodes it then selects the optimum one, in order to serve/accommodate the incoming flow.

For enabling Energy-Efficiency in the proposed framework a Backward Traffic Difference (BTD) estimation [19] methodology is used. The main additional contribution is that, in the proposed framework, the BTD estimation is bounded by the delay limitations of the transmission, whereas it takes into consideration the hop-by-hop link delay as well as the total end-to-end delay of the transmission. The later should satisfy the delay requirements of the transmission. The designed model guarantees the End-to-End availability of requested resources, while it significantly reduces the Energy Consumption and maintains the requested scheduled transfers, in a mobility-enabled communication. The innovation adopted in this scheme is that each secondary mobile node uses different assignment(s) of sleep-wake schedules based on the incoming traffic difference that each node receives through time. The Sleep-time duration is assigned according to the BTD scheme in a dissimilar manner in order to enhance node's lifetime, whereas it avoids mutation which, will result in network partitioning and resource sharing losses.

Assuming that a mobile secondary node has already used the depicted routing scheme of the previous section and established an End-to-End connection in order to transmit requested content/packets. Routing occurs on the End-to-End basis and each node separately runs the Traffic-aware mechanism using the BTD as is described in the following section. The mechanism measures the traffic that traverses each one of the nodes, where the BTD estimation through the assigned time-window frame will affect the Sleep-time duration and enable Energy Conservation onto nodes as conducted simulation experiments show.

3.2 Backward Traffic Estimation for Energy-Efficient Transmission

The scheme takes into consideration the incoming nodal traffic and estimates the Sleep-time duration of the node according to the Backward Traffic Difference (BTD), using a certain window frame-size. Traffic that is being traversed in a path is being forwarded on a hop-by-hop basis from one node to another until it reaches the destination. On one hand, if node is available and in Active-state, it receives the transfer (for example file), whereas if the node receiving the file is not the destination, it forwards the packet to the destination node via other neighbouring hop-nodes in the path[1]. On the other hand, if the next hop-node is not available to receive and process within a specified time-frame the transmission to the next-hop node, then promiscuous caching [10] of the transmitted packet occurs in the path. This is performed in order to buffer the packets that are intended for the destination node. Therefore the proposed traffic-based framework is focusing on the Traffic that is incoming for each node and for a specified time-window T. Packets will be sent from a source to a destination via intermediate nodes as expressed in the routing procedure earlier. The monofractality properties [10] were taken into consideration for modelling the energy schedules for a certain time-frame in order to enable Energy Conservation. The traffic and the monofractal characteristics of it were considered in order to enable greater associativity with the self-similar behaviour expressed in [10, 19], where the window of the traffic duration is t_w. In this work we consider the window to be

[1] The path is constructed according to the routing scheme explored in the previous section.

$$t_w(s,d) = \{\lim_{t\to k\cdot t} F_{n(t)} \in t(s,d) : R_N(t) \approx R_N(t-k\tau)\forall k < 2\} \quad (1)$$

where $t_w(s,d)$ is the time window measure for the multipath pair source-destination model and where the limit of it should be bounded into $k\tau$ time duration for the determined window size. *k* should be less than 2 in order to satisfy the monofractality property of the repetition index $R_N(t)$ of the incoming traffic [10].

When nodes in the transmission path are expecting traffic they keep their communication network interfaces in Active-state for time *t*. This means that if the transmission will delay with $d_t > t_{active}$ where t_{active} is the active time duration of the wireless nodal interface, then nodes may set their interfaces to an Energy Conservation state (Sleep-state). In this respect, the scheme enables the promiscuous caching [10] to be enabled. The packets that are destined for the certain node can be cached for a specified amount of time (as long as the *Node (i)* is in the Sleep-state) in the 1-hop neighbour node (*Node(i-1)* that is in Active state) in order to be recoverable when the referred node enters the Active state. When certain node has incoming traffic then the node remains active for prolonged time. As a showcase, this work takes the specifications of the WiMax IEEE 802.16e (specs v. 2005) [27] that are recommending the duration of the forwarding mechanism that takes place in a non-power saving mode lays in the interval *0.1 nsec < τ <1 psec*. This means that approx. 80 times in a msec the communication's triggering action between nodes may result a problematic end-to-end transmission reliability/accuracy. Adaptive Dynamic Caching [8] takes place and enables the packets to be "cached" in the 1-hop neighbouring nodes. Correspondingly, if node is no-longer available due to sleep-state in order to conserve energy (in the interval slot *T=0.08 μsec*), then the packets are cached into an intermediate node with adequate capacity equals to: $C_{t_f,k(s)}(t) > C_{t_f,i}(t)$, where $C_{t_f} > \alpha \cdot C_i$; where α_i is the capacity adaptation degree [10] based on the time duration of the capacity that is reserved on node N of C_k; where $C_{t_f,k(s)}(t)$ is the needed capacity where i is the destination node and k is the buffering node (a hop before the destination via different paths).

This work associates also the Backward Difference Traffic moments with the Sleep-time duration in order to tune the Active durations of a node according to the transmissions' activities and the expected traffic for the next time step. This is performed via the BTD estimation which enables the capacity of the traffic $C(t)$ that is destined for the Node *i* in the time slot (duration) *t*, and the traffic capacity $C_{N_i(t)}$ which is cached onto *Node (i-1)* for time *t*, to directly affect the Sleep-time of a node. The one-level Backward Difference of the Traffic is evaluated by estimating the difference of the traffic while the *Node(i)* is set in the Sleep-state for a period, as follows:

$$\begin{aligned}\nabla C_{N_i(1)} &= T_2(\tau) - T_1(\tau-1)\\ \nabla C_{N_i(2)} &= T_3(\tau-1) - T_2(\tau-2)\\ &\vdots\\ \nabla C_{N_i(n+1)} &= T_n(\tau-(n-1)) - T_2(\tau-(n-2))\end{aligned} \quad (2)$$

where $\nabla C_{N_i(1)}$ denotes the first moment traffic/capacity difference that is destined for *Node(i)* and it is cached onto Node (i-1) for time τ, $T_2(\tau)-T_1(\tau-1)$ is the estimated traffic difference while packets are being cached onto *(i-1)* hop for recoverability as in [19]. The Traffic Difference is estimated so that the next Sleep-time duration can be directly affected according to the following:

$$\delta(C(T)) = C_{total} - C_1, \forall C_{total} > C_1, T \in \{\tau-1, \tau\} \quad (3)$$

where the Traffic that is destined for *Node(i)*, urges the Node to remain active for $\frac{\delta(C(T))}{C_{total}} \cdot T_{prev} > 0$, T_{prev} is the previous Sleep-time duration ($\{\tau-1,\tau\}$) of the node. On the contrary with related approaches, this work measures the BTD within a certain transmission time-frame. This means that each transmission is bounded by a certain delay limitation (time-duration $t_w(s,d)$), which cannot be overtaken. When a node receives traffic, the traffic flow t_f, can be modelled as a stochastic process [19, 20] and denoted in a cumulative arrival form as $A_{t_f} = \{A_{t_f}(T)\}_{T \in N}$, where $A_{t_f}(T)$ represents the cumulative amount of traffic arrivals in the time space [0..T]. Then, the $A_{t_f}(s,T) = A_{t_f}(T) - A_{t_f}(s)$, denotes the amount of traffic arriving in time interval $(s < t]$. Hence the next Sleep-time duration for *Node (i)* can be evaluated as a function of the Traffic that traverses the *Node (i)* provided that the amount of traffic arriving in time interval $(s < t]$ is measured according to the total aggregated Traffic/Capacity that the channel can handle at time t. The next Sleep-time duration for *Node (i)*can be defined as:

$$L_i(n+1) = \frac{\delta(C(T) \mid A_{t_f}(s,T))}{C_{total}} \cdot T_{prev}, \forall \delta(C(T)) > 0,\ t_w(s,d) < 2\frac{\delta_{ij}}{\Delta_{\max}} \quad (4)$$

where δ_{ij} is the delay that the transmission experiences to reach destination *j*, $\Delta_{\max}$ is the max allowed delay-duration that the transmission cannot overtake. The aggregated traffic destined for *Node (i)* should satisfy the $\sup_{s \le T}\left\{\sum_{t_f=1}^{N} A_{t_f}(s,T) - C_{t_f}(T)\right\}$, for traffic flow t_f at time T and $C_{t_f}(T)$ represents the service capacity of the *Node(i-1)* for this time duration. Taking into consideration the above stochastic estimations, the

Energy Efficiency EE_{t_f} can be defined as a measure of the capacity of the *Node(i)* over the *Total Power consumed* by the *Node*, as:

$$EE_{t_f}(T) = \frac{C_{t_f}(T)}{TotalPower} \tag{5}$$

Equation 5 above can be defined as the primary metric for the lifespan extensibility of the wireless node in the system.

The basic steps of the proposed scheme can be summarized in the pseudocode of Table 1.

Table 1. Basic steps of the proposed BTD scheme with the bounded transmission delay

```
for Node(i) that there is C(t)>0 {
  while (C_{N_i(t)}>0) { //cached Traffic measurement
    Evaluate (∇C_{N_i(1)});
Calc( δ(C(T)) = C_total − C_1, ∀C_total > C_1, T ∈ {τ−1, τ} )
if (Activity_Period= δ(C(T))/C_total · T_prev >0)
//Measure Sleep-time duration
L_i(n+1) = δ(C(T) | A_{t_f}(s,T))/C_total · T_prev, ∀δ(C(T)) > 0, t_w(s,d) < 2 δ_ij/Δ_max
Sleep (L_i(n+1)); //sleep duration for the upcoming slot
 } //for
}//while
```

4 Performance Evaluation Analysis, Experimental Results and Discussion

A number of experimental tests were conducted, in order to validate the efficiency of the proposed routing protocol. Performance evaluation results were extracted, by conducting exhaustive simulation runs and experimentation using the NS-2 and the generated real traffic traces for implementing the proposed scenario. The energy consumption model used in the simulation for the calculation of the amount of energy consumed is based theoretically on the specifications of the WiMax IEEE 802.16e (ver. 2005) [27]. The extracted results are characterizing the trade-off issues between the performance in deploying the discussed scenario and the Energy consumption of each secondary CR node by using the proposed traffic-oriented scheme. Results also encompass comparisons with other existing schemes for the throughput, the reliability and the accuracy offered by the proposed framework as well as EC efficiency conveying an estimated confidence interval (CI) of approximately 3%<CI<5%. All confidence intervals were found to be less than 5% of the mean values of the certain examined parameters. The mobility model adopted in this work is based on the probabilistic mobility scenario derived by Fractional Random Walk. The probabilistic random walk mobility model was derived from the Brownian motion [28], where nodes are moving according to certain probabilities with respect to the location and the time.

According to such simulation scenario, a number of data flows are contending to pass through the same intermediate node, thus evaluation of delay metrics is crucial, for an efficient performance of the proposed routing protocol. In this context, a number of delay metrics [18] are evaluated, such as end to end delay, backoff delay, switching delay and queuing delay. End to end delay from the source node up to the destination node is computed as the overall sum of queuing delay and node delay:

$$D_{end\ to\ end} = D_{queuing} + D_{node} \tag{6}$$

Node delay at an intermediate node i is based on switching delay and backoff delay and is computed as follows:

$$D_{node} = \sum_{1}^{i}(D_{switching} + D_{backoff}) \tag{7}$$

Figure 3 depicts simulation results and performance comparison of mean end to end delay, while the number of simultaneous active flows is increasing. It is clear that when the proposed routing protocol incorporates the assigning mechanism, mean end to end delay is decreased in comparison to the performance of the simple version of the protocol. It has to be noted here that when the number of simultaneous active flows in the network is small, assigning mechanism does not show much advantage, comparing to the case where multiple flows are initiated in the network.

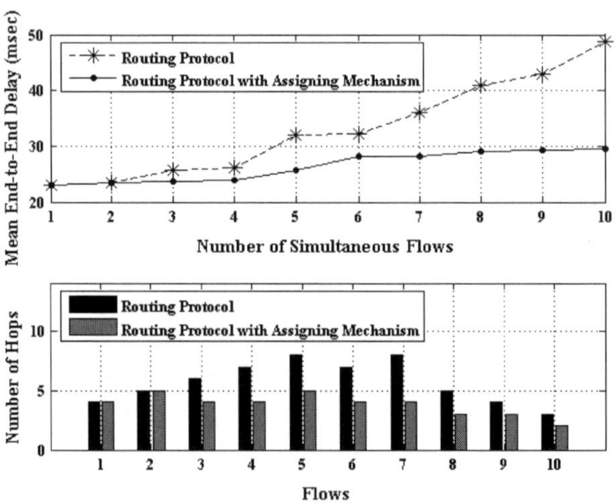

Fig. 3. Mean End-to-End delay

Furthermore, Figure 4 presents simulation results regarding the number of hops required, in order to make feasible ten simultaneous routing paths between source and destination nodes. This comparison results that the proposed routing protocol with the assigning mechanism performs better reducing the number of hops in the network. More specifically, the routing scheme with the traffic-aware activity and the Sleep-Wake schedules enables the extensibility of the lifespan of each secondary node in the system.

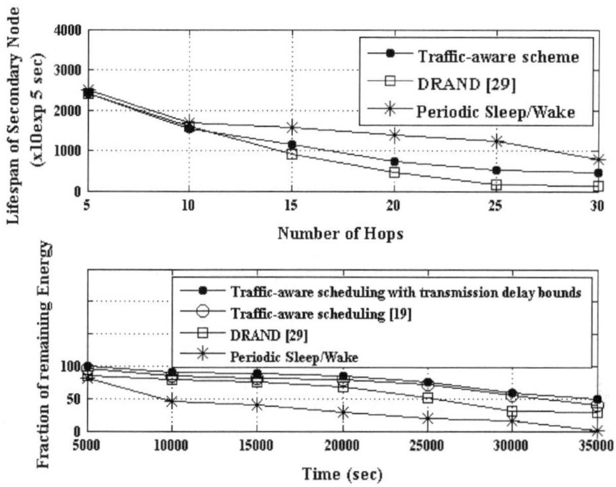

Fig. 4. Network Lifetime and Remaining Energy of each secondary node in the CR system

The comparative simulation conducted with other compatible schemes, like the DRAND [29] scheme, the periodic Sleep-Wake schedules and the single-moment Traffic-aware scheme adopted in [19], shows the efficiency of the proposed framework. In turn the fraction of the remaining energy of each secondary wireless device remains at relatively high levels, compared with the results extracted for different frameworks.

Figure 5 shows the Successful packet Delivery Ratio (SDR) with regards to the End-to-End streaming delay and the delay duration/total delay for k-hops in the

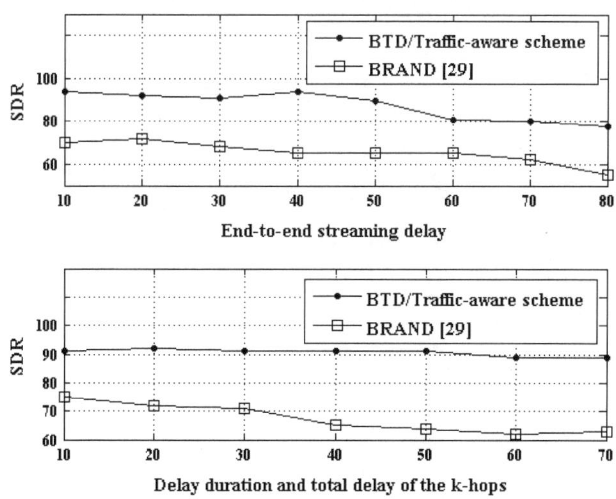

Fig. 5. Comparative results for the Successful packet Delivery Ratio (SDR)

communicating path, for mobile secondary nodes. The BTD scheme offers greater SDR in the End-to-End path even though the transmission delay in the path can be successively bounded to certain delay limitations.

Comparative results regarding the offered associated Throughput as well as the End-to-End latency for different fading and mobility models respectively are shown in Figure 6. As signal strength and the associated fading characteristics are posing a major factor for the End-to-End reliable connection, the proposed scheme behaves satisfactorily with respect to the fading Rayleigh model used. It is important to mention that the fading characteristics of the channels affect vertically the transmission rate of the channels, whereas the Throughput is also affected. In our conducted experiments the Rayleigh fading small scale fading of 2% with mean noise factor of 4dB causes minor Throughput degradation ranging from 4-31% -in worst case affection. The Throughput remained at significantly high levels due to the recoverability promiscuous caching methodology utilized. Moreover, the proposed scheme was evaluated for the End-to-End latency experienced by the secondary users since their initial request, for two mobility models, namely Uniform Random Mobility Pattern and Random Waypoint mobility. The latter, Random Waypoint model, uses for the movement of mobile users a certain likelihood based on the their location, velocity and acceleration change over time whereas, Uniform Mobility uses only uniform distribution model to denote the next mobility/movement.

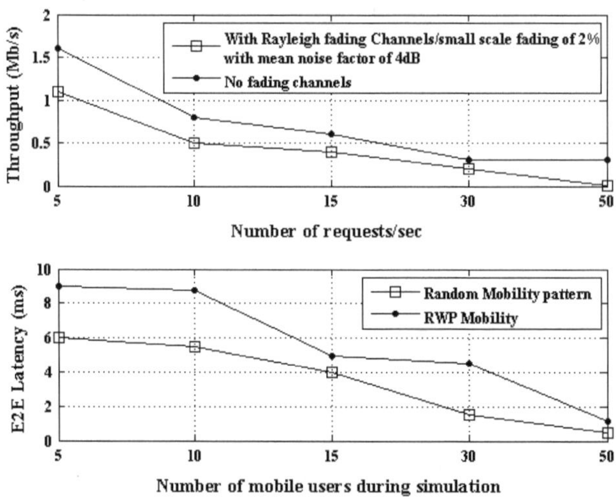

Fig. 6. Comparative results for the Throughput and the End-to-End latency

Figure 7 shows the Complementary Cumulative Distribution Function (CCDF) for the resource sharing reliability with the number of secondary users. Extracted measures include the fading characteristics where, with mean 4 dB noise factor using the Rayleigh fading, the CCDF remains at high levels. This occurs when the number of secondary users increases resulting to the awareness of an established path, which enables the secondary users to receive and transmit efficiently the requested data.

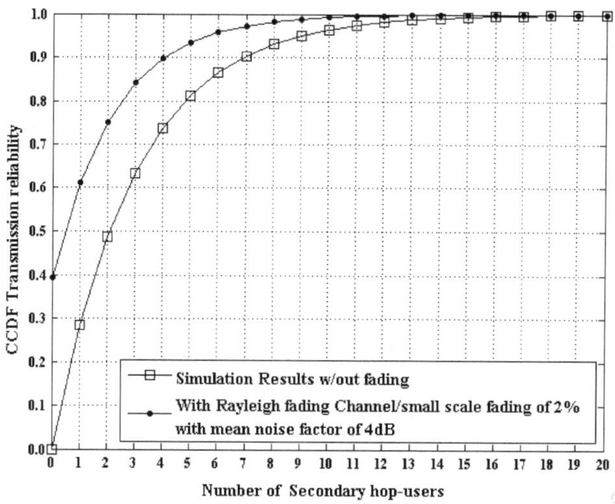

Fig. 7. CCDF Sharing Reliability with the Number of sharing secondary-users in the CR system

5 Conclusions and Further Research

This work proposes a routing mechanism, where in collaboration with the underlying BTD scheme, it enables energy conservation and reliable data flow among secondary communication nodes with heterogeneous spectrum availability in CR systems. The proposed routing scheme establishes an End-to-End optimal path whereas, secondary nodes in the CR system can efficiently and, in a collaborative manner, share requested data/resources. On the contrary with previous researches [30], [31], [32], [33], [34], this work measures the BTD within a certain transmission time-frame where, the bounded End-to-End delay of a transmission is taken into consideration for each secondary node. The performance evaluation through simulation shows that the proposed routing scheme in collaboration with the BTD mechanism, manipulates the energy consumption of each secondary node/device effectively. Moreover, the traffic-aware management scheme can significantly reduce the energy consumed and can keep the throughput response of the system at relatively high-levels. Comparative measurements with other similar schemes [35], [36], [37], [38] show that the proposed methodology can efficiently conserve the energy and can significantly extend the lifetime of each secondary node in the CR network.

References

1. Akyildiz, I.F., Lee, W.Y., Vuran, M.C., Mohanty, S.: NeXt generation/dynamic spectrum access/cognitive radio wireless networks: a survey. Computer Networks 50, 2127–2159 (2006)

2. Bourdena, A., Pallis, E., Kormentzas, G., Mastorakis, G.: A prototype cognitive radio architecture for TVWS exploitation under the real time secondary spectrum market policy. Physical Communications, Elsevier (2013), `http://dx.doi.org/10.1016/j.phycom.2013.11.003`
3. Bourdena, A., Pallis, E., Kormentzas, G., Skianis, C., Mastorakis, G.: Real-Time TVWS Trading Based on a Centralised CR Network Architecture. In: IEEE International Workshop on Recent Advances in Cognitive Communications and Networking, IEEE Globecom2011, pp. 994–999. IEEE Press, Texas (2011)
4. Bourdena, A., Pallis, E., Kormentzas, G., Mastorakis, G.: A radio resource management framework for TVWS exploitation under an auction-based approach. In: 8th IEEE International Conference on Network and Service Management (IEEE CNSM2012), pp. 204–208. IEEE Press, Las Vegas (2012)
5. Bourdena, A., Pallis, E., Kormentzas, G., Skianis, H., Mastorakis, G.: QoS provisioning and policy management in a broker-based CR network architecture. In: IEEE Globecom 2012, Anaheim, California, USA (2012)
6. Hossain, E., Niyato, D., Han, Z.: Dynamic spectrum access and management in cognitive radio networks. Cambridge University Press (2009)
7. Mavromoustakis, C.X.: Using Backward Traffic Difference Estimation for Efficient Energy Saving Schedules in Wireless Device. IEEE CommSoft E-Letters 1, 1–6 (2012)
8. Charalambous, M.C., Mavromoustakis, C.X., Bani Yassein, M.: A Resource Intensive Traffic-Aware Scheme for Cluster-based Energy Conservation in Wireless Devices. In: 14th IEEE International Conference on High Performance Computing and Communications (HPCC 2012) of the Third International Workshop on Wireless Networks and Multimedia (WNM 2012), pp. 879–884. IEEE Press, Liverpool (2012)
9. Shpungin, H.: Energy Efficient Online Routing in Wireless Ad Hoc Networks. In: 8th Annual IEEE Communications Society Conference on Sensor, Mesh and Ad Hoc Communications and Networks (IEEE SECON 2011). IEEE Press, Salt Lake City (2011)
10. Mavromoustakis, C.X., Zerfiridis, K.G.: On the diversity properties of wireless mobility with the user-centered temporal capacity awareness for EC in wireless devices. In: 6th IEEE International Conference on Wireless and Mobile Communications, ICWMC 2010, pp. 367–372. IEEE Press, Valencia (2010)
11. Khalife, H., Ahuja, S., Malouch, N., Krunz, M.: Probabilistic pathselection in opportunistic cognitive radio networks. In: IEEEGlobecom 2008. IEEE Press, New Orleans (2008)
12. Ma, H., Zheng, L., Ma, X., Luo, Y.: Spectrum aware routing for multihopcognitive radio networks with a single transceiver. In: 3rd International Conference on CognitiveRadio Oriented Wireless Networks and Communications (CrownCom 2008), pp. 1–6. IEEE Press, Singapore (2008)
13. Shiang, H.P., van der Schaar, M.: Delay-sensitive resource managementin multi-hop cognitive radio networks. In: 3rd IEEE Symposium on New Frontiers in Dynamic Spectrum Access Networks (DySPAN 2008). IEEE Press, Chicago (2008)
14. Xin, C., Ma, L., Shen, C.C.: A path-centric channel assignmentframework for cognitive radio wireless networks. Mobile NetworksApplications (Kluwer) 13, 463–476 (2008)
15. Chowdhury, K.R., Di Felice, M.: Search: a routing protocol for mobile cognitive radio ad-hoc networks. Computer Communication 32, 1983–1997 (2009)
16. Pal, R.: Efficient routing algorithms for multi-channel dynamic spectrum access networks. In: 2nd IEEE Symposium on New Frontiers in Dynamic Spectrum Access Networks (IEEE DySPAN 2007), pp. 288–291. IEEE Press, Dublin (2007)

17. Pefkianakis, I., Wong, S.H.Y., Lu, S.: SAMER: spectrum aware mesh routing in cognitive radio networks. In: 3rd IEEE Symposium on New Frontiers in Dynamic Spectrum Access Networks (DySPAN 2008). IEEE Press, Chicago (2008)
18. Chen, G., Liu, W., Li, Y., Cheng, W.: Spectrum aware on-demand routing in cognitive radio networks. In: 2nd IEEE Symposium on New Frontiers in Dynamic Spectrum Access Networks (IEEE DySPAN 2007), pp. 288–291. IEEE Press, Dublin (2007)
19. Mavromoustakis, C.X., Dimitriou, C.D., Mastorakis, G.: Using Real-Time Backward Traffic Difference Estimation for Energy Conservation in Wireless Devices. In: 4th International Conference on Advances in P2P Systems (AP2PS 2012), Barcelona, Spain, pp. 18–23 (2012)
20. Mavromoustakis, C.X., Karatza, H.D.: Real time performance evaluation of asynchronous time division traffic-aware and delay-tolerant scheme in ad-hoc sensor networks. International Journal of Communication Systems (IJCS), Wiley 23, 167–186 (2010)
21. Yu, J., Petropulu, A.P.: Study of the effect of the wireless gateway on incoming self-similar traffic. IEEE Trans. Signal Processing 54, 3741–3758 (2006)
22. Zhuang, X., Pande, S.: A scalable priority queue architecture for high speed network processing. In: 25th IEEE International Conference on Computer Communications (INFOCOM 2006), pp. 1–12. IEEE Press, Barcelona (2006)
23. Cao, Q., Abdelzaher, T., He, T., Stankonic, J.: Towards optimal sleep scheduling in sensor networks for rare-event detection. In: 4th International Symposium on Information Processing in Sensor Networks, Piscataway, NJ, USA (2005)
24. Cunqing, H., Tak-Shink, P.Y.: Asynchronous Random Sleeping for Sensor Networks. ACM Transactions on Sensor Networks (TOSN) 3 (2007)
25. Jawhar, I., Wu, J., Agrawal, P.: Resource Scheduling in Wireless Networks Using Directional Antennas. IEEE Transactions on Parallel and Distributed Systems 21, 1240–1253 (2010)
26. Perkins, C.E., Royer, E.M.: Ad hoc on-demand distance vector routing. In: IEEE Workshop on Mobile Computing Systems and Applications, pp. 90–100. IEEE Press, New Orleans (1999)
27. WiMAX Forum Specification. IEEE to WiMAX Forum (2011)
28. Camp, T., Boleng, J., Davies, V.: A Survey of Mobility Models for Ad Hoc Network Research. Wireless Communication & Mobile Computing (WCMC), Special Issue on Mobile Ad Hoc Networking: Research Trends and Applications 2, 483–502 (2002)
29. Rhee, I., Warrier, A., Min, J., Xu, L.: DRAND: distributed randomized TDMA scheduling for wireless ad-hoc networks. In: Mobile Ad Hoc Networking and Computing, pp. 190–201 (2006)
30. Mavromoustakis, C.X., Dimitriou, C.D., Mastorakis, G.: On the Real-Time Evaluation of Two-Level BTD Scheme for Energy Conservation in the Presence of Delay Sensitive Transmissions and Intermittent Connectivity in Wireless Devices. International Journal on Advances in Networks and Services 6(3 and 4), 148–162 (2013)
31. Mastorakis, G., Mavromoustakis, C.X., Bourdena, A., Pallis, E., Sismanidis, G.: Optimizing radio resource management in energy-efficient cognitive radio networks. In: Proceedings of the 2nd ACM Workshop on High Performance Mobile Opportunistic Systems, pp. 75–82. ACM (2013)
32. Mastorakis, G., Mavromoustakis, C.X., Bourdena, A., Pallis, E.: An energy-efficient routing scheme using Backward Traffic Difference estimation in cognitive radio networks. In: IEEE 14th International Symposium and Workshops on World of Wireless, Mobile and Multimedia Networks, WoWMoM 2013 (2013)

33. Mastorakis, G., Mavromoustakis, C.X., Bourdena, A., Kormentzas, G., Pallis, E.: Maximizing energy conservation in a centralized cognitive radio network architecture. In: 2013 IEEE 18th International Workshop on Computer Aided Modeling and Design of Communication Links and Networks (CAMAD 2013), pp. 175–179 (2013)
34. Mastorakis, G., Bourdena, A., Mavromoustakis, C.X., Pallis, E., Kormentzas, G.: An energy-efficient routing protocol for ad-hoc cognitive radio networks. In: Future Network and Mobile Summit (FutureNetworkSummit 2013) (2013)
35. Mavromoustakis, C.X., Dimitriou, C., Mastorakis, G., Pallis, E.: Real-Time Performance Evaluation of F-BTD scheme for optimized QoS Energy Conservation in Wireless Devices. In Proc. IEEE Globecom 2013, 2nd IEEE Workshop on Quality of Experience for Multimedia Communications (QoEMC 2013), Atlanta, GA, USA, December 09-13 (2013)
36. Dimitriou, C., Mavromoustakis, C.X., Mastorakis, G., Pallis, E.: On the performance response of delay-bounded energy-aware bandwidth allocation scheme in wireless networks. Paper Presented at the IEEE ICC 2013, Budapest, Hungary, June 9-13, pp. 9–13 (2013)
37. Bourdena, A., Mavromoustakis, C.X., Kormentzas, G., Pallis, E., Mastorakis, G., Yassein, M.B.: A Resource Intensive Traffic-Aware Scheme using Energy-efficient Routing in Cognitive Radio Networks, Future Generation Computer Systems, Elsevier (2014), `http://dx.doi.org/10.1016/j.future.2014.02.013`
38. Mavromoustakis, C.X., Mastorakis, G., Bourdena, A., Pallis, E.: Energy Efficient Resource Sharing using a Traffic-oriented Routing Scheme for Cognitive Radio Networks. IET Networks Journal (2013), doi:10.1049/iet-net.2013.0132

Spectrum Trading in Cognitive Radio Networks: Challenges and Solutions

Georgios Kormentzas

University of the Aegean, Department of Information and Communication Systems Engineering, Karlovassi, 83200, Samos, Greece
gkorm@aegean.gr

Abstract. Cognitive Radio (CR) paradigm has been introduced to overcome several problems related with the limited resources of radio spectrum in wireless networks. Towards this direction, CR network architectures exploit dynamic spectrum access mechanisms for an efficient usage of the available frequency slots. Such frequency slots include specific parts of radio spectrum, like the TV White Spaces (TVWS). In this context, this chapter elaborates on reviewing the related research approaches in the field of radio spectrum management and trading, considering both centralized and distributed network architectures. In addition, a radio resource management (RRM) and trading solution is presented to enable for the opportunistic exploitation of TVWS in a centralized CR network. Efficient administration of radio spectrum resources is achieved, by using a novel RRM framework, adopted in a spectrum broker that is in charge to effectively orchestrate the available wireless networking resources. Finally, fields for further research are proposed, concluding this chapter.

Keywords: Cognitive Radio, Spectrum Trading, Radio Resource Management, TV White Spaces.

1 Introduction

The current trend in emerging wireless network infrastructures is the realization of Cognitive Radio (CR) networks, researched and developed to overcome challenges related with the efficient exploitation of the available radio resources. Deployment of CR networks can satisfy/fulfill the increasing users' demand for bandwidth-hungry, QoS-sensitive mobile services, by enabling dynamic access to the available spectrum pool, along with on-demand utilization of radio resources. Cognitive radio networks use mechanisms to interact with the surrounding spectral environment [1] - [5] and dynamically identify locally unexploited frequency slots [6]. This capability gives the possibility to design dynamic spectrum access strategies/policies for reusing under-utilized spectrum at local level. Major obstacle, however, towards the wide-spread deployment and massive exploitation of CR networks is the current spectrum administration policy (i.e. Command-and-Control) that only allows licensed networks exploiting specific frequency bands, while prohibiting any other secondary

C.X. Mavromoustakis et al., *Resource Management in Mobile Computing Environments*, Modeling and Optimization in Science and Technologies 3,
DOI: 10.1007/978-3-319-06704-9_15, © Springer International Publishing Switzerland 2014

transmission within these frequencies. Such a case of licensed spectrum are the TV white spaces (TVWS) [7] (i.e. television frequencies that comprise VHF/UHF frequencies [8]). Therefore, introduction of CR networks in TVWS currently represents a disruption to the "command-and- control" paradigm of TV/UHF spectrum management and thus the exploitation of the CR technology is highly intertwined with the regulation models that would eventually be adopted. "Spectrum of Commons" enables the coexistence of secondary systems with primary ones in the same geographical location. In this case, there is no spectrum broker to manage resources allocation and Quality of Service (QoS) cannot be guaranteed QoS-sensitive applications. On the other hand, "Real Time Secondary Spectrum Management" (i.e. RTSSM) scheme enables primary networks to deal spectrum permission rights with secondary networks, towards establishing a spectrum market. Radio Resource Management (RRM) algorithms are utilized in this case to optimize the process of spectrum allocation.

In this context, the rest of this book chapter focuses on issues related to the efficient design of trading mechanisms that effectively allocate TVWS. More specifically, Section 2 presents related work and research motivation, as well as the potential technology of CR networks that can be utilized for optimum resource exploitation. Two main categories of CR topologies are presented, while potential implementation is also provided. In addition, a centralized network architecture that enables for radio resource management and trading is also presented in Section 3. The proposed architecture exploits a RRM and trading algorithm, that allocates the resources, either optimizing the spectrum usage or by increasing the spectrum broker utility through auction-based processes. Finally, Section 4 presents performance evaluation analysis and experimental results, while Section 5 concludes this work and proposes fields for possible future research.

2 Related Work and Research Motivation

2.1 Television White Spaces - TVWS

Transition from analogue to digital terrestrial television (i.e. Digital Switchover - DSO) [9] - [13] in EU countries [14] adopts DVB-T standard [15] for the provision of audio-visual content. The remaining available frequencies that will be completely unused (especially at local level [7]) due to frequency planning issues are known as "Interleaved Spectrum" [16], while frequencies that are released or remain "leftover" by the digital switchover process, are known as "Digital Dividend" [17] – [18]. Television White Spaces (TVWS) consist of Interleaved Spectrum, as well as Digital Dividend spectrum. TVWS usually sum up-to tenths of MHz at local/regional level, facilitate low cost and low power system design, provide superior propagation conditions and building penetration, while at the same time their sufficiently short wavelength allows the construction of resonant antennas. For instance, a potential exploitation of TVWS can be performed by a cognitive radio network, where in the coverage area between broadcasting services defines the white spaces. Several studies and results highlight that white spaces are present and fragmented.

2.2 Cognitive Radio Network Principles

Towards efficiently exploiting TVWS, state-of-the-art wireless technologies are required to allow opportunistic and dynamic exploitation of any available radio spectrum in the VHF/UHF bands, which are traditionally utilized in a static manner by primary/licensed systems (e.g. DVB-T/H). In this context, critical technical requirements among the envisaged schemes are the ability of these technologies for interference-free operation with incumbent transmissions (either by licensed users or other unlicensed ones). Towards this direction, recent research efforts and technological solutions resulted in the development of cognitive radio technology, i.e. a new communication paradigm for wireless communication systems, which aims to opportunistically exploit any available (in time and space) radio frequencies. The main attribute of a CR technology is that every action depends on its local observations. In this respect, CR networks require spectrum-aware operations, in order to adapt the transmission/reception mode according to the dynamic spectrum environment. In other words, CR networks access and exploitation of any locally available spectrum is a matter of a number of processes, interrelated and intimately tied-up each to other, altogether constituting the so-called cognitive/cognition cycle [19], [20], [21], [22]. The cognitive cycle mainly consists of four spectrum management functions: spectrum sensing, spectrum decision, spectrum sharing, and spectrum mobility [22].

In a cognitive radio network, two types of users exist, licensed or primary and unlicensed or secondary users. Primary users (e.g. DVB-T transmitters) are the incumbent systems of the spectrum band of UHF/VHF that exploit licensed operation, under the current spectrum policy (i.e. command-and-control). On the other hand, secondary users are able to operate in the absence of primary ones, or in cases that they do not cause interference to nearby licensed users. Thus, sophisticated spectrum sensing and management methods are required, as well as new spectrum policies have to be adopted, in order to permit the opportunistic operation of unlicensed users. A cognitive radio network enables to establish communication among cognitive radio nodes/users. Related communication parameters can be adjusted, according to change in the environment, topology, operating conditions, or user requirements [20].

2.3 Cognitive Radio Network Architectures

Architectural design issues categorise most cognitive radio networks as infrastructure-based or ad-hoc, according to their topology, or as single-hop or multi-hop according to the communication between a transmitter and a receiver [1]. For example, in a single-hop infrastructure-based CR architecture, communication among unlicensed users is achieved through a central controller (e.g. a base station [1]), which takes the responsibility of managing and coordinating secondary spectrum access (e.g. allocation of time, frequency band, and transmit power). On the other hand, in a multi-hop infrastructure-based CR architecture multiple base stations are utilised (i.e. relay nodes), enabling unlicensed users to exchange data even though they are not in the transmission range of each other (multi-hop communication) [1]. In an ad-hoc CR architecture, the unlicensed users communicate with each other directly (i.e. in a peer-to-peer mode) without requiring any base station assistance. The communication can be either single-hop or multi-hop [1]. It should be noted that for multi-hop communications, some unlicensed users can temporarily assume the role of relay stations [23]. Additionally to the above categorization, CR networks can be based, either on centralized architectures or distributed ones, depending on the place where the spectrum decision and access process is taking place.

2.3.1 Dynamic Radio Resource Management in Centralized Cognitive Radio Networks

A central controller (i.e. Spectrum Broker) is taking the decision on spectrum access in a centralized architecture, by collecting information about the spectrum usage by the licensed users. Based on this information, an optimal solution (e.g. one which maximizes spectrum utilisation while satisfies secondary users) on dynamic spectrum access can be obtained. The decisions of the central controller are communicated/ broadcasted to all unlicensed users in the network. However, information collection and exchange to/from the central controller can incur a considerable overhead. The centralized dynamic spectrum access schemes may be implemented following two approaches, i.e. based on decision-making optimization and based on auction theory. In short, an optimization-based approach emphasizes to the technical aspect, while an auction-based approach emphasizes to the economic aspect of dynamic spectrum access.

More specifically, in an optimization-based approach, the spectrum access by secondary systems/users is formulated as an optimization problem, which can be solved by a variety of techniques i.e. closed form solution, mathematical programming and integer/combinatorial programming. In such a formulation, the common objective is to maximize the spectrum efficiency (i.e. maximize spectrum utilization with as less as possible unusable spectrum fragments) while the common constraint is to maintain the interference at the lower tolerable limits to avoid causing interference to licensed users. The most common optimization implementations are the non-linear programming belonging to the field of mathematical programming, as well as the integer/combinatorial programming. The optimization process involving non-linear objective functions and constraints is called non-linear programming, where the key difference with the linear one is the inequality between the local optimum and the global optimum i.e. there can be more global optimums and a simple "climbing uphill" algorithm cannot solve the optimization problem. Popular solutions for solving a non-linear-programming problem are genetic algorithms, simulated annealing and the Monte Carlo method. On the other hand, the integer/combinatorial programming encompasses the optimization problems that involve parameters with integer values or parameters that are of combinatorial nature. These are multi-objective problems that can be solved only as a search for the optimal answer through the entire set of possible answers. The goal of the integer/combinatorial programming is shortening the search to a smaller subset of possibilities. The simplest solution is the backtracking algorithm that generates all possible spectrum allocations, by performing systematic/exact search. It should be noted, however, that the choice of the most appropriate decision-making optimization technique constitutes an application-driven approach, based on specific use-case scenarios, and by taking into account the corresponding implementation intricacies. Thereupon, metrics such as the complexity of the algorithm, the range of the possible solutions to be checked, the processing time and computational power required for obtaining the optimum solution have to be considered prior to choosing the most applicable technique.

Alternatively, centralized dynamic spectrum access schemes may be implemented by elaborating on economic issues [24]. A game theoretic perspective studies this process and provides solutions to the radio spectrum trading approach. The spectrum trading problem can be formulated as a game among two or more entities (i.e. spectrum broker and secondary systems) with incentives to conflict or to cooperate. In centralized architectures, where the spectrum broker orchestrates the operation of resource management, auction theory is also a well suited tool to elaborate on fairness, efficiency and valuation independence issues. Moreover, auctions assure truthfulness, which is an attribute that indicates the actual valuation of the bids that are sent by secondary systems to request access to the available radio spectrum. Based on the above, auction theory can be applied considering the TVWS as the commodity. The auction theory is capable not only to determine the spectrum access and allocation strategies, but also to determine the price, cost, and utility of spectrum access by the secondary systems.

Auction theory [25] has been exploited to determine the optimum allocation of resources, where sellers try to increase their revenue. Each secondary system in Auction theory sends its bids, reflecting the player's value. The most widely exploited Auction processes that have been used in such case are the single, the double and the combinatorial ones. Also, radio spectrum can be leased at the same time by using simultaneous auctions or sequentially through sequential auctions [26]. This last set includes the combinatorial auctions [27], [28].

2.3.2 Dynamic Radio Resource Management in Distributed Cognitive Radio Networks

In the case of a distributed decision process for dynamic radio-resource exploitation, the deployment of a central controller may not be feasible, thus the unlicensed/ secondary users can individually and autonomously decide on spectrum access. The communications overhead in this case would be smaller, but since each user has only local information, the optimal solution for spectrum access may not be achievable by all unlicensed users. The most common behaviour of a secondary user in a distributed cognitive radio network can be classified in collaborative–cooperative, collaborative–non-cooperative, and non-collaborative–non-cooperative.

In case of the cooperative behaviour, secondary systems try to obtain a common objective, even though this resource management decision may not provide the optimal one for individual purposes. In other words, a secondary user is concerned more about to maximize overall utility than its individual performance. For instance, a bargain game is usually adopted to formulate interaction between cooperative systems, where the cooperative behaviour can be either collaborative or non-collaborative. For instance, in a collaborative–non-cooperative case, secondary user may agree to exchange some information, but it makes a decision considering its own benefit. It is also obvious that in case a secondary system is collaborative–cooperative, it makes decisions considering the overall network performance. Alternatively, participants in non-cooperative, as well as non-collaborative, games behave selfishly (i.e. to maximize its own benefit), while individual players separately decide.

The choice of selecting a behaviour depends on both the type of the CR network and the secondary users. For instance, secondary users that belong to the same group have common objectives and typically have cooperative behaviour. In contrast, if the secondary users are independent and have different objectives, non-cooperative behaviour is the most appropriate. However in all cases of a distributed network architecture, the learning ability of CR users is vital in order to make intelligent and optimum decisions. In particular, secondary users have to observe and learn the network environment (e.g. the locally available radio spectrum), in order to get knowledge about the system, an attribute that drive to efficient decisions on spectrum exploitation. Also, knowledge can be either collaborate or non-collaborative.

2.3.3 Implementation of RRM in Centralized and Distributed Cognitive Radio Networks

Related research approaches have been considered for both architectures to solve the problem of spectrum management and trading. The choice of an architecture type is based on the main characteristic of the CR network that the researcher wishes to exam and present. For instance, authors in [29], [30], [31], present centralized architectures, where the unlicensed users share the resources under the spectrum of commons policy. In such cases, the authors elaborate on RRM schemes, considering the main attributes interference avoidance techniques for QoS guarantee, markov chains for licensed prioritization and admission control with power allocation. In this category, the optimization goal is to allocate the resources by protecting primary users. Moreover, authors in [32], [33], [34] implemented a centralised topology under the RTSSM policy that enables for coordinated spectrum trading and negations between the central controller and the unlicensed users. In such cases, the spectrum broker guarantees the protection of primary users. Additionally, authors in [35], [36], [37] solve the allocation and trading problem considering not only a coordinated allocation under the RTSSM but also a trading mechanism that exploit auctions. A number of auction schemes have been proposed such as synchronous, asynchronous, sequential, concurrent, multi-bid etc. Significant efforts also considered distributed topologies in both spectrum policies. Such topologies are selected to examine the behaviour of unlicensed users prior to obtain the resources. Authors in [38], [39], [40] implement a spectrum sharing model under the spectrum of commons policy, in order to efficiently allocate the resources at guaranteed QoS.

In this context, the next section presents a centralized architecture that enables for radio resource management and trading by exploiting the RTSSM policy for efficient spectrum usage and broker revenue [41], [42], [43], [44], [45], [46], [47], [48], [49], [50], [51]. The proposed architecture enables for QoS guarantees through the broker and interference avoidance through a geo-location spectrum database. The implemented RRM and trading algorithms, of the proposed architecture, allocate the resources either optimizing the spectrum usage or by increasing the spectrum broker utility through auction-based processes.

3 Overall Cognitive Radio Network Architecture

This section elaborates on a centralized CR network architecture that exploits both the optimization and auction approaches for obtaining spectrum sharing based on the RTSSM policy. A centralized topology approach has been followed, since QoS guarantee is important. In addition, this centralized approach enables the trading of radio spectrum, establishing secondary markets for leasing and auction. A spectrum broker entity in this network topology manages the available resources and power assigned to secondary systems, towards keeping the desired QoS and interference below the regulatory limits. The overall network architecture of such a centralized infrastructure-based CR network, is depicted in Figure 1. More specifically, this approach consists/comprises of a) a Spectrum Broker to manage TVWS access and b) a number of Secondary Systems (i.e. SS), like mobile network systems or wireless networks, each one accommodating users geographically adjacent to it, competing spectrum utilization.

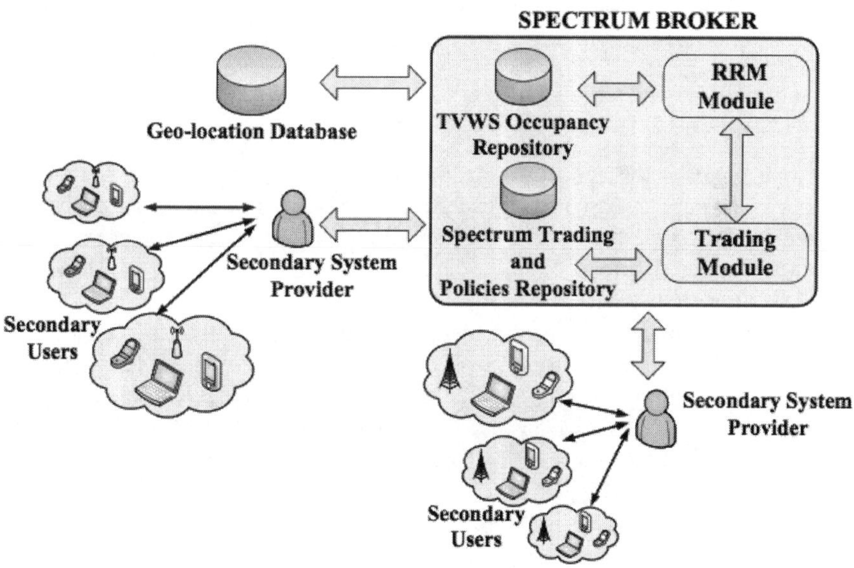

Fig. 1. Centralized CR network architecture

In particular, the Spectrum Broker consists of two sub-entities, a Dynamic TVWS Allocation Mechanism (RRM module), as well as a Trading and Price Discovery mechanism (Trading module), and a number of repositories. TVWS occupancy repository is provided with information from the Geo-location database (spectrum information supplier) that includes data, towards creating a pool of the available TVWS in specific geographical areas. The Geo-location database frequently updates TVWS occupancy repository, in order the latter to create a spectrum-portfolio.

More specifically, and according to this architecture, SSs' requests for TVWS access are communicated (e.g. via dedicated links) to the Spectrum Broker, where the Radio Resource Management module processes them as a matter of the Secondary System technical requirements (e.g. requested BW, transmission power, etc.). Prior to any spectrum allocation, the economics of TVWS transactions are also analysed/elaborated (Trading Module in Figure 1), considering the price per unit of spectrum.

All activities within the envisaged Real Time Secondary Spectrum Market are coordinated by the Broker, which is in charge to obtain the best-matching solution, taking into account parameters with integer values or combinatorial nature, such as the available TVWS, the number of secondary systems, the required bandwidth, the maximum allowable transmitted power, the spectrum-unit price, etc. Eventually, the anticipated best-matching solution (spectrum allocation scheme) will be the result of two alternative TVWS allocation mechanisms, based on a fixed-price or an auction-based policy. In case that a fixed-price policy is followed, an optimization algorithm (e.g. Backtracking, Simulated Annealing, Genetic Algorithm) obtains the best-matching/optimal solution. This can be performed by minimizing an objective function, as a matter of spectrum fragmentation and Secondary Systems prioritization. Alternatively, in the auction-based policy, the spectrum broker collects bids from the secondary systems to determine the allocation solution.

3.1 TV White Spaces Allocation Process

This section presents the spectrum allocation/leasing problem, following either the fixed-price or auction-based approaches. According to initial measurements conducted in [52], [53] frequencies availability comprises of ten TV channels (i.e. 8MHz each) scattered among digital terrestrial television radio spectrum. In case of fixed-price allocation process, the spectrum broker reaches to the most optimal allocation solution, by minimising an objective function *"C(A')"* (i.e. equation 1), as a matter of allowable transmission power (*P(i,f)*), requested bandwidth (*BW(i,f)*), spectrum fragmentation (*Frag(i,f)*), when a secondary system *"i"* is assigned to a specific frequency *"f"* and/or secondary systems prioritisation (*Pr(i)*) (e.g. in case that a number of secondary systems must be served before other ones, due to higher QoS level priority).

$$min\colon C\left(A'\right) = \sum_{i \in V} \sum_{f \in F} x_{if[P(i,f)+BW(i,f)+Frag(i,f)+\Pr(i)]} \qquad (1)$$

On the other hand, auction-based process is best suited in cases that total requirement for radio spectrum is considered more than the available radio resources (i.e. equation 2).

$$\sum_{i=1}^{i} s_i > S \qquad (2)$$

Furthermore, when the auction-based algorithm is followed, radio spectrum sellers are denoted as N = {1,2,…,n}. N is 1 in the proposed CR networking architecture, according to the simulation scenario (i.e. Spectrum Broker, leasing the available TVWS S = {1,2,…,s} to I = {1,2,…,i} secondary systems). Each buyer "i" is able to purchase x_i portions of radio spectrum for a specific time period t_i, by reporting a price $P_i^{(b)} = \{x_i, t_i\}$ (i.e. Bid Price of m for specific portion of spectrum in a specific time), while spectrum broker leases y_n portions of radio spectrum for a specific time t_i, by reporting a price $P_n^{(s)} = \{y_n, t_i\}$ (i.e. Asking Price of m for specific portion of spectrum in a specific time). Finally, $x_{i,n}$ is the quantity, which is leased by "i" secondary system from spectrum broker. The pair (i,s) in the pseudo-code of Table 1 represents possible combinations of solutions, regarding "s" TVWS to "i" Secondary systems. In case that spectrum broker benefit has to be maximized, an optimization problem is formulated as follows, based on linear programming (i.e. equation 3):

$$max: \sum_{i=1}^{i}\sum_{n=1}^{n} x_{i,n} t_i \left(P_i^{(b)} - P_n^{(s)}\right) \tag{3}$$

4 Performance Evaluation Analysis, Experimental Results and Discussion

Towards verifying the validity of the proposed approach, for different pricing models, several simulation tests were performed, while the available TVWS were based on real data [52], [53], [54]. Figure 2 depicts the average benefit/income of spectrum

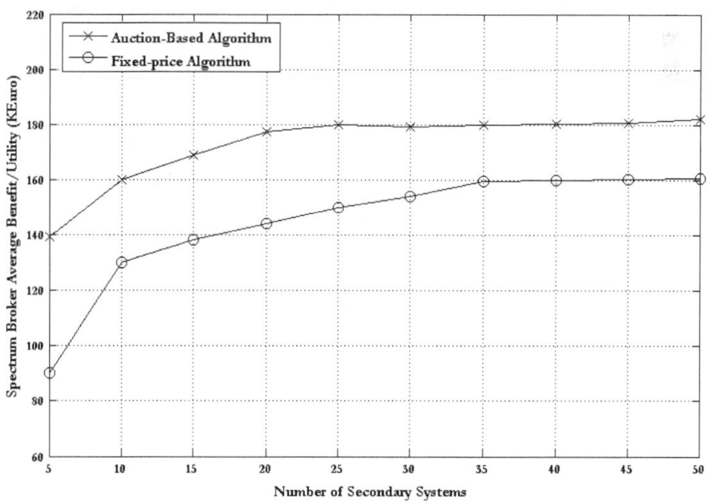

Fig. 2. Spectrum Broker Average Benefit

broker, under various numbers of bidders (i.e. LTE secondary systems). It is observed that the algorithm based on the auction process performs better providing higher revenue compared to the fixed-price algorithm.

Moreover, spectrum utilization was estimated as the percentage of the exploited bandwidth over the totally available TVWS, while spectrum fragmentation was estimated, by considering the unused parts of radio spectrum and the size for each individual fragment. Figure 3 and Figure 4 present evaluation results, regarding average spectrum utilization and spectrum fragmentation respectively. As spectrum utilization is increasing, resulting to an increased number of secondary systems that exploit TVWS, fragmentation is also increasing, getting worst. In case of the fixed-price algorithm, the most optimum solution is provided, based on minimizing spectrum fragmentation, resulting to lower levels of utilization. On the other hand, auction-based algorithm creates a more fragmented radio spectrum after the allocation process, when the number of secondary systems served, is higher in comparison to the fixed-price algorithm.

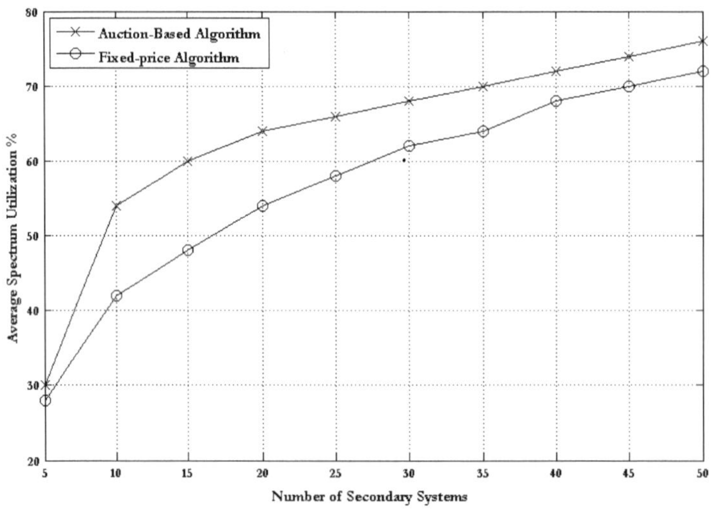

Fig. 3. Average Spectrum Utilization

Finally, probability of accessing TVWS is a metric that defines the possibility of a secondary system permitted to operate, exploiting radio spectrum resources. The auction-based algorithm provides a higher probability of using TVWS, in comparison to the fixed-price algorithm, as depicted in Figure 5. This implies that bidders are encouraged to participate in the auction-based process, increasing the possibility to access the available TVWS.

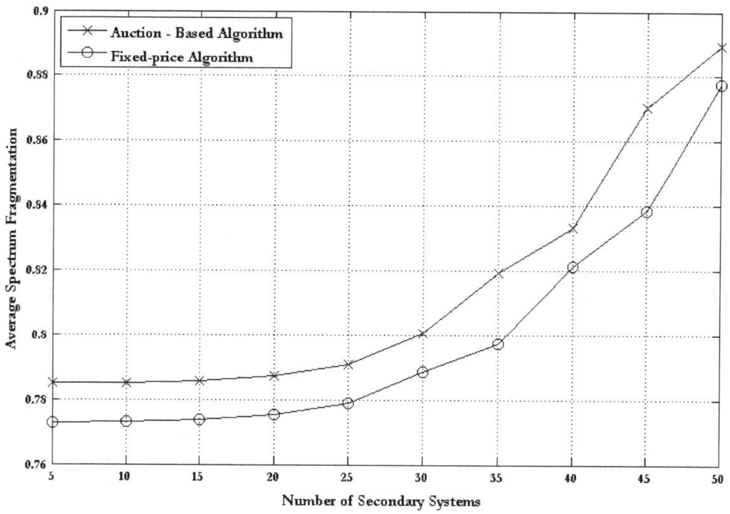

Fig. 4. Average Spectrum Fragmentation

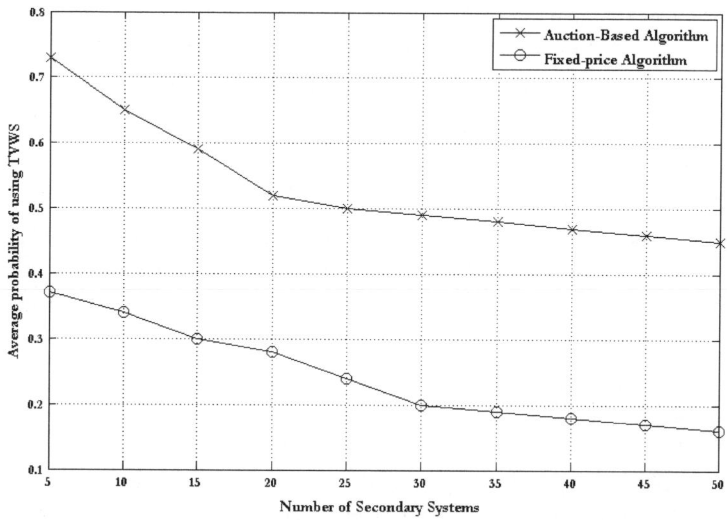

Fig. 5. Average Probability of accessing TVWS

5 Conclusion and Further Research

The work carried out within this book chapter elaborated on potential cognitive radio network architectures that allow efficient radio resource management and spectrum trading. These network architectures constitute the basis for the efficient and fast deployment/establishment of broadband metropolitan infrastructures. Also, a research work presented based on a centralized topology that exploits a prototype algorithm to provide both optimum spectrum exploitation and maximum possible revenue to the centralized entity (i.e. spectrum broker). However, in such architecture open issues have to be taken into account for future work. For instance, a potential field for future exploitation is the issue of energy efficiency in centralized cognitive radio network architectures. Measuring network lifetime is a critical design issue for uninterrupted information flow in CR networking architectures. CR networking nodes are energy 'hungry', operating on limited energy resources, whereas replacement or recharging of CR mobile terminals/devices is infeasible. Moreover, the fact that mobile devices change in time their location, there is no guaranteed QoS level during the resource exchange process, resulting to reduced performance responsiveness, under such conditions. Energy harvesting is important to be performed in CR networking systems, where network partition can frequently occur, since nodes move freely. This partitioning problem causes a part of data to be often inaccessible to a number of CR nodes. In turn, the energy consumed, by wireless devices, as well as the bandwidth allocation policy in collaboration with the power and frequency bandwidth measurements, pose a great challenge for both operators and wireless access technology planning developers, towards further exploiting the cooperative diversity and meeting QoS requirements of each service provided. As energy conservation is an important trade-off for high performance deployment in CR networks, the supporting energy schemes have to be reactive so that the energy levels of wireless nodes will be tuned, according to the associated parameters (i.e. capacity, traffic [55], [56] and remaining supporting energy of the nodes). In a general context, an energy-efficient scheme [57], [58], [59] has to take into consideration the bounded end-to-end delays of transmission and provide connectivity assurance, by exploiting technical measurements of the network lifetime that are closely related to the transmission characteristics and the underlying access policy used [60], [61], [62]. An efficient methodology should combine the temporal bandwidth-aware behaviour of the node [63], [64], [65] with the access policy used for achieving optimal energy in an end-to-end path, using both mobile peer-exchange mode and radio access mode.

References

[1] Hossain, E., Niyato, D., Han, Z.: Dynamic spectrum access and management in cognitive radio networks, 1st edn. Cambridge University Press (2009)

[2] Mitola III, J.: Software radio architecture: A mathematical perspective. IEEE J. Sel. Areas Communication 17(4), 514–538 (1999)

[3] Mitola III, J., Maguire, G.Q.: Cognitive Radio: Making Software Defined Radio More Personal. Paper Published at the IEEE Pers. Communication 6(4) (August 1999)

[4] Wyglynski, A.M., Nekovee, M., Hou, T.: Cognitive Radio Communications and Networks: Principles and Practice. Academic Press (2009)

[5] Thomas, R.-W., DaSilva, L.-A., MacKenzie, A.-B.: Cognitive Networks. Paper Presented at the IEEE DySPAN, First Symposium on Dynamic Access Networks, Baltimore, USA, November 8-11 (2005)

[6] Tu, S.-Y., Chen, K.-C., Prasad, R.: Spectrum sensing of OFDMA systems for cognitive radio networks. Paper Published at the IEEE Transaction Vehicular Technology 58(7), 3410–3425 (2009)

[7] OFCOM,Digital Dividend Review: geographic interleaved awards 470 - 550 MHz and 630 - 790 MHz – Consultation on detailed award design (June 2008)

[8] Unlicensed Operation in the TV Broadcast Bands, Final Rules (2009), `http://edocket.access.gpo.gov/2009/pdf/E9-3279.pdf` (accessed June 3, 2013)

[9] Pallis, E., Mantakas, C., Mastorakis, G., Kourtis, A., Zacharopoulos, V.: Digital Switchover in UHF: the ATHENA concept for broadband access. Paper Published at the European Transactions on Telecommunications 17(2), 175–182 (2006)

[10] Mastorakis, G., Pallis, E., Mantakas, C., Kormentzas, G., Skianis, C.: Exploiting Digital Switchover for Broadband Services Access in Rural Areas. Paper Published at the Journal of Communications 1(6), 45–50 (2006)

[11] Mastorakis, G., Kormentzas, G., Pallis, E.: A Fusion IP/DVB Networking Environment for Providing Always -on Connectivity and Triple-Play Services to Urban and Rural Areas. Paper Published at the IEEE Network Magazine 21(2), 21–27 (2007)

[12] Mastorakis, G., Pallis, E., Kormentzas, G.: A DVB/IP QoS aware Backhaul Networking Environment. Paper Published at the Wireless Personal Communications 52(3), 637–649 (2010)

[13] Mastorakis, G., Markakis, E., Sideris, A., Pallis, E., Zacharopoulos, V.: Experimental infrastructures for IP/DVB convergence: an actual substantiation for triple play services provision at remote areas. In: Paper Presented at the 18th Annual IEEE International Symposium on Personal, Indoor and Mobile Radio Communications (PIMRC 2007), Special Session on Advanced Experimental Infrastructures and Paradigms for NGNs, Athens, Greece, September 03-07 (2007)

[14] The Digital Dividend & the Future of Digital Terrestrial Television, DigiTAG (2009)

[15] ETSI EN 300 744,Digital Video Broadcasting (DVB); Framing structure, channel coding and modulation for digital terrestrial television. European Standard (Telecommunications series) (2004)

[16] The Economics of Spectrum management: A Review, Australian Communication and Media Authority, ACMA (2007)

[17] Analysys Mason, DotEcon, HOGAN & HARTSON, 'Exploiting the digital dividend' - a European approach, Report to the European Commission (August 14, 2009)

[18] CEPT REPORT 24. Technical considerations regarding harmonisation options for the Digital Dividend (July 2008)

[19] Haykin, S.: Cognitive radio: brain-empowered wireless communications. Paper Published at the IEEE Journal on Selected Areas in Communications 23(2), 201–220 (2005)

[20] Mitola III, J.: Cognitive Radio: An Integrated Agent Architecture for Software Defined Radio. PhD Dissertation Royal Institute of Technology, Stockholm, Sweden (May 2000)

[21] Fette, B.A.: Cognitive Radio Technology. Academic Press (2006)

[22] Akyildiz, I., Lee, W., Vuran, M., Mohanty, S.: NeXt generation/ dynamic spectrum access/cognitive radio wireless networks: A survey. Paper Published at the Computer Networks 50(13), 2127–2159 (2006)

[23] Sampath, A., Yang, L., Cao, H., Zheng, B., Zhao, Y.: High Throughput Spectrum-aware Routing for Cognitive Radio Based Ad-hoc Networks. Paper Presented at the Third International Conference on Cognitive Radio Oriented Wireless Networks and Communications (CROWNCOM) (May 2008)

[24] Niyato, D., Hossain, E.: Competitive pricing for spectrum sharing in cognitive radio networks: dynamic game, inefficiency of Nash equilibrium, and collusion. Paper Published at the IEEE Journal on Selected Areas in Communications 26(1), 192–202 (2008)
[25] Krishna, V.: Auction theory. Academic Press (2009)
[26] Sengupta, S., Chatterjee, M.: Sequential and concurrent auction mechanisms for dynamic spectrum access. Paper Presented at the International Conference on Cognitive Radio Oriented Wireless Networks and Communications (CrownCom) 2007 (July-August 2007)
[27] Shoham, Y., Leyton-Brown, K.: Multi-agent Systems: Algorithmic, Game-Theoretic, and Logical Foundations. Cambridge University Press (2009)
[28] Yoon, K.: The participatory Vickrey–Clarke–Groves mechanism. Paper Published at the Journal of Mathematical Economics 44(3-4), 3–4 (2008)
[29] Xing, Y., Mathur, C.-N., Haleem, M.-A., Chandramouli, R., Subbalakshmi, K.-P.: Dynamic spectrum access with QoS and interference temperature constraints. Paper Published at the IEEE Transactions on Mobile Computing 6(4), 423–433 (2007)
[30] Wang, B., Ji, Z., Liu, K.-J.-R.: Primary-prioritized Markov approach for dynamic spectrum access. Paper Presented at the IEEE International Symposium on New Frontiers in Dynamic Spectrum Access Networks (DySPAN), pp. 507–515 (April 2007)
[31] Zhang, L., Liang, Y.-C., Xin, Y.: Joint admission control and power allocation for cognitive radio networks. Paper Presented at the IEEE International Conference on Acoustics, Speech and Signal Processing (ICASSP), vol. 3, pp. 673–676 (April 2007)
[32] Kim, H., Lee, Y., Yun, S.: A dynamic spectrum allocation between network operators with priority-based sharing and negotiation. Paper Presented at the IEEE International Symposium on Personal, Indoor and Mobile Radio Communications (PIMRC), vol. 2, pp. 1004–1008 (September 2005)
[33] Subramanian, A.-P., Gupta, H., Das, S.-R., Milind, B.-M.: Fast spectrum allocation in coordinated dynamic spectrum access based cellular networks. Paper Presented at the IEEE International Symposium on New Frontiers in Dynamic Spectrum Access Networks (DySPAN), pp. 320–330 (April 2007)
[34] Kamakaris, T., Buddhikot, M.-M., Iyer, R.: A case for coordinated dynamic spectrum access in cellular networks. Paper Presented at the IEEE International Symposium on New Frontiers in Dynamic Spectrum Access Networks (DySPAN), pp. 289–298 (November 2005)
[35] Gandhi, S., Buragohain, C., Cao, L., Zheng, H., Suri, S.: A general framework for wireless spectrum auctions. Paper Presented at the IEEE International Symposium on New Frontiers in Dynamic Spectrum Access Networks (DySPAN), pp. 22–33 (April 2007)
[36] Kovacs, L., Vidacs, A.: One-shot multi-bid auction method in dynamic spectrum allocation networks. Paper Presented at the IST Mobile and Wireless Communications Summit, pp. 1–5 (July 2007)
[37] Sengupta, S., Chatterjee, M.: Synchronous and asynchronous auction models for dynamic spectrum access. In: Chaudhuri, S., Das, S.R., Paul, H.S., Tirthapura, S. (eds.) ICDCN 2006. LNCS, vol. 4308, pp. 558–569. Springer, Heidelberg (2006)
[38] Etkin, R., Parekh, A., Tse, D.: Spectrum sharing for unlicensed bands. Paper Published at the IEEE Journal on Selected Areas in Communications 25(3), 517–528 (2007)
[39] Berlemann, L., Hiertz, G.-R., Walke, B., Mangold, S.: Strategies for distributed QoS support in radio spectrum sharing. Paper Presented at the IEEE International Conference on Communications (ICC), vol. 5, pp. 3271–3277 (May 2005)

[40] Niyato, D., Hossain, E.: Cognitive radio for next generation wireless networks: an approach to opportunistic channel selection in IEEE 802.11-based wireless mesh. Paper Published at the IEEE Wireless Communications (2009)

[41] Bourdena, A., Pallis, E., Kormentzas, G., Mastorakis, G.: A prototype cognitive radio architecture for TVWS exploitation under the real time secondary spectrum market policy. Paper Published at the Special Issue on "Cognitive Radio for LTE Advanced & Beyond", Physical Communication (2013), http://dx.doi.org/10.1016/j.phycom.2013.11.003

[42] Bourdena, A., Pallis, E., Kormentzas, G., Mastorakis, G.: Opportunistic TV White Spaces Exploitation through Efficient Radio Resource Management Algorithms in Cognitive Radio Networks. Paper Published at the Transactions on Emerging Telecommunications Technologies, ETT (2013)

[43] Bourdena, A., Pallis, E., Kormentzas, G., Skianis, C., Mastorakis, G.: Real-Time TVWS Trading Based on a Centralised CR Network Architecture. Paper presented at the IEEE Globecom2011, IEEE International Workshop on Recent Advances in Cognitive Communications and Networking, Texas, Houston, USA, December 05-09, pp. 964–969 (2011)

[44] Bourdena, A., Pallis, E., Kormentzas, G., Mastorakis, G.: A centralised broker-based CR network architecture for TVWS exploitation under the RTSSM policy. Paper Presented at the 2nd IEEE Workshop on Convergence among Heterogeneous Wireless Systems in Future Internet (CONWIRE 2012), IEEE ICC 2012, Ottawa, Canada, June 10-15, pp. 5685–5689 (2012)

[45] Bourdena, A., Mastorakis, G., Pallis, E., Karditsis, E., Kormentzas, G.: A radio resource management framework for TVWS exploitation under the RTSSM policy. Paper Presented at the International Conference on Telecommunication & Multimedia, IEEE TEMU 2012, Heraklion, Crete, Greece, July 30-August 1, pp. 1–6 (2012)

[46] Bourdena, A., Mastorakis, G., Pallis, E., Arvanitis, A., Kormentzas, G.: A Dynamic Spectrum Management Framework for Efficient TVWS Exploitation. Paper Presented at the 17th IEEE International Workshop on Computer-Aided Modeling Analysis and Design of Communication Links and Networks, IEEE CAMAD, Barcelona, Spain, September 17-19, pp. 51–55 (2012)

[47] Bourdena, A., Pallis, E., Kormentzas, G., Mastorakis, G.: A radio resource management framework for TVWS exploitation under an auction-based approach. Paper Presented at the 8th International Conference on Network and Service Management, IEEE CNSM 2012, The Mirage, Las Vegas, USA, October 22-26, pp. 204–208 (2012)

[48] Bourdena, A., Mastorakis, G., Pallis, E., Mavromoustakis, C.X., Kormentzas, G., Karditsis, E.: A Radio Resource Management Framework for Opportunistic TVWS Access. Paper Presented at the1stACM Workshop on High Performance Mobile Opportunistic Systems, IEEE ACM HP-MOSys 2012, Paphos, Cyprus, October 21-25, pp. 33–38 (2012)

[49] Bourdena, A., Pallis, E., Kormentzas, G., Skianis, C., Mastorakis, G.: QoS provisioning and policy management in a broker-based CR network architecture. Paper Presented at the IEEE Globecom2012, IEEE International Workshop on Recent Advances in Cognitive Communications and Networking, Anaheim, San Francisco, USA, December 3-7, pp. 1841–1846 (2012)

[50] Bourdena, A., Pallis, E., Kormentzas, G., Mastorakis, G.: Radio Resource Management Algorithms for Efficient QoS Provisioning over Cognitive Radio Networks. Paper Presented at the IEEE ICC 2013, IEEE International Conference on Communications, Budapest, Hungary, June 9-13 (2013)

[51] Bourdena, A., Makris, P., Skoutas, D., Skianis, C., Kormentzas, G., Pallis, E., Mastorakis, G.: Joint Radio Resource Management in Cognitive Networks: TV White Spaces Exploitation Paradigm. Lagkas, T., Sarigiannidis, P., Louta, M., Chatzimisios, P. (eds.) Paper Published at the in the Book Evolution of Cognitive Networks and Self-Adaptive Communication Systems (2013)

[52] The European 7th Framework Programme project: COGEU (ICT-248560) - Deliverable 4.2 (2012), http://www.ict-cogeu.eu/pdf/COGEU_D4%201ICT_248460.pdf (accessed June 3, 2013)

[53] The European 7th Framework Programme project: COGEU (ICT-248560) - Deliverable 6.1 (2010) - Dynamic Radio Resource Management algorithms for an efficient use of TVWS (2010)

[54] The European 7th Framework Programme project: COGEU (ICT-248560) - Deliverable 6.4 - System level evaluation platform and simulation results (2012)

[55] Mavromoustakis, C.X., Dimitriou, C.D., Mastorakis, G.: Using Real-Time Backward Traffic Difference Estimation for Energy Conservation in Wireless Devices. Paper Presented at the 4th International Conference on Advances in P2P Systems (AP2PS 2012), Barcelona, Spain, September 23-28, pp. 18–23 (2012)

[56] Mavromoustakis, C.X., Zerfiridis, K.G.: On the diversity properties of wireless mobility with the user-centered temporal capacity awareness for EC in wireless devices. Paper Presented at the 6th IEEE International Conference on Wireless and Mobile Communications, ICWMC 2010, Valencia, Spain, September 20-25, pp. 367–372 (2010)

[57] Mavromoustakis, C.X., Dimitriou, C.D., Mastorakis, G.: On the Real-Time Evaluation of Two-Level BTD Scheme for Energy Conservation in the Presence of Delay Sensitive Transmissions and Intermittent Connectivity in Wireless Devices. International Journal on Advances in Networks and Services 6(3 and 4), 148–162 (2013)

[58] Mastorakis, G., Mavromoustakis, C.X., Bourdena, A., Pallis, E., Sismanidis, G.: Optimizing radio resource management in energy-efficient cognitive radio networks. In: Proceedings of the 2nd ACM Workshop on High Performance Mobile Opportunistic Systems, pp. 75–82. ACM (2013)

[59] Mastorakis, G., Mavromoustakis, C.X., Bourdena, A., Pallis, E.: An energy-efficient routing scheme using Backward Traffic Difference estimation in cognitive radio networks. In: IEEE 14th International Symposium and Workshops on World of Wireless, Mobile and Multimedia Networks, WoWMoM 2013 (2013)

[60] Mastorakis, G., Mavromoustakis, C.X., Bourdena, A., Kormentzas, G., Pallis, E.: Maximizing energy conservation in a centralized cognitive radio network architecture. In: 2013 IEEE 18th International Workshop on Computer Aided Modeling and Design of Communication Links and Networks (CAMAD 2013), pp. 175–179 (2013)

[61] Mastorakis, G., Bourdena, A., Mavromoustakis, C.X., Pallis, E., Kormentzas, G.: An energy-efficient routing protocol for ad-hoc cognitive radio networks. In: Future Network and Mobile Summit (FutureNetworkSummit 2013) (2013)

[62] Mavromoustakis, C.X., Dimitriou, C., Mastorakis, G., Pallis, E.: Real-Time Performance Evaluation of F-BTD scheme for optimized QoS Energy Conservation in Wireless Devices. In: Proc. IEEE Globecom 2013, 2nd IEEE Workshop on Quality of Experience for Multimedia Communications (QoEMC 2013), Atlanta, GA, USA, December 09-13 (2013)

[63] Dimitriou, C., Mavromoustakis, C.X., Mastorakis, G., Pallis, E.: On the performance response of delay-bounded energy-aware bandwidth allocation scheme in wireless networks. Paper Presented at the IEEE ICC 2013, Budapest, Hungary, June 9-13 (2013)

[64] Bourdena, A., Mavromoustakis, C.X., Kormentzas, G., Pallis, E., Mastorakis, G., Yassein, M.B.: A Resource Intensive Traffic-Aware Scheme using Energy-efficient Routing in Cognitive Radio Networks. Future Generation Computer Systems (2014), http://dx.doi.org/10.1016/j.future.2014.02.013
[65] Mavromoustakis, C.X., Mastorakis, G., Bourdena, A., Pallis, E.: Energy Efficient Resource Sharing using a Traffic-oriented Routing Scheme for Cognitive Radio Networks. IET Networks Journal (2014), doi:10.1049/iet-net.2013.0132

Energy and Resource Consumption Evaluation of Mobile Cognitive Radio Devices

George Mastorakis[1], Spyros Panagiotakis[2], Kostas Kapetanakis[2], Giorgos Dagalakis[2], Constandinos X. Mavromoustakis[3], Athina Bourdena[2], and Evangelos Pallis[2]

[1] Technological Educational Institute of Crete, Department of Business Administration, Lakonia, Agios Nikolaos, 72100, Crete, Greece
gmastorakis@staff.teicrete.gr
[2] Technological Educational Institute of Crete, Department of Informatics Engineering, Estavromenos, Heraklion, 71500, Crete, Greece
spanag@teicrete.gr, kapekost@ie.teicrete.gr, tp2730@edu.teicrete.gr, {bourdena,pallis}@pasiphae.eu
[3] University of Nicosia, Department of Computer Science, 46 Makedonitissas Ave., 2414 Engomi, Nicosia, Cyprus
mavromoustakis.c@unic.ac.cy

Abstract. This chapter proposes a Cognitive Radio network architecture that enables for the efficient operation of mobile devices over TV White Spaces. The proposed network architecture comprises of a Geo-location database and a spectrum broker that coordinates TV White Spaces access, by a number of 4G secondary communication systems, competing/requesting for the available radio spectrum. Furthermore, it introduces an innovative methodology for evaluation of energy and resource consumption in mobile cognitive devices that does not require any external metering device but exploits the advanced software and hardware features of modern smart phones to this end. In particular, the various APIs provided, by such operating systems for access to their functionality can be used for adequately auditing and reporting resource consumption on such mobile platforms. More specifically, we evaluate energy consumption and CPU utilisation in various communication scenarios via a number of experimental tests, carried out under controlled conditions. Network connectivity, calling and multimedia playback are some of the scenarios that are evaluated and presented here.

Keywords: Cognitive Radio Networks, Energy Consumption Evaluation, Resource Consumption, Mobile Devices.

1 Introduction

Ubiquitous provision of wireless services, rich in multimedia content, raises the need for increased radio spectrum availability and creates new challenges in cellular networks resources management and guaranteed Quality of Service (QoS). Radio spectrum utilization studies have resulted that most of licensed spectrum parts are

C.X. Mavromoustakis et al., *Resource Management in Mobile Computing Environments*, Modeling and Optimization in Science and Technologies 3,
DOI: 10.1007/978-3-319-06704-9_16,

under-utilized [1] and considerable units of it are available, when both space and time dimensions are taken into account. TV White Spaces (TVWS) is an example of under-utilized radio spectrum that comprises of VHF/UHF channels, released after analogue to digital terrestrial television switchover process and interleaved spectrum, which is available due to frequency planning issues [2]. TVWS include large parts of radio spectrum especially at local/regional level, enabling for low cost and low power systems design due to superior propagation conditions [3]. Therefore, TVWS are well suited for wireless networks services, provided by sophisticated cellular systems. Towards addressing the need for increased radio spectrum demand, a number of cutting-edge technologies may be exploited, such as the LTE standard [4] that provides flexible deployment, in terms of high spectral efficiency, effective bandwidth management, as well as different modulation/coding schemes. In addition, LTE systems can be designed to operate in alternative, un-used radio spectrum bands (e.g. TVWS), when both dimensions of space and time are considered [5], coexisting with other telecommunication systems.

LTE deployment over TVWS will enable cellular network operators to cover large geographical areas with less number of base stations, decreasing investment costs and providing cheaper cellular broadband services, especially in rural areas. Furthermore, this part of radio spectrum could be exploited to support peak data traffic in urban areas with increased bandwidth demands, while several schemes to share channels on a temporary basis of short or medium duration may be investigated, towards providing relief of crowded cellular networks that experience peak loads. Deployment of LTE systems requires a new radio spectrum management policy. Among the envisaged schemes [6], [7] the Real Time Secondary Spectrum Market (i.e. RTSSM) policy is the most appropriate solution, especially for deployments that require sporadic access to radio spectrum and for which QoS guarantees are important. RTSSM policy, adopts radio spectrum trading, by permitting spectrum license holder to run admission control algorithms that allow secondary systems to access wireless networking resources, only when QoS is adequate. Trading of secondary spectrum usage may occur through network management entities (e.g. spectrum broker), exploiting Radio Resource Management (RRM) algorithms [6], [8], in order to efficiently allocate the available resources to secondary systems [9], [10]. Secondary systems, in this case, dynamically request access, only when radio spectrum is needed and are charged based on channel utilization basis, as a matter of types of services, access characteristics and priority level requirements [11][12].

A vital enabler towards deployment of LTE systems over TVWS, considering the RTSSM policy, is Cognitive Radio (CR) technology/networks [13], [14]. CR networks enable dynamic access to radio spectrum from secondary systems, by avoiding to cause interference to primary ones. For this purpose, a centralized network architecture [15] is appropriate for LTE deployment based on RTSSM policy, rather than a distributed one, due to QoS provisioning requirements. Furthermore, the exploitation of a spectrum broker enables the orchestration of the available network resources, by collecting information about radio spectrum access usage stemming from primary systems, as well as information about the transmission requirements/demands from secondary systems. Based on this information, an optimal solution (e.g. solution that maximises spectrum utilisation) on dynamic spectrum access is obtained.

On the other hand, the evaluation of network nodes lifetime is a critical design issue for uninterrupted information flow in CR networking architectures. Especially as it concerns the modern mobile devices, smart phones or tablets, they are quite energy 'hungry' and operate on limited energy, hardware and network resources, whereas replacement or recharging is infeasible. This fact becomes really critical if we take into account that devices of this type are considered anymore as life companions and not just as means for occasional communication, as happened with mobile phones just a few years ago. This induces continuous and intensive use of them almost everywhere and at any time, hence they should assure their owners reliable and long-term operation under any circumstances, even when away from a power line for a long period. In that context, multimedia playback, gaming, browsing, calling and location-based services are included among the most popular, but energy- and resource- demanding, applications. The execution of such applications, sequentially or in parallel, consumes a great percentage of hardware and network resources (CPU, memory, battery and bandwidth) critically affecting device availability and degrading its performance. This inevitably affects the native and operational needs of such devices. For example, the fact that mobile devices update quite often their location with the network contributes to the absence of any guaranteed QoS level during the resource exchange process, resulting in reduced performance responsiveness, under such conditions. The same stands for the control plane signalling of communications. Despite the fact that each such system reserves resources for its operational needs, it is obvious that the more heavy applications are running, the more resources will be consumed and the more performance degradation the user will experience. Hence, efficient dynamic rescheduling of processes that would release resources and would lighten the execution load in the device would be really desirable. Wishing would be also a dynamic estimation of the available time an operation could be executed on a device under a given battery charge and execution load or of the device's performance degradation that is induced by the launching of an application. The first step towards such cognitive algorithms is the evaluation of various critical applications and operations in terms of resource demands including battery and CPU utilisation.

In this context, this chapter is making progress beyond the current state-of-the-art, by proposing a CR networking architecture, where its operation is controlled via a radio spectrum broker that receives information regarding the availability of TVWS from a geo-location database. Furthermore, it introduces an innovative methodology for evaluation of energy and resource consumption in mobile cognitive devices that does not require any external metering device since it exploits the advanced software and hardware features of modern smart phones to this end. In particular, the various APIs provided by such operating systems for access to their functionality can be efficiently used for adequately auditing and reporting resource consumption on such mobile platforms. Following this introductory section, the rest of this chapter is organised as follows: section 2 discusses the design of the spectrum broker, incorporated into the proposed CR network architecture and elaborates on energy and resource consumption issues of mobile devices. Section 3 details on the proposed methodology used to evaluate energy and resource consumption of the mobile CR mobile devices, providing experimental results on various communication scenarios, while section 4 concludes the chapter, by identifying fields for future research and experimentation.

2 Broker-Based Cognitive Radio Network Architecture

This section firstly presents a broker-based CR networking architecture for efficient TVWS exploitation, under the RTSSM policy. The overall networking architecture is depicted in Figure 1, comprising of a spectrum broker that coordinates TVWS management, a number of LTE based secondary systems, competing/requesting for TVWS access, as well as of a Geo-location database. According to this architecture, spectrum broker consists of five sub-entities, a TVWS occupancy repository, a RRM module for TVWS allocation, a spectrum trading and policies repository, a spectrum trading module and an energy controller. TVWS occupancy repository obtains information from the national database, namely as Geo-location database, which includes data, regarding the available TVWS in specific geographical locations and maximum allowable transmission power levels per channel, in order to avoid causing interference to primary systems. TVWS occupancy repository creates a spectrum-portfolio, including all the above mentioned information that is advertised to all LTE based secondary systems.

Moreover, the RRM module matches secondary systems requirements with available resources, allocating TVWS based on specific QoS requirements. This module adopts/implements RRM algorithms, by exploiting information stemming from the Geo-location database, in order to determine the available channels and maximum transmission power levels, at which a secondary system is allowed to operate, towards avoiding spectrum fragmentation, optimising RRM process and supporting QoS provision or guaranteed fairness to TVWS access. Furthermore, trading module is responsible to determine the revenue of spectrum broker, which aims to trade/lease radio spectrum with temporary exclusive rights to the most valuable secondary systems, while spectrum trading and policies repository possesses data, regarding the TVWS leasing process and the price per spectrum-unit that is vital during the phase of resources trading, towards creating a price-portfolio.

Radio spectrum broker entity adopted in the proposed CR networking architecture (see Figure 1) is in charge to manage/trade the available channels, among a number of secondary systems that participate to TVWS allocation process. It initially informs secondary systems, regarding spectrum portions available for leasing, as well as relevant information regarding maximum allowable transmission power thresholds. This information originated from the Geo-location database, is hosted within the TVWS occupancy repository. Spectrum broker advertises both a spectrum-portfolio and a price-portfolio to all secondary systems, informing them about transmission characteristics and call price of the available TVWS. After this stage, LTE systems provide their demand for the available spectrum portions, which is defined by the offered price. Spectrum broker firstly collects all radio spectrum requests/demands in the RRM module, which is in charge to analyse and process them, as a matter of all technical requirements by LTE systems and the available TVWS characteristics. For each spectrum portion/fragment, spectrum broker creates and maintains a list with the requests, namely as request-portfolio, in order to allocate each spectrum fragment to the most valuable LTE system that showed interest, respecting QoS requirements/ constraints (i.e. priority level).

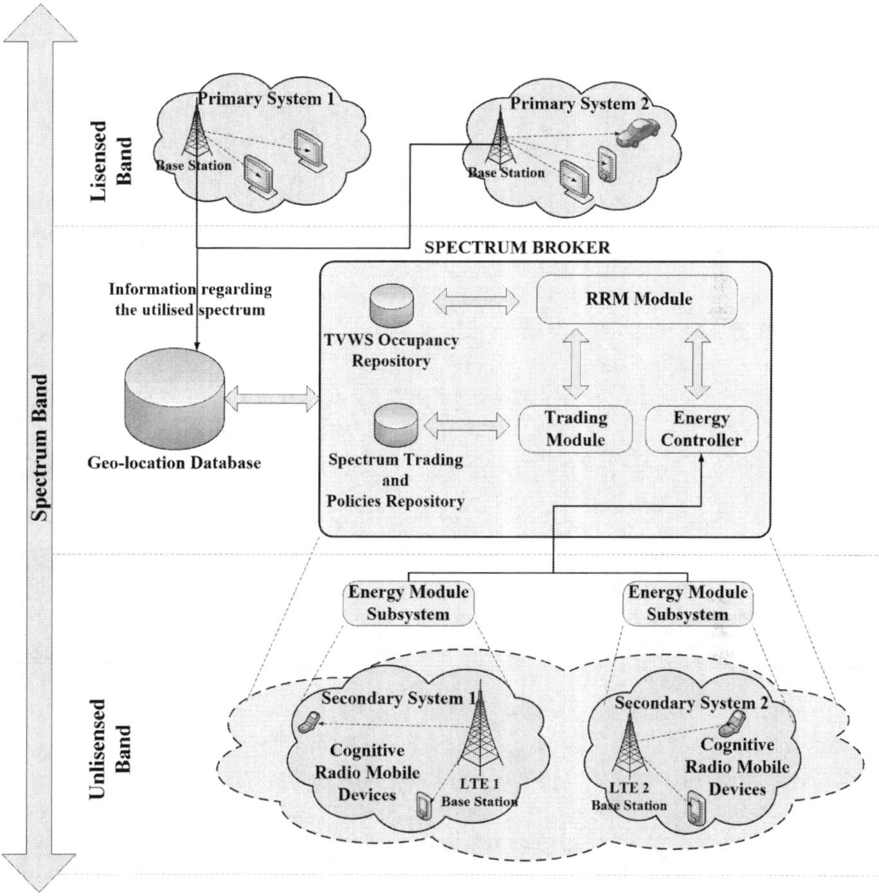

Fig. 1. Secondary communication nodes operating over heterogeneous TVWS

On the other hand, the evaluation of the network nodes lifetime is a critical design issue for uninterrupted information flow in CR networking architectures. More specifically, modern mobile devices, such as smart phones or tablets are quite energy 'hungry' and operate on limited energy, hardware and network resources, whereas replacement or recharging is infeasible. Hence, efficient dynamic rescheduling of processes, running in a device that would release resources and lighten the execution load, would be really desirable. The first step towards such a process is the evaluation of various critical applications and operations in terms of resource demands, including battery and CPU utilisation

In this respect, an Android application was developed as a tool for battery and resource measurement of mobile Cognitive Radio devices in the proposed network architecture above. Android is an open source mobile platform based on a Linux Operating System. Applications in Android (or apps) have an .apk extension (Android package), are developed in JAVA using the Android API and are executed on a Dalvik process virtual machine [16]. Every Android app runs in its own process with its own instance of the Dalvik virtual machine, isolated from the system and other applications. The Android apps are composed of one or more Activities. Activity is an application component used for every screen to be displayed. While the user navigates through an application using gestures, button tapping and other events, each Activity invokes another one. The communication between Activities is achieved using the Intents. Intent is an abstract description of an operation to be performed. Using Intents we can also start an application or a background process. In Android, the background processes can run with no user interface and are known as Services. Services are used to run log-running operations or to perform work for remote processes.

The Android API includes a Battery Manager object, which can be used to read the battery status, while the device is in use. Hence, we have developed a lightweight Java application to perform measurement tasks in Android smart phones. The application listens to the changes of the battery charging level. Whenever a change occurs, the process is triggered to collect the updated battery data. In particular, the remaining energy levels (in both millivolts and % of the total charge), along with the battery temperature, are captured and logged into a text file. When the battery reaches its lowest level, the process stops and closes the log file. Using MS Office Excel we can then plot consumption graphs for each test case. In [17] we introduced such an application for energy consumption evaluation on Android smart phones and we conducted a series of experiments to estimate energy consumption on scenarios including 3D graphics rendering.

In addition to the battery statistics, we also collect, here, data related to the CPU utilisation of the device and the signal strength of all connected networks. In particular to the former, we exploit common UNIX techniques to read system files and calculate the CPU utilisation. To this end, we have created a Java Object parsing the /proc/stat file that includes data related to processes [18]. To capture the signal strength for both 3G and Wi-Fi connections we use the native libraries of Android API. In particular, for cellular networks it is the Telephony package that contains the signal data and for Wi-Fi networks the WifiManager package.

The application is limited to logging the battery and the CPU status whenever they alter. Our intention was to minimize the resources' consumption due to the application's execution, thus we did not implement graphs plotting, event notifications and other graphic features that common monitoring applications provide [19]. The GUI is a simple Activity with two buttons, as it is depicted in Figure 2. The layout of the application is described in an XML file. An auto-generated class file named "R" holds the necessary JAVA code to build the fields in the application. Each field has a unique id. In the event listener of the main view we use the id of the buttons to detect which one is pressed and broadcast an Intent to start or stop our service (Figure 3). This service registers a receiver that can listen to our Intents. In Figure 4, the receiver is registered to listen for battery status Intents. When the registered event occurs, the data are delivered to the service. Then, the service writes the necessary data to a file as it is depicted in Figure 5.

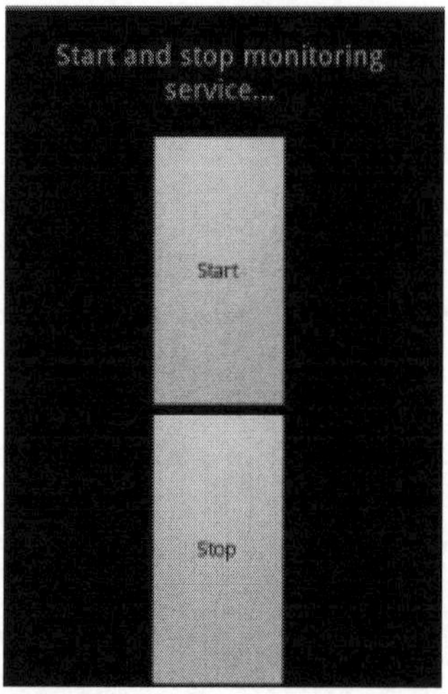

Fig. 2. GUI of the Android application for resources monitoring

```
public void onClick(View src) {
                    switch (src.getId()) {
                    case R.id.buttonStart:
                      Log.d(, "onClick: starting
srvice");
                      startService(new
Intent(this, monitorService.class));
                      break;
                    case R.id.buttonStop:
                      Log.d(, "onClick: stopping
srvice");
                      stopService(new Intent(this,
monitorService.class));
                      break;
                    }}
```

Fig. 3. Listening to button events to send an intent managing the service

```
this.registerReceiver(this.mBatInfoReceiver,
new IntentFilter(Intent.ACTION_BATTERY_CHANGED));
```

Fig. 4. Registering the receiver for listening to battery events

```
    try {
          SimpleDateFormat dateFormat = new
 SimpleDateFormat("HH:mm:ss");
                               Date date = new Date();
    myOutWriter.write("end: "+dateFormat.format(date)+"
 , "+ String.valueOf(voltage) + "\n");
        } catch (IOException e) {

             e.printStackTrace();
             }
```

Fig. 5. Dumping data to the file

3 Performance Evaluation Analysis, Experimental Results and Discussion

3.1 Evaluation of the Battery Measurement App

The associated Android Battery Manager was evaluated to assure reliability of our application. Hence, we set up a circuit by in parallel installing a Tektronix (TX3) measurement tool with 4.8 digits, as it is depicted in Figure 6, to continuously display the voltage of our smart phone battery. Our smartphone is a LG Optimus P970 with Android v4.0.4. At the same time, we let our application, installed and running in the smart phone, to capture battery's discharging as an android application executes. With this setting we tried to compare the recorded values by both measurement techniques. The results were more than satisfactory since the values were almost identical.

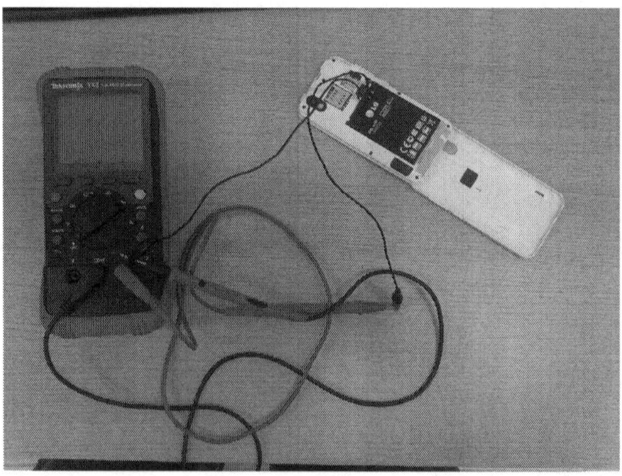

Fig. 6. Tektronix measurement tool installed in a parallel circuit with our LG (P970) smart phone

3.2 Experimental Scenarios and Methodology

Once we have assured that measurements taken by our application are reliable, we conducted a series of experiments to estimate resource consumption on a smart phone in various common communication scenarios. In particular, we have considered battery discharging and CPU utilisation in the following test cases:

1. without any network connection (3G and Wi-Fi are disabled).
2. with the device connected to a 3G network and the Wi-Fi disabled.
3. with the device connected to a Wi-Fi network and the 3G disabled.
4. with the device connected to both 3G and Wi-Fi networks.
5. with the device connected to both 3G and Wi-Fi networks and the GPS on.
6. during a Skype voice call via Wi-Fi, with the device already connected to a Wi-Fi network and the 3G disabled or enabled.
7. during a Skype video call via Wi-Fi, with the device already connected to both 3G and Wi-Fi networks.
8. during online video playback via Wi-Fi, with the device already connected to both 3G and Wi-Fi networks.
9. during online music playback via Wi-Fi, with the device already connected to both 3G and Wi-Fi networks.

Table 1 illustrates the specific configuration for each test case in terms of networking and running applications.

The selected scenarios include some of the most common data communication operations of a smartphone, thus they can be considered as representatives of its daily usage as a means for data communications. We note here, that only scenarios that include some form of data communication have been taken into account during our experimentation. During all experiments we have maintained the same settings as it concerns screen brightness (set to high) to assure it does not affect our measurements. To this end, before starting the measurements, we deactivated the light sensor to maintain stable levels of brightness. In addition, to assure trustworthy measurements the following configurations of our test-bed were set: a) all measurements were conducted over the same mobile device (a LG Optimus P970 with a Cortex-A8 processor at 1 GHz and 512 MB RAM, running Android v4.0.4). Our device is also equipped with a Li-Ion 1500 mAh battery, the fully charged voltage of which was measured at 4200 mV and its fully discharged voltage at 3200 mV. b) All experiments were conducted in the same room, over a short period of time and having the mobile device static to minimise the impact of mobility, environmental conditions and signal strength variation (especially for the experiments required 3G connectivity). In addition, all experiments were executed at least three times to avoid random behaviours. c) The mobile device was connected to Wi-Fi network situated in our local network to assure trustworthy connectivity and minimise unwanted propagation errors, d) The GSM / UMTS / Bluetooth / GPS antennas were deactivated whenever were not required, e) Besides our resource-aware application, only the factory default background applications were also running, f) Between measurements we let the phone to rest for several hours and the battery temperature to drop back to normal.

Table 1. Configuration for each test case

Case	3G/UMTS	Wi-Fi	GPS	Application
1	off	off	off	-
2a	On (weak signal)	off	off	-
2b	On (strong signal)	off	off	-
3a	off	On (weak signal)	off	-
3b	off	On (strong signal)	off	-
4	on	on	off	-
5	on	on	on	Google maps
6	on	on	off	Skype (voice call)
7	on	on	off	Skype (video call)
8a	on	on	off	YouTube Application
8b	on	on	off	Chrome Browser
9	on	on	off	Chrome Browser

We have also considered other similar scientific reports to assure accuracy of our work. In particular, in [20] and [21] the authors describe the Quality of Service (QoS) requirements of distributed multimedia streams. Based on this research we set our experiments according to an acceptable minimum level of QoS. In [19] Zhang et. al. present an Android app that provides dynamic power estimation for smartphones. With such a tool the end user is continuously informed about the remaining operational time of the device and developers may take into account these metrics for optimization purposes. Furthermore, researchers [22]-[26] have measured energy consumption on a variety of scenarios, which we have considered to set up our experiments and evaluate our results.

Hence, with experiment 1 we attempt to evaluate background battery consumption and CPU utilisation of the device when only its standard operations are executed and no connectivity is provided. This measurement is considered to be the reference measurement for the forthcoming ones. Then, with any next experiment we attempt to add just one additional operation to the previous settings, so we can evaluate its impact on resources consumption. Hence, with experiment 2 we attempt to evaluate the impact of 3G connectivity on battery and CPU consumption. With experiment 3 the impact of Wi-Fi connectivity. With experiment 4 the impact of simultaneous 3G and Wi-Fi connectivity. With experiment 5 the impact of GPS connectivity. With experiment 6 the impact of voice calls via Wi-Fi. With experiment 7 the impact of video calls via Wi-Fi, and with experiments 8 and 9 the impact of online video and music streaming under various options.

We would like to note here, that in our experimentations we have not considered the cases of voice calls via 3G and on line gaming, since the former does not concern data communications and the latter is very game-specific. We have not also considered data transfer via 3G due to the weak 3G signal's strength in our lab room, which would greatly affect the reliability of such measurements.

3.3 Results

3.3.1 Resources Consumption without Any Network Connection (3G and Wi-Fi Disabled)

In this experiment, after having fully charged the battery, we let the device in idle mode with the Wi-Fi and 3G features set to off. The device was set to maintain the screen always on with screen brightness to its highest value. This setting is maintained to all forthcoming experiments. The rest of the background processes remained as on device boot with only our resource-aware app also running. The experiment ran without any human interaction. The device was set on a horizontal position and was not moved until the end of the measurement. Figure 7 and Figure 8 illustrate battery discharging and CPU utilisation for this reference experiment. The horizontal axis is the Time axis and illustrates the total duration of measurements for this experiment.

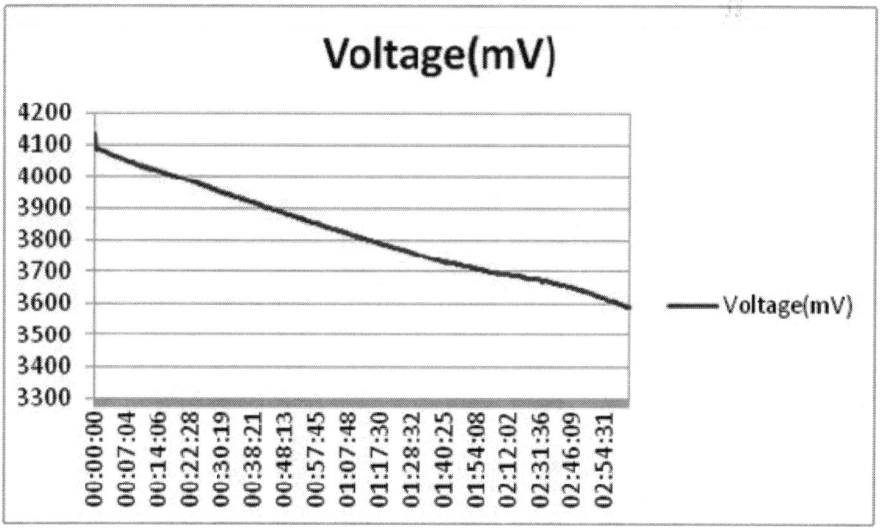

Fig. 7. Battery discharging for experiment 1

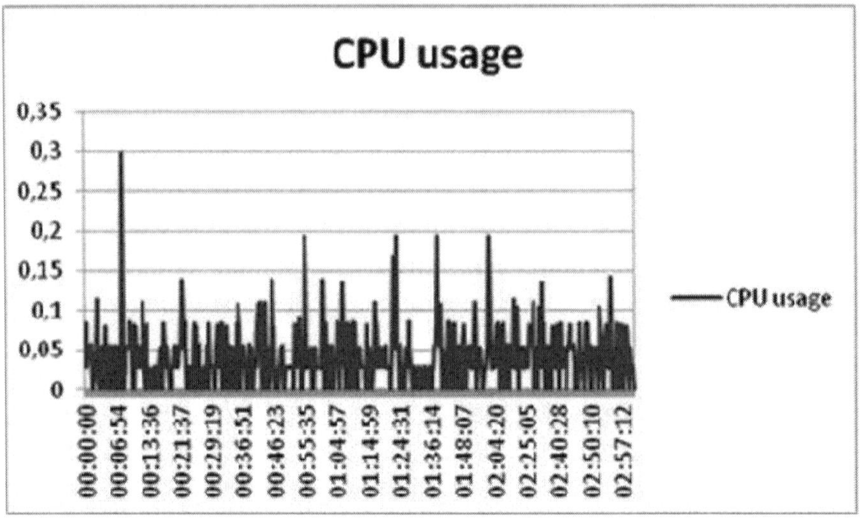

Fig. 8. CPU utilisation for experiment 1

3.3.2 Resources Consumption with the Device Connected to a 3G Network and the Wi-Fi Disabled

In this experimentation, after we fully recharged the battery, we let the device in idle mode with the Wi-Fi set to off and a 3G connection established. Screen brightness, background processes and device position were kept as on experiment 1. In

a. Weak 3G signal

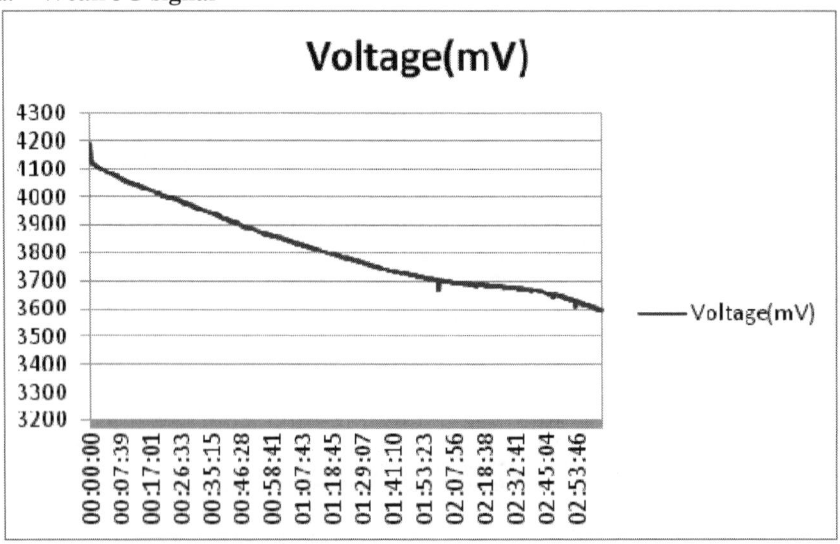

Fig. 9. Battery discharging for experiment 2a

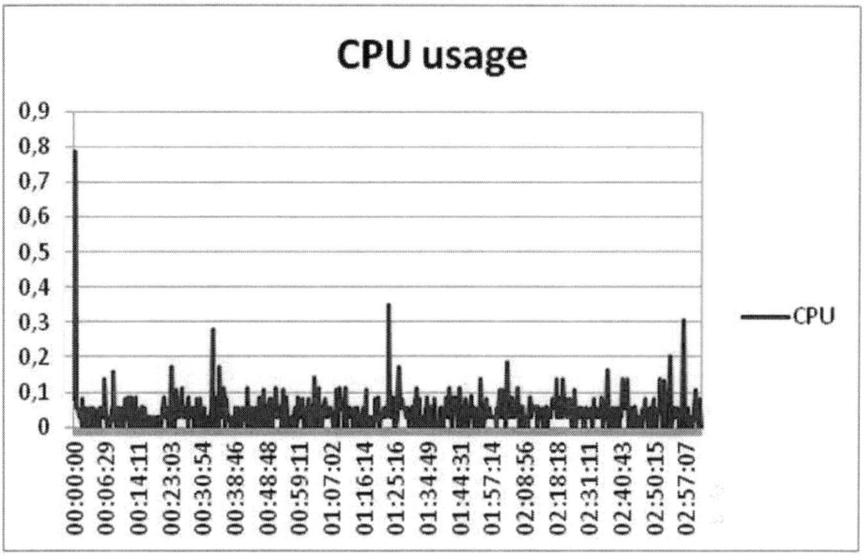

Fig. 10. CPU utilisation for experiment 2a

b. Strong 3G signal

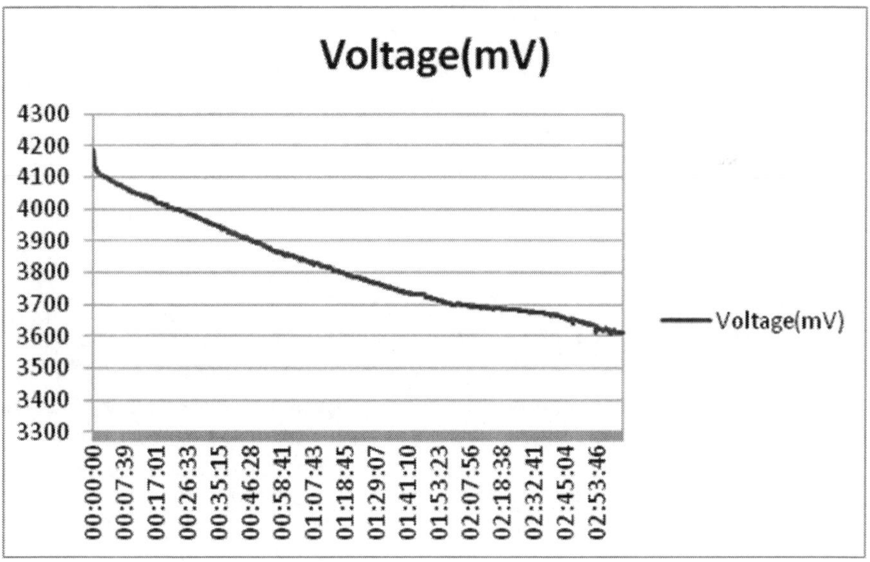

Fig. 11. Battery discharging for experiment 2b

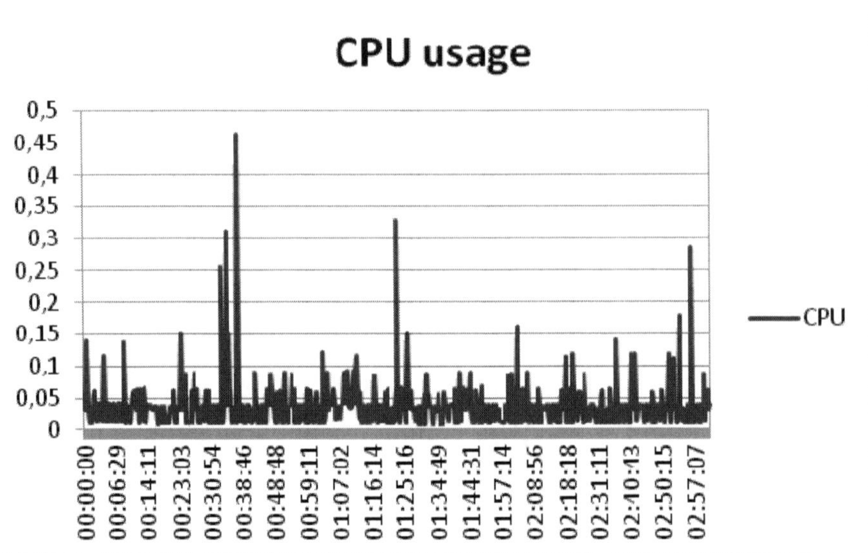

Fig. 12. CPU utilisation for experiment 2b

experiment 2a the signal strength from the 3G network was permanently weak during the measurement (3-5 ASU). In experiment 2b we chose a different location with permanently strong signal strength from the 3G network (27-30 ASU). Both experiments ran with no human interactions. Figure 9 and Figure 10 illustrate battery discharging, as well as CPU utilisation for experiment 2a. Figure 11 and Figure 12 present the results for experiment 2b.

3.3.3 Resources Consumption with the Device Connected to a Wi-Fi Network and the 3G Disabled

In this experimentation, after we fully recharged the battery, we let the device in idle mode with 3G set to off and a Wi-Fi connection established. In experiment 3a the signal strength from the Wi-Fi access point was intentionally maintained strong (-40 dBm) and in experiment 3b weak (-75 dBm). Screen brightness, background processes and device position were kept as on experiment 1. Both experiments ran with no human interactions. Figure 13 and Figure 14 illustrate battery discharging and CPU utilisation for experiment 3a. Figure 15 and Figure 16 present the results for experiment 3b.

a. Strong Wi-Fi signal

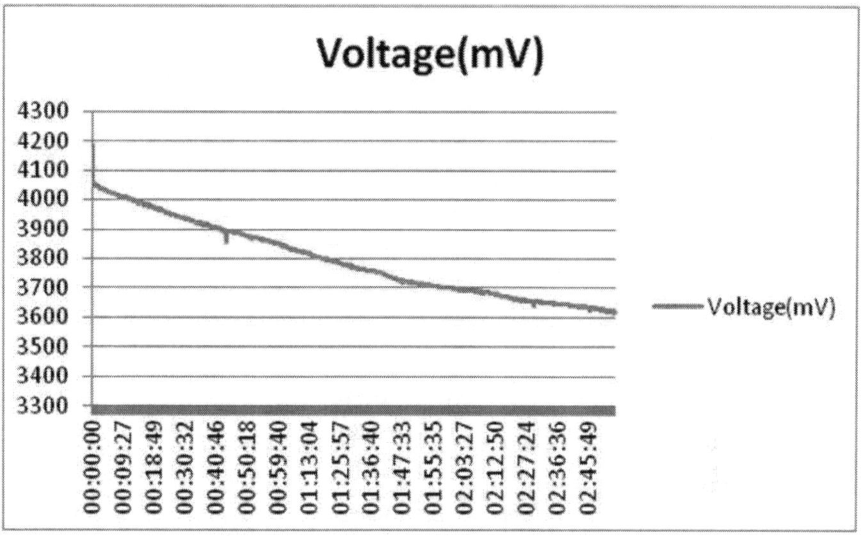

Fig. 13. Battery discharging for experiment 3a

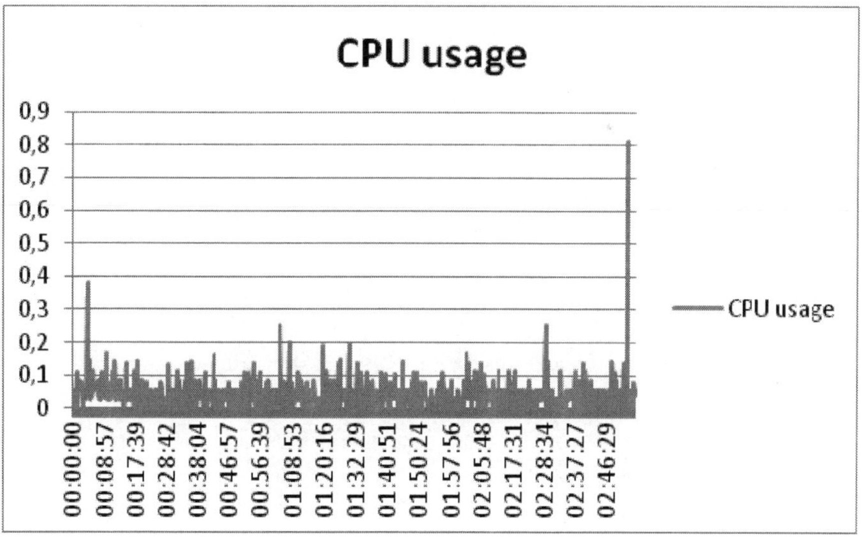

Fig. 14. CPU utilisation for experiment 3a

b. Weak Wi-Fi signal

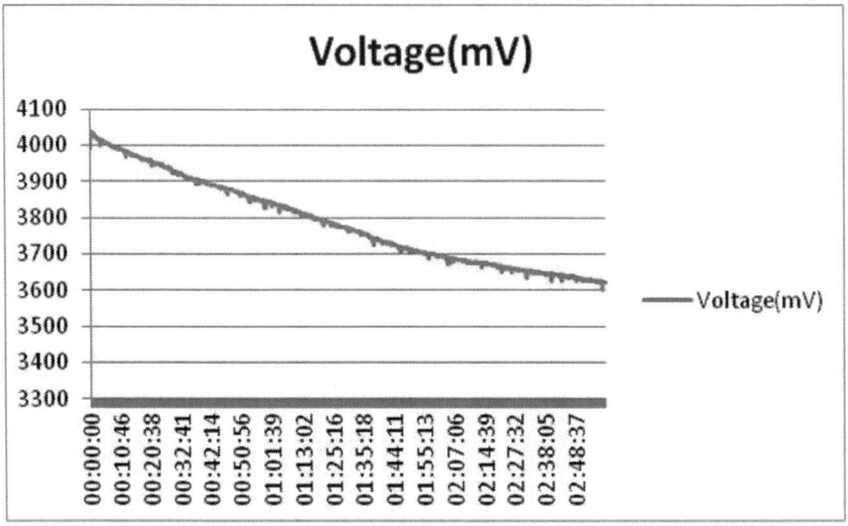

Fig. 15. Battery discharging for experiment 3b

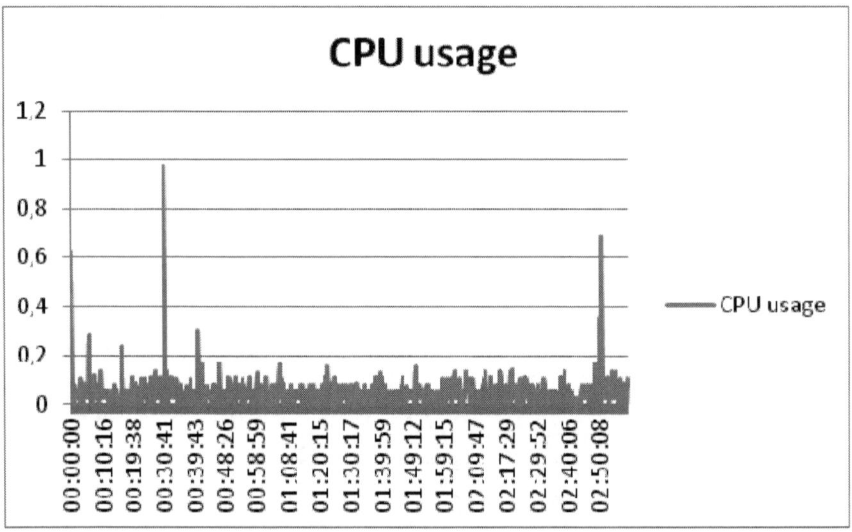

Fig. 16. CPU utilisation for experiment 3b

3.3.4 Resources Consumption with the Device Connected to Both 3G and Wi-Fi Networks

In this experiment, after we fully recharged the battery, we let the device in idle mode with 3G and Wi-Fi connections established. Screen brightness, background processes and device position were kept as on experiment 1. The signal strength from the 3G network was permanently weak during the measurement (3-5 ASU). The signal strength from the Wi-Fi access point was intentionally maintained strong (-40 dBm). The experiment ran with no human interactions. Figure 17 and Figure 18 illustrate battery discharging and CPU utilisation for experiment 4.

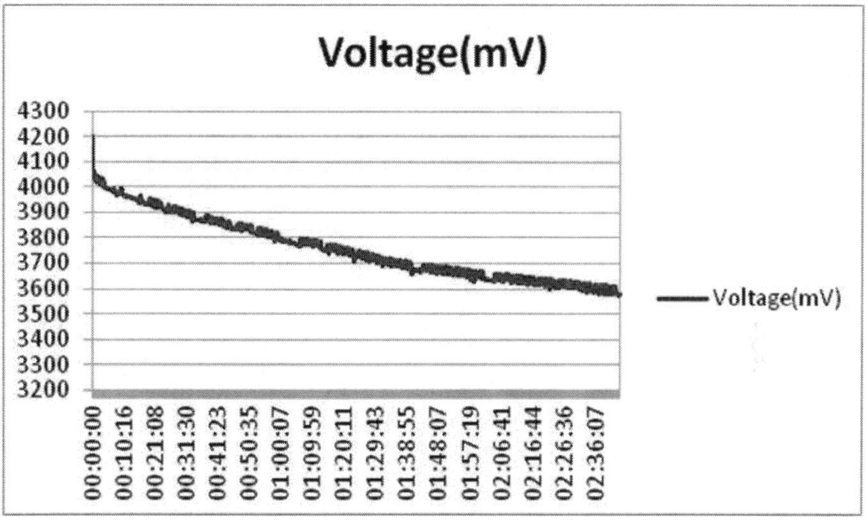

Fig. 17. Battery discharging for experiment 4

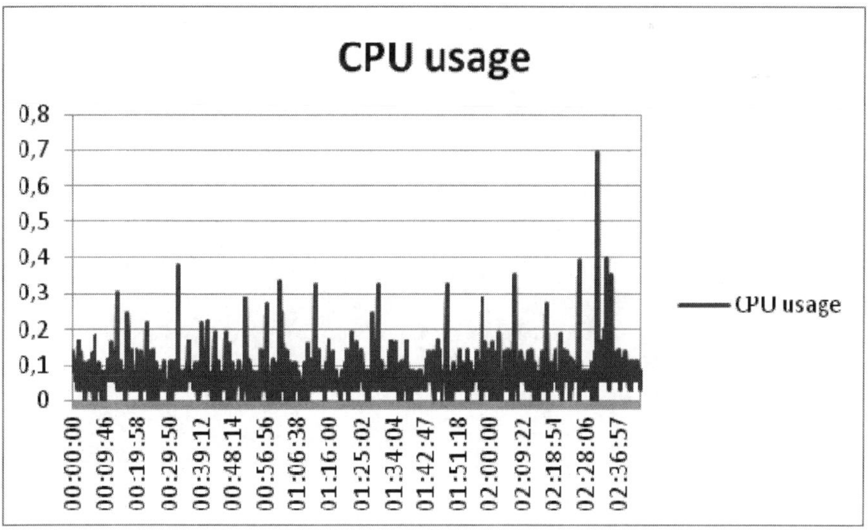

Fig. 18. CPU utilisation for experiment 4

3.3.5 Resources Consumption with the Device Connected to Both 3G and Wi-Fi Networks and the GPS on

In this experiment, after we fully recharged the battery, we let the device in idle mode with 3G and Wi-Fi connections established and the GPS sensor activated. In that context, we also used the Google Maps built in application (v. 6.11.1) to trigger the GPS antenna. Screen brightness, background processes and device position were kept as on experiment 1. The signal strength from the 3G network was permanently weak

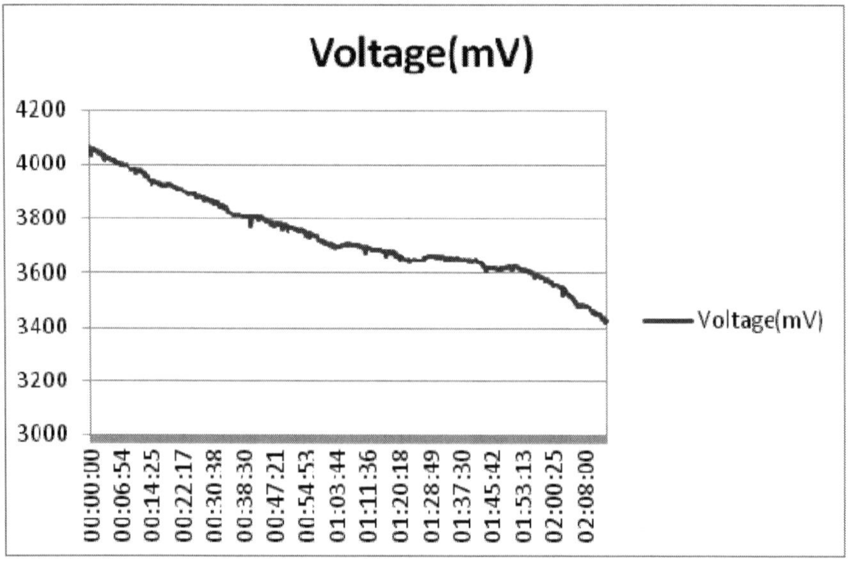

Fig. 19. Battery discharging for experiment 5

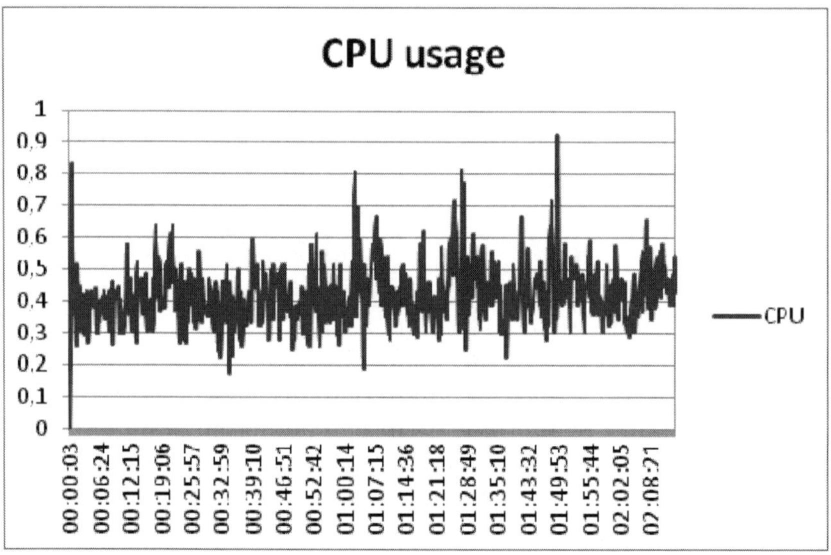

Fig. 20. CPU utilisation for experiment 5

during the measurement (3-5 ASU). The signal strength from the Wi-Fi access point was intentionally maintained strong (-40 dBm). The experiment ran with no human interactions. Figure 19 and Figure 20 illustrate battery discharging and CPU utilisation for experiment 5.

3.3.6 Resources Consumption during a Skype Voice Call via Wi-Fi, with the Device Already Connected to Both Wi-Fi and 3G Networks

In this experiment, after we fully recharged the battery, we let the device in idle mode with 3G and Wi-Fi connections established. Using Skype (v. 4.0.0.19550) we started a voice call towards a computer peer. The sound from the speaker was set to the

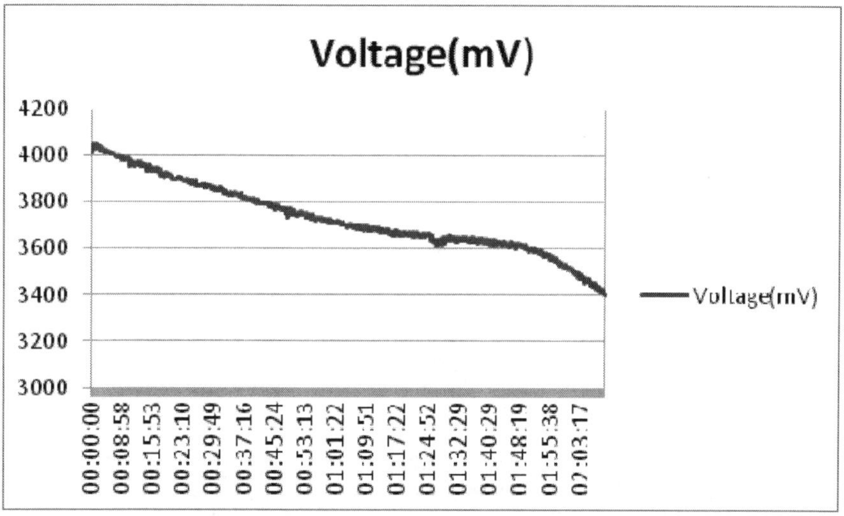

Fig. 21. Battery discharging for experiment 6

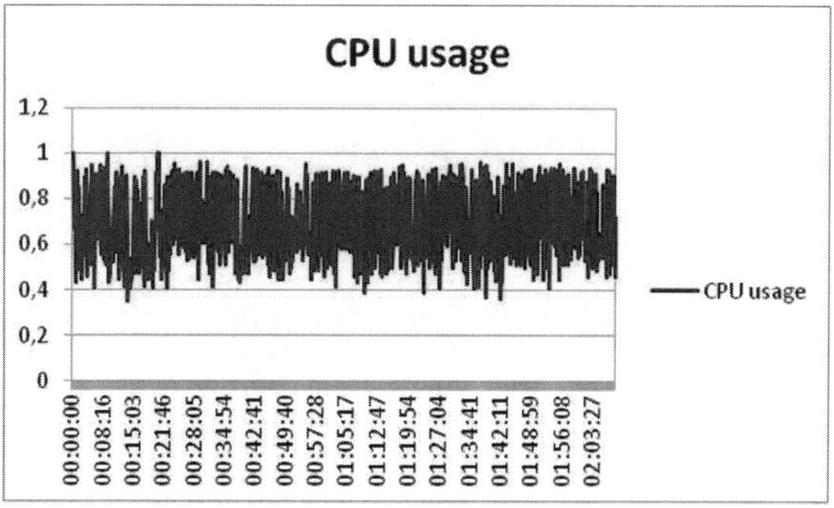

Fig. 22. CPU utilisation for experiment 6

highest level. Screen brightness, background processes and device position were kept as on experiment 1. The signal strength from the Wi-Fi access point was intentionally maintained strong (-40 dBm). The signal strength from the 3G network was permanently weak during the measurement (3-5 ASU). The experiment ran with no human interactions. Figure 21 and Figure 22 illustrate battery discharging and CPU utilisation for experiment 6.

3.3.7 Resources Consumption during a Skype Video Call via Wi-Fi, with the Device Already Connected to Both 3G and Wi-Fi Networks

In this experiment, after we fully recharged the battery, we let the device in idle mode with 3G and Wi-Fi connections established. Using Skype (v. 4.0.0.19550) we started a video call towards a computer peer. The video quality was set to its highest value. The sound from the speaker was set to the highest level. Screen brightness, background processes and device position were kept as on experiment 1. The signal strength from the 3G network was permanently weak during the measurement (3-5 ASU). The signal strength from the Wi-Fi access point was intentionally maintained strong (-40 dBm). The experiment ran with no human interactions. Figure 23 and Figure 24 illustrate battery discharging and CPU utilisation for experiment 7.

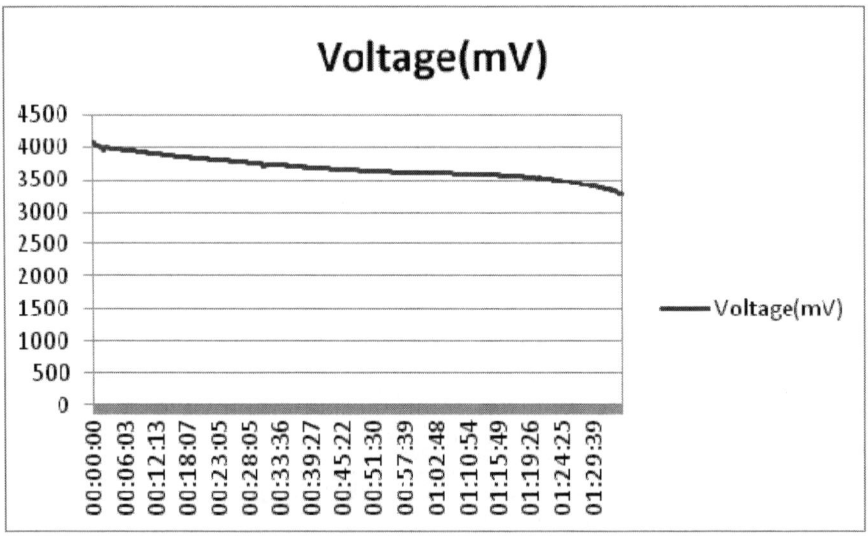

Fig. 23. Battery discharging for experiment 7

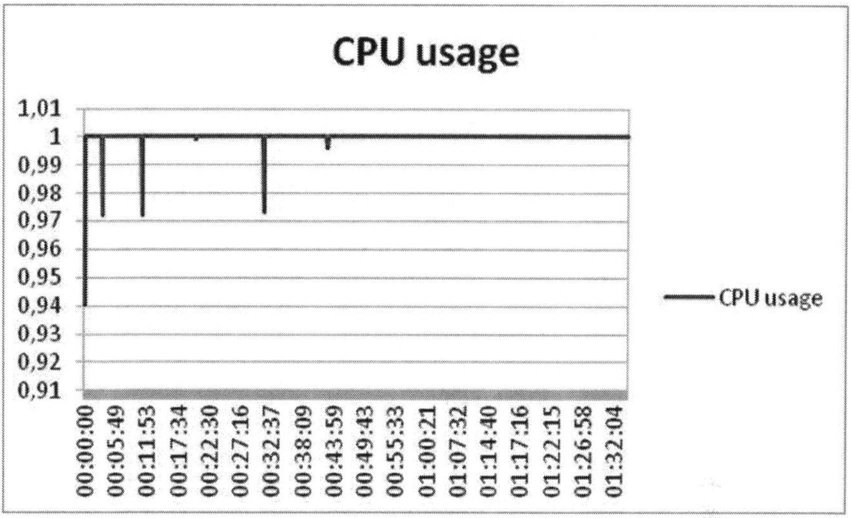

Fig. 24. CPU utilisation for experiment 7

3.3.8 Resources Consumption during Online Video Playback via Wi-Fi, with the Device Already Connected to Both 3G and Wi-Fi Networks

In this experimentation, after we fully recharged the battery, we let the device in idle mode with 3G and Wi-Fi connections established. In experiment 8a, using the YouTube application (v. 4.0.23) of our device we started a video stream at High Definition quality. The sound from the speaker was set to the highest level. Screen brightness, background processes and device position were kept as on experiment 1. The signal strength from the 3G network was permanently weak during the

a. Using a native Android app

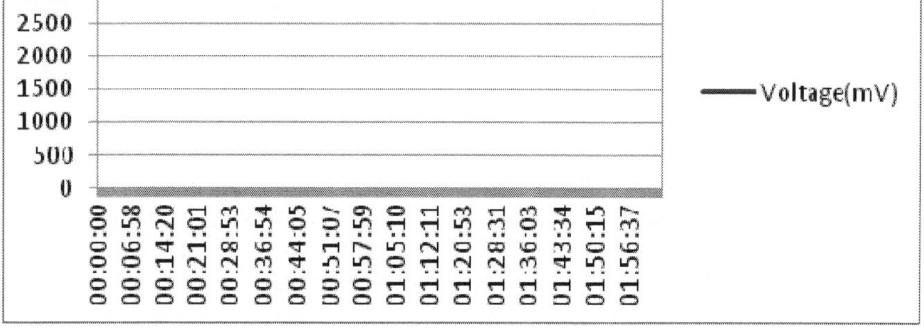

Fig. 25. Battery discharging for experiment 8a

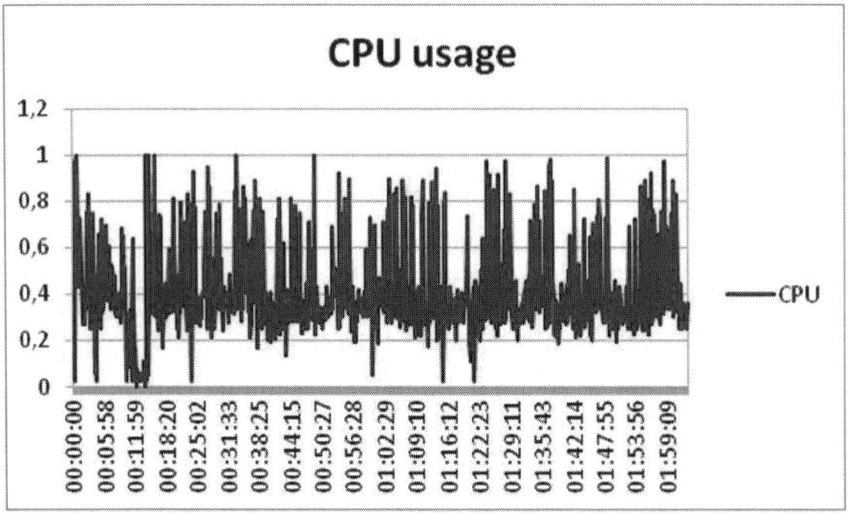

Fig. 26. CPU utilisation for experiment 8a

b. Using a browser

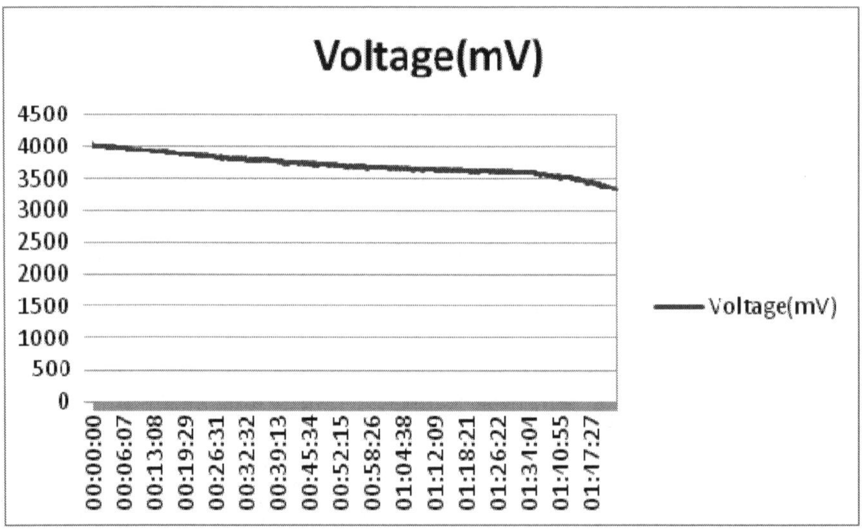

Fig. 27. Battery discharging for experiment 8b

measurement (3-5 ASU). The signal strength from the Wi-Fi access point was intentionally maintained strong (-40 dBm). The experiment ran with no human interactions. Following the completion of experiment 8a we repeated it as experiment 8b with exactly the same configuration but using this time the Chrome browser (v. 18.0.1025308) for the playback of the same online HD video instead of the YouTube application. Figure 25 and Figure 26 illustrate battery discharging and CPU utilisation for experiment 8a. Figure 27 and Figure 28 present the results for experiment 8b.

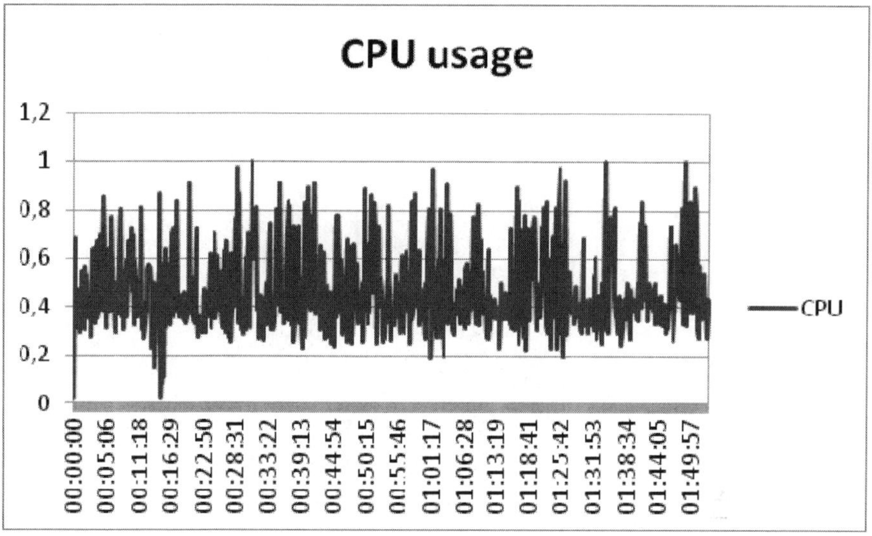

Fig. 28. CPU utilisation for experiment 8b

3.3.9 Resources Consumption during Online Music Playback via Wi-Fi, with the Device Already Connected to Both 3G and Wi-Fi Networks

In this experiment, after we fully recharged the battery, we let the device in idle mode with 3G and Wi-Fi connections established. Using the Chrome browser (v. 18.0.1025308) we started an e-radio music stream from http://www.fm1.teicrete.gr. The sound from the speaker was set to the highest level. Screen brightness,

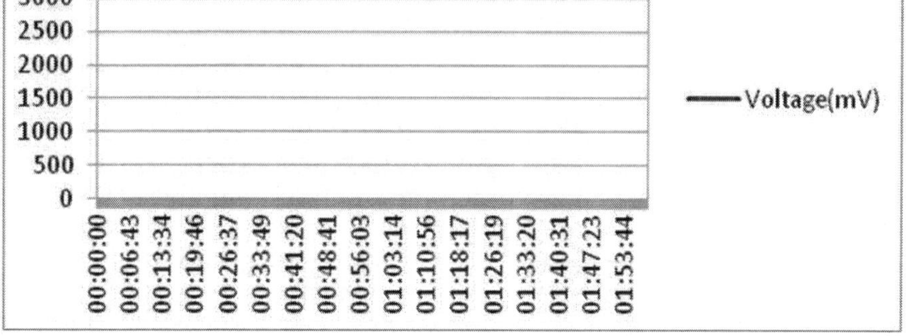

Fig. 29. Battery discharging for experiment 9

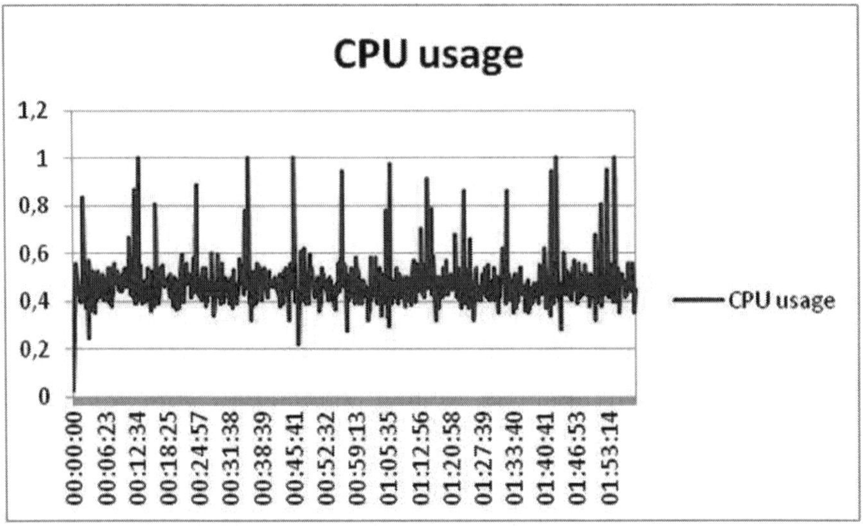

Fig. 30. CPU utilisation for experiment 9

background processes and device position were kept as on experiment 1. The signal strength from the 3G network was permanently weak during the measurement (3-5 ASU). The signal strength from the Wi-Fi access point was intentionally maintained strong (-40 dBm). The experiment ran with no human interactions. Figure 29 and Figure 30 illustrate battery discharging and CPU utilisation for experiment 9.

3.4 Discussion

Statistically processing the measurements from the nine experiments above, we can calculate for each experiment: a) the average discharging rate of our battery in millivolts per experiment's minute, b) the average CPU load during each experiment, and c) the operational time that our device could continuously execute each experiment with the specific battery. Table 2 summarises the results from statistical processing.

A first remark from our experiments is that there is not a direct relationship between CPU utilisation and battery consumption. Definitely, a high CPU utilisation induces high battery consumption but there is not a linear relationship between them. Thus, Skype video calling (experiment 7) and HD video play back via browser (experiment 8b) induce almost same battery discharging, although the former requires almost double CPU occupation. This result comes partially in contrast with [27] that designates screen and CPU as the most energy-demanding hardware components in a device. However, with experiment 1 we can assure the energy-demanding nature of screens, since with just the screen brightness on high and only the boot processes running, our device remains operational for just 5.5 hours.

Table 2. Statistical processing of measurements

Experiment	Configuration			Resources consumption and Time estimation		
	3G / UMTS	*Wi-Fi*	*GPS*	***Average Battery Discharging rate (millivolts per minute)***	*Average CPU load (%)*	**Operational time (hours)**
1	off	off	off	3.03	2.9	5.5
2a	on (weak signal)	off	off	3.30	4	5.05
2b	on (strong signal)	off	off	3.20	3.7	5.21
3a	off	on (weak signal)	off	3.13	3.5	5.32
3b	off	on (strong signal)	off	3.10	3.2	5.38
4	on (weak signal)	on (strong signal)	off	3.39	6	4.92
5	on (weak signal)	on (strong signal)	on	3.46	41.9	4.82
6	on (weak signal)	on (strong signal)	off	3.74	70	4.46
7	on (weak signal)	on (strong signal)	off	4.48	99.3	3.72
8a	on	on (strong signal)	off	4.15	44	4.02
8b	on	on (strong signal)	off	4.46	47	3.74
9	on (weak signal)	on (strong signal)	off	3.83	45	4.35

Taking into account that in our experimentation we gradually added complexity between experiments, we can now approximately compute the additional resources demand induced by each one operation tested. For example, juxtaposing experiments 1 and 2 we can compute the additional battery discharging rate and CPU load induced by the connection to a weak and a strong 3G network. The same derives for the connection to a strong and a weak Wi-Fi network by juxtaposing experiments 1 and 3. Table 3 includes the additional resources demand from the various operations.

From Table 3 we can assure outcomes from various independent studies [25],[29] that present 3G as a more energy-demanding networking technology than Wi-Fi, as well as that GPS is really battery-harvesting [30]. However, we notice that the most resource demanding operation is video processing, with video calls and video streaming to drain battery faster. This is also in line with various research reports such as [31], [32], [33].

As far as it concerns connectivity to weak or strong networks we observe that in 3G the mobility from an area with weak coverage to an area with strong signal releases about 0.1 mV per minute of operation while in Wi-Fi it releases about 0.03 mV/minute.

Table 3. Resources demand from the various operations

Operation	Additional Resources Demand	
	Average Battery Discharging rate (millivolts per minute of operation)	*Average CPU load (%)*
Connection to a 3G network (weak signal)	0.27	1.1
Connection to a 3G network (strong signal)	0.17	0.8
Connection to a Wi-Fi network (weak signal)	0.1	0.6
Connection to a Wi-Fi network (strong signal)	0.07	0.3
Enabling the GPS	0.07	0.1
Skype voice calling	0.35	0.46
Skype video calling	1.09	1.2
Online video play back using a native app	0.76	0.9
Online video play back using a browser	1.07	1.18
Browser operation in idle state	0.31	0.28
Online music play back using a browser	0.44	0.57

An analogous result is induced as it concerns online video streaming via a native Android application and a browser with the former to require 0.31 mV/minute less that the latter. Especially as it concerns experiment 8, by juxtaposing experiments 8a and 8b we can compute the average discharging rate induced by the operation of a browser in idle state (that is without running any webpage). This equals to 0,31 mV per minute of operation, an extremely high amount if we consider that on this should be added the energy cost from any running webpage.

Another remark is that CPU load due to connectivity to various networks is extremely low in comparison to the load induced by the execution of applications such as voice and video calling, video and music streaming, and GPS navigation. Thus, connection to both 3G and Wi-Fi networks enhances the CPU load at only 3% (from 3% to 6%), which equals to the activation of a browser, while a voice call via Skype enhances the CPU load at 64% (from 6% to 70%) and data streams (video or music) at about 40%.

4 Conclusions and Further Research

This chapter proposes a Cognitive Radio network architecture that enables for the efficient operation of mobile devices over TV White Spaces. The proposed network architecture comprises of a Geo-location database and a spectrum broker that coordinates TV White Spaces access, by a number of LTE based secondary systems, competing/requesting for the available radio spectrum. Furthermore, it introduces an innovative methodology for evaluation of energy and resource consumption in mobile cognitive devices that does not require any external metering device but exploits the advanced software and hardware features of modern smart phones to this end. In particular, the various APIs provided by Android (or other similar operating systems) are used here for adequately auditing and reporting resource consumption on such mobile platforms. More specifically, we evaluated energy consumption and CPU utilisation in various communication scenarios via a number of experimental tests, carried out under controlled conditions. Network connectivity, calling and online multimedia playback are some of the communication scenarios that were evaluated and presented here. In this respect, fields for future research include the exploitation of these results for the design of efficient cognitive algorithms providing dynamic rescheduling of processes running in a mobile device that would release resources and would lighten the execution load in the device. Desirables would be also cognitive algorithms for dynamic estimation of the available time an operation can be executed on a device under a given battery charge and execution load or of the device's performance degradation that is induced by the launching of an application. In addition, several approaches can be studied based on the research work published by the authors in [34], [35], [36], [37], as well as in [38], [39], [40], [41], [42] related with energy efficiency issues in cognitive radio networks.

References

[1] First Report and Order, Federal Communication Commission Std. (2010), http://fjallfoss.fcc.gov/edocs_public/attachmatch/FCC-10-196A1_Rcd.pdf

[2] The Economics of Spectrum management: A Review, Australian Communication and Media Authority, ACMA (2007)

[3] OFCOM, Digital Dividend Review: geographic interleaved awards 470 - 550 MHz and 630 - 790 MHz – Consultation on detailed award design (June 2008)

[4] 3GGPP LTE, http://www.3gpp.org/LTE

[5] Unlicensed Operation in the TV Broadcast Bands, Final Rules, http://edocket.access.gpo.gov/2009/pdf/E9-3279.pdf

[6] Bourdena, A., Pallis, E., Kormentzas, G., Skianis, C., Mastorakis, G.: Real-Time TVWS Trading Based on a Centralised CR Network Architecture. In: Proc. IEEE Globecom 2011, Texas, Houston, USA, pp. 964–969 (2011)

[7] Akyildiz, F., Lee, W.Y., Vuran, M.C., Mohanty, S.: A Survey on Spectrum Management in Cognitive Radio Networks. IEEE Coms. Mag. 46, 40–48 (2008)

[8] Cavalcanti, F., Andersson, S.: Optimizing Wireless Communications Systems. Springer (2009)

[9] Bourdena, A., Pallis, E., Kormentzas, G., Mastorakis, G.: A prototype cognitive radio architecture for TVWS exploitation under the real time secondary spectrum market policy. Physical Communication (2013), http://dx.doi.org/10.1016/j.phycom.2013.11.003
[10] Bourdena, A., Pallis, E., Kormentzas, G., Mastorakis, G.: A centralised broker-based CR network architecture for TVWS exploitation under the RTSSM policy. In: Proc. IEEE ICC 2012, Ottawa, Canada, June 10-15, pp. 7243–7247 (2012)
[11] Bourdena, A., Pallis, E., Kormentzas, G., Skianis, H., Mastorakis, G.: QoS provisioning and policy management in a broker-based CR network architecture. In: Proc. IEEE Globecom 2012, Anaheim, California, USA, December 03-07 (2012)
[12] Bourdena, A., Pallis, E., Kormentzas, G., Mastorakis, G.: Radio Resource Management Algorithms for Efficient QoS Provisioning over Cognitive Radio Networks. In: Proc. IEEE ICC 2013, Budapest, Hungary, June 09-13 (2013)
[13] Wyglynski, A.M., Nekovee, M., Hou, T.: Cognitive Radio Communications and Networks: Principles and Practice. Academic Press (2009)
[14] Bourdena, A., Pallis, E., Kormentzas, G., Mastorakis, G.: Efficient radio resource management algorithms in opportunistic cognitive radio networks. Transactions on Emerging Telecommunications Technologies (2013)
[15] Hossain, E., Niyato, D., Han, Z.: Dynamic spectrum access and management in cognitive radio networks, 1st edn. Cambridge University Press (2009)
[16] Dalvik Virtual Machine, https://code.google.com/p/dalvik/ (visited on August 12, 2013)
[17] Kapetanakis, K., Panagiotakis, S.: Efficient Energy Consumption's Measurement on Android Devices. In: Proceedings of the 2nd International Workshop on Mobile Device Software Development and Web Development, MDSD 2012 (in Conjunction with the 16th Panhellenic Conference on Informatics, PCI 2012), Piraeus, Greece, October 5-7 (2012)
[18] The kernel documentation as visited on August 14, 2013, https://www.kernel.org/doc/Documentation/filesystems/proc.txt
[19] Zhang, L., Tiwana, B., Qian, Z., Dick, R.P., Mao, Z.M., Yang, L.: Accurate online power estimation and automatic battery behavior based power model generation for smartphones. In: Proceeding of the Eighth IEEE/ACM/IFIP International Conference on Hardware/Software Codesign and System Synthesis (CODES+ISSS 2010), Scottsdale, AZ, USA, October 24-29 (2010)
[20] Malamos, A.G., Malamas, E.N., Varvarigou, T.A., Ahuja, S.R.: A Model For Availability of Quality of Service (QoS) in Distributed Multimedia Systems. Multimedia Tools and Applications Journal 16(3), 207–230 (2002)
[21] Malamos, A.G., Malamas, E.N., Varvarigou, T.A.: On The Definition, Modeling, And Implementation of Quality of Service (QoS) in Distributed Multimedia Systems. In: IEEE ISCC 1999, Egypt (July 1999)
[22] Carroll, A., Heiser, G.: An analysis of power consumption in a smartphone. In: Proceedings of the 2010 USENIX Conference on USENIX Annual Technical Conference, Boston, MA, June 23-25 (2010)
[23] Paul, K., Kundu, T.K.: Android on Mobile Devices: An Energy Perspective. In: IEEE International Conference on Computer and Information Technology (2010)
[24] Rynkiewicz, R.: Discharge and charge modeling of lead acid batteries. In: Proc. Appl. Power Electron. Conf. Expo (1999)

[25] Mittal, R., Kansal, A., Chandra, R.: Empowering developers to estimate app energy consumption. In: Proceedings of the 18th Annual International Conference on Mobile Computing and Networking, Istanbul, Turkey, August 22-26 (2012), doi:10.1145/2348543.2348583

[26] Ferreira, D., Dey, A.K., Kostakos, V.: Understanding Human-Smartphone Concerns: A Study of Battery Life. In: Lyons, K., Hightower, J., Huang, E.M. (eds.) Pervasive 2011. LNCS, vol. 6696, pp. 19–33. Springer, Heidelberg (2011)

[27] Shye, A., Scholbrock, B., Memik, G.: Into the wild: studying real user activity patterns to guide power optimizations for mobile architectures. In: Proceedings of the 42nd Annual IEEE/ACM International Symposium on Microarchitecture (MICRO 2009), New York, NY, USA, December 12-16 (2009)

[28] Kassinen, O., Harjula, E., Korhonen, J., Ylianttila, M.: Battery life of mobile peers with UMTS and WLAN in a Kademlia-based P2P overlay. In: Proceedings of the 20th Personal, Indoor and Mobile Radio Communications Symposium (PIMRC 2009), Tokyo, Japan, September 13-16 (2009)

[29] Balasubramanian, N., Balasubramanian, A., Venkataramani, A.: Energy Consumption in Mobile Phones: A Measurement Study and Implications for Network Applications. In: Proceeding of the 9th ACM SIGCOMM Conference on Internet Measurement Conference, IMC 2009, Chicago, Illinois, USA, November 4-6, pp. 280–293 (2009)

[30] Korhonen, K.: Predicting mobile device battery life. MSc. Thesis, Aalto University, Helsinki, Finland (2011)

[31] Shen, B., Tan, W.-T., Huve, F.: Dynamic Video Transcoding in Mobile Environments. IEEE Multimedia 15(1), 42–51 (2008)

[32] Šoštarić, D., Vinko, D., Rimac-Drlje, S.: Power Consumption of Video Decoding on Mobile Devices. In: Proceedings of the 52th International Symposium (ELMAR 2010), Zadar, Croatia, September 15-17, 2010, pp. 81–84 (2010)

[33] Lin, C.H., Liu, J.C., Liao, C.W.: Energy analysis of multimedia video decoding on mobile handheld devices. Elsevier Computer Standards & Interfaces 32(1-2), 10–17 (2010)

[34] Mavromoustakis, C.X., Dimitriou, C.D., Mastorakis, G.: On the Real-Time Evaluation of Two-Level BTD Scheme for Energy Conservation in the Presence of Delay Sensitive Transmissions and Intermittent Connectivity in Wireless Devices. International Journal on Advances in Networks and Services 6(3 and 4), 148–162 (2013)

[35] Mastorakis, G., Mavromoustakis, C.X., Bourdena, A., Pallis, E., Sismanidis, G.: Optimizing radio resource management in energy-efficient cognitive radio networks. In: Proceedings of the 2nd ACM Workshop on High Performance Mobile Opportunistic Systems, pp. 75–82. ACM (2013)

[36] Mastorakis, G., Mavromoustakis, C.X., Bourdena, A., Pallis, E.: An energy-efficient routing scheme using Backward Traffic Difference estimation in cognitive radio networks. In: IEEE 14th International Symposium and Workshops on World of Wireless, Mobile and Multimedia Networks, WoWMoM 2013 (2013)

[37] Mastorakis, G., Mavromoustakis, C.X., Bourdena, A., Kormentzas, G., Pallis, E.: Maximizing energy conservation in a centralized cognitive radio network architecture. In: 2013 IEEE 18th International Workshop on Computer Aided Modeling and Design of Communication Links and Networks (CAMAD 2013), pp. 175–179 (2013)

[38] Mastorakis, G., Bourdena, A., Mavromoustakis, C.X., Pallis, E., Kormentzas, G.: An energy-efficient routing protocol for ad-hoc cognitive radio networks. In: Future Network and Mobile Summit (FutureNetworkSummit 2013) (2013)

[39] Mavromoustakis, C.X., Dimitriou, C., Mastorakis, G., Pallis, E.: Real-Time Performance Evaluation of F-BTD scheme for optimized QoS Energy Conservation in Wireless Devices. In: Proc. IEEE Globecom 2013, 2nd IEEE Workshop on Quality of Experience for Multimedia Communications (QoEMC 2013), Atlanta, GA, USA, December 09-13 (2013)

[40] Dimitriou, C., Mavromoustakis, C.X., Mastorakis, G., Pallis, E.: On the performance response of delay-bounded energy-aware bandwidth allocation scheme in wireless networks. Paper Presented at the IEEE ICC 2013, Budapest, Hungary, June 9-13 (2013)

[41] Bourdena, A., Mavromoustakis, C.X., Kormentzas, G., Pallis, E., Mastorakis, G., Yassein, M.B.: A Resource Intensive Traffic-Aware Scheme using Energy-efficient Routing in Cognitive Radio Networks. Future Generation Computer Systems (2014), `http://dx.doi.org/10.1016/j.future.2014.02.013`

[42] Mavromoustakis, C.X., Mastorakis, G., Bourdena, A., Pallis, E.: Energy Efficient Resource Sharing using a Traffic-oriented Routing Scheme for Cognitive Radio Networks. IET Networks Journal (2014), doi:10.1049/iet-net.2013.0132

Efficient Resource Management Utilizing Content-Aware Multipath Routing

George Mastorakis[1], Evangelos Pallis[2], Constandinos X. Mavromoustakis[3], and Athina Bourdena[2]

[1] Technological Educational Institute of Crete, Department of Business Administration, Lakonia, Agios Nikolaos, 72100, Crete, Greece
gmastorakis@staff.teicrete.gr

[2] Technological Educational Institute of Crete, Department of Informatics Engineering, Estavromenos, Heraklion, 71500, Crete, Greece
{pallis,bourdena}@pasiphae.eu

[3] University of Nicosia, Department of Computer Science, 46 Makedonitissas Ave., 2414 Engomi, Nicosia, Cyprus
mavromoustakis.c@unic.ac.cy

Abstract. This chapter proposes a multipath routing approach that is adopted in content-aware networks, towards providing end-to-end Quality of Service (QoS) and an efficient exploitation/management of the resources. According to traffic identification mechanisms, this approach elaborates on effective packet distribution within several intra-domain paths. This is achieved through the capability of the edge nodes to recognize the content in a prototype network architecture. In addition, this chapter elaborates on content-awareness issues, regarding the utility estimation for a packet to select a particular path, based on different QoS-centric network criteria and requirements.

Keywords: Efficient resource management, multipath routing, content-aware networks.

1 Introduction

The most of the routing mechanisms in the current network infrastructures are based on algorithms that support single-shortest path approaches [1]. According to such approaches and taking into account that the packets have to be transferred as soon as possible, the same set of nodes are usually used to establish the shortest paths. Therefore, the routing algorithms that are based on the single-shortest path approaches, cause congestion in specific parts of the network, while other paths remain un-exploited. Towards minimizing this network congestion, multipath routing algorithms can be exploited for an optimum exploitation of the resources [2]. Multipath routing algorithms are used to route the packets of specific flows through multiple routes. This is achieved via the customization of routing mechanisms based on Quality of Service (QoS) requirements for different services, as well as via the

C.X. Mavromoustakis et al., *Resource Management in Mobile Computing Environments*, Modeling and Optimization in Science and Technologies 3,
DOI: 10.1007/978-3-319-06704-9_17,

improvement of end-to-end reliability and the dynamic sharing of the traffic through alternative routing paths. However, multipath routing is isolated from the characteristics of the services or the user-generated content that is provided within the network. Towards this direction, several modules are required to be adopted that should be able to analyze the content on-the-fly and identify the traffic. Then, a content-aware multipath routing approach can be adopted and applied on user-generated services transferred within the network.

In this respect, this chapter briefly discusses the multipath routing approach by elaborating on the description of a content-aware network architecture that exploits random utility theory. Section 2 describes the multipath routing approaches, while Section 3 elaborates on content-awareness issues, as well as traffic identification techniques. Section 4 presents a content-aware network architecture, where multipath routing is applied for efficient resource management. Finally, Section 5 is dedicated to conclude the chapter, by summarizing its main approach and highlighting open issues for further investigation.

2 Related Work and Research Motivation

Multipath routing algorithms are exploited to provide multiple paths among the source and the destination nodes. They are able to aggregate the available bandwidth through multiple paths, enabling for increased data transfer rates [3]. Several research approaches elaborate on extending the intra-domain routing algorithms (e.g. RIP and OSPF) to support the multipath approach [4], [5], [6]. The most known multipath algorithms include the Multiple Path Algorithm (MPA), the Multipath Distance Vector Algorithm (MDVA) [7], the Discount Shortest Path Algorithm (DSPA), the Capacity Removal Algorithm (CRA), the Equal-cost multi-path (ECMP) and the Multipath Partial Dissemination Algorithm (MPDA). ECMP is exploited to route packets through multiple paths of equal costs, while the load is equally distributed among the paths utilizing a round-robin distribution. MPA is also used to detect a number of paths to satisfy conditions for loop-freeness, while DSPA is exploited to minimize the delay, by considering the path quantity, as well as the path independence. CRA maximizes throughput depending on varying data traffic conditions, while MDVA provides multiple next-hop choices for every destination in the network. MPDA [8] is based on the Partial Dissemination Algorithm (PDA) to calculate the shortest distance up to the destination node, while QMPDA is an enhancement of MPDA algorithm, considering network failures, as well as network topology alterations. This is exploited for the provision of multiple Class of Services. The basic target is to aggregate the stream flows through the routing paths with flow classes. This is considered as a scalable technique that can be easily developed. In a general context, multipath routing is able to provide a scalable approach, towards supporting the provision of QoS. By considering all the above mentioned issues, this chapter elaborates on multipath routing issues based on random utility theory, as well as on Multinomial Logit Model (MNL) that is exploited for the prediction of individual's selection behaviour for a choice among a set of alternatives [9]. Random utility theory can be used to calculate the probability for each alternative to be chosen. Based on the multipath routing approach, a network of nodes interconnected by different links, can be represented as a directed graph [10]. In this case, network mechanisms are able to deliver packets under a more balanced approach through multiple paths, by exploiting algorithms to predict packet distribution among the alternative paths and indentifying the content provided.

3 Content-Awareness Concept

The traffic identification techniques can be exploited for the case that unclassified content is distributed over the network. Several such techniques like the ones proposed in [11], [12], are categorized based on the approach used to extract data from the traversing of network flows. For instance, several techniques exist using either data from the header of specific networking protocols and data extracted through the payload inspection [13], or data based on statistical flow information [14]. In a general context, no mechanism exists that can be uniformly adopted to support all possible cases for the user-generated services, while one technique that has been extensively exploited towards classifying data traffic, is the Deep Packet Inspection (DPI) method [15]. This method is used for the network flows to be inspected, towards extracting data for traffic classification. This approach is built according to two basic assumptions. According to such assumptions, third parties that are unaffiliated with either the source or the destination, are capable to inspect the payload of each packet and the classifier is aware of the relevant syntax of each packet payloads [16]. In this context, a library of protocol signatures, as well as the filter strings have to be developed. This library could be consulted, to enable for the accurate detection of the protocol. It has to be continuously updated as application protocols change or as new protocols appear. The most common algorithms that use DPI for signature analysis and mapping are summarized as follows:

- Automaton (Regular expression matching). It is used to track partially matched patterns in the data string by transition in either a Deterministic Finite Automaton (DFA) or Non-DFA implementation that accepts strings in the pattern set.

- Heuristics – A heuristic can be exploited to check a block of characters in the window suffix for its appearance in the patterns. It determines whether a match occurs and moves to the next window position if not.

- Filtering Based – This method is exploited to search the data string for necessary pattern features and quickly exclude the content not containing those features. A very common way of applying this approach for text filtering is using well-known Bloom filters.

Even though the DPI techniques can be used in data traffic classification of a network, they cannot be applied in several cases when encrypted traffic is not permitted to third parties. In such cases, it is a challenge for the classification mechanisms to properly classify the services, since specific encrypted data are invisible to DPI mechanisms. Therefore, statistical methods can be exploited to use data traffic characteristics, such as the packet inter-arrival time, the packet lengths, the flow idle time and the distribution of flow duration. These characteristics can be used to distinguish specific types of applications.

4 A Content-Aware Network Architecture to Apply Multipath Routing

The proposed approach can be applied in a network architecture, by utilizing content-awareness mechanisms at the edges of an autonomous system (AS). This network architecture is presented in Figure 1. It is able to classify and control the data traffic, as well as improve the network QoS. The proposed architecture is based on two basic approaches. The first one is associated with a novel service environment that includes sophisticated service management issues. This is based on an overlay of interconnected media-centric Home Gateways, as well as on an enhanced network approach that features Content Awareness. This approach will enable the development of virtual overlays networks (VCANs) that are associated with the media transport. In addition, this infrastructure consists of several edge routers (MANEs), as well as several core routers. The most difficult approach regarding the packet processing is related with the operation of the MANEs, since the core routers are mainly used to support the routing and forwarding procedures in the network.

There are several core network technologies, which are able to accommodate such functionality. The Carrier Ethernet family of standards and especially its application in metropolitan area networks provides among the others the capability of network virtualisation, using either MAC-in-MAC or Q-in-Q tagging and the appropriate network nodes. In some occasions, Carrier Ethernet is used either over already available SDH/SONET networks (ensuring high-availability) or over IP/MPLS networks (supporting Virtual Private Wire Service (VPWS) and Virtual Private Line Service (VPLS)) [17]. In optical networks, there is an increasing trend in the usage of Generalized Multi-Protocol Label Switching (GMPLS)[18], [19], which is derived from MPLS but adapted to the needs of these networks physical layer. GMPLS provides a common control plane for multiple layers, multiple vendors allowing greater service intelligence and efficiency. Based on the proposed approach of this chapter, this can be used to support the virtualisation by mapping VCANs in LSPs. In addition, MPLS can be combined with DiffServ mechanisms to assure QoS capabilities, while VCANs can be deployed at the edge routers and MPLS can be exploited at the core routers. In this respect, the MANE will be considered as a Label Edge Router (LER) that is improved with Content Awareness capabilities, by exploiting MPLS labels to the incoming flows. Also, the core routers will be considered as Label Switching Routers that are able to switch the packets.

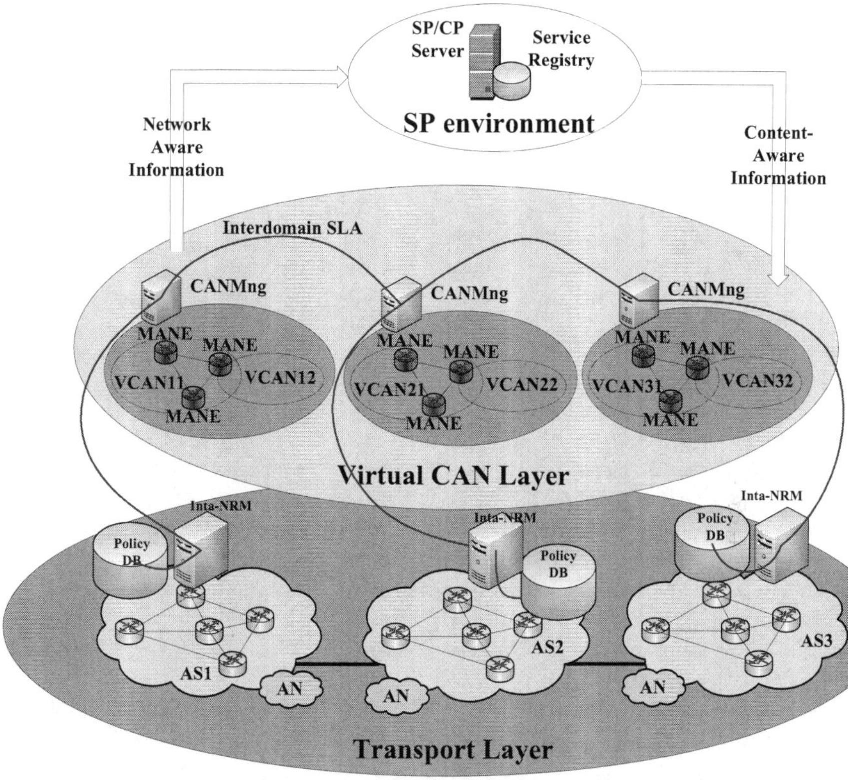

Fig. 1. Overall network architecture

5 Conclusions and Further Research

This chapter elaborates on a novel content-aware routing approach, by applying the random utility theory. The main traffic identification mechanisms have been described to realize content-awareness and a network architecture has been presented, where this concept may be applied. Future work based on the above mentioned issues, includes the investigation of methods to estimate the exponential value of the probability either with Monte Carlo simulations or with Taylor series approximations according to the alterations caused from the content-type information. In addition, several approaches can be studied based on the research work published by the authors in [20], [21], [22], as well as in [23], [24], [25], [26] related with novel routing protocols and algorithms.

Acknowledgments. This project is implemented through the Operational Program "Education and Lifelong Learning", Action Archimedes III and is co-financed by the European Union (European Social Fund) and Greek national funds (National Strategic Reference Framework 2007 - 2013).

References

[1] Awduche, D., Chiu, A., Elwalid, A., Widjaja, I., Xiao, X.: Overview and principles of Internet traffic engineering. RFC 3272 (2002)

[2] Banner, R., Orda, A.: Multipath routing algorithms for congestion minimization. In: Boutaba, R., Almeroth, K.C., Puigjaner, R., Shen, S., Black, J.P. (eds.) NETWORKING 2005. LNCS, vol. 3462, pp. 536–548. Springer, Heidelberg (2005)

[3] Suzuki, H., Tobagi, F.A.: Fat bandwidth reservation scheme with multi-link and multi-path routing in ATM networks. In: Proceedings of IEEE INFOOCOM (1992)

[4] Castineyra, I., Chiappa, N., Steenstrup, M.: The Nimrod Routing Architecture. IETF RFC 1992 (August 1996)

[5] Narvaez, P., Siu, K.Y.: Efficient Algorithms for Multi-Path Link State Routing. In: ISCOM 1999, Kaohsiung, Taiwan (1999)

[6] Vutukury, S., Garcia-Luna-Aceves, J.J.: A Simple Approximation to Minimum-Delay Routing. In: SIGCOMM 1999 (September 1999)

[7] Vutukury, S.: Garcia -Luna-Aceves, J.J.: MDVA: A Distance-Vector Multipath Routing Protocol. In: Proceedings of the INFOCOM (2001)

[8] Vutukury, S.: Multipath routing mechanisms for traffic engineering and quality of service in the Internet. Ph.D thesis, University of California, Santa Cruz (March 2001)

[9] Train, K.: Discrete Choice Methods with Simulation. Cambridge University Press (2003)

[10] Honma, Y., Aida, M., Shimonishi, H., Iwata, A.A.: New Multi-Path Routing Methodology Based on Logit Type Assignment. In: IEEE GLOBECOM Workshops (2009)

[11] Moore, A.W., Papagiannaki, K.: Toward the Accurate Identification of Network Applications. In: Dovrolis, C. (ed.) PAM 2005. LNCS, vol. 3431, pp. 41–54. Springer, Heidelberg (2005)

[12] Callado, A., Kamienski, C., Szabo, G., Gero, B., Kelner, J., Fernandes, S., Sadok, D.: A survey on Internet traffic identification. IEEE Communications Surveys & Tutorials 11(3), 37–52 (2009)

[13] Fernandes, S.F.L., Kamienski, C.A., Kelner, J., Mariz, D., Sadok, D.: A stratified traffic sampling methodology for seeing the big picture. Computer Networks 52(14), 2677–2689 (2008)

[14] Nguyen, T., Armitage, G.: A survey of techniques for internet traffic classification using machine learning. IEEE Communications Surveys (2008)

[15] Cascarano, N., Risso, F., Ciminiera, L.: Optimizing Deep Packet Inspection for High-Speed Traffic Analysis. Journal of the Network and Systems Management 19(1), 7–31 (2011)

[16] Dainotti, A., Pescapè, A., Claffy, K.: Issues and future directions in traffic classification. IEEE Network 26(1), 35–40 (2012)

[17] Sung, M., Xu, J.: IP Trace back-Based Intelligent Packet Filtering: A Novel Technique for Defending against Internet DDoS Attacks. IEEE Transactions on Parallel and Distributed Systems, 861–872 (September 2003)

[18] Mannie, E.: Generalized Multi-Protocol Label Switching (GMPLS) Architecture (October 2004)

[19] Berger, L., Fedyk, D.: Generalized MPLS (GMPLS) Data Channel Switching Capable (DCSC) and Channel Set Label Extensions (October 2010)

[20] Bourdena, A., Mastorakis, G., Kormentzas, G., Pallis, E.: A spectrum aware routing protocol for ad-hoc cognitive radio networks. In: 2012 IEEE 23rd International Symposium on Personal Indoor and Mobile Radio Communications (PIMRC 2012), pp. 1096–1101 (2012)

[21] Mastorakis, G., Bourdena, A., Mavromoustakis, C.X., Pallis, E., Kormentzas, G.: An energy-efficient routing protocol for ad-hoc cognitive radio networks. In: IEEE Future Network and Mobile Summit (FutureNetworkSummit) (2013)

[22] Mastorakis, G., Bourdena, A., Kormentzas, G., Pallis, E.: Spectrum aware routing in ad-hoc cognitive radio networks. In: IEEE Future Network & Mobile Summit (FutureNetworkSummit) (2012)

[23] Bourdena, A., Mavromoustakis, C.X., Kormentzas, G., Pallis, E., Mastorakis, G., Yassein, M.B.: A Resource Intensive Traffic-Aware Scheme using Energy-efficient Routing in Cognitive Radio Networks. Future Generation Computer Systems (2014), http://www.sciencedirect.com/science/article/pii/S0167739X14000387

[24] Mastorakis, G., Mavromoustakis, C.X., Bourdena, A., Pallis, E.: An energy-efficient routing scheme using Backward Traffic Difference estimation in cognitive radio networks. In: 2013 IEEE 14th International Symposium and Workshops on World of Wireless, Mobile and Multimedia Networks, WoWMoM 2013 (2013)

[25] Bourdena, A., Mastorakis, G., Pallis, E., Kormentzas, G.: A Routing Protocol for Enhanced Efficiency in Cognitive Radio Networks. Journal of Communication and Computer 10, 675–685 (2013)

[26] Bourdena, A., Mastorakis, G., Pallis, E., Arvanitis, A., Kormentzas, G.: A spectrum aware routing protocol for public safety applications over cognitive radio networks. In: 2012 IEEE International Conference on Telecommunications and Multimedia (TEMU), pp. 7–12 (2012)

Part IV

Resource and Power Management in Mobile Computing Systems

Scheduling for Efficient Telemedicine Traffic Transmission over Next Generation Cellular Networks

Theodora Kouskouli and Polychronis Koutsakis

Dept. of Electronic and Computer Engineering,
Technical University of Crete, Greece
ntora_kousk@hotmail.com, polk@telecom.tuc.gr

Abstract. Telemedicine traffic transmission has gained in importance during the past few years. Due to the fact that it carries critical information regarding the patients' condition, the expedited and errorless transmission of multimedia telemedicine traffic is of fundamental importance. The prioritized or guaranteed transmission of telemedicine traffic, however, can lead to the violation of the Quality of Service (QoS) requirements of regular traffic users and to the loss of guaranteed bandwidth in cases when it is left unused, due to the infrequent nature of telemedicine traffic. To resolve these problems, we propose a Multiple Access Control (MAC) protocol, for the integrated transmission of regular and telemedicine traffic transmission over next generation cellular networks and an adaptive bandwidth reservation scheme based on road map information and on user mobility and a fair scheduling scheme for telemedicine traffic transmission over cellular networks. The combination of the two schemes achieves high channel bandwidth utilization while offering full priority to telemedicine traffic over regular traffic. Three scheduling algorithms are evaluated, quantitatively and qualitatively, in terms of the QoS and the fairness they offer to different types of traffic.

1 Introduction

Telemedicine has been defined as using telecommunications to provide medical information and services [1], and is already being employed in many areas of healthcare.

The ultimate goal for all telemedicine applications is to improve the well-being of patients and bring medical expertise fast and at low cost to people in need [3, 4]. Thus, in addition to ambulance vehicles, it is also of critical importance for the provision of health care services at understaffed areas like rural health centers, ships, trains, airplanes, as well as home monitoring [3, 5]. Mobile healthcare (M-health, "mobile computing, medical sensor and communication technologies for healthcare" [6]) is a new paradigm that brings together the evolution of emerging wireless communications and network technologies with the concept of "connected healthcare" anytime and anywhere.

Next generation cellular networks will be able to provide voice, data and streamed multimedia to users on an "anytime, anywhere" basis. This will be achieved after

C.X. Mavromoustakis et al., *Resource Management in Mobile Computing Environments*, Modeling and Optimization in Science and Technologies 3,
DOI: 10.1007/978-3-319-06704-9_18, © Springer International Publishing Switzerland 2014

wired and wireless technologies converge and will be capable of providing very high data rates both indoors and outdoors, with premium quality and high security.

2 Description of System Model

In our work we consider the integration of regular and telemedicine traffic transmission. We consider voice, video, e-mail and short message service (sms) as representatives of regular traffic, and electro-cardiograph (ECG), X-ray, video and high-resolution medical still images for telemedicine traffic, in order to study the practical scenario of many different types of users simultaneously attempting to access the network and hence aggravating the access for telemedicine users. We describe below the characteristics of each traffic type.

2.1 Regular Multimedia Traffic

Four types of "regular" multimedia traffic are considered in our work: MPEG-4 video-conference, voice, email and mobile text messages (sms), which are the most common traffic types in cellular networks.

- Voice: The speech codec rate is 32 kb/s, and voice terminals are equipped with a voice activity detector (VAD) [9]. Voice sources follow an alternating pattern of talkspurt and silence periods (on and off), and the output of the voice activity detector is modeled by a two-state discrete time Markov chain (Figure 1). The mean talkspurt duration is 1 s and the mean silence duration is 1.35 s. The talkspurt to silence transition probability is PTS and the silence to talkspurt transition probability is PST. The talkspurt and silence periods are geometrically distributed with mean 1/PTS and 1/PST frames, respectively. Therefore, at steady state, the probability that a terminal is in talkspurt (speech activity), PT , or silence, PS, is obtained from the following equations:

$$P_T = P_{ST} / P_{ST} + P_{TS}$$
$$P_S = 1\text{-}P_T$$

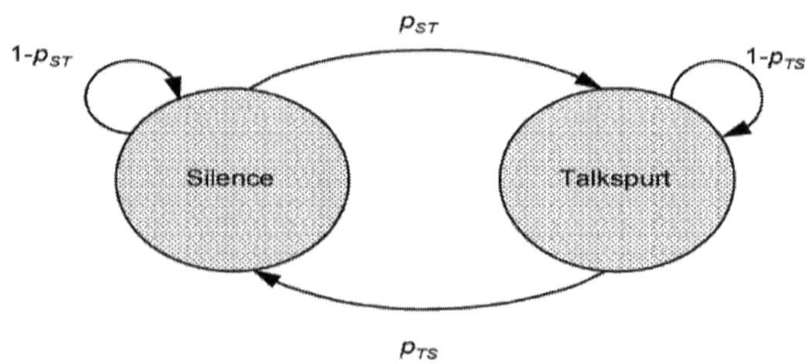

Fig. 1. The voice source activity model

The number of active voice terminals N in the system is assumed to be constant over the period of interest. All of the voice source transitions (e.g., talk to silence) occur at the frame boundaries. Reserved slots are deallocated immediately. The allowed voice packet dropping probability is set to 1%, and the maximum transmission delay for voice packets is set to 40 ms.

- Email: We adopt the data traffic model based on statistics collected on e-mail usage from the Finnish University and Research Network (FUNET) [8]. The probability distribution function f(x) for the length of the e-mail data messages of this model was found to be well approximated by the Cauchy (0.8, 1) distribution. The packet interarrival time distribution for the FUNET model is exponential, and the average e-mail data message length is 80 packets. A quite strict (considering the nature of this type of traffic) upper bound is set on the average e-mail transmission delay, equal to 5 s. The reason for this strict bound is that mobile users sending emails will be quite demanding in their QoS requirements, as they will expect service times similar to those of short message service traffic.
- SMS: Short Message Service (SMS) is a store-and-forward service that relies on a Short Message Service Center (SMSC). SMS messages are especially suitable for the transmission of small data bulks and for transmissions repeating in long time intervals (minutes to hours). The SMS payload is 140 bytes (including a header of 13 bytes) [24]. The message inter-arrival time distribution is considered exponential. In this work, in order to test our system under the strictest QoS requirements, we set an upper bound of two seconds in SMS transmission delay.
- Regular MPEG-4 Video Streams: The two MPEG-4 video streams used in our study have been extracted and analyzed from a camera showing the events happening within an office and a camera showing a lecture, respectively [10]. We have used the high quality version of the videos: one has a mean bit rate of 400 kb/s, a peak rate of 2 Mb/s, and a standard deviation of 434 kb/s, and the other one has a mean rate of 210 kbps, peak rate of 1.5 Mbps and standard derivation of 182 kbps. New video frames (VFs) arrive every 40 ms. We have set the maximum transmission delay for video packets to 40 ms, with packets being dropped when this deadline is reached. If a video packet does not arrive on time, the play out process will pause, which is annoying to the human eye. The allowed video packet dropping probability is set to 1% [11], as the loss of regular video packets is not of equally critical importance as that of telemedicine video packets for which the maximum allowed video packet dropping probability is 0.01%, as it will be explained in Section 2.2. The two video traces are chosen with equal probability by regular video users.

2.2 Telemedicine Traffic

Four types of telemedicine traffic are considered in our work: Electro-Cardiograph (ECG), X-ray, video and high-resolution medical still images.

- Electro-Cardiograph (ECG): We consider that ECG data is sampled at 360 Hz with 11 bits/sample precision. The arrival rate of ECG users is set to be λ_E users/frame following a Poisson distribution. The transmission of ECG traffic should be rapid and lossless, due to the critical nature of the data; additionally, we have set a strict upper bound of just 1 channel frame (12 ms) for the transmission delay of an ECG packet.
- X-Ray: We consider that a typical X-ray file size is 200 Kbytes [5] and that the aggregate X-Ray file arrivals are Poisson distributed with mean λ_X files/frame. The upper bound for the transmission delay of an X-Ray file, which again needs to be lossless, is set to 1 minute.
- Medical Images: Medical image files have sizes ranging from 15 to 20 Kbytes/image [2] and are Poisson distributed with mean λ_I files/frame. The upper bound for the transmission delay of a medical image is set to 5 seconds and the transmission needs to be lossless.
- Telemedicine Video: Since H.263 is the most widely used video encoding scheme for telemedicine video today, we use in our simulations real H.263 video-conference traces from [10] with mean bit rate of 91 Kbps, peak rate of 500 Kbps and standard deviation of 32.7 Kbps. The video frames arrive with constant rate (every 80 ms) with variable frame sizes. We have set the maximum transmission delay for video packets to 80 ms, with packets being dropped when this deadline is reached; i.e., all packets of a video frame must be delivered before the next video frame arrives. Due to the need for very high-quality telemedicine video, the maximum allowed video packet dropping probability is set to 0.01%.

3 Multimedia Integration Access Control (MI-MAC)

The Multimedia Integration Multiple Access Control (MI-MAC) protocol, introduced in [7] and based on Time Division Multiple Access with Frequency Division Duplex (TDMA-FDD), is one of the first works in the relevant literature for wireless picocellular networks that efficiently integrates voice (Constant Bit Rate, CBR, On/Off Traffic), bursty email, and sms traffic with either MPEG-4 or H.263 video streams (Variable Bit Rate, VBR) in high capacity picocellular systems with burst-error characteristics. The protocol was shown to be a good candidate for next generation cellular networks, as it outperformed (in simulation results and conceptually) other TDMA and Wideband Code Division Multiple Access (WCDMA)-based protocols when evaluated over a wireless channel with burst-error characteristics.

3.1 Channel Frame Structure

Within a picocell, spatially dispersed source terminals share a radio channel that connects them to a fixed base station (BS). The BS allocates channel resources, delivers feedback information, and serves as an interface to the mobile switching center (MSC). The MSC provides access to the fixed network infrastructure.

Our work focuses on the uplink (wireless terminals to Base Station) channel. The uplink channel time is divided into time frames of fixed length. The frame duration (12 ms accommodating 566 slots) is selected such that a voice terminal in talkspurt generates exactly one packet per frame. The packet size is considered to be equal to 53 bytes, 48 of which contain information. This choice was made in [7] in order to compare the protocol with other works in the literature; it is by no means restrictive, and does not influence the efficiency of our scheduling scheme. As shown in Fig. 2, which presents the channel frame structure, each frame consists of two types of intervals. These are the request interval and the information interval.

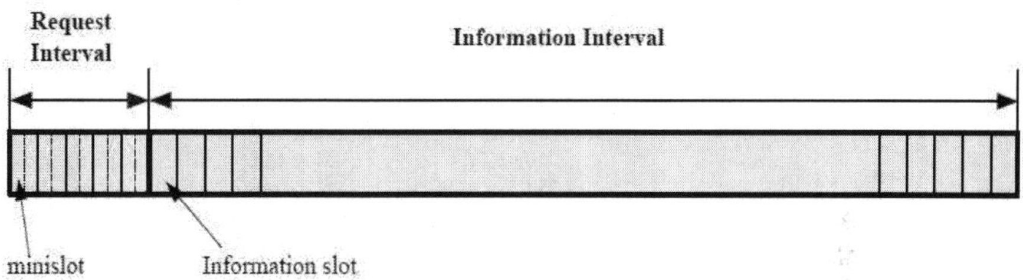

Fig. 2. Channel Frame Structure

By using more than one minislot per request slot, a more efficient usage of the available request bandwidth (in which users contend for channel access) is possible. We consider a 20 Mbps channel as a conservative estimation, given that next generation cellular networks are planned to have transmission rates exceeding 20 Mbps. We chose the number of minislots per request slot to be equal to 2, to allow for guard time and synchronization overheads, for the transmission of a generic request packet and for the propagation delay within the picocell. Each minislot accommodates exactly one fixed-length request packet. Within an information interval, each slot accommodates exactly one fixed-length packet that contains voice, video, or data information and a header. Any free information slot of the current channel frame can be temporarily used as an extra request (ER) slot to resolve the contention between requesting users. ER slots are again subdivided into two minislots. The function and operation of ER slots are exactly the same as those of the regular request slots.

3.2 Base Station Scheduling and Actions of Terminals

Terminals with packets, and no reservation, contend for channel resources using a random access protocol to transmit their request packets only during the request intervals. The Base Station broadcasts a short binary feedback packet at the end of each minislot, indicating only the presence or absence of a collision within the minislot [collision (C) versus non-collision (NC)]. Upon successfully transmitting a request packet the terminal waits until the end of the corresponding request interval to learn of its reservation slot (or slots). If unsuccessful within the request intervals of the current frame, the terminal attempts again in the request intervals of the next frame. A terminal with a reservation transmits freely within its reserved slot.

To resolve contention among all requesting users, different priorities were assigned to different types of users. The four types of telemedicine traffic are transmitted first; their priority order will be explained in Section 4. The four types of regular traffic follow, with priority order: video, voice, email, and sms. The above prioritization by isolating each type of traffic and letting it contend only with traffic of the same type is feasible due to the use of the two-cell stack reservation random access algorithm (by video and voice terminals) and the two-cell stack blocked access collision resolution algorithm [12] (by email and sms terminals) to resolve contention. To allocate channel resources, the Base Station maintains a dynamic table of the active terminals within the picocell. Upon successful receipt of a voice or data request packet, the Base Station provides an acknowledgment and queues the request. The BS allocates channel resources at the end of the corresponding request interval.

3.3 System State Transitions

As shown in Figure 3, an active terminal is described as being in one of four states: silent, contender, queued, or reserved. A silent terminal has no packet to transmit and does not require channel resources. Once the terminal has information to transmit, it enters the contender state and remains there until it either successfully transmits a request packet or drops all of its packets (in the case of video and voice terminals). Since the requests are queued at the BS, the terminal enters the queued state and remains there until it either receives a reservation or exits talkspurt. After receiving a reservation, the terminal enters the reserved state and transmits one (or more, in the case of video terminals) packet(s) per frame into its slot(s) until it exhausts its packets and returns to the silent state.

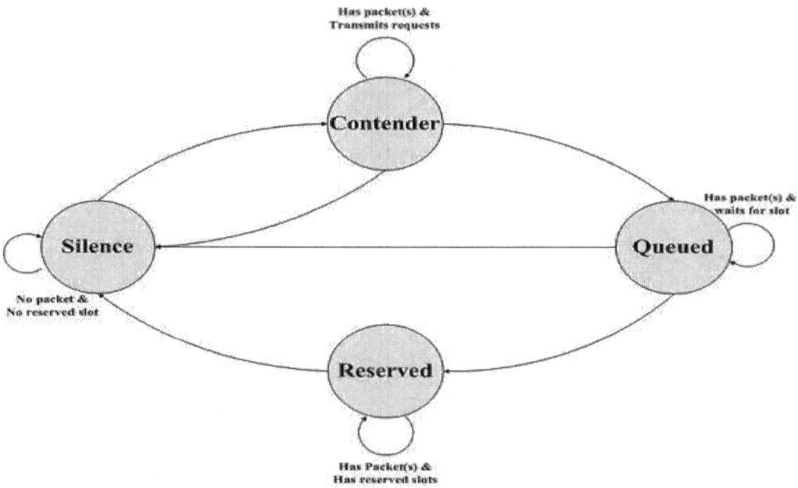

Fig. 3. State transition diagram for an active terminal

3.4 Channel Error Model

The most widely adopted wireless channel error model in the literature is the Gilbert-Elliot model [13, 14]. The Gilbert- Elliot model is a two-state Markov model where the channel switches between a "good state" (always error-free) and a "bad state" (error-prone). A better choice for a more robust error model for wireless channels is the one we adopt in our study, and which was proposed in [15]. This model, with the use of the short and long error bursts, makes more accurate predictions of the long-term correlation of wireless channel errors than the Gilbert-Elliot model. The error model consists of a three-state discrete-time Markov chain, where one state is the "good state" (error-free) and the other two states are the "bad states", the long bad and the short bad state, as we can see in Figure 4. A transmission is successful only if the channel is in the "good state" (G); otherwise, it fails. The difference between the long bad (LB) and short bad (SB) states is the time correlation of errors: LB corresponds to long bursts of errors, SB to short ones.

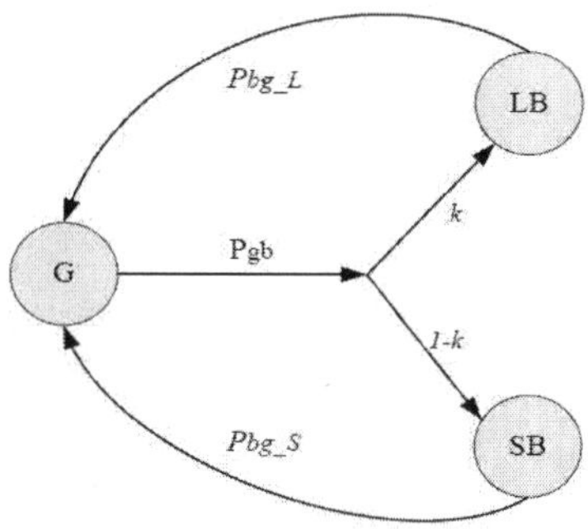

Fig. 4. Channel Error Model

4 The Proposed Scheduling Scheme

4.1 Introduction of the Scheduling Ideas

In our work the following design limitations have been adopted:

- A portion of the traffic arriving in each cell is handoff traffic from the other cells in the network. Handoff traffic is treated with full priority, with the use of the adaptive bandwidth reservation scheme which will be analyzed in Section 5.
- Video sources do not "live" permanently in the system, but have exponentially distributed sessions with a mean duration of five minutes [16]. This "relieves" a burden from the information interval of the channel; but adds a significant burden to the request interval, which has to compensate for the increase in contention as video users attempt to regain channel access. A small percentage of the bandwidth (4.4% in our work, i.e., 25 slots out of the 566 slots of the channel frame) suffices to be used for requests. This value

has been found via extensive simulations to provide a good tradeoff between allowing sufficient bandwidth for terminals to transmit their requests and allowing a large enough number of slots for terminals with a reservation to transmit their information packets.

- In reality, however small the picocell radius, the channel fading experienced by each user is different, since users are moving independently of each other; therefore, in the present work fading per user channel is considered. This new proposed mechanism aims at allocating as many of these slots as possible to other video terminals waiting for packet transmission, in order to decrease their transmission delay.

Still, the BS cannot know with certainty the type of channel state transition that takes place for a mobile terminal when it leaves the good state, i.e., if the terminal's channel has entered the SB state or LB state. Therefore, the BS can only make an estimation of each mobile video terminal's channel conditions, by monitoring the slots allocated to the terminal and checking whether the terminal is transmitting in them or not. If the total number of a terminal's failed transmissions within its allocated slots surpasses a given threshold, the BS in our scheme deduces that the terminal is in LB state, as the probability that it is in SB is very small given the high number of corrupted transmissions. Based on the channel error model it is easy to confirm by both analysis and simulation that the probability that a mobile terminal's channel is in SB when more than 6 slots have been wasted is 6.55%; hence we have set the threshold to be 6 subsequent transmission failures (choosing a higher threshold would result in a more accurate prediction of the channel condition, as the probability of a mistake in the prediction would be significantly lower; however, it would also lead to a higher number of lost slots while the BS is awaiting to make that prediction). When the BS determines that a mobile video terminal is in LB state, if that terminal has more reserved slots in the current channel frame, the BS deallocates these slots. Full priority for these slots is given to handoff telemedicine video terminals, followed by telemedicine video users originating from within the cell, then by hand-offed regular video users and finally by regular video users originating from within the cell; the allocation of the abandoned slots within each priority type is FCFS.

When the channel of the mobile terminal to which the slots were originally allocated returns to the good state, the terminal needs to inform the BS of this change, if it still has packets to transmit. This is done by transmitting a request packet. The terminal has to follow this procedure also in the case of a wrong estimation by the BS (i.e., if it was in SB state despite the long error burst).Therefore, in the (unlikely but not improbable) case of a wrong estimation, this does not directly influence the throughput achieved by our protocol in heavy traffic loads (slots are simply allocated to other telemedicine video and regular video users) but it results in an unnecessary increase of contention.

4.2 Scheduling Priorities and Contention Resolution

When resolving the contention among all requesting users, the BS needs to service the telemedicine traffic first, due to its urgency. To achieve this objective, we need to guarantee highest priority to telemedicine traffic. The priority order used by the BS in our proposed scheme is the following: ECG, X-Ray, telemedicine image, telemedicine video. The choice of priorities has been made based on the importance that each of these traffic types currently has for medical care. Highest priority is given to handoff telemedicine traffic, with telemedicine traffic originating from within the cell following in priority, in the same order (provided, of course, that the telemedicine users from within the cell have successfully transmitted their requests at the beginning of the frame request interval). Handoff regular traffic is transmitted next, with priority (video, voice, email, sms), based on the strictness of the QoS requirements for each traffic type (video and voice have the same QoS requirements of less than 1% packet dropping, but video traffic is much burstier, therefore it is granted priority over voice). Regular traffic originating from within the cell is transmitted last, with the same priority order.

We employ the two-cell stack reservation random access algorithm for telemedicine video, regular video and voice terminals, and the two-cell stack blocked access collision resolution algorithm to resolve the contention of ECG, X-ray, medical image, email and sms terminals.

4.3 Fair Scheduling

If users of the same type of traffic are served in a FCFS order once they are admitted into the network (as in [7] and in most relevant works on MAC protocols in the literature), the average performance evaluation metrics will give no insight on the QoS of each individual wireless subscriber; therefore, it could be the case that certain users have their QoS severely violated while others get exceptional QoS, which would give a seemingly acceptable average QoS over the total number of users. This approach is, however, unfair to users who arrive later in the network and hence are placed at the bottom of the BS service queue; the problem is especially significant in the case of telemedicine video and regular video users, where early arriving users may dominate the channel by being allocated large numbers of slots, allowing just a small number of resources to be available for users arriving later.

For this reason, we introduce the following Fair Scheduling scheme for telemedicine video and regular video users (the scheme is enforced separately among users of each of the two types of traffic, since telemedicine video users have higher priority). The BS allocates bandwidth by comparing the channel resources to the total requested bandwidth, currently, from all active video users. If the available bandwidth is larger than the total requested bandwidth, all users will be assigned as many slots as they have requested. If, however, the available bandwidth is smaller than the total requested bandwidth, then the available bandwidth will be shared among video users proportionally. More specifically, let M be the number of currently idle information slots in the frame and B_i the amount of bandwidth that will be assigned to video terminal i in every channel frame. B_i is given by:

$$B_i = M * (D_i / \Sigma_i D_i)$$

where D_i is the ith user's requested bandwidth and $\Sigma_i D_i$ is the total bandwidth requested by all of the video terminals at that moment.

It is intuitively clear, and it will also be shown from our results, that with the use of this formula the number of telemedicine and regular, respectively, video users whose

QoS is violated significantly decreases. The above scheme does not need to be implemented on any of the other types of traffic considered in our work, besides video, since they are allocated only one slot per frame.

4.4 FCFS – EDF – SJF Algorithms

As mentioned earlier, users of the same type of traffic are served in a FCFS order once they are admitted into the network. The First-Come-First-Served algorithm is the simplest scheduling algorithm. Processes are dispatched according to their arrival time on the ready queue. Jobs arriving are placed at the end of queue, the dispatcher selects the first job in the queue and this job runs to completion. The FCFS scheduling is fair in the human sense of fairness but it is unfair in the sense that long jobs make short jobs wait. For this reason, we wanted to compare its performance against two other well-known scheduling algorithms from the literature.

The Earliest Deadline First (EDF) algorithm operates on the logic that, among the users of one traffic class (e.g. voice), the user with the nearest deadline to transmit (a packet, message or video frame) will be accommodated first, instead of the user that arrived first (i.e. transmitted a request packet successfully in an earlier minislot). Earliest deadline first or "least time to go" is a dynamic scheduling algorithm used in real-time operating systems. It places processes in a priority queue. Whenever a scheduling event occurs (task finishes, new task released, etc.) the queue will be searched for the process closest to its deadline.

Shortest Job First (SJF), also known as Shortest Job Next (SJN), is a scheduling policy that selects the waiting process with the smallest execution time to execute next, and this process runs to completion. Shortest job next is advantageous because of its simplicity and because it maximizes process throughput (in terms of the number of processes run to completion in a given amount of time). We did not use SJF on voice users since we cannot predict a priori how long a conversation will last, while we know, e.g., the size of data messages or video frames.

4.5 Jain's Fairness Index

Aggregate performance evaluation metrics reveal very little, if any, information regarding the QoS of each wireless subscriber. Therefore, we should examine the impact of FCFS, EDF, SJF not only on commonly used performance metrics, such as throughput and delay, but also on fairness. We used Jain's Fairness Index [25] which is defined as follows:

$$J(x_1,\ldots,x_n) = \frac{\left(\sum_{k=1}^{n} x_k\right)^2}{n\sum_{k=1}^{n} x_k^2},$$

where n is the number of users and x_k is the throughput of user k. This index is continuous and bounded between 0 and 1, with 1 denoting maximal fairness. It is also very intuitive. If a ratio y of the users are treated fairly and (1 - y) are starved, then the resulting fairness index is y. We study fairness separately for telemedicine video users, regular video users and voice users. The throughput is measured in terms of the average number of allocated slots over 83 frames, which approximately corresponds to the average talkspurt duration (1 second); n is the number of users that were active during each 83-frame window (which is defined as a superframe).

5 Adaptive Bandwidth Reservation Based on Mobility and Road Information

5.1 Network and Mobility Models

It is a common assumption that the dissatisfaction of a wireless cellular subscriber who experiences forced call termination while moving between picocells is higher than that of a subscriber who attempts to access the network for the first time and experiences call blocking [17,18]. For this reason, it is important that the system is able at any point in time to accommodate newly arriving handoff calls in any cell of the network.

We consider an architecture of seven hexagonal cells which we place in two different topologies. In the first topology, we assume the "circular" case, where after leaving the last cell (G) a user enters cell (A) again. In the second, "open" topology, mobiles can get in or out of our system only via cell A and cell G, which are placed at the two edges of our network model [23]. So, all cells are connected in a straight line, as shown in Figure 5.

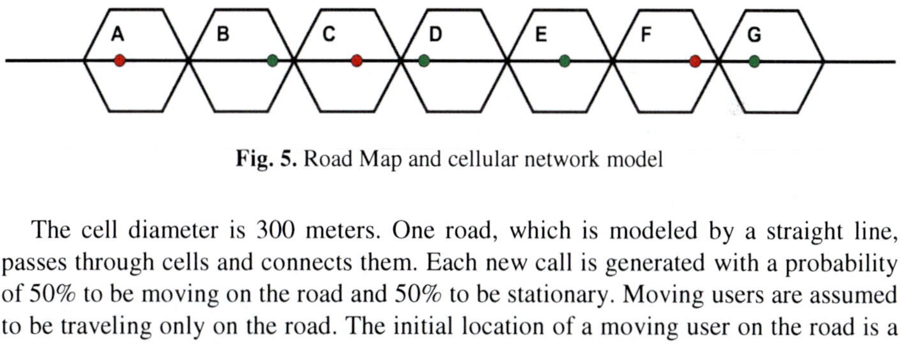

Fig. 5. Road Map and cellular network model

The cell diameter is 300 meters. One road, which is modeled by a straight line, passes through cells and connects them. Each new call is generated with a probability of 50% to be moving on the road and 50% to be stationary. Moving users are assumed to be traveling only on the road. The initial location of a moving user on the road is a uniform random variable between zero and the length of that road. During their call, stationary callers remain stationary and mobile users travel at a constant speed. Mobile users can travel in either of the two directions of a road with an equal probability, and with a speed chosen randomly in the range of [36, 90] Km/h. One traffic light is located randomly within each cell. A mobile user arriving at the traffic light of the cell might continue to go straight, or turn around with probabilities 0.9 and 0.1, respectively. If a mobile user chooses to go straight at the traffic light, it needs to stop there with probability 0.5 for a random time between 0 and 30 seconds due to a red traffic light, or else passes with probability 0.5. If the user chooses to turn around, it needs to stop there for a random time between 0 and 60 seconds due to the traffic signal [19, 20]. Each base station is loaded with the road map of its coverage area and its neighboring cells. Mobile stations report their position to the BS of their cell through a control channel. The position information includes the mobile user's exact location (cell), moving direction, and speed, and can be provided with an accuracy of 1m through GPS [19, 20, 21].

5.2 Adaptive Bandwidth Reservation Scheme

In our study, we adopt the idea that mobile stations only need to update their position information to the BS of their cell when they arrive at a traffic light. If the BS of the current cell of a mobile station predicts, based on the station's location (at a traffic light) and speed that the station is going to move to another cell, it sends a notification to the BS of that cell, including the current bandwidth used by the station and the estimated arrival time at the next traffic light. Hence, the proper amount of bandwidth is reserved for the station. For telemedicine video and regular video terminals, the bandwidth that is reserved in the next cell is equal to the remaining bandwidth that the terminal will need to complete its transmission (this bandwidth is declared in each video user's initial request to the BS). For all other types of users, the bandwidth that is reserved in the next cell is equal to their current bandwidth, so that they will seamlessly continue its transmission.

6 Results and Discussion

6.1 Simulation Setup

We use computer simulations to study the performance of our scheme. In our results, we use different traffic "combinations" from all types of traffic considered in our work, in order to test the system's performance in a large variety of cases. In this way, we try to produce results representative of different practical scenarios, where one type of telemedicine traffic might be more dominant than others in any given moment. Each simulation point presents the average result over 5 combinations which were used to create a specific traffic load, and for each combination we conducted 5 independent runs, each simulating 155000 frames (1/2 h of network operation), the first 5000 of which are used as warm-up period.

6.1.1 FCFS Results

We present below the first part of our results, in Figures 6-8. These figures include results derived with the use of the FCFS (First Come First Served) algorithm, for the "circular" topology.

In Figure 6, the improvement shown in the average video packet dropping is due to the proportionate allocation; our scheme prevents the case where a user whose transmission deadline is not imminent may dominate the channel, hence not allowing users with imminent transmission deadlines to transmit.

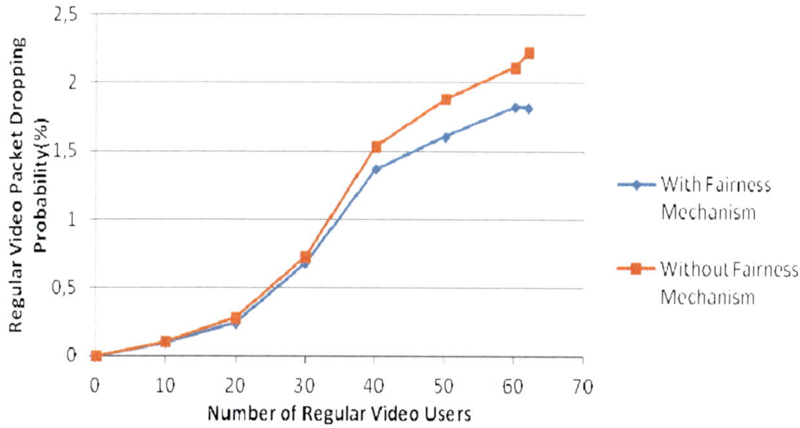

Fig. 6. Regular video packet dropping versus the number of regular video users

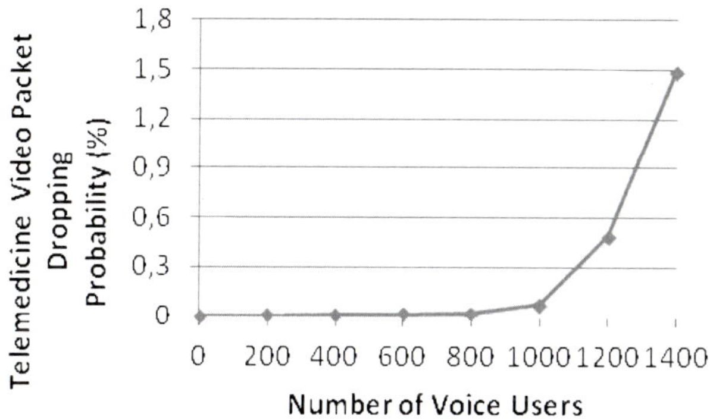

Fig. 7. Effect of regular voice traffic on telemedicine video traffic

Figure 7 shows the almost negligible effect on the QoS of telemedicine users that stems from the increase in the number of voice users. We present results on the telemedicine video packet dropping and it is clear that only in the case of very high voice loads there can be deterioration in telemedicine traffic QoS. The reason is that our combined scheduling and adaptive bandwidth reservation schemes guarantee full priority to all types of telemedicine traffic.

In Figure 8 the percentage of telemedicine video users whose QoS requirements for packet dropping (< 0.01%) are violated is much lower with the use of our fairness mechanism. The reason for this result is that our schemes are designed to offer maximum priority to telemedicine traffic; therefore, when telemedicine video packet dropping increases, the increase is almost "uniform" for all telemedicine video users.

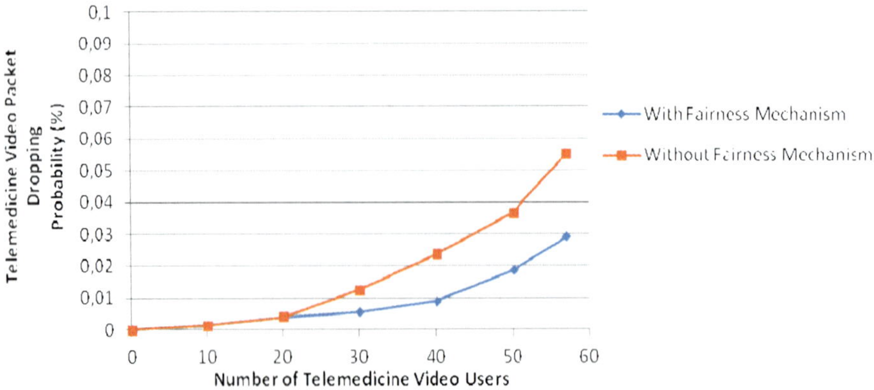

Fig. 8. Telemedicine video packet dropping versus the number of telemedicine video users

6.1.2 FCFS-EDF-SJF Results

Figures 9-10 present results derived with FCFS (First Come First Served), EDF (Earliest Deadline First) and SJF (Shortest Job First) algorithms. For these results we used the same "scenarios" for channel load as those that we used for the corresponding figures in the previous section.

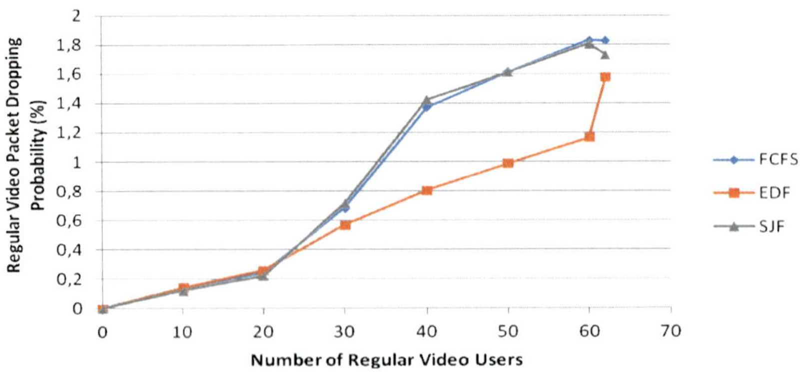

Fig. 9. Regular video packet dropping versus the number of regular video users

Figure 9 shows that the increase in the number of regular video users clearly affects that type of traffic, because the video packet dropping probability of regular video users significantly rises above the 1% acceptable upper bound. Initially, the three algorithms have almost the same efficiency. However, when the number of users in the system (i.e., over all seven cells) exceeds 20, the EDF algorithm produces significantly better video packet dropping probability results for regular video users, with FCFS being marginally better than SJF overall.

Figure 10 shows that the three queuing priority algorithms produce almost identical results for low and medium telemedicine video loads, in terms of regular and telemedicine video packet dropping. EDF excels once again in the case of high loads.

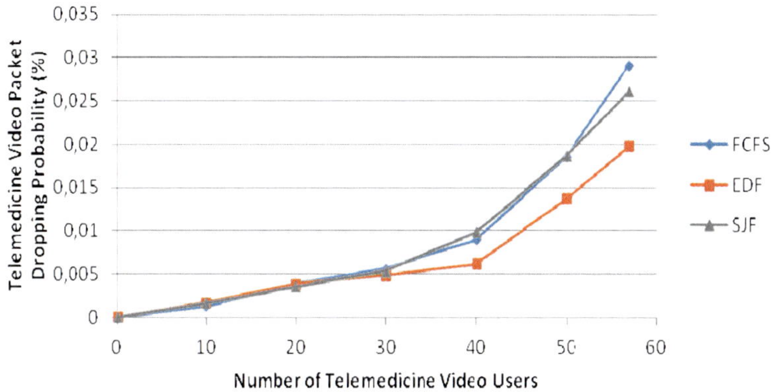

Fig. 10. Telemedicine video packet dropping versus the number of telemedicine video users

It is clear from the figures above that EDF excels in comparison to the other two algorithms. The reason is that it accommodates users based on their deadline, hence it manages to satisfy their QoS requirements in time. The FCFS and SJF algorithms base their respective policies on the arrival time and the size of new information, thereby ignoring the urgency of users' needs. The more aggressive policy implemented by EDF, of course, has, theoretically, the disadvantage that it can lead to unfairness in specific metrics; users with later deadlines can experience longer delays than they do, e.g., with the use of FCFS. For this reason, in Section 6.5 we will evaluate the fairness offered to regular and telemedicine users by each of the three scheduling algorithms. Finally, we need to note that, although FCFS and SJF have comparable results for almost all traffic loads, in very high load conditions FCFS performs always worse than SJF because it fails to offer any kind of priority based on the users' needs.

6.1.3 Open Topology Results

In this section we indicatively present results derived for the second topology that we studied. All cells are connected in a straight line and mobiles can get in or out of our system only via cell A and cell G, which are placed at the two edges of the network.

Figure 11 illustrates the increase of telemedicine video packet dropping probability, as we increase the number of telemedicine video users. It is useful to compare and contrast Figure 11 to Figure 10, where we considered the "circular" topology.

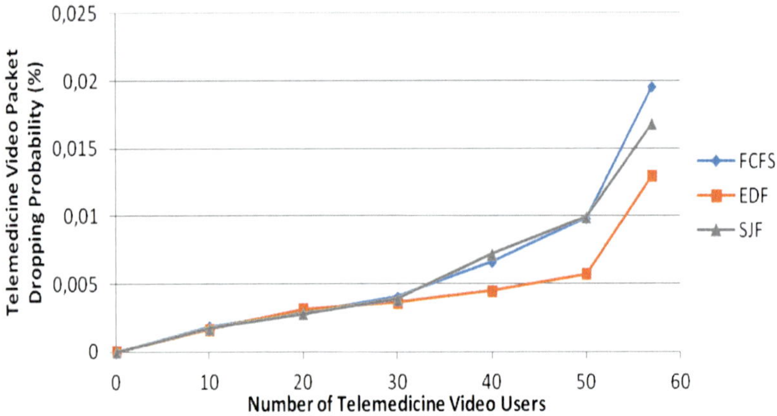

Fig. 11. Telemedicine video packet dropping versus the number of telemedicine video users

Comparing Figures 10 and 11 we note that the results are similar in nature, but the quantitative difference is large. Using the circular topology, average telemedicine video packet dropping probability was 0.03% with FCFS, 0.026% with SJF and 0.02% with EDF, while with the second map the corresponding values are 0.02%, 0.017% and 0.0128%.

Hence, in the case of the open topology, regular video packet dropping probability and telemedicine video packet dropping probability are much lower. The reason is that when the seven cells are connected in a circular fashion, users never leave the system; hence, their aggregate number steadily grows, considering that new users keep arriving in the system. On the contrary, when users can leave the system via cells A and G, the aggregate users' number fluctuates. This explains the worse QoS achieved by all algorithms in the case of the circular topology; this case actually represents a worst-case scenario for our system.

6.1.4 Jain's Fairness Results

Below, we present results derived with 3 different combinations of all loads of users, ranging from very low (9%) to very high (100%) and we present the throughput fairness results with the use of Jain's fairness index for voice, video and telemedicine video users.

Table 1. 9% traffic load

User's type/Algorithm	*FCFS*	*EDF*	*SJF*
Voice Fairness	0.9510	0.9513	-
Video Fairness	0.9978	0.9981	0.9976
Telemedicine Video Fairness	1.0000	1.0000	1.0000

Table 2. 51% traffic load

User's type/Algorithm	*FCFS*	*EDF*	*SJF*
Voice Fairness	0.9476	0.9470	-
Video Fairness	0.9868	0.9889	0.9859
Telemedicine Video Fairness	0.9963	0.9961	0.9965

Table 3. 100% traffic load

User's type/Algorithm	*FCFS*	*EDF*	*SJF*
Voice Fairness	0.9838	0.9925	-
Video Fairness	0.9876	0.9901	0.9866
Telemedicine Video Fairness	0.9920	0.9931	0.9902

The results presented in Tables 1-3 show that EDF outperforms FCFS and SJF in terms of fairness in throughput, something intuitively explained due to the nature of the algorithm (packets with imminent deadlines are transmitted first, therefore the smallest possible number of packets is dropped). FCFS outperforms SJF, which creates some unfairness by always servicing first the users with the least packets to transmit, and hence leading larger transmissions to possible packet dropping. Still, the use of throughput as a fairness metric does not suffice; delays and packet dropping are of equal importance, therefore we used them in our study and we present the results below.

We present two more cases, one with a high load (80%) and one corresponding to a traffic overload (110% of the channel capacity) and we evaluate fairness in terms of delay and packet dropping.

Table 4. 80% traffic load

	DELAY			*PDROP*		
User's type/Algorithm	*FCFS*	*EDF*	*SJF*	*FCFS*	*EDF*	*SJF*
Voice Fairness	0.9369	0.9327	-	0.9138	0.9094	-
Video Fairness	0.9799	0.9753	0.9226	0.9338	0.9203	0.8248
Telemedicine Video Fairness	0.9906	0.9893	0.9317	0.9893	0.9824	0.8465

Table 5. 110 % traffic load

	DELAY			*PDROP*		
User's type/Algorithm	*FCFS*	*EDF*	*SJF*	*FCFS*	*EDF*	*SJF*
Voice Fairness	0.8977	0.8813	-	0.8948	0.8723	-
Video Fairness	0.8858	0.8552	0.7912	0.6937	0.6779	0.6219
Telemedicine Video Fairness	0.9727	0.9788	0.8864	0.9551	0.9501	0.8065

For the results presented in Tables 4-5, we have kept constant the number of voice users and we increased the number of video and telemedicine video users. These Tables clearly show that as we increase the channel load, the fairness achieved by all three algorithms decreases significantly, especially when considering packet dropping as a metric. The traffic type that is influenced the most is regular video traffic, as voice is less demanding in bandwidth, and telemedicine video users have absolute priority in scheduling, therefore they are minimally affected. As intuitively expected and explained in Section 6.2.3, EDF does not outperform FCFS when delay and packet dropping are used as fairness metrics.

It is clear, however, from all the results presented in Tables 1-5, that our scheme achieves excellent fairness results for all traffic types and for all traffic loads that do not exceed the maximum channel capacity.

7 Conclusion and Future Work

This work has focused on the problem of scheduling integrated traffic transmissions from urgent types of traffic, like telemedicine, with regular wireless traffic over next generation cellular networks. We have extended a recent work [22] on a new MAC protocol by using three different scheduling algorithms over a simple network topology and by evaluating the algorithms' fairness based on a number of metrics. Our results have clearly shown that in terms of the provided QoS to wireless users, the EDF algorithm excels over SJF and FCFS, which has been widely used in the literature as it is the intuitively simplest choice and is marginally fairer than EDF.

In future work, we intend to experiment with more scheduling algorithms from the literature and to propose an algorithm of our own, which will incorporate the advantages of EDF and will provide increased fairness in comparison to FCFS. In order to achieve this, we believe that we will need periodic bandwidth reallocation from the BS to the wireless users, based on efficient traffic modeling.

References

1. Perednia, D.A., Allen, A.: Telemedicine Technology and Clinical Applications. The Journal of the American Medical Association (JAMA) 273(6), 483–488 (1995)
2. Chu, Y., Ganz, A.: A Mobile Teletrauma System Using 3G networks. IEEE Transactions on Information Technology in Biomedicine 8(4), 456–462 (2004)

3. Bhargava, A., Khan, M.F., Ghafoor, A.: QoS management in Multimedia Networking for Telemedicine Applications. In: Proceedings of the IEEE Workshop on Software Technologies for Future Embedded Systems, pp. 39–42 (2003)
4. Tulu, B., Chatterjee, S., Laxminarayan, S.: A Taxonomy of Telemedicine Efforts with Respect to Applications, Infrastructure, Delivery Tools, Type of Setting and Purpose. In: Proceedings of the 38th Hawaii International Conference on System Sciences, HICSS (2005)
5. Voskarides, S.C., Pattichis, C.S., Istepanian, R., Michaelides, C., Schizas, C.N.: Practical Evaluation of GPRS use in a Telemedicine System in Cyprus. In: Proceedings of the 4th IEEE International EMBS Special Topic Conference on Information Technology Applications in Biomedicine, pp. 39–42 (2003)
6. Garawi, S., Istepanian, R.S.H., Abu-Rgheff, M.A.: 3G Wireless Communications for Mobile Robotic Tele-Ultrasonography Systems. IEEE Communications Magazine 44(4), 91–96 (2006)
7. Koutsakis, P., Psychis, S., Paterakis, M.: Integrated Wireless Access for Videoconference from MPEG-4 and H.263 video coders with Voice, Email and Web traffic. IEEE Transactions on Vehicular Technology 54, 1863–1874 (2005)
8. Pang, Q., Bigloo, A., Leung, V.C.M., Scholefield, C.: Service Scheduling for General Packet Radio Service Classes. In: Proceedings of IEEE Wireless Communications and Networking Conference (1999)
9. Nanda, S., Goodman, D.J., Timor, U.: Performance of PRMA: A Packet Voice Protocol for Cellular Systems. IEEE Transactions on Vehicular Technology 40, 584–598 (1991)
10. Fitzek, F.H.P., Reisslein, M.: MPEG-4 and H.263 Video Traces for Network Performance Evaluation. IEEE Network 15(6), 40–54 (2001)
11. Tsai, C.-F., Tsang, C.-J., Ren, F.-C., Yen, C.-M.: Adaptive Radio Resource Allocation for Downlink OFDMA/SDMA systems. In: Proceedings of the IEEE International Conference on Communications, ICC (2007)
12. Cleary, A.C., Paterakis, M.: An investigation of Stack Based Algorithms for Voice Packet Transmission in Microcellular Wireless Environments. In: Proceedings of the IEEE International Conference on Communications, ICC (1995)
13. Gilbert, E.N.: Capacity of a Burst-noise Channel. Bell System Technical Journal 39, 1252–1265 (1960)
14. Elliot, E.O.: Estimates of Error Rates for Codes on Burst-noise Channels. Bell Labs Technical Journal 42, 1977–1997 (1963)
15. Bottigliengo, M., Casetti, C., Chiasserini, C.-F., Meo, M.: Short-term Fairness for TCP flows in 802.11b WLANs. In: Proceedings of IEEE Infocom (2004)
16. Baig, E.C.: Will Consumers Tune in to a Tiny TV in their hand (2006), `http://www.usatoday.com/tech/wireless/2006-08-17-mobile-tv_x.htm`
17. Yu, F.R., Wong, V.W.S., Leung, V.C.M.: A New QoS Provisioning Method for Adaptive Multimedia in Wireless Networks. IEEE Transactions on Vehicular Technology 57(3), 1899–1909 (2008)
18. Song, Q., Jamalipour, A.: A Negotiation-Based Network Selection Scheme for Next-Generation Mobile systems. In: Proceedings of the IEEE Globecom (2006)
19. Choi, S., Shin, K.: Adaptive Bandwidth Reservation and Admission Control in QoS-sensitive Cellular Networks. IEEE Transactions on Vehicular Technology 13, 882–897 (2002)

20. Lee, D.-S., Hsueh, Y.-H.: Bandwidth-reservation Scheme based on Road Information for Next-generation Cellular Networks. IEEE Transactions on Vehicular Technology 53, 243–252 (2004)
21. Soh, W.S., Kim, H.S.: A Predictive Bandwidth Reservation Scheme using Mobile Positioning and Road Topology Information. IEEE/ACM Transactions on Networking (TON) 14, 1078–1091 (2006)
22. Qiao, L., Koutsakis, P.: Adaptive Bandwidth Reservation and Scheduling for Efficient Wireless Telemedicine Traffic Transmission. IEEE Transactions on Vehicular Technology 60(2), 632–643 (2011)
23. Kwon, T., Choi, Y., Bisdikian, C., Naghsineh, M.: QoS Provisioning in Wireless/Mobile Multimedia Networks Using an Adaptive Framework. Wireless Networks 9(1), 51–59 (2003)
24. ETSI. Digital Cellular Telecommunications System (Phase 2+); Technical Realization of the Short Message Service (SMS); Point-to-Point (PP) (GSM 03.40)
25. Jain, R.: The Art of Computer Systems Performance Analysis: Techniques for Experimental Design, Measurement, Simulation, and Modeling. Wiley- Interscience, New York (1991)

An Energy-Efficient Contention-Aware Algorithm for Channel Selection in Cognitive Radio Networks

Agapi Mesodiakaki[1], Ferran Adelantado[2], Luis Alonso[1], and Christos Verikoukis[3]

[1] Technical University of Catalonia (UPC), Signal Theory and Communications Department, Castelldefels, Spain

[2] University Oberta de Catalunya (UOC), Barcelona, Spain

[3] Telecommunications Technological Centre of Catalonia (CTTC), Castelldefels, Spain

{agapi.mesodiakaki,luisg}@tsc.upc.edu, ferranadelantado@uoc.edu, cveri@cttc.es

Abstract. In recent years, due to the ever-increasing traffic demands and the limited spectrum resources, it is very likely that several cognitive radio ad hoc networks (CRAHNs) will coexist and opportunistically use the same primary user (PU) resources. In such scenarios, the ability to distinguish whether a licensed channel is occupied by a PU or by another CRAHN can significantly improve the spectrum efficiency of the network, while the contention among the CRAHNs already operating on the licensed channels with no PU activity, may further affect the performance of the network. Therefore, the proper selection of the frequency bands, already being used by other CRAHNs, could result in notable throughput and energy efficiency gains for the network under study. In this chapter, we propose a novel contention-aware channel selection algorithm, that: i) initially locates the spectrum holes by exploiting cooperative spectrum sensing, ii) categorizes the idle licensed channels based on their contention level (i.e., number of secondary users (SUs) belonging to other non-cooperating CRAHNs that are operating on the licensed channels), and iii) selects the less contended licensed channel to be accessed first by the CRAHN under study. The proposed algorithm is evaluated by means of simulations and it is shown that it can present gains up to 70% in throughput and up to 68% in energy efficiency.

1 Introduction

Cognitive radio (CR) is a promising technology proposed to cope with the spectrum scarcity problem, which has emerged as a result of the increased need for anywhere-anytime connectivity and the fact that the spectrum resources are limited. A CR, with its built-in intelligence and cognitive capabilities, can sense the radio spectrum, locate spectrum holes (i.e., underutilized or unused portions in the licensed spectrum) and opportunistically access them as long as the licensed users (also called primary users (PUs)) do not use the band [1].

As illustrated in Fig. 1, a typical duty cycle of a CR includes detecting the spectrum holes, selecting the best frequency bands, coordinating spectrum access with

C.X. Mavromoustakis et al., *Resource Management in Mobile Computing Environments*, Modeling and Optimization in Science and Technologies 3,
DOI: 10.1007/978-3-319-06704-9_19, © Springer International Publishing Switzerland 2014

other users and vacating the channel when a PU appears. Such a cognitive cycle is supported by the following functions: i) spectrum sensing, ii) spectrum analysis, and iii) spectrum decision. Through spectrum sensing and analysis, a CR can detect the spectrum holes and decide which ones it will utilize (spectrum decision). Through sensing a CR is also able to detect the change in the PU activity in the licensed channel that operates and thus to adapt its transmission so that no harmful interference is generated to the PU. Therefore, CR spectrum sensing is of high importance to guarantee PU protection.

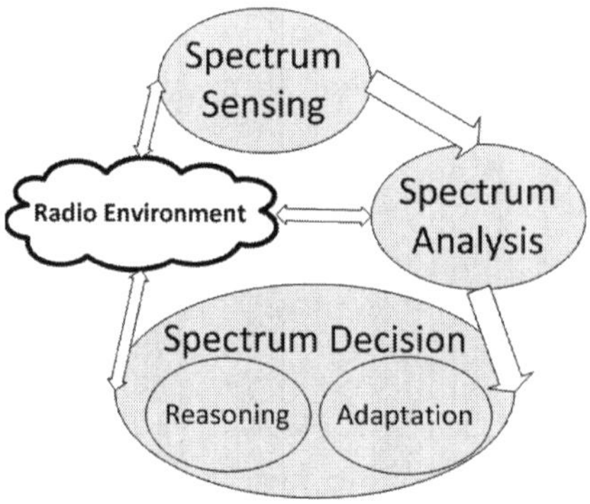

Fig. 1. Cognitive radio cycle

Spectrum sensing can be performed using several sensing techniques. The most important ones are energy detection and feature detection [2], [3]. Energy detection is the most commonly used technique because of its low computational and implementation complexity. However, it is susceptible to the noise power uncertainty [4]. On the other hand, feature detection enables the distinction between different types of signals (e.g., PUs' and secondary users (SUs)' signals) at the expense of higher complexity and longer sensing time. A specific type of feature detection is cyclostationary feature detection [3], which determines the presence of PU signals by extracting their specific features such as pilot signals, cyclic prefixes, symbol rate, spreading codes or modulation types from its local observation. Therefore, it requires prior information about the PU waveforms. However, notice that this is typically known for most standard technologies that operate on licensed channels [5].

To further improve the sensing accuracy, cooperative spectrum sensing is exploited, which benefits from the spatial and multiuser diversity and manages: i) to guarantee PU protection by decreasing the mis-detection probability, and ii) to utilize the idle[1] licensed channels more efficiently by reducing the false alarm probability [3], [6]–[7]. At this point we should mention that the mis-detection probability refers to the probability that a licensed channel that is busy will be erroneously sensed idle, while the false alarm takes place when a licensed channel that is idle is erroneously sensed busy.

---

[1] Please note that in CR, the characterization of a channel as busy or idle concerns exclusively the PU activity.

Moreover, the CR nodes may be battery-powered wireless devices and thus their energy efficiency plays a key role. At the same time, due to the increased need for anywhere-anytime connectivity and the fact that the spectrum resources are limited; the coexistence of CRAHNs that use the same licensed channels opportunistically has become a challenging topic. To that end, the motivation of studying the operation and performance of a CRAHN, in a scenario where other non-cooperating CRAHNs are also using the same PU resources opportunistically is twofold: First, the ability of detecting whether the activity in a particular band is caused by PUs or by secondary users (SUs) allows a more intensive reuse of the idle resources, which is the main aim of the CR paradigm. Second, the number of SUs belonging to other CRAHNs that contend for the same resources may have a significant impact both on the throughput and the energy efficiency of the CRAHN under study. Therefore, the proper selection of the frequency bands already being used by other CRAHNs can result in notable performance gains.

To that end, in this chapter, an energy-efficient contention-aware channel selection algorithm is proposed, in a scenario where the CRAHN under study coexists with other non-cooperating CRAHNs. The algorithm uses cooperative spectrum sensing, during which feature detection is combined with a technique that estimates the number of SUs in each licensed channel. Feature detection enables the distinction of the licensed channels into: i) those with PU activity, ii) those with SU activity and iii) those with no activity at all and will be used as the reference algorithm for the performance comparison of our approach; while the second technique estimates the contention in each licensed channel with SU activity. Specifically, it estimates the number of SUs belonging to the other non-cooperating CRAHNs that are already operating on the channel. Thereafter, the algorithm categorizes the idle licensed channels based on their contention level and it selects the less contended channel to be accessed first by the CRAHN under study.

The rest of the chapter is organized as follows: In Sections 2 and 3, the related work and the system model are respectively presented. In Section 4, a detailed description of the proposed algorithm is given and in Section 5, the simulation results are presented. Finally, Section 6 concludes the chapter.

2 Related Work

In CR literature, the research interest has been recently moved to energy-efficient cooperative spectrum sensing algorithms [8], [9]. In [8], the authors study how to choose an optimal sensing duration to strike a balance between energy consumption and system throughput, while in [9] the fusion rule threshold, the detector's thresholds at the SUs, the sensing time, and the number of cooperating SUs are optimized so that the energy efficiency of the CR network is maximized.

In parallel, several channel selection algorithms have been proposed for CRNs mainly aiming at maximizing the overall throughput. To that end, the authors in [10] propose an optimal channel sensing order policy based on previous observations of the channels, where apart from selecting the optimal channel to sense, the optimal sensing time that leads to maximized SU throughput is derived. In [11], a dynamic context-aware channel selection scheme that uses reinforcement learning to improve the throughput and delay performance is proposed, while in [12] the authors propose a MAC protocol for opportunistic spectrum access that uses two channel selection methods, a uniform and a spectrum opportunity-based channel selection. According to the first one, each SU chooses a channel among the available ones randomly, whereas the latter takes into account the different probabilities of spectrum availability in the channels. However, the authors assume that each SU can correctly estimate the spectrum availability probability (i.e., the number of active secondary flows) without justifying this assumption. Moreover, contrary to our approach, they consider a dedicated global common control channel.

All the aforementioned approaches consider only a single CR network, thus totally overlooking any coexistence issues among SUs. Nevertheless, as previously analyzed, due to the limited spectrum resources it is very likely to have several CRAHNs coexisting and sharing the same PU resources. To that end, in [13] a contention-aware channel selection strategy is proposed. However, this approach studies the trade-off between connectivity and contention, and it differs from our work as it aims at a different objective (i.e., to achieve high connectivity in the network).

3 System Model

We consider a set of M_s licensed channels; each one allocated to a PU, as depicted in Fig. 2, which can be opportunistically accessed by SUs as long as they remain unused. An unlicensed channel is also considered (e.g., the industrial, scientific and medical (ISM) band), which is highly congested. In addition, we consider a CRAHN, denoted as CRAHN under study, consisting of N SUs, which is able of operating both in the considered unlicensed and licensed channels. The CRAHN under study has a coordinator, whose role may be assigned to each one of the N SUs in a round robin way to achieve fairness in energy consumption [14]. Furthermore, each licensed channel, when being idle, can be accessed by CRAHNs, each one consisting of a number of SUs. We denote as N_{SUlic}, the maximum total number of SUs that can operate in a licensed channel and as N_{unlic} the number of users that operate in the unlicensed channel (this parameter does not include the N users of the CRAHN under study). At this point, it is worth mentioning that the N_{unlic} users that operate in the unlicensed channel may consist of a mixture of SUs belonging to other CRAHNs and users without cognitive radio capabilities that are only able to use the unlicensed channel.

The CRAHN under study does not use a dedicated global control channel for communication. This fact is very important, as a global common control channel could be a potential bottleneck for network performance and scalability and it is also prone to jamming attacks by malicious users, which is one of the main Denial of Service (DoS) attack threats in CR networks. Moreover, in-band signaling is considered (i.e., the channels are shared for both control and data transmission). However, notice that the CRAHN under study exploits explicitly the licensed channels for data transmission. In

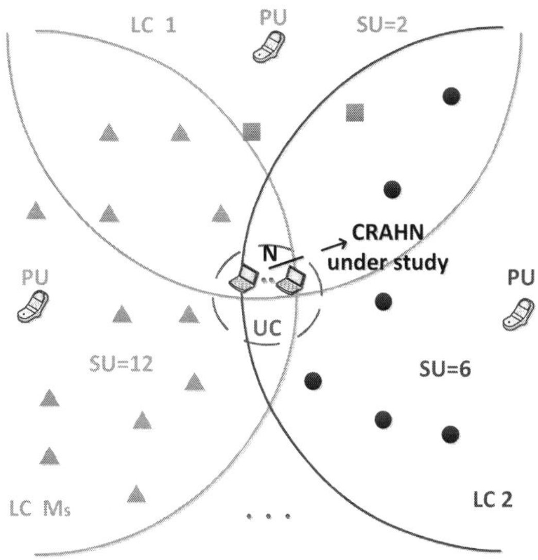

Fig. 2. System model example

the unlicensed band only control information exchange takes place during the initial set up of the network and in the case there is no licensed channel available and the CRAHN hops to the unlicensed channel for the recovery process.

Each SU is equipped with a single transceiver. Thus, even if it is capable of operating over multiple channels including the licensed ones, at a given time it can operate (either transmit or receive) only over a single channel. Obviously, the use of a single transceiver is less energy consuming and costly compared to CR nodes equipped with multiple transceivers and it has been already considered in some CR devices and prototypes [15].

Furthermore, we assume that the users of the CRAHN under study are capable of operating over the same set of channels, including the licensed ones and in addition, that they constitute a short range communication network, whose nodes are adequately close to each other, to be exposed to the same activity of accessible channels. However, notice that the sensing results of the SUs when sensing the same channel at the same time may be different due to the mis-detection and false alarm probability.

Moreover, the transmissions of the SUs both in the unlicensed and licensed channels follow the CSMA/CA method, as described in [16]. To that end, a node wishing to transmit data has to first listen to the channel for a predetermined amount of time (t_{DIFS}) to determine whether or not another node is transmitting. If no other node transmits, the node is permitted to begin the transmission process. Otherwise, the node defers its transmission for a random period of time t_{BO} (back-off time). It is worth noting here, that the PUs use their own access method, while accessing the licensed channels (e.g., SC-FDMA in the uplink and OFDMA in the downlink for LTE access).

4 Proposed Algorithm

The operation of the CRAHN under study can be divided into the operation in the unlicensed channel (UC) and the operation in the licensed channels (LCs), as depicted in the flowchart of Fig. 3. In particular, the CRAHN under study is assumed to be initially located in the highly congested unlicensed channel, where other N_{unlic} users are present. There, the coordinator triggers a sensing procedure, aiming at finding new spectrum opportunities in the licensed channels for the CRAHN under study to exploit.

As previously described in Section 3, in the unlicensed channel only sensing procedures take place at the following cases: i) at the initial setup of the CRAHN, and ii) at the recovery process, in the case that there is no licensed channel available. Thus, the unlicensed channel operation includes only control information exchange, as described in Section 4.1.

After the sensing procedure in the unlicensed channel has finished, a list is constructed that contains the sensed as idle licensed channels in the order defined by the applied channel selection algorithm (described in Section 4.3). At this point there are the following possible cases:

- *All the licensed channels have been sensed busy.* If the constructed list is empty, the CRAHN stays in the unlicensed channel and another sensing procedure is initiated.
- *There is at least one licensed channel that has been sensed idle.* If the list contains at least one licensed channel, the CRAHN under study hops to the first channel in the list and operates there following the procedure described in Section 4.2.

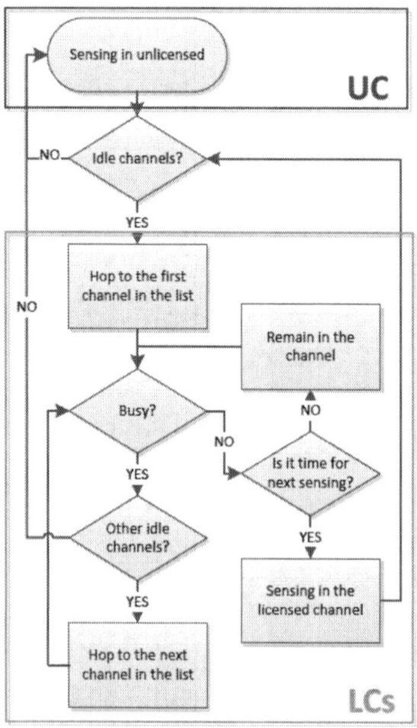

Fig. 3. Flowchart of the operation of the CRAHN under study

4.1 Operation in the Unlicensed Band

The operation time in the unlicensed channel can be divided into three time periods (t_1, t_2 and t_3), as depicted in Fig. 4. Specifically, during t_1 the sensing procedure is initiated, t_2 denotes the total sensing time and during t_3 the exchange of the sensing results takes place.

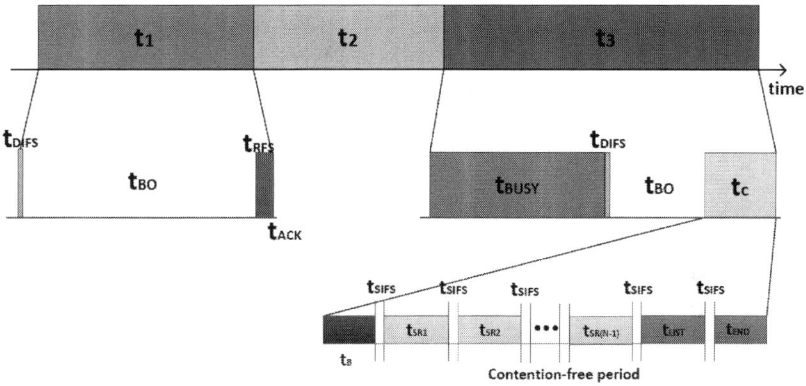

Fig. 4. Sensing procedure time diagram example

1. Time period t_1

During t_1, only the coordinator contends with the other N_{unlic} users to gain access to the unlicensed channel (i.e., the rest *N-1* SUs remain idle). Subsequently, after listening to the channel being idle for t_{DIFS} and having its back-off timer equal to zero, it broadcasts a RFS (Request For Sensing) packet.

In order to be reassured that the transmission of the RFS packet was successful, the algorithm used in [17] is applied. According to that, e.g., the node that is defined to send its sensing results first, acts as a leader or representative for the purpose of sending feedback to the coordinator. On erroneous reception of the RFS packet, the leader does not send an acknowledgment, prompting a retransmission. On erroneous reception of the packet at receivers other than the leader (i.e., at the rest *N-2* SUs of the CRAHN under study), the protocol allows negative acknowledgments from these receivers to collide with the acknowledgment from the leader and thus destroying the acknowledgment and prompting the coordinator to retransmit the packet. The role of the leader may also be assigned in a round robin way among the SUs (e.g., the leader of the current sensing procedure may be the coordinator of the next one).

The RFS packet includes the following information: i) which channels are going to be sensed by each SU of the CRAHN under study, ii) the order in which the SUs are going to report their sensing results to the coordinator and iii) how often the sensing procedure is going to be triggered, namely the time between two consecutive sensing procedures, denoted as T_S.

Specifically, the parameter T_S is constant and it is defined as the time period from the end of the sensing time until the moment that the coordinator has a new RFS packet to send. This value is closely coupled with the PU activity pattern. For quickly changing PU activity, the value should be kept low enough, so that the sensing procedure takes place more frequently, to keep the information for every channel updated.

2. Time period t_2

After having successfully received the RFS packet, the time period t_2 begins. Thus, all the SUs start sensing each one the licensed channels that were assigned to it. This assignment of licensed channels for sensing is equally performed among the SUs (that is, each SU senses the same number of channels), for the sake of fairness in energy consumption. In the case that a channel is sensed by more than one SU, cooperative spectrum sensing is applied using the OR fusion rule. The fusion rule refers to a predefined rule, according to which the central entity (the coordinator in our case) processes the SUs' reported sensing results to make a final decision about the state of the licensed channel. In particular, according to the OR fusion, when at least 1 out of k cooperating SUs detects that the PU is busy in the licensed channel, the final decision declares a PU is present. In particular, when applying the OR fusion rule the mis-detection (P_{md}) and the false alarm probability (P_{fa}) can be respectively expressed as:

$$P_{md} = 1 - \prod_{i=1}^{k} P_{md,i} \tag{1}$$

$$P_{fa} = 1 - \prod_{i=1}^{k} P_{fa,i} \tag{2}$$

where k is the number of cooperating SUs and $P_{md,i}$ and $P_{fa,i}$ denote the mis-detection probability and false alarm probability of the ith SU, respectively. Generally, the OR fusion rule presents low mis-detection probability and high false alarm probability. As a result, it provides higher protection to the PUs, at the expense of low reusability of the unoccupied licensed channels.

Furthermore, if the number of SUs of the CRAHN under study (N) is a multiple of the licensed channels (M_s), each channel is going to be sensed by the same number of users. Otherwise, a subset of licensed channels is sensed by more SUs, in order to achieve higher accuracy i.e., better mis-detection and false alarm probability.

During the sensing time that lasts for a constant period of time equal to t_2, cyclostationary feature detection is used [3]. As previously described, cyclostationary feature detection enables the distinction between PUs' and SUs' signals, at the expense of more complexity and longer sensing time. However, notice that its use is fundamental to such co-existence scenarios, as the use of a simpler technique (e.g. energy detection) would result in very low spectrum efficiency, as all the idle licensed channels being used by other CRAHNs, would have been considered busy and thus would have been avoided.

3. Time period t_3

After the sensing has finished, the CRAHN under study hops to the unlicensed channel for the exchange of the sensing results. The sooner this exchange takes place, the higher the accuracy in the results. Therefore, only the coordinator (from the CRAHN under study) contends again with the other N_{unlic} nodes to gain access to the unlicensed channel.

Once the coordinator is able to transmit a packet successfully, it broadcasts a beacon frame asking for the sensing results of the rest of the SUs. With this beacon a contention free period starts, as shown in Fig. 4, in which each SU of the CRAHN under study has to wait t_{SIFS} and then sends its sensing results to the coordinator in the previously defined (in the RFS packet) order. Thereafter, the coordinator constructs a list, which defines the order that the licensed channels that were sensed idle are going to be visited by the CRAHN under study. A detailed explanation of the construction of the list is given in Section 4.3. Subsequently, a packet containing the list (of transmission's duration t_{LIST}) is broadcasted by the coordinator and the contention-free period ends.

The duration of the contention-free period is constant and predetermined in the beacon frame, so that when the other nodes in the unlicensed channel hear the beacon to set their network allocation vector for time equal to this of the contention free period. As shown in Fig. 4, all the procedure of exchanging control information lasts for t_C.

At this point we have to clarify, that the constructed list contains only the sensed as idle licensed channels. However, notice that some of them may be erroneously sensed idle due to mis-detection probability.

4.2 Operation in the Licensed Band

While the CRAHN under study operates in licensed channels, the data exchange follows the CSMA/CA method, as previously mentioned. Specifically, the SUs that have a packet to send (belonging to the CRAHN under study or/and to the other co-existing CRAHNs) contend to gain access to the licensed channel. Hence, the operation time in the licensed channel consists of successful transmission, collision and idle slots. This normal CSMA/CA operation on the licensed channel is interrupted in the following cases:

- *The PU of the licensed channel remains idle and the time for the next sensing procedure to be initiated has come, namely T_S has elapsed from the previous sensing*. In this case, the coordinator, that has a new RFS packet to send to trigger a new sensing procedure, contends again to gain access to the licensed channel. At this point, we make the assumption that the PU activity does not change during the sensing procedure (that is, for $\overline{t_1} + t_2 + \overline{t_3}$, where $\overline{t_1}$ and $\overline{t_3}$ are the average values of the time periods t_1 and t_3, respectively, and t_2 is the constant duration of the time period t_2) and we will justify this assumption in the performance evaluation section (Section 5).

- *The state of the PU of the licensed channel changes to busy earlier than T_S.* In this case, the SUs have to leave the channel immediately in order not to interfere with the PU. We assume that the CRAHN under study can detect the PU activity after a period of time equal to t_r and then reacts by hopping to the next channel in the list of available ones. In case there is no other channel available in the list the CRAHN hops to the unlicensed channel and a new sensing procedure is triggered again from the beginning (recovery process).

4.3 List Construction

As the list defines the order in which the licensed channels are going to be visited by the CRAHN under study, it is of high importance. To that end, we propose the classification of the channels that were sensed idle, according to the number of contending SUs belonging to other CRAHNs that are already operating on them.

As a result, after each sensing procedure a list is constructed, containing all the sensed as idle licensed channels, sorted by the number of contending nodes in ascending order (that is, the channel with the least number of contending SUs takes the first place and thus higher priority). The main goal of our algorithm is to achieve throughput and energy efficiency gains by minimizing the time that the CRAHN under study spends in channels with high contention.

For the estimation of the number of competing SUs in each licensed channel, the following formula is used [18]-[19], given the assumption that all the SUs are in saturated conditions.

$$n_C = \frac{\log(1-p_C)}{\log\left(1-\dfrac{2(1-2p_C)}{(1-2p_C)(CW_{\min}+1)+p_C CW_{\min}(1-(2p_C)^m)}\right)} \tag{3}$$

According to [18]-[19], the estimated number of contending users (n_C), can be explicitly expressed as a function of the conditional collision probability (p_C), (that is, the probability that a packet being transmitted on the channel collides) and the known and constant back-off parameters CW_{min} and m, with $m = \log_2(CW_{max} / CW_{min})$.

When a SU senses a licensed channel with SU activity, it can efficiently measure p_C by simply monitoring the channel. Specifically, it is able to understand the collisions and the successful transmissions of other users by listening to their packet exchange [20]. Thus, p_C can be obtained by counting the number of slots that a successful transmission occurs (C_{succ}), as well as the number of slots that a collision of the other SUs occurs (C_{coll}), as in each of these slots a potential packet transmission of the SU (the one performing sensing) would have failed. Consequently, the conditional collision probability p_C can be expressed as:

$$p_C = \frac{C_{succ}+C_{coll}}{B} = 1-\frac{C_{idle}}{C_{succ}+C_{coll}+C_{idle}} \tag{4}$$

where B is the total number of observed slots, namely the sum of successful transmission (C_{succ}), collision (C_{coll}) and idle slots (C_{idle}).

The accuracy of the estimation is highly dependent on the observation time (B), and the number of contending users (n_C). Specifically, for a constant number of contending users on the channel, the longer the observation time, the more accurate their estimated value while for a given observation time, the more the contending users on the channel, the less accurate their estimated value. However, it is very important that the proposed algorithm performs well even if the estimated values are not very accurate, as the correct construction of the list depends only on the comparison between the estimated values of contending SUs in each channel and not on their exact value.

The process of the list construction is summarized in Algorithm 1.

Algorithm 1 List Construction

1. Each SU senses the LC/s that was/were assigned to it.
2. For the idle ones, it calculates the p_C and then the n_C from (2) and (1), respectively.
3. The SUs send their results to the coordinator.
4. The coordinator applies the OR fusion rule.
5. For each of the sensed as idle channels, it calculates the average of the estimations of n_C.
6. The list is constructed sorted by the estimation of n_C of each channel in ascending order.

Algorithm 1 List construction process

4.4 Operation Example

For the reader's convenience, a simple example of the operation of the CRAHN under study is depicted in Fig. 5. Specifically, we consider a CRAHN consisting of $N=3$ SUs and a set of three licensed channels (LC1, LC2 and LC3).

Initially, the CRAHN is located in the unlicensed channel (UC), where a sensing procedure takes place, as previously described. In particular, after the coordinator manages to successfully send a RFS packet, each SU senses a licensed channel and then they all hop to the unlicensed channel for the exchange of the sensing results. In the considered example the LC3 is found busy and thus it is not included in the list. For LC1 and LC2, the number of contending SUs that belong to other non-cooperating CRAHNs has been estimated to be 2 and 8 SUs, respectively. As a result, the LC1 will take the first place on the list, as it is the one with the least SU contention. After the coordinator has constructed and broadcasted the list, all the SUs of the CRAHN under study hop (at $t=\tau_1$) to the first channel on the list (i.e., LC1) and start communicating there following the CSMA/CA method, as previously described. At $t=\tau'$ the PU of LC1 becomes busy and thus the SUs have to evacuate the licensed channel as soon as possible in order not to interfere with the PU. As the CRAHN is able to detect the change in the PU activity after t_r, it reacts by hopping at $t=\tau_2$ to the next available channel in the list, i.e., LC2. While operating there, at $t=\tau''$ the PU of LC2 becomes busy and as there is no other licensed channel available in the list, the

CRAHN hops back to the unlicensed channel at $t = \tau_3$ for the recovery process. There the coordinator triggers a new sensing procedure. However, the updated list will contain this time the LC1 at the first place and the LC3 at the second place, as the LC1 was sensed free of both PU and SU activity and the LC3 having 12 SUs. Hence, at $t = \tau_4$ the CRAHN under study hops to the LC1 and a new cycle begins, as previously described.

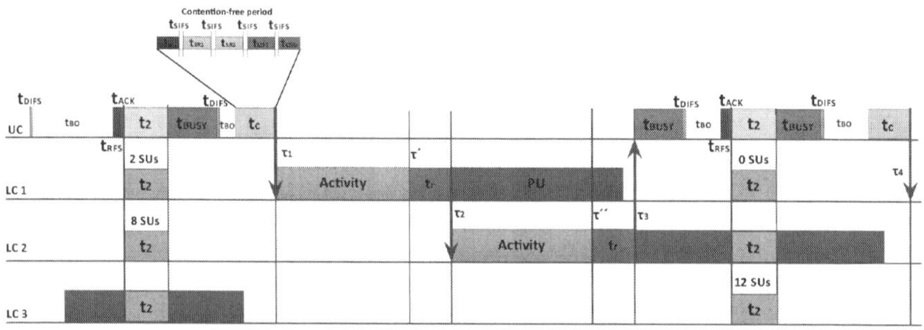

Fig. 5. Example of the operation of the CRAHN under study

5 Performance Evaluation

5.1 Simulation Scenario

In the simulations we executed in MATLAB™, we have considered a CRAHN consisting of N= 6, 12 and 18 SUs and a set of M_s=6 licensed channels. We assume that the PU activity on the licensed channels follows an exponential on-off traffic model, with the mean durations of on and off periods denoted by T_{on} and T_{off}, respectively. Moreover, it has been proved through extensive simulations, that the possibility that the PU activity changes during the sensing procedure in a licensed channel is lower than 3%, which accounts for our assumption that the PU activity does not change during the sensing procedure.

Furthermore, a highly congested unlicensed channel is considered with N_{unlic}=50 users. The false alarm and mis-detection probability of each SU are $P_{fa,i}=P_{md,i}=0.1$, respectively, whereas the back-off parameters are equal to $m=6$ and $CW_{min}=16$. All the users are assumed to be in saturated conditions (i.e., they always have a packet to transmit). The channel conditions are assumed ideal: no hidden terminals, no exposed terminals and no packet corruption are considered. These parameters would have the same impact on our proposal and the reference algorithm, and thus they are omitted to highlight the performance gains.

Moreover, the licensed channels are divided into four categories according to their number of contending SUs that are operating on the channel: the very low contended, the low contended, the medium contended and the high contended ones. The maximum number of SUs in a licensed channel has been considered equal to $N_{SUlic}=35$. Specifically, two very low contended channels (0-9 SUs), one low contended (9-18 SUs), one medium contended (18-27 SUs) and two high contended (27-35 SUs) are considered. The exact value of the actual number of SUs in a channel is randomly chosen between the interval of its category. At this point it is worth noting that a licensed channel may change category dynamically, depending on the choices of the other CRAHNs.

The MAC parameters have been selected according to IEEE 802.11g Standard [21], while the rest of the simulation parameters are summarized in Table 1.

In order to calculate the accuracy in the construction of the list we adopt the generalization of Kendall tau distance, K_D, which according to [22] is given by:

$$K_D(n_R) = \sum_{i<j} D_{ij}[n_R(i) > n_R(j)] \tag{5}$$

where D_{ij} denotes the distance between two licensed channels i and j that are erroneously inverted in the list, i.e., the position in the list of channel i is before j, even if its actual number of SUs, $n_R(i)$, is higher than j's, $n_R(j)$. The parameter D_{ij} is then equal to: $D_{ij} = n_R(i) - n_R(j)$.

Table 1. Simulation values

Parameter	Value
T_{on}, T_{off}	1 s
Time between two consecutive sensing periods (T_S)	200 ms
Time to react (t_r), Time to sense a channel	3 ms
Time to switch channel	9 ms
Slot time	9 μs
t_{SIFS}	10 μs
t_{DIFS}	28 μs
PLCP preamble and PHY header	20 μs
MAC Header	34 bytes
Payload	1000 bytes
ACK, Beacon packet, End packet length	14 bytes
SR packet, LIST packet	16 bytes
RFS packet length	20 bytes
Ctrl. Transmission Rate	6 Mbps
Data Transmission Rate	24 Mbps
Transmission Power, Power to switch channel	1900 mW
Receiving Power, Idle Power, Power to sense a channel	1340 mW

5.2 Simulation Results

In Fig. 6 and Fig. 7 the average throughput and generalized Kendall tau distance of the proposed algorithm versus the time to sense a channel are respectively depicted. As it can be noticed, devoting more than 3 ms to sense a channel (that is, the time needed for feature detection and the minimum possible for our algorithm), results in worse throughput even if better accuracy is achieved, namely even if the generalized Kendall tau distance gets closer to zero. In other words, for sensing time more than 3 ms, the loss in throughput due to the decrease of time available for data transmission outweighs the gain of better accuracy in the construction of the list.

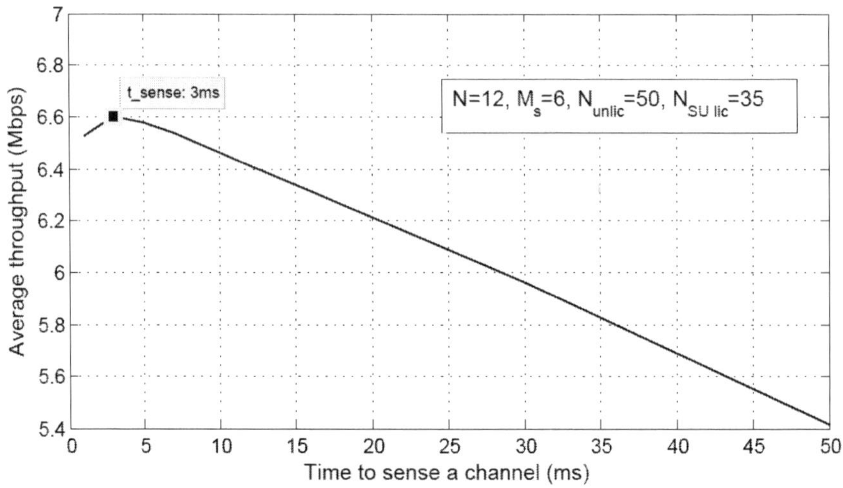

Fig. 6. Average throughput of the CRAHN under study vs. the time to sense a channel

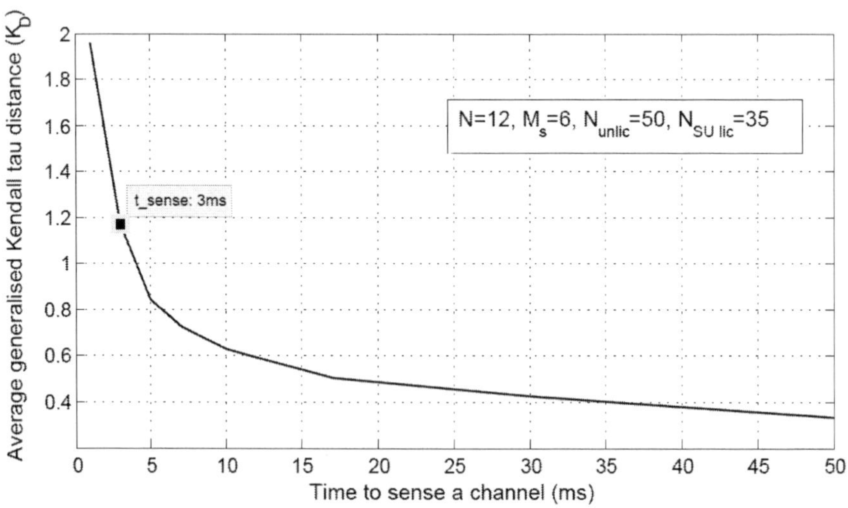

Fig. 7. Average generalized Kendall tau distance of the constructed list vs. the time to sense a channel

This is a general conclusion as this value has been proved through extensive simulations to be independent of the other simulation parameters. Therefore, later on we set the time to sense a channel equal to 3 ms for both our proposal and the reference algorithm. As a result, a very important outcome is that our proposal does not induce any significant overhead compared to the reference algorithm.

As previously explained, we consider as a reference algorithm a feature detection (FD) algorithm that uses no estimation technique for the number of contending SUs. Using FD, a SU is able to distinguish between the licensed channels that are busy and thus should be avoided, the ones with SU activity and others with no activity at all (neither PU nor SU). As a result, for a fair comparison, the list will include first the idle channels with no activity and then the channels with SU activity in a random order.

To that end, in Fig. 8 and Fig. 9 the average throughput and energy efficiency of the CRAHN under study are respectively depicted for the proposed algorithm in comparison with the reference algorithm vs. the number of users of the CRAHN under study (*N*). As it can be noticed in both algorithms, the more the SUs of the CRAHN under study, the higher their throughput until a point when an upper bound is reached, due to the saturation of the licensed channel (i.e., saturation throughput). On the other hand, the energy efficiency of the CRAHN is decreased with the increase of its SUs' number, as the energy consumption during sensing and accessing increases with higher rate than the successful bits transmitted by the CRAHN under study.

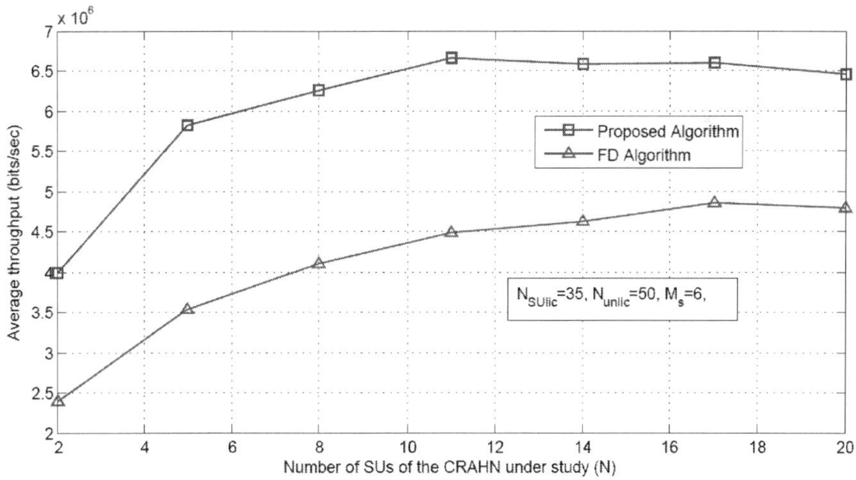

Fig. 8. Average throughput of the CRAHN under study for the proposed algorithm in comparison with the reference algorithm vs. its number of users (*N*)

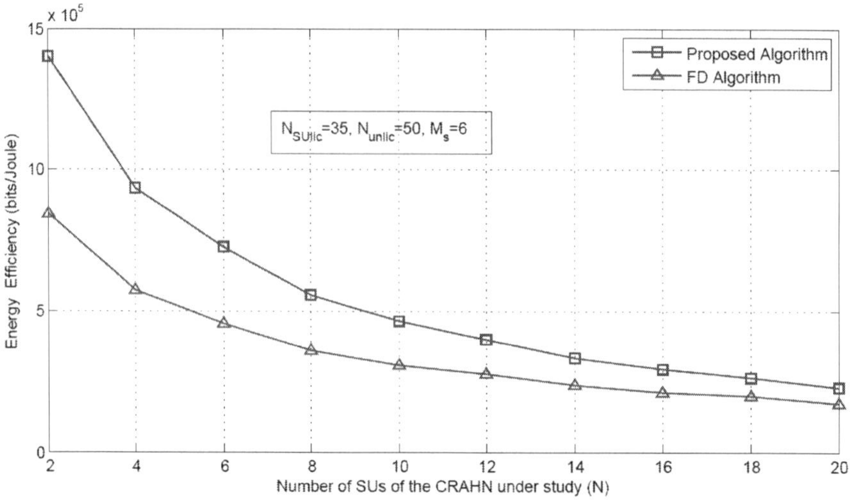

Fig. 9. Average energy efficiency of the CRAHN under study for the proposed algorithm in comparison with the reference algorithm vs. its number of users (N)

In comparison to FD, the proposed algorithm achieves higher throughput and energy efficiency, as there is more time spent in low contended channels, where the probability of a successful transmission is higher and thus more useful bits are sent. At the same time, due to lower contention in the visited licensed channels there is less energy consumption in idle and collision slots compared to FD. Moreover, notice that higher gain is achieved for smaller networks, by virtue of the fact that the more the users of the CRAHN under study, the less the effect that the contention can have in throughput and the more the energy consumption. In other words, in the case that the CRAHN under study consists of a large number of SUs, there will be high contention even in low contended channels, due to the contention among its users.

In Fig. 10 and Fig. 11, the average throughput and energy efficiency of the CRAHN under study are respectively depicted for the proposed algorithm in comparison with the reference algorithm vs. the maximum number of SUs in any licensed channel (N_{SUlic}). As it can be observed, the more the contention in the licensed channels (i.e., the higher the value of N_{SUlic}), the lower the throughput and the energy efficiency of the SN under study for both algorithms, as there are more collisions among the users and thus higher energy consumption.

Furthermore, notice that higher throughput and energy efficiency gains are achieved in comparison to FD for higher contention in the licensed channels, due to the contention-awareness of the algorithm. Specifically, the higher the contention effect, the higher its performance gains.

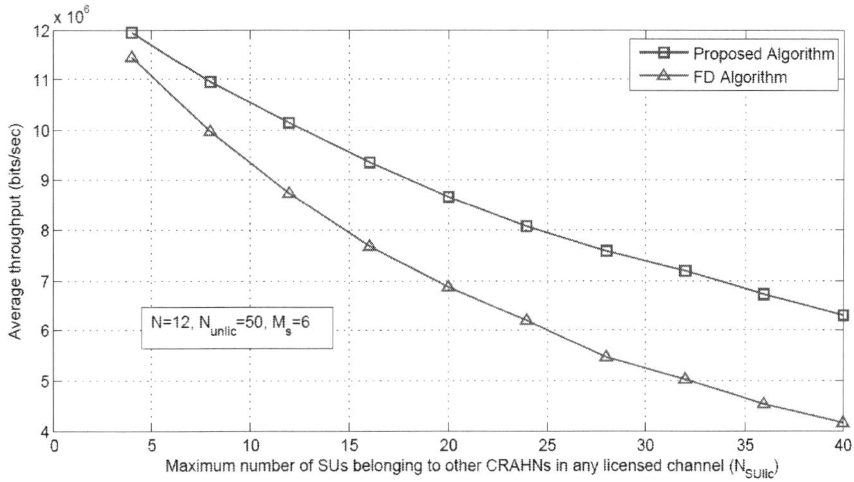

Fig. 10. Average throughput of the CRAHN under study for the proposed algorithm in comparison with the reference algorithm vs. the maximum number of SUs in any licensed channel (N_{SUlic})

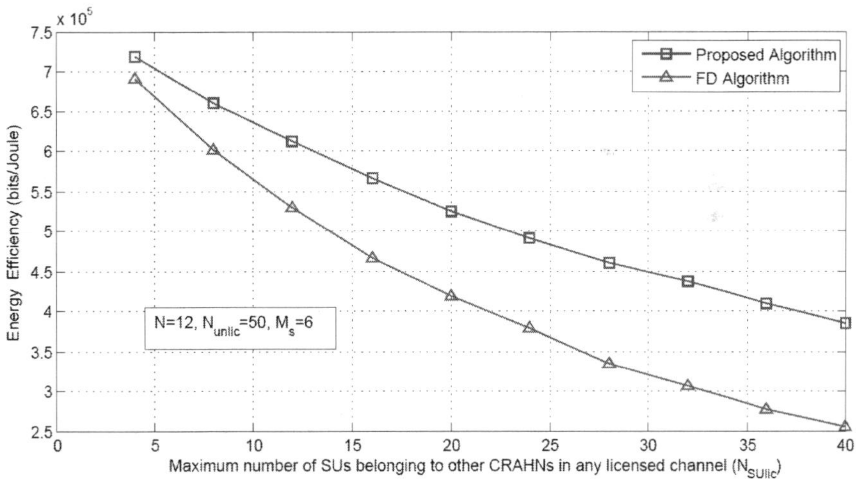

Fig. 11. Average energy efficiency of the CRAHN under study for the proposed algorithm in comparison with the reference algorithm vs. the maximum number of SUs in any licensed channel (N_{SUlic})

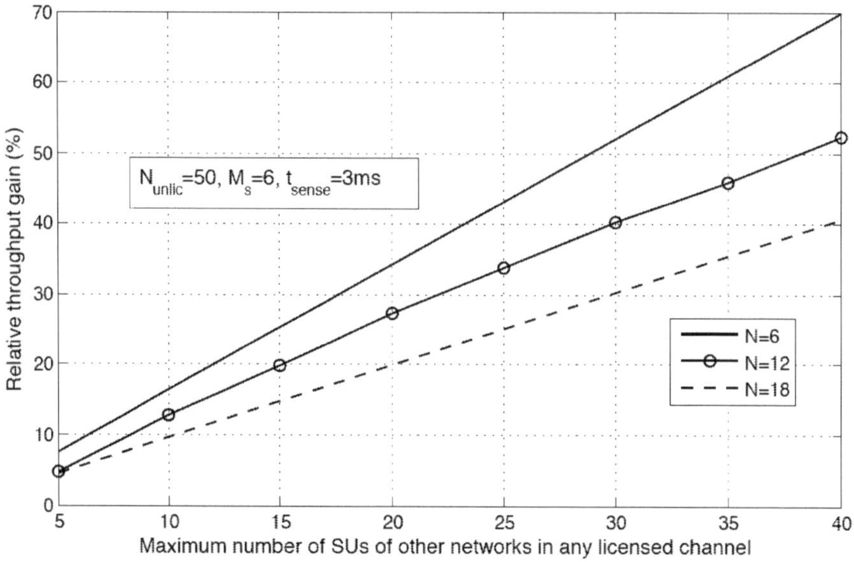

Fig. 12. Relative throughput gain of the proposed algorithm in comparison with the reference algorithm vs. the maximum number of SUs in any licensed channel (N_{SUlic})

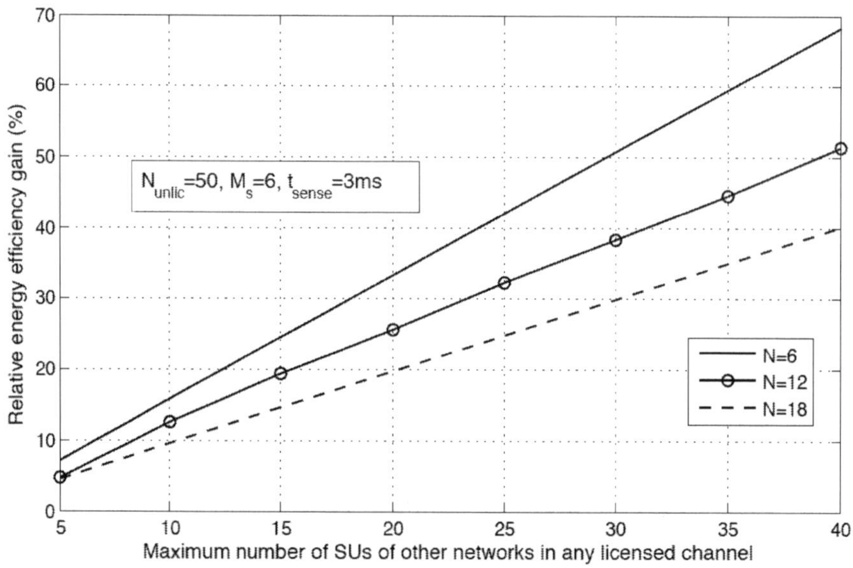

Fig. 13. Relative energy efficiency gain of the proposed algorithm in comparison with the reference algorithm vs. the maximum number of SUs in any licensed channel (N_{SUlic})

Finally, the relative gain in throughput and energy efficiency of our proposal compared to the FD algorithm vs. the maximum number of SUs in a licensed channel (N_{SUlic}), are respectively depicted in Fig. 12 and Fig. 13, for different numbers of SUs of the CRAHN under study (N). It can be noticed that as the contention in the licensed channels increases (N_{SUlic}), the relative gain in both throughput and energy efficiency increases. This stems from the fact that when the proposed contention-aware algorithm is applied, the CRAHN under study spends more time in low contended licensed channels and thus achieves better performance, as previously explained.

Moreover, as the number of users of the CRAHN under study (N) increases, the relative gain in both throughput and energy efficiency decreases, as previously analyzed. Thus, the proposed algorithm achieves the best relative gain under high contention and for a small CRAHN (i.e., for $N=6$ and $N_{SUlic}=40$) it can present up to 70% improvement in throughput and 68% in energy efficiency.

6 Conclusion

In this chapter, we have described the operation and studied the performance of a CRAHN in a scenario where other non-cooperating CRAHNs are also using the same primary resources. A novel energy-efficient contention-aware channel selection algorithm has been proposed, according to which the CRAHN under study selects the less contended idle licensed channels to access. The proposed algorithm uses cooperative spectrum sensing, during which feature detection is combined with another technique that estimates the number of SUs in each licensed channel. Moreover, it has been proved and explained that the proposed algorithm performs well even if the accuracy in the estimated values is low, as its performance only depends on the comparison between the estimated values and not on their exact value. Specifically, the proposed algorithm achieves its best performance for 3 ms, which is equal to the sensing time that feature detection requires. As a result, it has been shown that the proposed algorithm manages to achieve notable performance gains without inducing any significant additional overhead in comparison with the reference algorithm. In particular, it has been demonstrated that for high traffic in the licensed channels, it can present up to 70% gain in throughput and up to 68% in energy efficiency.

Acknowledgments. This work has been funded by the Research Projects GREENET (PITN-GA-2010-264759), CO2GREEN (TEC2010-20823), GREEN-T (CP8-006), GEOCOM (TEC2011-27723-C02-01) and COST-TERRA (IC0905).

References

1. Akyildiz, I.F., Lee, W.Y., Vuran, M.C., Mohanty, S.: Next generation/dynamic spectrum access/cognitive radio wireless networks: a survey. Comput. Networks 50(13), 2127–2159 (2006)

2. Yucek, T., Arslan, H.: A survey of spectrum sensing algorithms for cognitive radio applications. IEEE Commun. Surveys & Tutorials 11(1), 116–130 (2009)
3. Wang, B., Liu, K.J.: Ray Advances in Cognitive Radio Networks: A Survey. IEEE J. on Sel. Topics in Signal Proces. 5(1), 5–23 (2011)
4. Cabric, D., Mishra, S., Brodersen, R.: Implementation issues in spectrum sensing for cognitive radios. In: Thirty-Eighth Asilomar Conf. on Signals, Systems & Comput., pp. 772–776 (2004)
5. Lunden, J., Koivunen, V., Huttunen, A., Poor, H.V.: Collaborative cyclostationary spectrum sensing for cognitive radio systems. IEEE Trans. Signal Process. 57(11), 4182–4195 (2009)
6. Adelantado, F., Juan, A., Verikoukis, C.: Adaptive Sensing User Selection Mechanism in Cognitive Wireless Networks. IEEE Commun. Lett. 14(9), 800–802 (2010)
7. Akyildiz, I.F., Lo, B.F., Balakrishnan, R.: Cooperative spectrum sensing in cognitive radio networks: A survey. Phys. Commun. 4(1), 40–62 (2011)
8. Feng, X., Gan, X., Wang, X.: Energy-Constrained Cooperative Spectrum Sensing in Cognitive Radio Networks. In: IEEE Globecom 2011 (2011)
9. Peh, E.C.Y., Liang, Y.C., Guan, Y.L., Pei, Y.: Energy-Efficient Cooperative Spectrum Sensing in Cognitive Radio Networks. In: IEEE Globecom 2011 (2011)
10. Gong, S., Wang, P., Liu, W., Yuan, W.: Maximize secondary user throughput via optimal sensing in multi-channel cognitive radio networks. In: IEEE Globecom 2010 (2010)
11. Yau, K.L.A., Komisarczuk, P., Teal, P.D.A.: Context-Aware and Intelligent Dynamic Channel Selection Scheme for Cognitive Radio Networks. In: CROWNCOM 2009 (2009)
12. Le, L., Hossain, E.: OSA-MAC: A MAC Protocol for Opportunistic Spectrum Access in Cognitive Radio Networks. In: IEEE WCNC 2008 (2008)
13. Rehmani, M.H., Viana, A.C., Khalife, H., Fdida, S.: Toward Reliable Contention-aware Data Dissemination in Multi-hop Cognitive Radio Ad Hoc Networks. INRIA RR-0375 France (2009)
14. Nam, D., Min, H.: An Energy-Efficient Clustering Using a Round-Robin Method in a Wireless Sensor Network. In: 5th ACIS Int. Conf. (2007)
15. Harada, H., Tinnirello, I.: A Small-size Software Defined Cognitive Radio Prototype. In: IEEE PIMRC 2008 (2008)
16. Bianchi, G.: Performance analysis of the IEEE 802.11 distributed coordination function. IEEE J. Sel. Areas Commun. 18(3), 535–547 (2000)
17. Kuri, J., Kasera, S.K.: Reliable Multicast in Multi-access Wireless LANs. In: IEEE INFOCOM 1999 (1999)
18. Bianchi, G., Tinnirello, I.: Kalman Filter Estimation of the Number of Competing Terminals in an IEEE 802.11 network. In: IEEE INFOCOM 2003 (2003)
19. Zhao, L., Zhang, J., Zhang, H.: Using Incompletely Cooperative Game Theory in Wireless Mesh Networks. IEEE Network 22(1), 39–44 (2008)
20. Rayanchu, S., Mishra, A., Agrawal, D., Saha, S., Banerjee, S.: Diagnosing wireless packet losses in 802.11: Separating collision from weak signal. In: IEEE INFOCOM 2008 (2008)
21. (2012) Part 11: Wireless LAN Medium Access Control (MAC) and Physical Layer (PHY) Specifications. IEEE Std 802.11 (2012)
22. Kumar, R., Vassilvitskii, S.: Generalized Distances between Rankings. In: 19th WWW 2010 (2010)

Resource Management in Mobile Sink Based Wireless Sensor Networks through Cloud Computing

Yasir Saleem[1], Farrukh Salim[2], and Mubashir Husain Rehmani[3]

[1] National University of Sciences and Technology, Islamabad, Pakistan
yasirsaleem106@gmail.com
[2] NED University of Engineering and Technology, Karachi, Pakistan
engr_farrukhsaleem@yahoo.com
[3] COMSATS Institute of Information Technology, Wah Cantt, Pakistan
mshrehmani@gmail.com

Abstract. Wireless Sensor Network (WSN) is composed of distributed spatially connected sensor nodes with limited computing power and storage. However, Mobile Sink (MS) based Wireless Sensor Networks (WSN) showed great advantage over the traditional WSN for saving energy [36]. By adopting good movement strategy for mobile sink, routing, data gathering and communication policy, MS based WSN can more effectively utilize all the available resources as compared to traditional WSN. On the contrary Cloud Computing provides computing and storage resources typically through the Internet. Therefore, there is a recent trend to combine MS based WSN with the clouds so that both can get benefit from each other. In this chapter, we first give an introduction to Wireless Sensor Networks (WSN), Mobile Sink based WSN and Cloud Computing. After then, we give an overview of state-of-the-art work on Wireless Sensor based Cloud Computing (WSCC). Subsequently, integration of WSN and Cloud Computing is highlighted with some insights on how WSN and Clouds can both get benefits from each other. Applications of Wireless Sensors over the cloud are then described. Afterwards, we explain incorporation of mobile sink between WSN and Cloud. Finally, we discuss issues, challenges, and future directions in the realization of Wireless Sensor Network based Cloud Computing.

1 Introduction

WSN is composed of autonomous sensor nodes which are deployed in a region to observe physical or environmental conditions like motion, vibration, lightning condition, soil makeup, noise levels, pollutants, sounds, etc. A wireless sensor node is normally composed of a microcontroller, battery, communication element and sensor. Since wireless sensor nodes are deployed densely and in large number in the area under observation, therefore these are small in size and inexpensive which results in limitations on resources like energy, memory, computational speed and bandwidth. Wireless sensor nodes normally perform three main functions; i.e. sensing, processing and communication. Due to limited energy and memory of sensor nodes, lifetime of wireless networks is very insignificant.

C.X. Mavromoustakis et al., *Resource Management in Mobile Computing Environments*, Modeling and Optimization in Science and Technologies 3,
DOI: 10.1007/978-3-319-06704-9_20, © Springer International Publishing Switzerland 2014

However, with the widespread use of the WSN, the data storage and processing requirements also rose up. Usually the power consumption of storage and processing element of the wireless sensor node was assumed to be negligible. But with the increasing demand for the variety of information; the power consumed by the processing and the storage element can no more be ignored. To tackle such situation; cloud computing provides promising solutions for the storage and processing requirements of the wireless sensor nodes.

Traditionally in WSN, a static sink was deployed at the center of the network which fetches data from all wireless sensor nodes via multihop communication. Consequently, nodes near the sink drain their battery power quickly as nodes in vicinity of the sink have to send their own data as well as relay the data of other nodes. To overcome this problem, Mobile Sink (MS) was proposed in which sink was made to move. Thus mobile sink improves the overall lifetime of the network by minimizing number of transmissions by sensor nodes, eliminating redundant data as all sensor nodes have to send their own data only and by conserving energy because nodes do not have to relay the data of their neighboring nodes.

With the increasing trend of using cloud computing environment for data storage and processing requirements, the popularity of Cloud computing dramatically increases. By using cloud computing strategy, organizations and people can save power cost by reducing the number of servers required, thus runs their activities more productive and with ease. The services can be acquired via Internet and the requirement of services can be changed at any time without worrying about background physical management of servers. Moreover, some of the main features of cloud computing include abstraction, service on demand, elasticity of demand, network access, service measurement, resource pooling etc. [1].

This chapter is organized as follows. In section 2, models of cloud computing from deployment scenarios are presented. Then in section 3, integration of WSN with cloud computing is discussed and how each one is getting benefit from other is explained. After then, applications of WSN and cloud computing are elaborated in much details. Subsequently, the incorporation of mobile sink between WSN and cloud computing is explained in section 5 in which advantages of mobile sink, trajectories and their effects on network performance followed by mobile sink and challenges of mobile sink are discussed. In this section, it is also explained how WSN and cloud computing get benefits from mobile sink. In section 6, we discuss issues, challenges and future direction of WSN and cloud computing. Finally we conclude the chapter in section 7.

2 Cloud Computing Models

Most organizations are now moving towards cloud in-order to control operational costs and cut capital expenditures. However, in-order to adopt cloud, there are many challenges and security risks which could be more expensive to deal with [20]. Therefore organizations must consider them.

Primarily, four cloud models are recommended by [1] for different business scenarios and requirements. Therefore, organizations must understand their requirements before opting any cloud model. These deployment models are also shown in Fig 1.

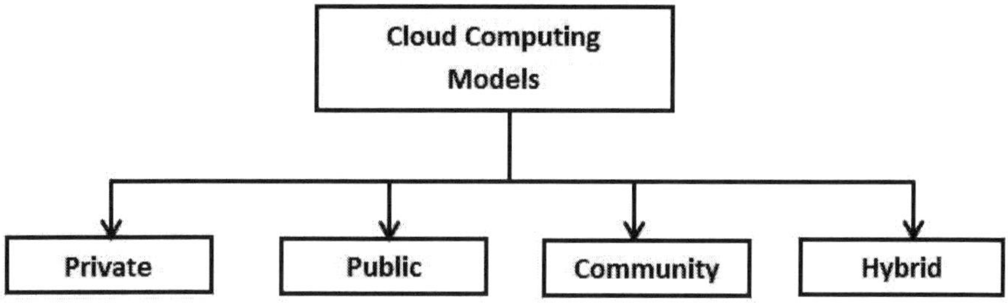

Fig. 1. Cloud Computing models from deployment scenario

2.1 Private Cloud

In this model, cloud infrastructure operates within a single organization and is managed by either the organization itself or by third party. The purpose of this model is to optimize and maximize the utilization of internal resources, providing data privacy, low data transfer cost, security etc. This model does not offer much cost efficiency but it provides security. Security concerns are addressed in this model through secure-access VPN.

Beside security, this model is also adopted by those organizations which have to conform to regulatory standards, e.g., Health Insurance Portability and Accountability Act of 1996 (HIPAA) [21], Statement on Auditing Standards (SAS) 70 [22] etc. where data privacy is of major concern. In-order to ensure data privacy, many cloud applications allow their clients to manage data on their own premises according to their business requirements. One of these applications is Amazon which provides Virtual Private Cloud.

2.2 Public Cloud

In this model, cloud infrastructure is provisioned for open use and is operated and owned by third parties. It is managed and existed on the premises of the cloud provider. In this model, same infrastructure is shared among all the customers with limited resources, configuration and security protections. This model reduces capital expenditures and operational costs [20]. The big example of public cloud is Google where vender provides service either free of charge or on pay per user base license policy.

2.3 Community Cloud

The cloud infrastructure in this model is created by multiple organizations in the community having same concerns and compliance considerations. Some policies for sharing of resources are developed which are followed by these organizations.

This cloud infrastructure is managed by either one or more of the community organizations or by any third party vendor. Since it is shared by larger group, therefore as compared to private cloud, it reduces operational IT cost. This model is suitable for those organizations which require access to same data. For example, many organizations requires access to same information, e.g., hospitals, roads, population etc.

2.4 Hybrid Cloud

This cloud infrastructure is the combination of private and public clouds. In this cloud model, third party cloud providers are being utilized by service providers, either fully or partially for increasing the flexibility of resources. By this model, organizations can secure their applications and host data on private cloud and in the meantime can also reduce their operational cost by keeping their applications and data on public cloud. This model is also used when private cloud is unable to handle load and requires support then its load is shared by public cloud hosting without affecting end users performance. The two major examples of hybrid cloud are Microsoft Azure and Force.com.

3 Integration of WSN with Cloud

In distributed computing environments, cloud computing is getting popular day by day [37]. As cloud computing offers different services over the Internet and the requirement of services can be changed any time without background physical management of servers, therefore, there is a recent trend of using cloud computing for security, storage and data processing needs.

On the other hand, wireless sensor networks are composed of autonomous sensor nodes, which are deployed in a region for monitoring different conditions. Sensor nodes are small in size and inexpensive and can be deployed in a region in large number. But these devices have limited resources in terms of energy, computational speed, storage, bandwidth etc. With the increasing demand of variety of information, the computational speed and storage of sensor nodes require much attention. Here cloud computing provides promising solutions to sensor nodes in terms of security, storage and processing requirements.

This section discusses the integration of WSN and Cloud Computing. More precisely, we discuss how WSN and Clouds can get benefits from each other. WSN and Cloud Computing can be integrated for communication, computing and network technologies. This integration overcomes the challenges of each and provides support to each other. We define how each one can get benefit from other below:

3.1 How WSNs Can Get Benefit from Cloud Computing?

Wireless sensor nodes are deployed geographically disparate to monitor the environment. With the passage of time, the contemporary data fetched from these autonomous WSNs is needed by many users to get up-to-date with the collected and instantaneous data. In many applications, there are complex computation and storage requirements and some applications also require security. For example, in weather forecasting application, it is necessary to convert raw data into useful information before providing it to users. Since wireless nodes have limited computational power, therefore they cannot process this raw data for converting into useful information. Similarly, in other applications such as transport monitoring, irrigation, disaster forecasting, volcanos monitoring systems etc., it is required to store large amount of data which is difficult for sensor nodes to store due to limited storage. Also, in military application, security is of major concern which cannot be efficiently provided by the sensor nodes.

Overwhelmingly, cloud solves these challenges of sensor nodes. It provides storage, processing power and security to wireless sensor networks. It enables the opportunity to share valuable information with the desired users. It assembles, store and disseminates valuable data among the anticipated users with security [4].

The sharing of WSNs data with cloud makes it possible to easily integrate and share different sensor networks data on one platform. It also empowers data and processing power scalability thus optimizing resources. It also facilitates worldwide sharing of data [5].

3.2 How Clouds Can Get Benefit from WSN?

Clouds also get many benefits from wireless sensor networks because there are many applications of WSNs such as military, weather forecasting, transport monitoring, irrigation, disaster forecasting, volcanos monitoring, forest fire detection, pipeline monitoring, mine-safety monitoring, target detection and tracking etc. In all such applications, wireless sensor network collect huge, real-time and up-to-date data from different domains and provide this data to cloud. The cloud then store and process the huge data by providing security. This huge data management of different applications increases the demand of cloud due to variety of data for storage, processing and security which results in promoting the cloud pricing for IT infrastructure [5].

We now discuss some applications of wireless sensor networks using cloud computing.

4 Applications of Wireless Sensor Networks Using Cloud Computing

There are many applications in which wireless sensor networks get benefit from Cloud Computing and also, Cloud Computing gets benefit from Mobile-sink based WSN. Some of these applications in which wireless sensor networks use cloud computing as backbone are described below and are also illustrated in Fig. 2.

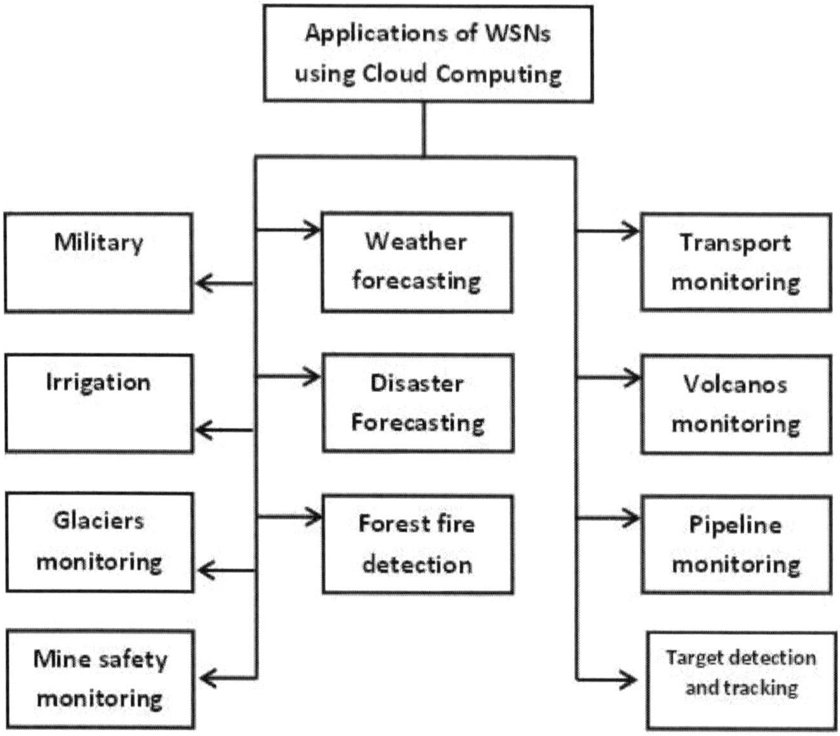

Fig. 2. Applications of Wireless Sensor networks using cloud computing

4.1 Military

In battlefields, sensor networks are used for military purposes for surveillance of enemy forces, targeting, monitoring of equipment, attacks detection etc. [1]. In such environment, mobile-sink plays a vital role in collecting data from sensor nodes. The monitoring forces send drones in the target areas where sensor nodes are deployed. These drones collect data from sensor nodes. Now due to security purpose, this collected data requires very high security therefore normal Internet connection cannot be used for transferring this data. Thus here, cloud computing plays its role by providing a secure infrastructure for military purposes only and help transferring this collected data.

4.2 Weather Forecasting

Weather forecasting is another example of WSN using cloud computing. These systems include data collection, data assimilation, numerical weather prediction and forecast presentation. Sensors nodes are used for sensing and collecting weather parameters which include speed, humidity, temperature, air pressure etc. This collected data is huge in size and much computation is required for computing complicated equations of weather forecast. Thus cloud computing provides storage and complex computation for WSNs [1].

4.3 Transport Monitoring

In transport monitoring system, sensors nodes are used for detection of vehicles, traffic signals lights, estimating vehicles speed and counting vehicle traffics. This data is used for automatic road enforcement. But it is huge in size and requires large computation cycles for data processing and generating traffic controlling data. Thus, the integration of WSN with cloud computing provides ease for storage and computation of large data and helps in generating data for traffic controlling [1].

4.4 Irrigation

Deployment of WSN in irrigation system through Cloud Computing [2] helps in precious agriculture techniques. In this system, sensor nodes are deployed underground in the soil. These sensor nodes collect information about soil moisture in the field and uses cellular phone network to send this collected data to the cloud. The cloud then convert this data into useful information and thus, the real-time data about soil moisture is readily available to farmers who can then take necessary actions according to the environmental conditions.

4.5 Disaster Forecasting System

Wireless Sensor Network in connection with Cloud Computing is used in disaster forecasting using neural network algorithm based on wavelet transform [3]. Cloud Computing is required in disaster forecasting because the forecasting of all types of disasters requires high storage and complex computation which cannot be performed by traditional servers. Therefore, in order to forecast all types of disaster, it is necessary to utilize cloud computing.

4.6 Volcanos Monitoring

WSN can also be used in monitoring volcanos eruptions and earth quakes. The designs of system volcanoes are illustrated in [12], [13] and [14] which are deployed at Tungurahua volcano, Ecuador in August 2007, Reventador volcano, Ecuador in August 2005 and Tungurahua volcano, Ecuador in July 2004 respectively. In these deployments, sensor nodes use event-detection algorithm and whenever any special volcanic activity occurs, they transfer data to the sink [13].

For such scenarios, reliable data collection is a major concern and a mechanism for achieving this goal is required. Thus mobile sink helps in these scenarios for reliable data collection by collecting data from sensor nodes and forwarding it to the cloud for processing and storage.

4.7 Glaciers Monitoring

Sensor networks for monitoring glacial environment are proposed in [15]. It also deployed a system for glaciers monitoring at Briksdalsbreen, Norway in August 2003. In such systems, sensor nodes are inserted in the glaciers which collect environmental data. Mobile sink then collects data from sensor nodes and forwards it to cloud for storage and processing.

4.8 Forest Fire Detection

In [16], WSNs are used for detection of forest fire. A system for detection of forest fire was deployed consisting of 3 sensor networks and 2 webcams in the Selway-Salmon Complex Fires of 2005. This system is called as FireWxNet which is a multi-tiered portable wireless system for monitoring environmental circumstances in forest fire environments. It provides fire and weather conditions within forest fires to fire-fighting community, who can then take necessary actions based on this information. In this system, sensor nodes are deployed in the forest whereas webcams are deployed at peaks. There are radios with directional antennas at each peak near base camp which relay data from sensor networks and webcams. But it is not a feasible system because if one radio is damaged due to fire then the data of sensor nodes cannot be reached to base camp. Thus it would be beneficial to use mobile sink, which will collect data from sensor nodes itself and then send it to cloud for storage. This data will then be utilized for extracting useful information.

4.9 PIPENET: Pipeline Monitoring

Pipeline monitoring is another example of WSN using cloud. A three tier system for detection and localization of leaks and failures in water transmission pipelines is proposed in [17] known as PipeNet. This system provides many advantages such as inexpensive installment, automatic detection of leaks/failures in pipelines, data collection with high frequency, programming environment based reusable data-flow. Authors claim that this system will improve the ability for monitoring and understanding large scale water supply systems.

4.10 Mine Safety Monitoring

Wireless Sensor Networks are used for Mine Safety Monitoring in [18]. It proposed a distributed heterogeneous hierarchical mine safety monitoring system known as HHMSM which monitors methane concentration and locate miner. For data collection, it also proposed over-hearing based adaptive scheme which reduce traffic and control overhead by exploiting redundancy and correlation of sample data. This system is easy and inexpensive to implement as compared to other schemes.

4.11 Target Detection and Tracking

The applications of wireless sensor networks for target detection and tracking are discussed in [19]. It deployed over 90 sensors nodes at MacDill Air Force Base in Tampa, Florida. This work designs a distributed and 2-dimensional surveillance system using sensor nodes. This model offers better detection and tracking along any arbitrary 2-dimensional path within an area as compared to 1-dimensional. In this model, sensor nodes process intrusion data locally and whenever something occurs, share data with their neighboring nodes and send their data to gateway. In such scenario, there are more chances of interference due to multi-hop communication therefore it is better to incorporate mobile sink for collecting data from sensor nodes and then forwarding this data to cloud.

4.12 Health Monitoring

Wireless sensor networks are widely in use for telemedicine applications [27]. They are used for monitoring patients in both clinics and at home. In hospitals, some sensors are connected with patient. Each sensor monitors different condition and collect user health information. Then this information is provided to doctors for personalized feedback. Health monitoring applications increase mobility, ease, comfort and reduce cost.

Other applications of wireless sensor networks using cloud computing are smart grid applications [24], measurement of rotational temperatures versus current variations [25], monitoring and controlling power distribution [26], homeland security (HLS) [28], structural health monitoring (SLH) [29], laser spectroscopic trace-gas [30], habitat monitoring [31], [32], wildlife tracking [33], underground communications [34], cane-toad monitoring [35] etc.

5 Incorporation of Mobile Sink between WSNs and Cloud Computing

The incorporation of Mobile Sink in WSN and Cloud provides many advantages. The most apparent advantage it provides to WSN is energy conservation for sensor nodes. Because with static sink, sensor nodes near the static sink have to send their data as well as relay the data of other sensor nodes in their neighborhood which are far away from static sink whereas on the contrary, mobile sink collects data from sensor nodes itself, thus there is no need for sensor nodes to relay the data of other sensor nodes. Additionally with mobile sink, there is no need of full network connectivity for data gathering. For cloud, it is not responsible to collect data from hundreds of sensor nodes because mobile sink performs this task. It collects data from sensor nodes and forwards this data to cloud. Thus cloud only has to store and process the data.

5.1 How WSN and Clouds Get Benefit from Mobile Sink

Mobile sink help WSNs in conserving their energy because mobile sink collect data from sensor nodes itself. Therefore, just unlike in the presence of static sink, sensor nodes only send their data to mobile sink instead of relaying other nodes data. Hence, this saves the energy of sensor nodes. Since sensor nodes have limited energy therefore this energy conservation in-turn increases the overall lifetime of the network. Also, in the presence of mobile sink, there is no need of full network connectivity because mobile sink itself collects data from sensor nodes, thus it creates a virtual fully connected network.

On the other hand, it is very difficult for cloud to collect data from hundreds and thousands of sensor nodes, thus mobile sink help clouds in collecting data from sensor nodes and then forwards this data to cloud. This is illustrated in Fig. 3. Also, if a

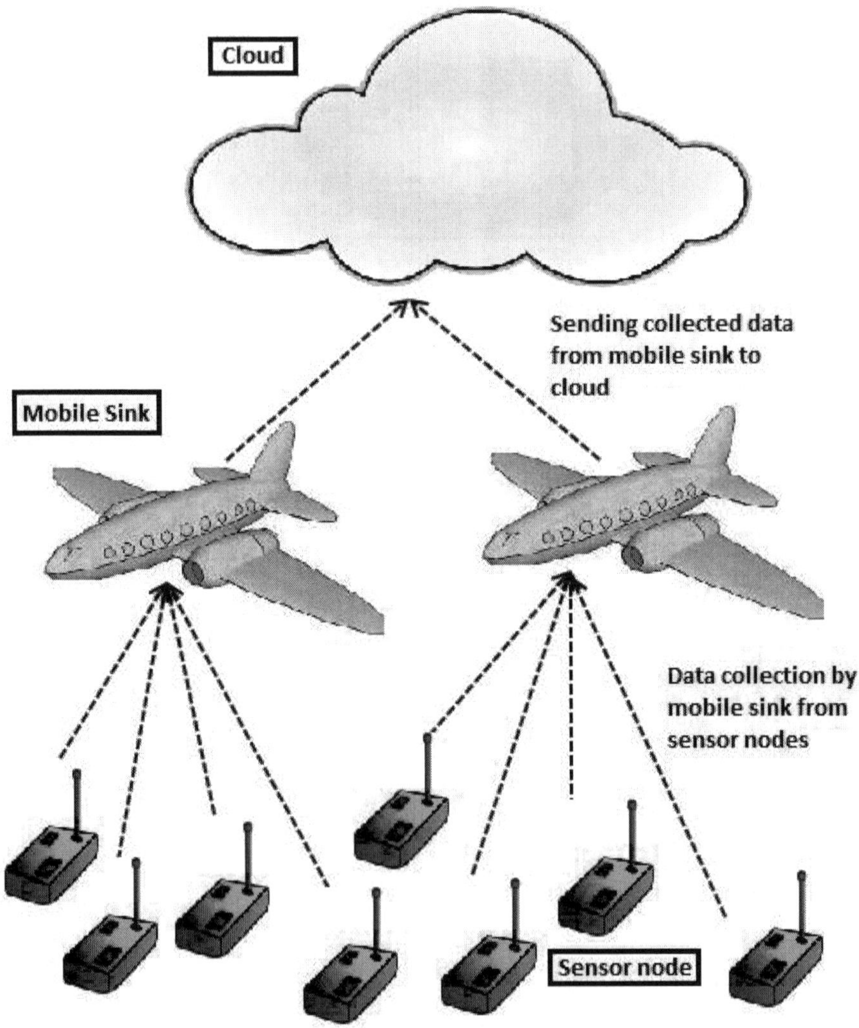

Fig. 3. Incorporation of Mobile Sink between WSN and cloud

network becomes disconnected, then static sink can only collect data from any one network with which it is connected and in this way, the data of other disconnected network could not be sent to clouds. But with mobile sink, there will be no effect of disconnected network for data collection because mobile sink itself collects data from sensor nodes by visiting them and sends it to cloud.

5.2 Advantages of Using Mobile Sink and Cloud Computing

There are many advantages of incorporating mobile sink for collecting data from sensor nodes and forwarding it to cloud. In the presence of mobile sink, it is not mandatory for the whole network to be connected for data collection. Also, sensor nodes do not have to find routes to sinks for transferring data due to which, generally hotspot problems occurs. Mobile sinks handle disconnected networks and increase throughput of the network. It also reduces the routing control overhead for the network. Since mobile sink collects data from sensor nodes itself, thus sensor nodes do not have to relay data of other nodes, therefore multi-hop routing is also not required. Also mobile sink can navigate through such areas where sensor nodes cannot operate, thus it increases the overall lifetime of the network.

The incorporation of mobile sink in WSN is very advantageous [6]. It reduces the overhead for cloud to collect data from sensor nodes. Some advantages are given below which show that how incorporation of mobile sink help cloud computing and WSNs.

5.2.1 Energy Conservation and Overall Network Lifetime Optimization

Sensor nodes have very limited battery and energy. Also, there are many scenarios, where batteries of sensor nodes neither be charged nor changed such as battlefields, where sensor nodes are deployed in enemies areas. In the presence of static sink, sensor nodes near the vicinity of sink have to relay the data of neighboring nodes, thus the battery of these nodes drain quickly due to high communication. While mobile sink collects data itself from all sensor nodes, thus nodes do not have to send data to multiple hops, which reduces the communication and energy utilization. Also, since mobile sink collects data itself by visiting sensor nodes, thus sensor nodes can reduce their transmission range to lowest possible value which also reduces energy utilization and in-turn increases the network overall lifetime.

5.2.2 Network Connectivity

By incorporating mobile sink, network is considered as fully connected because mobile sink itself collects data from sensor nodes by visiting them. Thus it does not matter if nodes are not connected with each other because they do not have to communicate with each other for data transfer rather they have to send their data to mobile sink when it visits them.

5.2.3 Hotspot Problem

In WSN when static sinks are used, hotspot problem occurs because sensor nodes in the vicinity of sinks drain their battery quickly due to relaying data of other nodes in the network. But with mobile sink, this hotspot problem is nearly completely removed because nodes do not have to relay data of other nodes.

5.2.4 Increased Throughput

Mobile sink reduces the number of communication in the network. This decreases the latency of the network and also reduces the routing control overhead. Thus due to these factors, the throughput of the network becomes increased.

5.2.5 Reduction in Transmission Errors and Collision

Transmission errors and collision occurs when multiple nodes try to send data at the same time in multi-hop network. Since there is no multi-hop routing in the presence of mobile sink, therefore, there will be much reduction in transmission errors and collision by incorporation of mobile sink.

5.2.6 Operational Cost

Since data is collected by mobile sink from sensor nodes and sensor nodes do not have to relay data of each other, therefore network connectivity among sensor nodes is not required for data collection.

Thus in-case of partition networks, there will be no operational cost for connecting these disconnected networks because mobile sink itself collects data from sensor nodes and make the network as virtually connected.

5.2.7 Enhancement in Security

In the presence of mobile sink, since there is no need of multi-hop routing, therefore data does not routed across the whole network. In this manner, security of the network is enhanced. Although, if an introducer can sniff data from any node, still then, the security of whole the network is preserved because he will not get information of whole network, rather will get information about that specific area only

5.3 Trajectories and Their Affect on Network Performance Followed by Mobile Sink

Mobile sink trajectories play a vital role in order to collect data from sensor nodes and forwarding this data to cloud in energy efficient manner. Many mobile sink trajectories have been proposed in the literature. The purpose of all these strategies is to optimize the energy consumption of sensor nodes while data collection and to forward data to cloud efficiently.

Generally, trajectories of mobile sink can be categorized into three types. In first category, mobile sink visits all nodes in the network, normally using one-hop communications. In second category, mobile sink visits some specific sensor nodes in the network as in hierarchical network, commonly using two-hop communication.

In the third category, mobile sink visits only single node in the network in-order to collect data. These categories are also presented in Fig. 4. In Fig. 4(a), mobile sink visits all nodes in-order to collect data. In Fig. 4(b), storage nodes (dark blue circles) collect data from non-storage nodes (empty circle) and mobile sink visits only storage node in-order to collect their and non-storage sensor nodes. In Fig. 4(c), there is only one storage node which is responsible for collecting data of all sensor nodes in the network. Mobile sink visits only this storage node in-order to collect data of whole the network. Note that it is important that this storage node must be provided with high storage and high energy, otherwise it will drain its battery quickly before transferring data to mobile sink.

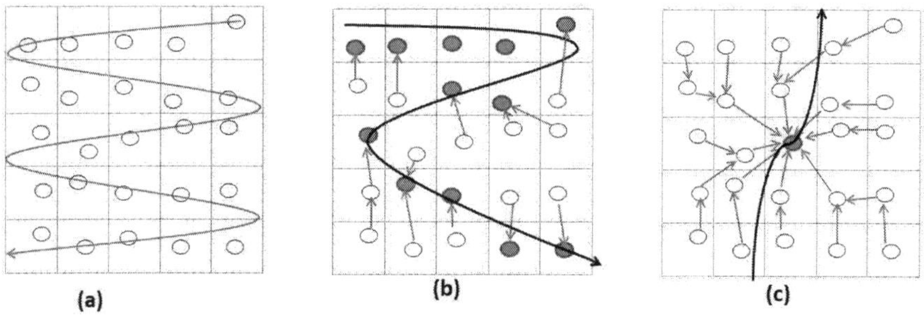

Fig. 4. Types of mobile sink trajectories

In first type of movement, energy is conserved extremely. But this type of trajectory is not feasible for time-sensitive environments and large sensing areas because visiting each sensor node by mobile sink will take considerable time. In second type of movement, mobile sink visits some specific nodes and these specific nodes collect data from their neighboring node and transfer this data along with their own data to mobile sink. In third type of movement, mobile sink collects data from one specific node in the network which is responsible for collecting data from all other sensor nodes. Since this specific node is responsible for much high communication, therefore it is equipped with high energy and battery power. Next we provide some mobile sink trajectories proposed in the literature.

The idea of making sink to be mobile was first given by [7]. This idea was proposed to avoid the hotspot problem. However, the visit time of mobile sink at each location was not discussed in this approach. In [8], the trajectory of mobile sink was considered using square grid topology. Based on energy consumption of sensor nodes, the visit/sojourn times for mobile sink for each grid cell were calculated. The location mobile sink was calculated using linear programming

Another trajectory of mobile sink is along network periphery proposed in [9]. The periphery path in this strategy is fixed, therefore periphery nodes have to relay the data of whole the network which creates overhead for these nodes. This trajectory is depicted in Fig. 5. In this figure, mobile sink visits along the periphery of the network in which nodes at the boundary collect data from other nodes and transfer this collected data along with their own data to mobile sink when it visits them.

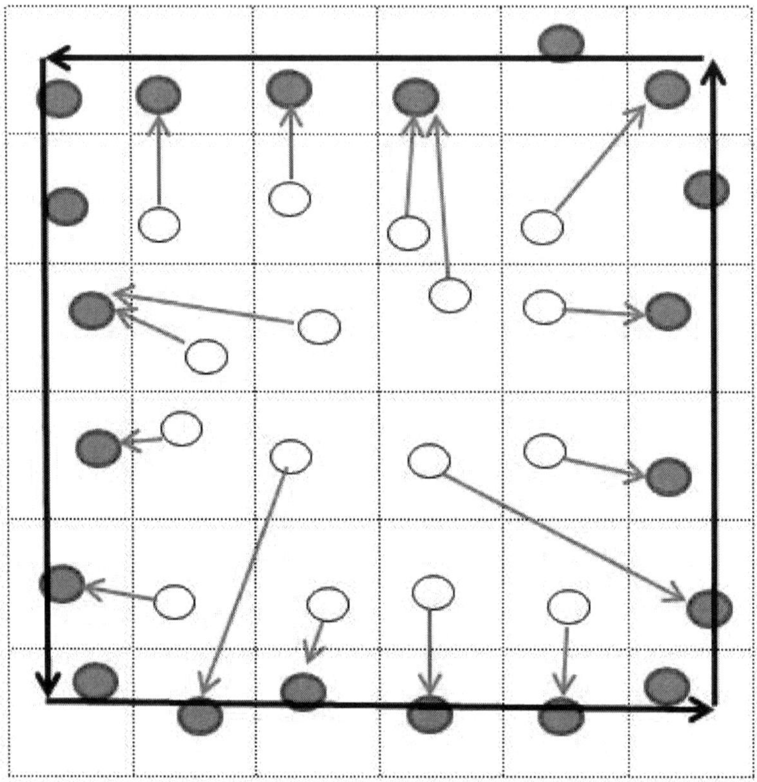

Fig. 5. Mobile sink peripheral movement

Self-directed movement strategy is another trajectory of mobile sink presented in [10]. In this strategy, mobile sink first gathers the energy information of all sensor nodes in data packets. Based on this information, node with minimum energy known as 'MoveDest' is chosen as next location of mobile sink. Also the sink gathers the energy information of sensor nodes from its neighboring nodes only. Thus it might be possible that nodes, which are not the neighbor of MoveDest have high energy threat, which needs to be visited earlier but mobile sink will not get their energy information. In this manner, the energy information of whole network is not obtained by mobile sink.

In [11], another trajectory of mobile sink based on hexagonal path is proposed in which multiple mobile sinks are considered. This trajectory is illustrated in Fig. 6. In this figure, multiple mobile sinks follow hexagonal path. In this strategy, all mobile sinks are synchronized with each other and follow hexagonal path. Sensor nodes only send data to that mobile sink that is at minimum distance. But the movement of mobile sink fails when there is no candidate location in the whole cluster.

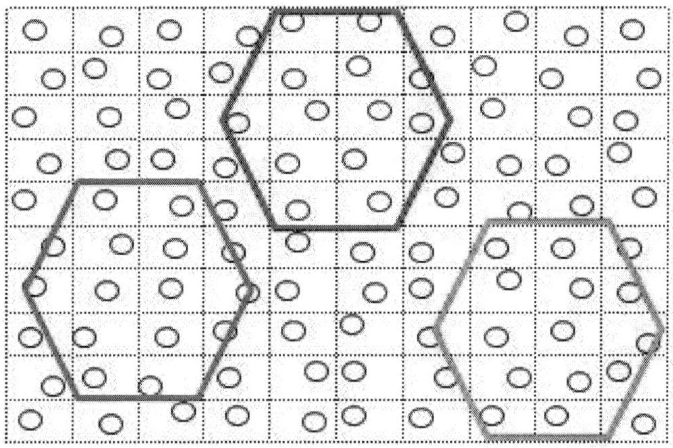

Fig. 6. Mobile sink hexagonal path movement

A data dissemination technique along with mobile sink movement is proposed in [6]. In this proposal, sensor nodes store information of other square-root N sensor nodes in the network and prove that by adopting good data dissemination technique, mobile sink can gather information of whole the network by visiting M nodes in the network. However, sensor nodes have to store the data of other nodes and there will be replication of data. Since they are energy constraints, therefore their battery will be drained quickly.

Another mobile sink trajectory is proposed in [23]. In this strategy, mobile sink first follow periphery movement along the network. After completing one cycle, its next periphery movement is some closer towards the center and form a spiral around the network. Similarly it keeps on moving in spiral movement until the center is reached and no further path is available to move in the spiral movement closer to center. Then it starts to move towards the boundary of the network, i.e., away from the center in the same spiral movement until it reaches to the boundary of the network. After then, it again starts its spiral movement towards the center. This movement is also illustrated in Fig. 7. In this trajectory, there is no much burden on any node because peripheral path is changed after each visit and its visit time is same for each node. But this trajectory does not consider the energy of each sensor nodes because in wireless sensor networks, sensor nodes might have different energies. Some might have greater energy and some might have low. Therefore it is necessary to visit nodes with low energy first so that their battery should not be drained until mobile sink visits them.

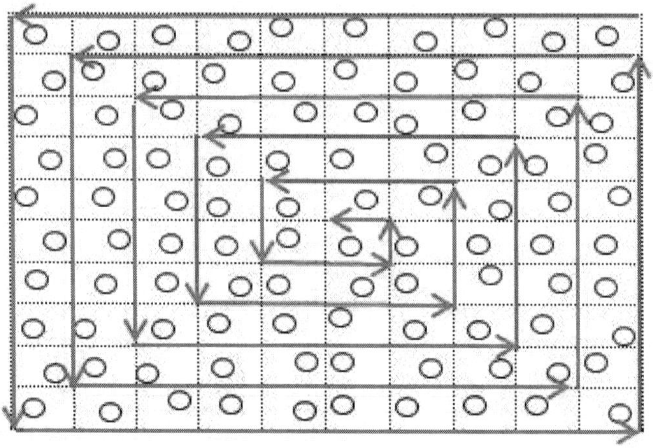

Fig. 7. Mobile sink spiral movement

5.4 Challenges of Mobile Sink When Incorporated between Sensor Nodes and Cloud Computing

The incorporation of mobile sink between WSNs and cloud computing incurs many challenges. Some of them are described below:

5.4.1 Speed of Mobile Sink for Data Collection

Speed of mobile sink for data collection is of much influential. If the speed of mobile sink is very fast, then there will be much packet loss. Otherwise, if the speed of mobile sink is too slow and the network is very large then it will take much time to collect data from all sensor nodes in the network. So, there is a need to investigate speed of mobile for data collection.

5.4.2 Mobile Sink Traversal Time

It is very critical that mobile sink will traverse the network in timely and efficient manner because if mobile sink fails to visit some sensor nodes then data of these nodes will not be collected by mobile sink. Thus it will result in data loss. Also during the traversal of mobile sink, if there is a gap then some nodes will be missed and their data will not be uploaded to sink which will again result in data loss. Therefore, mobile sink traversal time is also another big challenge.

5.4.3 Multiple Mobile Sinks

If multiple mobile sinks are incorporated in those schemes which are designed for single mobile sink then multiple sinks will not be exploited in those schemes and those schemes will not perform efficiently. Therefore, this is an open research issue to study the impact of multiple mobile sinks.

5.4.4 Security

Security is another most important issue of incorporation of mobile sink. Security is mainly required when sensor nodes are operated in the areas of enemies or in terrains. Thus it is very important to investigate that how mobile sink should collect data from sensor nodes.

6 WSN and Cloud Computing: Issues, Challenges, and Future Directions

Although wireless sensor networks based cloud computing with mobile sink are widely in used. But still there are many issues and challenges which are needed to be investigated. In this section, we describe several issues, challenges and future directions for wireless sensor network based Cloud Computing with mobile sink.

6.1 Data Storage on Cloud

When mobile sink collects data from sensor nodes in the network, it should be forwarded to cloud. It is still an open issue that how mobile sink will transfer data to cloud, what will be the time duration and what will be the frequency of transfer.

6.2 Speed of Mobile Sink

The speed of mobile sink for data collection from sensor nodes and forwarding it to cloud is another important issue. If the speed of mobile is too fast, then there will be high packet loss in data collection. Also, cloud will not be able to collect whole data from mobile sink due to high speed.

Also if the speed of mobile sink is very slow then it will cause delay which is harmful to time sensitive applications. Therefore, the speed of mobile sink should be investigated in details.

6.3 Multiple Mobile Sinks

If multiple mobile sinks are incorporated then it is necessary to revise the schemes and design them for catering multiple mobile sinks. Also, it is important that while forwarding data to cloud, there should be a mechanism that at one time, only one mobile sink should transfer its data to cloud otherwise if multiple mobile sinks try to forward their data to cloud simultaneously, then cloud will not be able to handle data from multiple sinks which will results in data loss.

6.4 Coordination between Multiple Mobile Sinks

When multiple mobile sinks are used, it is very important to initiate coordination between them. Because if there is no coordination between mobile sinks, then it is possible that multiple mobile sinks collect data from same nodes which will incurs duplication of data. Also it is possible that multiple mobile sinks try to either collect data from same sensor nodes or forward data to cloud at the same time which will result in data loss because it is not possible for mobile sink and cloud to collect data from multiple sources at the same time unless they are not equipped with multiple transceivers.

6.5 Security

Security is an essential issue in WSN based cloud with mobile sink and needs to be studied. There are three aspects of security in WSN based cloud with mobile sink. In the first aspect, it is necessary that sensor nodes should transfer their data only to reliable and authentic sink. Because in military applications, it is possible that enemies send their own mobile sink or drone in-order to collect data from sensor nodes in their neighboring terrains. Therefore, if security is not considered then it is possible that sensor nodes might transfer their data to enemy mobile sinks and thus secret and sensitive information will be collected by enemies.

The second aspect is that mobile sink should only transfer data to its own cloud. Because just like with sensor nodes, it might also be possible that enemies deploy their cloud in the region from where mobile sinks travels and in this case, the sensitive information will be collected by enemies. This is more severe than previous one. Because if enemies collect information from mobile sink, it means that information of whole network will be available to enemies. While in the previous case when information is collected from sensor nodes, then only the information of that specific region will be available.

In last, the third aspect is that cloud should also only collect information from authentic mobile sinks. It is also possible that an attacker should send their mobile sink with infectious data towards cloud and when cloud collect data from this unidentified mobile sink then it will infect and corrupt its whole data. Therefore it is important to ensure that cloud should only collect data from authentic mobile sinks.

6.6 Multiple Clouds

It is also possible to have more than one cloud. For example, it is possible that multiple clouds are deployed for collecting different information. In this case, mobile sinks should be intelligent enough about which information should be forwarded to which cloud. It is still an open issue and a lot of work can be done for incorporating this issue.

6.7 Mobile Sink Traversal Time

Mobile sink traversal time for collecting data from sensor nodes and forwarding it to cloud is very critical. A lot of work has been done in this dimension but still there are many issues. If mobile sink fails to visit some sensor nodes or there are some gaps in its traversal path, then data of all sensor nodes will not be available which will result in data loss. Also, traversal time is very important in forwarding data to cloud because mobile sink has to forward data of whole the network to the cloud. Thus it will take considerable time. On the other hand, it might be possible that energy of some sensor nodes is very low so it is also very important to collect data from them so that their data could be collected before draining their energy. Thus it is an open issue that how mobile sink should manage its traversal time in-order to collect data from all sensor nodes and forward it to cloud.

7 Conclusion

In this book chapter, the integration of wireless sensor networks and cloud computing and the need of mobile sink in cloud computing based wireless sensor networks for resource management is described. Wireless sensor networks are integrated with cloud computing for management of resources (storage and processing) due to drastic emergence in the applications of wireless sensor networks which require high storage and processing resources. But this integration imposes challenges of data collection from sensor nodes and forwarding this collected data to cloud for storage and processing. In-order to cope with these challenges, the need of mobile sink in cloud computing based wireless sensor networks is illustrated in this chapter. Mobile sink collects data from sensor nodes and forwards it to cloud using different trajectories. This incorporation optimizes resources of sensor nodes because in presence of mobile sink, sensor nodes do not have to relay data of other sensor nodes for providing it to the cloud, as mobile sink itself collects data from sensor nodes and forwards it to the cloud. But there are still some challenges and future directions in the incorporation of mobile sink based wireless sensor networks and cloud computing, which also needs to be explored further and are highlighted in this chapter.

References

[1] Dash, S.K., Mohapatra, S., Pattnaik, P.K.: A Survey on Applications of Wireless Sensor Network Using Cloud Computing. International Journal of Computer Science & Emerging Technologies 1(4) (December 2010)

[2] Tooker, J., Xin, D., Vuran, M.C., Irmak, S.: Connecting soil to the cloud: A wireless underground sensor network testbed. In: 9th Annual IEEE Communications Society Conference on Sensor, Mesh and Ad Hoc Communications and Networks (SECON), June 18-21 (2012)

[3] Jianhu, C., Tao, Y., Zhe, L., Song, J., Shaohua, L.: Developing a disaster surveillance system based on wireless sensor network and cloud platform. In: IET International Conference on Communication Technology and Application (ICCTA 2011), October 14-16, pp. 757–761 (2011)

[4] Rutakemwa, M.M.: From Physical to Virtual Wireless Sensor Networks using Cloud Computing. International Journal of Research in Computer Science 3(1), 19–25 (2013)

[5] Kurschl, W., Beer, W.: Combining cloud computing and wireless sensor networks. In: Proceedings of the 11th ACM International Conference on Information Integration and Web-based Applications & Services (iiWAS 2009), New York, NY, USA, pp. 512–518 (2009)

[6] Rehmani, M.H., Viana, A.C.: Decoupling Data Dissemination from the Mobile Sink Trajectory in Wireless Sensor Networks: Current Research and Open Issues. Masters Thesis, Laboratory of Signals and Systems, Supelec and University of Paris Sud- 11 (July 2008)
[7] Gandham, S.R., Dawande, M., Prakaish, R., Venkatesan, S.: Energy efficient schemes for wireless sensor networks with multiple mobile base stations. In: Proc. IEEE GLOBECOM, pp. 377–381 (December 2003)
[8] Wang, Z.M., Basagnii, S., Melachrinoudis, E., Petriolii, C.: Exploiting Sink Mobility for Maximizing Sensor Networks Lifetime. In: Hawaii International Conference System Sciences, HICSS 2005 (2005)
[9] Luo, J., Hubaux, J.: Joint mobility and routing for lifetime elongation in wireless sensor networks. In: INFOCOM, pp. 1735J–1746J (2005)
[10] Yanzhong, B., Limin, S., Jian, M., Na, L., Khan, I.A., Canfeng, C.: HUMS: An Autonomous Moving Strategy for Mobile Sinks in Data-Gathering Sensor Networks. EURASIP Journal on Wireless Communications and Networking (2007)
[11] Marta, M., Cardie, M.: Using sink mobility to increase wireless sensor networks lifetime. In: Proc. WoWMoM (2008)
[12] Werner-Allen, G., Lorincz, K., Welsh, M., Marcillo, O., Johnson, J., Ruiz, M., Lees, J.: Deploying a Wireless Sensor Network on an Active Volcano. IEEE Internet Computing 10(2), 18–25 (2006)
[13] Werner-Allen, G., Lorincz, K., Johnson, J., Lees, J., Welsh, M.: Fidelity and yield in a volcano monitoring sensor network. In: Proceedings of the 7th Conference on USENIX Symposium on Operating Systems Design and Implementation, Seattle, WA, November 06-08, vol. 7, p. 27. USENIX Association, Berkeley (2006)
[14] Werner-Allen, G., Johnson, J., Ruiz, M., Lees, J., Welsh, M.: Monitoring volcanic eruptions with a wireless sensor network. In: Proc. 2nd European Workshop on Wireless Sensor Networks (January-February 2005)
[15] Martinez, K., Hart, J.K., Ong, R.: Environmental Sensor Networks. Computer 37(8), 50–56 (2004)
[16] Hartung, C., Han, R., Seielstad, C., Holbrook, S.: FireWxNet: a multi-tiered portable wireless system for monitoring weather conditions in wildland fire environments. In: Proceedings of the 4th International Conference on Mobile Systems, Applications and Services, MobiSys 2006, Uppsala, Sweden, June 19-22, pp. 28–41. ACM Press, New York (2006)
[17] Stoianov, I., Nachman, L., Madden, S., Tokmouline, T., Csail, M.: PIPENET: A wireless sensor network for pipeline monitoring. In: 6th International Symposium on Information Processing in Sensor Networks, IPSN 2007, pp. 264–273. IEEE (April 2007)
[18] Niu, X.H., Xi, Z., Ze, Z., Yuhe, H., Changcheng Cui, L.: The Design and Evaluation of a Wireless Sensor Network for Mine Safety Monitoring. This Paper Appears in: Global Telecommunications Conference, GLOBECOM 2007, November 26-30, pp. 1291–1295. IEEE, Washington, DC (2007)
[19] Arora, A., Dutta, P., Bapat, S., Kulathumani, V., Zhang, H., Naik, V., ... Miyashita, M.: A line in the sand: a wireless sensor network for target detection, classification, and tracking. Computer Networks 46(5), 605–634 (2004)
[20] `http://www.cloudtweaks.com/2012/07/the-4-primary-cloud-deployment-models/`
[21] Annas, G.J.: HIPAA regulations—a new era of medical-record privacy? New England Journal of Medicine 348(15), 1486–1490 (2003)

[22] Nickell, C.G., Denyer, C.: An introduction to SAS 70 audits. Benefits Law Journal 20(1), 58–68 (2007)
[23] Salim, F., Jong, K.L.: Optimized Mobile Sink Trajectory for Efficient Energy Consumption in Wireless Sensor Networks. In: The 6th International Conference on Ubiquitous Robots and Ambient Intelligence (URAI), pp. 358–362 (2009)
[24] Erol-Kantarci, M., Mouftah, H.T.: Wireless Sensor Networks for smart grid applications. In: 2011 Saudi International Electronics, Communications and Photonics Conference (SIECPC). IEEE (2011)
[25] Nekahi, A., Farzaneh, M.: Rotational temperature measurement of an arc formed over an ice surface. IEEE Transactions on Dielectrics and Electrical Insulation 18(3), 755–759 (2011)
[26] Isaac, S.J., Hancke, G.P., Madhoo, H., Khatri, A.: A survey of wireless sensor network applications from a power utility's distribution perspective. In: AFRICON 2011, pp. 1–5. IEEE (September 2011)
[27] Zubiete, E.D., Luque, L.F., Rodriguez, A.V.M., González, I.G.: Review of wireless sensors networks in health applications. In: 2011 Annual International Conference of the IEEE Engineering in Medicine and Biology Society, EMBC, pp. 1789–1793. IEEE (August 2011)
[28] Lee, K.B., Reichardt, M.E.: Open standards for homeland security sensor networks. IEEE Instrumentation & Measurement Magazine 8(5), 14–21 (2005)
[29] Chintalapudi, K., Fu, T., Paek, J., Kothari, N., Rangwala, S., Caffrey, J., ... Masri, S.: Monitoring civil structures with a wireless sensor network. IEEE Internet Computing 10(2), 26–34 (2006)
[30] So, S., Koushanfar, F., Kosterev, A., Tittel, F.: LaserSPECks: laser SPECtroscopic trace-gas sensor networks-sensor integration and applications. In: Proceedings of the 6th International Conference on Information Processing in Sensor Networks, pp. 226–235. ACM (April 2007)
[31] Hakala, I., Tikkakoski, M., Kivela, I.: Wireless sensor network in environmental monitoring-case foxhouse. In: Second International Conference on Sensor Technologies and Applications, SENSORCOMM 2008. IEEE (2008)
[32] Cerpa, A., Elson, J., Estrin, D., Girod, L., Hamilton, M., Zhao, J.: Habitat monitoring: Application driver for wireless communications technology. ACM SIGCOMM Computer Communication Review 31(2 suppl.), 20–41 (2001)
[33] Juang, P., Oki, H., Wang, Y., Martonosi, M., Peh, L.S., Rubenstein, D.: Energy-efficient computing for wildlife tracking: Design tradeoffs and early experiences with ZebraNet. ACM Sigplan Notices 37(10), 96–107 (2002)
[34] Sicignano, D., Tardioli, D., Cabrero, S., Villarroel, J.L.: Real-time wireless multi-hop protocol in underground voice communication. Ad Hoc Networks (2011)
[35] Hu, W., Bulusu, N., Chou, C.T., Jha, S., Taylor, A.: The design and evaluation of a hybrid sensor network for cane-toad monitoring. In: Fourth International Symposium on Information Processing in Sensor Networks, IPSN 2005, pp. 503–508. IEEE (April 2005)
[36] Hamida, E.B., Chelius, G.: Strategies for data dissemination to mobile sinks in wireless sensor networks. IEEE Wireless Communications 15(6), 31–37 (2008)
[37] Aceto, G., Botta, A., De Donato, W., Pescapè, A.: Cloud monitoring: A survey. Computer Networks 57(9), 2093–2115 (2013)

Mobile Video Streaming Resource Management

Ram Gopal Lakshmi Narayanan

Senior Innovation Architect, Nokia Siemens Networks, California, USA

Abstract. Video traffic dominates internet, and expected to grow by many folds. Video sessions are longer, and demands both latency and bandwidth. On mobile networks during video play user experience frequent video interruption, often rate their user experience as poor quality of experience (QoE). Mobile network have constraints such as limited radio resource, time varying bandwidth, and finite battery power on the phones. All these factors must be taken into considering as part of video service resource management. Video service providers and operators have been trying out various resource management strategies to improve QoE. Several improvements have been made over the years to make streaming experience acceptable to users. In this chapter, we will analyse popular streaming techniques, and explore resource management techniques in mobile network architecture to support video streaming.

1 Introduction

Video traffic contributes towards major portion of internet traffic. Due to increase in penetration of smart phones and tablets, video traffic will grow by many folds [1]. User generated content (UGC) such as YouTube and premium traffic such as Netflix are major contributors in USA [2, 3]. Such a sudden explosion of video traffic makes network unmanageable. Lessons learnt from deploying Fourth generation mobile network such as Long Term Evolution (LTE) and Third generation mobile network such as Wideband Code Division Multiple Access (WCDMA) technologies have shown that video streaming does not perform well over mobile broadband networks. Major reasons for such poor performance are too many parallel developments in services, applications that are not network friendly and demand more data, and smart phones consuming energy than required. To manage network traffic, wireless operators move from flat rate billing to the tiered billing and hence usage of mobile network is restricted. Though such measure is considered temporary, but it clearly puts the requirement to improve the network infrastructure and service utilization.

Video streaming session is long duration, demands latency and bandwidth throughout the video play. Hence network resource must be made available throughout the video play. To satisfy these requirements, mobile operators have increased their server and network capacity. However, wireless signal strength varies with respect to location, time and environment, and delivering guaranteed bandwidth to video streaming application in such non-uniform wireless network condition becomes a challenge.

C.X. Mavromoustakis et al., *Resource Management in Mobile Computing Environments*, Modeling and Optimization in Science and Technologies 3,
DOI: 10.1007/978-3-319-06704-9_21,

Traditionally, to ensure guaranteed service to users, operators measure packet loss, jitter, delay, available bandwidth and other network related parameters as part of Quality of Service (QoS). Standard techniques and protocols to improve QoS are not widely deployed in internet, hence video streaming is delivered as best effort traffic over the internet. However, measuring network level information alone is insufficient to assess quality of video service. Therefore perceived user quality as QoE needs to be included when designing network and services [4]. For mobile networks, it has constraints such as energy consumption, limited radio spectrum, radio propagation characteristics, and interference. It is not possible to deliver video streaming data without adaptation at client or server or network. The goal of this chapter is to provide an overview of video streaming protocols, and various resource management techniques adopted during to improve QoE for users. Section 2 describes mobile broadband architecture, section 3 describes overview of video streaming protocols, section 4 and section 5 describes various network architecture to support video streaming, section 6 describes radio network impact and UE energy consumption due to video streaming and finally section 7 describes future research activities.

2 Mobile Broadband Architecture

Mobile broadband access network consists of Packet Core Network (CN), Radio Access Network (RAN) and Transport backhaul Network (TN). Simplified 3G and 4G mobile broadband network architecture is shown in Figure 1. Third Generation RAN shown in Figure 1(A) consists of NodeBs and Radio Network Controller (RNC). The functions of RAN include radio resource management (RRM), radio transmission and reception, channel coding and decoding, and multiplexing and de-multiplexing. Layer-2 radio network protocol messages are used to carry both control and user plane traffic from RAN to User terminal. RNC identifies signaling and user plan messages and forwards Layer-3 Mobility management messages towards core network HSS for authentication and authorization of users. Uplink user plane traffic from UE is received at RNC, and RNC performs GPRS Tunneling Protocol (GTP) operation and forward the IP packet towards GGSN. Similarly, downlink packets for UE are received at RNC, and GTP packets are terminated and inner IP packet is forwarded to UE.

3G packet core consists of Gateway GPRS Support Node (GGSN) and Serving GPRS Support Node (SGSN). The functions of Packet core include IP session management, Legal Interception functions, Policy based routing and charging functions, etc. User mobility and its associated sessions are handled by SGSN node. SGSN also acts as anchor point for ciphering and authentication of session between UE and wireless access networks.

Fourth Generation LTE radio network is shown in Figure 1(B) and it consist of eNodeB network element. LTE design is based on flat architecture with reduced number of network elements. The RAN functions are consolidated into single network element eNodeB. 3G and 4G RAN networks are not backward compatible as their physical layer technologies and radio network protocols are different. 4G uses Orthogonal Frequency Division Multiplexing (OFDM), whereas 3G uses WCDMA based technologies for wireless physical layer processing.

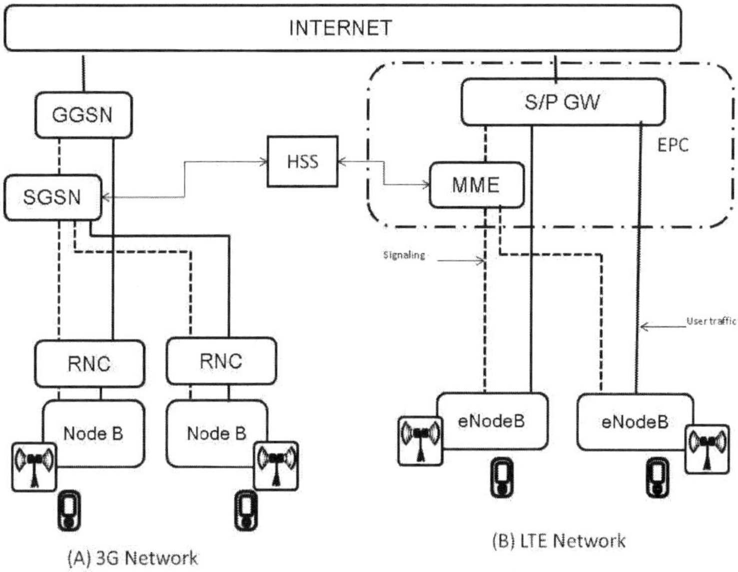

Fig. 1. Mobile broadband network

Similar to 3G, there is a logical separation of networks such as RAN, Packet Core and Transport network in 4G networks. Mobility Management Entity (MME) and Serving and Packet Gateway (S/P-GW) are part of Evolved Packet Core (EPC) in LTE. The functions of MME include Radio Signaling functions, and mobility management sessions maintenance. The EPC functions are similar to that of 3G GGSN, and contain improved IP mobility management functions.

3 Video Streaming Protocols

Three popular mechanisms exist in internet to deliver video content using standardized protocols. They are (a) user could download a video file from server using FTP service, and watch video locally in his device at later time, (b) video be streamed as video on demand (VOD) from a server, and (c) video could be delivered in real time. Internet Engineering Task Force (IETF) standards specified Real Time Protocol (RTP) and HTTP for transporting video content between server and client. RTP runs on top of User Datagram Protocol (UPD) and carries both video and audio streams [5, 6]. In RTP based implementation, server maintains states for each client, and client provides periodic feedback about its streaming experience. This state-full operation raises concern on server scalability. Moreover, most of the firewalls in internet do not allow RTP traffic and because of this, it has limited adoption. To overcome the weakness of RTP, HTTP which is pervasive in the internet has been chosen as transport protocol for video. Advantage of HTTP is that it is stateless protocol and support for HTTP services is already available in all Content Delivery Networks (CDN).

3.1 Pseudo Streaming

In Pseudo or Progressive streaming technique, video start to play as soon as client downloaded partial video content from server. In pseudo streaming architecture, video content is stored as one file at CDN servers. When user makes a video play request, the file is downloaded quickly on to client's video play out buffer. After partial download of file is complete, the client starts to play the video. In the background client continues to download the file from server and play the downloaded content from buffer concurrently. Figure-2 is a simplified view of YouTube video pseudo streaming architecture [7, 8]. Normally user performs sequence of operations such as visiting the website, then doing searching on video, and then selecting a video to be played. After a video is selected by user in YouTube namespace each video is uniquely identified by a video identifier. Client generates HTTP request containing video identifier to YouTube web server and is shown in message 1. YouTube web server acts like a lookup server, it does contain list of web servers where the actual video content is available. YouTube web server generates HTTP redirect response to client with list of servers where the actual video is available and is shown in Figure 2, message 2. In message 3, Client prepares request for video content based on received server list to video CDN. The Video CDN then streams video over the HTTP to the users as in message 4. The separation of video list that is maintained at YouTube web server and Video CDN is useful for load balancing, and for selecting appropriate server based on client location.

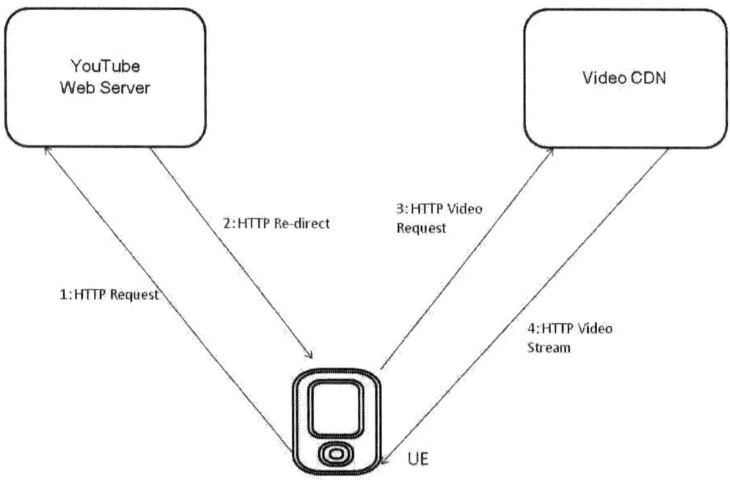

Fig. 2. YouTube pseudo streaming

Two key measurement parameters for video streaming are video start time and no video interruption during play. It has been observed that average user acceptable wait time for starting of video is between 6 and 9 seconds. To have interrupted video service available network bandwidth must be 1.2 times of video data rate. Video data rate depends on the type of codec used for generating a video file. Server sends video data to client at constant rate, during this period if the available bandwidth falls below the required data rate user will experience video interruption. Video interruption happens because the client has insufficient data in its play out buffer. Client will resume play once after it receives received sufficient data. This process is called re-buffering, and time to resume play after interruption is called re-buffering time.

It is not practical to expect a constant bandwidth for entire video session. Therefore, several possible streaming algorithms exist to avoid video stalls during play. We will discuss the common technique, and improvements to deliver video content. Early download is one possible streaming techniques, early download is possible when available bandwidth is much more than required bandwidth to play a video. Server instead of sending video content at constant rate sends data to available network bandwidth. With this approach, client receives video content very earlier than required and thus eliminate problem of video re-buffering. To have this implementation to work, client need to have sufficient buffer equal to that of video play duration. But, side-effect of this approach is that, more content is made available to client much earlier and if client prematurely terminates video session then downloaded video content is discarded and results in data wastage.

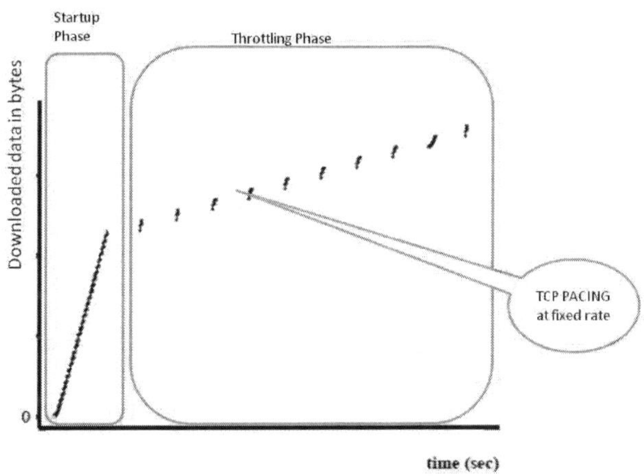

Fig. 3. YouTube pseudo streaming startup and throttling phase

To overcome data wastage, YouTube server implements two-phase approach namely start-up phase and throttling phase [9]. During startup phase, in order to start the video early, server fills client buffer at faster data rate up to certain seconds. As shown in figure 3, in order to start the video quickly the server sends at much higher data rate to fill in the client buffer. In throttling phase, data is throttled down and server sends at constant rate to the client until clients buffer gets filled and is shown in figure 4. During the throttling phase, server does constant TCP pacing and pumps data equal to 64Kbytes at constant rate. Client application consumes data from application buffer to play video to user, and when data in client buffer falls below the lower watermark, client requests for more video data. YouTube expects that available bandwidth is at least 1.2 times of required video rate. Under this condition, video data gets downloaded until client buffer is full, and stops downloading of data. Client will resume downloading of data only when data in client buffer falls below lower mark threshold. This behavior is observed as ON and OFF cycle in network traces and is shown in Figure (4). In this streaming approach, if the client buffer is kept low, then wastage of data can be low. Traces collected from internet shows that YouTube keeps client buffer size between 90 to 100 seconds of video data. In Figure 4, first ON-OFF duty cycle is 85-65 seconds respectively, and second ON-OFF duty cycle is 35-65 seconds respectively and third ON cycle is 35 cycles and last byte downloaded around 260 seconds for 6 minutes video. This approach improves resource utilization in mobile networks. Because, client pre-fetch only portion of the data, and then releases the network for other users. When there is a balance between available client buffer, and network capacity will minimize the data wastage and gives the notion of Just-In-Time (JIT) delivery.

Due to very high network fluctuations, it is difficult to achieve balance between available bandwidth, JIT, data wastage in pseudo streaming. For example, consider a scenario where user starts to play video at a location where there is good signal strength and bandwidth. When user starts to move to a new location and experiences lower bandwidth, then at the new location video play is interrupted frequently and eventually user may stop the video play. This changing network condition must to be adapted quickly to avoid video stalling.

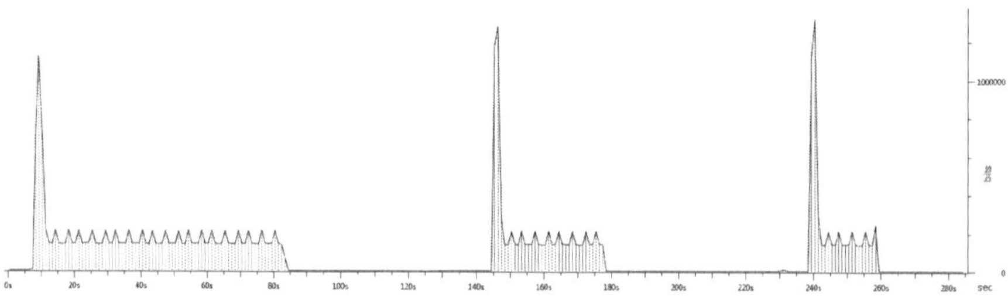

Fig. 4. YouTube 6 minutes pseudo streaming trace

3.2 Adaptive Bit Rate Streaming

Adaptive bit rate (ABR) streaming adapts quickly to changing network conditions and thus overcomes the problems of pseudo streaming. In ABR, client constantly monitors its network conditions, when available network bandwidth changes it makes new request that is suitable for the changed network condition bandwidth. This process of requesting to different video content based on available bandwidth and play without interruption is called stream switching. ABR is suitable for live and on-demand video content, and also minimize data wastage when client prematurely terminates a video session.

ABR streaming algorithms are not standardized, and the parameter that is being used by client is vendor dependent. Without going through a vendor specific implementation, we briefly describe ABR implementation. It consists of two steps namely ABR file preparation and ABR streaming. First a raw captured video is encoded at highest video quality rate (say for example HIGH_VIDEO_RATE). Then the encoded file is then coded into multiple different coding rates (say for example MEDIUM_VIDEO_RATE, LOW_VIDEO_RATE). Each of these different encoded files are further fragmented (or segmented) into multiple smaller file size called chunks. The size of chunk file is kept small to avoid data wastage and is usually between 3 and 8 seconds of video duration. After this process is completed, a Media Presentation and Description (MDP) is created which contains details about the number of bit rate supported such as HIGH, MEDIUM, and LOW etc, and intial file fragment, and location of server where these fragments are stored. MPD file along with all bit rate chunk files are kept at Video CDN and is now ready for client to access the files. ABR preparation process is also called as ingestion. Depending upon codec and adaption requirement number of bit rate supported may be varied.

Figure 5 describes the ABR message sequence between a client and video server. For illustration purpose the architecture is simplified, and shows only client and video server interaction. In reality, fragmented files could be kept in one or more servers, and it may be possible that there may be several path between client and server. Following are the message interaction between client and server.

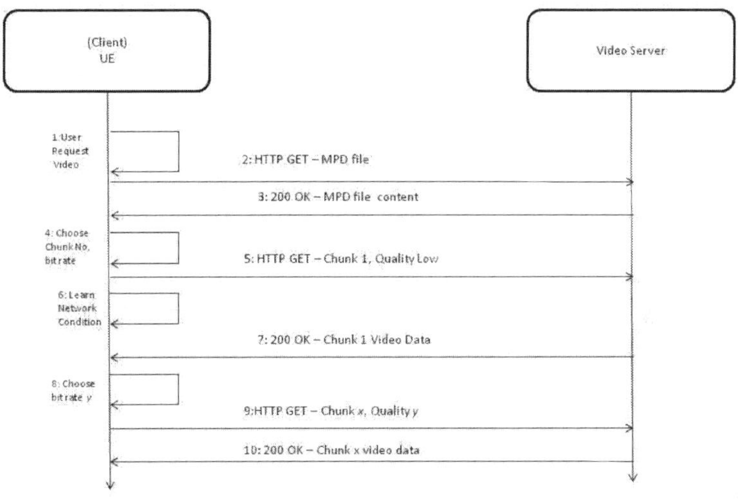

Fig. 5. Adaptive bit rate streaming

1. User selects a video for play.
2. Client application resolves the user request to a video server, and request MPD file.
3. The video server responds with MPD file, and MPD file contains details about list of supported bit rate, URL location for video chunks.
4. To start a video play most ABR client implementation starts with lowest bit rate and then quickly adapts by choosing appropriate bit rate for that network condition
5. Client requests first chunk with low video quality.

6. While downloading video data, client learns network conditions by combining several information including (a) either based on the delay of packets that it is receiving, or (b) based on the rate of TCP window size or (c) based on the TCP ACK it is generating towards to server.
7. Server responds with video data chunk.
8. Client after downloading first chunk of data, it has learnt available bandwidth for the future chunk request. Depending upon the rate adaptation algorithm, client makes future chunk request. For example, if available network bandwidth is more, then client will issue next chunk request with higher bit rate, and this process continues till it has reached a bit rate that is enough for the given bandwidth. This process is called *SWITCHING_UP*, and this normally happens during video start. How quickly client adapts to network condition from lowest supported bit rate is called initial adaptation time. After initial adaptation client continues to learn the network conditions, when there is a change in network bandwidth, client request chunks that is appropriate for the changed bandwidth. It may be possible that the bandwidth change is transient, and to avoid excessive *SWITCHING_UP* or *SWITCHING_DOWN*, client could wait for certain time before performing streaming switching.
9. Client choose chunk x of bit that it decides appropriate for the network conditions.
10. Video server responds with video data for chunk x.

There are three variations of ABR available today (a) Dynamic Streaming over HTTP from Adobe (b) Smooth Streaming from Microsoft, (c) HTTP Live Streaming from Apple [10]. Even though all these ABR streaming are conceptually similar, they differ in their semantics and hence makes them incompatible [11, 12]. MPEG standard organization has specified Dynamic Adaptive Streaming over HTTP (DASH) as standard to ensure interoperability and compatibility among the ABR technologies [13, 14, and 15].

4 Video Optimization

When same video content is delivered to all mobile users inside a mobile network user experience will be poor. Because, each mobile user experience different bandwidth variation from network, and there is heterogeneity in devices and operating system. To overcome this, mobile network video optimizer (VO) is deployed in the network. Function of VO is to transform one or more video properties to adapt to network conditions and device capabilities without noticeable degradation to QoE. Optimizer is a middle-box residing transparently between user terminal and video server. VO provides two functions namely media aware optimization such as transcoding, and network adaption based optimization, such as for instance traffic pacing, smart caching, and rate control [16]. This section illustrates how these techniques is applied in mobile networks to deliver video streaming in an effective manner while maintaining good QoE.

4.1 Video Transcorder

Video transcoding is an operation that modifies source video streams to target video stream based on transcoding inputs. In order to convert each source video stream, network adaptation function inside VO performs deep packet inspection (DPI) of client request, collects video request type, device preference, learns various network conditions and finally combines all these information to choose a appropriate input as transcoding parameters for transcoding. Transcoders are broadly classified as homogeneous and heterogeneous transcoder [17]. In homogeneous transcoders, transcoding happens without changing to video format, whereas in heterogeneous transcoders performs conversion of source to different target video formats. Video transcoding converts from one video stream to another video streams and includes functions such as Format conversion, bit rate conversion, spatial resolution reduction, temporal resolution reduction, and error resilience. To support transcoding functions, three types of transcoders architectures exist, namely cascaded-pixel transcoder, open-loop transcoder and closed loop transcoder.

- *Cascaded-pixel transoder*: In cascased-pixel transcoder, source video stream is fully decoded first, then encoded again to target video format. The Quality of the target video stream that is achieved by this transcoding operation is high and user is not be able to notice degradation. The operation of decoder-encoder is computationally intensive and not suitable for real-time streaming application.
- *Open-loop transcoder*: To overcome the computation complexity of cascaded-pixel transcoder, open-loop transcoder is used. In open-loop transcoder, on the fly information from the source video such as variable length is decoded to extract the variable length code words referring quantized DCR coefficients, along with macro blocks. Then it is translated to target video bit stream by remapping video motion vectors, requantizing to target bit rate, and applying variable length encoding. Remapping process involves predictive coding technique wherein the coded frame is predicted from previous frames and only the prediction error is coded, and the prediction error accumulates during the entire GOP period. When decoding is done at the client, it results in drift errors. This is the drawback of this approach.
- *Closed-loop transcoder*: It addresses drift error problem by taking approximation of cascaded-pixel transcoder and feeding it as feedback to compensate for initial prediction errors. Close loop transcoder provides balance between quality and computational load compared to cascaded-pixel domain and open-loop transcoders.

4.1.1 Non-adaptive Transcoder

Transcoder consists of set of algorithm and techniques that gets applied to video streams. Transcoding parameters are applied only once in the beginning of the video streams, and are not possible to change on the fly. Those transcoders are called Non-Adaptive transcoder.

4.1.2 Rate Adaptive Transcoder

Non-Adaptive transcoder is not suitable for mobile networks, where the available bandwidth keep changing. Therefore, Rate Adaptive transcoder is used, in this architecture transcoder parameter can be changed at any time during the video streaming. Normally Rate adaptive transcoder operation is combined with network aware streaming functions and is shown in Figure 6.

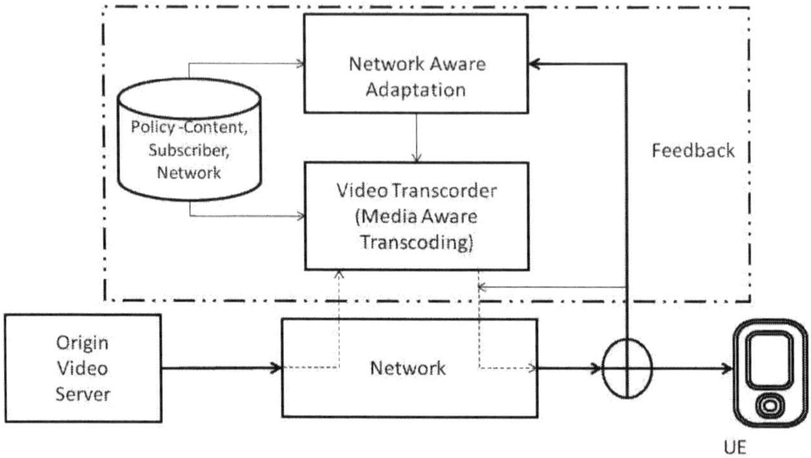

Fig. 6. Media and Network aware optimization

4.2 Network Adaptation

Network adaptation includes functions to monitor core network traffic and radio network traffic, and provides input to following functions:

1. *Transcoder Feedback*: Figure 6 describes feedback from network such as congestion information. With feedback information, rate adaptation transcoder dynamically adapt to changes for a given video streaming session, it needs following information as inputs:
    - Users available bandwidth: It depends on the degree of network congestion and user's subscription.
    - Device type: To decides screen size of the target stream.
    - Video format information either derived from the client request or by doing header inspection of source video stream streams.
    - Content policy: this depends on the agreement between mobile operator and content provider, this policy could include whether the content is allowed to be modified/cached etc.
2. *Media streaming functions*: For on-demand video sessions, even though the transoder is adapting the content, it still requires network level smoothing functions to minimize wastage of data and to ensure that the client do not experience re-buffering. Some of the popular techniques are (1) Pacing of packets are used to avoid excessive buffering of data in the client buffer, to avoid wastage of data. (2) Throttling the available bandwidth or delivering the data in shorter burst based on network conditions, (3) Re-routing the video traffic flows through alternate path when available etc.

VuClip studies show the effectiveness of VO in mobile network [18]. Thirty days of traces from video servers were collected, the traces reveals that there are more than 2000 device model type, tens of different video format resolutions, and large variation in number of supported operating system and native codec. It is observed that mean video duration of 162 seconds, and video traffic distribution follows Zipf like distribution pattern. VuClip performs transcoding depending upon user devices, and their capabilities, and it has been observed that around 73% of requested video content is HD quality which requires optimization.

4.3 Transcaching

To save bandwidth, on certain architecture transcoding and caching operations are performed together [19]. Depending on the design, either source video file or transcoded video file content or both are cached inside the network. When a client request for video stream, the data is fetched from the video server and transcoded to target bit stream on the fly, and at the same time there is an option to store either the source and/or target video bit stream locally to server future request from the clients. The advantage of caching the transcoded bit stream is to avoid repeated computation intensive transcoding operation. But this design approach shifts the problem of compute to storage of video content and we need to have algorithms to intelligently cache only popular content based on content popularity.

4.4 Deployment of Video Optimizers

Effectiveness of VO depends upon where it is positioned. VO can be inside the packet core network, radio network or outside of the operator network.

4.4.1 VO Integrated with GGSN

One possibility is to include VO functions as part of GGSN node itself, so that traffic that is flowing inside the wireless access network is optimized for both Transport backhaul network and Radio access network. As VO is more computationally intensive operations, it will be expensive to have integrated GGSN and VO as one unit. Alternatively architecture is to have cluster of VO nodes collocated with GGSN. The advantage of such separate nodes gives the advantage for scaling of both GGSN and VO independently. As discussed earlier, that network adaption must include both radio network and packet core network. When VO operations are performed at packet core, there must be an explicit notification to be made from RAN network about radio congestion to VO. 3GPP standards do not provide sufficient congestion notification for each IP flow towards packet core network from RAN network. Absence of such information will not make smooth VO adaptation and may result in re-buffering of video streams.

4.4.2 Application Aware RAN

One of the essential goals of Application aware RAN (ARAN) or Cloud RAN is to allow applications to work in radio network friendly manner. ARAN incorporates features to improve radio scheduler based on application requirements. When VO is performed in packet core or in internet it requires the knowledge of UE radio conditions to be made available for smooth streaming. ARAN provides following information (a) cell level congestion, (b) UE available throughput based on signal strength, (c) available throughput based on UE mobility.

ARAN provides information about each UE radio condition, and gives throughput guidance at regular interval to GGSN. To achieve this operation, ARAN must (a) parse each video HTTP request, (b) identifies video flows from HTTP flows (c) perform calculation and then summarized the collected data to GGSN via GTP tunnels. VO then uses these information and performs adaptation of video stream content on the fly.

Alternatively it is possible to host VO services as part of ARAN itself. Studies have shown that only 20% of cells contribute towards 80% of traffic volume [20]. In LTE ARAN architecture servers are part of RAN. When the traffic density is concentrated only to fewer base stations, and backhaul traffic is not constrained, proper dimensioning of users and usage could benefit both network operator and user. As of today this is a newer concept, and we need to see how content provider, operators and network vendors will get maximum benefit.

4.5 Concern Regarding VO Operations

Even though VO has been proved to be effective to deliver video streaming for various devices under dynamic network conditions, VO operations provide following concerns

- VO when deployed and operated by operator without involving content owner breaks net neutrality
- Video streaming servers perform Advertisement insertion, and caching of content inside of mobile network creates problems. It also breaks analytics and reporting operations
- When VO is performed transparently, and UE receives downgraded quality after transcoding, and if received quality is not acceptable to user, user will get the impression that video service provided by video service provider is poor rather than the middlebox that the operators have deployed.
- Video transcoding works well for plain content, if the content is encrypted then transcoder operation will not be performed.

To overcome some of the concerns of VO, operator hosted CDN based approaches are explored.

5 Mobile Content Delivery Network

Due to increase in traffic volume, it has become increasing important to consider CDN inside mobile network. The main reason for such solution is that, 70 msec is the latency between client and server in mobile networks and this is more than twice of fixed networks, and also studies have shown that geographical location dominates latency and throughput [21],[22].

5.1 CDN Enabled Mobile Packet Core

In this architecture CDN network is collocated with GGSN or GGSN is considered as part of CDN networks as shown in Figure 7(A). The advantage of this solution is that data can be pre-fetched before user makes a video request on to CDN. How data is pre-fetched involves combination of many factors including predictive analysis and machine learning technique. For example parameters that could be included are user preference, network load, location of user, video recommendation system etc. The effectiveness of such algorithms are still a question and hard to compare. However, from CDN perspective that the content could be pre-fetched because of higher bandwidth between Mobile packet core and CDN networks. The ingestion operation such as transcoding to different codec formats suitable for mobile devices; video data treatment for DASH including chunk/segmentation can be done at the video source, and distributed to this CDN.

CDN at Mobile core does not give substantial gain because radio network and transport network inside the mobile network are bandwidth constraint. When concurrent access is made to popular content, the content has to be delivered from the core network and this may result is congestion in transport network.

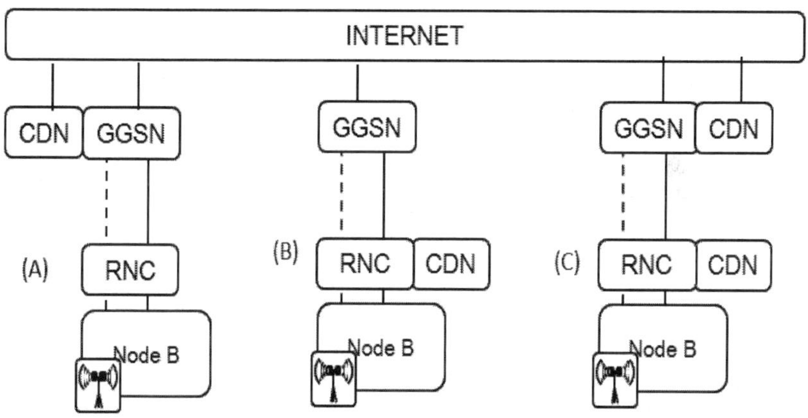

Fig. 7. Various CDN deployment inside Mobile Networks

5.2 Radio Access Networks CDN

When intelligence such as application aware processing are needed to be part of RAN for improving QoE of users, then ARAN or CRAN must be able to perform DPI for applicationflows. When CDN is collocated as part of ARAN and content being served from ARAN then transport backhaul congestion is reduced substantially. In these architecture both storage and processing capacity is limited therefore careful engineering principles must be adopted. Dimensiiong RAN CDN has couple of challenges:

1. Video data is huge and require storage capacity, and may be difficult to deploy in remote BTS sites where energy is not sufficiently available. Netflix one of the popular premium service provider prepares video in different format and then the content is made available for streaming [20, 23]. Content has to be prepared for different display type, screen resolution, different video bit rate, etc. Each of them increases the number of different files for each video. Few such combinations of video file itself will fillup the storage capacity in ARAN.
2. For small cell networks that are serving campus networks doing caching or CDN inside that network may not give cost benefit.
3. When network is having lot of transit users, then caching or providing CDN for those cells may not give higher benefit. Therefore when selecting RAN CDN it must be taken into account about the coverage, number of users accessing through the network, and their usage pattern.
4. In RAN CDN the content is made available locally to improve QoE, when a user moves from a network (say *H*) to another network (say *V*) and if the same content is not available in RAN network *V*, then those content must be fetched from internet. This could bring QoE difference to users but if ARAN concept is proven to be useful both for operators, content owners and users then eventually all RAN will be upgraded and this problem will dimish over time.

As of today concept of RAN CDN is still a research and experimental topic we need to wait for deployment and development of services.

5.3 Hierarchical Mobile Network CDN

Hierarchical CDN uses the strength of RAN CDN, and Core Network CDN and overcomes weakness of both architectures. It operates by combining both packet core CDN and RAN CDN. The CDN is organized in two hierarchical form namely 1st and 2nd level. Ist level CDN are part of RAN and 2nd level CDN are part of packet core. The operation of this network is simpler when user request a video content it is served by 1st level CDN when available. If the content is not present in 1st level then it is served from 2nd level CDN. Traces from mobile network reveals that (a) it not needed to deploy 1st level CDN at all eNodeBs, (b) It is not required to convert and make the different format files to be available at each eNodeBs. By doing predictive analysis, storage requirement can be reduced, and also availability of 1st level CDN is restricted to few eNodeBs. Intelligent protocols between 1st level and 2nd level will minimize the number of request and will minimize the transport backhaul network requirement.

Netflix encourages operators to deploy the CDN's based on their suitable architecture, so as to minimize the traffic crossing the operator network [24]. [25] have studied hierarchical CDN, and via simulation it was observed that there is 24% increase in cache hit ratio, and 45% savings in transport network capacity.

Both in RAN CDN and HCDN understanding of user behavior, and access pattern is critical to prove the effectiveness of CDN. Studies [26] reveals that traces collected from fixed network shows that distribution and long tail follows Pareto distribution. In mobile networks, placement of caching inside the mobile network is an important factor. As discussed hierarchical CDN and caching scheme will bring down the latency, but caching at small cell radio network may not benefit due to geographical coverage hence limited to fewer users. Nevertheless, power law distribution and video skewers are key findings and this helps in storing more popular content intelligently.

6 Impacts on Radio Resource Control and Energy

So far, we discussed placement of video servers inside the mobile network, and adaption of video content to stream the video data to improve users QoE. Another important aspect of mobile streaming is to consider how radio network signaling is performed and energy utilization of user devices. Already 3G and 4G networks has gone through several optimization radio network protocols to efficiently use radio resource and to increase UE battery lifetime. When an application that is not designed for mobile networks uses the radio resources inefficiently, battery drains quickly. Energy consumed by UE, and Radio state machine have direct relation for video streaming. In this section we will describe radio resource and UE energy impacts for video session.

Radio Resource Control (RRC) protocol is used to allocate and release radio resource for each UE. RRC protocol has internal states and is maintained both at UE and at RAN. Figure 8 describes 3G UMTS protocol state machine. In 4G only two states namely connected and disconnected state.

- *Idle mode:* When UE is ON, it will be in IDLE state and no radio resource is allocated to UE. UE consumes minimal energy in this state.
- *CELL_DCH (Dedicated Channel):* UE is in established state, and a dedicated channel is allocated by network that UE can use it exclusively for transferring uplink and downlink data. As radio resources are dedicated in this state, UE consumes maximum power even when no data is transferred with UE. Radio network has finite number of dedicated channel, when more UEs request for video streaming, it will exhaust radio resource quickly.
- *CELL_FACH (Forward Access Channel):* This is a shared channel, UE has established connection with the network, and network has allocated shared channel resource, and UE is in FACH state. Shared channel is used to transmit low volume of data say less than 20 Kbit/sec. This type of channel is ideal for application that require low data throughput rate. In this state, UE consumes half the power of compared to that of DCH.
- *CELL_PCH (Paging Channel):* This optional state offers the lowest current consumption of around 1-2 percent of the consumption in CELL_DCH state, because TX is off and RX is in stand-by. If there are downlink packets for the terminal, the terminal will be paged. In this state, the terminal is not able to send or receive packets, but the terminal will have to enter either the CELL_DCH or CELL_FACH state to send or receive.

State transitions are based on configured timers and amount of data being exchanged in each state. UE will be in FACH state if it has less data to exchange with the network, if more data needs to be transferred then UE moves to DCH state. To be on the same state UE must keep maintaining the data rate, when there is no data to send or receive, inactivity timer starts and upon expiry of inactivity timer UE moves back to low power state. For example, UE could start in IDLE, then it moves to FACH and then to DCH. After being in DCH, and no data is exchanged, when inactivity timer expires it moves to FACH or PCH state.

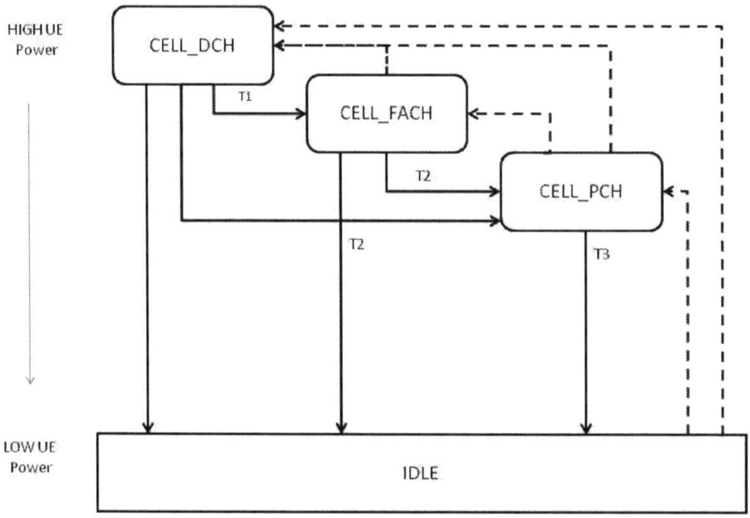

Fig. 8. UMTS Radio State Transition

Two factors that affect the Radio network protocol design are amount of signaling messages needed to move UE from one state to another and the amount of time UE spends on each state.

6.1 Fast Resumption

As per 3GPP specification for 3G PS call setup from IDLE state to operational state, it takes 30 RRC protocol messages [27- 28]. These messages exchange introduce latency for applications that needs to send or receive data quickly. Therefore, for video streaming it is not recommended to go to IDLE state after period of inactivity. Alternatively, UE can be made to park in PCH state for longer period of time upto 30 minutes. (a) when UE is moved from PCH->DCH state it takes around 12 messages, and (b) when UE is moved from PCH->FACH transition it takes only 3 messages. Therefore usage of PCH is recommended for always ON type application, and also more suitable for real-time ABR streaming. In ABR, data download happens in small chunks and no packet flows between two chunks. Therefore, UE instead of moving to IDLE state after inactive timer expiry, it moves to PCH state.

6.2 Synchronization of Radio States and Video Data Download

Figure 8 shows RRC states and possible state transition in 3G UMTS networks. There are several possible ways operator could configure the RRC state machine to transition. Decision to moving from low to high power state depends on the traffic usage. Following are possible ways that operator configure their network

(a) UE that is in IDLE state can make a transition from IDLE->FACH, then from FACH->DCH, DCH->FACH and FACH->PCH.

(b) Alternatively, operator could configure the transition from {IDLE ->FACH or IDLE->DCH], {FACH->PCH or DCH->PCH}.

Video streaming is longer duration session, and downlink traffic is more than uplink traffic. Following are the possible impact on radio state machine for different video streaming strategy.

- For pseudo streaming, when available network bandwidth is greater than required bit rate, then client makes single http request to download entire video data quickly at available network bit rate. Amount of data that is downloaded is large therefore most of the time UE will be DCH state. When UE is in DCH state it consume maximum power. As network is used for short period of time to download all content this mechanism may be power and radio efficient. But when user terminates video session prematurely data is wasted. It is difficult to achieve balance between radio state and data wastage.
- For YouTube like pseudo streaming as shown in Figure 3 and Figure 4. Even though network bandwidth is more than the required video data rate, but requires fixed bandwidth and uses TCP pacing techniques. The advantage of this mechanism is that, same amount of data is now spread out in time to avoid network congestion,. the side-effect is that UE needs to spend longer time in DCH and hence consume more UE energy. Also by having fixed TCP pacing during throttling phase as shown in Figure 3 it is not allowing UE to go to sleep. Therefore additional resource management steps needs to be in place to synchronize such activities. To achieve saving, UE can implement Discontinous Transmission (DTX) or Reception mechanism(DRX). DTX and DRX are time during which the UE transceiver will be switched OFF to save UE power. If the video server is performing TCP pacing and radio network is not aware of such pacing time, it not going to improve the RRC protocol efficiency.
- When ABR is being used, the video data is downloaded in chunks. There will be sufficient time gap between two consequentive chunks, and the gap between these chunks are not synchronized with radio state machine timers. For example, consider two consequtive chunks (*c1* and *c2*).
    - When content is adapted, and chunk *c1* is downloaded quickly, then UE will wait for *c1* to be downloaded first before start downloading next chunk *c2*. If this wait time is longer than the radio inactivity timer, then state machine will transition from DCH->PCH, and allocated resource for the UE will be used for other UEs.

  - If wait time to download chunk *c2* is less than inactivity timer, then radio state will remain in DCH, and there will be sufficient silence period between chunks download. When DTX and DRX timers are configured then those will save power. But time varying bandwidth will make both DTX and DRX timer statically configured for a network. We therefore need dynamic configuration to maximize DTX and DRX timers operation to save UE power.
- VO operations that we have discussed previously must consider UE power, and RRC timer into consideration for streaming the data. If there is no sychornization between Radio state, UE power and VO operation. It will waste UE energy and poor utilization of radio channels.

7 Further Research

There are many open problems yet to be addressed. In this section, a few problems are highlighted as future research studies.

- *Device-to-Device Communication:* Direct Device-to-Device (D2D), and Wi-Fi direct are new forms of communication methods being worked out in standards [29, 30]. Recently these types of communication approaches are getting more attraction in mobile environment to allow community and proximity based networking. D2D communication is being worked out between two or more mobile nodes with or without network assistance over licensed spectrum. When licensed spectrum are being allocated for D2D communication, then operators needs to ensure proper resources are being allocated for proper functioning of D2D based video streaming services. It is a new concept and touches on many areas including Peer-to-Peer streaming, cognitive networks.
- *Contextual analysis and streaming:* Factors that influence streaming include user preference, user devices configuration, available energy, user location, service subscription, recommendation from video recommender system etc. How to effectively use these factors to deliver mobile video streaming service has been explored as a research topic. To improve streaming service by using these factors involve machine learning approach and do predictive analysis. General principle behind machine learning is to pick up a known model and make predication, and let the model adapt as it learns from the data set. After initial learning phase, machine learning algorithm can give a good estimate to guide VO and CDN .
- *Streaming over Heterogeneous Networks:* Cellular networks gets congested while serving more users behind a cell, and offloading users to small cell such as Wi-Fi hotspots, and Femto cell are potential techniques. How such techniques impacts video streaming needs further study [31]. At this moment, the standards are being developed in parallel for Wi-Fi and LTE small cells, and the goal is to ensure seamless mobility for applications.

Acknowledgments. This work is supported by Nokia Siemens Networks. Opinions, findings or recommendations expressed in this paper are from the authors and do not reflect the views of Nokia Siemens Networks.

References

1. Cisco Visual Index, http://blogs.cisco.com/sp/cisco_visual_networking_index_forecast_annual_update/ (last accessed June 1, 2013)
2. YouTube, http://www.youtube.com
3. Netflix, http://www.netflix.com
4. ITU-T Recommendation P.910, SubjectiveVideo Quality Assessment Methods for Multimedia Applications. International Telecommunication Union, Geneva, Switzerland (1999)
5. Schulzrinne, H., Casner, S., Frederick, R., Jacobson, V.: RTP: A Transport Protocol for Real-Time Applications, RFC 3550, IETF (July 2003)
6. Huitema, C.: Real Time Control Protocol (RTCP) attribute in Session Description Protocol (SDP), RFC 3650, IETF (October 2003)
7. Wireshark, http://www.wireshark.org
8. Haddad, M., Altman, E., El-Azouzi, R., Jiménez, T., Elayoubi, S., Jamaa, S., Legout, A., Rao, A.: A Survey on YouTube. Streaming Service. In: VALUETOOLS (May 2011)
9. Ghobadi, M., Cheng, Y., Jain, A., Mathis, M.: Trickle: Rate Limiting YouTube Video Streaming. In: Usenix Annual Technical Conference (2012)
10. Akhshabi, S., Begen, A., Dovrolis, C.: An Experimental Evaluation of Rate-Adaptation Algorithms in Adaptive Streaming over HTTP. ACM Multimedia Systems, 157–168 (February 2011)
11. Riiser, H., Bergsaker, H.S., Vigmostad, P., Halvorsen, P., Griwodz, C.: A Comparison of Quality Scheduling in Commercial Adaptive HTTP Streaming Solutions on a 3G Network. In: Proceedings of the 4th Workshop on Mobile Video, pp. 25–30 (2012)
12. Muller, C., Lederer, S., Timmerer, C.: An evaluation of dynamic adaptive streaming over http in vehicular environments. In: Proceedings of the 4th Workshop on Mobile Video, pp. 37–42 (2012)
13. Stockhammer. T., Dynamic adaptive streaming over HTTP: Standards and design principles. In: Proc. of ACM MMSys, pp. 133–144 (February 2011)
14. 3GPP TS 26.234: Transparent end-to-end packet switched streaming service (PSS); Protocols and codecs, release 12
15. 3GPP TS 26.244: Transparent end-to-end packet switched streaming service (PSS); 3GPP file format (3GP)
16. Miller, R.B., Hearn, J.P., Purzynski, C., Cuervo, F., Scheutzow, M.: Mobile Video Delivery Using Network Aware Transcoding in an LTE Network. Bell Labs Technical Journal 16(4), 43–61 (2012)
17. Ishfaq, A., Wei, X., Sun, Y., Zhang, Y.: Video transcoding: an overview of various techniques and research issues. IEEE Transactions on Multimedia 7(5), 793–804 (2005)
18. Liu, Y., Li, F., Guo, L., Shen, B., Chen, S.: A Server's Perspective of Internet Streaming Delivery to Mobile Devices. In: Proc. IEEE Conference on Computer Communications INFOCOM (March 2012)

19. Maheshwari, A., Sharma, A., Ramamritham, K., Shenoy, P.: TranSquid: transcoding and caching proxy for heterogenous e-commerce environments. In: Proceedings of the Twelfth International Workshop on Research Issues in Data Engineering: Engineering E-Commerce/E-Business Systems, pp. 50–59 (2002)
20. Informa Report, Understanding Today's Smatphone User - Part I and Part II, http://www.informatandm.com/wp-content/uploads/2012/08/Mobidia_final.pdf (last accessed June 1, 2013)
21. Bahl, P., Han, R.Y., Li, L.E., Satyanarayanan, M.: Advancing the state of mobile cloud computing. In: Proceedings of the Third ACM Workshop on Mobile Cloud Computing and Services, MCS 2012, New York, USA (June 2012)
22. Leighton, T.: Improving performance on the internet. Communications of the ACM 52(2), 44–51 (2009)
23. Netflix Encoding types, http://vimeo.com/52637219 (last accessed June 1, 2013)
24. Temkin, D.: Netflix open connect. NANOG 55, CDN Panel (June 2012), http://www.nanog.org/meetings/nanog55/presentations/Tuesday/Temkin.pdf (last accessed June 2013)
25. Ahlehagh, H., Dey, S.: Video Aware Scheduling and Caching in the Radio Access Network. In: International Conference on Communication, pp. 7082–7087 (June 2012)
26. Cha, M., Kwak, H., Rodriguez, P., Ahn, Y.Y., Moon, S.: I Tube, You Tube, Everybody Tubes: Analyzing The Worlds Largest User Generated Content Video System. In: Proc. of ACM IMC (2007)
27. 3GPP TR 25.813: Radio interface protocol aspects (V7.1.0) (2006)
28. 3GPP TS 36.331: Radio Resource Control (RRC) (V10.3.0) (2011)
29. 3GPP TS 22.468 v0.3.0, Group Communication System Enablers for LTE (November 2012), http://www.3gpp.org/ftp/Specs/html-info/22468.htm
30. 3GPP TR 22.803, Feasibility study for Proximity Services (ProSe) (Release 12), http://www.3gpp.org/ftp/Specs/archive/22_series/22.803/
31. 3GPP TR 23.829, Local IP Access and Selected IP Traffic Offload, http://www.3gpp.org/ftp/Specs/html-info/23829.htm (last accessed June 1, 2013)

Using Traffic Diversities for Scheduling Wireless Interfaces for Energy Harvesting in Wireless Devices

Constandinos X. Mavromoustakis[1], Christos Dimitriou[1], George Mastorakis[2], Athina Bourdena[3], and Evangelos Pallis[3]

[1] University of Nicosia, 46 Makedonitissas Avenue, 1700 Nicosia, Cyprus
[2] Technological Educational Institute of Crete, Department of Business Administration, Lakonia, Agios Nikolaos, 72100, Crete, Greece
[3] Technological Educational Institute of Crete, Department of Informatics Engineering, Estavromenos, Heraklion, 71500, Crete, Greece
{mavromoustakis.c,dimitriou.cd}@unic.ac.cy, gmastorakis@staff.teicrete.gr, {pallis,bourdena}@pasiphae.eu

Abstract. The data traffic in many cases is specifying the expectations and the encouragement of new Mobile services/M-services (Location-based etc) and embosoms, through the user-centered awareness, to expose an innovative range of on-the-move applications. As today two distinct domains exist, the wireless world and the Internet world where, both can be met over a traffic-oriented framework providing reliable end-to-end users' connectivity and exchange of resources. The need to allocate and balance resources among different traffic classes to accomplish the best usage of network resources while maintaining the topology and the wireless connectivity of the users is today even more timely. The user's movements affect the type of connectivity, thus aggravating the degree of cooperation among users and degrading the reliability of communication. Traffic diversities are being considered in this chapter taking into consideration the traffic impact on the energy conservation of the nodes that are changing their location according to certain pattern as well as the consideration of the traffic as a feedback mechanism to prolong network's lifetime and nodes lifetime and communication duration extensibility. The chapter covers the primary traffic techniques and methodologies in order to show the direct dependencies between traffic and wireless interfaces' scheduling mechanisms as well as exposing the power-related parameters during the resource exchange process in order to enable the wireless communicating nodes to efficiently utilize their energy resources. Different variations of the proposed schemes are presented where the energy benefit is specified. The performance evaluations through conducted experiments were performed in real-time, through wireless sensor nodes, and through simulation presenting the effectiveness of the framework which efficiently maximizes the reliability of the resource exchange process of the nodes, while it minimizes the energy consumption and prolongs the system's lifetime.

Keywords: Energy Conservation; Scheduling management; Energy level self-control; Layered-based Energy Conservation; State-based Energy conservation; Mobile Peer to Peer Networks energy management scheme; High Resource Availability; Traffic management and composition; One-level Backward Traffic Difference scheme; Selective Two-level Backward Traffic Difference scheme; Traffic-oriented Energy Conservation; Traffic Volume and Capacity metrics.

C.X. Mavromoustakis et al., *Resource Management in Mobile Computing Environments*, Modeling and Optimization in Science and Technologies 3,
DOI: 10.1007/978-3-319-06704-9_22, © Springer International Publishing Switzerland 2014

1 Introduction

In many cases where traffic flow can be realised and understood in order to show the impact on various parameterised features of wireless devices, the reliability increases and similarly the fault significantly degrades. When the mechanisms that are associated with large scale network management involves several parameterised aspects for study like traffic engineering, resources' planning and provisioning, fault management and security, these parameters should be precisely elaborated through studies in order to improve the performance and the efficiency of the network .Considering the manipulation of the traffic in terms of the nodal traversal (incoming) traffic and the overall aggregated traffic in a subsystem [1] could improve the quality of service provision [2]. Moreover, there where the traffic can be bounded in a certain delay in order to be transmitted to any other host, the study of the behaviour of the traversal nodal traffic can be considered as a significant parameter for improving the quality and the reliability of the transmission. The traffic that traverses the wireless devices can be exploited targeting the minimization of the data delay trade-offs, and mitigate the energy consumption of each device. Each node should enable an energy conservation (EC) mechanism which could affect only the node according to the activity of the traffic that traverses the node [3]. This node-centric approach can be effective by considering the traversed nodal traffic and through a model that takes into consideration the volume of the traffic, it can then provide a feedback to prolong the nodal and/or network lifetime. This chapter describes some quantitative traffic-based approaches where energy-maximization mechanism takes place in order to enable each node to sleep according to the time duration of the active traffic. This traffic is based on the active nodal traffic that each node expects and experiences.

Different traffic tuning schemes have been proposed in the recent past addressing the improvements in terms of energy by traffic manipulation methodologies. Authors in [4] enable Sleep-proxies to regulate the activity of each node and use a cross-layered architecture to mirror any traffic changes onto different layers. However these schemes can be hosted by nodes which are using no delay characteristics in their transmissions. The approach is entirely different and the foundations can be disastrous, if the transmissions are considered as delay sensitive and are exposed to a bounded end-to-end delay -time duration. In the following sections this chapter will present the recent foundations in terms of optimizing the activity periods of the wireless devices by reducing or increasing the slots –where needed- in order to enable an improvement in the performance and the reliability of the transmitted data. Moreover the following sections present the approaches exposed by different researches using discrete mathematical methodologies like Backward Traffic Difference –aware techniques, in order to tune to an optimize duration each node to sleep and therefore provide the node with an energy-aware feedback to conserve energy.

1.1 Using Traffic-Based Schemes to Tune the Wireless Interfaces of the Nodes

Wireless communication traffic is expected to grow in the next 2 years reaching the 75% of the overall internet traffic. Therefore the consideration of schemes and efficient mechanisms that will ensure data availability is of vital importance. Moreover, since the mobile devices have limited lifetime in action, the energy consumption of these devices is becoming timely as the majority of application hosted onto wireless devices are dealing with resource-sharing and data exchanging activities. As these applications are rapidly gaining ground the traffic assessment and its impact onto the energy efficiency is considered of primary importance for enabling end-to-end wireless peer-communicating devices. In this regard, the underlying traffic-aware mechanisms should ensure that nodes transmit their information with the minimum energy consumption.

Mobility undermines several common assumptions for these devices: the intermittent connectivity is frequently occurring since users are moving and utilize applications involving physical mobility. In general, the computational context of the traffic is considered as unpredictable and cannot be predetermined. It is continuously changing in largely unpredictable ways that are usually affected by the users' mobility and resource sharing profile [5]. In resource-sharing applications the delay durations are non-guaranteed and cannot be predicted due to intermittent connectivity technical aspects of the nodes whereas, at the same time nodes participating in the resource sharing processes are consuming more energy [6]. This is due to the dramatic growth in the activity duration of the wireless nodes. Usually nodes that are hosting delay sensitive applications are experiencing more intermittent connections while on the move, causing in essence the nodes to be "energy-hungry" as more energy is needed for requesting resources and this result in collapsing. Notwithstanding, the mitigation of the resource manipulation mechanisms with the energy conservation schemes should be collaboratively balanced when utilizing delay sensitive applications. The resource exchange mechanism and the traffic-aware scheme should encompass the manipulation of resource sharing process in a middleware reflective manner, which will enable a direct impact on EC of each device. As in [4] this reflective scheme has to take into consideration the continuity of the incoming traffic, the possible intermittent diversities in the end-to-end connectivity and has to assess whether the communication (data and control packets) among peers within a period of time provides the expected energy conservation for the wireless devices.

1.2 Traffic-Aware Asynchronous Scheduling Scheme for Energy-Aware QoS Provision in Asymmetrical Wireless Devices

Many recent researches have shown that when considering delay sensitive streaming, satisfying some of the basic QoS metrics, the minimum transmission power that is required to achieve these transmissions in an end-to-end manner, reaches the optimal throughput performance. It can be extracted that the power control in wireless devices aims to determine the optimal transmit periods and the associated power level such that the energy consumed is steadily reduced. At the same time it aims to guarantee that the resource sharing process can be smoothly completed. Since nodes in wireless networks typically rely on their battery energy, the work done in [3] [4], propose mechanisms which host two variations of a traffic-aware scheme aim to reduce the energy consumption of wireless devices. The integrated middleware which merges the schemes in [3] and [4] is shown in figure 1. Figure 1 encapsulates the traffic schemes in a middleware and encompasses a feedback mechanism which evaluates the scheduled activity periods of each node. This mechanism is taking place in order to measure the scheduled time that each node can move into the sleep state in order to conserve energy. The framework adaptively enables the node which admits traffic to consider this traffic and uses a quantitative model based on the supervision of the Sleep-Proxy (a role-based node) which takes into consideration the cooperative caching and the resource exchange scheme for enabling Energy Conservation in the Cluster.

When delay-sensitive or -as called- streaming traffic is traversing mobile devices, the end-to-end path should be selected. However this path is not always the best effort in terms of energy. The streaming data correspond to audio or video applications' data and requires low packet loss and delay/jitter, whereas the elastic traffic is considered to be a "don't care" packet. This traffic usually corresponds to document transfers and have no deadlines in reaching the destination node. An efficient energy conservation scheme should be aware of the traffic pattern in order to enable higher QoS achievements by considering the class of the data that is being hosted, as well as by assessing the nodes that are needed to host these transmissions in the path. Therefore the associated QoS parameters have to be maintained below a given threshold and the packet loss probabilities should be minimized particularly in cases when users' mobility maximizes such likelihood [7]. Thus the energy conservation mechanism has to be closely collaborative with the incoming and outgoing traffic from/to a node, and with the routing protocol behavior used by nodes. Figure 1 shows the close dependencies regarding the delay sensitive transmissions hosted and being traversed by a mobile node.

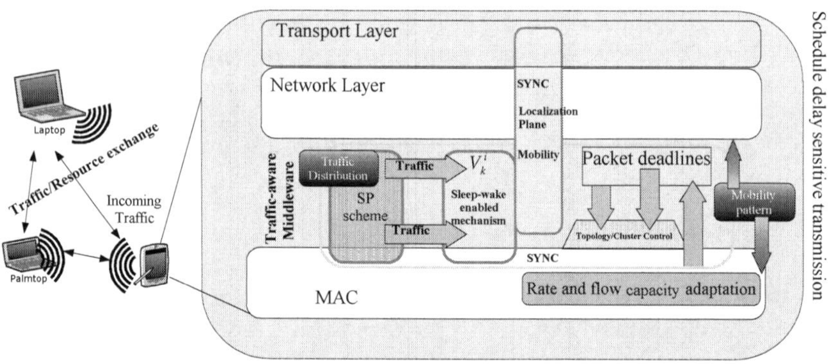

Fig. 1. The initial network formation in a peer-to-peer system according to node self-awareness

In Figure 1 the middleware which hosts the volume and the traffic changes and impacts directly the different layers, is shown. The cross layer interaction is performed via the traffic-awareness which is then passed to the layers that are manipulating the transmission characteristics in the end-to-end path. The real-time media traffic such as voice and video typically have high data rate requirements and stringent delay constraints for which the proposed middleware considers taking into account the deadlines of the packets. As wireless nodes have limited or momentarily connectivity the proposed traffic-aware reflective middleware enables the data packets to be traversed and manipulated through the utilized Data Link, Network, and Transport layers. These layers are considering the traffic awareness mechanism and the model for volume and capacity estimation to be reflected on these layers. The middleware of Figure 1 works as follows: the traffic-aware mechanism evaluates - after the bootstrap process- the estimated capacity that the incoming traffic is reserving onto a node (usually this traffic is the destination traffic of the node).

The traffic association with the mechanisms for affecting the energy conservation of the each node will be shown in the next section. The buffering capacity and traffic-aware mechanisms are considering the duty cycle schedules of each node so that the referred node can sleep using optimal schedule durations and in such a way to enable energy conservation. This is performed in order to tune the wireless interface of each device to sleep/wake according to the activity of each device. Activity of each device refers to the active slots of each node that is able to admit traffic by other nodes during the resource exchanging process. For this purpose a packet classification methodology was used -as in [8]- in order to mark the packets that are participating in the resource exchanging process as delay sensitive or not. On one hand if the resource exchange process considers the packets as delay sensitive, then strict deadlines are applied by the sender, according to the specifications set in the end-to-end resource exchanging path. On the other hand where no end-to-end delay deadlines are specified, then a recovery mechanism takes place using the promiscuous caching methodology explored in [9]. The promiscuous caching methodology enables the claimed and exchanged resources to be replicated in order to be available by the recipient node when needed. If the recipient node is in the sleep-state then the resource sharing reliability decreases dramatically. For facing the resource sharing degradation promiscuous caching methodology allows cached packets to be sent to the destination node, by buffering the packets to the 1-hop intermediate node. The promiscuous caching methodology is shown in Figure 2, and the quantitative analysis is presented in the next section.

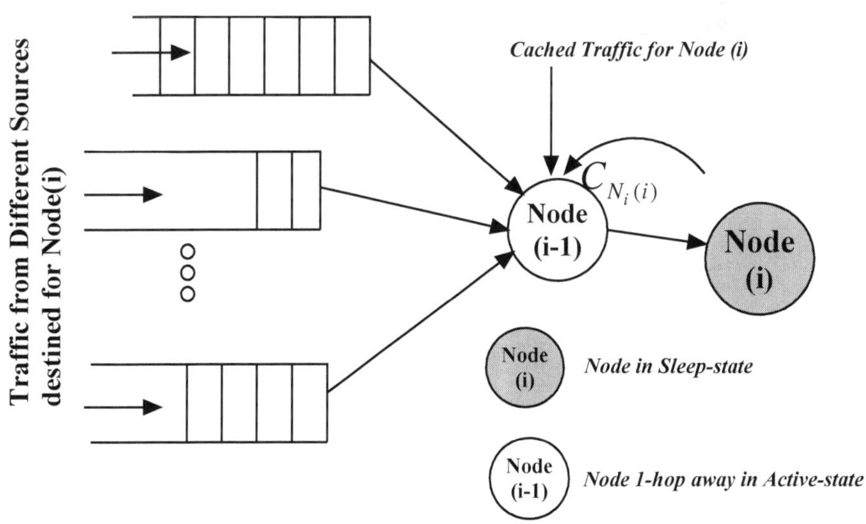

Fig. 2. A schematic diagram of the promiscuous caching mechanism addressed in this work

The activity period(s) of a node is primarily dependent on the nature and the spikes of the incoming traffic destined for this node [4]. If the transmissions are performed on a periodic basis then the nodes' lifetime can be forecasted and according to a model can be predicted and estimated [9]. Each node admits traffic while in the active state, whereas if the node is in the sleep-state it can cache the traffic onto the 1-hop neighbor node *(Node(i-1))* shown in Figure 2. As a showcase this work takes the specifications of the IEEE 802.11x that are recommending the duration of the forwarding mechanism that takes place in a non-power saving mode lays in the interval $1\ nsec<\tau<1psec$. This means that every ~0.125μsec (8 times in a msec) the communication triggering action between nodes may result a problematic end-to-end accuracy. Adaptive Dynamic Caching ([7], [13] and [16]), takes place and enables the packets to be "cached" in the 1-hop neighboring nodes. Correspondingly, if node is no-longer available due to sleep-state in order to conserve energy (in the interval slot $T=0.125\mu sec$), then the packets are cached into an intermediate node with adequate capacity equals to: $C_{t_f,k(s)}(t)>C_{t_f,i}(t)$, where $C_{t_f}>\alpha\cdot C_i$; where α_i is the capacity adaptation degree based on the time duration of the capacity that is reserved on node N of C_k; where $C_{t_f,k(s)}(t)$ is the needed capacity where i is the destination node and k is the buffering node (a hop before the destination via different paths).

By using the above estimations a new approach for an end-to-end energy-aware service provision is presented using a traffic-based layered asynchronous scheduling scheme. This new technique takes into account the asymmetrical spatiotemporal nature of the wireless devices, as well as capacity metrics that are responsible for optimizing the throughput response for the system. The traffic-aware mechanism mitigates the behavior between storage and bandwidth requirements and allows nodes to exchange resources in a reliable manner.

1.2.1 Traffic-Aware Asynchronous Scheduling Scheme

The main objective is to achieve a reduction in the rate of consumption by enabling an association of the active traffic with the sleep-time duration of each node. In Table 1 the Digital Radio Power States (DRPS) estimations provided by the specifications for some commercial transceivers. The rate of consumption in the "receiving" state could be less 50% of that consumed in "transmitting" state. It is important to appropriately determine when and at what power level a mobile host should attempt transmission or retransmission of packets[1]. The Traffic-aware asynchronous scheduling scheme is based on the traffic traces which show structural likeness for a wide range of time aggregations as in [4]. Node's incoming traffic is associated with the next time slot time duration. This suggests that examining the traffic Fractality principle and behavior, defines a way to mitigate the energy aware slots with the activity periods of each node. Different duration of sleep/wake schedules can be applied instead of the periodic ones in order to enable nodes to conserve energy by minimizing to the least possible and to the best effort communication the activity of their communication

[1] Node's transceiver should be powered off when not in use.

interfaces. Figure 3 shows the dissimilar sleep/wake schedules, based on node's incoming sleep-history traffic and the associated discrete time gain in terms of overall energy.

Table 1. Digital Radio Power States for different commercial transceivers

RADIO INTERFACE IEEE 802.11 Interfaces (2.4GHz)	TRANSMIT	RECEIVE	IDLE	SLEEP MODE	Mbps
Lucent Silver	1.3 W	0.90 W	0.74 W	.048 W	11
WaveLAN Turbo Card	285mA	185mA	185mA	9mA	11
Aironet PC4800	1.4-1.9 W	1.3-1.4 W	1.34 W	0.75W	11

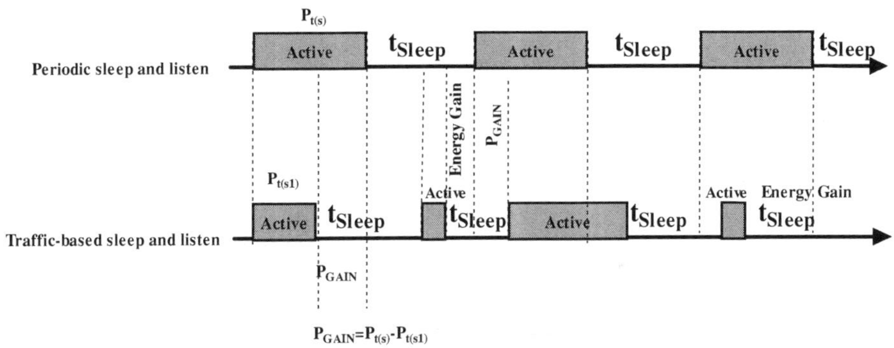

Fig. 3. Dissimilar duration of sleep/wake schedules, based on node's incoming sleep-history traffic and the associated discrete time gain (or power loss) in terms of overall energy

The capacity of the cached data is associated with the time duration $t_i\left(t-t_D\right)$ which corresponds to the next sleep duration of the destination node. Sleeping time for destination node is high enough (according to [4]) when the node during the previous time sleep/inactive slot, that is $t_i\left(t-t_{D(\tau-1)}\right)$, did not receive any packets. Hence it stands that for time τ:

$$T_{C(\tau),N} < T_{C(\tau-1),N} \text{ then } t_{D(\tau+1)} \geq t_{D(\tau)} \text{ for node N, and } \tau = 1,2,3,4...m. \tag{1}$$

where the traffic is $T_{C(t),N}$ where T is the incoming traffic, and $C(t),N$ is the capacity in the certain time interval t for node N. it is of considerable interest to see that if t_D

is high then the next sleep duration of node N will be in turn higher than the previous one (due to inactivity of the node in the $t_{D(\tau-1)}$). In this way each node evaluates dissimilar sleep and active states based entirely on each node's incoming sleep-history traffic (in-sleep history) as shown in Figure 3. By applying the simple superposition theorem, the power gain aggregation is significant. In the case of delay sensitive packets, the sleep time duration can be measured using the following expressions:

$$\Pi_{MM} = (1 - \frac{cached\_MM\_packs}{Total\_MM\_packs\_destined\_for\_D})^{T \cdot N} \tag{2.1}$$

$$S(t_{new})_D = S(t)_{D(\tau-1)} - S(t)_{D(\tau-1)} \cdot \left[\frac{Packs\_cached}{Total\_packets\_destined\_for\_D} \right] \tag{2.2}$$

where Π_{MM} is the streaming factor based on each multimedia stream, T is the time that passed since the beginning of MM packet transmission from source node, and N is the number of hops in the path from source node to destination. $S(t_{new})_D$ stands for the new time duration[2] depending on the traffic history of the route destined for D, and $S(t)_{D(\tau-1)}$ is the previous time step (slot) duration. Then the new sleep time duration for the destination node is evaluated in:

$$S(t_{new})_D = S(t_{new})_{D(\tau-1)} \cdot \Pi_{MM} \tag{3}$$

where Π_{MM} is the streaming factor previously mentioned. The capacities of each node as well as the durations of the delays of the transmitted packets are expected to reach the associated destinations, are *normalized* as in [9]. PROC packets are purely responsible for informing the node for any limitations. If a node does not receive any packets for a long period due to the stochastic nature of incoming traffic, the sleep time will increase at a level which affects system performance. Thus PROC packets by using a specified field PROC_tLIM place a limit on the sleep time of each node to avoid increased latency, which results in network partitioning. Also in order to prevent a large number of nodes to enter into the sleep state, a maximum number of nodes that are allowed to be in the sleep state is set. In this way the traffic aware scheduling scheme avoids the potential network partitioning.

Figure 3 shows the state transition diagram of each node and the transitions from the 'sleep' state to the 'active' and 'idle' states. If a node is set in the active state its interface can transmit or receive data at any time. This causes an increment in the energy consumption. If incoming traffic is continuously traversing a node as a destination or as intermediate), then this node is kept in the active state. The referred closed "loop" is not energy efficient and it is intermitted if the traversal traffic is

[2] New time duration cannot exceed twice the corresponding value.

reduced. In order to face this issue, node changes into listening mode for a specified amount of time (Equations 2-3). At $t_{D(\tau)}$ the sleep time is further reduced, because the incoming traffic for the time duration $t_{D(\tau-1)}$ is negligible. In this way at the time $t_{D(\tau)}$ node has significantly reduced the sleep time duration. The sleep mode of a node is only corrupted if serious latency limitations exist, and/or sleep period expires. Figure 4 shows the sequence diagram regarding the wireless interface scheduling for a single multimedia stream. At the initial state S_0, a node sends route request (r_{REQ} [3]) to a candidate next hop node. In turn the candidate node answers to source node with a determined and specified path, and source node decides whether the route is suitable. Delay sensitive packets have transmission expiration time and a TTL tag. At the S_0 state, a node knows at any time the candidate next hop node N_{i+1} by sending continuously r_{REQ}. Then the source node sends back to N_{i+1} the number of MM packets that are destined for N_{i+d} node (d is the number of hops to the κ destination, $\kappa = i + d$). Indicative simulation experiments are shown in Figures 5 and 6 where the specifications of the WaveLAN PC/Card energy consumption were used found in the study by Feeney and Nilsson [10]. According to [10] these characteristics can be summarized in Table 2, where the currents drawn by the card at 4.74 Volts power supply are measured at 285, 190, 156, and 10 mA for transmit, receive, idle, and sleep modes, respectively.

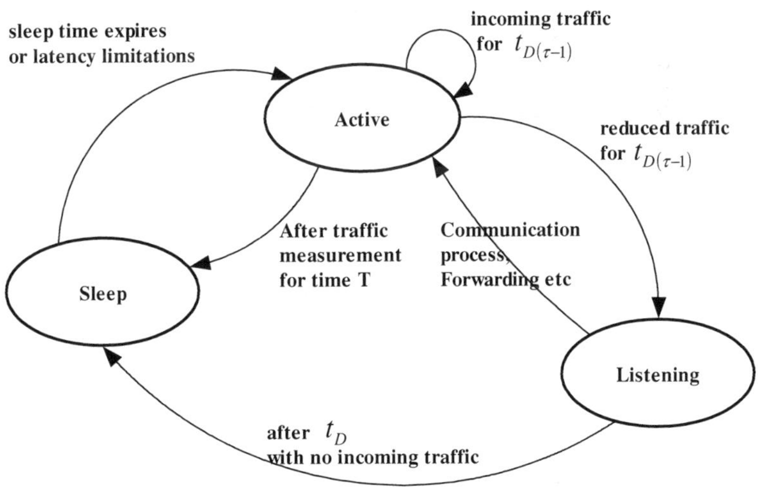

Fig. 4. Node operation cycle as a state transition diagram

[3] Calculates the route request and the end-to-end delay of the requested path.

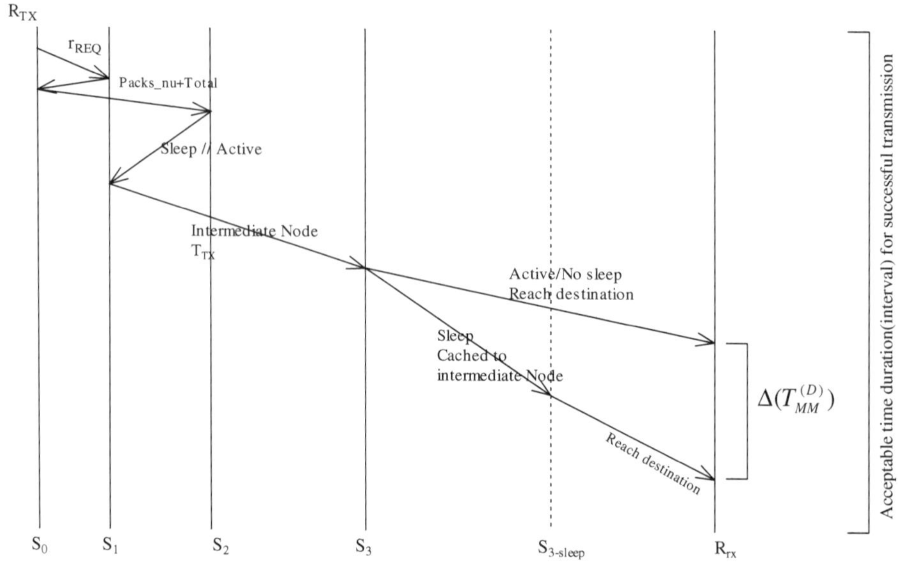

Fig. 5. State-Time (sequence) scheduling diagram for a single multimedia transmission

Table 2. Energy consumption characteristics of WaveLAN PC/Card (average currents)

Mbps	SLEEP MODE	TRANSMIT	RECEIVE	IDLE
11	10mA	285mA	190mA	156mA

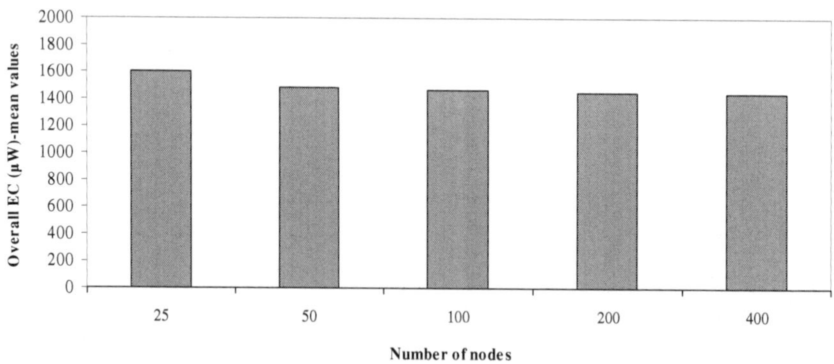

Fig. 6. Network's overall EC (μW) with the number of nodes

Figure 6 shows that by increasing the number of nodes, the network's overall EC in (μW) remains almost the same or fluctuates between a small range of values whereas Figure 7 depicts that this scheme can handle in an efficient manner manage the energy on a 'self-tuned' mode based on the quantitative measures of the buffered (cached) traffic onto intermediate nodes.

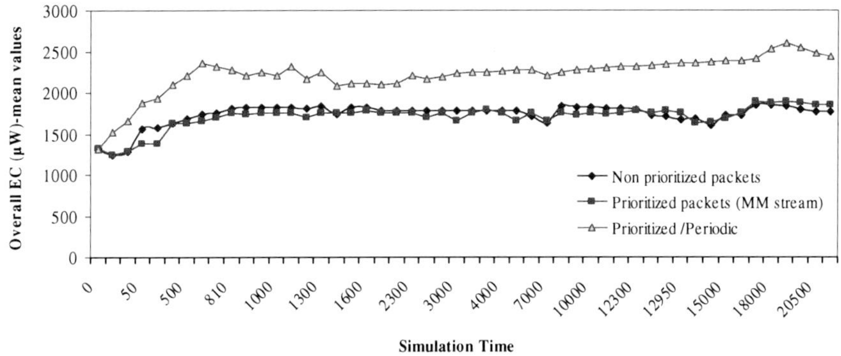

Fig. 7. Mean Energy consumption comparison for periodic and non periodic using MM and non prioritized packets

1.2.2 Using Backward Traffic Difference Asynchronous Scheduling Scheme

Appropriate and efficient mechanisms are required to estimate the optimal activity periods of each one of the nodes. Therefore the traversed traffic can be the parameterized input which, in order to evaluate the next sleep-time or the active periods, it considers the terminal's transmission and reception durations. This can be achieved through the Backward Traffic Difference evaluation. The network lifetime in the presence of symmetric-duration and periodic sleep-wake periods, will be significantly decreased [4] due to frequent and redundant activity periods where no traffic traverses the node. The Backward Traffic Difference (BTD) estimates the traversed traffic of a node and associates these estimations with the previous moments, in order to conserve energy. Each wirelessly communicating node uses different assignment(s) of sleep-wake schedule estimation, based on the traffic difference through time. The traffic difference can be estimated as the difference of the incoming traffic to a certain node for time t, with the difference of the incoming traffic of the same node for the time instant t-1(previous time moment). The traffic can be seen as a renewal process [11] that has aggregation characteristics [12] from different sources.

Let $C(t)$ be the capacity of the traffic that is destined for the Node *i* in the time slot (duration) *t*, and $C_{N_i(t)}$ is the traffic capacity that is cached onto *Node (i-1)* for time *t*. Then, the one-level Backward Difference of the Traffic is evaluated by estimating the difference of the traffic while the *Node(i)* is set in the Sleep-state for a period, as follows:

$$\begin{aligned}
\nabla C_{N_i(1)} &= T_2(\tau) - T_1(\tau - 1) \\
\nabla C_{N_i(2)} &= T_3(\tau - 1) - T_2(\tau - 2) \\
&\vdots \\
\nabla C_{N_i(n+1)} &= T_n(\tau - (n-1)) - T_2(\tau - (n-2))
\end{aligned} \tag{4}$$

where $\nabla C_{N_i(1)}$ denotes the first moment traffic/capacity difference that is destined for *Node(i)* and it is cached onto Node (i-1) for time τ, $T_2(\tau)-T_1(\tau-1)$ is the estimated traffic difference while packets are being cached onto *(i-1)* hop for recoverability. Equation (4) depicts the BTD estimation for one-level comparisons, which means that the moments are only being estimated for one-level ($T_2(\tau)-T_1(\tau-1)$). The Traffic Difference is estimated so that the next Sleep-time duration can be directly affected according to the following:

$$\delta(C(T)) = C_{total} - C_1, \forall C_{total} > C_1, T \in \{\tau-1, \tau\} \tag{5}$$

where the Traffic that is destined for *Node(i)*, urges the Node to remain active for $\frac{\delta(C(T))}{C_{total}} \cdot T_{prev} > 0$.

According to the work done in [11], the Long-Range Dependence of self-similar incoming traffic can be measured using the probability density function of the Pareto distribution and the corresponding mean value, whereas, the load generated by one source is mean size of a packet train divided over mean size of packet train and mean size of inter-train gap or it is the mean size of ON period over mean size of ON and OFF periods as follows:

$$L_i = \frac{\overline{ON_i}}{\overline{ON_i} + \overline{OFF_i}} \tag{6}$$

Figure 8 shows the traffic aggregation and the difference with the Figure 3 when each node considers the ON and OFF durations of the cached traffic.

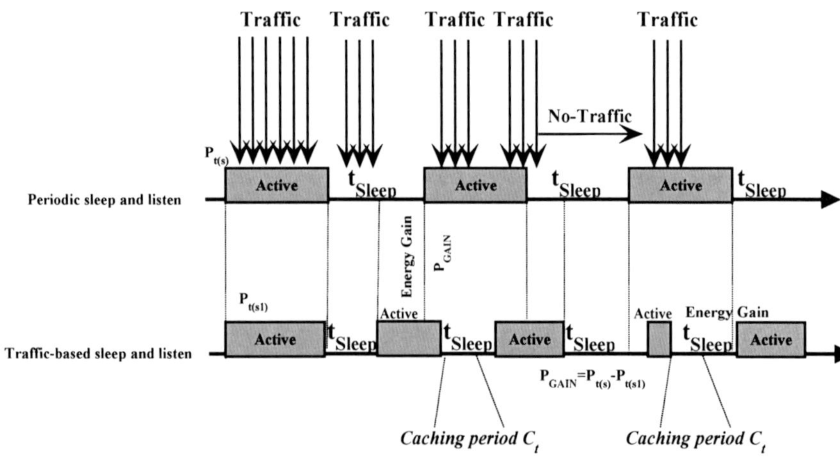

Fig. 8. ON and OFF periodic durations of a Node with the associated cached periods

Considering the above estimations in Equations 4-5, the next Sleep-time duration for *Node (i)* can be evaluated as:

$$L_i(n+1) = \frac{\delta(C(T) \mid A_{t_f}(s,T))}{C_{total}} \cdot T_{prev}, \forall \delta(C(T)) > 0 \tag{7}$$

and for the case that the $\delta(C(T)) < 0$ it stands that $\delta(C(T)) = C_{total} - C_1, \forall C_{total} < C_1, T \in \{\tau-1, \tau\}$, and $\frac{\delta(C(T))}{C_{total}} \cdot T_{prev} < 0$, $\forall T_{prev} > T_{prev}(\tau-1)$, and $A_{t_f}(T)$ represents the cumulative amount of traffic arrivals in the time space [0..T]. Then, the $A_{t_f}(s,T) = A_{t_f}(T) - A_{t_f}(s)$, denotes the amount of traffic arriving in time interval $(s\text{<}t]$. The total active duration of a node increases gradually according to the:

$$T_{sleep} = T(\tau - t_1) - (-C_{N_i}) = T(\tau - t_1) + T_{C_{N_i}} \tag{8}$$

where the $T_{C_{N_i}}$ is the estimated duration for the capacity difference for $C_{N_i} < 0$, whereas the Sleep-time duration decreases accordingly with Equations (6) and (7), iff the $C_{N_i} < 0$. The basic steps of the proposed scheme can be summarized in the pseudocode of the Table 3.

Table 3. The main steps of the BTD scheme

```
for Node(i) that there is C(t)>0 {
  while (C_{N_i(t)}>0) { //cached Traffic measurement
    Evaluate (∇C_{N_i(1)});
Calc(δ(C(T)) = C_total − C_1, ∀C_total > C_1, T ∈ {τ−1, τ})
if (Activity_Period= δ(C(T))/C_total · T_prev >0)
//Measure Sleep-time duration
L_i(n+1) = δ(C(T) | A_{t_f}(s,T))/C_total · T_prev, ∀δ(C(T)) > 0
else if (δ(C(T)) < 0)
   T_sleep = T(τ−t_1) − (−C_{N_i}) = T(τ−t_1) + T_{C_{N_i}};
   Sleep (T_sleep);
    } //for
}//while
```

The main observations of the optimizations offered by the BTD scheme are shown in Figure 9-11. For the mobility of the users a Fractional Random Walk model was used allowing nodes to move from a certain location with a randomly selected speed in a randomly selected direction as real time mobile users. These results include

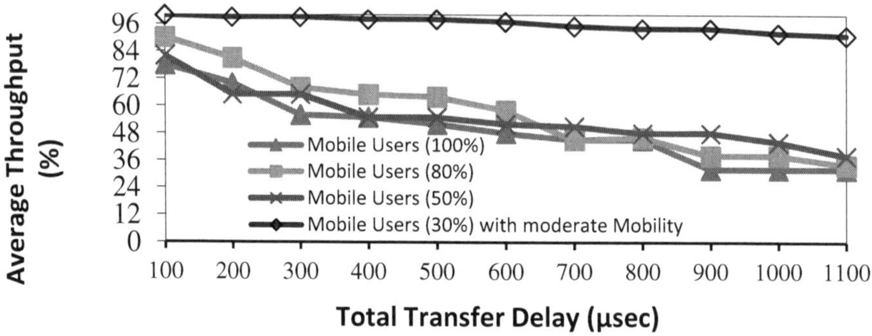

Fig. 9. The Average Throughput with the Total Transfer Delay (μsec)

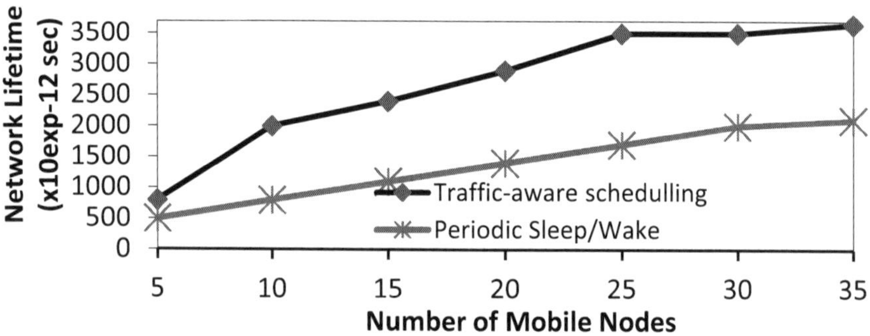

Fig. 10. Network Lifetime with the Number of Mobile Nodes

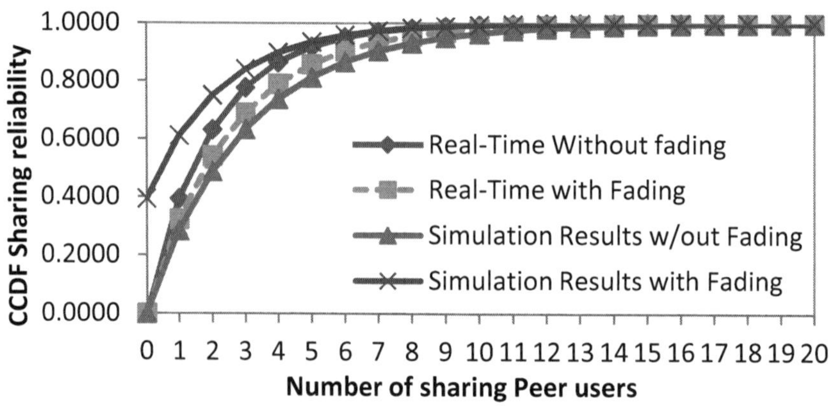

Fig. 11. CCDF Sharing Reliability with the Number of sharing Peer-users

comparative evaluations of the BTD in real time with the BTD performance extracted by simulation experiments.

In Figure 9 the Average Throughput with the Total Transfer Delay in (μsec) is presented, for different mobility models. Figure 9 shows the benefit and losses for the Average Throughput responses with the mobility. The Figure depicts the different Throughput responses for the number of mobile users in the system, for full node mobility, moderate and low (30%) mobility. Figure 10 shows the network lifetime with the number of Mobile Nodes whereas, Figure 11 shows the associated CCDF (Complementary Cumulative Distribution Function (CCDF or simply the tail distribution)) Sharing Reliability with the Number of sharing Peer-users. The extracted results for CCDF Sharing Reliability with the Number of sharing Peer-users, were examined for both Simulation experiments and Real-Time estimations. In this way the benefit and the convergence or divergence of the extracted results can be shown and a comparison between the simulated results and the results extracted from Real-Time Traffic and experiments can be drawn.

2 Conclusions and Open Ended Research Issues

This chapter has provided an overview of the up-to-date traffic-based schemes that are used to optimize the energy consumption and related aspects in wireless devices. Efficient energy conservation techniques encompass many different parameterized issues which in most cases they affect the resource sharing whereas, they enable high energy consumption. The properties of individual wireless nodes pose additional challenges to the development of energy efficient communication protocols. These schemes and the supported protocols are mainly tailored to provide high energy efficiency whereas, they need to ensure the correct transmission and reception during the resource exchanging process. To this end, traffic plays a catalytic role in the resource exchanging process and a major role in the energy consumption and the power scaling levels of the wireless devices. In an end-to-end communication it is u6ndoubtedly true that high energy costs and rare or limited resources prevent the establishment of the end-to-end reliability. Localized reliability and traffic-aware mechanisms are necessary, instead. Since wireless nodes are limited in terms of processing, storage, and energy consumption, traffic-based schemes and methodologies aim to exploit the collaborative capabilities of the wireless nodes and minimize the potential weaknesses in the communication process to a more Physical level communication.

Many open ended issues remain to be addressed in this context. Firstly the exploitation of the traffic pattern to affect the regional capacity. Secondly the mobility pattern in contrast to the nodal capacity and the impact of this limited capacity to the exchange of information. Third, the exploitation of different infrastructures and systems in the resource exchange process as in [13] and [14] where the Cognitive Radio configuration was used. There should be a traffic-aware policy for enabling efficient and reliable communication whereas the moving nodes should be able to consider their motion characteristics to this evaluation. Different nodal motion patterns will expose different characteristics for the traffic which in turn will impact the scheduling management of the wireless node. This comprises of an open research issue for dynamically changing sensor systems where, the nodal traffic and the related motion can be modeled to increase the efficiency and the performance of the system.

References

[1] Mavromoustakis, C.: Using Backward Traffic Difference Estimation for Efficient Energy Saving Schedules in Wireless Device. IEEE CommSoft E-Letters 1(1), 1–6 (2012)

[2] Mavromoustakis, C., Karatza, H.: Quality of Service Measures of Mobile Ad Hoc Wireless Network using Energy Consumption Mitigation with Asynchronous Inactivity Periods. Simulation: Transactions of the Society for Modelling and Simulation International 83, 107–122

[3] Mavromoustakis, C., Karatza, H.: A tiered-based asynchronous scheduling scheme for delay constrained energy efficient connectivity in asymmetrical wireless devices. The Journal of Supercomputing 59(1), 61–82 (2012)

[4] Charalambous, M.C., Mavromoustakis, C.X., Muneer, B.Y.: A Resource Intensive Traffic-Aware Scheme for Cluster-based Energy Conservation in Wireless Devices. In: Proceedings IEEE 14th International Conference on High Performance Computing and Communications (HPCC 2012) of the Third International Workshop on Wireless Networks and Multimedia (WNM 2012), to be Held in Conjunction, June 25-27 (2012)

[5] Mavromoustakis, C., Karatza, H.: Performance evaluation of opportunistic resource sharing scheme using socially-oriented outsourcing in wireless devices. The Computer Journal 56(2), 184–197 (2013)

[6] Mavromoustakis, C.: Mitigating file-sharing misbehavior with movement synchronization to increase end-to-end availability for delay sensitive streams in vehicular P2P devices. International Journal of Communication Systems (2013) (accepted)

[7] Younis, O., Fahmy, S.: Distributed clustering in ad-hoc sensor networks: a hybrid, energy-efficient approach. In: Proceedings of IEEE INFOCOM 2004, Hong Kong, China (2004)

[8] Mavromoustakis, C.: On the impact of caching and a model for storage-capacity measurements for energy conservation in asymmetrical wireless devices. In: 16th International Conference on Software, Telecommunications and Computer Networks (SoftCOM 2008), IEEE Communication Society (COMSOC), September 25 & 26, pp. 243–247 (2008); "Dubrovnik", September 27, Split and Dubrovnik

[9] Mavromoustakis, C., Karatza, H.: A Gossip-based optimistic replication for efficient delay-sensitive streaming using an interactive middleware support system. IEEE Systems Journal 4(2), 253–264 (2010)

[10] Feeney, L., Nilsson, M.: Investigating the energy consumption of a wireless network interface in an ad hoc networking environment. In: Proc. of IEEE InfoCom, vol. 5(8) (2001)

[11] Yu, J., Petropulu, A.P.: Study of the effect of the wireless gateway on incoming self-similar traffic. IEEE Trans. Signal Proceesing 54(10), 3741–3758 (2006)

[12] Sheluhin, O., Smolskiy, S., Osin, A.: Self-Similar Processes in Telecommunications, p. 334 (2007) ISBN: 978-0-470-01486-8

[13] Mastorakis, G., Mavromoustakis, C.X., Bourdena, A., Pallis, E.: An Energy-Efficient Routing Scheme using Backward Traffic Difference Estimation in Cognitive Radio Networks. In: Third IEEE Workshop on Convergence among Heterogeneous Wireless Systems in Future Internet (CONWIRE 2013) in Conjunction, IEEE WoWMoM 2013, The Fourteenth International Symposium on a World of Wireless, Mobile and Multimedia Networks, Madrid, Spain, June 4-7 (2013)

[14] Mastorakis, G., Bourdena, A., Mavromoustakis, C., Pallis, E., Kormentzas, G.: An Energy-efficient Routing Protocol for Ad-hoc Cognitive Radio Networks. In: Cunningham, P., Cunningham, M. (eds.) Future Network &MobileSummit 2013 Conference Proceedings, p. 10. IIMC International Information Management Corporation (2013) ISBN: 978-1-905824-36-6

Unified Platform for M2M Virtual Object Interoperability

Nikolaos Zotos[1,2], Charalampos Stergiopoulos[2], Konstantinos Anastasopoulos[2], Georgios Bogdos[2], Evangelos Pallis[3], Charalabos Skianis[1], George Mastorakis[4], and Constandinos X. Mavromoustakis[5]

[1] Department of Information & Communications Systems Engineering, University of the Aegean, Karlovasi, 83200, Samos, Greece
nzotos@icsd.aegean.gr, cskianis@aegean.gr

[2] Research Department, Future Intelligence Ltd., Agia Paraskevi, 15341, Athens, Greece
{nzotos,csterg,kanastasopoulos,gbogdos}@f-in.gr

[3] Department of Informatics Engineering, Technological Educational Institute of Crete, Estavromenos, Heraklion, 71500, Crete, Greece
pallis@pasiphae.eu

[4] Department of Business Administration, Technological Educational Institute of Crete, Lakonia, Agios Nikolaos, 72100, Crete, Greece
gmastorakis@staff.teicrete.gr

[5] Department of Computer Science, University of Nicosia, 46 Makedonitissas Ave., 2414 Engomi, Nicosia, Cyprus
mavromoustakis.c@unic.ac.cy

Abstract. Sensor networks contribute to the interconnection of a large variety of devices (i.e. transducers, sensors, actuators) thus, enable monitoring and control processes. While new wireless technologies are emerging, a major issue of interoperability has to be addressed in terms of data communications, controlling and interfacing in order to confront the heterogeneity of networks and connected devices and enable end-to-end communication, as well as efficient resource management. A proposed solution to this problem is the provision of a unified service access architecture, which will support common interfaces for data communications, as well as device management and control that will be based on open standards. The concept of object virtualization and IP based networking is introduced for resolving interoperability issues. In this context, the architecture can serve applications that are in line with Machine-to-Machine (M2M) and Internet-of-Things (IoT) concepts.

Keywords: Wireless sensor networks, information management, object virtualization, Low power IP networking, IEEE 1451.x, Machine-to-Machine, Internet-of-Things.

1 Introduction

Sensor networks provide the means for the monitoring, control and the communication of connected devices. These devices may include computers, home appliances,

C.X. Mavromoustakis et al., *Resource Management in Mobile Computing Environments*, Modeling and Optimization in Science and Technologies 3,
DOI: 10.1007/978-3-319-06704-9_23, © Springer International Publishing Switzerland 2014

sensors, actuators, transducers, mobile devices etc. Dedicated platforms comprised of such devices may constitute a monitor and control layer, which is deployed over a system that serves a specific purposes (e.g. an industrial manufacturing system). Wireless sensor networks (WSNs) support actual measurements, data communication and data processing, using multi-hop, as well as ad-hoc networking all combined with sensible cost and low power consumption [1]. A large number of applications of various fields can be benefitted by WSNs like environment monitoring, health care, home automation, industrial control etc. The conventional perception of sensor networking is characterizing by a limited number of features that do not allow pervasive sensing. The major challenge that has to be addressed is the heterogeneity in terms of device, data formatting and networking [2]. The sensor devices are using various communication and networking protocols, incompatible interfaces and data structures. This kind of heterogeneity becomes a serious problem for the design, implementation and deployment of a sensor network. The integration of various types of sensors into a common platform is not favored as long as interoperability cannot be ensured in a consistent and widely accepted manner. A promising solution is the proposal of a unified service access architecture that resolves all the above mentioned issues and enables efficient resource management. This architecture will be based on open standards and will support common interfaces for data communications, device management and control. Based on this architecture an integrated platform could be designed and implemented. This platform will address the issue of objects interoperability, in terms of networking and communication. A scalable middleware will interconnect and manage network objects and their attached devices (sensors, actuators etc.). Specially designed interfaces both software and hardware based on open standards would be implemented. The IEEE Instrumentation and Measurement Society Technical Committee on Sensor Technology has already been defined in IEEE 1451, a family of smart transducer interface standards, in order to standardize the sensor network interface, implement interchanges and operations between sensors [3]. An essential feature that will contribute to the concept of interoperability is the object virtualization, meaning a software equivalent of real objects through which data management and control will be possible in a consistent way. IP based networking using standardized technologies could be used, in order to resolve multiple network heterogeneity. During the design of such an architecture issues like low power consumption, scalability and ease of deployment should be considered for following the trend line of machine to machine (M2M) and internet-of-things (IoT) concept.

2 Machine-to-Machine (M2M)

The Machine-to-Machine (M2M) concept refers to the communication between computers, embedded systems, sensors, actuators and mobile devices with little or no human intervention [4]. This concept is trying to cover a basic need for interconnecting objects in order to implement more sophisticated and intelligent applications. During the near future, M2M will mark its presence as already predicted [5]. According to these predictions made by market players and researchers, by 2014, 1.5 billion devices will work without human intervention and will be wirelessly connected. A number of application in various fields have begun emerging (e.g. home automation, smart grids, health care, industry control etc.). The components of M2M systems are usually connected using wireless sensor networks (WSNs). This interconnection when combined with intelligent functionalities, decision making and control can convert such systems into smart ones. M2M is one of the hottest topics found in the agenda of ICT applications to the industry domain. M2M services and wireless communication technologies have begun bringing considerable profits. A very active and emerging market is involved in this business trying to leverage these technologies, by implementing modern applications. A significant factor for a business, which wants to be a successful player on this field is the simplicity of platforms, services and potential applications offered to the customers. An added value to these products and services will be its cost effectiveness. The business model of M2M is based on the mass market use and deployment. This is the reason why any solutions offered must be built on standards. Standardization will contribute the development of horizontal rather than vertical solutions that are tailor made for specific applications and thus, avoiding complex integration with existed platforms. Moreover, it will guarantee the interoperability between M2M systems in terms of communications and data transfer. Major standardization bodies have acknowledge the need and the importance of introducing standards for the M2M industry. Specifically, the European Telecommunications Standards Institute (ETSI) has published the first release of its Machine to Machine (M2M) service standards. These standards are introduced by a platform having the role of managing the complexity of multiple M2M services and technologies [6]. The ETSI basic M2M architecture is shown in Figure 1 below.

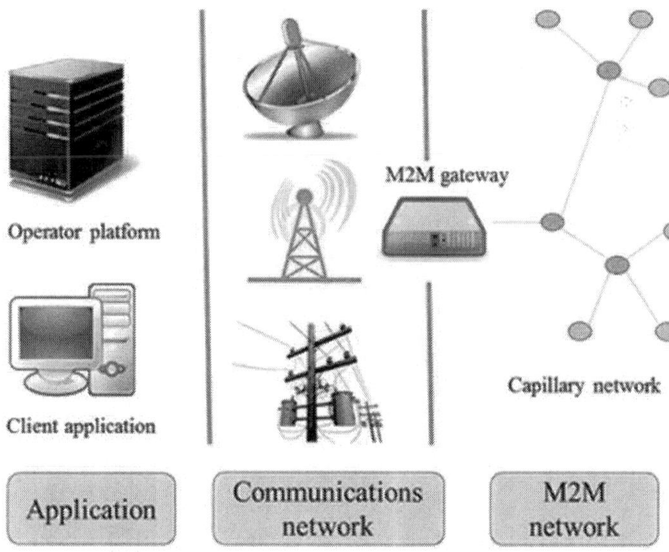

Fig. 1. ETSI M2M architecture domains

According to this architecture, the M2M network is divided into three distinct domains that are: The application for management and monitoring, the M2M area network and the communications network. The M2M network domain is consisted of a number nodes along with a gateway. Each node is a device that has sensing and communication features (i.e. a mobile wireless device that has is equipped with sensors for real-time monitoring). A node, after collecting data from the sensing devices, will make the decision of transmitting them through the network in a single-hop or multi-hop manner, towards the gateway. The gateway is responsible for sending these data through the communication network to the back-end application supported by a platform. Many existing technologies could effectively play the role of the communication network and provide cost-effective and reliable solutions. These technologies could be wired networks (PLC, xDSL etc.) or wireless ones (WiFi, WiMAX, GPRS etc.). The application domain will include a server that is responsible for collecting and storing the data coming from the M2M network as well as for monitoring and control the M2M devices and network.

The ETSI M2M specifications lay out the service requirements for M2M devices on the M2M network, which cover the management, security and addressing of the system [7]. Ubiquitous networking along with its services and properties is used by M2M platforms, thus implementing features like Quality of Service (QoS), different communication modes (multicast, unicast etc.) and IP-based networking (IPv4, IPv6). The ETSI standardized M2M is using for the data transfer the representational state transfer (REST) architecture. M2M networks integrate these ubiquitous technologies in such a way as to facilitate reliable, simple and secure global information transfer across an abstracted network for a multitude of applications. M2M use network technologies and resources for the communication of remote machines with a remote application infrastructure, which serves as a monitoring and control mechanism. Considering the M2M nodes as smart objects, their interconnection is covered by the Internet-of-things concept, thus merging data coming from the physical world by sensing devices with those coming from the digital world in terms of communications and networking.

3 Internet-of-Things (IoT)

Internet-of-Things (IoT) is defined as "a network in which a variety of things interconnected on a global scale can uniquely address and access each other using standard communication protocols" [8]. The distinctive characteristic is that an IP addressing based networking system enables the communication and the interwork of entities that are characterized by autonomous and smart features. The term "IoT" is said to have been introduced by researchers working in the Auto-ID Labs of Massachusetts Institute of Technology (MIT) on the topic of RFID tags [9]. An RFID system is based on a small reader, which is able to receive a low power radio signal and tags that are actually the signal transmitters. RFID tags can be passive, having no power source or active with a power source included. These tags are attached to things and transmit information about the ID, the properties, the condition and status of a thing. A sensor network is also a paradigm of IoT. A sensor network is most of the times a wireless one and has numerous sensor nodes that are transmitting through the network. Two basic features of these networks are their energy efficiency, in terms of low power usage and reliability in communication. Nodes in a sensor network can be classified into parent nodes and child nodes, according to the roles they play in terms of data routing and networking as shown in Figure 2. A parent node plays the role of a sink. This sink node collects information gathered by the child nodes, most of the times in multi-hop way as data are forwarded to their destination through a route created by multiple nodes.

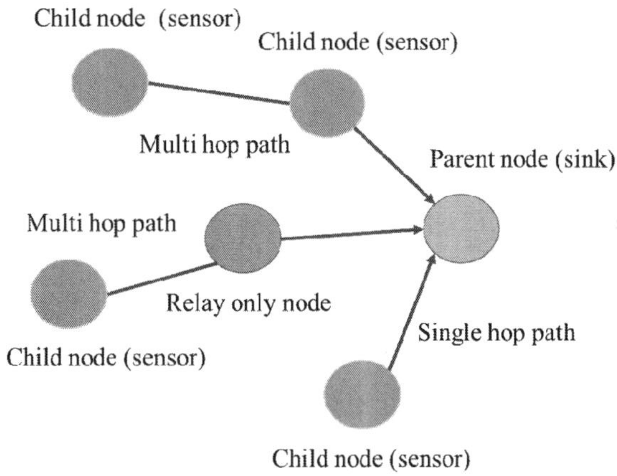

Fig. 2. Conceptual diagram of a sensor network

The variety of hardware and software platforms that can potentially form a wireless sensor network (WSN) constitutes in many cases a problem for deploying and control such networks because of the inherent heterogeneity. The issue of heterogeneity could be solved by the introduction of a middleware that will serve the WSN.

4 Heterogeneity in Wireless Sensor Networks (WSNs)

The issue of heterogeneity referring to Wireless Sensor Networks (WSNs) prevents from the incorporation of sensor platforms into a common application. Many differences may exist in terms of communication interfaces, programming languages, operating systems and deployment mechanisms. A comparison of commonly-used WSN technologies can be found in the bibliography [10]. The problem of heterogeneity is further prolonged, if we consider a vast set of devices that incorporate many different types of sensors. A classic example is the one of a modern smart phone, which includes GPS, accelerometer, camera etc. In order to collect data from these different platforms, a common way of data capturing and reading is needed, which will be based on widely accepted standards. Different data types and communication models of a WSN are studied from standardization bodies, in order to remove the barriers of heterogeneity. The Sensor Web Enablement (SWE) from the Open GeoSpatial Consortium (OGC) is one of the initiative for resolving heterogeneity issues, especially when WSNs are connected to the Internet. The proposal of SWE is the usage of a data model, which will be standardized by the introduction of an XML schema. this schema will be formatted having in mind that the encoded data are coming from measurements. Following this concept, a Sensor Model Language (SensorML) is also created for describing properties, capabilities and information related to a sensor. SWE also describes a number of service protocols for accessing the sensors: Sensor Observation Service (SOS) and Sensor Planning Service (SPS) provide access mechanisms for sensor data retrieval. The Sensor Alert Service (SAS) is actually an alert and notification mechanism. Finally, Web Notification Services (WNS) are a set of tools for asynchronous access to SAS and SPS web services.

The OGC SWE is a remarkable effort for trying to resolve heterogeneity of sensors and introduce interoperability. Nevertheless the SWE standards present challenges for their implementation and operation [11]. For better understanding and description of different systems, a common ontological approach is needed for a standardized description. Efforts have been made to describe the SWE standards using semantics [12]. Other attempts for creating an ontological system for virtual sensor representation exist [13]. The World Wide Web Consortium (W3C) has created a working group (W3C Semantic Sensor Network (SSN) Ontology) that also aims to create a standardized sensor ontology.

A main characteristic of a pervasive system is the context, which includes information for the state of an object or for an action that will be useful for the user of the application. Context in terms of data may be gathered from various sources, both physical and non-physical. For example, if referring to a WSN, the source of context data will be captured and transmitted from sensors that are measuring environmental parameters. An additional source of context may be the World Wide Web (WWW). The current WWW encompasses diverse documents and services that produce and/or maintain real-time, as well as static information relating to a wide range of users, places, topics, and concepts as shown in Figure 3. The new concepts of Machine-to-Machine (M2M) and Internet-of-Things (IoT) are based on the availability and data exchange of context data. So, sources that provide these data, physical and non-physical that are interconnected are expected to be rapidly increased in the near future.

A physical sensor is a device that actually measures and monitors a parameter (i.e. temperature) found in the physical environment. A virtual sensor is a surrogate of the physical object in the digital world and represents a physical sensor or a group of them. Soft sensors can also exist as special software components that access data from non-physical resources. The continuously increased number of these components, software and hardware are part of data exchange network. Their communication, monitoring, control as well as the need for data fusion is an active area of research.

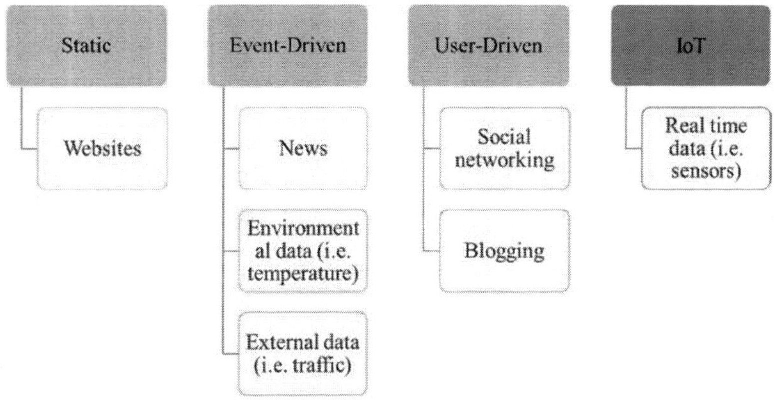

Fig. 3. A generic taxonomy of context data sources

5 The Role of a Middleware in M2M

The increased demand for efficient resource management and control of connected objects demonstrates the need that a software platform must exist, which will be able to provide a set of services, as well as object abstraction and virtualization in a non-physical domain as software entities. This platform or framework used in M2M and IoT can be defined as a middleware. The middleware will address interoperability issues caused by heterogeneous devices, adaptation problems, context awareness, device control and management. A middleware will also contribute to the pervasive computing and ubiquitous networking [14] and will act as an adaptation layer which will support a plug-and-play mode [15].

A framework acting as a middleware is essential for IoT, M2M systems and applications mainly for the following reasons:

- The need of a common way of communication between networked sensor devices.
- The existence of a layer that can "absorb" heterogeneity of components.
- An abstraction/adaptation layer in order to address applications of diverse domains.
- The provision of an Application Provision Interface (API) for the communication and offered services that encapsulates the details of diversity.

In order to address the above mentioned reasons a middleware must have the following features: interoperability, device discovery and management, context detection and awareness, data management, data security and privacy.

Interoperability can be described as the information exchange across diverse domains of applications over different communication interfaces. The categories that interoperability has to be applied using an appropriate middleware are network, syntactic and semantics [16]. Network interoperability refers to basic connectivity issues that correspond to almost all OSI layers (physical, data-link, network, transport, session and application). Syntactic interoperability deals with the format and structure of the information encoding exchanged among components. Semantic interoperability defines a set of rules for interpreting in a common way the information content by creating a semantic model.

Device discovery and management is the feature that enables any network component been aware of its neighboring devices. This component also make its presence known to the network. A proper ontological model can contribute to the device discovery and management especially in the case of heterogeneous devices. A middleware for IoT and M2M must be context aware as one of its purpose is to enable and serve smart environments. For this reason a knowledge database must be designed and implemented for effectively support of context-aware systems.

Context awareness is based on two features: context detection and context processing. During the process of context detection data are detected and collected while identifying any factor that might have been important for a prompt response. Context processing is related to the decision making process which follows the extraction and processing of data.

Security and privacy are responsible for confidentiality, authenticity, and non-repudiation. Managing data volumes must be an essential feature of the middleware. Having in mind the number of interconnected devices that is continuously increased and the volume of data coming for different sources physical and non-physical, efficient and reliable mechanisms should exist for data management. In this case the challenges involve in querying, indexing, process modelling, and transaction handling.

In Figure 4 the architecture of a middleware is shown. The architecture is based on the division of the system to layers. The basic layers are interface protocols, device abstraction responsible for providing interoperability, device abstraction, and the management module, which is a core component as it includes the functionalities if device discovery management and context detection. Application abstraction module provides the interface for the communication between remote applications and the local one. Local applications mainly run as event driven services. The other components like context analysis, knowledge database may reside in remote system essentially generating a need of distributed architecture, which can potentially be hosted in a cloud computing environment.

As M2M and IoT demands a complex distributed architecture with numerous different kind of components, ubiquitous and smart computing must be obtained. This facts points out the significance of a middleware system for IoT and M2M that will act as a bond joining the heterogeneous domains of applications communicating over heterogeneous interfaces.

A middleware can also be an intermediate layer of virtualization, which will contribute to resolve the issue of interoperability. The term "virtualization" is used here for describing a layer of the system that allows the creation, hosting and management of virtual objects. A virtual object may be a software equivalent of a physical object (e.g. a sensor device) that has inherited all its properties. This way a platform or a middleware can efficiently manage the attached resources through

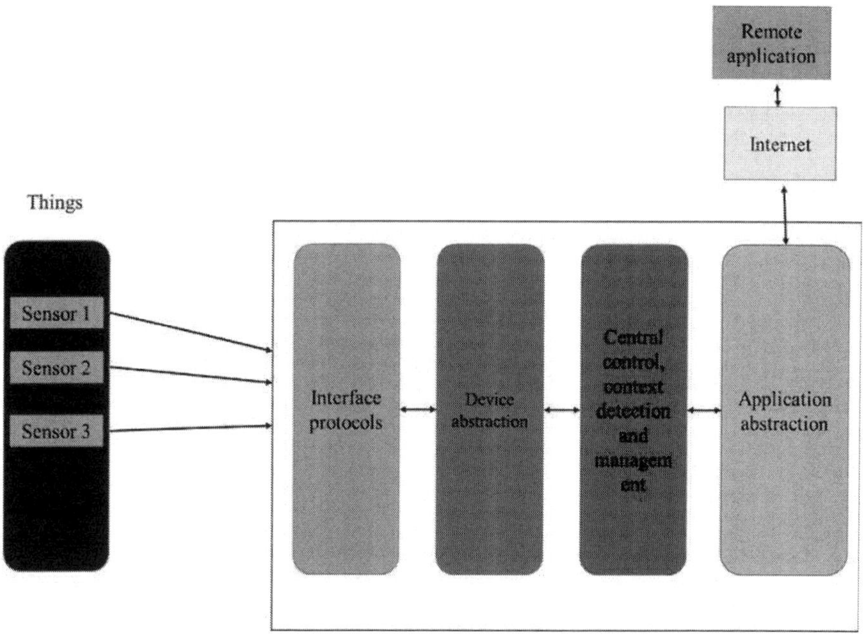

Fig. 4. Components of a middleware architecture

virtual objects that help the context awareness and data manipulation for monitoring and control purposes. Virtual objects do not necessarily correspond to physical equivalents. They may correspond to groups of physical objects (i.e. a group of sensors that measure temperature in a building) or be custom made virtual objects that gather properties from different physical objects, according to the management policies of each system.

In order to exploit the potential of a middleware, proper communication protocols have to be used for transferring the data from an end point (sensor) to the application through the middleware. Once again, these communication technologies have to be bases on open standards that are widely used.

6 End-to-End Communication Solution

The development of heterogeneous wireless technologies, such as IEEE 802.11 (WiFi), Bluetooth, WiMAX or IEEE 802.15.4 has been performed very fast. During the last years, these technologies were used for the deployment of wireless sensor networks and transducers that were used in IEEE 1451 [17]. Most of the solutions did not support end-to-end IP based architecture, thus maintaining the problem of interoperability. Currently, the development of IPv6 over Low power Wireless Personal Area Networks (6loWPAN) [18] using on top the IEEE 1451.5 standard allows the design of smart sensors fully end-to-end IPv6 based architecture using the physical layer IEEE 802.15.4 with low power options and limited computation.

The IEEE 1451.5 Standard for a Smart Transducer Interface for Sensors and Actuators provides a standardized communication protocol for transducers (i.e. sensors, actuators etc.), which incorporates network connection interfaces. A transducer interface module (TIM) is a module that contains a transducer as well as logic to implement an interface and can interact with a network via a network-capable application processor (NCAP). The IEEE 1451.0 standard also states that a compliant TIM must have Transducer Electronic Data Sheet (TEDS) for each transducer. In addition, it describes the structure of many of these TEDSs. A TIM may have many transducers (either sensors or actuators, or a combination of the two). IEEE 1451.5 extends this standard to wireless networks for these TIMs, referenced as wireless transducer interface modules, or WTIMs.

6LoWPAN is a very recent technology, which provides a solution that features low-power, low-transmission-rate and energy-efficiency. At the same time it supports Internet Protocol Version 6 (IPv6) using IP packet and header compression mechanisms, which enable data packets to be transferred over IEEE 802.15.4 hardware. Sensors that are part of the WSN using 6LoWPAN can have direct communication with the Internet because of the IP- based technology.

7 Description of the Unified Platform

The main problem currently faced by various industries is the existence of multiple different systems from different manufacturers and with different functions. The basic requirement is the unified resource management of these systems or parts of these systems at both technical and operational level. The collection and management of various information from a single platform would result in better and more effective management of problems / processes, minimizing human errors and providing decision making processes. For addressing these issues, a platform for interoperable management of heterogeneous sensors and other similar devices is proposed, connecting them with the use of network objects implementing M2M and IoT technologies in a simple manner, which is transparent to the user.

The features of the proposed platform are:

- The system should allow a wide range of heterogeneous sensors, actuators and data loggers that are scattered to get interconnected using the platform.
- Interfacing will be performed by the use of hardware and software objects equipped with suitable connection interfaces based on widely used standards. An interface will be used for the connection of sensors and other devices to the platform by the help of a special designed hardware, while another one will be used for the connection to a data network in order to facilitate the data transmission from the sensor to the application layer.
- The platform will provide the means of connectivity with existing systems, through recognized standards and the development of appropriate interfaces for connection.
- The management of networking objects and their extensions extension (sensors connected to these devices) will be performed by the use of a middleware, which will introduce the concept of virtual objects with related information coming from physical objects like sensors. The middleware will be considered as a virtualization layer, in terms of hosting virtual objects.

- The middleware will have the ability of creating custom virtual objects, by grouping multiple objects and heterogeneous sensors into new virtual object (i.e. all the lamp posts of a lighting system in a particular street, all the temperature sensors of a building, etc.). This way group of physical objects can be monitored and controlled as groups.
- A communication network will be deployed based on IPv6 protocol, in order to provide unique address to network objects. Networking over IPv6 network will be based on the 6LowPAN standard for connecting and managing a large number of networked objects wirelessly.
- The platform architecture will be based on clearly defined services in order to achieve better flexibility and structure (modularity) combining appropriate mechanisms and intelligent data transfer maximum scalability.
- Automatic discovery and registration of new objects in an ad-hoc way will be enabled.
- The architecture including hardware, software and middleware will be designed and implemented giving emphasis on data security through authentication and secure connections to ensure the protection and privacy of data.

8 The Platform Architecture

Four different hardware components will be the main parts of the platform besides any software implantation. Their role is specific and dedicated to certain processes. Their operation will be based on embedded microprocessors and software giving them smart features and capabilities. The overall system architecture is shown in Figure 5.

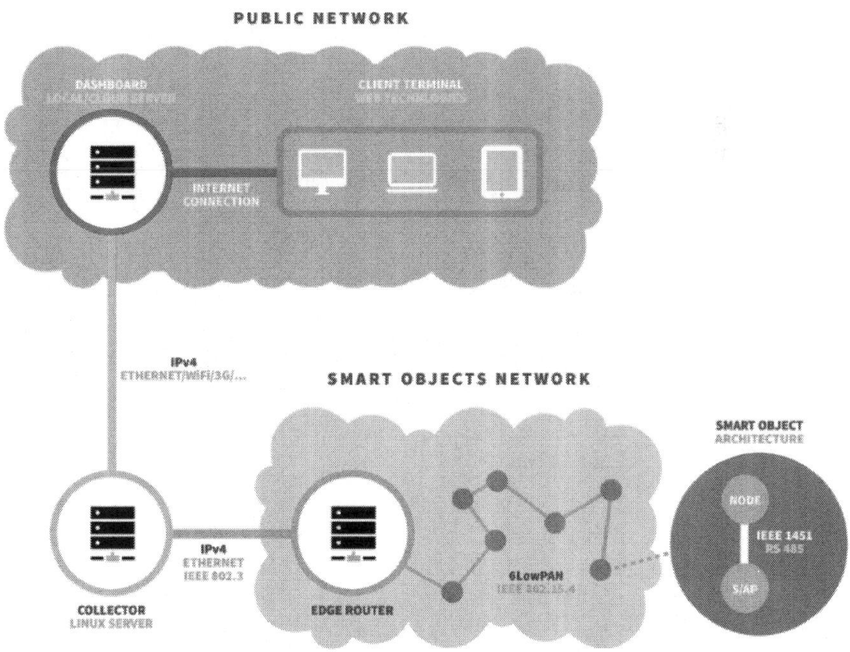

Fig. 5. Overall system architecture

A hardware device called the S/AP (Sensor/Actuator Peripheral) is a key component for the connection of the sensors to the system. Its main role is to provide proper interfaces to the peripheral devices (sensor, actuators etc.). Through the connection the data collected from these devices can be transmitted to the next level above using the network connection.

The key parts of the S/AP are:

- A microcontroller processor RISC 8bit of low consumption for analog and digital peripherals.
- An Interface communication bus that connects the S/AP to the next level through a wired communication. It is compatible with the industry standard ANSI/TIA/EIA-485 [19].
- The device is easy to install near the sensor. It will be inside a case that will be compatible to the standard EN 50022 (DIN-Rail) for being easily adapted to the infrastructure and the premises where the system is located.

Another hardware device named as Node will have the task of interlinking the individual S/APs on the network data. At this layer of the system the requirement is the gathering of any data coming from the sensors in a structured way, in order to be easily read and interpreted after sending them to the next layer using the communication network. A special device called "Node" will be responsible for these processes, which will be performed by the use of special software, while the operating system that will run on this level, will be based on Contiki [20]. An enhanced version of the Node with multiple interfaces will be able to support through its connections more than one S/Aps acting as a local hub for the connection to the communication network. Much like the S/AP, the Node will have the ability for easy installation and deployment as its casing will be compatible with standard EN 50022 (DIN-Rail). The S/AP and the Node together constitute an object within the network. The main parts of the Node are:

- A unit microcontroller processor RISC 16bit of low consumption for analog and digital peripherals.
- Wireless communication is the core of the circuit wireless connectivity. It will have an integrated circuit compatible with IEEE 802.15.4, low-power, transmitting at the frequency band of 2.4GHz. Each Node will have a unique address in the IPv6 network.
- An interface communication bus that interconnects the Node with S/AP via wired communication, according to the standard ANSI/TIA/EIA-485.
- Compatibility with standard management sensors IEEE 1451 [21] and IEEE 1451.5.
- An Interface - compatible with RS232 that will be used for programming and maintenance procedures of the unit.

The third device that will be developed is an Edge router. Its tasks are: The interconnection of Nodes and the network configuration between nodes, promoting data to the next level placed inside the middleware. The Edge Router just like the Node will run a Contiki based operating system, interfacing directly with the level of data collection using web sockets. An Edge router will be also easily installed as its casing is compatible with standard EN 50022 (DIN-Rail). The key parts of the Edge router are:

- A unit microcontroller processor RISC 16bit, low consumption for analog and digital peripherals.
- Regional wireless communication by the use of an integrated circuit compatible with IEEE 802.15.4 [22], low-power, frequency 2.4GHz. The interconnection of Nodes will be in full mesh topology, allowing the development of a scalable network of Nodes without the need for direct connection of each Node with the Edge Router. Some of the Nodes will have the role of expanding the network without being connected to S/AP.
- Ethernet interface (IEEE 802.3) for transferring data to the collection level.
- Interface - compatible with RS232 support for programming and maintenance procedures of the unit.

At the level of collection, a device named Collector will have the role of collecting and configuring the information derived from the level below. More specifically:

- It will accept connections, sending data from the lower levels (via web sockets)
- It will perform appropriate synchronization operations with objects (question / answer opportunities objects, finding / tracking objects in the network)
- It would impose rules of behavior of objects in the network
- It maintains a table possible S/AP, which will be displayed in the network
- It will include a properly configured database for managing information from heterogeneous sensors.
- It has a properly configured software, which will define the level of virtualization of data flows from lower levels to aid in enforcing rules, and automation.
- It is compatible with standard IEEE 1451 sensor management.
- It provides the ability to interface with other systems utilizing technology SensorML based on OGC (xml based structure).
- It will maintain the level of implementation that will be provided to the user through an appropriately structured user interface (HMI) and the ability to manage the objects of an environment easily accessible (webportal).

The device named the Collector is a PC based on x86 or ARM architecture and may be a real or a virtual machine. A cloud infrastructure hosts the platform as its architecture favors its distributed nature capable of collecting data from multiple places. Such a Collector installed in appropriate infrastructure can support multiple applications and networking items installed in independent and far apart places. The Collector will be based on open-source software (linux based) and different levels of middleware daemons and applications implemented using appropriate programming languages (Python, Perl, C, Java, scripting languages, etc) can play management

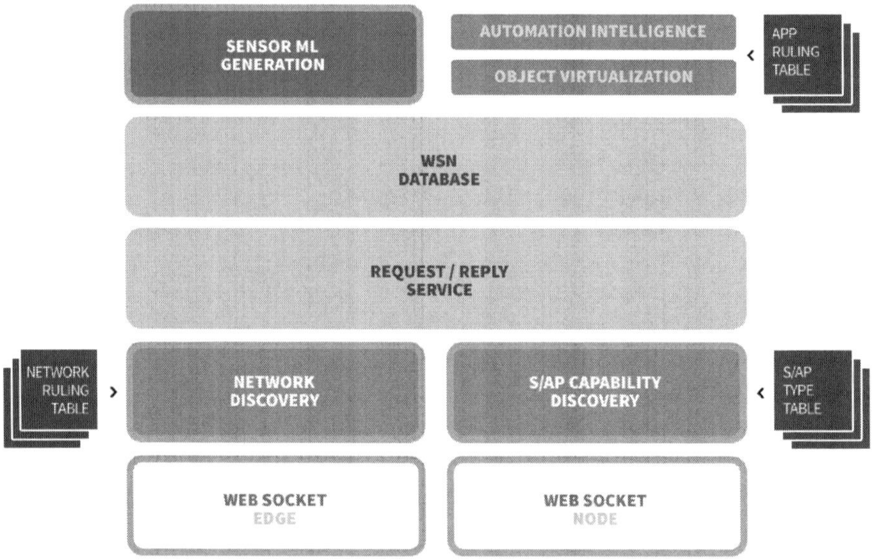

Fig. 6. Collector architecture

roles, in terms of scalability. The data management subsystem will address appropriate databases to manage large flows of sensor data. Finally, the platform interoperability is ensured by the use of open standards, which are supported by existing hardware and/or software, as well as by those that will appear at the market in the near future. For ensuring interoperability, the platform will provide a modeled markup language based n SensorML for handling data from heterogeneous sensors and objects. Moreover, virtual objects as mentioned in previous sections will play an important part for addressing interoperability issues. The latter option offers the opportunity to develop entire ecosystem of applications that can utilize the data obtained from this platform. The Collector will support the functions of the middleware is characterized in the context of the architecture as a virtualization layer, as it will host the virtual objects.

Finally, the platform will support end-to-end security. To ensure the operation of the wireless medium, a mechanism of symmetric encryption AES [23] will be applied. Moreover, due to the use of IPv6 addressing security, the protocol stack protection IPsec [24] will be used.

9 Conclusions

The increased demand for interconnecting objects and the introduction of concepts, such as the M2M and IoT have created the need for efficient resource management, as well as monitoring and control in an intelligent way. A proposed solution to this problem is the provision of a unified service access architecture, which will support common interfaces for data communications, as well as device management and control based on open standards. A platform based on this architecture will involve a middleware, as well as software and hardware components that will play their role for bringing solutions to address the problems of heterogeneity, in terms of devices and communications, as well as to ensure the interoperability between systems.

References

1. Akyildiz, I.F., Su, W., Sankarasubramaniam, Y., Cayirci, E.: A survey on sensor networks. IEEE Communications Magazine 40(8), 102–114 (2002)
2. O'Grady, M.J., Murdoch, O., Kroon, B., Lillis, D., Carr, D., Collier, R.W., O'Hare, G.M.: Pervasive Sensing: Addressing the Heterogeneity Problem. Journal of Physics: Conference Series 450(1), 012044 (2013)
3. Yongping, X., Yun, Z., Zhijia, Y.: A Middleware Framework Based on the IEEE 1451 Standards. In: International Conference on Wireless Communications, Networking and Mobile Computing, WiCom 2007, pp. 3119–3122 (2007)
4. Watson, D.S., Piette, M.A., Sezgen, O., Motegi, N., Ten Hope, L.: Machine to machine (M2M) technology in demand responsive commercial buildings (2004)
5. Fadlullah, Z.M., Fouda, M.M., Kato, N., Takeuchi, A., Iwasaki, N., Nozaki, Y.: Toward intelligent machine-to-machine communications in smart grid. IEEE Communications Magazine 49(4), 60–65 (2011)
6. Cackovic, V., Popovic, Z.: Cloud based service for M2M communication. In: 2012 IX International Symposium on Telecommunications (BIHTEL), pp. 1–6. IEEE (2012)
7. Potter, C.H., Hancke, G.P., Silva, B.J.: Machine-to-Machine: Possible applications in industrial networks. In: 2013 IEEE International Conference on Industrial Technology (ICIT), pp. 1321–1326 (2013)
8. Atzori, L., Iera, A., Morabito, G.: The internet of things: A survey. Computer Networks 54(15), 2787–2805 (2010)
9. Ashton, K.: That 'Internet of Things' Thing. RFiD Journal 22, 97–114 (2009)
10. Johnson, M., Healy, M., van de Ven, P., Hayes, M.J., Nelson, J., Newe, T., Lewis, E.: A comparative review of wireless sensor network mote technologies. In: 2009 IEEE Sensors, pp. 1439–1442 (2009)
11. Rouached, M., Baccar, S., Abid, M.: Services (SERVICES). In: 2012 IEEE Eighth World Congress, pp. 65–72 (2012)
12. Broring, A., Foerster, T., Jirka, S.: Interaction patterns for bridging the gap between sensor networks and the sensor web. In: 2010 8th IEEE International Conference on Pervasive Computing and Communications Workshops (PERCOM Workshops), pp. 732–737 (2010)
13. Stevenson, G., Knox, S., Dobson, S., Nixon, P.: Proceedings of the 1st Workshop on Context, Information and Ontologies CIAO 2009, pp. 91–98. ACM, New York (2009)
14. Yang, D., Liu, F., Liang, Y.: A Survey of the Internet of Things. In: ICEBI 2010, Advances in Intelligent Systems Research (2010) ISBN, 978, 90-78677
15. Bandyopadhyay, S., Sengupta, M., Maiti, S., Dutta, S.: Role of middleware for Internet of things: a study. International Journal of Computer Science & Engineering Survey (IJCSES) 2 (2011)
16. Terziyan, V., Kaykova, O., Zhovtobryukh, D.: Ubi-Road: Semantic Middleware for Cooperative Traffic Systems and Services. International Journal on Advances in Intelligent Systems 3(3 and 4), 286–302 (2010)
17. Higuera, J., Polo, J.: Understanding the IEEE 1451 standard in 6loWPAN sensor networks. In: 2010 IEEE Sensors Applications Symposium (SAS), pp. 189–193 (2010)

18. Montenegro, G., Kushalnagar, N., Hui, J., Culler, D.: Transmission of IPv6 packets over IEEE 802.15. 4 networks. Internet proposed standard RFC, 4944 (2007)
19. Neilson, C., Donaldson, S.: IPv6 Maintenance Working Group K. Lynn, Ed. Internet-Draft Consultant Intended status: Standards Track J. Martocci Expires: Johnson Controls (September 13, 2012)
20. Dunkels, A., Gronvall, B., Voigt, T.: Contiki-a lightweight and flexible operating system for tiny networked sensors. In: 29th Annual IEEE International Conference on Local Computer Networks, pp. 455–462 (2000)
21. Song, E.Y., Lee, K.: Understanding IEEE 1451-Networked smart transducer interface standard-What is a smart transducer? IEEE Instrumentation & Measurement Magazine 11(2), 11–17 (2008)
22. Callaway, E., Gorday, P., Hester, L., Gutierrez, J.A., Naeve, M., Heile, B., Bahl, V.: Home networking with IEEE 802.15. 4: a developing standard for low-rate wireless personal area networks. IEEE Communications Magazine 40(8), 70–77 (2002)
23. Babu, R., Abraham, G., Borasia, K.: A Review On Securing Distributed Systems Using Symmetric Key Cryptography, arXiv preprint arXiv:1303.0351 (2013)
24. Ferguson, N., Schneier, B.: A cryptographic evaluation of IPsec. Counterpane Internet Security, Inc., 3031 (2000)

Part V

Performance Evaluation of Mobile Computing Systems

Computational Resource Management for Video Coding in Mobile Environments

Guilherme Correa[1], Pedro Assuncao[2], Luciano Agostini[3], and Luis A. da Silva Cruz[1]

[1] Instituto de Telecomunicações, University of Coimbra, Coimbra, Portugal
[2] Instituto de Telecomunicações, Polytechnic Institute of Leiria, Leiria, Portugal
[3] GACI, Federal University of Pelotas, Pelotas, Brazil

Abstract. The increase of computational resources in mobile devices and the availability of reliable communication infrastructures provide support for acquisition, display, coding/decoding and transmission of high-resolution video in a broad set of equipment such as tablets and smartphones. Nevertheless, real-time video encoding and decoding is still a challenge in such computing environments, especially when considering the amount of computational resources required by state-of-the-art video coding standards. Moreover, battery technologies did not evolve as much as desired, which makes power consumption minimization an important issue for the mobile devices industry and users. Therefore, in current mobile systems, the available computational resources along with battery-life are responsible for imposing significant limitations on mobile real-time multimedia communications. This chapter presents an overview of the state-of-the-art research on management of computational resources for video encoding systems in mobile communications equipment. A review on computational complexity analysis of both H.264/AVC and HEVC video coding standards is presented, followed by a description of current methods for modelling, reducing and controlling the expenditure of computational resources on these video codecs. Finally, future trends on computational complexity management for video codecs implemented on power-constrained devices are lined out.

1 Introduction

The advances of semiconductor technologies in the last decades have fostered a considerable growth in the market of multimedia-capable mobile devices, such as mobile TVs, tablets and smartphones. The increase of computational resources and the availability of reliable communication infrastructures allowed a broad set of equipment such as tablets and smartphones to receive and display high-resolution digital video. Very common are also portable devices with the capability of capturing and transmitting digital video through wired and wireless networks. Furthermore, the current trend to embed increasingly higher resolution cameras into these portable devices together with the relatively small transmission bandwidth of wireless

C.X. Mavromoustakis et al., *Resource Management in Mobile Computing Environments*, Modeling and Optimization in Science and Technologies 3,
DOI: 10.1007/978-3-319-06704-9_24,

channels, calls for the addition of high-resolution video encoding and decoding capabilities to these devices.

Despite this recent evolution in portable devices, particularly regarding communications technology and computational power, video encoding and decoding in real time is still a challenge in mobile computing environments, especially when considering the amount of computation required by state-of-the-art video coding standards, such as H.264/AVC [1] and High Efficiency Video Coding (HEVC) [2]. Besides the real time transmission problem, power consumption is also one of the most important issues in the mobile devices industry, since limited battery capacity still imposes major constraints in multimedia applications demanding for high computational power. Experiments have shown that video encoding is responsible for about 2/3 of the total power for video communications over Wireless LAN [3]. It has also been noticed that this ratio tends to increase if the required video distortion is lower [4], which mandates higher transmission bitrates over the wireless interface.

These two different problems are actually strongly connected, since they are both effectively addressed by methods to manage the use of computational resources by controlling the complexity of video encoding/decoding tasks. While academy and industry keep researching and developing new and more efficient solutions for increasing processing speed and battery life, video coding experts try to decrease encoding/decoding time and power consumption by wisely managing computational resources usage by their algorithms and systems.

In a generic video encoder, major signal processing units such as motion estimation/compensation, forward and inverse transform and quantisation, and entropy coding are responsible for the largest share of calculations. As most of these units can operate in multiple modes and setting points, the choice of the operating modes and points for each unit that achieves the best overall coding trade-off between computational resource use and compression efficiency has become an important design challenge.

Until recently, H.264/AVC has been considered the state-of-the-art video compression standard. It introduced several parameter choices that specify the encoding/decoding operating modes, including the number of reference frames, the use of variable block sizes for motion estimation/compensation, and the resolution of motion vectors. The broad number of encoding possibilities allowed the standard to achieve bit rate savings of about 50% in comparison to MPEG-2 for the same video quality [5]. Currently, the High Efficiency Video Coding (HEVC) standard is gradually being introduced to industry and is expected to substitute H.264/AVC during the next years as the state-of-the-art video coding standard. This new standard is able to reduce bitrates by about 40% over H.264/AVC through the use of several new tools and operating modes that complement or improve those already used by H.264/AVC [6]. Obviously, as more operating modes are made available, the computational complexity involved in the encoding process of HEVC becomes higher than that of H.264/AVC and previous standards. As a result, efficient computational resource management becomes even more important for next-generation multimedia portable devices, which will certainly include HEVC codecs.

This chapter presents an overview of the state-of-the-art research on computational resource management for video encoding systems in mobile computing environments. Section 2 presents an introduction to video compression and an overview of the H.264/AVC and HEVC video coding standards, including a brief analysis of their computational complexity. Section 3 reviews current works on complexity modelling for H.264/AVC and HEVC. Section 4 and 5 present a set of proposed methods for complexity reduction and control on both standards, respectively. Future trends and challenges on the field are outlined in section 6 and conclusions are presented in section 7.

2 Video Compression and Computational Complexity

A generic video compression system is based on the following signal and data processing operations: (i) spatial and temporal prediction, (ii) decorrelating transform, (iii) quantisation and (iv) entropy coding. The prediction step is usually followed by the transform and quantisation of prediction residues, which is then followed by entropy coding, as shown in Fig. 1, where the basic components of an encoder are shown.

Spatial prediction is also known as intra-frame prediction and is responsible for reducing redundancy between regions within the same frame. Temporal prediction, also known as inter-frame prediction, which usually also involves motion compensation, reduces the temporal redundancy between time-sequentially close frames. They are both lossless methods (i.e., no information is lost when these techniques are applied), but usually require a large number of operations to be performed, specially the motion compensation which involves a costly motion estimation procedure. The difference between the original and the predicted frame, the residual data, is transformed and quantised in order to reduce spatial redundancy. Even though the transformation step does not incur in data loss, quantisation is a lossy compression technique. Finally, entropy coding, which is a lossless method, processes the transformed and quantised data in order to reduce quantised data redundancy.

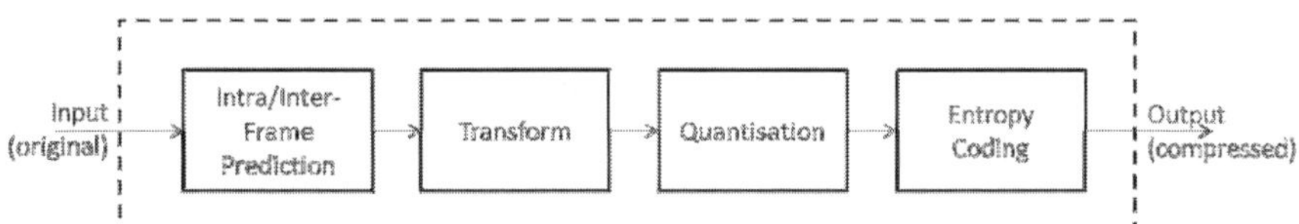

Fig. 1. Generic video compression system

Many methods for intra/inter-frame prediction, data transformation, quantisation and entropy coding are used in the existent video coding standards. Most of them allow different operating modes, so that the encoding process can be adjusted to the video characteristics in order to better exploit the nonstationarities of the video source. Choosing correctly the operating mode for each tool in an encoder is essential for obtaining high compression efficiency levels.

The two important parameters for measuring compression efficiency are the coding bit rate (R) and the picture quality or distortion (D). For video transmission applications the coding bit rate is determined by the communication channel bandwidth. In the case of storage applications usually the coding bit rate is decided based on available storage capacities or the retrieval link speed. The picture quality can be measured according to several different metrics, such as the Peak Signal-to-Noise Ratio (PSNR), the Structural Similarity (SSIM) and the Mean Squared Error (MSE).

The Rate-Distortion Optimisation (RDO) method has been proposed in [7] for minimizing distortion in video compression algorithms given a certain bit rate allocation. Each set of encoding parameters is tested resulting in an operating Rate-Distortion (R-D) point, which represents the mean bit rate and distortion achieved by the coding using that parameter set. When performing RDO, the encoder searches the parameter set space looking for the points that are as close as possible to the optimal R-D curve, as shown in Fig. 2.

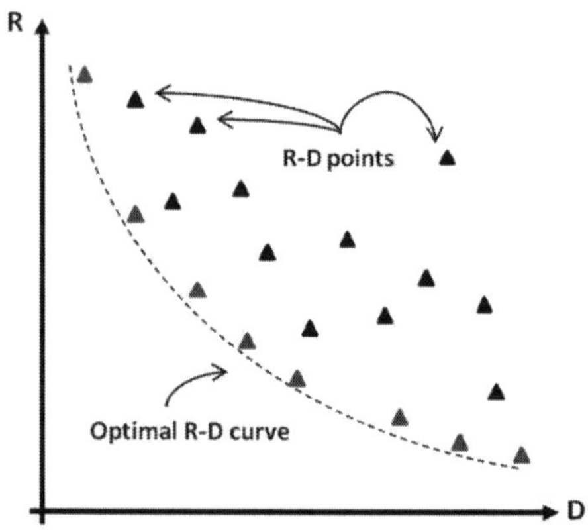

Fig. 2. Rate-Distortion points and optimal curve

The R-D optimisation task can be solved using Lagrangian optimisation, in which the distortion term is weighted against the bit rate term. The minimisation problem is represented as in (1), where J is the R-D cost to be minimized and λ is the Lagrange multiplier [5, 8].

$$\min\{J\}, where\, J = D + \lambda \cdot R \tag{1}$$

The RDO technique is very effective, yielding the best possible choice among all the encoding parameter sets. However, if every combination of operating modes is tested and evaluated by exhaustive search over the parameter and mode space, the computational complexity involved in the task becomes a limiting factor. In practical applications, a number of operating modes must be ignored in the RDO process to comply with the limitations imposed by available computational resources, but the selection of such modes is not a trivial task. Understanding the operations of video coding standards is the key to design a video coding system which manages efficiently the computational resources.

The next subsections will detail the operations of the current state-of-the-art video coding standards, H.264/AVC and HEVC. An overview of the computational complexity of both standards is also presented.

2.1 H.264/AVC

The H.264/AVC video coding standard was developed by the Joint Video Team (JVT) with the main target of doubling compression rates in comparison to any other existing standard [5]. Several innovations were introduced in the encoding process, such as the intra-frame prediction, the Hadamard transform, the use of multiple reference frames for inter-frame prediction and the mandatory use of a deblocking filter. A significant gain in terms of compression was achieved in H.264/AVC in comparison to previous standards at the cost of increased computational complexity.

In H.264/AVC each video frame may comprise one or more slices partitioned into macroblocks (MB) with 16x16 luminance samples and two chrominance blocks which dimensions depend on the colour sub-sampling scheme. Slices are self-contained in the sense that they are coded independently from other slices. The syntactic elements representing the coded data can be parsed from the bitstream and the image samples can be correctly decoded without reference to data from other slices provided that the reference pictures are identical at the encoder and the decoder [5].

Type I slices contain only I MBs, which are those encoded using intra-frame prediction. Type P slices may contain both I and P MBs, which are those encoded using the inter-frame prediction method. Type B slices contain I, P and B MBs. MBs of type B are also encoded with inter-frame prediction, but they allow the use of two reference frames instead of only one to form predictions. There is another important type of MBs, allowed in P or B slices, called *SKIP*. When it is used, no information is encoded for that MB, except for a *SKIP* mode signalling flag. Usually, type I slices require much fewer computational resources to be encoded and decoded than type P and B slices, since the prediction is performed only based on the pixels of the current frame. Besides that, fewer memory resources are required when encoding I slices, because no reference frames need to be stored for temporal prediction. On the other hand, the compression levels achieved with P and B slices are usually much higher than with I slices.

The kernel of the H.264/AVC encoder is composed by the following modules: intra-frame prediction, inter-frame prediction, direct and inverse transforms (T and IT), direct and inverse quantisation (Q and IQ), deblocking filter and entropy coding, as shown in Fig. 3. Each one of these modules uses different techniques and encoding modes.

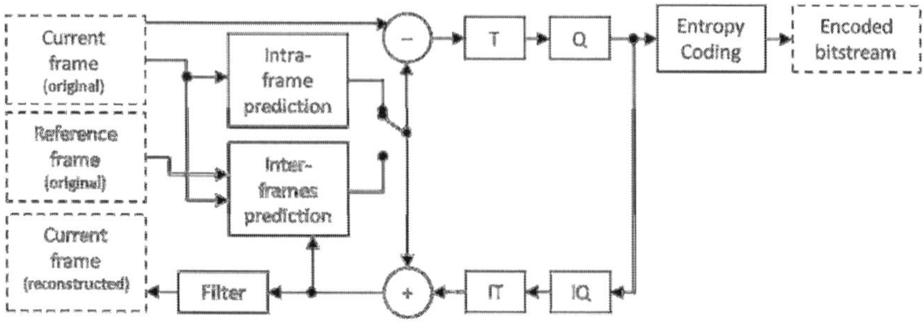

Fig. 3. H.264/AVC encoder

The intra-frame prediction in H.264/AVC operates on pixels located in the same frame. Each block is predicted from pixels in the neighbouring blocks according to 9 different modes for 4x4 blocks or 4 different modes for 16x16 blocks. The inter-frame prediction comprises the Motion Estimation (ME) and the Motion Compensation (MC) modules. During motion estimation each block in the frame is compared to blocks in the reference frames (up to 16 reference frames are allowed in H.264/AVC) and the best match is used for the motion compensated prediction. Different block sizes can be used in the inter-frame prediction (16x16, 16x8, 8x16, 8x8, 8x4, 4x8, or 4x4). A motion vector is computed to indicate the reference block position in the reference frame in relation to its position in the frame being encoded.

The encoder must take several decisions during the encoding process: whether intra-frame or inter-frame prediction is used, which is the best block size (16x16 or 4x4) and the best mode for the intra-frame prediction, which block size should be employed in inter-frame prediction, whether or not the *SKIP* MB and the bi-prediction are used, which one is the best reference block for prediction in the reference frames, and so on. If the RDO method is used, every possibility is considered and thus all combinations are tested and compared in terms of Rate-Distortion efficiency. The amount of resources such as processing and storage elements required to perform all these decisions is a major preoccupation of researchers, designers and engineers working on H.264/AVC systems. The next section presents a brief overview on H.264/AVC computational complexity.

2.2 H.264/AVC Computational Complexity

Several works have been published addressing the problem of complexity analysis of video encoders and decoders. Processing time has been largely used to measure computational complexity, but other approaches such as theoretical analysis, direct analysis of reference code and power consumption evaluation have also been frequently applied to measure complexity.

The computational complexity of H.264/AVC is evaluated in [9] by comparing its encoding speed to that observed in the H.263 and H.263+ encoders. The computational complexity was evaluated under seven different configurations, in which the number of possible block sizes was incremented from 1 to 7. It was observed that the H.264/AVC encoder is 1.2 times more complex than the H.263 encoder when only 1 block size is allowed and 2.8 times when 7 block sizes are allowed. The computational complexity of operations composing the Motion Estimation was also evaluated, since it is the most time-consuming task of the H.264/AVC encoder.

Several tools of the H.264/AVC encoder are analysed in [10] in terms of computational complexity and compression efficiency. Comparisons between H.264/AVC and MPEG-4 Part 2 are also presented. The authors analysed 18 possible configurations, comparing all of them in terms of computational complexity versus PSNR in order to identify the best ones. The *Atomium* software profiler was used to obtain the complexity data by executing the instrumented code with representative test stimuli. The obtained results show that even though a compression efficiency increase of up to 50% is achieved with H.264/AVC, the computational complexity increases by a factor of 2 in the decoder and by more than one order of magnitude in the encoder, when compared to MPEG-4 Part 2.

2.3 High Efficiency Video Coding

The High Efficiency Video Coding standard (HEVC), recently approved as international standard, is based on the same classic hybrid video coding scheme of H.264/AVC, also being based in inter-frame/intra-frame prediction and transform coding [11]. The encoding structure of HEVC is similar to the coder structure presented in Fig. 3, even though the actual operations performed by each module are not exactly the same in both standards. For instance, in HEVC the *Filter* module of Fig. 3 is composed by the Deblocking Filter and the Sample Adaptive Offset, as explained later.

Nevertheless, essential modifications were introduced in HEVC [12]. One of the most important changes is the use of quadtree-based frame partitioning structures, which are broadly discussed in [13]. In HEVC, each video slice is divided into a number of square blocks of equal size called treeblocks (CTB) or largest coding units (LCU), which are used as the roots of each coding tree. Each leaf of the coding tree is called a coding unit (CU) and its dimensions can vary from 8x8 to 64x64 luminance samples (or up to the CTB size), depending on the tree depth. Fig. 4 shows an example of how CUs are organised in the form of a coding tree. In order simplify the visualisation, Fig. 4 shows only the divisions for the first CU at each depth level. The CU at the first depth is a LCU (a 64x64 CU, in the example), which may be divided into four smaller CUs (32x32). This process is repeated until the maximum coding tree depth allowed is reached (4th CU depth, in the example). This is where a significant amount of computational resources are consumed, since the whole encoding process must be performed for each possible coding tree configuration if a RDO-based decision method is used.

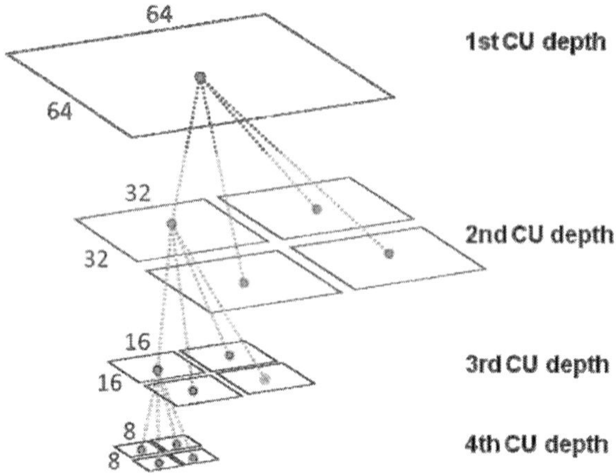

Fig. 4. Quadtree-based structure for coding units (CUs)

For inter-frame prediction, each CU can be divided into two prediction units (PU) of up to five different shapes (called *2Nx2N*, *2NxN*, *Nx2N*, *2Nx(N/2+3N/2)* and *(N/2+3N/2)x2N*). For intra-frame prediction, the CU can be split into four PUs (the *NxN* shape) if it is located at the last level of the coding tree. Otherwise, it is always predicted using the *2Nx2N* shape.

When transform coding the prediction residue, each CU is assumed to be the root of another quadtree-based structure called residual quadtree (RQT). Each leaf of the RQT is known as a transform unit (TU) and can take on sizes from 4x4 to 32x32 luma samples.

Intra-frame prediction in HEVC is an extension of intra-frame prediction in H.264/AVC, supporting larger block sizes (from 4x4 to 64x64) and allowing more prediction directions: 16 different luma angular modes for 4x4 PUs, 2 luma angular modes for 64x64 PUs and 33 luma angular modes for the other PU sizes. Besides these directional prediction modes, a DC prediction mode similar to DC in H.264/AVC and a planar prediction mode designed to reconstruct smooth image segments through bilinear interpolation from the PU borders may also be applied to any PU.

Similarly to H.264/AVC, HEVC inter-frame prediction relies on block-based Motion Compensation (MC). Motion compensation uses inter-frame motion information obtained by the motion estimation (ME) step, which can use up to 16 reference frames in HEVC. The ME is based on Advanced Motion Vector Prediction (AMVP), which includes a Motion Vector Competition scheme [14] along with a Motion Merge [15] technique. The former defines a set of motion vector predictor candidates comprising motion vectors of spatially and temporally neighbouring PUs. The best motion vector predictor is selected from the set of candidates through RD optimisation. Motion merging is based on a similar principle, but in this case the motion parameters are not explicitly transmitted. They are instead derived at the decoder based on the prediction mode and merge mode parameters.

Transform and quantisation follow similar principles to those used in H.264/AVC, even though several new features were introduced in HEVC. Besides larger transform block sizes (4x4, 8x8, 16x16, and 32x32), mode dependent coefficient scanning, last significant coefficient coding, improved significance flag context modelling, multilevel significance maps, and sign data hiding are some of these features [16]. An integer approximation to the Discrete Cosine Transform (DCT) is applied to the prediction residue and a mode dependent integer function based on Discrete Sine Transform (DST) is alternatively specified for residuals from 4x4 intra-frame prediction [12].

Regarding the entropy coding, HEVC supports only Context-Adaptive Binary Arithmetic Coding (CABAC), which was already used in H.264/AVC. Even though the CABAC structure is basically the same as it was in H.264/AVC, HEVC aims at decreasing the computational burden associated with binarisation, context update and access by reducing the number of syntax elements to binarise and the number of contexts, and also by increasing throughput by enabling parallel processing of quantized transform coefficients [17, 18].

The HEVC deblocking filter (DF) is applied to boundaries belonging to CUs, PUs, and TUs larger than 4x4 pixels [19]. Vertical and horizontal edges are filtered according to a border filtering strength, which varies from 0 (no filtering) to 4 (maximum filtering strength) and also depends on the border characteristics. In addition to the DF, HEVC employs the Sample Adaptive Offset (SAO) [20]. SAO processes the entire picture as a hierarchical quadtree, classifying the reconstructed pixels into different categories and then reducing distortion by adding an offset to the pixels in each category. This classification is performed taking into consideration the pixel intensity and edge properties.

Internal bit depth increase may also be used in some encoder configurations in HEVC for increased calculation accuracy. In such cases, a 2-bit extension is applied to each sample for additional calculation precision, which means that each sample is multiplied by 4 before further calculations.

If the RDO method is used, the encoder must take a huge number of decisions during encoding time, such as: the best coding tree configuration, the best PU shape division, the best RQT structure, the best intra-frame and inter-frame block shapes, the best intra-frame mode for luma and chroma samples, the best inter-frame mode and reference block in the reference frames, the best strength for the DF and classifications for the SAO, and so on. Therefore, both the encoding structures and coding tools used are responsible for an enormous share of the overall amount of computational resources used in HEVC, as explained in the next section.

2.4 HEVC Computational Complexity

In [21], the computational complexity of the modules composing the HEVC Test Model (HM) encoder was extensively analysed using the *Intel VTune Amplifier XE 2011* software profiler [22]. A set of 16 different encoding configurations was used to investigate the impact of each tool, varying the encoder parameter set and comparing the results with a baseline HEVC encoder and with an H.264/AVC encoder. It was observed that HEVC presents a computational complexity increase varying from 9% up to 502% in comparison to H.264/AVC High Profile, depending on the HEVC encoding configuration. The work also shows that some tools such as the Hadamard Motion Estimation and the Filters yield a better R-D performance than other tools and therefore should be first enabled in a complexity-constrained system.

Even though CUs, PUs and TUs only represent groups of samples to be encoded, predicted and transformed together as a single unit, the definition of their size is a very complex task. If the RDO process is used, the process of defining the optimal combination of CUs, PUs and TUs would involve encoding the image using all possibilities allowed in HEVC, comparing their RD costs and finally choosing the best one. An analysis of the computational complexity for defining coding tree structures is presented in [23]. Fig. 5 shows average results from [23] in terms of percentage of computational complexity involved for encoding CUs at each depth of the tree structure. Each CU depth is presented in the y-axis in Fig. 5 and the computational complexity is presented in the x-axis. As previously explained, the nature of the data structures used in HEVC leads to nested encoding loops, such that CUs at large tree depths are encoded inside CUs at smaller tree depths. This is shown in Fig. 5 as follows: for each CU depth (vertical axis), the computational complexity is divided in three components: (1) the complexity of performing inter-frame prediction for CUs at the current depth (in grey); (2) the complexity of coding the same CU as sub-CUs at a larger depth (in white); and (3) the complexity of other operations (in black).

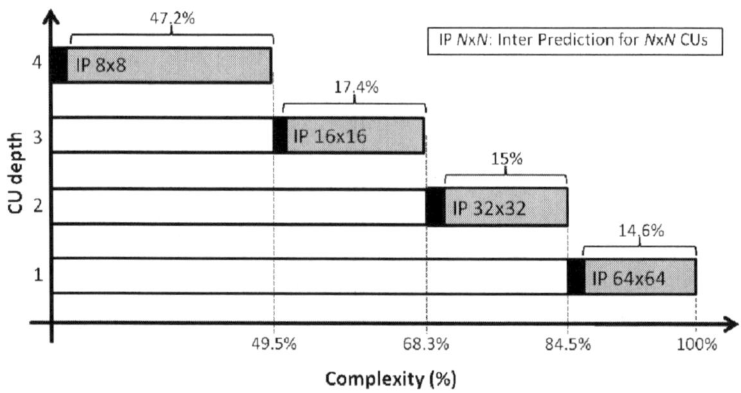

Fig. 5. Computational complexity associated to each coding tree depth [23]

It can be observed in Fig. 5 that larger CU depths are responsible for most of the encoding computational complexity (especially in *IP 8x8*). For example, although compressing CUs in the third depth represents 68.3% of the overall computational complexity, in fact only 17.4% is the real complexity associated to this CU depth, whereas almost all the remaining complexity is dedicated to inter prediction for 8x8 CUs (*IP 8x8*) and other operations (49.5%). These results provide relevant insight about the distribution of complexity over CUs of different sizes and take into account the nested coding structure used in HEVC.

The HM encoder and decoder complexity and implementation were analysed in [24]. The times spent by the main modules were measured with a software profiler, as in [21]. It was observed that in the *All Intra* encoder configuration the most time-consuming modules are the Transform and Quantisation (24.4%), due to the Rate-Distortion Optimised Quantisation technique, the Intra Prediction (16.6%) and the Entropy Coding (2.2%). In the *Random Access* encoder configuration, the Motion Estimation takes up a significant portion of encoding time (38.8% for SAD calculations and 19.8% for fractional pixel search refinement). The decoding computational complexity is dominated by the Inverse Transforms (15.9%) and the Filters (12.9%) in the *All Intra* configuration, and by the Motion Compensation (24.8%) and the Filters (12.4%) in the *Random Access* configuration.

According to [24], HM encoding time is around 5000 larger than the real-time encoding requirements. Decreasing its computational complexity is still a serious challenge for researchers and engineers, who are expected to develop heuristics for simplifying HEVC compatible encoder implementations in the coming years.

This section and the previous one addressed the importance of computational complexity management in video coding systems, especially in those cases in which computational resources are not abundant and battery resources are scarce. Besides analysing computational complexity, several works have also focused on modelling, reducing and controlling the computational complexity of video encoding and decoding operations. The next sections will review these works.

3 Modelling Computational Resource Use

The key for designing complexity-aware video coding applications which carefully manage computation resources use is to be able to detect when the encoding operations will exceed the system's available computational resources. The idea is to detect such cases without the need of performing the actual operations. Modelling computational complexity of video codecs is an important task, since it allows estimating the amount of computational resources that a certain operation will require prior to its execution.

Nevertheless, it is not a trivial task because video encoders are composed of several interdependent tools with parameterised functionalities that can deeply affect the level of computational complexity. Besides, according to [25], in the case of a complex multimedia system an overall view may not be enough for a complete comprehension of it; even though detailed information about all the constitutive blocks and their theoretical complexity may be available, the overall codec behaviour and complexity are very difficult to be understood, because they depend on the input data characteristics.

In the recent past modelling the computational complexity of video encoders and decoders was investigated in several works. In [26-28], the authors focus on modelling the computational complexity of inter-frame prediction operations at the

H.264/AVC decoder due to their large share in the overall complexity. Since different inter-prediction modes correspond to different complexities, this was taken into account to estimate the overall decoding complexity in [26].

The number of motion vectors was used in [27] to estimate the decoding complexity of a MB. According to the authors, the decoding computational complexity tends to decrease with the number of motion vectors in a MB. In [28], the number of interpolation filter operations was used for the complexity estimation. A MB with fewer interpolation filter operations should have lower decoding complexity.

The computational complexities of both intra-frame and inter-frame prediction are modelled in [29] for the H.264/AVC decoder. The MC complexity is modelled as a linear function of the number of cache misses, the number of interpolation filters and the number of motion vectors per MB. The intra-frame prediction decoding complexity is modelled as a function of the number of prediction directions for a specific intra type.

There are some approaches, which deal with the computational complexity of video encoders and decoders based on their basic data structures. In H.264/AVC, these structures are blocks, MBs, frames and Group of Pictures (GOP). In [30], the computational complexity of the H.264/AVC encoding process for a frame is estimated based on the computational complexity of the previous frame. A frame is divided into multiple groups of MBs, which are further used by the complexity control algorithm, which assigns different complexity budgets for each group. The complexity of the reference frame is computed by simply summing up the complexity of each group of MBs, which in turn is computed by summing up the complexity of each MB.

Some works have also proposed computational complexity models for video encoders, which are dependent upon configuration parameters used for complexity control. The H.264/AVC encoder complexity was modelled in [3] by first dividing it into three major parts: *ME* (Motion Estimation), *PRECODING* (Transforms, Quantisation and picture reconstruction) and *EC* (Entropy Coding). The *ME* complexity is given by (2), where C_{SAD} is the computational complexity of a SAD operation and λ_{ME} is the number of SAD calculations per frame, which is the complexity control parameter for the *ME* module. The *PRECODING* complexity is given by (3), where C_{NZMB} is the *PRECODING* computational complexity of one nonzero MB and λ_{PRE} is the number of nonzero MBs in a frame, which is the complexity control parameter for the *PRECODING* module. Here, a NZMB is a MB with nonzero DCT coefficients after the quantisation. Finally, the *EC* computational complexity is given by (4), and it is proportional to the bitrate R. In (4), C_{BIT} is the per bit complexity associated to *ENC* and S is the size of the picture, which is needed because R represents the coding bitrate in the unit of bits per pixel. The total computational complexity C of the video encoder, is given by (5), where λ_F is the encoding frame rate. The model represents a complexity-scalable architecture for video encoding, for which the computational complexity is controlled by the parameters λ_{ME}, λ_{PRE} and λ_F.

$$C_{ME} = \lambda_{ME} \cdot C_{SAD} \tag{2}$$

$$C_{PRE} = \lambda_{PRE} \cdot C_{NZMB} \tag{3}$$

$$C_{ENC} = S \cdot R \cdot C_{BIT} \tag{4}$$

$$C(R; \lambda_{ME}, \lambda_{PRE}, \lambda_F) = \lambda_F \cdot (\lambda_{ME} \cdot C_{SAD} + \lambda_{PRE} \cdot C_{NZMB} + S \cdot R \cdot C_{BIT}) \tag{5}$$

Similarly, in [31] the computational complexity of the H.264/AVC encoder is also modelled by dividing it into three modules: *ME*, *PRECODING* and *ENC*. However, when the RDO method is enabled the encoder executes the ME, Direct Transform, Direct Quantisation and Entropy Coding modules for each MB mode tested. Therefore, two models can be used: the first considers MB modes in the complexity estimation for the cases in which RDO is enabled, while in the second model RDO is off. The first model is shown in (6), where C_{RDOon} represents the encoding complexity of the frame and C^i_{16x16} - C^i_{I4x4} represent the complexity of each mode at the ith MB. C^i_M is the complexity of Direct Transform, Direct Quantisation and Entropy Coding of each mode for the ith MB. C^i_{PRE} and C^i_{ENC} are the complexity of the *PRECODING* and *ENC* modules after the mode decision process at the ith MB. The second model is shown in (7) for C_{RDOoff}. Since in this case the RDO method is off, the mode decision is only based on the results of the *ME* process, so that C^i_M is removed from (6).

$$\begin{aligned} C_{RDOon} = \sum_{i=0}^{M-1} (C^i_{16x16} + C^i_{16x8} + C^i_{8x16} + C^i_{8x8} + C^i_{I16x16} + C^i_{I4x4} + C^i_M) \\ + \sum_{i=0}^{M-1} (C^i_{PRE} + C^i_{ENC}) \end{aligned} \tag{6}$$

$$\begin{aligned} C_{RDOoff} = \sum_{i=0}^{M-1} (C^i_{16x16} + C^i_{16x8} + C^i_{8x16} + C^i_{8x8} + C^i_{I16x16} + C^i_{I4x4}) \\ + \sum_{i=0}^{M-1} (C^i_{PRE} + C^i_{ENC}) \end{aligned} \tag{7}$$

In [32], the computational complexity of H.264/AVC encoders is modelled based on the complexity of each prediction mode (*SKIP*, *Inter 16x16*, *Inter 16x8*, *Inter 8x16*, *Inter 8x8*, *Inter 8x4*, *Inter 4x8*, *Inter 4x4*, *P8x8*, *I4x4*, *I16x16*). The model for the full computational complexity of a P frame is calculated as in (8), where W^{RD}_i indicates the relative R-D computational cost for mode *i*, W^{ME}_i denotes the factor of single-reference ME for the inter modes, N_{MB} is the number of MBs in the frame, N_{RF} is the number of reference frames, W^O is the complexity consumed by the remaining encoder modules.

$$C_{Frame}^{Full} = N_{MB} \cdot [W_0^{RD} + \sum_{i=1}^{7} (W_i^{RD} + W_i^{ME} \cdot N_R) + W_8^{RD} + W_9^{RD} + W_{10}^{RD} + W^O] \tag{8}$$

The computational complexity of the H.264/AVC decoding process is modelled in [33] for complexity control, using an approach where each decoding module (DM) is separately modelled. Entropy Decoding complexity is modelled as the product of bit decoding complexity and the number of bits involved, as shown in (9), where C_{vld} is the DM complexity, k_{bit} is the average number of cycles required for decoding one bit and n_{bit} is the number of bits for a frame. Side information preparation, including MB sum/clip and deblocking filter strength calculation, is modelled as in (10), where C_{sip} is the DM complexity, k_{MBsip} represents the average clock cycles for side information preparation per MB and n_{MB} is the number of MBs in a frame. Inverse Transform complexity is modelled in (11), where $k_{MBitrans}$ is a constant that describes the complexity of MB de-quantisation and inverse transform and n_{nzMB} is the number of nonzero MBs per picture. Intra-frame prediction is modelled in (12), as the product between the average complexity of intra-frame prediction in one intra-coded MB ($k_{intraMB}$) and $n_{IntraMB}$ is the number of intra MBs per frame. The MC computational complexity model is given in (13), where k_{half} is the average complexity required to conduct a 6-tap Wiener filtering and n_{half} is the number of filterings to decode a frame. The Deblocking Filter is modelled as in (14), where k'_{α} and k'_{β} are the complexities to perform strong and normal filterings, respectively, and n_{α} and n_{β} are the numbers of strong and normal filterings, respectively. Finally, the total complexity required to decode a frame is expressed as in (15), where $k_{CU(DM)}$ is the complexity to decode a basic coding unit (a MB, for example) in a particular module, $N_{CU(DM)}$ is the number of basic coding units to be decoded by that module.

$$C_{vld} = k_{bit} \cdot n_{bit} \tag{9}$$

$$C_{sip} = k_{MBsip} \cdot n_{MB} \tag{10}$$

$$C_{itrans} = k_{MBitrans} \cdot n_{nzMB} \tag{11}$$

$$C_{Intra} = k_{IntraMB} \cdot n_{IntraMB} \tag{12}$$

$$C_{mcp} = k_{half} \cdot n_{half} \tag{13}$$

$$C_{dblk} = k'_{\alpha} n_{\alpha} + k'_{\beta} n_{\beta} \tag{14}$$

$$C_{frame} = \sum_{DM} C_{DM} = \sum_{DM} k_{CU(DM)} \cdot N_{CU(DM)} \tag{15}$$

4 Computational Resource Saving Strategies

Low-complexity algorithms for video encoders and decoders have been proposed in the last years as an attempt for enabling their use in complexity constrained platforms. As the encoder's computational complexity is much higher than the decoder's, most works focus on its modules and features.

4.1 H.264/AVC

As explained in section 2.2, ME and Mode Decision are the most complex tasks in the encoding process and this is the reason why most works focus on them to decrease the encoding computational complexity. The next sub-sections will present the most important works found in the literature for ME and MD complexity reduction.

4.1.1 Motion Estimation

Block matching motion estimation (BMME) is used to determine motion vectors (MV) representing the displacement of pixel blocks in the current frame in relation to the best matching block in the reference frame. Several algorithms have been proposed for searching candidate blocks and the algorithmically simplest but the most demanding in terms of computation needs is the Full Search (FS). In FS, the best match is found by searching all possible candidate blocks in the search window (SA), which leads to the optimal result. However, the computational complexity involved in this process is very high and faster approaches are used.

Several Fast Motion Estimation (FME) techniques have been proposed in the literature. They may be classified into two categories: (a) those which decrease the number of candidate blocks in the SA and (b) those which decrease the computational complexity required to compare such blocks to the current block (i.e., distortion measure computation, such as SAD or MSE).

Examples of the first category are the Three-Step Search (TSS) [34], the Block-Based Gradient Descent Search (BBGDS) [35], the New Three-Step Search (NTSS) [36], the Diamond Search (DS) [37, 38], the Hexagon-Based Search (HEXBS) [39], the One at a Time Search (OTS) [40], the Dual Cross Search (DCS) [41], and so on. All of them are sub-optimal algorithms which present a rate-distortion performance equal or less than that obtained with the Full Search, with the advantage of significantly decreasing the ME computational complexity.

Methods from the second category include techniques such as sub-sampling [42], Partial Distortion Search (PDS) [43], Normalized PDS [44, 45] and the Successive Elimination Algorithm (SEA) [46]. Even though these methods are capable of maintaining a good rate-distortion performance, their speedup gain is limited.

There are still other methods, which reduce ME computational complexity by applying different techniques. In [47], the authors propose a strategy based on sorting the MVs and coding modes such that the decision process is stopped when the bit rate for a MV or coding mode exceeds the average bit rate of any previous MV or mode tested, skipping the evaluation of some cases. In [48], a controller is added to the encoder in order to extract from the motion search results on the input signal statistics and use them to dynamically configure the ME parameters, such as the number of reference frames, valid block modes and search area for each 16x16 block and its sub-partitions. A method for estimating which MBs can be encoded in *SKIP* mode without trying the other modes is proposed in [49]. The authors propose a Bayesian framework using the R-D cost difference between coding and skipping a MB as the single decision feature. Savings in processing time of more than 80% for low-motion sequences are claimed by the authors at a cost of small quality decrease.

4.1.2 Mode Decision

The Mode Decision process involves different operations from various encoder modules. If RDO is enabled, then every possible mode is computed and their results in terms of R-D performance are compared. Many works based on heuristics have been proposed to decrease the number of modes to be tested during both intra-frame and inter-frame prediction.

In [50-53], the authors propose fast algorithms to select only some intra-frame prediction modes to be compared in the mode decision stage. The authors in [54] and [55] claim that not all intra modes need to have the rate-distortion performance evaluated, especially those with a high Sum of Absolute Transformed Differences (SATD) value, since they tend to result in high encoding bit rates. This way, only those modes with an SATD value lower than a threshold are evaluated and the remaining modes are discarded, saving computations.

In [56], the transforms are applied to the intra-frame prediction residue and the transformed coefficients are analysed in order to detect which is the best mode to be used. In [57] the low-frequency transformed coefficients for each MB are analysed in order to decide the best intra-frame block size (4x4 or 16x16).

A two-step algorithm for intra-frame mode decision was proposed in [58] based on heuristics. The first step consists in performing intra-frame prediction for every mode in both 16x16 and 4x4 prediction block sizes and computing the distortion between the original and predicted blocks. The mode leading to the smallest distortion is used as the best mode for its corresponding block size. In the second step, the MB homogeneity is computed in order to decide whether the 16x16 or the 4x4 block size is used. A pruned DCT is applied to the original MB in order to compute a subset of coefficients, which are enough for detecting its homogeneity level. Accordingly, homogeneous areas of the image must be encoded with large blocks, while heterogeneous areas must be encoded using small blocks.

In [59], a hierarchical fast inter-frame mode decision was proposed based on the calculation of stationarity, homogeneity and block border strength. The method is divided into three steps. In the first one, stationarity is detected by calculating the distortion between a MB and its co-located MB in the reference frame. If stationarity is detected, the MB is encoded as *SKIP* and the MD process is terminated. Otherwise, the second step takes place and the homogeneity of the MB is computed as in [58] in order to decide if the ME is performed with large or small blocks. In the third and final step, the intensity variation of luminance samples along the edges of possible blocks is calculated in order to detect the block shape which best fits the image for inter-frame prediction, eliminating the need of testing all the remaining block shapes. This method was integrated with the intra-frame decision heuristics proposed in [58], forming a complete mode decision scheme for H.264/AVC in [60].

According to [61], only 3% of the modes chosen in P frames are intra prediction modes, on average. This way, the evaluation of inter-frame modes prior to the evaluation of intra-frame modes is proposed in this work as a way of reducing the necessary computational resources without significant decrease in image quality or bit rate. In [62] a set of heuristics is presented for speeding-up the complete mode decision process (i.e., both intra and inter-frame predictions). The authors observe that homogeneous and static regions are mostly encoded with *P16x16* and *SKIP* modes, so that a Sobel Operator is used to detect homogeneity and the SAD calculation is used to detect stationarity. The proposed algorithm reduces the encoding complexity through RDO early termination when one of the specific conditions (homogeneity or stationarity) is true. If neither is true, then all the remaining prediction modes are evaluated.

In [63] the inter-frame mode decision complexity is decreased by analysing the spatial continuity of the motion field, which is generated by ME using 4x4 pixel blocks. A set of experiments led to the conclusion that video regions with high motion continuity present a higher probability of being encoded with *P16x16*, *P16x8* and *P8x16* modes, while regions with low motion continuity are mostly encoded with *P8x8* and smaller block sizes. As in [62], this work also uses the Sobel Operator to identify the borders of objects in an image.

4.2 HEVC

Even though the HEVC standard has just been finalized, there is already a small number of works published in the literature presenting methods to reducing the encoder's computational complexity. Most of these works aim at decreasing the computational complexity involved in the definition of the new quadtree-based structures, namely the Coding Units, the Prediction Units and the Transform Units. Other works, as in H.264/AVC, focus on decreasing the complexity of the most resource demanding functions, such as the Motion Estimation and Mode Decision.

4.2.1 Coding Unit Size and Depth Decision

A fast splitting and pruning method for intra coding is proposed in [64]. It was found that the main contributor to the computational complexity of intra prediction in HM is the calculation of full RD costs (J_{FRD}) for deciding intra prediction modes of CUs at all possible coding tree depths using RDO. In order to reduce this, the method allows skipping the J_{FRD} calculation for determined CUs and allows terminating subsequent CU splitting and pruning process. A low-complexity RD cost calculation (J_{LRD}), which does not require the whole encoding process to be completed, is used to determine the best intra prediction mode in each possible CU. The early splitting test evaluates the possibility of each possible CU being split into sub-CUs based on a Bayesian decision rule with the J_{LRD} costs. If it is determined that the CU is to be split into its sub-CUs, the computation of J_{FRD} is not necessary. The idea behind the early splitting test is that a CU with sufficiently large J_{LRD} is likely to be split into its sub-CUs. When the CU is not split based on the result of the early splitting test, the early CU pruning test is performed based on the J_{FRD} value of the best prediction mode for that CU. The idea is that a CU with sufficiently small J_{FRD} value is likely to be a leaf CU and no further tests in lower depths are necessary. The statistical parameters of the Bayesian tests are periodically updated online so as to adapt to the changing characteristics of the video sequence. Experimental results have shown that the method is able to decrease the intra coding computational complexity by 50% with a *Bjøntegaard Delta* [65] bit rate (BD-rate) increase of 0.6%.

Temporal correlation in neighbouring frames is exploited in [66] in order to reduce the number of quadtree splitting decisions. Based on the tree depth used in the co-located CU in the previous frame and its neighbouring CUs, the encoder decides whether or not to split the current CU into sub-CUs. Additionally, an early *SKIP* mode decision at the prediction stage is performed to further reduce the computational complexity. Experimental results have shown that a computational complexity reduction varying from 20% to 33%, depending on the encoder configuration, was achieved with this method. BD-rate increases varied from 0.1% to 0.47%.

Feature extraction to assist fast decisions of CU splitting is performed in [67]. The CU size decision is based on a Bayesian rule to avoid RDO search on all possible CU sizes and its modes. Predicting CU split or not is formulated as a two-class classification problem and the features include information such as variance of prediction error, SATD between original and predicted pixels, MV magnitude, RD costs and others. An average complexity reduction of 41.4% was achieved with this method, in comparison to the original HM encoder. Average BD-rate increase is around 1.88%.

A fast CU depth decision based on spatial and temporal correlation is proposed in [68]. The authors claim that since usually successive frames are strongly correlated, especially with the high frame rates lately used in video sequences, the final depth information or split structure of co-located coding trees in neighbouring frames is also highly correlated. The algorithm first determines the depth search range according to the similarity degree between neighbouring CTBs. Three classes of similarity are defined (high, medium, low). Then, once the depth range is settled, the final searching depths can be derived by selecting depths with high probability of occurrence and excluding low probability depths. Experiments have shown that the method is able to reduce the encoding computational complexity in 25% with an increase of 0.16% in bit rate.

In [69], a fast coding unit decision algorithm which operates in both frame level and coding unit level is proposed for computational complexity reduction of the HEVC encoder. The algorithm focus only on the inter mode early selection, since intra mode is essential for encoding high resolution video. The main idea of the frame level part of the method is to skip those depths which are rarely used in the reference frame. The number of CUs encoded at a certain tree depth in a frame is compared to a threshold in order to detect its level of usage. The coding unit part of the method relies on the fact that motion and texture detail of one particular part of the image tends to stay the same from one frame to another. By checking the spatial and temporal neighbouring CUs of a certain CU, the candidate CU depth can be predetermined. An average decrease of 45% in the ME computational complexity was achieved with this method. Bit rate increases remained under 0.3%.

4.2.2 Prediction Unit and Transform Unit Size Decision

A heuristic method to reduce the computational complexity of deciding the Prediction Unit (PU) size is proposed in [70]. The Motion Vectors Merging (MVM) merges *NxN* Prediction Units (PUs) in order to compose larger ones instead of performing ME for every possible PU partition. The method is applied to decide all PU sizes larger than *NxN*. When certain conditions are met, *NxN* partitions are merged into *2NxN*, *Nx2N* or *2Nx2N* partitions without performing the ME operations for each one of them. Initially, ME is performed only for the four *NxN* partitions in a CU. If the MV for the four partitions is the same, they are merged into one single *2Nx2N* PU. If they are not the same, the rectangular shapes are tested in a similar manner. Experimental results point out an average computational complexity reduction of 34% with an average bit rate increase of 1.37% and a PSNR drop of 0.08 dB.

In [71] a two-stage PU size decision algorithm is proposed to speed up the intra coding process in HEVC. In the first stage, before intra-frame prediction starts, texture complexity of CTBs and its sub-blocks are measured in order to filter out unnecessary PU sizes. The threshold for filtering PU sizes is calculated dynamically according to the content of the video sequence and to pre-defined coding parameters.. The frame texture complexity is calculated by down-sampling each 64x64 CTB to a 16x16 block and then computing its variance. In the second stage, which takes place during the intra-frame prediction, the PU sizes of neighbouring 32x32 blocks are analysed in order to skip small PUs. The average computational complexity reduction for the intra coding process achieved in this work varies from 28.8% to 44.9%, depending on the video resolution. Average bit rate increases and PSNR decreases stayed under 0.47% and 0.02 dB, respectively.

An early Transform Unit (TU) decision method for HEVC encoders is proposed in [72]. By analysing the residual information, it was possible to detect that the determined TU sizes and the number of nonzero coefficients are strongly correlated. This way, the number of nonzero coefficients is used in this method as a threshold to stop further R-D cost evaluation. A computational complexity decrease of 61% in the TU processing was achieved in this method with little losses in terms of compression efficiency.

4.2.3 Mode Decision and Rate-Distortion Optimisation

An early *SKIP* mode detection scheme is proposed in [73] for complexity reduction of the HEVC encoding process. According to the authors, *SKIP* mode is chosen in about 83% of CUs and detecting its occurrence would allow ignoring all the remaining modes in the RDO process. The proposed method pre-detects *SKIP* mode using the differential motion vector (DMV) and coded block flag (CBF) information of inter *2Nx2N* mode. The encoder first searches the best inter *2Nx2N* mode (i.e., chooses between competition mode and merging mode) and after selecting the one with the minimum RD cost, it checks the DMV and the CBF of it. If they are both equal to zero, then the best mode is determined as the *SKIP* mode and the remaining PU modes are not tested. The method reduced computational complexity by about 35% with a BD-rate increase of 0.5%.

Two fast RDO techniques for HEVC are proposed in [74] aiming at saving computational resources at small RD performance loss. The first method, the *Top Skip*, avoids checking the RD cost for large block partitions when they are unlikely to be chosen. A starting CTB depth is selected based on the minimum depth of the co-located CTB in the reference frame. The second method, the *Early Termination*, avoids checking smaller blocks unlikely to be selected. The algorithm stops the CU splitting process if the best RD cost is already lower than a given threshold. The threshold is adaptively computed based on the standard deviation of RD costs relative to the CUs at spatially and temporally neighbouring CTBs. When both methods are integrated to the RDO process, a computational complexity reduction of 40% is achieved for the whole encoding process with an average BD-rate increase of 1.9%.

In [75], a low complexity RDO coding scheme based in three different solutions is proposed. The method reduces the number of available candidates for evaluation in the intra-frame prediction mode decision, in the reference frame selection and in the CU splitting. Specifically for intra prediction, the direction information of the neighbouring blocks is used to speed up the decision. In the reference frame selection, the spatial and temporal correlations among neighbouring frames and CUs are used to decrease the number of reference frames candidates. For inter CU decision, correlation between the energy of prediction residuals and the CU splitting is used to accelerate the CU splitting termination process. Experiments have shown that the scheme yields an average computational complexity reduction of 30% in the encoding process with an average BD-rate increase of 0.8%.

5 Computational Resources Use Management

To efficiently manage the available computational resources, it is not enough to implement low complexity methods. It is also necessary to design a control system in which computational complexity can be adaptively adjusted according to specific conditions, such as the device's battery status, time limitations imposed by the transmission environment and user preferences. Therefore such systems generally have multiple operation modes, which can be chosen dynamically by accepting the option made by a user, by sensing environment changes or even by predicting a user's preference.

In the last years, dynamic control of computational complexity in video encoding has been a very active research field. The ideal solution for a complexity management system would be to extend the original RDO problem to a third dimension such as Rate-Distortion-Complexity Optimisation (RDCO) or Power-Rate-Distortion Optimisation (PRDO). This solution would find the best encoder configuration leading to the optimal visual quality under predefined rate and computational complexity constraints. However, joint Rate-Distortion-Complexity analysis is an extremely complex task due to the huge number of possible combinations of modes and encoding parameters. As exhaustive search is infeasible in complexity constrained environments, several heuristic solutions have been proposed to dynamically adjust the computational complexity of video encoding in order to manage the available computational resources.

5.1 H.264/AVC

Due to its high computational complexity, the ME process is used by many methods to control the encoding computational complexity of H.264/AVC and other previous standards. A complexity control scheme for the MPEG-4 encoder was presented in [76] based on adjusting ME parameters and using the search options of Full Search, the Three-Step Search and the Spiral Search algorithms. The remaining parameters are the search window size, the SAD threshold for early termination of the Spiral Search algorithm, the use of pixel sub-sampling and the number of bits used to represent a pixel.

In [77], a classification-based method is proposed for the ME process. MBs are classified into different categories according to their importance in the frame and a complexity controlled ME scheme applies different operations to each MB according to its class. Initially, a total computation budget for the video sequence is divided and a computation budget is allocated for each frame. Then, the frame budget is divided into three independent sub-budgets, which are assigned for each class of MBs. When performing ME, each frame is classified into one of the three classes and a computation budget is allocated for each MB according to its class. Finally, according to the MB computation budget, the encoder allocates more or less computations into each ME step for it.

Three other adjusting parameters for ME are proposed in [30]: partial cost evaluation for fractional motion estimation (FRME), block size adjustment for FRME, and search range adjustment for integer motion estimation (IME). By combining these parameters, 12 configurations were defined and used in the tradeoff between compression efficiency and computational complexity. Based on the complexity measure from the previous frames, the complexity control algorithm allocates a certain budget for each group of MBs in an image and then for each MB within the group.

A two-stage complexity control method is proposed in [31] based on adjusting the ME operation. In the first stage, an encoding time control algorithm is applied. It consists of encoding the whole frame only using the *P16x16* mode and then, based on encoding time information for this mode, estimating the total encoding time, the target encoding time and the parameters to be used in the second stage for complexity control. In the second stage, the number of *SKIP* MBs in the frame is used to adjust the encoder computational complexity to a determined target complexity level.

As in other methods aiming just to reduce the computational complexity, many complexity control methods are based on the adjustment of MD and ME operations. In [78] a scheme to control complexity is proposed which is based on adjusting parameters that affect the aggressiveness of an early stop criteria for ME, the number of prediction modes tested in the MD, and the accuracy of ME steps for inter modes.

Before deciding the number of tested modes, they are ordered based on the statistical frequency of the optimal modes for a given type of video, so that the first modes tested are those which most probably yield the best R-D performance. The computational complexity is controlled by adjusting one single parameter, which is mapped to the algorithmic parameters based on a rule tuned by a training process that uses several typical video sequences.

The approach of sorting modes was also used in [79] to control the MD computational complexity. A ranking with the most popular ones, including intra and inter modes, compose a subset which is tested in the RDO process, while the remaining modes are suppressed from the tests. Initially, a few MBs are randomly selected for the frequency distribution analysis. Each mode is then associated with a frequency of occurrence and a computational complexity. Then, based on the target complexity to encode a complete image, the dominant mode set is chosen and used to encode the next frame.

In [80], the computational complexity control is divided into two problems: the first one consists in how to allocate the available computational resources to different frames and encoding modules, while the second consists on how to optimally use the allocated computational resources by adjusting the encoding parameters. To tackle the first problem, a computation allocation model is proposed to divide available resources among the video frames. The second problem is solved by using a complexity-adjustable ME and a complexity-adjustable MD. The ME complexity is adjusted by allowing more or less operations (such as fractional pixel ME, searching point refinement, etc.) to be executed, while the MD complexity is adjusted by allowing more or less modes to be tested in the RDO process. The list of tested modes is sorted according to their occurrence frequency in the spatial and temporal neighbouring MBs.

The MD complexity was controlled in [32, 81] by an adaptive computational allocation method at the MB level. The computational cost of ME in 16x16 blocks is taken as the basic computational unit and the costs for the remaining modes were obtained through practical simulations and represented as weighting factors of the basic unit. A target complexity is calculated based on the total computational complexity estimated for a frame, which depends on the number of MBs in the frame and on the computational weighting factors previously defined. The computational budget allocated for each MB is computed based on the target complexity, on the complexity consumed by the previously encoded MBs in the frame and on the number of MBs already coded. The computational budget for a determined MB is then used adaptively by choosing which modes are to be tested and which are not.

A MD early termination method is used in [82] to decrease the encoding process computational complexity. By calculating the difference between the cost of encoding a MB as *SKIP* and an estimated coding cost, the encoder is able to stop the MD evaluation process just after encoding the MB as *SKIP*. A threshold calculated using conditional probability estimates of skipped and not skipped MBs is used in the early termination decision. The authors also propose a complexity control method which aims at maintaining a target level of complexity through a feedback algorithm which updates probability models so that small rate-distortion performance losses are noticed.

Other parameters which control other operations besides ME and MD are explored in some works. In [83], an empirical study on the controllability of parameters for complexity control on video encoding cloud services is presented. The work shows experimental results in terms of encoding time, bit rate and objective quality when varying the number of B-frames, the level of refinement for sub-pixel ME and the operations performed in trellis quantisation.

In [84], the number of reference frames, the method used for sub-pixel ME, the partition sizes allowed for intra-frame and inter-frame prediction and the quantisation approach are used as the parameters to adjust computational complexity. Considering all possible combinations among these four parameters would result in a total of 3360 possible parameter settings, which is infeasible for real time applications. Two fast algorithms are devised for finding the parameter settings which leads to high distortion-complexity performance. The algorithms are based on the generalised *Breiman, Friedman, Olshen and Stone* (GBFOS) algorithm [85] and use training sequences to find the best parameter settings.

The number of motion search positions and the frame rate were the two parameters used in the method proposed in [86]. By using the Adaptive Critic Design (ACD) technique, a class of approximate dynamic programming methods, an online complexity control scheme was developed based on neural networks.

A two-level method is proposed in [87]. In the frame-level algorithm, the encoding process is not changed in order to maintain acceptable image quality, but frames are dropped when necessary to decrease the amount of computations. In the per-frame algorithm, computational complexity is controlled for each frame in order to achieve the target coding time. The frame-level algorithm calculates a target encoding time for each frame in the video sequence based on the total delay experienced by the frame in the input buffer, as if in a real time case. The target time is then used by the per-frame algorithm, which adjusts computational complexity in a frame by increasing or decreasing the number of MBs encoded as *SKIP*, similarly to [82].

A dynamic framework which consists of a set of optimised core components is proposed in [88]. The ME, DCT, quantisation and mode decision processes can be configured to achieve a desired computation-performance tradeoff in the encoder. These modules can all be assembled to form an H.264/AVC encoder with various degrees of computational complexity, which is able to adapt itself according to the available computational resources. Eleven parameters are used to adjust the computational effort applied to the different modules, such as the number of ME search points and whether or not DCT is applied to the residual MB. As determining the best combination of parameters of each video through exhaustive optimisation is quite computationally demanding, a simpler, suboptimal greedy optimisation method is used.

Rate-Complexity-Distortion Optimisation (RDCO) and other similar approaches have also been proposed in several works, which treat the problem as an extension of Rate-Distortion Optimisation (RDO). In [3] and [89], a Power-Rate-Distortion (PRD) analysis framework was developed in order to build a parametric video encoding architecture which controls computational complexity of its modules by varying encoding parameters. Based on the rate-distortion behaviour of these parameters and on their associated computational complexity, a PRD model was created and used to determine the best configuration of parameters according to the available power supply level of the device in which the encoder is implemented and on the target bit rate. The same authors propose in [90] an operational approach for offline PRD analysis and modelling based on a wide set of training data. Based on the models developed, a control database for online resource allocation and energy minimization is proposed.

Game theoretical analysis is used in [91] to model the power consumption in video encoders. The encoder is divided into modules, which are treated as players competing for the use of a computational resource on a limited budget aiming at maximizing its efficiency. In [92], the complexity dimension was added to the RDO strategy. For each particular encoder setup, the total bit rate (*R*), PSNR (*D*) and ratio between the time spent to encode a training sequence and the time spent by the full-featured encoder (*C*) are calculated. The RDC points are then plotted to 2D charts (for constant bit rate) and a lookup table is built from the convex hull points in order to provide optimal starting RDC points. The trellis quantisation, the level of refinement in ME and the number of partitions allowed are the parameters adjusted.

5.2 HEVC

Even though recent works have explored computational complexity reduction for HEVC, methods for computational complexity control are still rare. Some methods developed for H.264/AVC may be adapted for HEVC, especially those which use ME parameters to adjust computational complexity. However, there are still no works found in the literature besides those in [23, 93-95], which aim at dynamically controlling the HEVC encoder computational complexity by adjusting the maximum tree depth allowed for coding tree blocks.

5.2.1 Complexity Control Using Co-located Maximum Tree Depth

As already mentioned in regard to Fig. 5, the major contributors to the overall computational complexity in HEVC are the inner CUs in a coding tree [23]. The coding tree behaviour along the temporal domain was also investigated. Experiments have shown that the maximum coding tree depth is kept constant during long periods of the sequence, which means that once a tree depth is used in a determined area of the video, it tends to be used for a long period in co-located areas of adjacent frames. Based on these observations, a method aiming at constraining the maximum number of possible CUs tested in the RDO process was developed. The goal is to avoid processing CUs at large tree depths when computational resources are limited. To achieve this, the algorithm encodes periodically some frames without constraining the tree depths used in the CUs. The largest depths used in this unconstrained encoding process are then used as maximum depths allowed in the next frames. These two types of frames are called unconstrained (*Fu*) and constrained (*Fc*) frames, respectively. Each *Fu* frame is followed by a number of constrained frames (*Nc*), as shown in Fig. 6.

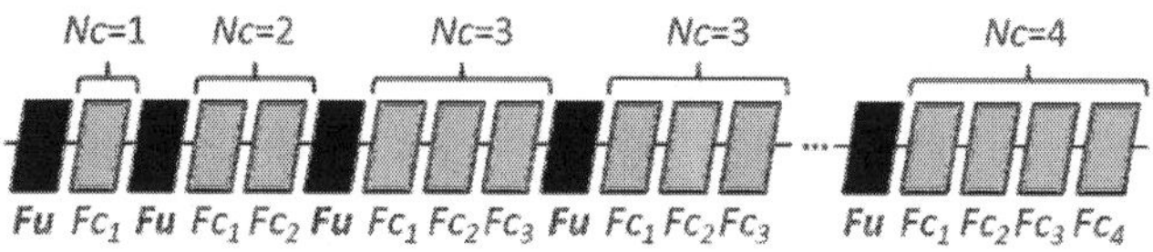

Fig. 6. Operation of the complexity control algorithm. Fu: unconstrained frames; Fc: complexity controlled (constrained) frames [23].

When a *Fu* is encoded, the maximum tree depth used in each CTB is defined by the RDO process and stored into a matrix of size *n*x*m* (*MTDM[n][m]*), where *n*, *m* are the number of CTBs in the horizontal and vertical axis of a frame, respectively, and MTDM stands for *Maximum Tree Depth Map*. When *Fc* frames are being encoded, the RDO process uses the information in *MTDM[n][m]* and limits the maximum tree depth tested in each CTB to the value saved in the corresponding MTDM position.

A target complexity is defined as a percentage of the maximum possible complexity that can be used to encode a predefined temporal segment of the video sequence. This is given by the maximum possible processing time that can be used to encode in real time all frames of the temporal segment. In portable devices, the target computational complexity can be either derived from a user-defined parameter or directly computed according to the battery energy level. The calculation of *Nc* is based on this target computational complexity *Tt*, on the complexity already spent in the encoding process *Td*, and on a prediction of the encoding complexity for the remaining frames within the temporal segment of the video sequence *Tp*.

When encoding a *Fc* frame, if the previously predicted complexity *Tp* plus the complexity already spent in previous frames is smaller than the target *Tt*, then more computation effort can be allocated and less frames need to be constrained. Thus, *Nc* can be decreased. Otherwise, *Nc* is increased. A new value for *Tp* is calculated after coding each group of *Nc*+1 frames according to

$$Tp = \frac{Te \cdot (Nt - Nd)}{(Nc + 1)},$$

where *Te* is the encoding complexity of the last *Fu* frame plus the encoding complexity of the last *Nc* frames of type *Fc*, *Nt* and *Nd* are the total number of frames in the video temporal segment and the number of frames processed so far, respectively.

The experimental results show that the computational complexity can be scaled down to 40% of the case in which no complexity control is used. A PSNR drop lower than 0.8 dB and a bit rate increase lower than 5.7% were noticed for the worst case (i.e., when the target complexity was set to 40% of the uncontrolled case).

5.2.2 Dynamic Complexity Control Using Co-located Maximum Tree Depth

In [93], the method proposed in [23] was extended in order to adjust the number of constrained frames (*Nc*) faster. Instead of incrementing and decrementing the value o *Nc*, it is recalculated after each group of *Nc* constrained frames according to the difference between the predicted complexity for the remaining frames within the temporal segment and the target complexity defined by the user or by the system in which the encoder is implemented. Fig. 7 illustrates the behaviour of the complexity control algorithm.

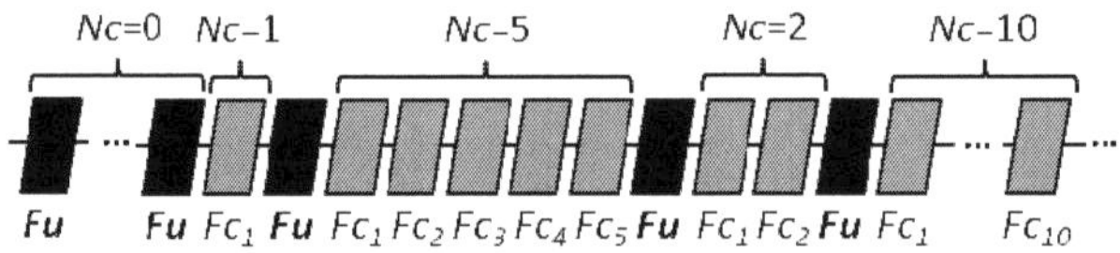

Fig. 7. Operation of the complexity control algorithm [93]

Experimental results have shown that the encoding computational complexity can be scaled down to 60% of the case in which no complexity control is applied. In the worst case, the average PSNR drop observed is lower than 0.1 dB and the average bit rate increase is under 4%.

5.2.3 Motion Compensated Depth Limitation for Complexity Control in HEVC

Even though the algorithms proposed in [23] and [93] provide an efficient way of controlling the computational complexity of HEVC, they assume that the maximum computational complexity is the same during the whole video sequence, which may not be true as fast motion and scene changes can induce broad variations in the computational complexity of the encoding operations. Besides, in such cases image areas may move away from the position where they were located in the last *Fu* frame before another *Fu* frame is encoded and the MTDM is updated. As a result, the *Fc* frames occurring between two consecutive *Fu* frames are encoded using a MTDM that may not be well matched to the current frame content with the mismatch becoming worse towards the end of the group of *Fc* frames. This problem is more severe in cases where the target complexity is small and thus *Nc* large.

To solve these problems, the maximum encoding complexity can be updated after each video segment and the MTDM is updated after encoding each *Fc* frame according to the average motion of each CTB, effectively motion compensating the MTDM [94]. As explained in 5.2.1, the maximum tree depths for each treeblock are saved in the MTDM when encoding a *Fu* frame. Simultaneously, a weighted average motion vector (MV) for each treeblock is computed and stored in another $n \times m$ matrix ($TBmotion[n][m]$), where the weights are proportional to each PU size. When the first *Fc* frame which follows an *Fu* frame is encoded, the values stored in $MTDM[n][m]$ are offset according to the motion vectors stored in $TBmotion[n][m]$. Each treeblock in *Fc* is then encoded with this updated MTDM. New average motion vectors for each treeblock are then computed and stored in $TBmotion[n][m]$ while encoding the *Fc* frame to be used in the next *Fc* frame.

Fig. 8 shows an example in which the maximum depth of a treeblock is updated according to the derived motion direction. The average motion vector (*MV* in Fig. Y) of a determined treeblock (in black) belonging to an Fc_i frame is computed. Assuming that the motion of that texture area is constant, the most probable location for this treeblock in the next Fc_{i+1} frame is indicated by the opposite of the average motion vector previously computed (*MV'* in Fig. 8). The maximum coding tree depth of that treeblock stored in the MTDM is thus moved from its current position to the one pointed by *MV'*. If the average motion vector points to a frame which was encoded before Fc_{i-1}, their horizontal and vertical components are scaled inversely to the number of frames between the current and the reference frames before deriving the value of *MV'*.

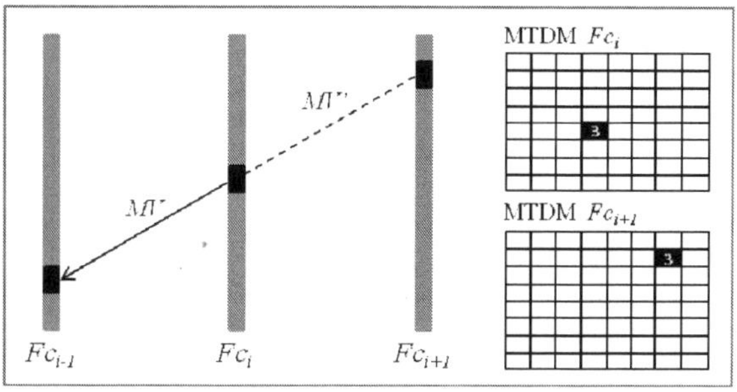

Fig. 8. MTDM motion compensation [94]

Experimental results for the [94] work show that the computational complexity can be scaled down to 60% of the uncontrolled case. The average PSNR drop observed in the worst case is lower than 0.1 dB and the average bit rate increase is around 3%.

5.2.4 Coding Tree Depth Estimation for Complexity Control in HEVC

The algorithm proposed in [95] aims at decreasing rate-distortion performance losses introduced in the previous versions of the work [23, 93, 94] by exploring both spatial and temporal correlation among treeblocks. Even though the core of the complexity control method is the same in [95], this work proposes a different way of deciding the maximum coding tree depth to be used in constrained frames.

The maximum coding tree depth decision used in [23] relies on the fact that the maximum coding tree depth tends to be constant during long periods of the sequence for co-located frame areas. However, this is not true for fast motion scenes and this problem was solved in [94] with the introduction of the motion-compensated MTDM. In [95], the video spatial correlation was also explored by considering the tendency of neighbouring CTBs to be encoded with the same or similar maximum tree depths. A set of experiments revealed that more than 80% of CTBs are encoded with the maximum coding tree depth equal to or smaller than its neighbouring CTBs. These experiments were performed considering the top, left and top-left CTBs, since they are the available neighbours when encoding a certain CTB *y*.

Based on these observations and on the algorithm proposed in [94], in this new proposal the maximum coding tree depths allowed for each CTB are decided taking into consideration the type of frame (*Fc* or *Fu*) as well as the maximum depths used in the temporal and spatial neighbouring CTBs and the current *Nc* value.

Let $CTB^{k}_{i,j}$ be a CTB located at position *i*, *j* of frame *k*. If *k* is a *Fu* frame, the CTB is encoded with no complexity limitation, which means that the maximum coding tree depth possible is allowed. If *k* is a *Fc* frame, the maximum coding tree depth allowed is defined to be the largest of the maximum coding tree depths used at the:

- Left side neighbouring tree block – $CTB^{k}_{(i\ 1,j)}$,
- Top neighbouring tree block – $CTB^{k}_{(i,j\ 1)}$,
- Top-left neighbouring tree block – $CTB^{k}_{(i\ 1,j\ 1)}$,
- Co-located tree block in the previous frame – $CTB^{k\ 1}_{(i,j)}$,
- Motion compensated tree block in the previous frame – $CTB^{k\ 1}_{(o,p)}$.

In cases with very small target complexities, the maximum coding tree depth allowed is decreased by one additional unit, if the maximum value for *Nc* is already achieved but complexity is still above the target. The *Nc* limit is defined as half the frame rate of the video.

Experimental results have shown that the computational complexity can be scaled down to 60% of the uncontrolled case with this method. In the worst case (target complexity equal to 60%), the observed PSNR drop is around 0.07 dB and the average bit rate increase is around 1.4%.

6 Challenges and Future Trends

As explained in the previous sections, computational resource management for video encoding through reduction and control of algorithm complexity had many research advances in recent years. However, an optimal solution is still far from being found, which is especially true when one takes into consideration how fast are mobile devices evolving in terms of computational power and display technologies.

Challenges come from different sides. The increasing screen resolution of current mobile devices and cameras allow higher resolution video sequences to be played, recorded and transmitted. Video content with higher resolutions require greater computational efforts be processed and transmitted, increasing power consumption in such devices. In order to reduce the number of bits and decrease the energy spent on transmission, more efficient compression methods must be used, which in turn increases even more the computational complexity and power consumption. It is easy to perceive that there is no way to avoid or ignore the computational complexity increase incurred by the latest mobile technology advances.

Methods for computational complexity analysis, reduction and control for HEVC are currently being developed by researchers. As discussed in the previous sections, many encoding parameters can be employed as the reduction and control variables for the optimisation of mobile multimedia systems. Large amounts of encoding parameters incur in a very complex analysis of the encoder's Rate-Distortion-Complexity efficiency, but it also allows finding better ways of reducing and controlling computational complexity. The HEVC Test Model (HM) reference software includes less encoding parameters than H.264/AVC, which on one side reduces the amount of Rate-Distortion-Complexity analysis to be performed, but on the other side reduces the number of encoding possibilities to be considered when developing a complexity controlled system. This is a challenge to be overcome by identifying in the HEVC standard which tasks should be parameterised in order to be made complexity scalable. Also, as different video sequences with different contents have different Rate-Distortion-Complexity results, developing efficient and accurate content-aware Rate-Distortion-Complexity analysis models is also another important open issue is to be solved.

Some of the methods proposed for previous standards can also be applied to HEVC after some adjustments, especially those which focus on reducing complexity of ME and MD. However, the overall complexity reduction would most probably not be the same for HEVC, since more tools and compression mechanisms have been added to the new standard, increasing the complexity of the remaining modules. The same problems stand out for other future video coding standards or current extensions of H.264/AVC and HEVC, such as the 3D, Scalable Video Coding (SVC) and Multiview Video Coding (MVC) extensions, all of which increase even more the demand for computational resources on the encoders through the inclusion of more tools and coding modes.

Real-time video coding can also be achieved if parallel computing solutions are explored. It is now common to find handheld devices equipped with multicore general processors and graphics processing units (GPUs), which can be used to increase encoding speed through the divide-and-conquer approach. Co-exploration between algorithm and architecture (CEAA) for parallel computing in HEVC has been exposed in [96]. For a detailed overview on the use of multicore graphics processors for video coding, the reader is referred to [97].

The use of computational resources on the decoder side will also become more important with the introduction of more complex video codecs and videos with higher spatial resolution, such as ultra-HD. Techniques for resources management on video decoders will need to be further investigated, especially for those devices with fewer computational capabilities. Furthermore, the heterogeneity of mobile devices, varying from those with fast multicore processors and GPUs to those with slower single-core processors, will also call for methods that allow scaling computational complexity on the decoder side according to the target platform constraints. The authors in [98] propose a model that allows an encoder to perform different encoding operations, so that the generated bitstream is suitable for a determined receiver platform with a certain constraint in terms of computational resources. Solutions on this range aiming at HEVC decoders are still absent.

7 Conclusions

This chapter has presented the main issues on computational resources management for video coding systems in resource constrained mobile environments. The operations of the H.264/AVC and High Efficiency Video Coding (HEVC) standards and their computational complexity analysis have been discussed based on the main works found in the literature. A study on the main works focusing on computational complexity modelling, reduction and control for H.264/AVC and HEVC was also presented in this chapter. As the HEVC standard draft has just been completed, special attention was devoted to the few existing complexity reduction and control methods for its encoder that have been published to date. Open research challenges to be dealt in the next years were finally discussed.

Acknowledgments. The authors acknowledge financial support from the National Council for Scientific and Technological Development (Brazil) and the Instituto de Telecomunicações (Portugal).

References

1. International Telecommunication Union, ITU-T Recommendation H.264 (May 2003): advanced video coding for generic audiovisual services (2003)
2. ISO/IEC-JCT1/SC29/WG11, High Efficiency Video Coding (HEVC) text specification draft 10, Geneva, Switzerland (2013)
3. Zhihai, H., Yongfang, L., Lulin, C., Ahmad, I., Dapeng, W.: Power-rate-distortion analysis for wireless video communication under energy constraints. IEEE Transactions on Circuits and Systems for Video Technology 15, 645–658 (2005)
4. Ates, H.F., Altunbasak, Y.: Rate-Distortion and Complexity Optimized Motion Estimation for H.264 Video Coding. IEEE Transactions on Circuits and Systems for Video Technology 18, 159–171 (2008)
5. Wiegand, T., Sullivan, G.J., Bjontegaard, G., Luthra, A.: Overview of the H.264/AVC video coding standard. IEEE Transactions on Circuits and Systems for Video Technology 13, 560–576 (2003)
6. ISO/IEC-JCT1/SC29/WG11, Comparison of Compression Performance of HEVC Working Draft 4 with AVC High Profile, Geneva, Switzerland (2011)
7. Sullivan, G.J., Wiegand, T.: Rate-distortion optimization for video compression. IEEE Signal Processing Magazine 15, 74–90 (1998)
8. Ortega, A., Ramchandran, K.: Rate-distortion methods for image and video compression. IEEE Signal Processing Magazine 15, 23–50 (1998)
9. Hallapuro, A., Lappalainen, V., Hamalainen, T.D.: Performance analysis of low bit rate H.26L video encoder. In: 2001 IEEE International Conference on Acoustics, Speech, and Signal Processing, vol. 2, pp. 1129–1132 (2001)
10. Saponara, S., Denolf, K., Lafruit, G., Blanch, C., Bormans, J.: Performance and Complexity Co-evaluation of the Advanced Video Coding Standard for Cost-Effective Multimedia Communications. EURASIP Journal on Applied Signal Processing 2004, 220–235 (2004)
11. Joint Collaborative Team on Video Coding (JCT-VC) (2013), `http://www.itu.int/en/ITU-T/studygroups/com16/video/Pages/jctvc.aspx`
12. Sullivan, G.J., Ohm, J., Woo-Jin, H., Wiegand, T.: Overview of the High Efficiency Video Coding (HEVC) Standard. IEEE Transactions on Circuits and Systems for Video Technology 22, 1649–1668 (2012)

13. Marpe, D., Schwarz, H., Bosse, S., Bross, B., Helle, P., Hinz, T., et al.: Video Compression Using Nested Quadtree Structures, Leaf Merging, and Improved Techniques for Motion Representation and Entropy Coding. IEEE Transactions on Circuits and Systems for Video Technology 20, 1676–1687 (2010)
14. Laroche, G., Jung, J., Pesquet-Popescu, B.: RD Optimized Coding for Motion Vector Predictor Selection. IEEE Transactions on Circuits and Systems for Video Technology 18, 1681–1691 (2008)
15. Helle, P., Oudin, S., Bross, B., Marpe, D., Bici, M.O., Ugur, K., et al.: Block Merging for Quadtree-Based Partitioning in HEVC. IEEE Transactions on Circuits and Systems for Video Technology 22, 1720–1731 (2012)
16. Sole, J., Joshi, R., Nguyen, N., Tianying, J., Karczewicz, M., Clare, G., et al.: Transform Coefficient Coding in HEVC. IEEE Transactions on Circuits and Systems for Video Technology 22, 1765–1777 (2012)
17. Sze, V., Budagavi, M.: High Throughput CABAC Entropy Coding in HEVC. IEEE Transactions on Circuits and Systems for Video Technology 22, 1778–1791 (2012)
18. Sze, V., Budagavi, M.: Parallelization of CABAC transform coefficient coding for HEVC. In: 2012 Picture Coding Symposium, pp. 509–512 (2012)
19. Norkin, A., Bjontegaard, G., Fuldseth, A., Narroschke, M., Ikeda, M., Andersson, K., et al.: HEVC Deblocking Filter. IEEE Transactions on Circuits and Systems for Video Technology 22, 1746–1754 (2012)
20. Chih-Ming, F., Alshina, E., Alshin, A., Yu-Wen, H., Ching-Yeh, C., Chia-Yang, T., et al.: Sample Adaptive Offset in the HEVC Standard. IEEE Transactions on Circuits and Systems for Video Technology 22, 1755–1764 (2012)
21. Correa, G., Assuncao, P., Agostini, L., da Silva Cruz, L.A.: Performance and Computational Complexity Assessment of High Efficiency Video Encoders. IEEE Transactions on Circuits and Systems for Video Technology 22, 1899–1909 (2012)
22. VTune™ Amplifier XE 2011 from Intel (2013), `http://software.intel.com/en-us/intel-vtune-amplifier-xe`
23. Correa, G., Assuncao, P., Agostini, L., da Silva Cruz, L.A.: Complexity Control of High Efficiency Video Encoders for Power-Constrained Devices. IEEE Transactions on Consumer Electronics 57, 1866–1874 (2011)
24. Bossen, F., Bross, B., Suhring, K., Flynn, D.: HEVC Complexity and Implementation Analysis. IEEE Transactions on Circuits and Systems for Video Technology 22, 1685–1696 (2012)
25. Mattavelli, M., Brunetton, S.: Implementing real-time video decoding on multimedia processors by complexity prediction techniques. IEEE Transactions on Consumer Electronics 44, 760–767 (1998)
26. Valentim, J., Nunes, P., Pereira, F.: Evaluating MPEG-4 video decoding complexity for an alternative video complexity verifier model. IEEE Transactions on Circuits and Systems for Video Technology 12, 1034–1044 (2002)
27. van der Schaar, M., Andreopoulos, Y.: Rate-distortion-complexity modeling for network and receiver aware adaptation. IEEE Transactions on Multimedia 7, 471–479 (2005)
28. Wang, Y., Shih-Fu, C.: Complexity Adaptive H.264 Encoding for Light Weight Streams. In: 2006 IEEE International Conference on Acoustics, Speech and Signal Processing, vol. 2, p. II (2006)
29. Szu-Wei, L., Kuo, C.C.J.: Motion Compensation Complexity Model for Decoder-Friendly H.264 System Design. In: 2007 IEEE Workshop on Multimedia Signal Processing, pp. 119–122 (2007)

30. Rhee, C.E., Jung, J.S., Lee, H.J.: A Real-Time H.264/AVC Encoder With Complexity-Aware Time Allocation. IEEE Transactions on Circuits and Systems for Video Technology 20, 1848–1862 (2010)
31. Kim, W., You, J., Jeong, J.: Complexity control strategy for real-time H.264/AVC encoder. IEEE Transactions on Consumer Electronics 56, 1137–1143 (2010)
32. Xiang, L., Wien, M., Ohm, J.R.: Rate-Complexity-Distortion Optimization for Hybrid Video Coding. IEEE Transactions on Circuits and Systems for Video Technology 21, 957–970 (2011)
33. Zhan, M., Hao, H., Yao, W.: On Complexity Modeling of H.264/AVC Video Decoding and Its Application for Energy Efficient Decoding. IEEE Transactions on Multimedia 13, 1240–1255 (2011)
34. Koga, T., Iinuma, K., Hirano, A., Iijima, Y., Ishiguro, T.: Motion compensated interframe coding for video conferencing. In: National Telecommunications Conference, pp. G5.3.1–G5.3.5 (1981)
35. Lurng-Kuo, L., Feig, E.: A block-based gradient descent search algorithm for block motion estimation in video coding. IEEE Transactions on Circuits and Systems for Video Technology 6, 419–422 (1996)
36. Renxiang, L., Bing, Z., Liou, M.L.: A new three-step search algorithm for block motion estimation. IEEE Transactions on Circuits and Systems for Video Technology 4, 438–442 (1994)
37. Jo Yew, T., Surendra, R., Ranganath, M., Kassim, A.A.: A novel unrestricted center-biased diamond search algorithm for block motion estimation. IEEE Transactions on Circuits and Systems for Video Technology 8, 369–377 (1998)
38. Shan, Z., Kai-Kuang, M.: A new diamond search algorithm for fast block-matching motion estimation. IEEE Transactions on Image Processing 9, 287–290 (2000)
39. Ce, Z., Xiao, L., Chau, L.P.: Hexagon-based search pattern for fast block motion estimation. IEEE Transactions on Circuits and Systems for Video Technology 12, 349–355 (2002)
40. Richardson, I.E.: Video Codec Design: Developing Image and Video Compression Systems. John Wiley & Sons, Inc. (2002)
41. Xuan-Quang, B., Yap-Peng, T.: Adaptive dual-cross search algorithm for block-matching motion estimation. IEEE Transactions on Consumer Electronics 50, 766–775 (2004)
42. Lin, Y.-L.S., Kao, C.-Y., Kuo, H.-C., Chen, J.-W.: VLSI Design for Video Coding: H.264/AVC Encoding from Standard Specification to Chip. Springer Publishing Company, Inc. (2010)
43. Eckart, S., Fogg, C.E.: ISO-IEC MPEG-2 software video codec. In: SPIE Conf. Visual Communications and Image Processing, pp. 100–109 (1995)
44. Chok-Kwan, C., Lai-Man, P.: Normalized partial distortion search algorithm for block motion estimation. IEEE Transactions on Circuits and Systems for Video Technology 10, 417–422 (2000)
45. Lengwehasarit, K., Ortega, A.: Probabilistic partial-distance fast matching algorithms for motion estimation. IEEE Transactions on Circuits and Systems for Video Technology 11, 139–152 (2001)
46. Li, W., Salari, E.: Successive elimination algorithm for motion estimation. IEEE Transactions on Image Processing 4, 105–107 (1995)
47. Moecke, M., Seara, R.: Sorting Rates in Video Encoding Process for Complexity Reduction. IEEE Transactions on Circuits and Systems for Video Technology 20, 88–101 (2010)

48. Saponara, S., Casula, M., Rovati, F., Alfonso, D., Fanucci, L.: Dynamic control of motion estimation search parameters for low complex H.264 video coding. IEEE Transactions on Consumer Electronics 52, 232–239 (2006)
49. Bystrom, M., Richardson, I., Zhao, Y.: Efficient mode selection for H.264 complexity reduction in a Bayesian framework. Signal Processing: Image Communication 23, 71–86 (2008)
50. Pan, L.-J., Yo-Sung, H.: A Fast Mode Decision Algorithm for H.264/AVC Intra Prediction. In: 2007 IEEE Workshop on Signal Processing Systems, pp. 704–709 (2007)
51. Chun-Hao, C., Jia-Wei, C., Hsiu-Cheng, C., Yao-Chang, Y., Jinn-Shyan, W., Jiun-In, G.: A Quality Scalable H.264/AVC Baseline Intra Encoder for High Definition Video Applicaitons. In: 2007 IEEE Workshop on Signal Processing Systems, pp. 521–526 (2007)
52. De-Wei, L., Chun-Wei, K., Chao-Chung, C., Yu-Kun, L., Tian-Sheuan, C.: A 61MHz 72K Gates 1280x720 30FPS H.264 Intra Encoder. In: 2007 IEEE International Conference on Acoustics, Speech and Signal Processing, vol. 2, pp. 801–804 (2007)
53. An-Chao, T., Jhing-Fa, W., Jar-Ferr, Y., Wei-Guang, L.: Effective Subblock-Based and Pixel-Based Fast Direction Detections for H.264 Intra Prediction. IEEE Transactions on Circuits and Systems for Video Technology 18, 975–982 (2008)
54. Yu-Ming, L., Yu-Ting, S., Yinyi, L.: SATD-Based Intra Mode Decision for H.264/AVC Video Coding. IEEE Transactions on Circuits and Systems for Video Technology 20, 463–469 (2010)
55. Hyungjoon, K., Altunhasak, Y.: Low-complexity macroblock mode selection for H.264-AVC encoders. In: 2004 International Conference on Image Processing, vol. 2, pp. 765–768 (2004)
56. Wang, F., Fan, Y., Lan, Y., Liu, W.: Fast intra mode decision algorithm in H.264/AVC using characteristics of transformed coefficients. In: 2008 International Conference on Visual Information Engineering, pp. 245–249 (2008)
57. Yu-Ming, L., Jyun-De, W., Yinyi, L.: An improved SATD-based intra mode decision algorithm for H.264/AVC. In: 2009 IEEE International Conference on Acoustics, Speech and Signal Processing, pp. 1029–1032 (2009)
58. Correa, G., Diniz, C., Bampi, S., Palomino, D., Porto, R., Agostini, L.: Homogeneity and distortion-based intra mode decision architecture for H.264/AVC. In: 2010 IEEE International Conference on Electronics, Circuits, and Systems, pp. 591–594 (2010)
59. Correa, G., Palomino, D., Diniz, C., Agostini, L., Bampi, S.: SHBS: A heuristic for fast inter mode decision of H.264/AVC standard targeting VLSI design. In: 2011 IEEE International Conference on Multimedia and Expo, pp. 1–4 (2011)
60. Correa, G., Palomino, D., Diniz, C., Bampi, S., Agostini, L.: Low-Complexity Hierarchical Mode Decision Algorithms Targeting VLSI Architecture Design for the H. In: 264/AVC Video Encoder, vol. 20, pp. 1–20 (2012)
61. Jeyun, L., Byeungwoo, J.: Fast mode decision for H.264. In: 2004 IEEE International Conference on Multimedia and Expo, vol. 2, pp. 1131–1134 (2004)
62. Wu, D., Pan, F., Lim, K.P., Wu, S., Li, Z.G., Lin, X., et al.: Fast intermode decision in H.264/AVC video coding. IEEE Transactions on Circuits and Systems for Video Technology 15, 953–958 (2005)
63. Liquan, S., Zhi, L., Zhaoyang, Z., Xuli, S.: Fast Inter Mode Decision Using Spatial Property of Motion Field. IEEE Transactions on Multimedia 10, 1208–1214 (2008)
64. Cho, S., Kim, M.: Fast CU Splitting and Pruning for Suboptimal CU Partitioning in HEVC Intra Coding. In: IEEE Transactions on Circuits and Systems for Video Technology (2013) (accepted for publication)

65. Bjontegaard, G.: Calculation of average PSNR differences between RD-curves, doc. VCEG-M33 (2001)
66. Tai, S.-C., Chang, C.-Y., Chen, B.-J., Hu, J.-F.: Speeding Up the Decisions of Quad-Tree Structures and Coding Modes for HEVC Coding Units. In: Pan, J.-S., Yang, C.-N., Lin, C.-C. (eds.) Advances in Intelligent Systems & Applications. SIST, vol. 21, pp. 393–401. Springer, Heidelberg (2012)
67. Xiaolin, S., Lu, Y., Jie, C.: Fast coding unit size selection for HEVC based on Bayesian decision rule. In: 2012 Picture Coding Symposium, pp. 453–456 (2012)
68. Zhang, Y., Wang, H., Li, Z.: Fast Coding Unit Depth Decision Algorithm for Interframe Coding in HEVC. In: 2013 Data Compression Conference, vol. 1, pp. 53–62 (2013)
69. Jie, L., Lei, S., Ikenaga, T., Sakaida, S.: Content Based Hierarchical Fast Coding Unit Decision Algorithm for HEVC. In: 2011 International Conference on Mutimedia and Signal Processing, vol. 1, pp. 56–59 (2011)
70. Sampaio, F., Bampi, S., Grellert, M., Agostini, L., Mattos, J.: Motion Vectors Merging: Low Complexity Prediction Unit Decision Heuristic for the Inter-prediction of HEVC Encoders. In: 2012 IEEE International Conference on Multimedia and Expo, pp. 657–662 (2012)
71. Guifen, T., Goto, S.: Content adaptive prediction unit size decision algorithm for HEVC intra coding. In: 2012 Picture Coding Symposium, pp. 405–408 (2012)
72. Kiho, C., Jang, E.S.: Early TU decision method for fast video encoding in high efficiency video coding. Electronics Letters 48, 689–691 (2012)
73. Jaehwan, K., Jungyoup, Y., Kwanghyun, W., Byeungwoo, J.: Early determination of mode decision for HEVC. In: 2012 Picture Coding Symposium, pp. 449–452 (2012)
74. Cassa, M.B., Naccari, M., Pereira, F.: Fast rate distortion optimization for the emerging HEVC standard. In: 2012 Picture Coding Symposium, pp. 493–496 (2012)
75. Ma, S., Wang, S., Wang, S., Zhao, L., Yu, Q., Gao, W.: Low Complexity Rate Distortion Optimization for HEVC. In: 2013 Data Compression Conference, Snowbird, Utah, vol. 1, pp. 73–82 (2013)
76. Jain, P., Laffely, A., Burleson, W., Tessier, R., Goeckel, D.: Dynamically Parameterized Algorithms and Architectures to Exploit Signal Variations. The Journal of VLSI Signal Processing 36, 27–40 (2004)
77. Weiyao, L., Panusopone, K., Baylon, D.M., Ming-Ting, S.: A Computation Control Motion Estimation Method for Complexity-Scalable Video Coding. IEEE Transactions on Circuits and Systems for Video Technology 20, 1533–1543 (2010)
78. Akyol, E., Mukherjee, D., Yuxin, L.: Complexity Control for Real-Time Video Coding. In: 2007 IEEE International Conference on Image Processing, vol. 1, pp. 77–80 (2007)
79. da Fonseca, T.A., de Queiroz, R.L.: Macroblock sampling and mode ranking for complexity scalability in mobile H.264 video coding. In: 2009 IEEE International Conference on Image Processing, pp. 3753–3756 (2009)
80. Li, S., Yan, L., Feng, W., Shipeng, L., Wen, G.: Complexity-Constrained H.264 Video Encoding. IEEE Transactions on Circuits and Systems for Video Technology 19, 477–490 (2009)
81. Xiang, L., Wien, M., Ohm, J.R.: Medium-granularity computational complexity control for H.264/AVC. In: 2010 Picture Coding Symposium, pp. 214–217 (2010)
82. Kannangara, C.S., Richardson, I.E., Bystrom, M., Yafan, Z.: Complexity Control of H.264/AVC Based on Mode-Conditional Cost Probability Distributions. IEEE Transactions on Multimedia 11, 433–442 (2009)

83. Li, X., Cui, Y., Xue, Y.: Towards an Automatic Parameter-Tuning Framework for Cost Optimization on Video Encoding Cloud. International Journal of Digital Multimedia Broadcasting (2012)
84. Vanam, R., Riskin, E., Ladner, R., Hemami, S.: Fast algorithms for designing nearly optimal lookup tables for complexity control of the H.264 encoder. In: Signal, Image and Video Processing, pp. 1–13 (2012)
85. Breiman, L.: Classification and regression trees. Chapman & Hall, New York (1984)
86. Sun, Z., Xi, C., Zhihai, H.: Adaptive Critic Design for Energy Minimization of Portable Video Communication Devices. In: 2008 International Conference on Computer Communications and Networks, pp. 1–5 (2008)
87. Kannangara, C.S., Richardson, I.E., Miller, A.J.: Computational Complexity Management of a Real-Time H.264/AVC Encoder. IEEE Transactions on Circuits and Systems for Video Technology 18, 1191–1200 (2008)
88. Ismaeil, I.R., Docef, A., Kossentini, F., Ward, R.K.: A computation-distortion optimized framework for efficient DCT-based video coding. IEEE Transactions on Multimedia 3, 298–310 (2001)
89. Yongfang, L., Ahmad, I.: Power and Distortion Optimization for Pervasive Video Coding. IEEE Transactions on Circuits and Systems for Video Technology 19, 1436–1447 (2009)
90. Zhihai, H., Wenye, C., Xi, C.: Energy Minimization of Portable Video Communication Devices Based on Power-Rate-Distortion Optimization. IEEE Transactions on Circuits and Systems for Video Technology 18, 596–608 (2008)
91. Ji, W., Liu, J., Chen, M., Chen, Y.: Power-efficient video encoding on resource-limited systems: A game-theoretic approach. Future Gener. Comput. Syst. 28, 427–436 (2012)
92. da Fonseca, T.A., de Queiroz, R.L.: Complexity-Constrained Rate-Distortion Optimization for H.264/AVC Video Coding. In: 2011 IEEE International Symposium on Circuits and Systems, pp. 2909–2912 (2011)
93. Correa, G., Assuncao, P., Da Silva Cruz, L.A., Agostini, L.: Adaptive coding tree for complexity control of high efficiency video encoders. In: 2012 Picture Coding Symposium, pp. 425–428 (2012)
94. Correa, G., Assuncao, P., Agostini, L., da Silva Cruz, L.A.: Motion compensated tree depth limitation for complexity control of HEVC encoding. In: 2012 IEEE International Conference on Image Processing, pp. 217–220 (2012)
95. Correa, G., Assuncao, P., Agostini, L., da Silva Cruz, L.A.: Coding Tree Depth Estimation for Complexity Reduction of HEVC In. In: 2013 Data Compression Conference, vol. 1, pp. 43–52 (2013)
96. Kiho, C., Jang, E.S.: Leveraging Parallel Computing in Modern Video Coding Standards. IEEE Multimedia 19, 7–11 (2012)
97. Ngai-Man, C., Xiaopeng, F., Au, O.C., Man-Cheung, K.: Video Coding on Multicore Graphics Processors. IEEE Signal Processing Magazine 27, 79–89 (2010)
98. Szu-Wei, L., Kuo, C.C.J.: Complexity Modeling of Spatial and Temporal Compensations in H.264/AVC Decoding. IEEE Transactions on Circuits and Systems for Video Technology 20, 706–720 (2010)

Resource Allocation and Scheduling for Video Transmission over LTE/LTE-A Wireless Systems

Nabeel Khan and Maria G. Martini

Kingston University, London
{n.khan,m.martini}@kingston.ac.uk

Abstract. Multimedia applications contribute a major portion of applications in current and emerging wireless networks. In particular, video streaming is one of the multimedia applications that must be supported efficiently by the Long-Term Evolution (LTE) standard. Quality of Service (QoS) requirements for video streaming are quite stringent and must be met for all active flows. Hence, resource allocation and scheduling at the MAC layer of an LTE system becomes extremely important in determining the overall system performance. Different strategies can be adopted to share the resources among the users. Some of these assume knowledge of channel information and allocate more resources to users experiencing better channel conditions in order to maximize the cell throughput. However, for real-time applications such schedulers are incapable of providing fairness among users. Achieving a good trade-off between efficiency and fairness is critical for scheduling strategies. This chapter presents the different aspects of scheduling strategies for multimedia applications in the downlink of LTE networks, highlighting how these tackle the aforementioned trade-off, and introduce solutions proposed by the authors to address it.

1 Introduction

The expectation from a cellular network is now to provide multi-play applications of Voice over IP (VoIP), video, data to a continuously growing number of cellular users. The scarcity of the available radio spectrum coupled with the unique traffic handling and Quality of Experience (QoE) requirements of the converged services poses a huge challenge to network operators. The evolution of mobile networks to high speed, IP-based infrastructure has put onus on the network operators to provide quality services to users running differentiated services on mobile devices. The solution of over-provisioning the network by increasing the amount of bandwidth is not economical, hence in addition to adding network capacity, network operators must devise a policy for efficiently partitioning the network resources. The diverse Quality of Service (QoS) requirements of different traffic types are shown in Table 1 [35].

As shown in the table, each service has different QoS requirements in terms of packet delay, packet loss rate and bit rate. Furthermore, each traffic type exhibits different bit rate characteristics. For instance, video traffic exhibits variable bitrate characteristic along with high bandwidth requirements. On the other hand, VoIP traffic exhibits constant bitrate characteristic with low bandwidth needs. Similarly the traffic characteristics of web browsing, File Transfer Protocol (FTP), and gaming are different. The heterogeneity in the bitrate in different traffic classes and strict QoS requirements have to be

C.X. Mavromoustakis et al., *Resource Management in Mobile Computing Environments*, Modeling and Optimization in Science and Technologies 3,
DOI: 10.1007/978-3-319-06704-9_25,

Table 1. Comparison of Quality Expectation and Performance Requirements by Service Type [35]

Services	Quality Expectation	Performance Attributes
Internet	Low – best-effort	Variable bandwidth consumption, Latency and loss tolerant
Enterprise/Business Services	High - critical data	High bandwidth consumption, Highly sensitive to latency, High security
Peer-To-Peer	Low - best effort	Very-high bandwidth consumption, Latency and loss tolerant
Voice	High - Low latency and jitter	Low bandwidth per call, Highly sensitive to latency
Video	High - low jitter and extremely-low packet loss	Very-high bandwidth consumption, Very sensitive to packet loss
Gaming and Interactive	High - low packet loss	Variable bandwidth consumption, Very sensitive to packet loss, Highly sensitive to latency

dealt with a dynamic policy management. The policy management depends upon the network ability to respond to congestion by gracefully degrading the service quality so that a minimum level of service quality is maintained by the network. Although adding capacity to the wireless network by increasing the spectrum is an important step which must be taken by the operators to handle the continuous growth of mobile data, any increase in system capacity by increasing the bandwidth would eventually be consumed by bandwidth hungry applications such as video and peer to peer traffic.

Advanced Radio Resource Management (RRM) procedures are one of the key features of 4G wireless systems which must be exploited by the operators. For instance, one option is the maximization of the system throughput by allocating the resources to the most appropriate users (users with the best channel quality). On the other hand, meeting the service quality requirements of each and every user is also critical. These two objectives are conflicting and there is a risk in achieving one at the expense of the other. Therefore, the trade-off between these objectives has to be addressed according to the policy rules of the operators. Packet scheduling is one of the most important functions of RRM and plays a key role in distributing radio resources among different users with different service needs.

LTE is a major advancement in terms of physical layer technology in the cellular paradigm with respect to previous releases of 3GPP, with the adoption of OFDM (Orthogonal Frequency Division Multiplexing) as the main new concept in cellular technology; LTE takes benefit from this technology to achieve its design goals, reported in Table 2. Some of its salient features include fully switched packet core, only one node in the radio access network, called eNodeB, which links the mobile stations, called User Equipment (UE), to the core network which achieves very low latency in user plane as well as control plane. LTE (like previous 3GPP releases) relies heavily on adaptive

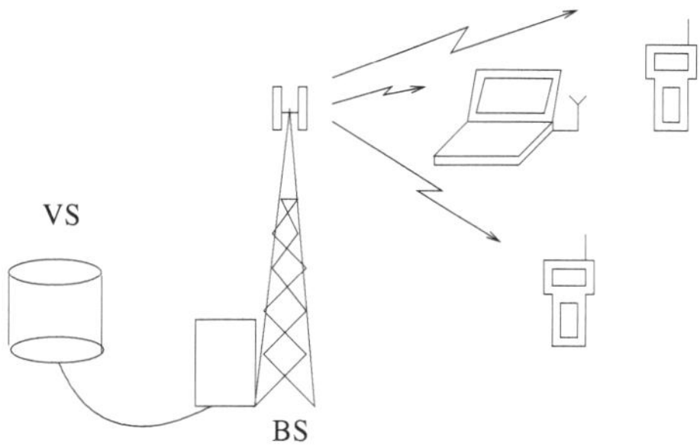

Fig. 1. Reference scenario

modulation and coding (AMC) and Hybrid Automatic Retransmission reQuest (HARQ) to achieve gains in throughput.

Similar as in [22] [16] [15] [17] [18] [12] [21], we consider in this chapter the scenario depicted in Figure 1 where a video server (VS) is directly linked to the base station (BS) / eNodeB, which transmits different video streams to multiple users. Different users can be characterized by specific user requirements, different channel conditions and by the reception of video sequences with different characteristics in terms of video activity.

Efficiently serving video streaming sessions over a multiple access wireless channel with shared communications resources is a critical problem that can be afforded at different layers: dynamic rate control strategies optimized across the users can be considered at the application layer in order to allocate resources according to users requirements and transmission conditions. Packet scheduling schemes across multiple users can be considered below in the protocol stack, for instance taking into account information about the channel conditions experienced by each user. Content-aware scheduling can also be considered [24].

Channel aware schedulers take channel information into account and maximize the cell throughput [34]. Maximization of cell throughput for real-time applications does not provide a fair system. Efficient scheduling schemes must provide a fair service and all the active flows in the system must get a minimum guaranteed service by efficiently utilizing the spectrum. AMC and HARQ are the basis which provide a framework for an efficient channel aware scheduling strategy.

Information about the content of the multimedia data transmitted can be exploited in content-aware scheduling approaches [24] [22] [25] [15], where priority is given to the most important information from the user's perspective.

The remainder of this chapter is organized as follows. Section 2 presents the main features of LTE systems. A basic review of the most used scheduling strategies is then presented, together with performance evaluation results in a simulated LTE system. We then present a novel strategy for scheduling delay sensitive applications over the downlink of LTE / LTE-A systems and compare its performance with state-of-the art solutions. QoE-based strategies are also briefly addressed, and recent research trends are highlighted.

2 The LTE Standard

The major design goals for 3GPP release 8 are reported in Table 2 [10].

Table 2. LTE design goals [10]

Feature	Value
Scalable Bandwidth	1.4, 3, 3.2, 5, 10, 15 and 20 MHz
Peak Data Rate	Up to 100 Mbps for 20 MHz in downlink for uplink up to 50 Mbps.
Antenna Configuration	Tx (no. of transmit antenna) * Rx (no. of receive antenna) for Downlink: 4*2, 2*2, 1*2 and 1*1. For Uplink: 1*2 and 1*1.
Spectrum Efficiency	3 to 4 times with respect to HSDPA in Downlink and 2 to 3 times in Uplink
Latency	C-plane less than 50-100 ms, U-plane less than 10 ms
Mobility	High performance at speed up to 120 km/hr and maintain link speed up to 350 km/hr
Coverage	Optimal performance up to 5 km radius and slight degradation up to 5-30Km, operation up to 100 km may be supported.

The remainder of this section describes the main functionalities of the different layers addressed in the standard.

2.1 Packet Data Convergence Protocol (PDCP)

The main function of this layer is to compress the header in the incoming IP packet [5]; as there is no circuit switching, the entire architecture is packet switched, hence compression (especially for VoIP) becomes very important. Robust Header compression (ROHC) is the protocol used for compression; ROHC can reduce the IP header size from 40 bytes to approximately 1 to 4 bytes. PDCP is also responsible for the security of user plane data, RRC (Radio Resource Control) data and NAS (Non access stratum) data. ROHC is performed before the security operation (encryption, decryption, integrity protection) as ROHC cannot compress encrypted packet.

2.2 Radio Link Control (RLC)

In 3GPP release 7, the main RLC functions (segmentation, concatenation, in sequence delivery through ARQ) were performed by a separate node called radio network controller (RNC). In LTE there is only one node called eNodeB in the radio access architecture, hence RLC is located in the eNodeB [4]. Apart from segmenting and concatenating the compressed IP packets, RLC provides reliability through ARQ operation. RLC is operated in two modes: Acknowledged Mode (AM) and Unacknowledged Mode (UM). In the first one, RLC requests retransmission of the missing protocol data units (PDUs) from the transmitting entity; this mode is mainly used for TCP-based applications where reliability is more important. For delay sensitive applications (VoIP or video) UM is used, where missing PDUs are not requested for retransmission. It is important to note that almost all the errors due to the dynamic nature of the wireless channel are handled by a much more efficient and fast (ideal for delay sensitive application) retransmission mechanism called HARQ (Hybrid Automatic Repeat Request). Hence a question arises about the retransmission mechanism (ARQ) at RLC. MAC (Medium Access Control) based HARQ can transfer erroneous packets to the RLC in the event of ACK/NAK (HARQ Acknowledge/Non-Acknowledge) corruption, for instance when NAK (due to noise) is interpreted as ACK. The possibility of this event to occur is low, approximately 1% [10], which is however still high when considering that the maximum data rate which LTE can support is 100 Mbps. According to [23] for TCP based applications the probability that a packet may be lost should be less than 10^{-5}. Hence ARQ at RLC becomes extremely important for TCP based traffic. Since RLC and MAC layer resides in one node in LTE, there is a tighter interaction between RLC and HARQ, hence RLC retransmissions are faster in LTE than the previous releases of 3GPP.

2.3 MAC (Medium Access Control)

The MAC layer performs HARQ retransmissions, scheduling (both uplink and downlink) [7], and handles control and traffic channels [6]. For details on scheduling in LTE refer to section 3. Traffic channels carry user data. Control channels, also called logical channels, are extremely important for the operation of LTE. From the MAC layer, the RLC layer uses services in the form of logical channels, as reported below.

Control Channels. Broadcast Control Channel (BCCH): Important system information such as operating bandwidth, cell ID and other important configuration information are transmitted through this channel. UE accesses this information before entering the system and the eNodeB continuously broadcasts all the configuration information on this channel. When a UE wants to access the system resources, it acquires all the configuration information broadcasted by the eNodeB.

Paging Control Channel (PCCH): In the event of unknown UE location, this control channel is used to page the UE so that cell level location of the UE is known to the network. Paging message is sent to the UE, which returns its cell location to the network.

Dedicated Control Channel: This channel is used to individually configure each UE; information such as handover message is carried over this channel. Hence whenever there is a need for an individual configuration of UE, this control channel is used.

Multicast Control Channel: This channel is used to control multicast information carried over Multicast Traffic channel. Control information related to MBMS (Multimedia Broadcast and Multicast Service) is transmitted over this channel.

Multicast Traffic Channel: All the multicast transmissions (MBMS from eNodeB to UEs) are carried over this channel. MBMS services, first introduced in Release 6 of 3GPP, make it possible to transmit the same content across multiple users, thus paving the way for Mobile TV services. MBMS is further divided into broadcast and multicast services. In broadcast, part of the radio resources in a cell is reserved and all UEs (subscribed to this service) supposed to receive the transmitted signal. It is important to note that there is no need to track the UEs (as done in unicast transmission) in the

RAN (Radio Access Network). A user subscribed to this service simply receives the content without informing the network. On the other hand in multicast services a group is formed, called a multicast group, where each UE joins this group by notifying the network so that the appropriate amount of radio resources can be assigned.

Dedicated Traffic Channel: Actual user data is transmitted over this channel.

Transport Channels. In order to offer services to the MAC layer, the physical layer provides transport channels. Transport channels carry user data, and the format and the characteristics with which the data is transmitted over the radio interface is specified by the MAC layer. The MAC layer (MAC scheduler) decides the type of modulation, coding, the size of the transport block (number of used PRBs) and also antenna mapping when MIMO (Multiple Input Multiple Output) mode is used. These characteristics specify the transportation format of the TB (Transport Block) and are called Transport Format. One TB (two in MIMO mode) is transmitted over the radio interface in each Transmission Time Interval (TTI). For information on TTI in LTE see section 2.6.

Broadcast Channel (BCH): It is used to transport information on the BCCH channel. The format with which this information is transmitted is fixed, generally the most robust modulation and coding scheme is used, *i.e.* the Transport Format is always fixed so that there is a high probability of acquiring the broadcast information over the dynamic channel.

Paging Channel: It is used to transport information on the PCCH channel. In the LTE standard paging time is kept fixed, so that mobile terminals remains in the sleep mode (in the event of no transmission) and wake up only (according to the predefined time) in the event of paging time instant. This feature saves the battery power of the mobile terminal.

Downlink Shared Channel (DL-SCH): Downlink data is transported over this channel. The MAC layer specifies the Transport Format according to which modulation scheme and coding rate are specified for each user. All the important features like HARQ, spatial multiplexing and link adaptation are supported by this channel. It is important to note that this channel transmits user data in the unicast mode only, i.e., broadcast and multicast data are not supported.

Uplink Shared Channel (UL-SCH): Similar to the DL-SCH instead of downlink, all the features are used for uplink.

Multicast Channel: This channel is used to transport multicast and broadcast data within a cell or from one cell to another. All the MBMS data is transported on this channel.

2.4 Physical Layer

The main operations of the physical layer include coding, HARQ operation (the physical layer performs the soft combining operation in which redundancy information is increased in each retransmission), modulation and multi antenna processing in case of MIMO (Multiple Input and Multiple Output). The PRB, shown in Figure 3, is the basic transmission unit. PRBs are controlled by the MAC layer and these are assigned to different UEs according to certain criteria (such as channel conditions and buffer status of each user). After MAC layer scheduling, a decision is taken about the assignment of PRBs to different UEs. These assigned PRBs are then processed by the physical layer which performs Cyclic Redundancy Check (CRC). The CRC bits are attached to each PRB, then FEC (Forward Error Correction) through Turbo coding is applied to each PRB. After FEC, modulation is performed. It is important to note that modulation scheme and coding rate are determined by the scheduler at the MAC layer. UE receives the transmitted signal, demodulates and decodes the signal, and informs the HARQ at the MAC layer about the status of the transmission. In the event of erroneous transmission, the erroneous packet is kept at the HARQ buffer and NAK is sent to the MAC layer scheduler at the eNodeB.

2.5 LTE Transmission Schemes

Physical layer OFDM and MAC layer OFDMA offer considerable improvements over 3GPP's previous technologies such as Universal Mobile Telecommunication System (UMTS) and (High Speed Packet Access (HSPA).

The use of OFDM is a novel concept in cellular systems: previously single carrier modulation was employed, resulting in a considerable amount of Inter Symbol Interference (ISI). ISI is the result of the delay spread in a signal due to the multipath effect. In single carrier systems if the data rate increases the symbol time decreases. In the event of considerable increase in the data rate one symbol can spill into the adjacent symbol, thus ISI imposes an upper limit on the data rate in a single carrier system. In the frequency domain, multipath propagation causes different amounts of distortion (phase shift) and the signal arrives at the receiver out of phase which further limits the capacity of the system. In order to combat multipath propagation, channel inversion or rake equalizers are employed, which cause a complex channel equalizer implementation. The complexity exponentially increases above 5 MHz. Contrary to the single carrier system, OFDM utilizes narrow band subcarriers which results in long symbol duration and thus an increase in the data rate can be achieved by simply increasing the number of subcarriers (increase in the parallel data stream). All the subcarriers are tightly spaced and are orthogonal to the adjacent subcarrier; this causes an efficient and flexible use of bandwidth. Another remarkable feature of OFDM is the use of cyclic prefix that further combats the ISI. Cyclic prefix is simply a fixed duration of guard band.

The only disadvantage of OFDM is a considerable increase in the signal's peak-to-average power ratio (PAPR). The single carrier system utilizes GMSK (Gaussian minimum shift keying) and PSK (phase shift keying) which produces a constant envelope modulation which results in a linear operation of power amplifiers. In OFDMA larger PAPR causes the power amplifier to be operated in the clipping regions (maximum and minimum amplitude of the signal) which decreases the efficiency of the RFPA (Radio Frequency Power Amplifier). Due to the reduced efficiency, higher RFPA are required which results in more power consumption at the base station. Due to the limited power capability of the mobile terminal, Single Carrier-Frequency Domain Multiple Access (SC-FDMA) is employed. For more information on SC-FDMA the reader can refer to [10].

This chapter mainly focuses on the downlink, hence the remaining discussion will focus on LTE downlink transmission.

2.6 LTE Frame Architecture

In the LTE standard two frame structures are defined, one for TDD (Time Division Duplexing), referred as frame structure Type 2, and the other one used for both TDD and FDD (Frequency Division Duplexing), referred as frame structure Type 1. The total duration of a downlink frame is 10 ms, which comprises 20 slots of 0.5 ms each, as shown in Figure 2. A frame is further divided into 10 subframes; the duration of each subframe is 1 ms, i.e., two slots equal one subframe, which is also called a Transmission Time Interval (TTI). The Physical Resource Block (PRB) is a basic time-frequency resource which comprises a set of 12 contiguous sub-carriers over one slot as shown in Figure 3. This is the basic resource unit allocated by the scheduler, i.e., resources are allocated in units of PRB. The number of PRBs available depends upon the system bandwidth, ranging from 6 in 1.4 MHz to 100 in 20 MHz bandwidth.

There are seven OFDM symbols in one slot, with each symbol separated by a guard interval called cyclic prefix (normal cyclic prefix); this makes a total of 84 resource elements (7 OFDM symbols $\times$ 12 sub-carriers). When the extended cyclic prefix is used, there are six OFDM symbols per slot, which makes a total of 72 resource elements as compared to 84 resource elements with the normal cyclic prefix. Therefore, the extended cyclic prefix has more overhead (large guard intervals), but such mode is more robust in an environment where delay spread is an issue. In the remainder of this chapter, the normal cyclic prefix mode is considered and frame structure Type 1 (FDD duplexing) is used.

The remainder of this chapter discusses scheduling strategies used or proposed for LTE wireless systems.

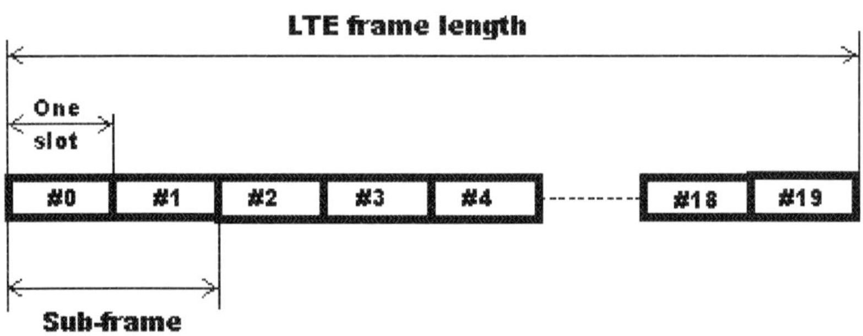

Fig. 2. LTE radio frame structure [37]

3 Scheduling in the Downlink of LTE Systems

The time-frequency resource is dynamically assigned among the users in LTE, whereas in HSPA the resources are divided in terms of time and channelization code. In shared channel transmission the LTE scheduler plays a key role in dividing the resources among the users. The scheduler decides which user will get what amount of resources in terms of PRBs. Scheduling decisions are taken at 1 ms intervals. By exploiting multiuser

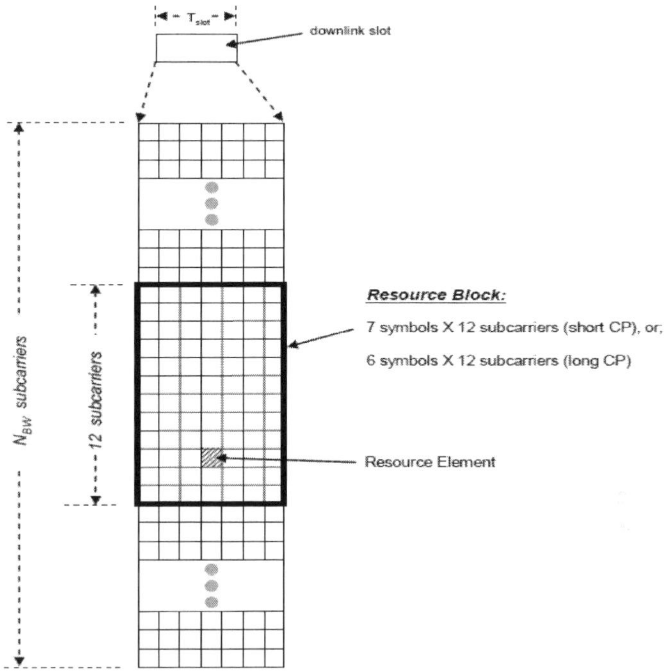

Fig. 3. Basic PRB structure [37]

channel diversity, considerable gain can be achieved, hence the throughput achieved by the scheduler can be highly increased if the scheduler takes channel conditions into account. The same concept was employed in HSPA, where the scheduler selects the users whose channel conditions are better, thus maximizing the cell throughput. Because of the OFDMA technology, LTE utilizes not only time domain variations but also frequency domain variations. Each UE periodically or aperiodically (depending on the system configuration) sends instantaneous channel quality information (CQI) to the base station (eNodeB). Based on this channel quality information, the scheduler allocates the appropriate PRBs which face less attenuation for each UE. There are different feedback granularities in the standard, for instance the UE can send just a single Channel Quality Indicator or a separate CQI for each PRB. The scheduler then carries out the averaging, *i.e.*, it maps the signal to interference noise ratio experienced by the PRBs into one (equivalent) Additive White Gaussian Noise (AWGN) SNR. One of the mostly used averaging methods is the MIESM (Mutual Information Effective SNR Mapping) [36] [11]. After resource allocation, the scheduler uses the Mutual Information Effective SNR Mapping (MIESM) method to calculate a single Channel Quality Indicator (CQI) value to be used for all the allocated PRBs of a particular UE.

Scheduling strategies proposed and adopted for several systems, including LTE wireless systems, are listed below.

3.1 Round Robin Scheduler

The Round Robin scheduler is the simplest scheduling algorithm; this scheduler assigns all the system resources to a flow in a Round Robin fashion. For example if there are 6 PRBs in the system (considering 1.4 MHz bandwidth), all the PRBs are assigned to one UE with appropriate AMC according to the CQI feedback for a period of 1 TTI. Time is shared in equal intervals among all the users. For wireless communications this basic scheduler is ineffective, since it neither provides higher system throughput nor it provides fairness. UEs with a low CQI will have an extremely low throughput since the total number of bits transmitted in the corresponding TTI is extremely small because of the necessity to use a robust AMC mode due to the low CQI value. On the other hand UEs with better channel quality utilize the channel only for 1 TTI and then their turn will come after serving all the UEs in the cell.

3.2 Best CQI Scheduler

The most common channel dependent scheduler is the Best CQI scheduler. This scheduler exploits channel variations between users to achieve maximum cell throughput: the higher the difference in the channel quality between users, the higher is the gain in throughput with respect to channel-unaware schedulers. However, users experiencing bad channel quality may not be served. For example users at the cell edges or users experiencing deep fades may not be scheduled compared to users experiencing good channel quality. Thus user i maximizing the rate $R_{i,\varphi}^{(n)}$ on a PRB (φ) will be assigned the PRB at scheduling epoch n.

3.3 Proportional Fair Scheduler

A Proportional Fair Scheduler [14] schedules users based on the following criterion: $i^*(n) = \arg\max \frac{R_{i,\varphi}^{(n)}}{R_{(i,\text{ave})}^{(n)}}$, where $i^*(n)$ represents the index of the user to be served at the scheduling instant n, $R_{(i,\text{ave})}^{(n)}$ is the average rate achieved over a moving average window. PF compromises cell throughput so that a minimum amount of fairness can be achieved among the different competing flows. UEs having good channel quality can still maximize throughput but not at the expense of depleting users with bad channel quality. Hence there is a trade off between maximum system throughput and fairness among different competing flows.

3.4 MAXMIN Scheduler

The target of a MaxMin scheduler is to maximize the minimum of the users' throughputs. This scheduling strategy is not throughput optimal but maintains a very high level of fairness in terms of throughput: this scheduler is Pareto optimal, i.e., the rate of one UE cannot be increased without decreasing the rate of another UE with a lower rate than the one considered.

3.5 Resource Fair Scheduler

The resource fair scheduler evenly distributes the resources among all the competing flows. For instance in 1.4 MHz bandwidth, 6 PRBs are available in each scheduling instant (1 ms), hence if there are 3 UEs competing for accessing radio resources, a resource fair scheduler will assign 2 PRBs to each of the UEs. RF also tries to maximize the throughput by maximizing the sum rate of all the users. This scheduler ensures minimum fairness through fair distribution of PRBs, *i.e.*, $(\frac{N}{I})$ where N is the number of PRBs available and I is the number of UEs. When the number of PRBs is not an integer multiple of the number of UEs some UEs will be assigned $\lfloor \frac{N}{I} \rfloor$ and others $\lceil \frac{N}{I} \rceil$ in order to make up the available resources. In order to ensure fairness a UE assigned $\lfloor \frac{N}{I} \rfloor$ in a current TTI will get $\lceil \frac{N}{I} \rceil$ in the next TTI through uniform randomization.

3.6 Performance Comparison

Simulation Set-Up. A C++ based simulation platform in MATLAB is used for the performance analysis of an LTE-like system. All the LTE physical layers features such as HARQ soft combining, Turbo coding, AMC and Cyclic Redundancy Check (CRC) are implemented in the C++ based simulator in MATLAB. Furthermore some of the physical layer features such as Fast Fourier Transform (FFT) and Inverse Fast Fourier Transform (IFFT) are readily available in the communications toolbox of MATLAB. Therefore, in order to investigate the performance of the scheduling algorithms discussed above, a link-level simulator built on MATLAB's object oriented features is selected as the simulation platform with all the basic features of an LTE MAC and physical layer.

Table 3 shows all the CQI values used in the LTE standard. Each CQI value corresponds to a particular coding rate and modulation scheme. CQI '1' corresponds to the most robust set of coding and modulation used by AMC when channel quality is extremely poor, whereas CQI '15' is associated to the least robust modulation and coding set, used when channel conditions are at its best so that the maximum rate can be achieved.

According to the LTE standard one CQI value is used for all the PRBs assigned to an UE, therefore the scheduler, after assigning all the PRBs, performs Signal to Interference Noise Ratio (SINR) averaging, i.e., the SINR values experienced on the assigned PRBs are mapped into a single (equivalent) SNR (Signal to Noise Ratio). MIESM (Mutual Information Effective SNR) specified by [36] [11] is the mostly used technique to carry out SINR averaging. It has been proved in [31] [30] that the MIESM based wideband feedback strategy produces optimum results in terms of throughput. Therefore MIESM is used as the averaging method throughout the simulation. All the simulation parameters are shown in Table 4.

Throughput vs. Fairness Comparison. In the following simulations we consider $I = 6$ users (UEs) with different channel conditions, hence the performance of the scheduler for different channel qualities for each UE is studied. The simulation parameters are shown in Table 4. In order to analyse the performance of each scheduler a full buffer situation is assumed. For the proportional fair scheduler an averaging window of 100

Table 3. LTE Channel Quality Indicators

CQI	MODULATION	Coding rate * 1024	Efficiency
1	QPSK	78	0.1523
2	QPSK	120	0.2344
3	QPSK	193	0.3770
4	QPSK	308	0.6016
5	QPSK	449	0.8770
6	QPSK	602	1.1758
7	16QAM	378	1.4766
8	16QAM	490	1.9141
9	16QAM	616	2.4063
10	64QAM	466	2.7305
11	64QAM	567	3.3223
12	64QAM	666	3.9023
13	64QAM	772	4.5234
14	64QAM	873	5.1152
15	64QAM	948	5.5547

Table 4. LTE Simulation Parameters

PARAMETERS	**VALUE**
No. of UE	6
Subcarrier Spacing	15 kHz
Bandwidth	1.4 MHz
Number of subcarriers per slot	72
Cyclic Prefix	Normal
Slot Duration	0.5 ms
Scheduling Time (1 TTI)	1 ms
HARQ model	Incremental Redundancy.
Total HARQ processes	8
Total Retransmissions	3 (not including the original)
Schedulers	Best CQI, PF, MAX-MIN and RF
Mode	Tx = 1 and Rx=1 or SISO mode (single input single output)
Uplink delay	1 ms (ACK/NAK and CQI reporting rate)
CQI Feedback Granularity	1 CQI value for all PRBs
Scheduler Assignment	Dynamic (Adaptive Modulation and Coding)
Channel model	3GPP-TU
Fading model	Block Fading
Receiver	ZF (Zero Forcing)

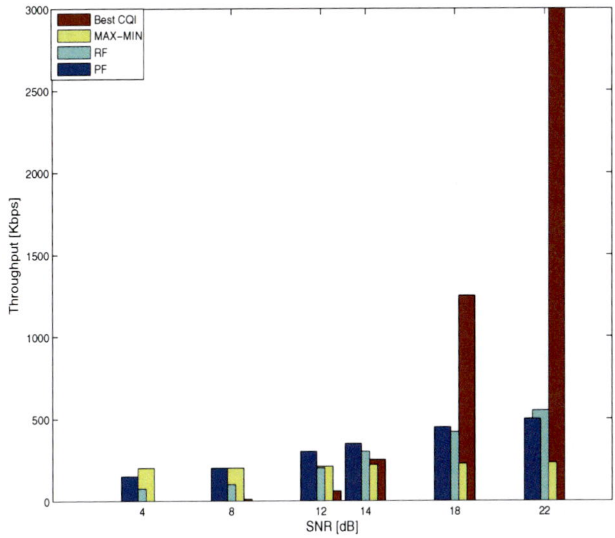

Fig. 4. Throughput (Kbps) achieved for different users with different schedulers

subframes (100 ms) is considered, i.e., an average rate of 100 ms is taken into account. For the Best CQI schedulers if more than one UE have the same channel quality then PRBs are assigned randomly among the users having the same CQI. Figure 4 reports the throughput for different users, each characterised by a different value of signal-to-noise ratio, for different scheduling strategies. It is clear from Figure 4 how efficiently the Best CQI scheduler utilizes channel variations and produces maximum cell (system) throughput, as also shown in Figure 5. The MAX-MIN scheduler produces the minimum cell throughput but achieves a fairer system in terms of throughput as shown in Figure 4 and in Figure 6.

In order to measure the fairness of a systems, the Jain's fairness index [13] can be used. It is calculated as shown below:

$$J = \frac{\left(\sum_{i=1}^{I} T(i)\right)^2}{I \sum_{i=1}^{I} \left(T(i)\right)^2} \tag{1}$$

where *T(i)* is the throughput achieved by each user *i*. If all the UEs have the same throughput, the fairness index is 1. Fairness decreases as the differences between throughput increase.

In Figure 6, the Jain's fairness index is used to analyze the fairness performance. The performance of the *PF* scheduler is better in terms of fairness as well as system throughput, as shown in Figure 5. PF is a better choice than RF when a level of fairness in terms of throughput is required, as PF also produces better system (cell) throughput, as shown in Figure 5.

The basic scheduling strategies discussed in this chapter exploit multi-user channel diversity by utilizing the AMC feature. The exploitation of the HARQ procedure further increases the system throughput. AMC and HARQ are the most important features

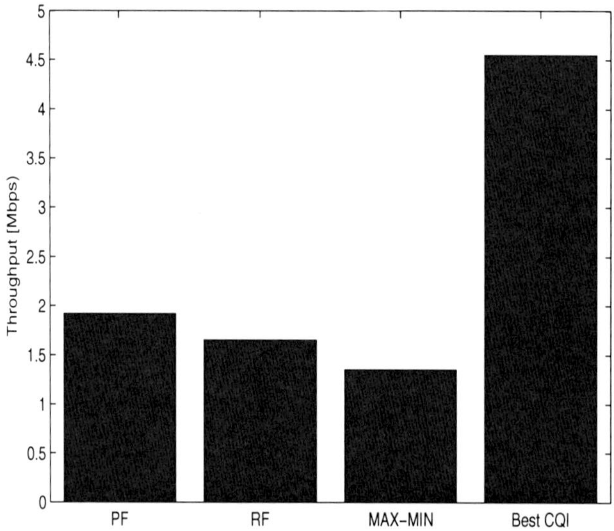

Fig. 5. Total System or Cell throughput (Mbps) for different schedulers

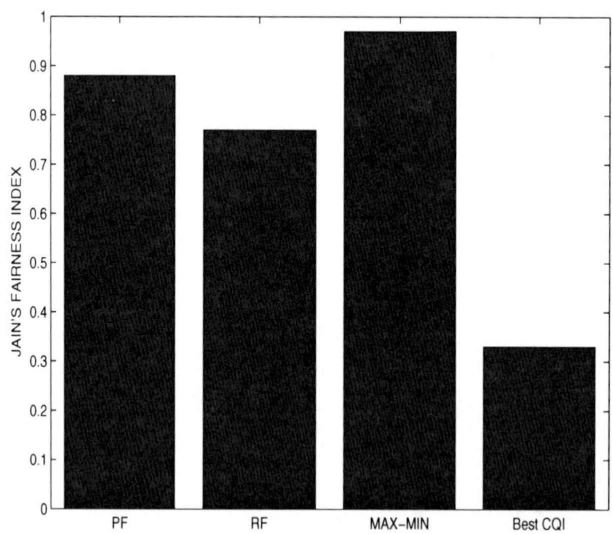

Fig. 6. Jain's fairness index for different schedulers

in optimum exploitation of the link capacity. The channel aware scheduling strategies discussed in this chapter form the basis of designing a QoS aware scheduling rules. When spectral efficiency is considered as the performance measure, then the Best CQI scheduler is the best scheduling rule as it maximizes the system throughput. Other issues such as fairness and QoS provisioning can be addressed at the expense of reduced system throughput. The MAX-MIN scheduler achieves fairness (in terms of throughput) at the expense of system throughput. An effective scheduling strategy achieves a good trade-off between fairness and efficiency, such as the PF scheduler discussed above. This scheduler finds a good trade-off between fairness and efficiency in terms of throughput.

We considered in this section a full buffer scenario, where the queue size of each flow is considered infinitely large. However, in order to design QoS-aware strategies, the consideration of parameters such as the packet's waiting time in the queue becomes extremely important.

4 QoS-Aware Scheduling

The main goal of a packet scheduler is to avoid QoS violations in terms of packet delay and loss rate. As shown in Table 1, different traffic classes have different QoS parameters. A QoS-aware scheduler must guarantee a limit on packet's delay and loss rate according to its Quality Class Indicator (QCI). The QCI is the mechanism determining the QoS requirement of each flow (also called radio bearer). The increase in network traffic above the system capacity causes the scheduler to increase the packet delay of best-effort traffic while maintaining the delay of VoIP traffic at a constant level. Since best-effort traffic can tolerate higher packet delays, its QoE satisfaction level decreases marginally with the increase in latency and the scheduler maintains satisfactory service even under network congestion. QoS-aware scheduling strategies are interesting, promising and simple in terms of implementation.

However, this type of schedulers require strict admission control as fairness is an inherent feature of QoS-aware schedulers. Any increase of the incoming traffic above the system capacity would violate the QoS performance of flows already in the network. Therefore, with a QoS-aware scheduler in place, the admission control policy must limit the incoming traffic of the real-time flows according to the system capacity. Video traffic exhibits a highly variable rate (high peak to average rate ratio). Considering a scenario where all the video traffic flows are admitted according to average rate requirements and the admission control blocks further flows from entering the network, the network resources may get under utilized when video traffic is low or may result in QoS violations in peak traffic conditions.

One of our proposals for QoS-aware scheduling is reported in the following.

5 Opportunistic Packet Loss Fair Scheduling for Delay-Sensitive Applications over LTE Systems

In this section, we propose a QoS-aware scheduling strategy aimed at guaranteeing QoS requirements such as delay bound and packet loss rate thresholds of different real-time flows.

Besides the basic schedulers presented in the previous sections, there are many scheduling algorithms designed for single carrier systems to accommodate real time traffic, such as the Modified Largest Weighted Delay First (M-LWDF) algorithm [9]. This scheduler serves the flow maximizing the product $\gamma_i H_i^{(n)} R_i^{(n)}$, where $H_i^{(n)}$ is the Head of Line (HOL) packet delay for queue i, $R_i^{(n)}$ is the rate available at scheduling instant n according to the instantaneous channel condition; γ_j is a constant whose value is adjusted to account for different delay requirements for different flows. There are two main reasons why M-LWDF is used for delay sensitive traffic: one is the fact that it is throughput optimal (analytically), proved in [8], and the second is that it is relatively simple to implement this algorithm, since only time stamping is required for the incoming packets. Due to its simplicity, this algorithm has been widely used in single carrier systems for real time applications. The use of M-LWDF has been adapted for the LTE system (which is based on Orthogonal Frequency Division Multiple Access (OFDMA)) in [26]; a comparison is provided among well known packet scheduling algorithms designed for single carrier systems, such as Proportional Fair (PF) [14], M-LWDF [9], maximum throughput [34], Exponential Proportional Fair (EXP/PF) [28] [27]. According to this paper, M-LWDF, in terms of efficiency and fairness, outperforms the PF, maximum throughput, and EXP/PF scheduling algorithms.

Some delay based scheduling algorithms have also been proposed for multicarrier systems. In [19] an Opportunistic Scheduling (OS) algorithm is proposed. This scheduling algorithm is quite complex due to the fact that it performs resource assignment and resource allocation in different steps, resulting in a more complex scheduling algorithm. The algorithm operates in two steps, where the first step consists of allocating subcarriers to users with a good channel by exploiting multiuser diversity; if during the first step some of the subcarriers remain unused, then those subcarriers are allocated to users having higher HOL delay to incorporate fairness. Hence, the second step consists of a subcarrier assignment algorithm which assigns the best subcarriers to users which have suffered from higher HOL delay violations. This scheduling algorithm lacks satisfactory fairness, since there is a high probability that the scheduler allocates all the subcarriers to good channel users, thus inhibiting the fairness step. Moreover, there is a need for a joint scheduling and resource allocation step which takes all the necessary information into account before assigning the resources, in order to allow a tighter control over the resources with lower complexity. Thus the signalling cost is high with such an approach. OS performance over LTE is analyzed in [29], where a simple scheduling algorithm, called Delay Prioritized Scheduler (DPS), is also designed. The DPS algorithm takes only packet delay information into account and assigns the best PRB to the users whose packets have remained in the buffer for a longer period of time. The best PRB for each user is determined by taking instantaneous Signal to Noise Ratio (SNR) in to account. The DPS algorithm calculates the packet delay of all users and assigns the best PRB to the user with the highest packet delay; in the next iteration the same process is repeated until all the PRBs are assigned. This delay prioritized scheduling algorithm outperforms the OS algorithm in terms of system throughput and also achieves very low packet loss rate (PLR) even under loaded conditions. However, in the DPS algorithm the only scheduling criterion is the packet delay information, hence users with bad channel conditions will force the PRB allocation towards themselves, thus limiting the system capacity. In [32] a Packet Loss Fair (PLF) scheduling rule is proposed for OFDMA systems in order to provide QoS for diverse real time traffic; the PLF rule provides short term as well as long term fairness, but this scheduling algorithm lacks the exploitation of statistically independent multiuser frequency selective fading.

We propose here a scheduling strategy overcoming some of the limits of the preceding scheduling algorithms. This is an Opportunistic Packet Loss Fair (OPLF) scheduling algorithm, based on calculating a simple dynamic priority function depending on HOL packet delay, PLR, and achievable instantaneous downlink rate of each user.

The remainder of this section is organized as follows. The system model considered is described in subsection 5.1, whereas subsection 5.2 presents the proposed scheduling strategy. Simulation set-up and results are presented in subections 5.3 and 5.6, respectively.

5.1 System Model

The system we consider consists of an OFDM single antenna Single Input Single Output (SISO) multiuser LTE system. We focus here on the downlink. In SISO systems a PRB can only be assigned to one user at any scheduling instant and hence there is no overlapping in PRB allocation. We consider a single cell scenario where the serving LTE base station (eNodeB) is at the center of the cell. The serving eNodeB's MAC scheduler controls all the available PRBs by allocating them to active flows competing for resources. Each user is assigned a buffer at the eNodeB. When a packet reaches the serving eNodeB, the buffer management system time stamps and queues each packet in a First In First Out (FIFO) order. At the start of each scheduling instant, *i.e.*, before the multiuser scheduling decision, the HOL packet delay for each user's packet is calculated by subtracting the arrival time of the packet from the current time. If the HOL packet delay is above the considered threshold $D_{\max}$, depending on the QoS requirements for the application, the packet is discarded by the buffer management system.

In the network, users report their instantaneous channel quality by means of CQI. After resource allocation, the scheduler uses the MIESM method to calculate a single CQI value to be used for all the allocated PRBs of a particular user.

5.2 Opportunistic Packet Loss Fair Scheduler

Although delay sensitive applications often tolerate losses, they typically require that the PLR is kept below a threshold. Hence we propose an Opportunistic Packet Loss Fair (OPLF) scheduling algorithm, based on calculating a simple dynamic priority function which depends on the HOL packet delay, the PLR and the achievable instantaneous downlink rate of each user.

The packet scheduler exploits channel information in order to achieve multiuser diversity gain, by assigning PRBs to the users experiencing lower attenuation. However, it is important to note that relying only on the diversity gain can lead to an unfair treatment of users experiencing lower average channel quality. Our proposed scheduling strategy addresses this issue by providing fairness and at the same time exploiting multiuser frequency diversity.

The available resource blocks are allocated to users through an iterative process, where the total number of iterations is equal to the total number of PRBs available at each Transmission Time Interval (TTI) and at each iteration only one PRB is allocated to the user which maximizes a priority function.

The proposed priority function of user i (OPLF rule) at scheduling epoch n is:

$$\mathrm{PRF}^{(\mathrm{n})}{}_{i} = \frac{\overline{R}_i^{(n)}\, H_i^{(n)}\, \mathrm{plr}^{(\mathrm{n})}{}_{i}}{H_{\max}\, \mathrm{plr}_{\mathrm{thr}_i}} \tag{2}$$

with

$$H_i^{(n)} = n - n_{\mathrm{enter}}(i) \tag{3}$$

and

$$\mathrm{plr}_i^{(n)} = \frac{\sum_{m=n-t_w}^{n} P_{\mathrm{drop}_i}^{(m)}}{\sum_{m=n-t_w}^{n} \left(P_{\mathrm{transmit}_i}^{(m)} + P_{\mathrm{drop}_i}^{(m)}\right)}. \tag{4}$$

$H_i^{(n)}$ is the HOL packet delay of user i at current scheduling instant n, $n_{\mathrm{enter}}(i)$ is the time at which a packet of user i enters the buffer of the eNodeB and is time stamped by the buffer manager. $\mathrm{plr}_i^{(n)}$ is the packet loss rate of user i at scheduling instant n calculated over the moving average transmission window t_w, $P_{\mathrm{transmit}_i}^{(m)}$ and $P_{\mathrm{drop}_i}^{(m)}$ are the number of transmitted and dropped packets over the moving average transmission window t_w. $\mathrm{plr}_{\mathrm{thr}_i}$ is the maximum PLR tolerated for user i. $\overline{R}_i^{(n)}$ is the instantaneous rate of user i averaged over all the unallocated PRBs:

$$\overline{R}_i^{(n)} = \frac{1}{|\Phi_{\mathrm{URB}}(n,k)|} \sum_{\varphi \in \Phi_{\mathrm{URB}(n,k)}} R_{i,\varphi}^{(n)} \tag{5}$$

where $\Phi_{\mathrm{URB}}(n,k)$ denotes the set of unallocated PRBs during iteration k at scheduling instant n and $|\Phi_{\mathrm{URB}}(n,k)|$ its cardinality; $R_{i,\varphi}^{(n)}$ is the instantaneous rate of user i on PRB φ.

The priority function above is high when the instantaneous PLR of user i is high with respect to the threshold (hence the user should get more resources to reduce it) and also when the HOL delay is high with respect to the maximum tolerated delay. $R_i^{(n)}$ improves the system efficiency by exploiting the statistically independent multiuser frequency selective fading. The priority function tends to favor the user with the highest HOL delay and the highest PLR; however, if some of the PRBs of such a user are under a deep fade, the factor $R_i^{(n)}$ ensures that those PRBs are not allocated to such user.

The OPLF scheduling strategy is described as a pseudo-code in Algorithm 1.

Algorithm 1. Opportunistic Packet Loss Fair Scheduler

repeat
  Calculate $H_i^{(n)}$, for all i, according to (3)
  Calculate $\text{plr}_i^{(n)}$, for all i, according to (4)
  while $|\Phi_{\text{URB}}(n,k)| > 0$ **do**
    Calculate $\overline{R}_i^{(n)}$, for all i, according to (5)
    for $i = 1$ to $N_{\text{active flows}}$ **do**
      Calculate $\text{PRF}_i^{(n)}$ according to (2)
    end for
    $i^* = \arg\max \text{PRF}_i^{(n)}$
    $\varphi^* = \arg\max R_{i^*,\varphi}^{(n)}$ where $\varphi \in \Phi_{URB}(n,k)$
    $\Phi_{\text{PRB},i^*}(n,k+1) = \Phi_{\text{PRB},i^*}(n,k) + \{\varphi^*\}$
    $\Phi_{\text{URB}}(n,k+1) = \Phi_{\text{URB}}(n,k) - \{\varphi^*\}$
    if packet of user i^* is scheduled **then**
      $H_{i^*}^{(n)} = n - n_{\text{enter}}(i^*)$
    end if
    If a user is scheduled on more than one PRB, then MIESM is applied to calculate the average CQI.
  end while
  TTI = TTI + 1
until END OF SIMULATION

$\Phi_{\text{PRB},i^*}(n,k)$ denotes the set of PRBs allocated to user i^* which maximizes the priority function $PRF_i^{(n)}$ by iteration k at scheduling instant n. It is important to note that at each iteration only one PRB is allocated to the user which maximizes the priority function, *i.e.*, the total number of iterations is equal to the total number of PRBs available at each TTI. In order to fully utilize the available resources, if more than one PRB have the same instantaneous CQI, then the PRB which experiences the maximum fading for other users, or in other words has the lowest quality for the other users, is selected. In this way, multiuser frequency diversity is exploited. We highlight here that the priority function is dynamic, since when a best quality PRB is allocated to the user, in the next iteration the factor $\overline{R}_i^{(n)}$ will be changed as user i^*, which maximizes $\text{PRF}_i^{(n)}$, is allocated its best remaining PRB φ^*. Our proposed scheduling algorithm is different from the M-LWDF rule, which also exploits HOL delay but relies on the the proportional fair rule in order to provide fairness. Our proposed scheduler provides fairness by exploiting multiuser frequency diversity, since in frequency selective fading the same PRB for different users undergoes a statistically independent fading, hence a PRB which is under a deep fade might be the best PRB for another user. The $\text{plr}_i^{(n)}$ term in the priority function equation provides a degree of fairness for users having poor average channel condition. Simulation results as seen in Section 5.6 confirm that our proposed scheduler provides better fairness and throughput than the M-LWDF and the PLF rule which uses proportional fair characteristics.

We summarize in the following the positive features of the proposed algorithm.

1. The bandwidth is utilized efficiently, as each user is scheduled on its best remaining PRBs and the scheduler will not assign resources to a user whose channel is under a deep fade, as indicated by the priority function.
2. Real-time traffic with diverse QoS requirements can be accommodated, and only information on the delay threshold $H_{\max}$ and on the packet loss threshold PLR_{thr} is required.
3. QoS parameters are used in the scheduling decisions, which ensures that users will get a minimum proportion of resources even if the average channel condition is low.
4. For real time applications fairness is guaranteed when the current packet loss rate is distributed proportionally equal among all the users competing for resources. In our context we aim to ensure fair PLR distribution over the moving average transmission window of size t_w. Therefore we achieve short term fairness; as highlighted in [20] short term fairness guarantees long term fairness, but not vice versa.

5.3 Simulation Environment

The assumptions we considered in the simulations are reported below.

- The channel quality of each user remains constant during the subframe period of 1 ms, although it changes from subframe to subframe.
- CQI feedback from UE to the eNodeB is error free. The error free assumption of the feedback channel is satisfied by using efficient and heavily coded feedback stream as is anyway customary for LTE system.
- It is assumed that equal downlink transmit power is allocated on each PRB.
- It is assumed that, at any time instant, pathloss is fixed on each PRB. Multipath induced fading is represented by a tapped delay line model (Typical Urban). The propagation model used follows the guidelines in [1][3].

Simulation parameters are reported in Table 5.

Video Traffic Model. In order to analyze the performance of channel aware schedulers on real-time traffic, a Near Real Time Video (NRTV) model [2] is used. The parameters of the NRTV model are given in Table 6. In order to model the variability in size and inter-arrival time between packets, a Truncated Pareto Distribution is considered, with probability density function (pdf) reported below:

$$f_x(x) = \begin{cases} \dfrac{ak^a}{x^{a+1}} & k \leq x < m, \\ (\dfrac{k}{m})^a & x = m \end{cases} \tag{6}$$

The values considered for the parameters of the distribution a, m and k are reported in Table 6.

Table 5. Simulation parameters - Downlink LTE scheduling for real time applications

PARAMETERS	VALUE
Bandwidth	3 MHz
Carrier frequency	2.1 GHz
No. of PRBs	15
No. of users	Variable (40 to 60)
UE distribution	Uniform
Cell radius	2 km
Application	NRTV
Admission Control	No Admission control
Mode	Tx = 1 and Rx = 1 (SISO mode)
Channel	3GPP-TU (Typical Urban)
Pathloss model	Hata-Cost-231 model (urban pathloss model)
HARQ	Synchronous retransmissions (Up to 3 retransmissions)
Channel Fading	Block Fading (1 ms)
UE speed	15 to 100 km/h (users moving independently at variable speed)
H_{max}	50 ms
PLR_{thr_i}	1%
t_w	1s

The total number of slices in a frame is deterministic and each slice corresponds to one packet; in the scenario we consider eight slices (packets) per frame. The total duration of one frame is also deterministic, considered here as 50 ms, which results in 20 fps. Based on these parameters, packets are produced and streamed into the user buffer at the eNodeB. Before entering the buffer of each user, packets are time stamped and served in FIFO order. According to the parameters below, the average rate of video streaming is approximately 120 kbps. QoS requirements are assumed the same for each flow. The maximum tolerated delay $H_{\max}$ is set to 50 ms and the packet loss rate threshold $\mathrm{plr}_{\mathrm{thr}_i}$ is set to 1%. When scheduled and transmitted successfully through the air interface, packets are assumed to be played. A packet is assumed to be lost if, due to retransmissions, the deadline of the packet is reached.

5.4 Benchmark Scheduling Strategies

The priority functions of the PLF, M-LWDF, and PF rules, considered in the following for comparison, are reported below:

PLF Scheduling Rule

$$\mathrm{PRF}_{i,\varphi}^{(n)} = \left(\frac{R_{i,\varphi}^{(n)}}{R_{i,\mathrm{ave}}^{(n)}}\right) * \left(\frac{\mathrm{plr}_i^{(n)}}{\mathrm{plr}_{\mathrm{thr}} * H_{\max}}\right) \tag{7}$$

Table 6. Video traffic model parameters

Streaming Information	Distribution	Parameters
Inter-arrival time between frames	Deterministic	20 fps: duration of one frame is 50 ms
Total packets or slices in one frame	Deterministic	8
Inter-arrival time between packets	Truncated Pareto distribution	Min time k=4 ms Max time m= 8 ms a=1.2
Packet (slice) size	Truncated Pareto distribution	Min size k=65 bytes Max size m= 150 bytes a=1.2

M-LWDF Scheduling Rule

$$\mathrm{PRF}_{i,\varphi}^{(n)} = \gamma_i * \frac{R_{i,\varphi}^{(n)}}{R_{i,\mathrm{ave}}^{(n)}} * H_i^{(n)} \tag{8}$$

PF Scheduling Rule

$$\mathrm{PRF}_{i,\varphi}^{(n)} = \frac{R_{i,\varphi}^{(n)}}{R_{i,\mathrm{ave}}^{(n)}} \tag{9}$$

In the equations above $R_{i,\mathrm{ave}}^{(n)}$ and γ_i are respectively the moving average of the rate achieved over a transmission window size t_w for user i at time t, and a constant whose value is adjusted to account for different delay requirements of different flows.

It is important to note that all the scheduling rules are implemented in a way that the priority functions of all active flows are calculated on each PRB and the scheduler allocates the PRB to the user which maximizes the priority function. Hence, all these rules exploit multiuser frequency diversity and we have a fair comparison among the scheduling rules.

5.5 Performance Metrics

The performance of the proposed algorithm is evaluated and compared with the benchmark scheduling strategies above. The comparison is performed in terms of cell PLR and its standard deviation, system throughput, and average system delay.

The cell throughput or system throughput represents the total amount of packets successfully transmitted through the air interface from eNodeB to all the active flows in the time unit. The cell PLR is the ratio of the total number of lost packets to the total number of packets produced. The first two metrics evaluate the efficiency of each algorithm.

In order to analyze the fairness of the different scheduling strategies, the standard deviation of the PLR of each user is also calculated. The lower the standard deviation, the higher the level of fairness among the users.

The average packet delay corresponds to the average amount of time packets reside in the buffer. The characteristics of all the considered schedulers are summarized in Table 7.

Table 7. Characteristics of different schedulers

Scheduler	Channel aware	Delay aware	Packet loss aware
PF	yes	no	no
M-LWDF	yes	yes	no
PLF	yes	no	yes
OPLF	yes	yes	yes

5.6 Simulation Results

The system throughput performance of the proposed algorithm and the reference ones is shown in Figure 7. It is clear from the figure that OPLF outperforms the M-LWDF, PF and PLF scheduling rules. When the load is above 45 users, the PLF rule outperforms the M-LWDF rule, mainly because the PLF rule does not incorporate HOL delay information, but uses only PLR and the proportional fair rule, therefore users having higher PLR will get higher priority irrespective of the packet's deadline, whereas the proportional fair term in the PLF rule ensures that bad channel users do not affect the system efficiency. The performance of the PF algorithm is the worst mainly because this rule

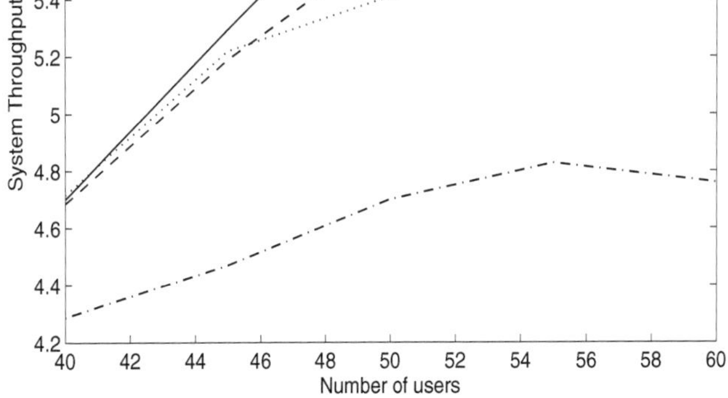

Fig. 7. System throughput vs. number of users

neither takes HOL delay nor PLR information into account in the scheduling decision. It is important to note that the system throughput of all the schedulers decreases after 55 users as this is an extreme load condition. This means that the QoS of most of the users is affected as more users are added and, since fairness is incorporated in all the algorithms, the effective throughput decreases due to the limited number of resources available, which results in more deadline violations.

The cell PLR performance of all the algorithms is shown in Figure 8. It is clear from the figure that OPLF outperforms the M-LWDF, PF and PLF rules also with this metric. Note that the PLR threshold considered for this application is 1%. It is important to note that when the number of users is 45, M-LWDF is slightly better than the PLF scheduling rule, however under high load PLF performs better than the M-LWDF rule which does not exploit PLR information in the priority function.

Figure 9 can be considered for evaluating the fairness performance of the algorithms. Fairness is an important criterion as algorithms may provide higher cell throughput at the expense of lower throughput (and hence quality) for users with bad channel conditions. In order to analyze the fairness performance of all the algorithms, we consider here the standard deviation of the PLR of all the users; a higher standard deviation implies that some users receive good service whereas some receive bad service, hence fairness is low. Our proposed scheduler achieves lower standard deviation in comparison with the other considered algorithms. When a user suffers from HOL packet delay violation, such an event is characterized by a higher priority function value in the case of PLF and OPLF schedulers. Therefore more resources are assigned to such a user so that further delay violations are avoided. This is not the case in M-LWDF, delay aware and packet loss blind, and PF, delay and packet loss blind, schedulers. The fairness performance of the OPLF scheduler is better than the PLF scheduler mainly because OPLF

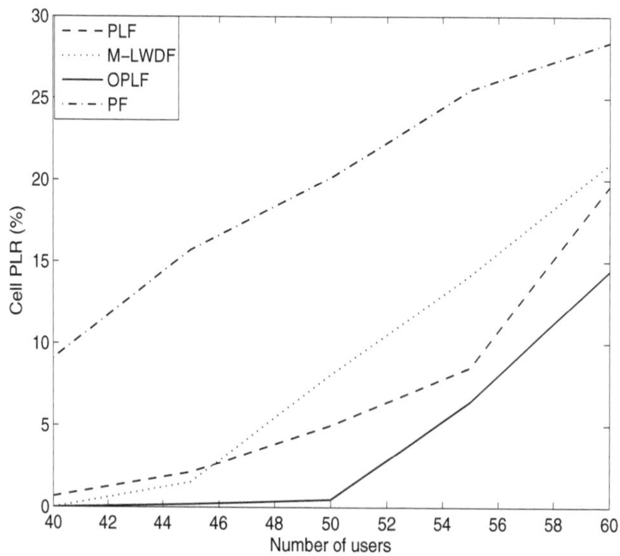

Fig. 8. Cell packet loss ratio (%) vs. number of users

is deadline aware. If users have the same PLR violation, OPLF adapts to this situation by giving more priority to the user having a higher HOL packet delay, therefore further delay violations are avoided. It is important to note that, although M-LWDF results in a slightly better system throughput than the PLF scheduling rule when the number of users is 45, PLF outperforms M-LWDF in terms of fairness, as the scheduling decision in PLF is mainly based on the PLR. The fairness performance of the PLF rule is better and close to the OPLF rule under heavy load.

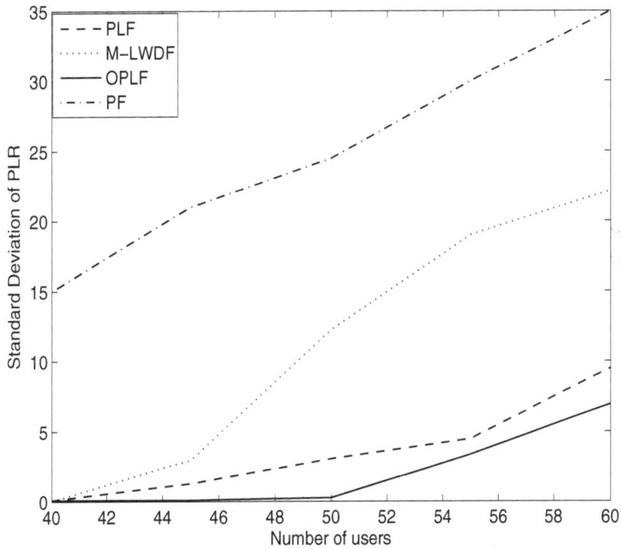

Fig. 9. Standard deviation of PLR vs. number of users

6 QoE-Aware Scheduling

Video traffic contributes a major proportion of network traffic. QoE-aware packet scheduling strategies are specifically designed for video streaming traffic. By considering the content of the video traffic, the scheduler aims to maximize the perceived video quality. The information on the contents of different video traffic is provided through cross-layer signaling. Under congestion, the scheduler exploits the redundancy in the video traffic by either dropping the redundant quality layers present in the video stream or sending a feedback signal to the video server to perform timely rate adaptation. These type of schedulers consider different objective functions based on video quality in the scheduling decision, the main goal of the scheduler being to maximize the video quality of the users. When the network is heavily congested, the scheduler can exploit the quality, temporal and spatial redundancy in the video stream, and drops packets having little contribution towards the overall video quality. Different information can be provided to the scheduler such as the decoding deadline of a video frame, packet's dependency on other packets, packet's contribution to the overall video quality.

Optimizing video traffic by designing a specific scheduling function has been gaining importance mainly because most of the mobile network traffic will be video in the near future. There are many content aware scheduling strategies designed for video traffic. Examples include [24] [22] [25] [15] [33].

However, content aware scheduling strategies pose a limitation from an implementation point of view because of the extensive cross layer signaling requirements. Complex application layer information such as distortion associated with each video packet, video frame size and its dependency level, rate and quality of each video layer are required by the scheduling function. Operators intending to use video quality based scheduling strategies must introduce new network elements which can parse a video packet and provide information to the scheduler regarding the reduction in video quality if the packet is not successfully scheduled and other information related to the video quality.

Research is ongoing to investigate novel strategies for content-aware scheduling requiring less signaling information and providing the best trade-off between efficiency and fairness in terms of quality among the users.

7 Conclusion

After an introduction on LTE wireless systems and on scheduling over the downlink of LTE systems, we presented our proposed OPLF algorithm for downlink scheduling of delay sensitive traffic at the MAC layer of wireless systems based on OFDMA. This algorithm outperforms state-of-the art algorithms, such as M-LWDF, PF, and PLF, in terms of throughput, packet loss rate and fairness, also keeping packet delay below a fixed threshold. With respect to existing algorithms, the proposed algorithm will thus enable the allocation in a cell of a higher number of users served with satisfactory quality. QoE-aware scheduling is finally presented, with a discussion on the main achievements and open issues.

Acknowledgments. The authors wish to thank the colleagues in DOCOMO Euro-labs for the fruitful discussions and the support.

References

1. Tech. rep. m.1225: Guidelines for evaluation of radio transmission technologies for IMT-2000, Geneva, Switzerland (1997)
2. 1xev-dv evaluation methodology- addendum (v6). 3GPP2 WG5 Evaluation AHG (2001)
3. Technical Specification Group GSM/EDGE Radio Access Network, Radio transmission and reception, Annex C.3 Propagation Models, 3GPP. Tech. Rep. TS 05.05 V.8.20.0 (Release 1999) (2005)
4. 3GPP - TS 36.322 Radio Link Control (RLC); protocol specification (Release 10) (2011)
5. 3GPP - TS 36.323 Packet Data Convergence Protocol (PDCP) specification (Release 10) (2011)
6. 3GPP - TS 36.331 Medium Access Control (MAC); protocol specification (Release 10) (2011)
7. 3GPP - TS 36.331 Radio Resource Control (RRC); protocol specification (Release 10) (2011)

8. Andrews, M., Kumaran, K., Ramanan, K., Stolyar, A., Vijaykumar, R., Whiting, P.: CDMA data QoS scheduling on the forward link with variable channel conditions. Bell Labs Technical Memo No. 10009626-000404-05TM (2000)
9. Andrews, M., Kumaran, K., Ramanan, K., Stolyar, A., Whiting, P., Vijayakumar, R.: Providing Quality of Service over a Shared Wireless Link. IEEE Communications Magazine 39(2), 150–154 (2001)
10. Dahlman, E., Parkvall, S., Skold, J., Beming, P.: 3G Evolution: HSPA and LTE for Mobile Broadband. Academic Press (2007)
11. He, X., Niu, K., He, Z., Lin, J.: Link layer abstraction in MIMO-OFDM system. In: International Workshop on Cross Layer Design (IWCLD), Jinan, China (2007)
12. Hewage, C.T.E.R., Martini, M.G., Khan, N.: 3D medical video transmission over 4G networks. In: International Symposium on Applied Sciences in Biomedical and Communication Technologies (ISABEL), Barcelona, Spain (2011)
13. Jain, R., Chiu, D., Hawe, W.: A quantitative measure of fairness and discrimination for resource allocation in shared computer systems. Tech. Rep. TR-301 (1984)
14. Jalali, A., Padovani, R., Pankaj, R.: Data Throughput of CDMA-HDR a High Efficiency-High Data Rate Personal Communication Wireless System. In: IEEE Vehicular Technology Conference (VTC), Tokyo, Japan (2000)
15. Khan, N., Martini, M., Bharucha, Z.: Quality-aware fair downlink scheduling for scalable video transmission over LTE systems. In: IEEE Conf. on Signal Processing Advances in Wireless Communications (SPAWC), Cesme, Turkey (2012)
16. Khan, N., Martini, M.G., Bharucha, Z., Auer, G.: Opportunistic Packet Loss Fair Scheduling for Delay-Sensitive Applications over LTE Systems. In: IEEE Wireless Communications and Networking Conference (WCNC), Paris, France (2012)
17. Khan, N., Martini, M.G., Staehle, D.: Opportunistic Proportional Fair Downlink Scheduling for Scalable Video Transmission over LTE Systems. In: IEEE Vehicular Technology Conference (VTC), Las Vegas, USA (2013)
18. Khan, N., Martini, M.G., Staehle, D.: Opportunistic QoS-Aware Fair Downlink Scheduling for Delay Sensitive Applications using Fuzzy Reactive and Proactive Controllers. In: IEEE Vehicular Technology Conference (VTC), Las Vegas, USA (2013)
19. Khattab, A.K.F., Elsayed, K.M.F.: Opportunistic Scheduling of Delay Sensitive Traffic in OFDMA-based Wireless Networks. In: International Symposium on World of Wireless, Mobile and Multimedia Networks (WoWMoM), Buffalo, NY, USA (2006)
20. Koksal, C., Kassab, H., Balakrishnan, H.: An Analysis of Short Term Fairness in Wireless Media Access Protocol. In: ACM SIGMETRICS, California, USA (2000)
21. Martini, M., Hewage, C., Nasralla, M.: 3-D Robotic Tele-Surgery and Training over Next Generation Wireless Networks. In: Proc. IEEE Engineering in Medicine and Biology Conference (EMBC 2013), Osaka, Japan (2013)
22. Martini, M.G., Tralli, V.: Video quality based adaptive wireless video streaming to multiple users. In: IEEE International Symposium on Broadband Multimedia Systems and Broadcasting, Las Vegas, NV (2008)
23. Padhye, J., Firoiu, V., Towsley, D.F., Kurose, J.F.: Modelling TCP Reno performance, "a simple model and its empirical validation". IEEE Transactions on Networking 8(2), 133–145 (2000)
24. Pahalawatta, P.V., Berry, R., Pappas, T.N., Katsaggelos, A.K.: Content-aware resource allocation and packet scheduling for video transmission over wireless networks. IEEE Journal on Selected Areas in Communications (JSAC) 25(4), 749–759 (2007)
25. Ralalage, H.A., Martini, M.G., Hewage, C.T.: Channel and content aware 3D video scheduling with prioritized queuing. In: Proc. of IEEE Wireless Advanced 2012, London, UK (2012)

26. Ramli, H.A.M., Basukala, R., Sandrasegaran, K., Patachaianand, R.: Performance of Well Known Packet Scheduling Algorithms in Downlink 3GPP LTE System. In: IEEE Malaysia International Conference on Communications (MICC), Kuala Lumpur, Malaysia (2009)
27. Rhee, J.H., Holtzman, J.M., Kim, D.K.: Scheduling of Real and Non-real Time Services: Adaptive EXP/PF Algorithm. In: IEEE Vehicular Technology Conference (VTC), Jeju Island, Republic of Korea (2003)
28. Rhee, J.H., Holtzman, J.M., Kim, D.K.: Performance Analysis of the Adaptive EXP/PF Channel Scheduler in an AMC/TDM System. IEEE Communications Letters 8(8), 497–499 (2004)
29. Sandrasegaran, K., Ramli, H.A.M., Basukala, R.: Delay-Prioritized Scheduling (DPS) for Real Time Traffic in 3GPP LTE System. In: IEEE Wireless Communications and Networking Conference (WCNC), Sydney, Australia (2010)
30. Schwarz, S., Wrulich, M., Rupp, M.: Calculation of the Spatial Preprocessing and Link Adaption Feedback for 3GPP UMTS/LTE. In: IEEE Wireless Advanced, London, UK (2010)
31. Schwarz, S., Wrulich, M., Rupp, M.: Mutual Information based Calculation of the Precoding Matrix Indicator for 3GPP UMTS/LTE. In: IEEE Workshop on Smart Antennas, Bremen, Germany (2010)
32. Shin, S., Ryu, B.H.: Packet Loss Fair Scheduling Scheme for Real-Time Traffic in OFDMA Systems. ETRI Journal 26(5), 391–396 (2004)
33. Thakolsri, S., Cokbulan, S., Jurca, D., Despotovic, Z., Kellerer, W.: QoE-driven cross-layer optimization in wireless networks addressing system efficiency and utility fairness. In: IEEE Workshop on Multimedia Communications and services (GLOBECOM Workshops), Houston, USA (2011)
34. Tsybakov, B.S.: File transmission over wireless fast fading downlink. IEEE Transactions on Information Theory 48, 2323–2337 (2002)
35. V11.7.0, G.T.: Technical specification, policy and charging control architecture (release 11) (2012)
36. Wan, L., Tsai, S., Almgren, M.: A fading-insensitve performance metric for a unified link quality model. In: IEEE Wireless Communications and Networking Conference (WCNC), Las Vegas, USA (2006)
37. Zyren, J., Coy, W.M.: Overview of the 3GPP long term evolution physical layer. White paper, Freescale Semiconductors (July 2007)

Performance Analysis of Epidemic Routing for Delay-Tolerant Networks

Qingshan Wang[1] and Zygmunt J. Haas[2]

[1] School of Mathematics, Hefei University of Technology, Hefei 230009, China
qswang@hfut.edu.cn

[2] School of Electrical and Computer Engineering, Cornell University, Ithaca, NY 14853, USA
haas@ece.cornell.edu

Abstract. Epidemic routing is considered one of the most effective routing schemes for delay-tolerant networks; indeed, epidemic routing ensures the shortest routing delays. In this chapter, first the epidemic routing is introduced and then the network and mobility models are presented. Finally, we analyze the performance of epidemic routing in two different ways. Most of the analytical work in the technical literature concentrates on the performance metric equation from a macroscopic view, for example, delivery delay, which is the time from packet generation to delivery at the destination. However, insight into the temporal nature of the epidemic routing process can add significant value; e.g., in designing new variations of the epidemic routing protocol. This chapter focuses on the second method of performance analysis. We investigate the probability distribution function of the number of infected nodes, the average number of infected nodes, and the probability of all nodes becoming infected, all as functions of time.

1 Introduction

In traditional communication networks, the notion of "store-and-forward" routing is employed, where it is assumed that a path exists between the source node and the destination node before communication commences. However, in some communication environments, such as low node density, short communication range mobile networks, links between nodes are established only for short periods of time. Thus, owing to the intermittent nature of links between nodes, a complete path between the source and destination nodes is unlikely to exist at any particular time. Thus, various works have proposed a new routing scheme called "store-carry-forward", in which a packet carrier stores a packet in its buffer, carries the packet, and forwards it to other nodes. This scheme exploits opportunistic connectivity and leads to significant routing delays. Such a routing scheme may be useful only if the network (and the underlying applications) can tolerate the resulting delays— hence the name, delay tolerant networks (DTNs). DTNs have many applications, for example, wildlife tracking and habitat monitoring networks [1], vehicular DTNs [2], and underwater networks [3].

C.X. Mavromoustakis et al., *Resource Management in Mobile Computing Environments*, Modeling and Optimization in Science and Technologies 3,
DOI: 10.1007/978-3-319-06704-9_26, © Springer International Publishing Switzerland 2014

Epidemic routing (ER) [4], adopting the "store-and-forward" paradigm, has been proposed as an efficient routing scheme for DTNs. Analogous to the spread of an infectious agent [5], in ER a node that carries a packet passes on a packet copy to every other node with which it can communicate. Owing to mobility, links are sporadically created and copies of the packet spread from the source node to all the other network nodes. Nodes that carry the packet are referred to as *infected*, while nodes that are yet to receive a copy of the packet are called *susceptible*. In particular, the packet will be received by the destination node if the destination node encounters an infected node.

Various works have been focused on reducing redundant transmissions and improving the performance of the epidemic routing. As the creation of each new copy of a packet in the network requires energy to transmit the packet to minimize the energy required for the routing operation, some schemes have been proposed that reduce the number of copies of a packet. Typically, such studies involve models based on Markov Chains ([6], [7]). SWIM (the Shared Wireless Infostation Model) [8] introduces the notion of an "anti-packet", which is a small packet created by the sink, and which propagates within the network deleting obsolete packets from network nodes. Social feature-based multi-path forwarding [9] has also been proposed.

In this chapter, we investigate performance analysis of epidemic routing in DTNs. The method in [10] calculates the packet delivery delay in a two-hop multiple-copy protocol and an unrestricted multiple-copy protocol, while obtaining the number of packet copies of a packet at the time of delivery to the destination node. In both forwarding protocols, the source node replicates the packet to all other nodes without the packet. However, whereas in the unrestricted multi-copy protocol an infected node can transmit the packet to any susceptible node if there is a communication opportunity between them, in the two-hop multiple-copy protocol an infected node can only forward the packet to the destination node. Furthermore, a rigorous, unified framework based on ordinary differential equations [11] has been used to investigate various forwarding and recovery schemes, which delete unnecessary packets from the networks to obtain the formulas for packet delivery delay, number of copies sent, and buffer occupancy. The authors in [12] proposed (p, q)-epidemic routing using the "anti-packet" notion introduced in [8], and derived the distribution of delivery delay. In (p, q)-epidemic routing, when a susceptible node encounters the source, it accepts a packet copy with probability $q(0 \le q \le 1)$. When a susceptible node encounters an infected node (excluding the source), it obtains a packet copy with probability $p(0 \le p \le 1)$. When the destination node encounters an infected node for the first time, the packet is forwarded to it with probability 1. Compared with the above works, the study in [15] investigated the temporal aspects of the process of packet propagation in epidemic routing. The novel contribution of this paper is the use of random graph theory [16] together with epidemic modeling to study the temporal packet spread in a DTN. In particular, we develop an analytical iterative formulation that can be used to derive *temporal* functions for the probability of infection, the distribution function of the number of infected nodes, the average number of infected nodes, and the probability of all nodes being infected. Finally, the theoretical models are verified through simulations.

2 Epidemic Routing

In ER [4], a node copies all of its packets to any node that it encounters. Fig. 1 illustrates how ER routing works in a DTN. Node S has a packet for delivery to node D in Fig. 1(a). However, owing to low node density, a fully connected path does not exist between nodes S and D. In Fig. 1(b), node S copies the packet to node A because they are within transmission range of each other. As a result of node mobility, node A meets node D and copies the packet to it, as shown in Fig. 1(c). Thus, the packet is delivered to its final destination. ER uses every possible forwarding opportunity for quick distribution of copies of the packet, resulting in the shortest possible delivery delay and maximal delivery ratio.

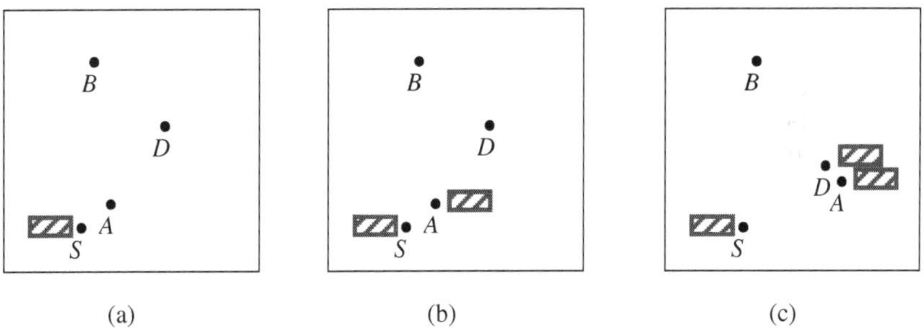

Fig. 1. Epidemic routing in DTNs

3 Network and Mobility Models

We assume that the network consists of N nodes, randomly placed within a square L[m]-by-L[m] network, which is assumed to be closed (i.e., forming a toroid), so that no "boundary effects" need be considered. Nodes move with some mobility pattern; for our analytical study, we do not specify a particular mobility model, but only require that the mobility pattern is random, independent, unconstrained, and stationary. In other words, we assume that the velocity (speed and direction) of a node is chosen randomly from some fixed distribution and independently of that of the other nodes; that a node can move freely to any location within the network (i.e., there are no "exclusion zones" within the network area); and that the mobility parameters (e.g., the spatial and temporal distribution) remain constant over time. We further assume that all nodes move with the same mobility pattern (although this restriction could be lifted), and that the transmission range of a node is fixed at r[m], where $r \ll L$.

Given this network model, we postulate that the inter-meeting times of any two particular nodes can be modeled as an exponentially distributed random variable [10, 17] with rate β, where the inter-meeting time is defined as the time between two consecutive meeting instances of any two specific nodes. For the random direction (RD) mobility model and the random waypoint (RWP) mobility model in particular, estimates for the exponential rate β are found and given by the following lemma.

Lemma 1: The exponential rates β for the RD and RWP can be represented as:

$$\beta_{RD} \approx \frac{2r\mathrm{IE}[V^*]}{L^2} \quad \text{and} \quad \beta_{RWP} \approx \frac{2\omega r\mathrm{IE}[V^*]}{L^2}$$

respectively, where $\mathrm{IE}[V^*]$ is the expected relative speed between two nodes, and $\omega \approx 1.3683$ is a constant for the RWP. $\beta_{RD} \approx \frac{8rv}{\pi L^2}$ and $\beta_{RWP} \approx \frac{8\omega rv}{\pi L^2}$ if the two nodes have an identical and constant speed v.

Because of the memory-less property of the exponential distribution, the probability of two nodes meeting within the next Δt is given by:

$$p = 1 - e^{-\beta \cdot \Delta t}, \tag{1}$$

where we assume that Δt is sufficiently small, so that the probability of more than one encounter between the two specific nodes is negligibly small.

There are two special nodes in the network, the *source* node and the *destination* node, where the source node generates a message that is intended for delivery to the destination node. Because the union of the transmission areas of the nodes is significantly smaller than the entire network area, i.e. $N\pi r^2 \ll L^2$, the average number of neighboring nodes is significantly smaller than one. Therefore, typically, when the message is generated, there is no end-to-end path between the source and the destination nodes.

For simplicity, we consider "propagation" of a single packet only and we assume that such copying occurs at a "very high" transmission data rate, so that any encounter between two nodes allows each node to copy the packet from its memory to the other node's memory (see [8] for treatment of the case where this assumption does not hold). Such propagation continues uninterrupted until the destination node or all network nodes have received the packet.

4 Performance Analysis of Epidemic Routing

ER has been proposed as an efficient approach for routing in DTNs. Several works [8, 11-15] have evaluated the performance of ER, including delivery delay, number of copies, and delivery probability. These works can typically be divided into two groups: those focusing on the performance metric equation from a macroscopic perspective, for example, delivery delay from packet creation time until its delivery to the destination node, and those that obtain insight into the temporal nature of ER.

4.1 Performance Metric Equation from a Macroscopic Perspective

4.1.1 Ordinary Differential Equations Model

A rigorous, unified framework based on ordinary differential equations (ODE) [11] has been presented to model the delivery delay of a packet in ER. A Markovian chain is constructed, as illustrated in Fig. 2, before delivery of a copy to the destination. Let

$n_I(t) \in [1,N]$ be the number of infected nodes at time t. Thus, the transmission rate from state n_I to state $n_I + 1$ is given by

$$r_N(n_I) = \beta n_I(N-1-n_I) \overset{\lambda=\beta(N-1)}{=} (N-1)\lambda \frac{n_I}{N-1}(1-\frac{n_I}{N-1}). \qquad (2)$$

Based on Theorem 3.1 in [18], it can be proved that, as N increases, the fraction of infected nodes $n_I(t)/(N-1)$ converges asymptotically to the solution of the following equation

$$i^{'}(t) = \lambda i(t)(1-i(t)) \text{, for } t \geq 0 \qquad (3)$$

with the initial condition $i(0) = \lim_{N\to\infty} n_I(0)/N$. We obtain a differential equation for the average number of infected nodes $I(t)$ thus:

$$I^{'}(t) = \beta I(t)(N - I(t) - 1) \qquad (4)$$

with the initial condition $I(0)=1$. Therefore, $I(t) = (N-1)/(1+(N-2)e^{-\beta(N-1)t})$.

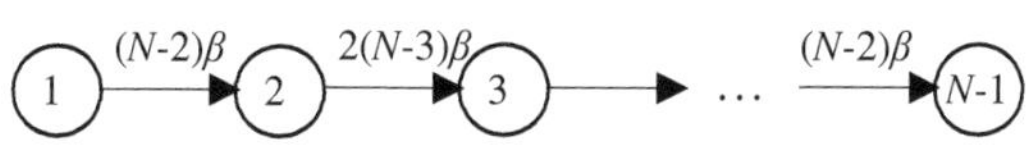

Fig. 2. Markovian chain giving the number of copies before delivery of a copy to the destination

Let the packet delivery delay be denoted by T_d and its cumulative distribution function be $P_N(t) = \text{Prob}(T_d < t)$. Then, we have

$$P_N(t+dt) - P_N(t+dt) = \text{Prob}\{t \leq T_d \leq t+dt\}$$

$$= \text{Prob}\{\text{ destination meets an infected node in } [t,t+dt] | T_d > t\} \times \text{Prob}\{T_d > t\}$$

$$= \text{Prob}\{\text{ destination meets one of the } n_I(t) \text{ infected nodes in } [t,t+dt]\} \times (1-P_N(t))$$

$$= \text{E}[\text{Prob}\{\text{ destination meets one of the } n_I(t) \text{ infected nodes in } [t,t+dt] | n_I(t)\}] \times (1-P_N(t))$$

$$\approx \text{E}[\beta n_I(\text{t})\text{dt}](1-P_N(t))$$

$$= \beta \text{E}[n_I(\text{t})](1-P_N(t))dt$$

$$= \lambda \text{E}[n_I(\text{t}) / (\text{N-1})](1-P_N(t))dt$$

Thus, we define the following equation:

$$\frac{dP_N(t)}{dt} = \lambda \text{E}[n_I(\text{t}) / (N\text{-}1)](1-P_N(t)) . \qquad (5)$$

With N increasing, $\mathrm{E}[n_I(\mathrm{t})/(\mathrm{N}-1)]$ converges to $i(t)$, and $P_N(t)$ converges to the solution of the following equation:

$$P'(t) = \lambda i(t)(1-P(t)) . \quad (6)$$

With the initial condition $P(0)=0$, we obtain $P(t)=1-(N-1)/(N-2+e^{\beta(N-1)t})$. Therefore, the average delivery delay can be expressed as

$$E[T_d] = \int_0^\infty (1-P(t))dt = \ln(N-1)/(\beta(N-1)) . \quad (7)$$

Moreover, the ODE model can be used to model more complex variants of epidemic routing with a "recovery process". When a node forwards a packet to the destination node, it should delete the copy so as not to infect other susceptible nodes. At the same time, to avoid infection again, an anti-packet [8] is created. This recovery scheme is called the IMMUNE scheme. If the anti-packet is also propagated to infected nodes and susceptible nodes, the corresponding schemes are called the IMMUNE_TX and VACCINE schemes, respectively. The number of copies and the average storage requirements under these recovery schemes are also investigated.

4.1.2 Continuous Time Markovian Chain Model

In [12], (p, q)-epidemic routing using the notion of "anti-packet" is presented. This approach implements the delivery delay by exploiting a continuous time Markovian Chain (CTMC) with an absorption state. When a susceptible node encounters the source, it accepts a packet copy with probability $q(0 \le q \le 1)$. Moreover, when a susceptible node encounters an infected node (excluding the source), it accepts a packet copy with probability $p(0 \le p \le 1)$. When the destination node meets an infected node for the first time, it is infected with a probability of 1.

The following four special situations are included in (p, q)-epidemic routing: direct source-destination delivery when $p=q=0$, two-hop forwarding when $p=0$, $q=1$, probability forwarding when $0<p=q<1$, and epidemic routing when $p=q=1$.

The (p, q)-epidemic scheme can be modeled as a CTMC $\{X(t);t \ge 0\}$ with N states from 0 to N-1 as shown in Fig. 3. $X(t)=k(1 \le k \le N-1)$ indicates that there are k packet copies, and state 0 is the only absorbing state. Thus, the packet delivery delay T_d is the first passage time to the single absorbing state. $a(k)$ in Fig. 3 is given by

$$a(k)=(N-k-1)((k-1)p+q) . \quad (8)$$

The transition rate from state k to state k+1 $(1 \le k \le N-2)$ is equal to $a(k)\beta=(N-k-1)((k-1)p+q)\beta$ since there are k infected nodes (including the source) and $N-k-1$ susceptible nodes (excluding the destination). The transition rate from state $k(1 \le k \le N-1)$ to state 0 is $k\beta$, which means that the destination has received a copy of the packet. Based on the CTMC in Fig. 3, $E[T_d]$ is given by

$$E[T_d]=\sum_{k=1}^{N-1} r_k \frac{1}{k\beta}, \tag{9}$$

where r_k is the probability of the absorption from state k to state 0, and is expressed as

$$r_k=\left(\prod_{m=1}^{k-1}\frac{a(m)}{a(m)+m}\right)\frac{k}{a(k)+k}. \tag{10}$$

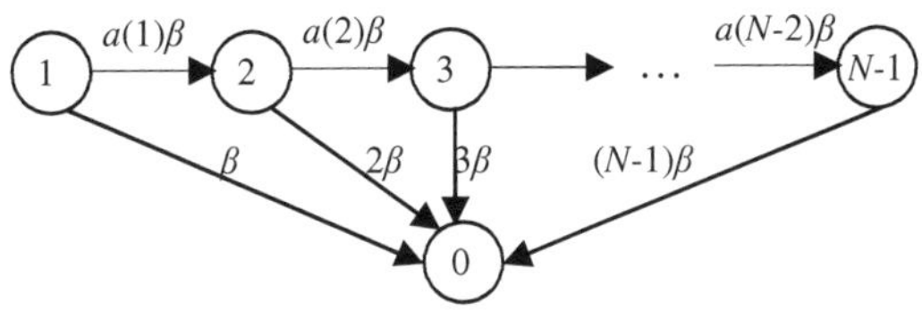

Fig. 3. Transition rate diagram of the absorbing Markovian chain

By analogy, a continuous time Markov framework is introduced to calculate the packet delivery probability [13] for both two-hop and probabilistic epidemic forwarding. In two-hop routing, the source can copy the packet to any susceptible nodes (excluding the destination node) with probability $p(t)\in[0,1]$, and forward the packet to the destination node with probability 1 at time t. However, other nodes can only forward the packet to the destination with probability 1. In probabilistic epidemic forwarding, an infected node can replicate the packet to any susceptible nodes, except the destination node, with probability $p(t)\in[0,1]$ at time t. However, infected nodes forward the packet to the destination with probability 1.

$X(t)$ is also modeled as a CTMC. Let $F(t)$ be the delivery probability that the packet has been delivered to the destination node at time t, given by

$$F(t)=1-e^{-\beta\int_0^t E(X(s))ds}, \tag{11}$$

where $F(0)$=0, and the expectations $X(t)$ of two-hop forwarding and probabilistic epidemic forwarding are $N-(N-X(0))e^{-\beta\int_0^t p(\varsigma)d\varsigma}$, and $\dfrac{NX(0)\mathrm{e}^{\beta N\int_0^t p(\varsigma)d\varsigma}}{N-X(0)+X(0)\mathrm{e}^{\beta N\int_0^t p(\varsigma)d\varsigma}}$, respectively.

4.2 The Temporal Nature of the Packet Spreading

4.2.1 Example

The second kind of performance analysis studies the temporal packet spread process [15] in DTN by using the random theory. As an example, Fig. 4 shows the probability of packet infection as a function of time. The results were obtained by 1000 simulation runs of a network of size 1000[m]-by-1000[m], with 30 nodes that are moving with random direction mobility pattern ([12]). Rather intuitively, the probability first increases while the number of susceptible nodes is still large; but although the number of infected nodes continues to increase, after achieving its maximum, the probability of infection decreases due to the decreasing number of susceptible nodes. Thus, the graph in Fig. 4 is a result of the two trends: increase in the number of infected nodes and decrease in the number of susceptible nodes. The location and the value of the maximal probability, the average time of infection, and the spread of the infection time are all functions of the network parameters, such as the nodes' mobility pattern(s) and the number of network nodes.

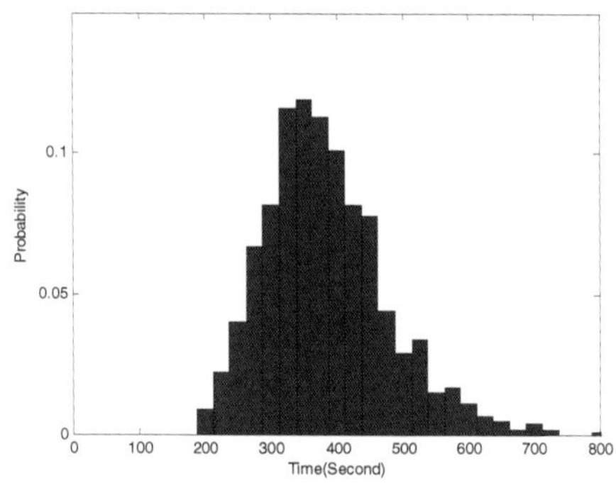

Fig. 4. Probability of a node infection as a function of time

4.2.2 Performance Evaluation

We are interested in understanding how the propagation of the packet occurs within the network as a function of time. Thus we establish the following main metrics: the probability function of the number of infected nodes as a function of time, the average number of infected nodes at a particular time, and the probability of all nodes being infected by a particular time.

To carry out our study, we use random graph theory. The network is modeled as a graph, where the nodes are modeled as vertices, and the encounter between any two nodes as an edge between the two corresponding vertices in the graph. Initially there exists one infected node (vertex), while all the remaining N-1 nodes are susceptible. Finally, time is slotted in slots of size Δt.

Let p_k be the probability that a particular node meets exactly k different nodes out of the other N-1 network nodes within the next slot of Δt. That is the degree of the node is k in the corresponding graph. Thus, because of the independent mobility of the nodes, p_k is given by

$$p_k = \binom{N-1}{k} p^k (1-p)^{N-1-k} . \tag{12}$$

The probability generating function is applied to the above result, and we obtain the following lemma.

Lemma 2: The probability generating function (PGF) of the degree of a vertex in the graph is:

$$G_0(x) = E(X^K) = \sum_{k=0}^{N-1} (p_k \cdot x^k) \text{ and } z_1 = \sum_{k=1}^{N-1} (k \cdot p_k), \tag{13}$$

where z_1 is the average degree of the nodes in the graph.

Our analysis is based on the derivation of the probability generating function of the number of infected nodes as a function of time. We assume that time is slotted and that every slot is of duration Δt (i.e., first slot is $(0, \Delta t]$, second slot is $(\Delta t, 2\Delta t]$, …, and the mth slot is $((m-1)\cdot\Delta t, m\cdot\Delta t]$).

We first introduce two basic definitions:

Definition 1: $Ge_m(x)(m = 0,1,2,\ldots)$ is the generating function of the number of infected vertices at time t, where $t = m\cdot\Delta t$.

Definition 2: $s_m(m = 0,1,2,\ldots)$ is the average number of infected vertices at time t, where $t = m\cdot\Delta t$.

Of course, $s_m = Ge_m'(1)(m = 1,2,\ldots)$.

At time zero, there is only one infected node, i.e., the source node. Thus $Ge_0(x) = x$, and $s_0 = 1$. During the time period of $(0, \Delta t]$, the single infected node (the source node) will infect new nodes based on the probability generating function of the degree of the infected vertex $G_0(x) = \sum_{k=0}^{N-1} p_k \cdot x^k$. Therefore, the probability generating function of the number of infected vertices at time $t = \Delta t$ can be expressed as:

$$Ge_1(x) = G_0(x) \cdot x = \sum_{k=0}^{N-1} p_k \cdot x^k \cdot x = \sum_{k=0}^{N-1} p_k \cdot x^{k+1} = \sum_{k=1}^{N} p_{k,1} \cdot x^k, \tag{14}$$

where $p_{k,1} = p_{k-1}$.

Based on Def. 2 and Eq. 14, we can obtain the current average number of infected vertices as:

$$s_1 = Ge_1'(1) = 1 + z_1, \tag{15}$$

where $z_1 = \sum_{k=1}^{N-1} k \cdot p_k$.

We define the set of infected nodes at time $t = m \cdot \Delta t$ as $s_{cluster}(m)$, and when not ambiguous we drop the time index and simply refer to the set as $s_{cluster}$. (Note that s_m represents the average size of $s_{cluster}(m)$; i.e., $s_m = |s_{cluster}|$.)

Naturally, we should now move on to calculate $Ge_m(x)$ for $m > 1$. However, we realize that there is a difficulty with the extension for large m due to the following two scenarios. First, as for $m > 1$ the set of infected nodes, $s_{cluster}$, will typically be larger than 1. Second, within the same time-slot, for $m > 1$ it is possible that within Δt a node will be infected more than once (i.e., by more than one encounter, each with a different infected node). In both cases, not every encounter of an infected node leads to a new infection. In other words, for $|s_{cluster}| > 1$ the probability generating function of the degree of a vertex, $G_0(x)$, overestimates to the number of new infections in a time-slot. We discuss this problem next in more details.

More specifically, when the number of infected vertices is larger than 1, we consider an encounter event where two already infected nodes meet with each other within a time-slot. We refer to this type of an event as a *repeated infection* or an *RI* event. The probability of an *RI* event is provided by the following theorem:

Theorem 1: Suppose that the number of infected vertices equals s_m, and the probability of one particular vertex doesn't meet any other infected vertices is labeled as $p(s_m)$. Then, the probability of two infected vertices meeting within a time-slot is

$$1.0 - p(s_m) \times p(s_m - 1) \times ... p(2) \quad \text{where} \quad p(s_m) = \sum_{k=0}^{N-s_m} p_k \times \frac{\binom{N-s_m}{k}}{\binom{N-1}{k}}.$$

To proceed now with our calculation of the probability generating function of the number of infected nodes as a function of time, we make a simple assumption that, for large networks and for small values of *t*, the likelihood of an *RI* event is negligible. In particular, when the size of the set of infected vertices is very small relative to the total number of vertices; i.e., $|s_{cluster}| \ll N$ based on Theorem 1, the chance of an *RI* event can be neglected.

Thus, at the beginning of the second slot, (Δt, $2\Delta t$], the number of infected vertices is *k* with the probability of $p_{k,1}(k = 1, 2, \cdots, N)$. Therefore, based on the property of probability generating functions ([18]), the distribution of the number of infected vertices at the end of the second slot is $(G_0(x))^k \cdot x^k$ with the probability of $p_{k,1}(k = 1, 2, \cdots, N)$. Therefore we obtain the probability generating function of the number of infected vertices at time $t = 2\Delta t$ as:

$$Ge_2(x) = \sum_{k=1}^{N} p_{k,1}(G_0(x))^k \cdot x^k = \sum_{k=1}^{N} p_{k,2} x^k, \tag{16}$$

and the average number of infected vertices is:

$$s_2 = Ge_2'(1) = (1 + z_1)^2. \tag{17}$$

In general, for $t = m \cdot \Delta t$ (m= 1,2, ...) the probability generating function of the number of infected vertices can be written as:

$$Ge_m(x) = \sum_{k=1}^{N} p_{k,m-1}(G_0(x))^k \cdot x^k = \sum_{k=1}^{N} p_{k,m} x^k, \tag{18}$$

where $p_{k,m} = \sum_{u=1}^{k} (p_{u,m-1} \sum_{\substack{u_0+u_1+\cdots u_{N-1}=u, \\ u_1+2u_2+\cdots(N-1)u_{N-1}=k-u,}} \frac{u!}{u_0 !u_1 !\cdots u_{N-1} !} p_0^{u_0} p_1^{u_1} \cdots p_{N-1}^{u_{N-1}})$, where $u_0, u_1, \cdots, u_{N-1}$ are integers in [0, u].

Based on Eq. 18, the probability distribution of the number of infected vertices at time $t = m{\cdot}\Delta t$ (m = 1, 2, …) is given by $p_{k,m} (1 \le k \le N)$, and the average number of the infected vertices is given by the follow theorem.

Theorem 2: The average number of the infected vertices at time $t = m{\cdot}\Delta t$ (m = 1, 2, …) is $s_m = (1+z_1)^m (m=1,2,\cdots)$ when $|s_{cluster}| \ll N$.

Using Eq. 12 for large N and small p, we obtain that $p_k = \binom{N-1}{k} p^k (1-p)^{N-1-k} \approx (\prod_{i=1}^{k} \frac{\lambda}{i}) e^{-\lambda}$ where $\lambda = (N-1)\cdot p \approx N \cdot p$. Thus, the random graph has a Poisson distribution of vertex's degrees and $z_1 = N \cdot p$. Consequently, the mean size of the set of the infected vertices is given by:

$$s_m = (1+z_1)^{m-1} = (1+Np)^m . \tag{19}$$

Finally, the probability of all nodes being infected at time $t = m \cdot \Delta t (m = 1,2,\ldots)$ can be obtained from Eq. 18 by calculating $p_{N,m}$.

We now consider our assumption that the probability of an *RI* event is negligible. Clearly, as m increases, the validity of our assumption diminishes and, based on Theorem 1, the error introduced in Eq. 18 becomes more significant. We choose some probability P_{th} and we find the largest time $t= m_{\max} \cdot \Delta t$ at which the probability of an *RI* event is still below P_{th}. Using Theorem 1 and Eq. 19, we obtain:

$$m_{max} = \left\lfloor log_{(1+Np)} (P_{th} \cdot (N-1)) \right\rfloor . \tag{20}$$

Therefore, we state that Eq. 18 is accurate for $m \le m_{\max}$ within the probability error of P_{th}.

For completeness of discussion, we also present an alternative approach, which is valid for $t = m \cdot \Delta t, m \ge m_{max} + 1$. This approach will allow us to fully compare our analytical results with simulation in following sections.

In this approach, we assume that the time-slot is small enough so that the probability of more than one encounter between any two nodes within a time-slot is negligible. We emphasize that this includes encounters between infected and susceptible nodes.

Let q_m be the number of infected vertices at time $t = m \cdot \Delta t, m \ge m_{max} + 1$, and $p(q_m = k)$ be the probability that the number of infected vertices is $k (k = 1,2,\ldots)$. We define $p_1(k)$ and $p_0(k)$ as the probabilities of an infection and of no infection within a time-slot, respectively, when the number of infected vertices is k. We obtain $p_0(k) = ((1-p)^k)^{N-k}$, because any of the N-k susceptible vertices don't meet any of the

k infected vertices. Based on the assumption that there is at most one infection within the time-slot, we conclude that $p_1(k) \approx 1 - p_0(k)$.

Using the above definitions and the Markovian model in Fig. 5, we write down an iterative equation for $p(q_m = k)$ $(k \geq 2)$, as:

$$p(q_m = k) = p(q_{m-1} = k)p_0(k) + p(q_{m-1} = k-1)p_1(k-1) \tag{21}$$

and $p_0(k)$ is as defined above. The initial condition for Eq. 21 could be the value of $p(q_{m_{max}} = k) = p_{k,m_{max}}$, where $p_{k,m_{max}}$ $(k = 1, 2, \cdots, N)$ is calculated based on Eq. 18 at time $t = m_{max} \cdot \Delta t$.

We note that the iteration of Eq. 21 could also be started from $k = 2$ in which case we use the fact that $p(q_0 = 1) = 1$ and $p(q_m = 1) = [p_0(1)]^m$.

The average number of infected nodes by time $t = m \cdot \Delta t$ can then be written as follows:

$$s_m = \sum_{k=1}^{N} p(q_m = k) \cdot k . \tag{22}$$

Finally, the probability of all nodes being infected at time $t = m \cdot \Delta t, m \geq m_{max} + 1$ can be obtained from Eq. 21 by calculating $p(q_m = N)$.

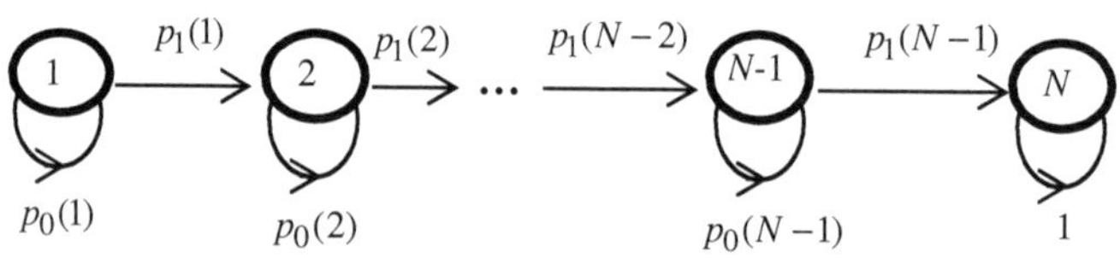

Fig. 5. Markov Chain of the number of infected nodes [15]

The size of the time-slot, Δt, is important to ensure the validity of our analytical model; e.g., if the slot-size is too large, the probability of more than one event with the time-slot cannot be neglected, and our model would introduce an error by underestimating the number of encounters. In a DTN with N nodes, the infection rate is maximal when there are approximately $N/2$ infected vertices and $N/2$ susceptible nodes. We can obtain the size of the time-slot $\Delta t = -\ln[1 - \sqrt{\varepsilon} / (N/2)^2] / \beta$ [15].

4.2.3 Simulation Results

We start with simulation to verify our analytical model. Nodes are randomly placed in a square area of 4000×4000[m^2], and are moving according to the random direction (RD) mobility ([20]). The speed and travel time of a node are chosen uniformly from the interval [0, 10[m/sec]] and [0, 120[sec]], respectively. The network area wraps around ([21]), so a node that hits the area boundary reenters the network at the opposite boundary. The transmission range of a node r is set to be 25[m]. To estimate the encounter rate β, we run our simulation model with only two nodes, and we record the time at which the two nodes firstly meet. The simulation results are obtained from an average of is on N_{top} runs.

For comparison purposes, we assume in our analytical model that ε and P_{th} equal 1.0×10^{-2} and 0.001, respectively. Unless otherwise stated, N = 150 and N_{top} = 1000. We choose the basic metric of the probability distribution of the number of infected nodes as a function of time in Section 4.2.2 as the performance metric.

In order to compare the analytical and the simulation results, we use the χ^2 (Chi-Square) statistical test ([22]). In particular, for the performance metric of distribution of the number of infected nodes as a function of time, the range of integers from 1 to N is divided into k bins, where a bin is assigned a range of integers that correspond to the number of infected nodes at a particular time instance. The ith bin counts the number of simulation runs, O_i, in which the number of infected nodes fits within the range assigned to the bin. Furthermore, we define E_i as the expected number of runs based on the proposed model, whose number of infected nodes belongs to the ith bin. In particular, we choose the k-1 integers $n_i (1 \le i \le k-1)$ to partition the integers' range from 1 to N, as to define the k bins. We determine n_i to be the smallest integer, so that the number of simulation runs whose number of infected nodes is expected to fall within the [1, n_i] range is at least $N_{top} \times i / k$, where N_{top} is the number of simulation runs. The χ^2 statistical test is defined by the following formula:

$$\chi^2 = \sum_{i=1}^{k} \frac{(O_i - E_i)^2}{E_i}, \tag{23}$$

and the hypothesis that the simulation results agrees with the analytical is rejected if:

$$\chi^2 \ge \chi_\alpha^2 (k-1), \tag{24}$$

where α is the significance level, and $\chi_\alpha^2 (k-1)$ is the critical value of the test. Otherwise, the hypothesis is accepted. In our evaluation, we assume that $\alpha = 0.01$. k is set to 4. Thus, the critical value is $\chi_\alpha^2 (k-1) = \chi_{0.01}^2 (3) = 11.35$ ([22]).

The probability distribution of the number of infected nodes as a function of time provides an important insight into the process of the packet spread. Fig. 6 shows the probability distribution of the number of infected nodes at the times of 2600[s], 3400[s], 4200[s], and 5000[s], when N_{top} = 10000. Note that Fig. 6(b) is a zoomed in version of Fig. 6(a). For the probability distribution of infected nodes at 2600[s], the χ^2 statistic in Fig. 6, as compared with the analytical results, is 3.55, which confirms a very good match between analysis and simulation. As a point of reference, the maximal gap between our analytical model and simulation at the above four time instances is less than 2.9%. From the results presented in this section, we conclude that our analytical model faithfully represents the infection-based spreading process of packets in a DTN.

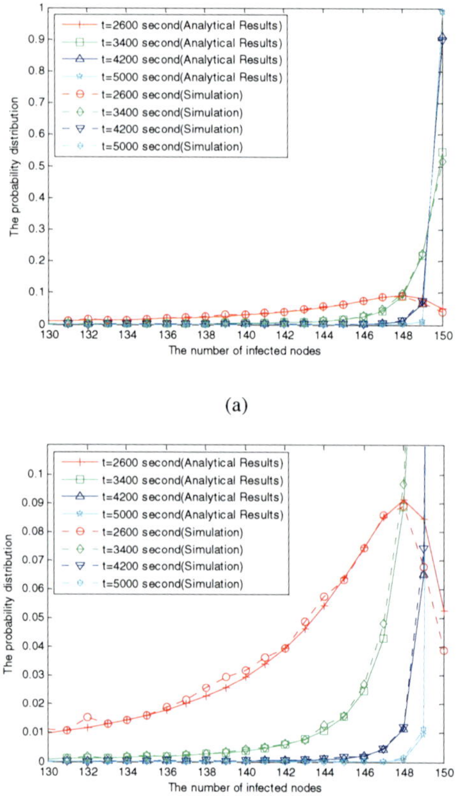

(b)

Fig. 6. Probability distribution of the number of infected nodes [15]

5 Summary

In this chapter, we first presented an overview of epidemic routing in DTNs. We then introduced the network and mobility models. We also carried out a performance analysis of epidemic routing, including obtaining a performance metric equation from a macroscopic perspective, for example, delivery delay, and gaining an understanding of the temporal aspects of the process of packet spread in DTNs. By emphasizing the latter analysis, we derived three metrics: the probability distribution of the number of infected nodes, the average number of infected nodes, and the probability of all nodes being infected, all as functions of time.

References

[1] Ehsan, S., Bradfor, K., Brugger, M., Hamdaoui, B., Kovchegov, Y.: Design and analysis of delay-tolerant sensor networks for monitoring and tracking free-roaming animals. IEEE Transactions on Wireless Communications 11(3), 1220–1227 (2012)

[2] An, N., Riihijarvi, J., Mahonen, P.: Studying the delay performance of opportunistic communication in VANETs with realistic mobility models. In: Proceedings of IEEE VTC Spring, pp. 1–5 (2009)

[3] Ribeiro, F.-J.-L., Pedroza, A.-D.-C.-P., Costa, L.-H.-M.-K.: Deepwater monitoring system in underwater delay/disruption tolerant network. IEEE Latin America Transactions 10(1), 1125–1134 (2012)

[4] Vahdat, A., Becker, D.: Epidemic routing for partially-connected ad hoc networks. Technical Report, Duke University CS-200006 (2000)

[5] Brauer, F., Castillo-Chávez, C.: Mathematical models in population biology and epidemiology. Springer, New York (2001)

[6] Yoon, S.-K., Haas, Z.-J., Kim, J.-H.: Efficient tradeoff of restricted epidemic routing in mobile ad-hoc networks. In: Proceedings of IEEE MILCOM, pp. 1–5 (2007)

[7] Zhu, H., Chang, S., Li, M., Naik, K., Shen, S.: Exploiting temporal dependency for opportunistic forwarding in urban vehicular networks. In: Proceedings of IEEE INFOCOM, pp. 2192–2200 (2011)

[8] Haas, Z.-J., Small, T.: A new networking model for biological applications of ad hoc sensor networks. IEEE/ACM Trans. on Networking 14(1), 27–40 (2006)

[9] Wu, J., Wang, Y.: Social feature-based multipath routing in delay tolerant networks. In: Proceedings of IEEE INFOCOM, pp. 1368–1376 (2012)

[10] Groenevelt, R., Nain, P., Koole, G.: The message delay in mobile ad hoc networks. Performance Evaluation 62(1-4), 210–228 (2005)

[11] Zhang, E., Neglia, G., Kurose, J., Towsley, D.: Performance modeling of epidemic routing. Computer Networks 51(10), 2867–2891 (2007)

[12] Matsuda, T., Takine, T. (p, q)-Epidemic routing for sparsely populated mobile ad hoc networks. IEEE Journal on Selected Areas in Communications 26(5), 783–793 (2008)

[13] Li, Y., Jiang, Y.-R., Jin, D.-P., Su, L.: Energy-efficient optimal opportunistic forwarding for delay-tolerant networks. IEEE Trans. on Vehicular Technology 59(9), 4500–4512 (2010)

[14] Fan, J., Chen, J.-M., Du, Y., Gao, W., Wu, J., Sun, Y.-X.: Geocommunity-based broadcasting for data dissemination in mobile social networks. IEEE Trans. on Parallel and Distributed System 24(4), 734–713 (2013)

[15] Wang, Q.-S., Haas, Z.-J.: Analytical model for epidemic routing in delay-tolerant networks. In: Proceedings of the 1st ACM Workshop on High Performance Mobile Opportunistic systems (HP-MOSys), pp. 1–8 (2012)

[16] Bollobás, B.: Random graphs, 2nd edn. Cambridge University Press, New York (1999)

[17] Groenevelt, R., Nain, P., Koole, G.: Message delay in MANET. ACM SIGMETRICS 33(1), 412–413 (2005)

[18] Kurtz, T.-G.: Solutions of ordinary differential equations as limits of pure jump markov processes. Journal of Applied Probabilities, 49–58 (1970)

[19] Wilf, H.-S.: Generating functionology, 2nd edn. Academic Press, London (1994)
[20] Nain, P., Towsley, D., Liu, B., Liu, Z.: Properties of random direction models. In: Proceedings of IEEE INFOCOM, pp. 1897–1907 (2005)
[21] Haas, Z.-J.: The routing algorithm for the reconfigurable wireless networks. In: Proceedings of ICUPC, pp. 562–566 (1997)
[22] Greenwood, P.-E., Nikulin, M.-S.: A guide to chi-squared testing. Wiley, New York (1996)

Author Index

Subject Index